PREVENTIVE MATERIALS, METHODS, AND PROGRAMS, VOL 4

THE AXELSSON SERIES
ON PREVENTIVE DENTISTRY

The world-renowned authority on preventive and community dentistry presents his life's work in this five-volume series of clinical atlases focusing on risk prediction of dental caries and periodontal disease and on needs-related preventive and maintenance programs.

Volume 1 An Introduction to Risk Prediction and Preventive Dentistry
Provides a general overview of current and future trends in risk prediction, control, and nonaggressive management of caries and periodontal disease; preventive dentistry methods and programs; and quality control.

Volume 2 Diagnosis and Risk Prediction of Dental Caries
Includes a comprehensive discussion of the etiology, pathogenesis, diagnosis, risk indicators and factors, individual risk profiles, and epidemiology of caries.

Volume 3 Diagnosis and Risk Prediction of Periodontal Diseases
Presents a comprehensive discussion of the etiology, pathogenesis, diagnosis, risk indicators and factors, individual risk profiles, and epidemiology of periodontal diseases. Considers periodontal diseases as a possible risk factor for systemic diseases and presents current and future trends in the management of periodontal diseases, including nonaggressive debridement and preservation of the root cementum.

Volume 4 Preventive Materials, Methods, and Programs
Discusses self-care and professional methods of mechanical and chemical plaque control, use of fluorides and fissure sealants, and integrated caries prevention. Addresses needs-related preventive programs based on risk prediction and computer-aided epidemiology analysis for quality control and outcome.

Volume 5 Nonaggressive Treatment, Arrest, and Control of Periodontal Diseases and Dental Caries
Details current and future trends in nonaggressive treatment methods that seek to preserve the root cementum; surgical versus nonsurgical periodontal therapy; repair and regeneration of periodontal support; management of furcation-involved teeth; restricted use of antibiotics; arrest of noncavitated enamel, dentin, and root caries lesions; nonaggressive mini-preparations; esthetic and hygienic aspects of restorations; and management of erosions. Focuses on needs-related maintenance programs to ensure the long-term success of treatment and to prevent recurrence of periodontal disease and dental caries.

PREVENTIVE MATERIALS,
METHODS, AND PROGRAMS, VOL 4

Per Axelsson, DDS, Odont Dr

Professor and Chairman
Department of Preventive Dentistry
Public Dental Health Service

Karlstad, Sweden

DATE DUE

quintessence
books

Quintessence Publishing Co, Inc
Chicago, Berlin, Tokyo, Copenhagen, London, Paris, Milan, Barcelona,
Istanbul, São Paulo, New Delhi, Moscow, Prague, and Warsaw

To my wife Ingrid, my daughter Eva, and my son Torbjörn

Library of Congress Cataloging-in-Publication Data

Axelsson, Per, D.D.S.
      Preventive materials, methods, and programs / Per Axelsson.
            p. ; cm. – (The Axelsson series on preventive dentistry ; v. 4)
      Includes bibliographical references and index.
      ISBN 0-86715-364-4 (hardcover)
      1. Dental prophylaxis. 2. Preventive dentistry 3. Dental
plaque–Prevention. 4. Dental caries–Prevention. 5.
Fluorides–Therapeutic use. I. Title. II. Series: Axelsson, Per, D.D.S.
Axelsson series on preventive dentistry ; v. 4.
      [DNLM: 1. Dental Caries–prevention & control. 2. Dental
Deposits–prevention & control. 3. Dental Prophylaxis–methods. 4.
Fluorides, Topical–therapeutic use. 5. Oral Hygiene. 6. Preventive
Dentistry–methods. WU 250 A969p 2004]
      RK60.7.A943 2004
      617.6'01–dc22
                                                            2003016826

© 2004 Quintessence Publishing Co, Inc
Quintessence Publishing Co, Inc
551 Kimberly Drive
Carol Stream, IL 60188
www.quintpub.com

Questions and lecture or course requests may be directed to the author by fax at +46 (0) 54-52 56 54 or by e-mail
at per.axelsson@karlstad.mail.telia.com.

Editor: Kathryn O'Malley
Production: Thomas Pricker

Printed in Slovakia

CONTENTS

PREFACE

The etiology of dental caries and periodontal diseases is well understood, and we now have developed efficient methods for prevention, treatment, arrest, and control of these diseases. For example, over the last 25 years in the county of Värmland, Sweden, large-scale implementation of our preventive programs for children has led to an increase in the percentage of caries-free 3-year-old children from 30% to 97%, as well as a reduction in caries prevalence in 12-year-old children from an average of 25 to less than 0.5 decayed or filled surfaces, with 85% of children in this age group caries free.

In our 30-year longitudinal study in adults, the mean number of lost teeth was only about 0.5 per subject per 30 years, and the periodontal attachment level was unaltered irrespective of age. Large-scale implementation of the study's methods in the preventive programs for our adult population has led to an increase of more than 15% in the number of remaining teeth in 65-year-old adults, as well as a reduced loss of periodontal support by more than 20% during only a 10-year period and a reduction in the percentage who were edentulous from 17% to 7%.

According to the principles of *lege artis*, all members of our profession are obliged to offer treatment based on the most current scientific and clinical knowledge available. We must there-fore continue to concentrate our efforts on preventing, controlling, and arresting the development of dental caries and periodontal diseases. However, needs-related preventive programs must be cost-effective and should be based on information derived from comprehensive diagnoses, histories, and risk predictions at group, individual, and tooth-surface levels. For quality control and evaluation of such programs, computer-aided analytical epidemiology, using relevant variables, should be introduced.

The aim of this book, the fourth of a five-volume series of textbooks and atlases, is to provide the reader with updated information about various preventive materials and methods, needs-related preventive programs, and analytical epidemiology for quality control. Each chapter contains clear, concise text as well as numerous illustrations describing how to use the materials and methods addressed, supplemented with schedules showing how they should be used on a needs-related basis, according to the state of the art and based on evidence from recent research. Thus, the volume should act as a preventive dentistry "cookbook."

The first chapter focuses on the rate and pattern of plaque formation, which represents the etiology of dental caries and periodontal diseases. Because some microorganisms are more cariogenic and aggressive periopathogens than others, daily mechanical cleaning of *all* tooth surfaces is the most rational and efficient method for prevention and control of periodontal diseases and dental caries. Thus, chapter 2 presents the most cost-effective approach, mechanical toothcleaning by self-care, describing in detail the associated materials and methods, as well as principles for successful establishment of needs-related self-care habits based on self-diagnosis. Chapter 3 presents efficient materials and methods for professional mechanical toothcleaning (PMTC), as well as evidence-based effects of PMTC with and without oral hygiene education on dental caries, gingivitis, and periodontal diseases.

Combinations of mechanical and chemical plaque control by self-care, supplemented with needs-related intervals of PMTC and professional application of chemical plaque control agents, are the most successful methods for prevention and control of periodontal diseases, as well as dental caries. Therefore, chapter 4 provides a supplementary comprehensive presentation of chemical plaque control agents for self-care and professional application, as well as a schedule for needs-related use of such products based on risk prediction.

A unique chemical caries-preventive material, fluoride has been shown to exert almost 100% of its effect posteruptively. The caries preventive mechanisms of fluoride, as well as materials and methods for rational use of fluorides by self-care and professionals are presented in chapter 5.

Almost 100% of all occlusal caries lesions are initiated in the distal and central fossae and related fissures of the permanent molars during the extremely long period of eruption (12 to 18 months). Without functional wear, cariogenic plaque biofilms reaccumulate extremely quickly and remain undisturbed on such key-risk surfaces. Therefore, if fissure sealants and other methods of intensified caries prevention are used until full eruption, the risk of developing fissure caries can be avoided altogether, as discussed in chapter 6.

Because of the multifactorial nature of dental caries, several caries-preventive methods must be integrated in needs-related caries-preventive programs. Chapter 7 discusses principles for integrated caries prevention, as exemplified by a 20-year longitudinal needs-related caries-preventive program for children and young adults.

Preventive dental programs also must be related to the oral health status and socioeconomic conditions of the population, as well as available dental care resources. In countries with high prevalence of dental diseases, low socioeconomic standards, and limited dental care resources, so-called population-based preventive programs should be implemented. However, in industrialized countries with high socioeconomic standards and well-developed dental care resources, more individualized preventive programs based on predicted risk for dental caries and periodontal disease are recommended from a cost-effectiveness point of view. New principles for oral health promotion and needs-related preventive programs are presented and discussed in chapter 8. In addition, schedules for preventive materials and methods related to predicted caries and periodontitis risk in different age groups are presented.

Finally, chapter 9 presents the rationale and methods for periodical evaluation of the effect of such programs by relevant diagnostic variables in computer-aided analytical epidemiology at clinic and population levels.

This project could not have been completed without the assistance and support of my family, friends, and colleagues. I offer my deepest thanks to my wife, Ingrid, and my daughter, Eva; my son, Torbjörn; and their families, as well to all my other relatives and friends, for their patience and understanding throughout the last 7 years, in which I have spent almost every night, weekend, and vacation preparing the material for these five volumes. I also wish to thank my wonderful staff at the Department of Preventive Dentistry, Public Dental Health Service, county of Värmland, for all their services, and particularly my assistant, Pia Hird Jonasson, who typed most of my manuscript. I owe special thanks to Art Director Fredrik Persson and Dr Jörgen Paulander, for their excellent support with computer-aided illustrations.

I am grateful to all my colleagues and friends around the world and to several publishers (Munksgaard International, The American Academy of Periodontology, S. Karger Medical and Scientific Publisher, FDI World Dental Press, and WHO Oral Health Unit), who have generously permitted me to use their illustrations (about 30% of the total). Last but not least, the excellent cooperation of the publisher is gratefully acknowledged.

CHAPTER 1

PLAQUE CONTROL:
THE KEY TO PREVENTIVE DENTISTRY

Plaque control is the key to prevention of gingivitis, periodontitis, and dental caries. It is causally directed toward the sole etiologic factor in these diseases: the pathogenic microflora that colonize the tooth surfaces and form dental plaque (biofilm). Both animal and human studies have confirmed the role of plaque; germ-free animals frequently fed with sugar do not develop caries until they are infected by cariogenic microflora, which colonize the tooth surfaces (Orland et al, 1954). Sugar is not an etiologic factor in dental caries, but an external modifying risk factor, as demonstrated in classic studies by von der Fehr et al (1970) and Löe et al (1972). Under extreme experimental conditions (the absence of oral hygiene, ie, unrestricted plaque accumulation, and nine daily sucrose rinses), gingivitis and enamel caries were induced in healthy young adults within 3 weeks. When the same research team carried out the same study but introduced chemical plaque control, ie, twice daily mouthrinsing with 0.2% chlorhexidine solution, no gingivitis or caries developed (Löe et al, 1972).

Figure 1, from a subject in the "experimental gingivitis" study by Löe et al (1965), shows the accumulated plaque and the resultant inflamed gingival margin, particularly at maxillary sites. Once adequate oral hygiene resumes, the gingival inflammation subsides within a week (Fig 2). The thickness of the gingival plaque gradually increases during the 3-week experimental period (Fig 3). For the first few days, this plaque is composed of gram-positive cocci and rods, representing the indigenous microflora of the tooth surface. After 4 to 5 days, filamentous organisms and gram-negative cocci and rods "infect" the gingival plaque; nonattaching spirochetes gradually appear in the gingival sulcus; and the assortment of microorganisms in the gingival biofilm increases continuously. As a consequence, the first clinical signs of gingivitis develop within 2 to 3 weeks. When accumulated plaque is mechanically removed and daily oral hygiene is reestablished, the gingivae heal within about 1 week.

These findings have since been confirmed in many human and animal studies. Egelberg (1964) found subclinical symptoms of gingival inflammation, in the form of an exudate from the gingival sulcus, as early as 4 days after free accumulation of plaque. This experimental gingivitis model is frequently used as a standard for development of gingivitis in studies designed to evaluate the effect of different chemical plaque control agents on gingivitis.

Figure 4 illustrates the mean Löe and Silness (1963) Gingival Index (GI) for all gingival units in subjects cleaning every 12, 48, 72, or 96 hours over an observation period of 6 weeks (Lang et al,

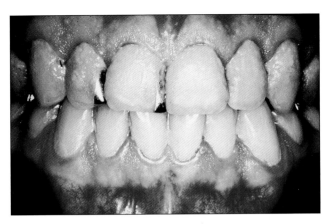

Fig 1 Plaque accumulation in a subject in an experimental gingivitis study. Note the relationship between the plaque and inflammation of the gingival margin. (From Löe et al, 1965. Reprinted with permission.)

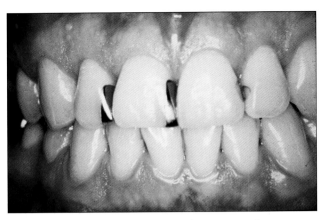

Fig 2 Resolution of the gingival inflammation shown in Fig 1, within 1 week of resumption of adequate oral hygiene. (From Löe et al, 1965. Reprinted with permission.)

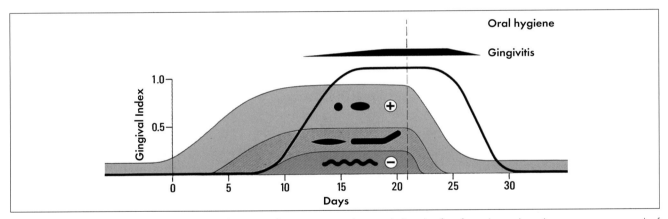

Fig 3 Increase in gingival plaque over the 3-week experimental period. For the first few days, this plaque is composed of gram-positive (+) cocci and rods, the indigenous microflora of the tooth. After 4 to 5 days, filamentous organisms and gram-negative (–) cocci and rods "infect" the plaque. Gradually, nonattaching spirochetes appear in the sulcus, while the assortment of microorganisms in the gingival biofilm increases continuously. (Modified from Löe et al, 1965 with permission.)

1973). Clinical signs of gingivitis developed only when plaque was allowed to accumulate for 72 or 96 hours, and not in subjects cleaning every 12 or 48 hours, although the mean Silness and Löe (1964) Plaque Index (PI) was almost identical after 48 and 72 hours. That is because plaque accumulation of more than 48 hours may have greater pathogenic potential than less mature plaque. This might be attributable to the dramatic change that occurs between day 2 and days 3 and 4 of de novo plaque accumulation: a massive increase in the thickness of the plaque and thereby the total number of plaque microorganisms (Fig 5). The study by Lang et al (1973) clearly demonstrated

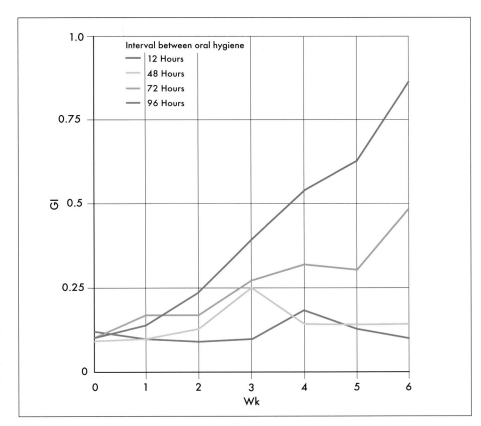

Fig 4 Mean Löe and Silness Gingival Index score for all gingival units in subjects cleaning at various intervals during a 6-week study. (From Lang et al, 1973. Reprinted with permission.)

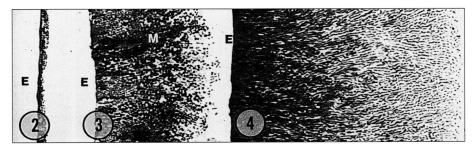

Fig 5 Thickness of freely accumulated gingival plaque after 2, 3, and 4 days. (E) Tooth enamel; (M) Microbial plaque. (From Listgarten, 1976. Reprinted with permission.)

that meticulous plaque control at intervals of 48 hours is compatible with the absence of clinical signs of gingival inflammation; at greater intervals, gingivitis develops.

In another study, Bosman and Powell (1977) induced gingivitis experimentally and then randomly allotted the subjects to one of four test groups using chemical plaque control. Participants rinsed with 0.2% chlorhexidine solution twice a day or every second, third, or fourth day. Within 7 to 10 days, gingivitis was eliminated in all subjects rinsing twice a day and every second day but persisted among those rinsing only every third or fourth day. This study confirmed the find-

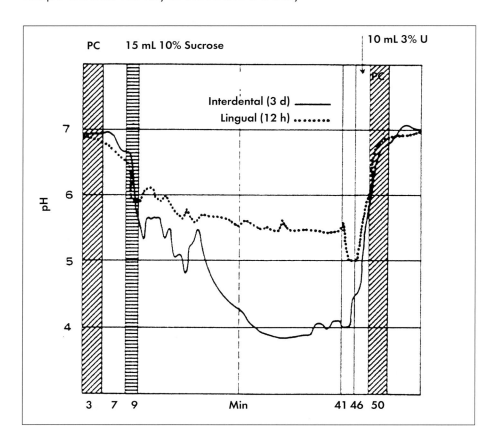

Fig 6 Telemetric measurement of the response to rinsing with 10% sucrose solution. The drop in pH in 12-hour-old lingual plaque is limited compared to the critical drop in pH beneath 3-day-old interdental plaque. (PC) Paraffin chewing; (3% U) 2-minute rinse with 3% urea solution. (Courtesy of T. Imfeld.)

ings by Lang et al (1973) that plaque control at least every second day eliminates or prevents the development of gingivitis.

Experimental studies in animals have shown that untreated, plaque-induced gingivitis can eventually progress to periodontitis (Lindhe et al, 1975; Saxe et al, 1967). In humans, although gingivitis is very common, progressive periodontitis develops in only a minority of individuals and at isolated sites. At the First European Workshop on Periodontology (Lang and Karring, 1994), there was consensus that periodontitis is always preceded by gingivitis: Prevention of gingivitis should also therefore prevent periodontitis.

The telemetric method developed by Graf and Mühlemann (1966) allows in vivo measurement of the "true" pH on the tooth surface beneath the undisturbed plaque. The importance of the age, amount, and composition of plaque as well as different concentrations of sugar can thereby be evaluated. Using the telemetric method, Imfeld (1978) showed that rinsing with 10% sucrose solution caused a dramatic drop in pH to below 4 in 3-day-old interdental plaque. Such plaque is typical for the approximal surfaces of the molars and premolars in a toothbrushing population. In contrast, the fall in pH in immature lingual plaque (12 hours old) was very limited (Fig 6).

Firestone et al (1987) used the same telemetric test in vivo, measuring the drop in pH on molars after subjects rinsed with 10% sucrose solution. Four different sites with approximal plaque were compared to plaque-free approximal surfaces (Fig 7). The authors concluded that "removing plaque from interdental surfaces significantly reduced the exposure of the surfaces to plaque acids following sucrose rinse. This further supports mechanical removal of plaque from interdental surfaces as a means of reducing dental caries."

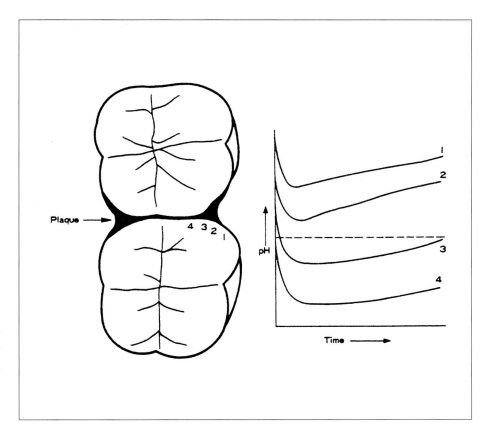

Fig 7 Drop in pH exhibited by molars with approximal plaque at four different sites after subjects rinsed with sucrose solution. (From Firestone et al, 1987. Reprinted with permission.)

In toothbrushing populations, for those who have an established habit of using a toothbrush and fluoride toothpaste daily, dental plaque more than 2 days old is located mainly on the approximal surfaces of the molars and premolars, partly subgingivally. Access with a toothbrush to the wide approximal surfaces is limited by the buccal and lingual papillae. In European countries, although daily toothbrushing with a fluoride dentifrice is an established oral hygiene habit, special aids to approximal oral hygiene, such as dental floss, dental tape, toothpicks, and interdental brushes, are used daily by less than 10% of the population (Kuusela et al, 1997). This explains why caries, gingivitis, and marginal periodontitis are much more prevalent on the approximal surfaces of the molars and premolars than on the buccal and lingual surfaces of the dentition.

Two conclusions may be drawn from these studies:

1. Prevention of gingivitis and marginal periodontitis must be based on plaque control.
2. Prevention of dental caries should be based on plaque control.

Plaque disclosure illustrates the close correlation between the sites of gingivitis and caries and the key etiologic factor: the dental plaque biofilm (Figs 8 and 9). Clean teeth will never decay.

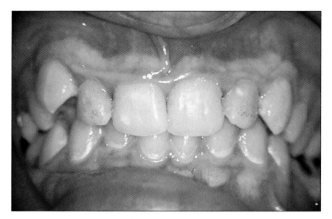

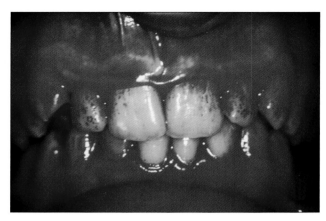

Fig 8 Anterior teeth of a 12-year-old boy with gingivitis at the following sites: the mesiobuccal (MB) surface of tooth 13 (Fédération Dentaire Internationale [FDI] tooth-numbering system); the distobuccal (DB) surface of tooth 12; the DB surface of tooth 21; the MB and DB surfaces of tooth 22; the MB surface of tooth 23; the MB surface of tooth 43; the DB surface of tooth 42; the DB surface of tooth 32; and the MB surface of tooth 33. Enamel (incipient) caries is located at the MB surface of tooth 13; the MB surface of tooth 43; the buccal surface of tooth 42; the DB surface of tooth 32; the MB surface of tooth 33; and the MB surface of tooth 34. In addition, a cavity is present on the distal surface of tooth 22.

Fig 9 Erythrosin-stained dentition of the patient in Fig 8. Note the close correspondence between areas of plaque and areas affected by gingivitis and caries.

ROLE OF THE ORAL ENVIRONMENT

Macroenvironment of the oral cavity

In certain aspects, the oral cavity may be regarded as a single microbial ecosystem. The oral environment is dominated by the flow of saliva, at the rate of approximately 0.4 mL/min under resting conditions and 2.0 mL/min under stimulated conditions; the average flow rate during waking periods is 0.6 mL/min. Approximately 15% of the adult population have stimulated flow rates of less than 0.7 mL/min, which may be related to medication with anticholinergic drugs, irradiation, or disease. The total daily secretion has been estimated to be 600 to 700 mL. The resting volume of saliva is approximately 7.0 mL and is replenished about 10 times per minute during waking hours. Salivary flow almost ceases during sleep, when the salivary flow rate may be as low as 0.25 mL/min.

In addition to saliva, the oral fluid contains tissue fluid, originating predominantly from the gingival sulci or crevices. The volume of crevicular fluid is dependent on the health of the periodontal soft tissues.

The composition of the oral fluid is very complex. It contains both organic and inorganic components. The organic components include a variety of proteins, carbohydrates, and enzymes. Each component plays an important role in controlling the oral environment. For example, secretion of immunoglobulin A (IgA), agglutinins, antimicrobial enzymes (lysozyme, lactoperoxidase, and lactoferrin), and glycosyl transferase significantly influences the number and variety of oral microflora as well as the plaque formation rate (Axelsson, 1987, 1991).

Although saliva is not a good medium for supporting the growth of large numbers of bacteria, 1 mL of whole saliva may contain more than 200 million microorganisms, representing more than 400 different species. Several periodontal pathogens can be cultured from saliva (Asikainen et al, 1991; Petit et al, 1993; van Winkelhoff et al, 1988). It is likely that organisms shed from intraoral reservoirs find transient residence in saliva. Investigation of untreated subjects with suppurating periodontitis revealed that *Porphyromonas gingivalis* and *Prevotella intermedia* could be detected in saliva at levels exceeding 10^6 cells/mL (van Winkelhoff et al, 1988), implying that saliva facilitates intraoral spread of these organisms. This is supported by the observation that trypsin-like activity, a characteristic of *P gingivalis*, can be detected in the saliva of patients with periodontitis before treatment (Zambon et al, 1985).

It is also likely that microbes attached to desquamated epithelial cells spread via saliva to different tooth surfaces; these microbes typically colonize sheltered regions: interproximal spaces, gingival margins, and occlusal fissures (Saxton, 1975).

Microenvironments of the oral cavity

In the oral cavity, there are several major and minor compartments, each constituting a separate microenvironment not easily affected by major events in the oral cavity. These environmental factors significantly influence the access and release of antimicrobial agents from toothpastes and particularly from delivery systems, such as mouthrinses and gels. For example, suspensions of charcoal in water placed in the vestibule of one side of the mouth of an individual who is not talking or chewing will spread, within about 5 minutes, to the dorsum of the tongue and the hard palate on the same side of the mouth.

However, there is no spread to the other side of the mouth, even after prolonged conversation. If charcoal is placed initially under the tongue, the movement is such that the entire dorsum will be covered within 1.5 minutes, but the hard palate on both sides will not be covered for at least 4 minutes (Jenkins and Krebsbach, 1985). Thus, even the major subcompartments within the oral cavity, eg, the sublingual space and vestibule, are not in immediate communication.

There are also many discrete compartments, such as the interproximal spaces, gingival crevices, gingival pockets, occlusal fissures, the papillae of the dorsum of the tongue, and the crypts and fissures of the tonsils, each of which creates a special microenvironment within the open system of the oral cavity as a whole.

Microbial colonies

All the surfaces of the oral cavity are colonized by microorganisms. Facultatively anaerobic streptococci constitute an essential part of the microflora that constantly colonize the mucous membranes and the teeth. Microorganisms are regularly swallowed with saliva. The amount within the oral cavity fluctuates, simply because the microbial deposits building up on mucous membranes and in particular on tooth surfaces grow and multiply and thus provide a reservoir for the oral environment. Fluctuations occur during sleeping and waking hours and also as a result of activities such as eating and drinking and oral hygiene procedures.

The microflora in mixed saliva comprises mainly microorganisms shed from the oral surfaces and to some extent reflects the gross composition of the microbial deposits on the various oral surfaces.

Specific environments for microbial deposits or plaque

A specific area that supports a bacterial flora is termed a *habitat*. The flora of a habitat develops through a series of stages, collectively called *colonization*. Colonization is a complex process because it involves not only interactions between bacteria and their environment but also interactions among bacteria.

The first important aspect of colonization of a habitat by bacteria is access. Because the organisms must be able to enter the habitat, there must be a means of transmission from one habitat to another. For example, mothers can serve as reservoirs of oral bacteria that colonize their children's mouths, and, within a single host, bacterial reservoirs can aid the survival of the organism. In the human mouth, the tongue and tonsils as well as the oral mucosa may serve as reservoirs for bacteria, which, under favorable conditions, may colonize periodontal pockets: Several investigators have observed that black-pigmented species, including *P gingivalis* and *P intermedia*, can be cultured from the dorsum of the tongue (van Winkelhoff et al, 1988), the tonsils (van Winkelhoff et al, 1988), and the oral mucosa (van Winkelhoff et al, 1986). It has also been demonstrated that *Actinobacillus actinomycetemcomitans* may attach to the teeth, tongue, and buccal mucosa (Asikainen et al, 1995; Müller et al, 1990; Slots et al, 1980).

Comparison of the microbiota on the tongue and tonsillar area in subjects with periodontitis and those without periodontitis indicates a correlation between disease and the presence in these areas of *P intermedia* and motile bacteria (van Winkelhoff et al, 1986). Similarly, in a group of untreated patients with *A actinomycetemcomitans*–associated periodontitis, the subgingival load of *A actinomycetemcomitans* has been associated with the number of *A actinomycetemcomitans* on cheek mucosa (Müller et al, 1990).

The dorsum of the tongue is the main reservoir for *Streptococcus salivarius*, a very potent cariogenic (acidogenic) bacteria. However, in one study, a higher number of *Streptococcus mutans* was repeatedly found on the dorsum of the tongue after five thorough scrapings with a tongue scraper than was found prior to scraping, indicating the tongue is also an important reservoir for *S mutans* (Axelsson et al, 1987b). Lindquist et al (1989b) found a significant correlation between the prevalence of *S mutans* in saliva and the prevalence on the dorsum of the tongue.

These data support the inclusion of the dorsum of the tongue in oral hygiene procedures, at least in patients highly infected by periopathogens and cariogenic bacteria such as *S mutans*, as part of the so-called complete-mouth disinfection principle (Axelsson et al, 1987b; Mongardini et al, 1999; Quirynen et al, 1995). The tongue-cleaning procedure should include initial thorough mechanical removal of deposits from the dorsum, which is normally covered by thick deposits that impede penetration of the antimicrobial solution. Studies using DNA fingerprints have revealed reservoirs, or fugitive habitats, where, for example, *S mutans* and periopathogens can survive chlorhexidine treatment (Kozai et al, 1991).

The oral cavity presents two types of surfaces for colonization by bacteria: the soft tissues, modified to some extent by a coating of saliva, and the hard tooth surfaces, modified by a pellicle that is formed by adsorption of salivary components. A distinct and important difference between the two types of surfaces is that the soft tissue surfaces desquamate, while the hard tissues do not. Bacteria colonizing the soft tissue surfaces are lost when cells are shed: For survival in this habitat, readherence is essential. In contrast to the hard surfaces, which will support heavy deposits of bacteria in dental plaque, soft tissue surfaces do not form complex layers of bacteria (biofilms).

In the oral cavity, the physical dimensions of a habitat do not fall within specific limits. The whole oral cavity, an occlusal tooth surface, or even a defined area on the occlusal surface may be considered a habitat. These habitats, together with their physiologic characteristics, are referred to as *microenvironments*. In oral microbiology, changes in the flora of a habitat such as the mouth may indicate, for example, patients at risk of caries, while changes in tooth surface microenvironments can identify a surface at risk of disease. Thus, although general definitions of habitats can be made, studies of the oral microflora should always include careful definition of the habitats being examined.

Generally, the oral cavity includes local habitats, such as the mucosa, tongue, saliva, and tooth surfaces. Included in the microenvironments are local areas where hard surfaces and soft tissues are juxtaposed, such as the gingival sulcus and periodontal pockets.

FORMATION OF PLAQUE BIOFILMS

Role of saliva in formation of pellicle and plaque

The saliva contains many different proteins, some of which are small organic proteins that together protect the oral cavity (the soft tissues as well as the teeth) from frictional wear, dryness, erosion, pathogenic bacteria, and other insults. Almost all salivary proteins are glycoproteins; ie, they contain variable amounts of carbohydrates linked to the protein core. Glycoproteins are often classified according to their cellular origin and subclassified on the basis of their biochemical properties.

A characteristic feature of glycoproteins is that many occur in multiple forms, constituting families, which may, however, exhibit remarkable functional differences. The mucins are produced by the minor salivary glands in the palate, and they provide a nonfrictional, lubricant layer, protecting the soft tissues from wear and tear and facilitating the swallowing of food. Because the mucins have a strongly negative charge, other negatively charged molecules, such as those contained in the cell walls of many oral bacteria, are repelled from the mucin-coated oral mucosa. Among other properties, the mucins also bind water and thereby protect the oral mucosa from drying out.

Serous glycoproteins have a much lower molecular weight than do mucins and contain less than 50% carbohydrate. Many belong to a group called *proline-rich glycoproteins* (PRPs). These proteins are secreted from the parotid and submandibular glands. *Glycoprotein*, the collective name for all carbohydrate-linked proteins, encompasses a large, heterogenous group. Most salivary proteins, such as secretory IgA, lactoferrin, peroxidases, and agglutinins, belong to this group.

Because human saliva is supersaturated with most calcium phosphate salts, some proteins are necessary to inhibit the spontaneous precipitation of these salts in the salivary glands and their secretions. Such proteins include statherin and PRPs. Statherin is present in both submandibular and parotid saliva. Proline-rich proteins form a complex group with large numbers of genetic variants, some of which also have the ability to inhibit spontaneous precipitation of calcium phosphate salts.

A newly proposed biologic function of adsorbed acidic PRPs is their ability to selectively mediate bacterial adhesion on tooth surfaces. The negative charge of the PRPs binds electrostatically to calcium on the tooth surfaces, while the outer ends, consisting of proline and glutamine amino acids, attract and bind very strongly to the harmless, protective normal microflora of the teeth (*Streptococcus oralis*, *Streptococcus sanguis*, and *Streptococcus mitis*). This may explain early electron micrographs by Lie (1978) that show the differences in the ways that gram-positive pioneer colonizers and gram-negative bacteria attach to the pellicle-covered tooth surface (Figs 10 and 11).

This primary colonization of the protective normal microflora occurs during the first 24 hours after cleaning. However, recent research has shown that so-called secondary colonization by other, more pathogenic microorganisms (gram-positive as well as gram-negative) is strongly related to the binding between galactose amine structures on the surfaces of the normal microflora as well as the secondary colonizers. The

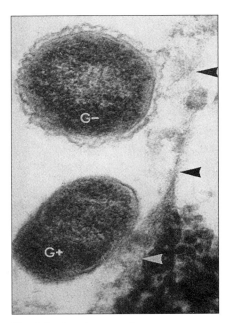

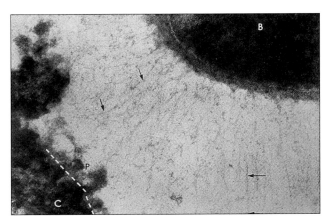

Fig 11 Closeup of the attachment (*arrows*) between a gram-positive bacterium (B) and the pellicle (P). (C) Hydroxyapatite crystals (original magnification ×150,000). (From Lie, 1978. Reprinted with permission.)

Fig 10 Attachment of a gram-positive (G+) pioneer colonizer to the pellicle-covered tooth surface (*white arrow*) in contrast to the relationship between a gram-negative bacterium (G-) and the pellicle (*black arrows*). (From Lie, 1978. Reprinted with permission.)

production and the individual structures of acidic PRPs and galactose amines are genetically related and may partly explain individual variations in plaque formation rates. This is a field of ongoing research (Strömberg, 1996).

As described earlier, saliva plays a significant role in maintaining an appropriate balance within the ecosystem associated with tooth surfaces. This fact is of great importance in the control of the oral microflora, because saliva enhances the ability of some bacteria to survive and reduces the competitiveness of others. This control over the oral flora is achieved by salivary components that may be present constantly or activated by a specific host response.

The major antimicrobial proteins are listed in Box 1. Although most of these proteins can inhibit the metabolism, adherence, or even the viability of pathogenic microorganisms in vitro, their role in vivo is largely unknown. It seems that they are important for the control of microbial overgrowth in the mouth, but their selectivity against pathogens has not been determined.

Lysozyme in whole saliva is derived from the major and minor salivary glands, gingival crevicular fluid, and salivary leukocytes. Salivary lysozyme is present in newborn babies at levels equal to those of adults, suggesting a preeruptive antimicrobial function. The classic concept of the antimicrobial action of lysozyme is based on its muramidase activity, ie, the abilty to hydrolyze the bond between *N*-acetyl muramic acid and *N*-acetyl glucosamine in the peptidoglycan layer of the bacterial cell wall. Gram-negative bacteria are more resistant to lysozyme because of the protective function of the outer lipopolysaccharide layer. In addition to its muramidase activity, lysozyme is strongly cationic and can activate bacterial autolysins, which can destroy the components of the cell wall.

Lactoferrin is an iron-binding glycoprotein secreted by the serous cells of the major and minor salivary glands. Polymorphonuclear leukocytes (PMNLs) are also rich in lactoferrin and release it into gingival fluid and whole saliva. The biologic function of lactoferrin is attributed to its high affinity for iron and its consequent expropriation of this essential metal from pathogenic microorganisms. If the lactoferrin molecule is saturated with iron, this bacteriostatic effect is lost, a factor that should be taken into account in areas where the drinking water is rich in iron. In its iron-free state, lactoferrin (apo lactoferrin) also has a bactericidal, irreversible effect on a variety of microorganisms.

Salivary peroxidase is produced in the acinar cells of the parotid and submandibular glands but not in the minor salivary glands. Salivary peroxidase systems have two major biologic functions: (1) antimicrobial activity and (2) protection of host proteins and cells from hydrogen peroxide toxicity.

Salivary agglutinins are glycoproteins that have the capacity to interact with unattached bacteria, resulting in clumping of bacteria into large aggregates that are more easily flushed away by saliva and swallowed; the term *aggregation* is therefore often used synonymously with *agglutination*. Box 1 lists salivary proteins with agglutinating capacity.

The secretory immunoglubulins, most notably sIgA, act by aggregating bacteria. They target specific bacterial molecules, such as adhesins, or enzymes, such as glucosyl transferase. Saliva also contains IgG and IgM, which are derived from serum and produced locally in the gingival tissues.

Formation and functions of pellicle

Saliva is seldom in direct contact with the tooth surface, but separated from it by the acquired *pel-*

Box 1 Major antimicrobial proteins of human whole saliva*

Nonimmunoglobulin proteins
- Lysozyme
- Lactoferrin
- Salivary peroxidase system (enzyme-SCN^--H_2O_2)
- Myeloperoxidase system (enzyme-SCN^-/halide-H_2O_2)
- Agglutinins
  - Parotid saliva glycoproteins
  - Mucins
  - $Beta_2$-microglobulin
  - Fibronectin
- Histidine-rich proteins (histatins)
- Proline-rich proteins

Immunoglobulins
- Secretory immunoglobulin A
- Immunoglobulin G
- Immunoglobulin M

*From Tenovuo and Lagerlöf (1994).

---

licle, defined as an acellular layer of salivary proteins and other macromolecules approximately 2 to 10 µm thick, adsorbed onto the enamel surface. Pellicle forms a base for subsequent adhesion of microorganims, which, under certain conditions, may develop into dental plaque biofilms. The pellicle layer, although thin, has an important role in protecting the enamel from abrasion and attrition, but it also serves as a diffusion barrier.

The undisturbed pellicle is formed in different layers (Fig 12). There are many nonattaching bacteria close to the outer surface of the pellicle. Because of abrasion from, for example, toothbrushing, the thickness will vary between 2 and 10 µm, depending on the intervals between brushing.

Saxton (1976) showed that complete removal of the pellicle requires about 5 minutes' pumicing with a rotating rubber cup. Figure 13 shows a groove in the pellicle down to the enamel surface, made by a knife. Such grooves were earlier thought to be abrasive defects in the enamel surface resulting from abrasive toothpaste. In Fig 14 the pellicle was removed by intensive cleaning for about 5 minutes. The pellicle-free enamel surface

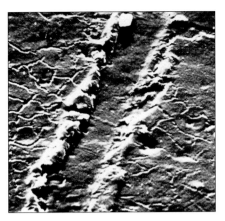

Fig 13 Groove in the pellicle, down to the enamel surface. The groove was made with a knife. (From Saxton, 1976. Reprinted with permission.)

Fig 12 Undisturbed pellicle in cross section *(left)*. It is formed in different layers. Many nonattaching bacteria *(right)* are close to the outer surface of the pellicle. (From Saxton, 1976. Reprinted with permission.)

was partly covered with nail varnish; the outer part was exposed to saliva in vivo for several hours. Figure 14 shows the thickness of the new pellicle compared to the naked enamel surface after removal of the nail varnish.

In the pellicle, movement of molecules by forces other than diffusion is low compared to most other parts of the salivary film. The relatively undisturbed layer of liquid in the pellicle will certainly influence the solubility behavior of the enamel surface. Adsorption of macromolecules, usually originating from the saliva, to the enamel is selective; ie, certain macromolecules show a higher affinity for the mineral surface than do others.

In the normal oral pH range, the enamel surface has a negative net charge because of the structure of hydroxyapatite, in which phosphate groups are arranged close to the surface. Counterions (of opposite charge), eg, calcium, are attracted to the surface, forming a hydration layer of unevenly distributed charges. The exact composition of this layer will be determined by several factors, eg, pH, ionic strength, and the types of ion present in the saliva. Because calcium predominates, the resulting net charge of the enamel surface with its hydration layer is positive, implying that the hydration layer will attract negatively charged macromolecules.

Negative charges on macromolecules are found in acidic side chains with end groups of phosphate or sulfate. These side chains have a high affinity for the tooth surface. Recent research has shown that the bulk of the pellicle consists of salivary micellelike structures of great importance for reducing diffusion through the pellicle and reducing friction between the teeth and other oral tissues. Pellicle formation is rapid during the first hour, then decreases. It seems likely that adsorption of the first layer of molecules to a clean surface is instantaneous. The formation rate varies

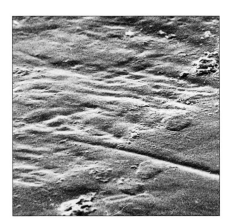

Fig 14 The groove shown in Fig 13 was removed by 5 minutes' intensive cleaning. The pellicle-free enamel surface was partly covered with nail varnish (*left*), while the other part (*right*) was exposed to saliva for several hours. The surface is shown after removal of the nail varnish. (From Saxton, 1976. Reprinted with permission.)

among individuals, probably because of differences in salivary composition.

Formation and composition of supragingival plaque biofilms

Dawes et al (1963) described dental plaque as "the soft tenacious material found on tooth surfaces which is not readily removed by rinsing with water." Nolte (1973) defined dental plaque as "the non-mineralized microbial accumulation that adheres tenaciously to tooth surfaces, restorations, and prosthetic appliances, shows structural organization with predominance of filamentous forms, is composed of an organic matrix derived from salivary glycoproteins and extracellular microbial products, and cannot be removed by rinsing or water spray."

The most readily discernible plaque on the smooth surfaces of the teeth, which occurs along the gingival margin, is termed *dentogingival plaque*. Dentogingival plaque on the approximal surfaces, apical to the contact points, is called *approximal dental plaque*. Plaque occurring below the gingival margin, in the gingival sulcus or in the periodontal pocket, is known as *subgingival plaque* (Theilade and Theilade, 1976). So-called occlusal or fissure plaque may also form, particularly in erupting molars.

It is estimated that only 1 mm^3 of dental plaque, weighing about 1 mg, contains more than 200 million bacteria. Other microorganisms, such as mycoplasma, "yeasts," and protozoa, also occur in mature plaque; sticky polysaccharides and other products form the so-called plaque matrix, which constitutes 10% to 40% by volume of the supragingival plaque. Studies of the composition and structure of dental plaque describe how the components, predominantly bacteria, are organized and interrelated. Recent studies have shown that mature dental plaque is a well-organized society. The dentogingival plaque is therefore regarded as a *biofilm*, defined as "matrix-enclosed bacterial populations adherent to each other and/or to surfaces or interfaces" (Costerton et al, 1994).

Phases of plaque growth

The first condition for plaque formation is a solid surface. In the oral cavity, only the pellicle-covered tooth surfaces offer such conditions. Although microorganisms may attach to the outer surface of the epithelial cells of the soft tissues of the oral cavity, plaque cannot form, because the surface layers of epithelial cells are continuously desquamated. At best, the microorganisms on desquamated epithelial cells may be transported by the oral fluid, eventually attaching to a solid tooth surface; only then can colonization occur and plaque formation begin.

Figure 15 is a scanning electron micrograph of the buccal enamel surface of a newly erupted and

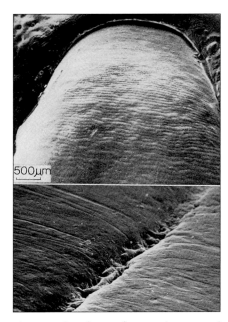

Fig 15 (*top*) Facial enamel surface of a newly erupted and cleaned permanent central incisor. The perikymata are still discernible. (*bottom*) Increased magnification of the view. (From Saxton, 1975. Reprinted with permission.)

Fig 16 Formation of bacterial colonies (original magnification ×4,000). (From Saxton, 1975. Reprinted with permission.)

cleaned permanent central incisor tooth. The perichymata are still discernible. The first cells to adhere to pellicle on tooth surfaces or other solid surfaces are coccoid bacteria, epithelial cells, and polymorphonuclear leukocytes; the bacteria occur singly or as aggregates, either on or within the pellicle. Larger numbers of microorganisms may be carried to the tooth surface by epithelial cells, as described earlier. During the first few hours, bacteria that resist detachment from the pellicle may start to proliferate, forming small colonies of morphologically similar organisms (Fig 16).

Plaque growth may also be initiated by microorganisms harbored in minute irregularities, such as grooves in tooth surfaces, the margins of restorations, the cementoenamel junction, and the gingival sulcus, where they are protected from natural cleaning of the tooth surface. As early as 1975, Saxton showed that gingival crevicular fluid (GCF) released from inflamed gingiva increases reaccumulation of bacterial colonies and plaque

immediately after toothcleaning because GCF is rich in nutrients not readily available in the saliva. Figure 17 shows bacterial colonies and seepage of GCF from the gingival sulcus. The largest colonies are located close to the GCF (Fig 18).

Four hours after cleaning, there are 10^3 to 10^4 bacteria per 1 mm^2 of tooth surface (Nyvad and Kilian, 1987), predominantly streptococci and actinomyces. Within a day, the number of bacteria increases 100- to 1,000-fold, mainly because of the growth of streptococci.

The initial bacteria are called *pioneer colonizers*, because they are hardy and successfully compete with the other members of the oral flora for a place on the tooth surface (Gibbons and van Houte, 1980). The deposition of these pioneer species is not a chance occurrence, but the result of an exquisitely sensitive interaction between protein adhesins on the surface of the colonizing bacteria and carbohydrate receptors on the salivary components adsorbed to the tooth surface.

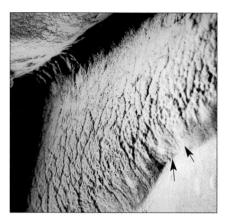

Fig 17 Relationship between bacterial colonies (*arrows*) and seepage of gingival crevicular fluid from the sulcus (original magnification ×200). (From Saxton, 1975. Reprinted with permission.)

Fig 18 The largest bacterial colonies (*arrows*) are located close to the gingival crevicular fluid (original magnification ×4,000). (From Saxton, 1975. Reprinted with permission.)

After initial deposition, clones of pioneer colonizing bacteria, *S sanguis*, begin to expand away from the tooth surface to form columns that move outwardly in long chains of pallisading bacteria. These parallel columns are separated by uniformly narrow spaces. Plaque growth proceeds by deposition of new species into these open spaces (Listgarten et al, 1975) (Fig 19).

These newly deposited species attach to pioneer species in a specific molecular lock-and-key manner. Expansion of existing species in a lateral direction causes the interbacterial spaces to merge. It is hypothesized that, once the spaces are close enough, a starter substance is secreted by bacteria within the plaque matrix, stimulating a growth spurt in the surrounding bacteria. The tooth surface adjacent to the gingiva is rapidly covered by intermeshed bacteria. New bacteria, derived from saliva or surrounding mucous membranes, now sense only the bacteria-laden landscape of the tooth surface and attach by a bonding interaction to bacteria already attached to the plaque. These associations, called *intergeneric coaggregations*, are mediated by specific attachment proteins that occur between two partner cells. All this activity occurs within the first 2 days of plaque development and for descriptive

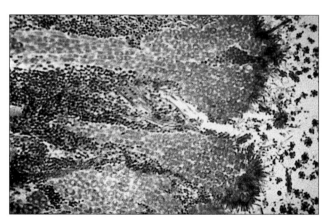

Fig 19 Cross section of columns of colonizing bacteria, separated by open spaces. (From Listgarten et al, 1975. Reprinted with permission.)

purposes is called *phase I* of plaque formation (Theilade et al, 1976).

After 24 to 48 hours, continuous plaque has formed along the gingival margin (Fig 20). The plaque is dominated by streptococci and a few rods. Figures 21 to 23 show 2 days' free plaque accumulation in a caries-free young adult without gingivitis. Even in such a healthy mouth, there has been continuous plaque accumulation at so-

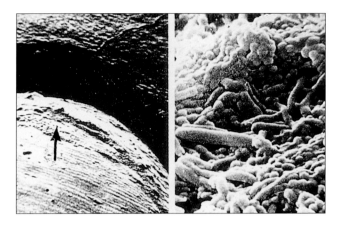

Fig 20 *(left)* Plaque formation along the gingival margin after 24 to 48 hours *(arrow)* (original magnification ×100). *(right)* Increased magnification revealing that the plaque is dominated by streptococci and a few rods (original magnification ×4,000). (From Saxton, 1973. Reprinted with permission.)

Figs 21 to 23 Two days' free plaque accumulation in a caries-free young adult without gingivitis. Plaque has accumulated in stagnant zones.

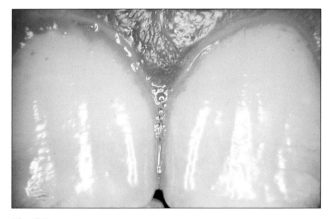

Fig 21

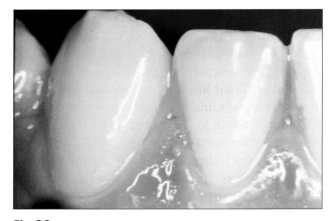

Fig 22

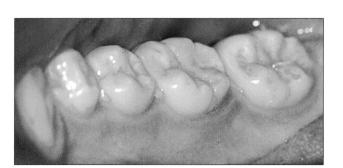

Fig 23

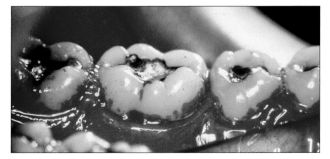

Fig 24 Gingivitis at mandibular lingual and interproximal sites in another young adult. Much more plaque has accumulated in less than 2 days in this individual than in the patient shown in Figs 21 to 23.

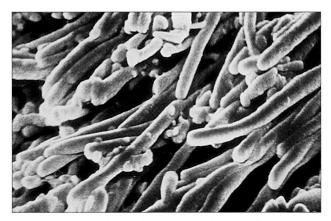

Fig 25 Outer surface of plaque in phase II of plaque development, covered by gram-positive tall rods (original magnification ×8,000). (From Saxton, 1973. Reprinted with permission.)

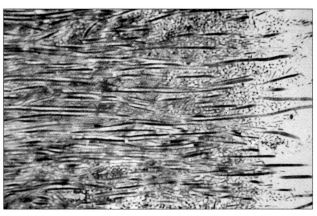

Fig 26 Filamentous bacteria on the surface of predominantly coccoid plaque during phase II of plaque development (original magnification ×1,500). (From Listgarten et al, 1975. Reprinted with permission.)

called stagnant zones (interdentally and along the gingival margins), ie, sites at which plaque can accumulate undisturbed by friction from chewing and tongue movements. Figure 24 shows gingivitis at mandibular lingual and interdental sites in another young adult, in whom much more plaque reaccumulated in less than 2 days.

During the first 2 days of plaque formation, the dentogingival plaque is dominated by the relatively "harmless" normal microflora of the tooth surface, consisting of facultatively anaerobic gram-positive streptococci (*S sanguis* and *S mitis*) and a minority of gram-positive rods (*Actinomyces* species), which may impede infiltration of more pathogenic microorganisms.

In phase II, from around day 3, filamentous bacteria can be found on the surface of the predominantly coccoid plaque (Figs 25 and 26).

In phase III, competitive growth among the predominantly coccoid microbial colonies continues for about 1 week. Filamentous bacteria then begin to penetrate the coccoid plaque from the surface, and it gradually becomes predominantly filamentous. The process may continue for about 2 more weeks. The columnar microbial colonies are replaced by a dense mat of filamentous bacteria, oriented roughly perpendicular to

the colonized surface (Figs 27 and 28). Several species of coccoid bacteria are able to aggregate with some of the filamentous bacteria to produce "corncob" formations that are detectable on the plaque surface in this phase, particularly along the gingival margin (Figs 29 and 30). These may persist on the plaque surface along the gingival margin and can be identified in supragingival plaque samples known to be 3 weeks old.

In phase IV, 1 to 2 weeks after initiation, the diversity of the flora has increased to include motile bacteria, spirochetes, and vibrios, as well as fusiforms. Attached gingival plaque fills the gingival sulcus, while spirochetes and vibrios move around in the outer and more apical regions of the sulcus (Fig 31). Gradually the gingival margin becomes inflamed, and there is an increase in volume of GCF. In the depths of the plaque that fills the sulcus, minerals from the GCF accumulate, forming black calculus (Fig 32). In addition, the anaerobic environment in the depths of such thick, mature plaque leads to a shift toward gram-negative anaerobic microflora, ie, the creation of growth conditions favorable to most periopathogens.

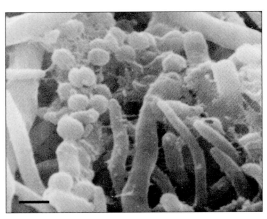

Fig 27 Dense mat of filamentous bacteria, oriented roughly perpendicular to the colonized surface, in phase III of plaque formation. (From Listgarten et al, 1975. Reprinted with permission.)

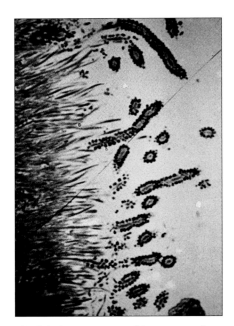

Fig 28 Surface of phase III plaque (bar = 5μm). (From Adriaens et al, 1988b. Reprinted with permission.)

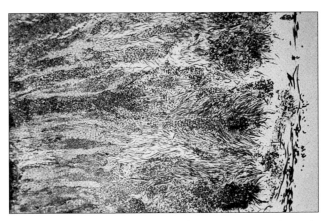

Fig 29 Corncob formation of coccoid and filamentous bacteria in phase III of plaque formation (original magnification ×8,000). (From Saxton, 1973. Reprinted with permission.)

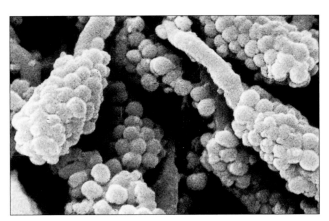

Fig 30 Cross section of the outer surface of gingival plaque, containing several corncob formations (original magnification ×1,000). (From Listgarten, 1976. Reprinted with permission.)

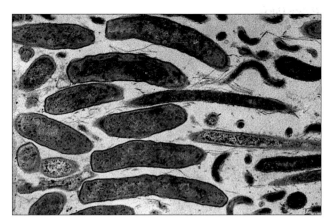

Fig 31 Cross section of gingival plaque filling the gingival sulcus while spirochetes and vibrios move around in the outer and more apical regions of the sulcus during phase IV of plaque development (original magnification ×12,000). (From Listgarten, 1976. Reprinted with permission.)

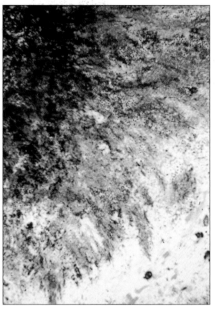

Fig 32 Accumulation of minerals in the deeper part of the plaque, resulting in calculus formation (original magnification ×400). (From Listgarten, 1976. Reprinted with permission.)

The biofilm concept

Mature dental plaque is a microbial biofilm. Although first described some years ago, more recent research into their molecular organization, physiochemical properties, and growth characteristics has led to the concept of biofilms as ecological communities that evolved to permit survival of the community as a whole (Costerton et al, 1995). Evidence from a variety of sources is consistent with the concept of coevolution of a mixed microbial community to live in proximity to the tooth surface, above and below the gingival magin. This is supported by the presence of a primitive circulatory system and metabolic cooperation, but there is as yet no published documentation of mutational analysis and structure-function analysis, in vitro or in vivo (Fig 33).

Although the capacity to attach plays a key role in the variety of different microbial species that constitute the dental plaque biofilm, bacterial growth is the primary determinant of the relative proportions of the bacteria. Plaque doubling times are rapid in early development and slower in more mature films because of the structure of the biofilms, which contain areas of high and low bacterial biomass, interlaced with aqueous channels of different sizes (see Fig 33). It is believed that these channels transport nutrients and metabolic waste products within the colony. For example, dissolved oxygen in living biofilms was measured to be at nearly anoxic values within microcolonies but at significant concentrations at all levels in cell-free aqueous channels (Costerton et al, 1995). This structure provides a means by which the different bacterial species can benefit from their juxtaposition, facilitating a type of physiologic cooperation not seen in mixed populations of planktonic organisms. The unique conditions of physiologic cooperation in biofilms may influence the occurrence of bacterial blooms (periods of rapidly accelerated growth of specific species or groups of species) described in periodontal plaque. Further investi-

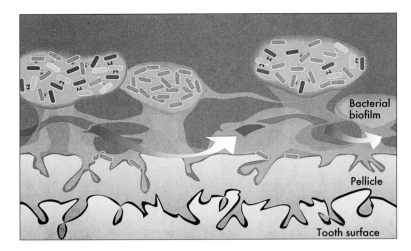

Fig 33 Schematic illustration of a plaque biofilm. The pellicle and biofilm extracellular matrix are depicted as firmly embedded in each other, contributing to the well-known recalcitrance of dental plaque. The unique shape of biofilms are believed to facilitate growth and symbiosis among the microbiota. The *large arrows* depict solvent flow that occurs through both large and small aqueous channels that are believed to carry nutrients and metabolic products to different members of the community. (From Darveau et al, 1997. Reprinted with permission.)

gation is required to define the factors that influence growth of selected microcolonies in dental plaque biofilms.

Amount of Plaque

Measurement of plaque

The presence and/or the amount of supragingival plaque is usually measured by one of the established plaque indices, which have been used for many years to record oral hygiene standards in epidemiologic studies, clinical trials, and clinical practice. Assessment may be based on the presence or absence of plaque at certain sites, the area or thickness of plaque, or gravimetric measurement. Recordings are usually made from the exposed buccal and/or lingual surfaces of the teeth, although interproximal determinations can be attempted by supplementary probing. Disclosing agents may be applied to reveal the colorless plaque to the examiner and, particularly, to the pa-

tient. Although simplified indices are often used in epidemiologic studies, complete-mouth recordings are strongly recommended in clinical trials and clinical practice.

Plaque Index

The plaque index most extensively applied in clinical trials during the last few decades is that originally described by Silness and Löe (1964), which is based on estimated measurements of plaque by examination of the whole or parts of the dentition. It has been applied to studies in children as well as adults and is reliable for evaluating both mechanical and chemical plaque control. Each of the four gingival areas of the tooth is given a score from 0 to 3; this is the Plaque Index (PI) for the area. The scores from the four sites on the tooth may be added and then divided by 4 to give the PI for the tooth. The scores for individual teeth (incisors, premolars, and molars) may be grouped to designate the PI for groups of teeth. The Plaque Index for the individual is obtained by adding the

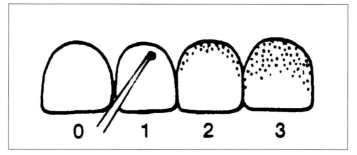

Fig 34 Silness and Löe Plaque Index (1964): 0 = The tooth surface is clean; 1 = The tooth surface appears clean, but dental plaque can be removed from the gingival third with a sharp explorer; 2 = Visible plaque is present along the gingival margin; 3 = The tooth surface is covered with abundant plaque.

area scores for each tooth and dividing by the number of teeth examined.

Each gingival area is scored as follows (Fig 34):

- Score 0 = The gingival area of the tooth surface is literally free of plaque. The tooth is dried, and a sharp probe is run across the tooth surface at the entrance to the gingival crevice. If no soft matter adheres to the point of the probe, the area is considered clean.
- Score 1 = No plaque can be observed in situ by the naked eye, but plaque is visible on the point of a probe drawn across the tooth surface at the entrance to the gingival crevice. Disclosing solution was not used in the original investigation but may be useful for recognizing this film of plaque.
- Score 2 = The gingival area is covered by a thin to moderately thick layer of plaque, visible to the naked eye.
- Score 3 = Heavy accumulation of soft matter fills the niche produced by the gingival margin and the tooth surface. The interdental area is full of soft debris.

The PI scores record only differences in the thickness of the soft deposits on the gingival area of the tooth surfaces and not the coronal extent of the plaque. Plaque formation on calculus de-

posits, restorations, and crowns is assessed. A major criticism of the Silness and Löe index is subjectivity in estimating plaque, which becomes apparent when several examiners are participating in a study. It is therefore recommended that a single examiner be trained and used with each group of patients throughout a clinical trial. However, the Plaque Index is not linear, and nonparametric methods are therefore necessary for analysis of the data. It is essential that not only the actual plaque scores be recorded, but also the changes in the scores through the course of the study be monitored.

The Silness and Löe PI has frequently been used in longitudinal plaque control studies to evaluate the correlation between PI and Gingival Index (GI) according to Löe and Silness (1963). Strong correlations have been found at group, individual, and tooth surface levels (Ainamo, 1970; Axelsson and Lindhe, 1974, 1977).

Plaque control record

A very simple and therefore reliable method for evaluating oral hygiene procedures was proposed by O'Leary and coworkers (1972). The disclosed plaque accumulations on all teeth are scored dichotomously. Four or six surfaces per

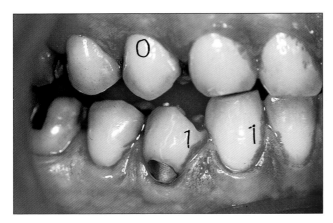 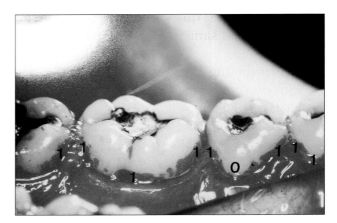

Figs 35 and 36 Examples of scoring with the Plaque Control Record (O'Leary et al, 1972). Plaque is scored as present (1) or absent (O) on four or six surfaces per tooth. The number of positively scored surfaces is divided by the total number of surfaces evaluated, and the result is multiplied by 100 to obtain a percentage.

tooth are recorded. The number of positively scored units is divided by the total number of tooth surfaces evaluated, and the result is multiplied by 100 to express the index as a percentage. With this method, the topographic distribution of plaque throughout the dentition can be readily assessed. In daily practice, repeated scoring facilitates evaluation of the efficacy of oral hygiene programs.

Figures 35 and 36 exemplify some scores according to O'Leary et al (1972). This index is particularly useful in clinical practice for monitoring a patient's standard of oral hygiene and as a basis for individual education in self-care. However, it only reveals areas that the patient has failed to clean effectively, despite a special effort on the day of the dental appointment; it does not indicate the rate at which plaque forms in the individual, or the oral hygiene status 1 week before or after the dental appointment (for reviews, see Barnes et al, 1986; Fischman, 1986; and Lang, 1998).

Pattern of remaining plaque in toothbrushing populations

The pattern and quantity of plaque remaining after people "clean their teeth" will vary widely among populations and individuals, depending on factors such as the oral hygiene aids being used, differences in instruction in self-care, and socioeconomic conditions. The habit of brushing daily is generally well established, whereas interproximal cleaning is not (Kuusela et al, 1997).

Specific individual patterns of plaque accumulation have been recognized in young individuals who have had no special instruction in home care (Cumming and Löe, 1973) and have been attributed to patterns in the quality of local plaque control. Although the pattern was unique to each of the subjects in the study, a general pattern for the whole group emerged. Although some regions of the dentition were sometimes plaque free and sometimes covered by plaque, other areas in the mouths of all subjects were either consistently clean or consistently covered by plaque. The pattern of remaining plaque, based on the Silness and Löe (1964) PI, is shown in Fig 37.

The lowest frequency of plaque deposits was recorded for the facial surfaces. Plaque was slightly more common on mandibular teeth than on maxillary teeth and far more frequent on molars than on incisors and premolars. The pattern was repeated for mandibular surfaces: Remaining plaque was observed more frequently on lingual than on facial aspects. Interproximal surfaces harbored the greastest quantities of plaque, and the distribution was similar to that of the other surfaces.

Similar findings were made in a study of the patterns of plaque removal in students and faculty members of a dental school (Lang et al, 1977): Oral cleanliness, PI, and gingival health (GI) were assessed in 150 dental students and 101 faculty members. Analysis of plaque distribution showed that posterior teeth consistently had more plaque than did anterior teeth, and the heaviest deposits were interproximal. Furthermore, a positive relationship was found between oral hygiene habits and the subjects' academic status and involvement in clinical dentistry.

Although mean plaque scores observed by examining a number of representative tooth surfaces might identify the need for improved oral hygiene in general, they fail to identify specific potential problem regions in subjects with better-than-average oral hygiene standards. Because each individual appears to perform plaque control in specific patterns, the clinician has to identify inadequacies in personal oral hygiene programs by serial registrations of plaque deposits on single surfaces. Generally, in undisturbed plaque accumulation, the heaviest deposits seem to form interproximally and on the lingual mandibular surfaces. Because this is also the case in the average patient without special home care instruction (Cumming and Löe, 1973), it is obvious that special attention has to be paid to interdental cleansing when patients are being instructed in self-care.

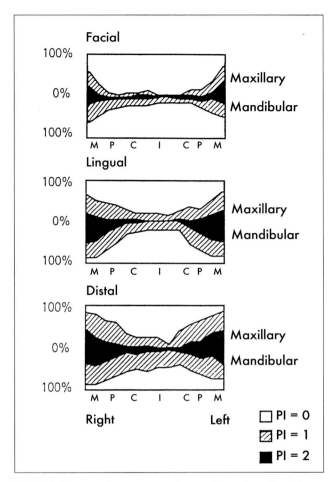

Fig 37 Pattern of remaining plaque, based on the Silness and Löe PI. The frequency of plaque deposits is lowest on the facial surfaces and greatest at interproximal sites. (M) Molar; (P) Premolar; (C) Canine; (I) Incisor. (Modified from Cumming and Löe, 1973.)

FORMATION RATE AND REACCUMULATION OF PLAQUE

Measurement of formation rate

The amount and location of plaque recorded at clinical examination discloses only where the patient has been unsuccessful in cleaning, despite extra effort on the day of the dental appointment. It does not reveal the age of the plaque, the rate of accumulation, or the patient's usual standard of oral hygiene. It is therefore important to appreciate the difference between plaque indices, which are static, and the plaque reaccumulation rate, which is dynamic. An understanding of plaque formation rates and patterns is an essential foundation of successful strategies for primary and secondary prevention of periodontal diseases as well as dental caries.

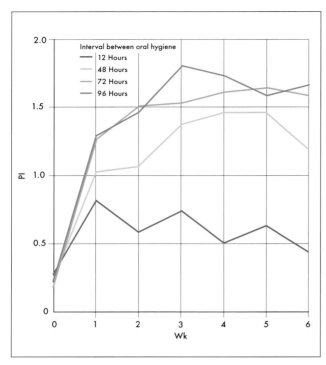

Fig 38 Levels of plaque accumulation in dental students performing oral hygiene measures at differing intervals. The Silness and Löe (1964) PI was significantly lower after 12 hours than it was after any of the other time intervals. (From Lang et al, 1973. Reprinted with permission.)

cumulated on 70% to 100% of the approximal surfaces and 90% to 100% of the lingual surfaces of the mandibular molars and premolars, to a visible plaque score of 2 on the Silness and Löe (1964) scale (Lang et al, 1973).

The aim of the clinical study by Lang et al (1973) was to assess the rate and pattern of plaque development in 32 dental students with excellent oral hygiene and clinically healthy gingival conditions. Four groups of volunteers performed complete plaque removal every 12, 48, 72, or 96 hours. During a period of 6 weeks, PI (Silness and Löe, 1964) was recorded immediately before cleaning. Figure 38 illustrates the different levels of plaque accumulation in the four groups: The mean level of undisturbed plaque accumulation after 12 hours was significantly lower than that after 48, 72, or 96 hours.

Plaque Formation Rate Index

The quantity of plaque that forms on clean tooth surfaces during a given time represents the net result of interactions among etiologic factors, many internal and external risk indicators and risk factors, and protective factors (Box 2). This observation has been the rationale for the con-

According to the nonspecific plaque hypothesis, mechanical removal of dental plaque, being causally directed, is a rational method for prevention and control of periodontal diseases and dental caries. However, for cost effectiveness, cleaning should be related to the pattern of plaque reaccumulation and otherwise predicted risk. In the past three decades, questions regarding the rate of plaque growth and its pattern of development on the dentition have been addressed by several investigators.

After mechanical removal, plaque slowly reaccumulates along the gingival margins of the teeth over the following 2 days. After the second day, the thickness increases dramatically, reaching a maximum after 7 days (Lang et al, 1973; Listgarten et al, 1975; Löe et al, 1965; Saxton, 1973). Forty-eight hours after toothcleaning, plaque has reac-

Box 2 Factors affecting plaque reaccumulation

- The total oral bacterial population
- The quality of the oral bacterial flora
- The anatomy and surface morphology of the dentition
- The wettability and surface tension of the tooth surfaces
- The salivary secretion rate and other properties of saliva
- The intake of fermentable carbohydrates
- The mobility of the tongue and lips
- The exposure to chewing forces and abrasion by foods
- The eruption stage of the teeth
- The degree of gingival inflammation and volume of gingival exudate
- The individual's oral hygiene habits
- The use of fluorides and other preventive products, such as chemical plaque-control agents

struction of the Plaque Formation Rate Index (PFRI) by Axelsson (1987, 1991). It includes all but the occlusal tooth surfaces and is based on the amount of plaque freely accumulated (de novo) in the 24 hours following professional mechanical toothcleaning (PMTC), during which period subjects refrain from all oral hygiene practices. In a pilot study on 50 adult subjects, adherent plaque was disclosed on 5% to 65% of the total number of tooth surfaces (for details on materials and methods, see Axelsson 1987, 1991). On the basis of this study, the following five-point scale was constructed for the PFRI:

- Score 1 = 1% to 10% of surfaces affected: very low
- Score 2 = 11% to 20% of surfaces affected: low
- Score 3 = 21% to 30% of surfaces affected: moderate
- Score 4 = 31% to 40% of surfaces affected: high
- Score 5 = More than 40% of surfaces affected: very high

In 1984, PFRI was evaluated in a cross-sectional study of 667 schoolchildren aged 14 years in the city of Karlstad, Sweden. The subjects were followed over a 5-year period, up to the age of 19 years (Axelsson, 1987, 1991).

Many indicators and factors possibly related to PFRI were also evaluated, including caries prevalence and caries incidence; gingival inflammation; PI; dietary intake during the 24 hours of free plaque accumulation; salivary levels of *S mutans* and glucosyl transferase; agglutinin levels in resting saliva; and oral hygiene, dietary, and fluoride habits.

Figure 39 shows the frequency distribution of PFRI scores among the 14-year-old schoolchildren. Most were low (score 2 = 48%) or moderate (score 3 = 27%) plaque formers. Among schoolchildren in Karlstad, the standard of oral hygiene is very high, and, as a consequence, the gingival health status is excellent and the caries prevalence is low.

Among other observations from the study were:

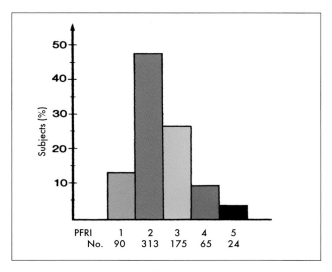

Fig 39 Stratification of PFRI in 14-year-old schoolchildren in Karlstad, Sweden. (From Axelsson, 1987, 1991. Reprinted with permission.)

1. Individuals with a PFRI score of 4 or 5 had considerably higher scores for gingival bleeding than did those with a score of 1 or 2.
2. An initially high PI score usually correlated with PFRI scores of 3 to 5.
3. There was no significant correlation between different salivary *S mutans* levels and PFRI scores.
4. The level of salivary glucosyl transferase was lower in individuals with a PFRI score of 4 or 5 than in those with a score of 1 or 2, probably because glucosyl transferase had already accumulated in the matrix of the plaque in the high and very high plaque formers.
5. The scores for individuals with very low and low PFRI (scores 1 and 2, respectively) tended to remain constant over the 5-year period, but scores tended to vary in some individuals with scores of 3 to 5, increasing or decreasing by 1 unit.

This final observation indicates that in subjects with a score of 4 or 5, plaque formation rates can be reduced. Thorough evaluation of such subjects should identify the factors contributing to the rapid plaque formation. Needs-related preventive measures could then be introduced.

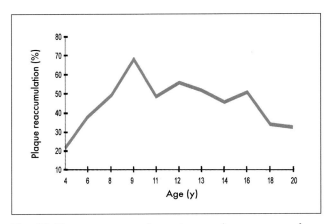

Fig 40 Plaque reaccumulation by age (percentage of surfaces with plaque). (From Cunea and Axelsson, 1997. Reprinted with permission.)

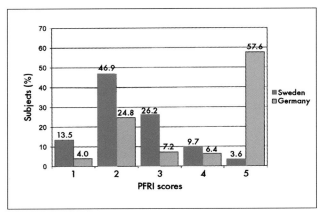

Fig 41 Comparison of PFRI scores of German and Swedish children. (From Cunea and Axelsson, 1997. Reprinted with permission.)

For example, there is a strong correlation among plaque formation rate, the severity of gingival inflammation, and the volume of gingival exudate (Axelsson, 1987, 1991; Quirynen et al, 1991; Ramberg et al, 1994, 1995a; Saxton, 1973, 1975). In individuals with a PFRI score of 4 or 5 and high GI scores, initially intensive, frequent mechanical and chemical plaque control is indicated to heal all inflamed sites as soon as possible and thereby reduce the plaque formation rate.

If the high plaque formation rate is associated with inadequate salivary secretion, frequent plaque control (before every meal) should be supplemented with salivary stimulation by using fluoride chewing gum immediately after every meal.

A high intake of fermentable carbohydrates, particularly sucrose, will result in sticky plaque, rich in polysaccharides, and an increased plaque formation rate (Carlsson and Egelberg, 1965). Needs-related prevention for individuals with a PFRI score of 4 or 5 and a frequent intake of sugar-containing products should, therefore, emphasize not only frequent plaque control but also a reduction in frequency of sugar intake. In the above study, it was observed that some individuals with a PFRI of 5 had reported a high consumption of bananas during the 24-hour period

of free plaque accumulation. Many other factors are also related to plaque formation rate.

The PFRI has recently been applied in studies on different populations and age groups. From more than 1,000 residents of Karlstad, Sweden, aged 17 to 19 years, those who were among the 30% with the highest GI score were selected to participate in a 4-month double-blind mouthrinse study. At baseline, most of the subjects had PFRI scores of 3 (more than 40%) or 4 (about 25%) (Axelsson et al, 1994b). Subjects with the highest GI scores also had the highest PFRI scores. In addition, sites with gingival inflammation had significantly higher plaque formation rates than did healthy gingival sites (Ramberg et al, 1995a).

The caries prevalence in Brazil is among the highest in the world. In São Paulo, a 3-year caries-preventive study based on self-diagnosis and self-care was conducted in 12- to 15-year-old schoolchildren. The PFRI was used as a tool for self-diagnosis and establishment of needs-related oral hygiene habits. At baseline, almost 100% of the 12-year-old schoolchildren had a PFRI score of 5. The mean percentage of surfaces with reaccumulated plaque was more than 70%, probably because of the extremely high caries prevalence, a high GI, and the presence of erupting permanent teeth. At reexamination 3 years later, the PFRI had dropped significantly: Most of the 15-year-old students had scores of 3 or 4. The main contributing factors were improvement in oral hygiene habits and gingival health and the fact that all teeth were now fully erupted (Albandar et al, 1994a; Axelsson et al, 1994a; Buischi et al, 1994).

In Duisburg, Germany, the PFRI was evaluated in different age groups of children: preschool children, children with mixed dentitions and erupting permanent teeth, and children with fully erupted teeth. Children with erupting teeth had the highest PFRI scores (Fig 40). According to the World Health Organization's Data Bank (1993), caries prevalence in German 12-year-old children is high, and the German children generally had higher PFRI scores than did Swedish children of comparable age with very low caries prevalence, excellent gingival conditions, and oral hygiene habits (Fig 41; Axelsson, 1991; Cunea and Axelsson, 1997).

Pattern of plaque reaccumulation

As discussed earlier, plaque formation rate is influenced by factors such as the anatomy and surface morphology of the teeth; the stage of eruption and functional status of the teeth; the wettability and surface tension of the tooth surfaces (both intact and restored surfaces); and gingival health and volume of gingival exudate. The pattern of plaque reaccumulation will also be influenced by these factors, but may differ somewhat on tooth surfaces exposed to chewing forces; abrasion from foods; and friction from the dorsum of the tongue, the lips, and the cheeks compared with the pattern on less accessible areas, such as approximal sites, along the gingival margin, and in irregularities such as occlusal fissures, particularly erupting molars. These areas are often designated "stagnation areas" for plaque.

In the 6-week study by Lang et al (1973), plaque reaccumulation was registered in four groups of dental students who carried out oral hygiene procedures (mechanical toothcleaning by self-care) with different frequencies: twice daily or every second, third, or fourth day. Figure 42 shows the pattern of reaccumulated plaque according to the Silness and Löe (1964) PI (scores 0 to 3). After only 12 hours of free plaque reaccumulation, there was visible plaque on some of the approximal surfaces of the molars and the lingual surfaces of the mandibular molars (score 2). After 48 hours, almost 100% of these surfaces and most of the remaining approximal surfaces had scores of 2 or 3. The pattern of visible plaque after 2 and 3 days seems to be similar, except for the facial surfaces.

According to Listgarten (1976), freely accumulated plaque is about five times thicker after 3 days than it is after 2 days (see Fig 5). This explains why gingivitis developed in the group of students cleaning only every third or fourth day but not in those who cleaned their teeth at least every second day.

Figure 43 presents the percentage of freely reaccumulated (de novo) plaque, 24 hours after PMTC, in 667 children aged 14 years in the city of Karlstad (Axelsson, 1987, 1991). Plaque reaccumulation was greatest on the mesiolingual and distolingual mandibular surfaces (33%), particularly on the molars, followed by the mesiobuccal and distobuccal surfaces on both maxillary and mandibular teeth, particularly the molars. There was almost no plaque reaccumulation (3%) on the palatal surfaces of the maxillary teeth, mainly because of friction from the rough dorsum of the tongue.

Figures 44 and 45 illustrate the percentage of de novo plaque on maxillary and mandibular tooth surfaces, respectively, 24 hours after PMTC, in young German subjects (Cunea and Axelsson, 1997). The highest percentages are found in 6- to 14-year-old subjects with many erupting teeth, on distobuccal and mesiobuccal surfaces of molars, and on distolingual and mesiolingual surfaces of mandibular molars.

In another recent study (Furuichi et al, 1992), the pattern of de novo plaque formation after 1,

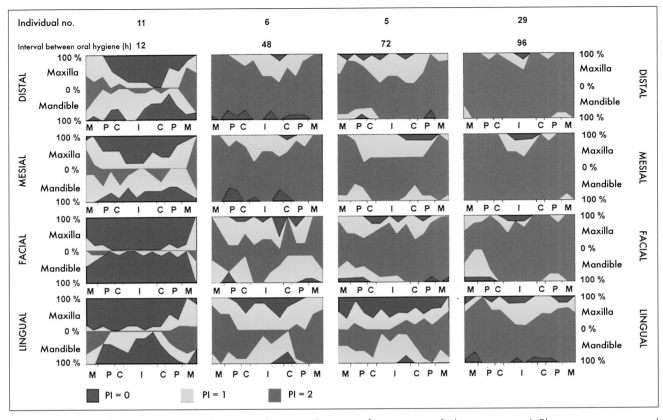

Fig 42 Patterns of plaque reaccumulation in subjects with various frequencies of plaque removal. Plaque was measured with the Silness and Löe (1964) PI. (Modified from Lang et al, 1973. Reprinted with permission.)

4, 7, and 14 days of undisturbed plaque accumulation was studied in 10 subjects, aged 24 to 29 years. At the beginning of the study, the subjects received a thorough PMTC and oral hygiene instruction. At the end of a 2-week preparatory phase, the subjects were examined to ensure that the gingivae were healthy at baseline. Mechanical toothcleaning was then discontinued, and plaque accumulation was recorded as PI (Silness and Löe, 1964). During a 2-week period without oral hygiene, most plaque formed during the first 4 days. The amount of plaque was greater on mandibular than on maxillary teeth, greatest on the approximal surfaces, and least on the palatal surfaces. These differences, observed on day 4, persisted throughout the 2-week monitoring period. These findings verified the patterns of plaque formation described by Lang et al (1973) and Axelsson (1987, 1991).

In a microbiologic study, Mombelli et al (1990) analyzed subgingival plaque samples from maxillary and mandibular right canines, premolars, and first molars (distal, midbuccal, and lingual) in 10 healthy subjects who had refrained from oral hygiene procedures for 4 days. The samples were examined by darkfield microscopy. Distobuccal samples contained more bacteria than did buccal samples, and buccal samples contained more than did lingual samples. Bacterial counts were higher in samples from posterior sites than they were in more anterior samples and significantly higher in maxillary samples than they were in mandibular samples (Fig 46). This microscopic study also appeared to confirm the distinct pattern of plaque development on clean tooth surfaces.

These findings are supported by two studies on the role and pattern of approximal mutans

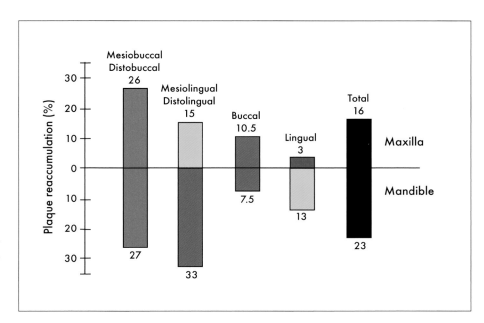

Fig 43 The pattern of plaque formation rate in 14-year-old subjects 24 hours after PMTC. (From Axelsson, 1987, 1991. Reprinted with permission.)

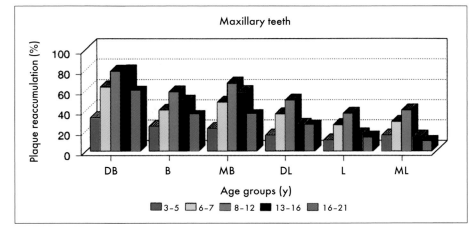

Fig 44 Percentage of new plaque accumulation related to age on distobuccal (DB), buccal (B), mesiobuccal (MB), distolingual (DL), lingual (L), and mesiolingual (ML) maxillary tooth surfaces 24 hours after PMTC. (From Cunea and Axelsson, 1997. Reprinted with permission.)

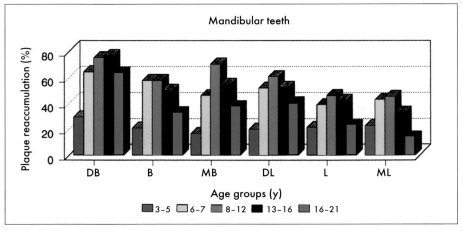

Fig 45 Percentage of new plaque accumulation related to age on distobuccal (DB), buccal (B), mesiobuccal (MB), distolingual (DL), lingual (L), and mesiolingual (ML) mandibular tooth surfaces 24 hours after PMTC. (From Cunea and Axelsson, 1997. Reprinted with permission.)

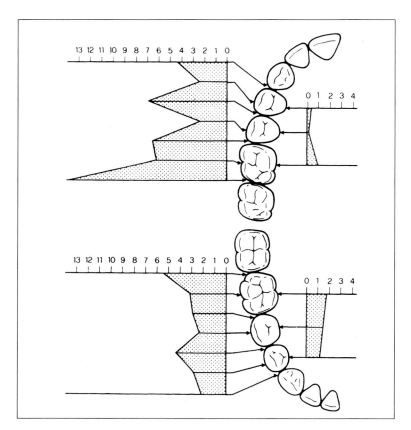

Fig 46 Bacterial counts in subgingival plaque from various intraoral sites. Numbers represent ×10^6 organisms/mL. (From Mombelli et al, 1990. Reprinted with permission.)

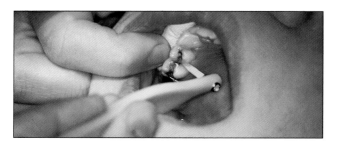

Fig 47 Interproximal samples of mutans streptococci obtained with a sterile wooden toothpick.

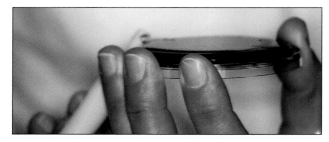

Fig 48 Sides of the toothpick contaminated with mutans streptococci are pressed on agar plates.

streptococci colonization in a 13-to-16-year-old toothbrushing population (Axelsson et al, 1987b; Kristoffersson et al, 1984). During a 30-month period, mutans streptococci was studied on all the approximal surfaces in 187 subjects, aged 13 to 16 years, with more than 1 million colony-forming units of mutans streptococci per 1 mL of saliva, selected from a population of 720 subjects aged 13

years. Every 6 months, mutans streptococci was sampled from saliva, the dorsum of the tongue, and every approximal tooth surface. Interproximal samples were obtained, as described by Kristoffersson and Bratthall (1982), with a sterile, wooden triangular pointed toothpick (Fig 47). The contaminated sides of the toothpick were then pressed directly against selective (mitis-

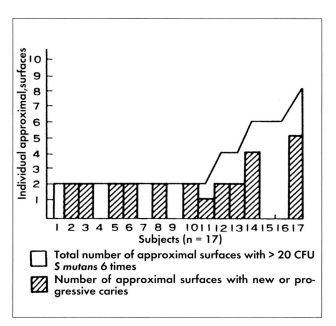

Fig 49 Development of caries lesions on surfaces highly colonized with mutans streptococci. (From Axelsson et al, 1987b.)

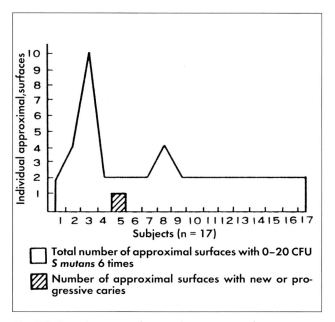

Fig 50 Development of caries lesions on surfaces sparsely colonized with mutans streptococci. (From Axelsson et al, 1987b.)

salivarius-bacitracin) agar plates (Fig 48). After incubation, the number of colonies formed was evaluated for every approximal surface.

In 17 subjects, who consistently had a minimum of one surface highly colonized with mutans streptococci and a minimum of one mutans streptococci–negative or lightly colonized surface, about 50% of the highly colonized surfaces developed caries (Fig 49). Only 3% of the mutans streptococci–negative or sparsely colonized surfaces developed caries (Fig 50) (Axelsson et al, 1987b).

In a prior study of 14-year-old children with more than 1 million colony-forming units of mutans streptococci per 1 mL of saliva, the surfaces most heavily colonized with mutans streptococci were the approximal surfaces of the molars and the second premolars (Fig 51) (Kristoffersson et al, 1984). In fact, the previously mentioned study of more than 600 subjects aged 14 years showed that the same surfaces also had the highest PFRI scores (see Fig 43). These observations explain why, in toothbrushing populations, the highest prevalence of decayed, missing, or filled surfaces

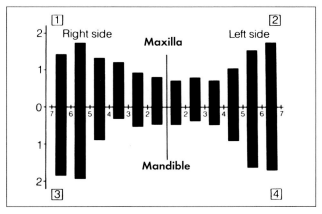

Fig 51 Colonization by mutans streptococci in 14-year-old children. The approximal surfaces of molars and second premolars are the most highly colonized, according to approximal mutans streptococci scores 0 to 3. (From Kristoffersson et al, 1984.)

is recorded on these approximal surfaces. Figure 52, data from 12-year-old residents of the county of Värmland, Sweden, shows the pattern of decayed, missing, or filled surfaces in 1964, 1974,

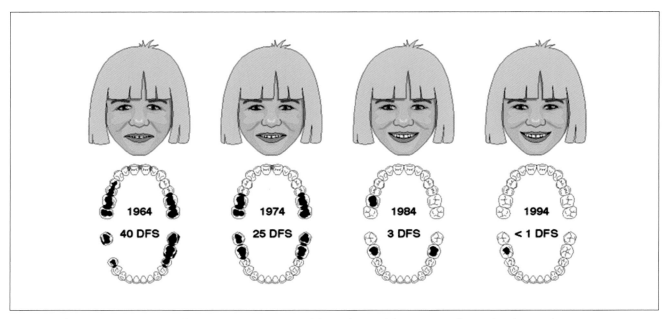

Fig 52 Change in caries prevalence among 12-year-old residents of the county of Värmland, Sweden, 1969 to 1994. (DFS) Decayed or filled surface. (From Axelsson, 1998. Reprinted with permission.)

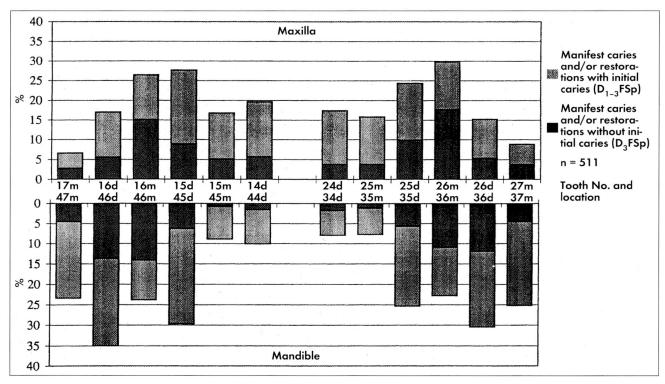

Fig 53 Mean pattern of manifest caries or restorations with or without initial caries (enamel caries) included on the posterior approximal surfaces of 19-year-old subjects (FDI tooth-numbering system). (D_1, D_2) Enamel caries lesion; (D_3) Dentinal caries; (FS) Filled surface; (p) Posterior; (m) Mesial; (d) Distal. (From Forsling et al, 1999. Reprinted with permission.)

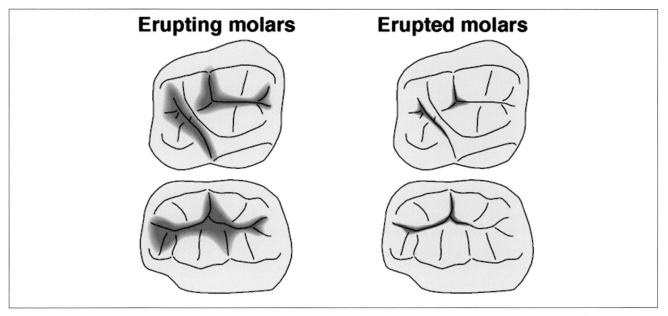

Fig 54 Reaccumulation of plaque (*shaded areas*) at 48 hours in partially and fully erupted molars. (Modified from Carvalho et al, 1989. Reprinted with permission.)

1984, and 1994. For comparison, Fig 53, data from a randomized sample of 19-year-old residents of other counties of Sweden, shows the mean pattern of manifest caries and restorations with or without incipient (enamel) caries lesions on the approximal surfaces of the posterior teeth. For optimal caries prevention in such populations, plaque control and topical application of fluorides should target these key-risk surfaces.

Carvalho et al (1989) studied the pattern and amount of de novo plaque, 48 hours after PMTC, on the occlusal surfaces of partially and fully erupted first molars. Figure 54 illustrates the heavy plaque reaccumulation, particularly in the distal and central fossae of the erupting maxillary and mandibular molars, in contrast to reaccumulation in the fully erupted molars, which are subjected to normal chewing friction. Abrasion from normal mastication significantly limits plaque formation; this explains why almost 100% of occlusal caries in molars begins in the distal and central fossae during the eruption period of 14 to 18 months.

It is important to differentiate between plaque indices and the plaque reaccumulation rate (ie, PFRI). For successful strategies for primary and secondary prevention of dental caries and periodontal diseases, an understanding of plaque formation rates and patterns is essential. Mechanical removal of dental plaque according to the nonspecific plaque hypothesis is a rational method for prevention and control of periodontal diseases as well as dental caries because it is directed toward the cause (etiology) of these diseases. However, for cost effectiveness, the program should be related to the rate and pattern of plaque reaccumulation, PFRI, and otherwise-predicted risk (for reviews on plaque formation, see Axelsson, 1994, 1998; Lang et al, 1997; Listgarten, 1994; and Straub et al, 1998).

CONCLUSIONS

Classic experimental studies in humans have shown that:

1. In the absence of oral hygiene, gingivitis develops within 2 to 3 weeks.
2. In the absence of oral hygiene, and with frequent sugar intake (rinsing with sucrose solution nine times a day), incipient enamel caries develops within 3 weeks.
3. With twice-daily plaque control, rinsing nine times a day with sucrose solution does not lead to the development of enamel caries within 3 weeks.
4. Cleaning at least once every second day prevents gingivitis, whereas cleaning only every third or fourth day results in the development of gingivitis within 6 weeks.
5. Gingivitis will heal within 1 week if plaque control measures are carried out at least once every second day, but not if plaque control occurs only every third and fourth day.
6. Rinsing with 10% sucrose solution does not result in a critical drop in pH in 12-hour-old plaque, while it does in 3-day-old plaque. In toothbrushing populations, such mature plaque remains only on the approximal surfaces of the posterior teeth.

Thus, plaque control should be the basis of prevention and control of gingivitis, periodontitis, and dental caries. Meticulous cleaning of all tooth surfaces twice a day should prevent the development of gingivitis and dental caries.

Role of the oral environment

Although the oral cavity may in some ways be regarded as a single microbial ecosystem, it is important to recognize that there are several major and minor compartments, each constituting a separate microenvironment, not easily affected by major events in the oral cavity; for example, the dorsum of the tongue, each tooth surface, particularly the fissures of molars, and the approximal surfaces of the posterior teeth, rough tooth surfaces, and restorations can be considered microenvironments. Each such microenvironment offers specific conditions for plaque formation, composition of the microflora, and accessibility for plaque removal.

Formation of plaque biofilms

The entire oral cavity, including the tooth surfaces, gingivae, tongue, and oral mucosa, is covered by a glycoprotein layer (pellicle), derived from the saliva. The pellicle (1 to 10 μm thick) acts as a non-friction layer and protects the teeth and the soft tissues from abrasion. Studies have shown that complete removal of the pellicle from the enamel surface requires the application of a rotating rubber cup and pumice for about 5 minutes. Normal toothbrushing with a medium abrasive toothpaste does not remove the pellicle from the enamel surfaces but temporarily reduces its thickness.

The first requirement for plaque formation is a solid surface. In the oral cavity, only pellicle-covered tooth surfaces offer such conditions. Although microorganisms may attach to the outer surface of the epithelial cells of the soft tissues, no plaque can form because these cells are continuously desquamated.

It is estimated that 1 mm³ of dental plaque, weighing about 1 mg, contains more than 200 million bacteria. The tooth surfaces are rapidly recolonized, and plaque re-forms, particularly in so-called stagnant zones: along the gingival margin, on the approximal surfaces of the posterior teeth, and in the fissures of erupting molars. Only 4 hours after cleaning, there are 10^3 to 10^4 bacteria per 1 mm² on such surfaces. Within a day, the number of bacteria increases 100- to 1,000-fold, mainly by growth of the harmless normal microflora, predominantly facultatively anaerobic streptococci. After 24 to 48 hours, continuous plaque has formed along the gingival margin. After 3 to 4 days, the thickness of de novo accumulated plaque increases dramatically and reaches its maximum after about 7 days. Further accumulation is limited by friction from chewing and so on. The range of microorganisms in the plaque increases, gram-negative and gram-positive long rods predominate, and a well-organized biofilm is formed.

Amount of plaque

The amount of plaque on the tooth surfaces is usually expressed by specific plaque indices. The most common are the Plaque Index (PI) by Silness and Löe (1964), and the Plaque Control Record, based on the presence or absence of disclosed plaque (O'Leary et al, 1972).

In toothbrushing populations, most remaining plaque is localized on the linguoapproximal surfaces of the mandibular posterior teeth and the buccoapproximal surfaces of the maxillary posterior teeth. However, the amount and location of plaque recorded at clinical examination discloses only where the patient has been unsuccessful in cleaning, despite extra effort on the day of the dental appointment. It does not reveal the age of the plaque, the rate of accumulation, or the patient's usual standard of oral hygiene. It is therefore important to appreciate the difference between plaque indices, which are static, and the plaque reaccumulation rate, which is dynamic.

Formation rate and reaccumulation of plaque

The quantity of plaque that forms on cleaned tooth surfaces during a given time represents the net result of interactions among etiologic factors (the number and quality of the oral microflora), many internal and external modifying risk indicators and risk factors, and protective factors. This observation has been the rationale for the construction of the Plaque Formation Rate Index (PFRI) (Axelsson, 1987, 1991). It includes all but

the occlusal tooth surfaces and is based on the amount of plaque freely accumulated (de novo) in the 24 hours following professional mechanical toothcleaning, during which period subjects refrain from all oral hygiene.

Studies have shown that PFRI scores are low in populations with low caries prevalence and high standards of oral hygiene, and high in populations with high caries prevalence and poor oral hygiene. The recommended frequency of mechanical plaque control by self-care should be based on the subject's plaque formation rate.

The pattern of plaque formation is influenced by factors such as the anatomy and surface morphology of the teeth; the stage of eruption and functional status of the teeth; the wettability and surface tension of intact and restored tooth surfaces; gingival health and volume of exudate; exposure to chewing forces; abrasion from foods; friction from the dorsum of the tongue, the lips, and the cheeks; and the presence of less accessible areas.

Most plaque reaccumulation has been shown to occur on the linguoapproximal surfaces of the mandibular posterior teeth, the buccoapproximal surfaces of the maxillary molars, and in the fissures of erupting molars. Almost no plaque reaccumulates on the lingual surfaces of the maxillary teeth (mainly because of friction from the rough dorsum of the tongue). Mechanical plaque control should target the surfaces on which most plaque reaccumulates.

Studies have also shown that in toothbrushing populations, the greatest number of microorganisms, including the cariogenic mutans streptococci, are found on the approximal surfaces of the molars and premolars, and these surfaces predominate in scores for decayed, missing, or filled surfaces. Toothbrushing should therefore be supplemented by approximal oral hygiene aids on these surfaces. (For reviews on plaque formation and plaque control, see Axelsson, 1981, 1994, 1998; Frandsen, 1986; Garmyn et al, 1998; Hotz, 1998; Jepsen, 1998; Kinane, 1998; Kuusela et al, 1997; Lang et al, 1997; Lang, 1998; Listgarten, 1994; Mayfield et al, 1998; Mombelli, 1998; Quirynen, 1986; Saxton, 1975; and Straub et al, 1998.)

CHAPTER 2

MECHANICAL PLAQUE CONTROL
BY SELF-CARE

Plaque control can be achieved mechanically or chemically by self-care or by professionals (dentists or dental hygienists). Plaque control programs based on needs-related combinations of these methods are to date the most successful for prevention of gingivitis, marginal periodontitis, and dental caries. The effectiveness of plaque control by self-care depends on the patient's motivation, knowledge, oral hygiene instruction, oral hygiene aids, and manual dexterity. A huge range of assorted oral hygiene aids are available: The dentist and the dental hygienist should assess the individual needs of the patient and recommend appropriate aids (Fig 55).

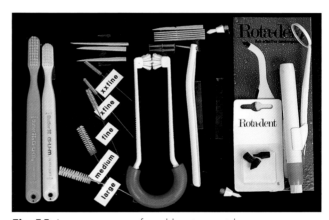

Fig 55 An assortment of oral hygiene aids.

MATERIALS AND METHODS

Manual toothbrushes

Toothbrushing is the most widespread mechanical means of personal plaque control in the world. Figure 56 shows the percentage of children aged 11 years from 23 European countries who brush their teeth more than once a day. Denmark and Sweden exhibit the highest percentage (83%), followed by Germany (76%). The lowest percentage is found in the Baltic countries (about 40%).

As early as 1973, a randomized Swedish survey showed that no fewer than 99.5% of all dentate adults used a toothbrush. However, only 70% of men older than 30 years used a toothbrush daily, compared with 85% to 90% of women (Håkansson, 1978). A more recent study in randomized samples of 35-, 50-, 65- and 75-year-old subjects in the county of Värmland, Sweden, showed that 93% of the 35-, 50- and 65-year-old subjects and 79% of the 75-year-old subjects brushed their teeth ≥ 2 times per day (Axelsson et al, 2000).

Enthusiastic use of the toothbrush is not, however, synonymous with a high standard of oral hy-

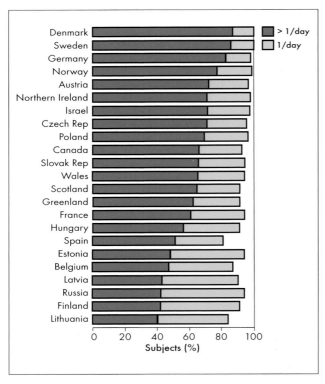

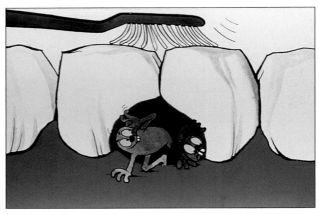

Fig 57 In members of toothbrushing populations, dental caries is concentrated in the approximal surfaces of the posterior teeth. Therefore, supplementary plaque control methods should be performed on these high-risk surfaces. (Illustration by J. Waerhaug, courtesy of the University of Oslo.)

Fig 56 Percentage of children who brush their teeth more than once a day in different countries. (From Kuusela et al, 1997. Reprinted with permission.)

giene. The toothbrush has very limited access to the wide approximal surfaces of the molars and premolars (Fig 57). Clinical, visual assessment of plaque removal by toothbrushing does not mean that all bacteria on the tooth surfaces have been removed (Fig 58).

Role of toothbrush design

There are numerous designs for manual toothbrushes, and, in the past, claims have been made of superiority for plaque removal by individual brands. Although world workshops on plaque control and oral hygiene practices have consistently concluded that there is insufficient evidence that any one toothbrush design is superior (for review, see Frandsen, 1986), novel designs continue to be introduced, supported by studies

Fig 58 Toothbrush bristles removing dental plaque (original magnification ×20). (Courtesy of L. Nilsson.)

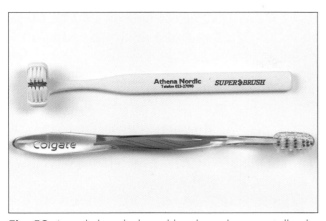

Fig 59 A triple-headed toothbrush and a specially designed conventional toothbrush.

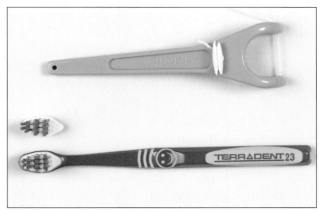

Fig 60 A toothbrush with disposable heads and a dental tape holder for supplementary approximal toothcleaning.

that demonstrate their superiority. For many of these new products, there is only one published report (Benson et al, 1993; Hanioka et al, 1995; Soparkar et al, 1991; Yankell et al, 1993). Other recent studies have failed to demonstrate that new toothbrush designs are superior to conventional designs for plaque removal (Claydon and Addy, 1995; Emling and Yankell, 1994; Stabbe et al, 1988; Yankell and Emling, 1994).

Buccal surfaces and anterior teeth are usually brushed most thoroughly (Bay et al, 1967; Rugg-Gunn and Mac Gregor, 1978), perhaps because these are the most accessible surfaces and because patients place special emphasis on esthetics. Mandibular lingual surfaces are much less obvious and much less accessible, especially in the molar areas, where the tongue may obstruct access to the gingival margins. Several studies have indicated that a double- or triple-headed toothbrush may improve plaque control by parents in preschool children, especially lingually. Figure 59 shows such a triple-headed toothbrush and a specially designed conventional toothbrush that facilitates cleaning of the lingual-approximal sulcus.

Differences in plaque removal arising from specific variations in toothbrush handle design

have been evaluated in several studies. Davies et al (1988) found that, during supervised brushing, toothbrushes with long, contoured handles performed significantly better than did those with short, uncontoured handles. Generally, it is important that the length and width of the handle be adequate. For preschool children, the handle should be big enough for the parent's grasp, and the head of the toothbrush should be small enough for the child's mouth.

Unfortunately, the trend in recent years has been toward toothbrushes with intricate, exclusive, expensive handles and less durable bristles. An optimally functional, exclusive handle design would be acceptable if the heads were disposable with well-designed bristles of high quality (Fig 60). This would appeal to consumers and benefit the environment.

Bergenholtz et al (1969) compared the plaque-removing ability of four standardized toothbrushes that differed in the stiffness and density of their bristles (hard and soft, multitufted and space-tufted); in a single-use study in dental students, the researchers found no significant differences. Finkelstein and Grossman (1984), after comparing seven toothbrushes of different designs, concluded that design features other than bristle texture accounted for the differences in the mechanical cleaning efficiency. In contrast, Pretara-Spanedda et al (1989) demonstrated that a single brushing with brushes that had a higher bristle density removed significantly more plaque.

In addition, Bergenholtz et al (1984) found no differences in plaque removal by unsupervised brushing with straight, multitufted, or V-shaped brushes. When used professionally, the V-shaped toothbrush was better than the straight brush for interproximal plaque removal. Chong et al (1985) also reported better plaque removal by a V-shaped toothbrush, although the difference was not statistically significant. No differences were found when a new cross-tuft filament layout was compared with a standard vertical-tuft toothbrush (Apiou et al, 1994).

Van Swol et al (1996) evaluated the effect of a small, imperceptible electric current on established dental plaque and gingivitis during manual toothbrushing in 64 adults. From baseline to 6 months, the Quigley and Hein Plaque Index scores showed significantly greater improvement for teeth cleaned with the test brush than for those cleaned with the uncharged control brush.

Role of toothbrushing method

To systematize the toothbrushing procedure, different methods have been recommended. These methods can be categorized according to the pattern of motion of the brush:

1. Roll: Rolling stroke; modified Stillman
2. Vibratory: Stillman; Bass
3. Circular: Fones
4. Vertical: Leonard
5. Horizontal: Scrubbing

For the past decade, the Bass method (Bass, 1954) has been the most frequently recommended. Unless instructed otherwise, however, people tend to use the horizontal scrubbing method. The different techniques have been compared in a num-

ber of studies. In children (Sangnes, 1974) and in adolescents (Rugg-Gunn et al, 1979), the horizontal scrubbing technique had a somewhat better plaque-removing potential than did the roll technique.

Arai and Kinoshita (1977) compared six toothbrushing methods, using several kinds of toothbrushes, and found that the Fones and scrubbing methods were the most effective with respect to plaque-removing ability.

Hansen and Gjermo (1971) studied the plaque-removing effect of four toothbrushing methods (roll, horizontal scrub, Charters, and interbrush methods). After 3 days of undisturbed plaque accumulation, a dental hygienist brushed the patients' teeth with brushes recommended for the different methods. The roll method left the most residual plaque.

Gibson and Wade (1977) compared plaque removal by the Bass and the roll techniques in 38 dental students after a study period of 2 weeks. The Bass technique was superior to the roll method in cleaning adjacent to the gingival margins on the lingual and facial aspects, but no statistically significant differences were found in overall effectiveness. Despite detailed instruction at a single visit and renewed instruction within 3 weeks, the participants failed to achieve an adequate standard of cleanliness.

Bergenholtz et al (1984) compared plaque removal by professional toothbrushing in which dental assistants used the Bass, roll, circular, and horizontal scrubbing methods. The methods were randomly assigned to four quadrants in the mouths of 24 adult patients. Compared with the other methods, the roll technique was less effective on buccal surfaces, and the Bass method was more effective on axiolingual surfaces.

In a 1-week study in 39 patients, Bastiaan (1984) found that the combined action of the modified Bass and roll techniques was less efficient than the modified Bass technique alone. Experimentally, it has been shown that proper use of the Bass method, three times a week, will prevent subgingival plaque formation on buccal surfaces accessible to the toothbrush and that plaque

Fig 61 Buccal gingival plaque before toothbrushing. (Illustration by J. Waerhaug, courtesy of the University of Oslo.)

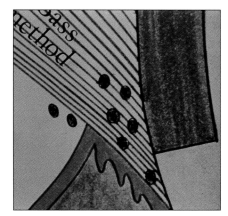

Fig 62 The effect of the Bass method of toothbrushing. (Illustration by J. Waerhaug, courtesy of the University of Oslo.)

can be removed to at least 1 mm subgingivally (Waerhaug, 1981) (Figs 61 and 62).

Because of varying experimental conditions, studies on the efficiency of brushing methods are difficult to compare. To date, no method of toothbrushing has been shown to be clearly superior to others, but the results from the aforementioned and other controlled studies (Frandsen et al, 1970; McClure, 1966; Rodda, 1968) indicate that the roll method is inferior to most other methods.

In recent years, no further studies comparing different toothbrushing methods have been published. As early as 1986, Frandsen commented that "The reason for the low research priority given to this area is similar to that suggested for toothbrushes. Researchers have realized that improvement in oral hygiene is not as dependent upon the development of better brushing methods as upon improved performance by the persons using any one of the accepted methods."

Toothpastes should not be regarded merely as sophisticated vehicles for the delivery of fluoride and chemical plaque control agents. In toothbrushing studies standardized with respect to brushing time, frequency, and pressure on the buccal surfaces, it has been shown that a medium-abrasive toothpaste will increase plaque removal by about 50% (Badersten and Egelberg, 1972).

Role of toothbrushing frequency

Because the frequency of plaque removal required to prevent dental disease is not known, there is no consensus as to the optimum frequency of toothbrushing. However, for cost effectiveness, the frequency of plaque removal should be individualized on the basis of plaque formation rate (see chapter 1) and predicted risk for dental caries and periodontal disease.

A prospective study by Lang et al (1973), discussed in chapter 1, analyzed the relationship between toothbrushing frequency and the development of gingivitis. Two similar investigations were conducted to determine the minimum brushing frequency necessary to prevent the development of gingivitis. The study populations, comprising dental students or young dental faculty members with healthy gingivae, were assigned to groups with different cleaning frequencies over study periods of 4 to 6 weeks. Brushing and interdental cleaning once a day or even once every second day prevented the development of gingivitis, whereas cleaning only every 3 days did not. In 1977, Bosman and Powell evaluated the minimum frequency of cleaning required to reverse experimental gingivitis: Cleaning once every day or every second day was sufficient, but cleaning only every 3 days failed to reverse the gingivitis.

Caution should be exercised in extrapolating the results of the aforementioned studies in dentally aware subjects to the average patient. It appears that few patients are able to achieve total plaque control at each cleaning. In fact, de la Rosa et al (1979) studied the pattern of plaque accumulation and removal with daily toothbrushing during a 28-day period following professional mechanical toothcleaning (PMTC) in 180 teenage boys. The results of this study suggested that the average person does not clean very effectively, leaving large quantities of residual plaque, despite daily brushing. On average, about 60% of the plaque remained after brushing and promoted rapid regrowth (for a review of the pattern of remaining plaque, see Cumming and Löe, 1973).

There are no published reports of prospective, experimental studies of different frequencies of toothbrushing in subjects susceptible to caries and periodontal disease. Such trials cannot be conducted, not only because of the long observation periods needed for disease progression to occur but also for ethical reasons. The in vivo telemetric studies by Imfeld (1978) and Firestone et al (1987), discussed in chapter 1, indicated that to prevent critical drops in pH (pH < 5) after sugar intake, plaque removal from all tooth surfaces once or twice a day should be adequate in most subjects. If, however, 3-day-old plaque is present on the approximal surfaces of the molars, the pH will fall to less than 4 after a rinse with 10% sugar solution. The following procedures should be recommended for high–caries-risk patients with high or very high plaque formation rates (Plaque Formation Rate Index of 4 or 5):

1. Clean all tooth surfaces immediately before, instead of after, meals to prevent drops in pH during meals and abrasion of surfaces previously attacked by acids from cariogenic plaque and from the diet (eg, fruit acids).
2. Use a fluoride chewing gum for 15 to 20 minutes after meals: Salivary stimulation and fluoride release will arrest microdemineralization.

Other attempts to clarify the relationship between brushing frequency and caries or periodontitis are based on voluntary reports from subjects in cross-sectional studies of disease exposure in various populations. Contradictory results are reported for periodontal disease (Dale, 1969; Hansen and Johansen, 1977; Hansen et al, 1990; Murtomaa et al, 1984). In a recent study, Lang et al (1995) reported that plaque scores and observed brushing thoroughness were better linked to periodontal conditions than was frequency of brushing. The effect of mechanical plaque control by self-care on caries and periodontal diseases is discussed further at the end of this chapter.

In summary, cross-sectional studies on the association between cleaning and caries and periodontal disease indicate that the quality of cleaning, rather than the frequency, may be related to disease. For practical and esthetic reasons, and for the perception of oral freshness, the generally recommended practice of brushing twice daily may be appropriate. However, there are inadequate scientific data to support this practice as an effective measure against disease (for reviews on manual toothbrushing, see Bergenholtz, 1972; Frandsen, 1986; and Jepsen, 1998).

Iatrogenic effects of incorrect toothbrushing

Cervical abrasion, gingival irritation, and gingival recession may occur as sequelae to vigorous toothbrushing, incorrect technique, or the use of an inadequate toothbrush (Gillette and Van House, 1980). In most instances, toothbrushing trauma has been reported in epidemiologic studies and case reports.

In a study of uninstructed children and young adults, most right-handed subjects started brushing the buccal surfaces of the anterior teeth or the left side. The mean brushing time was about 50 seconds, only 10% of which was spent on the lingual surfaces (McGregor and Rugg-Gunn, 1979). Accordingly, the most severe gingival recession and abrasion defects are found on the buccal sur-

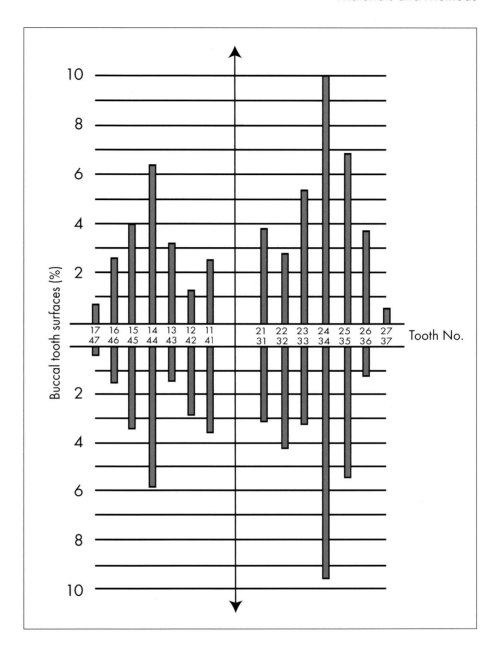

Fig 63 Percentage of the occurence of iatrogenic toothbrushing lesions on individual buccal tooth surfaces. (Fédération Dentaire Internationale [FDI] tooth-numbering system). (Modified from Sangnes, 1976 with permission.)

faces on the left side (Källestål and Uhlin, 1992; Sangnes, 1976).

Figure 63, from a study by Sangnes (1976), shows the extent of abrasion on the individual buccal surfaces. In this toothbrusing population, comprising about 90% right-handed individuals, the most advanced abrasion is clearly on the left first premolars. Figures 64 to 67 show right-handed, self-taught subjects, who have brushed with

the horizontal scrubbing method since their teens. On the left side of the dentition, abrasion of varying severity is present on the buccal surfaces of both natural and crowned teeth.

Most self-taught subjects begin by scrubbing the buccal surfaces, especially the anterior teeth, and rarely proceed to the lingual surfaces. They do not clean interproximally. Most oral hygiene brochures recommend that cleaning begin, with

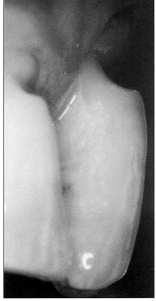

Fig 64 Advanced iatrogenic defect caused by abrasion of toothbrushing in a right-handed, self-taught position.

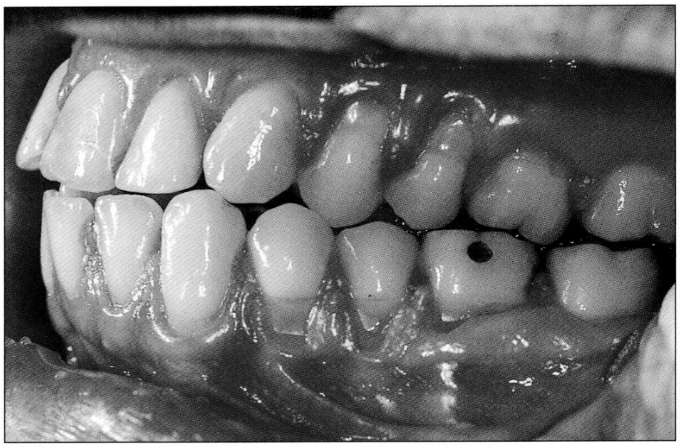

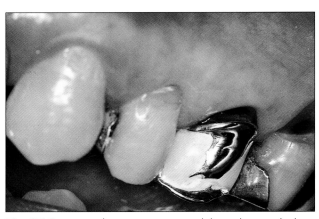

Fig 65 Typical pattern of iatrogenic defects on the left side in a right-handed, self-taught toothbrushing patient.

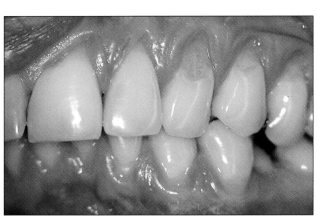

Fig 66 According to the findings by Sangnes (1976) (see Fig 63), iatrogenic abrasions from the toothbrush most frequently are found on the maxillary left canines and premolars and mandibular left first premolars in right-handed subjects.

Fig 67 Even acrylic crowns may exhibit advanced abrasion defects.

toothbrush and paste, on the maxillary teeth, first on the buccal surfaces and then on the palatal surfaces. The mandibular teeth are cleaned next: first the buccal surfaces and then the lingual surfaces. Only then is interdental cleaning recommended. There is no scientific basis for this routine.

The patient's level of ambition is always greatest at the beginning of cleaning. Moreover, there is more toothpaste on the brush in the initial phase, and the bristles are most rigid. In the buccal region, the alveolar bone is very thin and may be totally absent over the canine teeth. On the lingual and the palatal aspects of the teeth, however, the alveolar bone is normally very dense. It is evident that there is a high risk of damage if toothbrushing begins on the buccal surfaces of the maxillary teeth. Moreover, after instruction, many subjects tend to revert from a correct Bass method to horizontal scrubbing, again increasing the risk of trauma to the most prominent buccal surfaces.

As a result, there is a strongly positive correlation between brushing frequency and gingival recession and the development of abrasive lesions on the buccal surfaces. The greatest loss of periodontal attachment is generally found on the buccal surfaces in subjects aged 35 years and younger, mainly because of toothbrushing. In cases where the bifurcations on the buccal surfaces of the molars have been exposed, there is an indirect risk of complications and eventually the need for root separation or extraction.

Sangnes and Gjermo (1976) studied the prevalence of hard and soft tissue lesions related to mechanical toothcleaning procedures: of 533 subjects, 45% had one or more wedge-shaped cervical defects; the frequency was higher in subjects who brushed more than twice daily or who had good oral hygiene (Fig 68).

In another study, Bergström and Lavstedt (1979) found an evident relationship in 818 subjects between abrasion in the form of wedge- or saucer-shaped depressions and the frequency with which the subjects' toothbrushes wore out, especially with a horizontal brushing technique; factors such as bristle stiffness and dentifrice abrasivity were much less important. Similar results were observed in another investigation (Bergström and Eliasson, 1988) that found cervical abrasion in 85% of 250 subjects, aged 21 to 60 years; the abrasion was most likely related to toothbrushing. The prevalence and severity of the abrasion increased with the patient's age and the presence of defects caused by periodontal disease, suggesting a poor correlation between aggressive toothbrushing and prevention of periodontal disease. Lussi et al (1991) also reported the presence of wedge-shaped cervical defects, not necessarily caused by toothbrushing, in 47% of 391 subjects

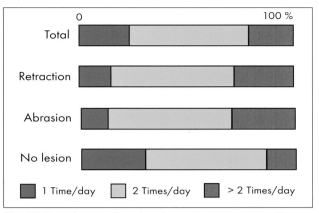

Fig 68 Prevalence of hard and soft tissue iatrogenic lesions related to toothbrushing frequency. (From Sangnes and Gjermo, 1976. Reprinted with permission.)

(aged 26 to 30 and 46 to 50 years), most of whom experienced hypersensitivity at these sites.

It is generally accepted that most cases of abrasion are attributable to the combined effect of brushing and the use of toothpastes that contain abrasives (Sangnes, 1976), but the role of the latter is probably minor (Axelsson et al, 1997b; Volpe et al, 1975). Although the dentin and enamel abrasion values for commercial toothpastes vary widely, brushing force and the hardness of the bristles may be of more importance than the abrasiveness of the toothpaste (Harte and Manly, 1975). In one investigation in dental students (Nordbo and Skogedal, 1982), the average rate of abrasion after two years of toothbrushing with different toothpastes was 0.2 mm, but no difference in effect was observed between conventional and nonabrasive dentifrices.

In an animal model, the rate of enamel abrasion was accelerated in the presence of erosion (Attin et al, 1997). In fact, tooth erosion (the progressive destruction of enamel and dentin by chemical means) can be observed even in young populations (Milosevic et al, 1997) and should be considered a risk factor associated with tooth abrasion.

Electric toothbrushes may have lower potential for dentinal abrasivity than conventional toothbrushes (McLey et al, 1997; Schemehorn and Zwart, 1996; Wilson et al, 1993), probably because less force is applied to the tooth surface (Van der Weijden et al, 1996c). Localized dentinal hypersensitivity is often associated with cervical abrasion, possibly as an effect of vigorous toothbrushing (Chabanski et al, 1996). The prevalence of dentinal hypersensitivity is high, around 15% (Murray and Roberts, 1994), although certainly less than the prevalence of abrasion.

With recession of the gingival margin, the root surface is exposed. This is a common condition associated with periodontal disease, inadequate toothbrushing (Björn et al, 1981), and repeated periodontal instrumentation (Löe et al, 1992). Among precipitating factors, vigorous toothbrushing would be the most significant, often in combination with predisposing factors such as inadequate attached gingiva and prominent roots. Another significant factor is the characteristics of the gingiva. Although in childen the amount of both keratinized and attached gingiva increases over time, the labial gingiva is thin, and resistance to mechanical irritation may be reduced. Although it has been shown that gingival recession in young children often diminishes over time, in these patients toothbrushing must be performed carefully and should be supervised. In an adult population, subjects with a thin periodontium may be more susceptible to gingival recession (Olsson and Lindhe, 1991).

The prevalence of gingival recession increases with age, in terms of both the number of individuals and the number of tooth surfaces affected. Several studies have demonstrated the role of inadequate toothbrushing as a causal factor: The methods (Björn et al, 1981; Paloheimo et al, 1987), direction (Sangnes, 1976), frequency (Ainamo et al, 1986; Sangnes and Gjermo, 1976; Vehkalahti, 1989), and magnitude of brushing are all important factors. The hardness and material of the bristles and the shape of the bristle tips should also be considered. Electric toothbrushes are generally considered safe for the gingival tissues.

In cases of gingival recession related to toothbrushing, the loss of soft tissue is mostly observed at the buccal surfaces of maxillary first molars, premolars, and canines (Björn et al, 1981; Paloheimo et al, 1987); approximal surfaces do not tend to show recession, except in cases of early periodontitis (Clerehugh et al, 1988; Källestål et al, 1990). Buccal recession is particularly common in young individuals (Källestål and Uhlin, 1992; Löe et al, 1992) with good oral hygiene (Björn et al, 1981). In older individuals, gingival recession is more prevalent and tends to have a generalized pattern, perhaps as the combined consequence of loss of attachment because of periodontal disease, the presence of calculus, and toothbrushing trauma (Serino et al, 1994).

The relationship between periodontal disease and inadequate toothbrushing related to gingival recession warrants further investigation. As stated by Joshipura et al (1994), more research is needed to develop a brushing method that does not cause recession but maintains good oral hygiene. This is discussed further later in this chapter (for reviews on iatrogenic abrasion and recession, see Axelsson et al, 1997b and Echeverria, 1998).

Electric toothbrushes

Effectiveness

In 1986, an international workshop on oral hygiene concluded that, to date, powered toothbrushes removed no more plaque than manual brushes, regardless of method (Löe and Kleinmann, 1986). At that time, only conventional electric toothbrushes were available; they had a head design similar to that of a manual toothbrush and a combined horizontal and vertical motion. Subsequently, different designs were introduced, eg, the Rota-dent (Pro-Dentec, Batesville, AR), a rotary-action single-tufted brush with small bristles that reach only one surface per tooth. It comes with three head designs: short and pointed, elongated, and hollow-cup brush tips (Figs 69 and 70).

In a short-term study of dental students, Walsh and Glenwright (1984) compared removal of 3- to

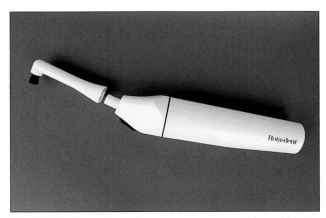

Fig 69 The Rota-dent electric rotating pointed toothbrush.

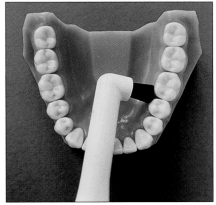

Fig 70 Rotating pointed interspace toothbrush (Rota-dent), useful for cleaning approximal surfaces in open approximal spaces and the fissures of erupting molars.

4-day-old plaque by the Rota-dent and a manual brush and found no statistically significant difference. In contrast, Glavind and Zeuner (1986) found that the improved Rota-dent was as effective as a combination of manual toothbrushing, flossing, and toothpicks in a test group of periodontal patients. In both the Rota-dent and the control groups, the plaque level had dropped at the 3-month examination. Boyd et al (1989a), in a 12-month study of a group of maintenance patients, reported that the Rota-dent was just as effective as the oral hygiene kit used in the study by Glavind and Zeuner (1986).

Other short-term studies have indicated improved interproximal plaque removal (Müller et al, 1987; Preber et al, 1991). Silverstone et al (1992) conducted a 6-week study in 30 subjects, comparing the Rota-dent and the Oral-B 40 (Gillette, Boston, MA) soft toothbrush, and reported no differences in gingival inflammation between the two groups. The Rota-dent brush is appropriate for periodontitis patients with "open" interproximal spaces, for "semiprofessional" mechanical cleaning of the fissures of erupting second and third molars, and for parents cleaning the fissures of erupting first molars.

The Interplak (Dental Research Corp, Tucker, GA) electric toothbrush was the next innovative toothbrush design, introduced in the mid-1980s. It has a rectangular head, with six to eight bristle tufts that individually rotate counterclockwise (Fig 71). Baab and Johnson (1989) assessed the ability of the Interplak to remove plaque under professional supervision. Subjects using the electric brush had lower plaque scores, attributable to increased effectiveness of the brush in the interproximal regions. In a 12-month study, Wilson et al (1993) found a larger reduction in plaque with the Interplak than with the Butler Gum 311 (Butler, Chicago, IL) manual toothbrush. No differences with respect to gingivitis were observed.

The Braun/Oral-B Plak Control (D5, D7, and D9) electric brush (Gillette) was introduced in 1991. It has a small circular head with an oscillating-rotating movement (Figs 72 and 73a). The Plak Control head has been shown in clinical trials to be superior to that of a conventional electric toothbrush and more effective than a manual toothbrush (Van der Weijden et al, 1993a, 1995a). Stoltze and Bay (1994) compared the Braun D5 to a manual toothbrush during a 6-week period. The electric toothbrush was more effec-

Fig 71 The Interplak counter-rotational electric toothbrush with separate bristle tufts that individually rotate counterclockwise.

Fig 72 Electric toothbrush (Braun/Oral-B) with a round brush head and 70-degree back-and-forth movement.

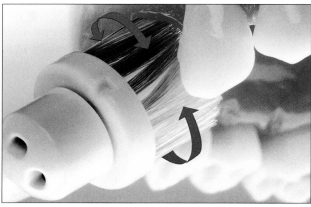

Fig 73a Close-up of the head of the Braun/Oral-B Plak Control.

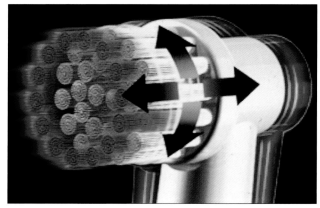

Fig 73b Close-up of the head of the new Braun/Oral-B 3D Plaque Remover showing its three-dimensional action.

tive in removing plaque, mainly on the interproximal surfaces.

In an 8-month preventive program, the Braun D5 was compared to a manual toothbrush (Butler Gum 311) in a group of gingivitis subjects (Van der Weijden, 1994). Plaque, gingivitis, gingival abrasion, and calculus were assessed. At the end of the trial, differences in plaque scores and gingival bleeding favored the Braun/Oral-B Plak Control.

In 1996, with the introduction of the D9 model, the frequency of movement was increased from 2,800 (D5) to 3,800 strokes per minute (D9). In addition, in the newer model, the angle of rotation was diminished from 70 to 60 degrees. A comparative study disclosed no significant difference between the D7 and D9 versions (Van der Weijden et al, 1996a). The Braun/Oral-B Plak Control electric toothbrush also seems to have high ac-

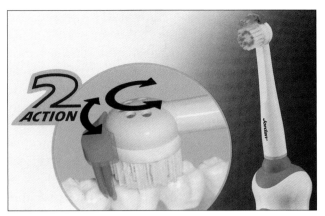

Fig 74 The Philips HP 510 oscillating-rotating electric toothbrush.

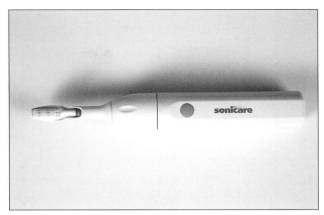

Fig 75 The Sonicare electric toothbrush.

ceptance in patients of all ages (Heasman et al, 1999).

Recently, the Braun/Oral-B 3D Plaque Remover (D15) (Gillette) was introduced. This novel electric toothbrush, developed from the oscillating/rotating Braun/Oral-B Ultra Plaque Remover (D9) (Gillette), incorporates an additional pulsating action to facilitate deeper cleaning in approximal areas and features a brush head with three-dimensional action (Fig 73b). The established oscillating/rotation action at a frequency of 63 Hz is retained, but the brush head has an additional pulsating action in the direction of the long axis of the bristles at the frequency of 170 Hz and an amplitude of approximately 0.15 mm.

In a recent single-blind, 6-month, longitudinal study on 48 periodontal maintenance patients, Haffajee et al (2001) compared the effect of this new Braun/Oral-B 3D Plaque Remover (D15) with a manual toothbrush on clinical variables. Both the manual and D15 reduced pocket depth, plaque index, and bleeding on probing. However, only D15 significantly reduced the mean gingival index and probing attachment loss. The greatest benefit of D15 was at the mandibular and lingual surfaces.

Dörfer et al (2001) compared the effect of D15 with a new high-speed, "microtation," electric-powered brush with an additional "microbrush clip" (Rowenta Dentasonic MH921S; Rowenta,

Germany) on 48 hours de novo accumulated plaque on 82 healthy subjects in a split-mouth study. The results showed that both brushes were able to remove a significant amount of plaque, but D15 was significantly more effective compared to Dentasonic. The additional microbrush clip for the proximal embrasures on the Dentasonic failed to improve plaque removal from these tooth surfaces compared with the D15 alone.

Recently, the Philips HP 510 (Philips, The Netherlands) was introduced. It is an oscillating-rotating electric toothbrush that has a circular head, similar in design to the Braun/Oral-B Plak Control, but with an active tip at the end of the head, which makes a small sweeping motion (Fig 74). Data from the manufacturer indicate that the efficacy is similar to that of the Braun oscillating-rotating toothbrush (De Jager et al, 1998).

The Sonicare electric toothbrush (Philips Oral Healthcare, Snoqualmie, WA) (Fig 75) was introduced in 1993. It has a rectangular head with bristles arranged in a sawtooth design. The side-to-side movement operates at a high frequency of 250 Hz. In a 4-week study in adults, the Sonicare proved to be more effective than a manual toothbrush (Oral-B 30) in removing plaque from the lingual and interproximal surfaces (Johnson and McInnes, 1994). In a recent 12-week study, the Sonicare was more effective than the manual toothbrush in removing plaque but comparable to it in reducing inflammation (Tritten and Armitage, 1996).

The Sonex ultrasonic brush (Salton, Lake Forest, IL) is designed with a piezoelectric transducer, operating at 1.6 MHz and located in the handle of the toothbrush. It is claimed that these vibrations pass from the handle along the head and down the bristles. A short-term study (Terezhalmy et al, 1995a) showed that the Sonex significantly reduced the bleeding and gingival indices. However, at the end of a 6-month study (Terezhalmy, 1995b), no difference was observed between the test group using the Sonex ultrasonic brush and a control group using a manual toothbrush.

The efficacy of most of the aforementioned electric toothbrushes is attributable mainly to improved interproximal cleaning, and it may be argued that subjects who use a manual toothbrush supplemented by some form of interproximal cleaning device would not benefit from an electric toothbrush. However, as discussed by Axelsson (1994), interproximal cleaning is not an established oral hygiene routine in Europe and is practiced regularly by, at most, 15% to 20% of the population of industrialized countries. By improving interproximal cleaning, the electric toothbrush therefore has a potentially important role in the prevention of periodontal diseases.

Factors that influence effectiveness

Several studies have indicated that, to achieve maximum benefit from an electric toothbrush, professional oral hygiene instruction is important. In studies of toothbrushing by patients with no prior professional instruction or with taped instructions, few or no differences were found (Barnes et al, 1993; Stoltze and Bay, 1994). Significantly better results were achieved for the electric toothbrush in studies that included professional brushing instruction (Grossman et al, 1995; Van der Weijden et al, 1993a, 1994). In the study by Ainamo et al (1997), although instruction was given only at the outset and not re-

peated during a 12-month study, the electric toothbrush was superior to the manual.

A few studies have investigated stain removal as well as plaque removal. Using an experimental model in which stain was induced by rinsing for 4 days with an intense chlorhexidine tea regimen (no other form of oral hygiene was allowed), Grossman et al (1996) showed that electric toothbrushes were more effective than manual brushes in removing extrinsic dental stain. This confirmed in vitro findings by Schemehorn and Henry (1996). When the chlorhexidine-induced stain model was used for an extended period, up to 4 weeks, both the Braun Plak Control and Sonicare were superior to a manual toothbrush for stain removal (McInnes et al, 1994; Moran and Addy, 1995).

Patients are generally unwilling to spend the time recommended by dental professionals for brushing and interdental cleaning. Electric toothbrushes are potentially more efficient than manual brushes, and this could improve plaque control in most adults (Boyd, 1997). Most patients brush their teeth for less than 1 minute instead of the recommended 5 minutes (Hawkins et al, 1986). Clinically, Preber et al (1991) demonstrated that the Rota-dent takes only half as long as a manual brush to remove the same amount of plaque.

Two studies have addressed the relationship between duration of brushing and efficacy of plaque removal (Van der Weijden et al, 1993b, 1996a); results indicated that, in the same period of brushing, a manual toothbrush removes less plaque than does an electric toothbrush. After 6 minutes' brushing by a professional, the manual toothbrush removes only 75% of the plaque; this level of cleanliness is achieved after only 1 minute with the electric toothbrush. With increasing duration of brushing, efficacy increases up to 6 minutes, but, for electric toothbrushes, 2 minutes' use appears to be optimum. Compared to manual toothbrushes, electric toothbrushes seem to result in less abrasion and recession, mainly because, as discussed earlier, less force is applied.

It is well documented that plaque removal increases with duration of brushing. Although the optimal time is at least 2 minutes for electric toothbrushes (Van der Weijden et al, 1993b, 1996b), most individuals brush for only 60 seconds (Huber et al, 1985; Van der Weijden et al, 1993b, 1996b). Instruction should therefore emphasize an increase in the duration of brushing. Most currently available electric toothbrushes are equipped with a timer, heightening patient awareness of the importance of duration of brushing in achieving an optimal mechanical effect. An electric toothbrush removes more plaque in the same time as a manual brush, and this ease of use is an important aspect of patient acceptance (Van der Weijden et al, 1993b).

Different electric toothbrushes have been compared in a number of studies (eg, Bader and Williams, 1997; Grossman et al, 1995; Robinson et al, 1997; Van der Weijden et al, 1993b, 1996b, 1996c), but the findings are inconclusive. To date, the data indicate that the Rota-dent, counter-rotational, and oscillating-rotating electric toothbrushes, are all very efficient.

Selection of an appropriate brush should also take into account aspects other than plaque removal efficacy. Tritten and Armitage (1996) compared the sonic electric toothbrush (Sonicare) and a traditional manual toothbrush for efficacy in removing supragingival plaque and reducing gingival inflammation in a 12-week, single-blind clinical trial. Although both types of brush were effective in removing supragingival plaque, the sonic brush was better at removing subgingival plaque, particularly in posterior teeth. Both devices were equally effective in reducing gingival inflammation.

The pressure with which a toothbrush is applied may be important. Using a standardized brushing machine, Sarker et al (1997) compared the cleaning efficiency of four automatic toothbrushes (Rota-dent, Braun/Oral-B, Interplak, and Sonicare) and a manual toothbrush at pressures typically applied in vivo. The rotary action of the Rota-dent provided the most efficient combination of low abrasion and high cleaning efficiency.

Bader and Williams (1997) compared the efficacy of two powered brushing instruments (Rota-dent and Interplak) for control of plaque and gingivitis in interproximal spaces and furcations. The results indicated that the Rota-dent was significantly more effective than the Interplak; there were clinically relevant differences in all indices measured.

Patient acceptance of the electric toothbrush is an important aspect in selection of an appropriate brush, as illustrated by a 2-month trial comparing the Braun electric toothbrush with a sonic toothbrush. The volunteers using the Braun/Oral-B Plak Control wished to continue with the toothbrush, whereas 25% of the Sonicare group disliked the device and discontinued its use (Grossman et al, 1995). This was also confirmed by Van der Weijden et al (1996b). In a study in which the Plak Control was compared with the Philips HP 500, which has an action similar to that of a conventional toothbrush, subjects were allowed to keep one toothbrush at the end of the study (Van der Weijden et al, 1995b). The majority of the panelists preferred the oscillating-rotating toothbrush (Braun/Oral-B Plak Control). Baab and Johnson (1989) conducted a telephone survey 6 months after their investigation and learned that most subjects were not using the Interplak twice a day as they had done during the study period.

Patients who may benefit from use of an electric toothbrush

Electric toothbrushes may be of more benefit to some types of patients than to others, eg, poorly compliant periodontal maintenance patients, small children, adolescent orthodontic patients, the elderly, and the disabled. In periodontal maintenance patients, supragingival plaque control is an important factor in preventing periodontal breakdown. Patients with suboptimal plaque control usually need more frequent maintenance visits and are more likely to develop loss of attachment (Lindhe and Nyman, 1984).

It is well established that the use of electric toothbrushes is of particular advantage in controlling plaque accumulation in patients who comply poorly with respect to oral hygiene. Hellstadius et al (1993) studied a group of such patients, referred to a specialist for periodontal treatment. Despite previous extensive instruction in the use of manual oral hygiene aids over periods of up to 40 months, the percentage of tooth surfaces exhibiting plaque remained unacceptable at 48%. A change from a manual to an electric brush resulted in a reduction to a mean of 12%, and this level was maintained for the observation period of up to 3 years. Yukna and Shaklee (1993) reported that, in a comparable patient group, the electric toothbrush proved to be a useful adjunct in maintaining reduced plaque levels and favorable gingival conditions.

Most published studies on the use of electric toothbrushes in children have described electric toothbrushes developed in the 1960s. In one early study, Lefkowitz et al (1962) compared the use of an electric toothbrush with that of a manual brush in two groups of children, aged 7 to 9 and 10 to 12 years. The electric brush removed more plaque in both age groups.

A recent study compared the plaque control efficacy of a new electric toothbrush (oscillating-rotating), designed specifically for use by children, with a children's manual brush (Grossman and Proskin, 1997). In this population, aged 8 to 12 years, the electric brush achieved significantly greater plaque removal. When parents brush their children's teeth, electric toothbrushes can be of great benefit, particularly for cleaning the fissures in erupting first molars. In studies in which brushing was done by a professional, efficacy was high (Van der Weijden et al, 1993a, 1993b, 1996b).

Especially for children, a small brush head should be available. For children who do their own brushing, a small, light handle is more suitable. A small head is also practical for adults, because posterior teeth are difficult to reach. De Jager et al (1998) compared two oscillating-rotating toothbrushes, the Braun and the Philips. Although no overall difference was found, the Philips was more effective in the molar area. This was attributed to the reduced height of the brush head. The Braun was more effective in the anterior region, probably for reasons other than the height of the brush head.

In adolescent orthodontic patients, fixed appliances often impede effective plaque control (Boyd, 1997). Both Yankell et al (1985) and Wilcoxon et al (1991) have shown the Interplak brush to be more efficient than a manual toothbrush in controlling gingivitis and plaque in orthodontic patients. In a 1-month study in adolescent orthodontic patients with existing gingivitis, the Sonicare toothbrush, after oral hygiene instruction, resulted in greater reduction in both plaque and bleeding than did a manual brush (Ho and Niederman, 1997). Although another 3-month study by White (1996) concluded that the Sonicare may help orthodontic patients to improve their oral health, this was not a blind study. Jost-Brinkman et al (1994) reported that three electric toothbrushes (Interplak, Braun/Oral-B Plak Control, and Rota-dent) showed efficacy comparable to that of a manual system comprising a toothbrush, floss, and an interspace toothbrush.

The only long-term clinical trial evaluating the effectiveness of an electric brush on the periodontal health of orthodontic patients is an 18-month study by Boyd et al (1989b), showing that the Rota-dent can be more effective than conventional toothbrushing.

Recently, Sicilia et al (2002) presented a systematic qualitative review of randomized controlled trials on the effect of power-driven toothbrushes compared to manual toothbrushes in terms of gingival bleeding or inflammation resolution in the treatment of gingivitis and chronic periodontitis. A total of 21 studies in adults were selected. In 10 studies, power-driven toothbrushes were more efficient than manual toothbrushes in reduction of gingival bleeding and inflammation. This effect was related to the capacity of plaque reduction and was more evident in counter-rotational and oscillating-rotating toothbrushes. No evidence was found that sonic toothbrushes have a greater effect than manual toothbrushes. However, in short-term studies, no significant differences were found between power-driven toothbrushes and manual toothbrushes if PMTC was carried out after the initial examination, regardless of the type of power-driven toothbrushes tested.

Although it is suggested in the literature that electric toothbrushes are especially useful for the disabled patient (eg, Cancro and Fischman, 1995), there are few controlled clinical studies. Two studies have shown that electric toothbrushes are valuable for mentally disabled children and for disabled children with poor manual dexterity (Kelner, 1963; Smith and Blankenship, 1964). The few available recent studies have shown that electric toothbrushes are also valuable for disabled adults (Blahut, 1993; Bratel and Berggren, 1991). In a study by Martin et al (1987) of institutionalized elderly patients with limited manual dexterity, subjects were given no oral hygiene instruction. Oral cleanliness and gingival health were assessed both before and after the study, and the results suggested that the increased efficacy of the electric toothbrush would aid the maintenance of oral hygiene in this category of patients.

Finally, it should also be observed that "semi-professional" and professional brushing with the electric toothbrush is highly effective (Van der Weijden, 1993a, 1993b, 1996a). In cases where a caregiver is responsible for oral hygiene, the electric toothbrush can be a useful tool (for reviews on electric [automated] toothbrushes, see Hancock, 1996; Van der Weijden et al, 1998; and Sicilia et al, 2002).

Interdental cleaning aids

In toothbrushing populations, the approximal surfaces of the molars and premolars are the predominant sites of residual plaque; plaque reaccumulation; decayed or filled surfaces (DFSs); and periodontal treatment needs (see chapters 1 and 9). Cumming and Löe (1973) found that, in a toothbrushing population of teenagers, most residual plaque is found on the approximal surfaces of the molars and premolars, followed by the lingual surfaces of the mandibular molars (see Fig 37 in chapter 1). Most de novo plaque accumulates on the approximal surfaces of molars and premolars after 12 and 48 hours (Lang et al, 1973) (see Fig 42 in chapter 1) and 24 hours (Axelsson, 1987, 1991) (see Fig 43 in chapter 1).

In subjects who refrained from cleaning for 4 days (Mombelli et al, 1990), by far the highest number of bacteria was found in subgingival plaque from the approximal surfaces of molars (particularly maxillary molars) (see Fig 46 in chapter 1). In a large group of 13- to 14-year-old toothbrushing children, colonization by mutans streptococci was also greatest on the approximal surfaces of the molars (see Fig 51 in chapter 1) (Kristoffersson et al, 1984).

Because of this pattern of residual plaque and plaque reaccumulation, in toothbrushing populations of 12- to 20-year-old subjects, DFSs comprise mainly the approximal surfaces of the molars and premolars and the occlusal surfaces of the molars (see Figs 52 and 53 in chapter 1). The approximal surfaces of the posterior teeth are also the sites of greatest periodontal treatment needs. These findings highlight the tremendous potential for improving prevention and control of periodontal disease and dental caries by supplementing toothbrushing with regular gingival plaque removal from the approximal surfaces of the molars and premolars.

Frequency of approximal cleaning

Special aids for approximal cleaning, such as dental floss, dental tape, toothpicks, interdental brushes, and electric brushes, should be considered in the context of the subject's (1) age; (2) susceptibility to caries, gingivitis, or periodontitis; (3) caries and periodontal disease experience; (4) manual dexterity; and (5) knowledge and motivation in oral hygiene practices. Interdental cleaning is different in adults and children because in adults the spaces are generally bigger (particularly in patients with periodontitis) and caries incidence is generally much lower.

The previously mentioned study by Kuusela et al (1997) showed that the use of dental floss in 11-year-old children was uncommon. More Canadian schoolchildren flossed their teeth daily than did children in other countries. Children in Hungary, Finland, and the Slovak Republic rarely flossed (Table 1). Girls used dental floss more frequently than did boys; there were statistically significant differences between girls and boys using dental floss daily in Canada (30% and 20%, respectively), Norway (20% and 13%, respectively), and Northern Ireland (18% and 11%, respectively).

It may be argued that supplementary approximal toothcleaning—except in high-caries-risk children—is not necessary until the molars and premolars are fully erupted, ie, from the age of about 12 years in girls and 12.5 to 13 years in boys. According to an earlier survey by Honkala et al (1991), during the 1990s, there was an increase not only in the use of dental floss but also in toothbrushing more than once per day in 11-year-old children in most European countries.

In a recent, large-scale English study (n = 4,142) of 12- to 15-year-old subjects, MacGregor et al (1998) found that only about 8% to 9% used dental floss daily. However, there was a positive correlation between frequency of flossing and other factors such as the frequency of washing their hands after going to the lavatory, frequency of bathing, time since their most recent dental appointment, and having a current friend of the opposite sex.

In 1973, a Swedish national dental survey disclosed that wooden toothpicks were used sporadically by approximately 45% of the adult population and daily by only 12%. Dental floss was used irregularly by 12% and daily by only 2%. Scandinavian adults use triangular, pointed toothpicks more frequently than they do dental floss or tape. Among adults in the United States, dental floss or tape is the most frequently used interproximal oral hygiene aid.

A more recent Swedish national survey, performed by Eureka Research in 1994, showed that almost half of the adult Swedish population regularly uses toothpicks (48%) or dental floss or dental tape (41%). In a randomized sample of more than 600 subjects aged 50 to 55 years in the county of Värmland, Sweden, as many as 64% used a toothpick daily, and 35% used dental tape daily. About 35% used interdental brushes daily as a supplementary aid in some individual interproximal spaces (Axelsson and Paulander, 1994). In a more recent analytic epidemiologic study of randomized samples of 35-, 50-, 65-, and 75-year-old subjects in the county of Värmland, Sweden, it was shown that 20%, 51%, 85%, and 66%, respectively, used toothpicks daily; 26%, 29%, 53%, and 28%, respectively, supplemented with daily use of dental tape; and 4%, 15%, 25%, and 45%, respectively, used interdental brushes daily (Axelsson et al, 2000). However, in most countries of the world, no interproximal cleaning is practiced.

Table 1 Percentage of children who floss their teeth daily (standardized for gender) in different countries*

Country	Children flossing daily (%)
Canada	25
Norway	17
Northern Ireland	15
Greenland	12
Israel	10
Denmark	9
Spain	9
Austria	7
Sweden	6
Czech Republic	6
Slovak Republic	4
Finland	3
Hungary	2

*From Kuusela et al (1997). Reprinted with permission.

Use of dental floss and tape

Use of flat dental tape combined with a fluoride dentifrice is recommended for cleaning the approximal surfaces of molars and premolars in children and young adults. With a "rubbing" motion, holding the tape by hand or in a special holder (Fig 76), patients can remove plaque from the ap-

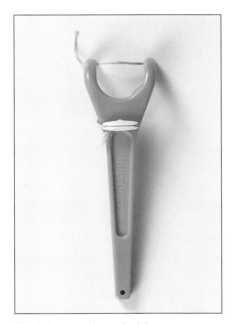

Fig 76 Dental tape holder.

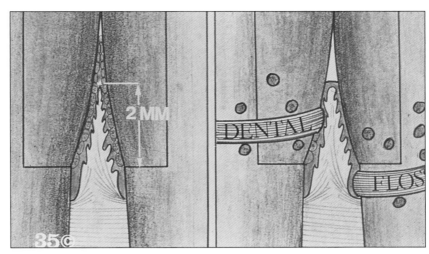

Figs 77 and 78 Plaque removal from the approximal surfaces of molars. Flat, fluoridated dental tape, used in a rubbing technique, can remove 2 mm of subgingival plaque. (Illustrations by J. Waerhaug, courtesy of the University of Oslo.)

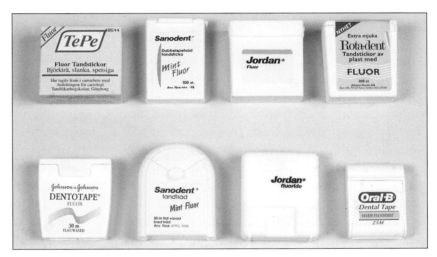

Fig 79 Assortment of fluoridated wooden toothpicks (top) and dental tape (bottom).

proximal surfaces of the molars up to 2 mm subgingivally (Figs 77 and 78). Several brands of fluoridated dental tape and fluoridated toothpicks are now commercially available (Fig 79).

The techniques for dental floss and dental tape are quite different. The thin, circular floss is inserted between the approximal tooth surface and the papilla. The plaque is removed in a coronal direction. The use of dental tape, which involves a rubbing action, is easier to learn.

Unlike dental floss, dental tape can be combined with the application of a medium-abrasive fluoride toothpaste to provide supplementary cleaning and fluoridation effect on the approximal surfaces. Applied with a toothbrush to the buccal, lingual, and occlusal surfaces, fluoride toothpaste is unquestionably beneficial; when applied with approximal oral hygiene aids to the approximal surfaces of the molars and premolars (the key-risk surfaces), it should provide an even greater beneficial effect. Fluoride toothpaste can be applied interdentally with a toothbrush, a finger, or a cotton roll. The best method of application is by syringe, as in PMTC procedures (see chapter 3).

When toothbrushing is supplemented by flossing or the use of tape, more plaque is removed from the proximal surfaces than is accomplished by toothbrushing alone (Bergenholtz et al, 1974; Kiger et al, 1991). Graves et al (1989), in a 2-week supervised clinical trial of patients with gingivitis, found that interdental bleeding was reduced by about 67% with flossing and brushing; only a 35% reduction was achieved by toothbrushing alone. Subsequent trials using comparable protocols have recorded similar reductions (Kinane et al, 1992). A study by Reitman et al (1980), showed that, compared with toothbrushing alone, supplementary flossing increased plaque-free surfaces by 30% to 80%.

Dental tape and particularly dental floss require greater manual dexterity than do wooden toothpicks and take almost twice as long (Gjermo and Flötra, 1970). However, toothpicks are contraindicated in children and young adults because the interdental space is filled by a normal papilla. Waxed and unwaxed floss do not differ appreciably in cleaning efficacy (Bergenholtz and Brithon, 1980; Gjermo and Flötra, 1970). Dental tape and waxed or unwaxed dental floss also appear to be similarly effective in reducing interdental gingival bleeding (Graves et al, 1989). Dental tape and floss have been shown to be superior to wooden toothpicks, especially in removing plaque from the lingual aspect of proximal surfaces (Bergenholtz and Brithon, 1980; Gjermo and Flötra, 1970).

A holder may facilitate the use of dental tape or floss. In a recent study, the use of a special individually threaded floss holder, was compared with manual use of similar unwaxed dental floss. A group of 30 volunteers used either the floss holder or finger flossing for 30 days, followed by a 14-day washout period, in which they brushed but used no floss. They then used the other product for 30 days. Both groups demonstrated improved gingival health, but no intergroup differences were noted.

Use of triangular-tipped toothpicks

Figure 80 is a typical bitewing radiograph from a Scandinavian adult, aged about 50 years. The patient has approximal restorations with subgingival overhangs, secondary caries, and some loss of alveolar bone. It has been shown in vivo and at autopsy that a triangular-tipped toothpick, inserted interproximally, can maintain a plaque-free region 2 to 3 mm subgingivally (Fig 81) (Mörch and Waerhaug, 1956). The resilience of the gingival papilla allows cleaning apical to the subgingival margins of restorations (ie, surfaces at risk for recurrent caries). Thus, a more appropriate term for plaque removal by a triangular-tipped toothpick, particularly on the approximal surfaces of the posterior teeth, is *gingival*, rather than *supragingival, plaque control.*

For prevention and control of periodontal diseases on these key-risk surfaces, removal of subgingival plaque is probably more important than supragingival plaque control. For open interproximal surfaces, common among normal adults, toothpicks are most appropriate. In restoration of approximal surfaces, a triangular-shaped wedge ensures a correctly shaped embrasure, readily accessible by the aids normally used for interdental oral hygiene. The bitewing radiograph can act as a guide in selecting a suitable oral hygiene aid for interdental cleaning. The best time to apply fluoride toothpaste, for optimal cleaning and delivery of fluoride, is just as the gingival papilla is depressed (see Fig 81).

In subjects aged 40 years or younger, approximal plaque distal to the canines is often difficult to detect, even after application of a disclosing agent, because the gingival papillae completely occupy the interproximal spaces, obscuring the disclosed plaque. The few exceptions are those patients younger than 40 years with approximal surfaces exposed by advanced periodontitis. Because the posterior interproximal spaces normally have wider lingual embrasures, the most effective means of plaque removal is by inserting triangular-tipped toothpicks from the lingual aspect.

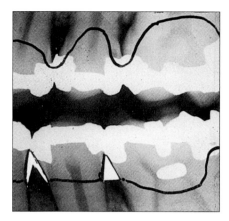

Fig 80 Bitewing radiograph of a 50-year-old Scandinavian. The location of the gingival margin and papillae is marked. Placement of two pointed (wedge-shaped and triangular) toothpicks is illustrated.

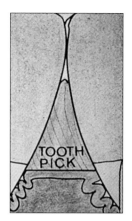

Fig 81 Triangular pointed toothpick inserted interdentally. Because of the resilience of the papillae, plaque biofilms can be removed 2 to 3 mm subgingivally. Delivery of fluoride from toothpaste is also enhanced by depression of the gingival papilla. (Illustration by J. Waerhaug, courtesy of the University of Oslo.)

Fig 82 The EVA-H holder with EVA-7 wooden pointed toothpick cleaning interproximally between the mandibular first and second molars (key-risk surfaces in a right-handed person) from a lingual position.

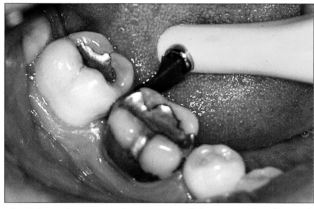

Fig 83 A close-up with EVA-H and the triangular-shaped, pointed EVA-123 tip in the same position as in Fig 82.

Insertion is more comfortable and effective if the toothpick is fixed in a handle, eg, the EVA-H (Dentatus, Hägersten, Sweden) (Axelsson, 1969) (Figs 82 to 85). Figure 84 shows triangular-tipped, wooden toothpicks (EVA-7; Dentatus) with the white surface to be used against the gingival papilla. If the papilla is inflamed, the surface will be discolored by bleeding, an indication of ineffective approximal plaque control. The V-shaped, pointed, flexible, plastic-tipped EVA-123

(Dentatus) can also be used in the handle (see Figs 80 and 83).

In adults, needs-related interproximal cleaning should start with the mandibular teeth, approached from the lingual aspect of the embrasures; this is followed by the approximal surfaces of the maxillary molars and premolars, also approached from the lingual aspect (Fig 86).

Bergenholtz et al (1974) demonstrated that, in open interdental spaces, round or rectangular

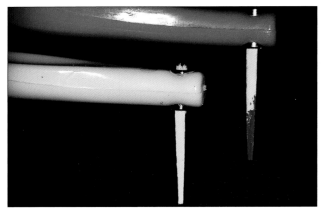

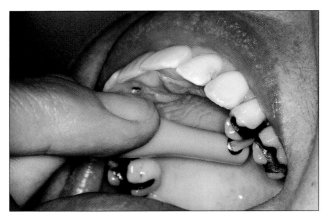

Fig 84 Blood on the back side (ie, the side that is painted white and is placed against the papilla) of the EVA-7 wooden tip indicates that the papilla is still inflamed and interproximal cleaning must be improved.

Fig 85 Interproximal cleaning with EVA-H and EVA-7 between the maxillary second premolar and first molar from a palatal position.

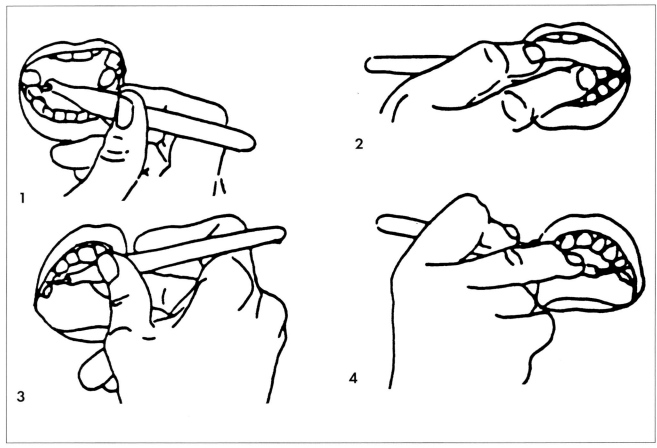

Fig 86 Interproximal cleaning from lingual and palatal positions with EVA-H and EVA-7 in ranking order: (1) Mandibular right side; (2) Mandibular left side; (3) Maxillary right side; (4) Maxillary left side.

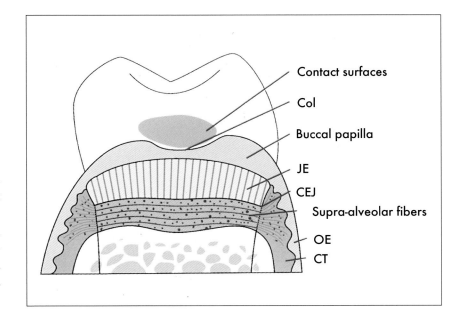

Fig 87 Cross section of hard and soft tissues in the approximal area of mandibular first molar contact surface. (JE) Junctional epithelium, which attaches to the tooth surface during healthy conditions; (CEJ) Cementoenamel junction; (OE) Oral epithelium; (CT) Connective tissue.

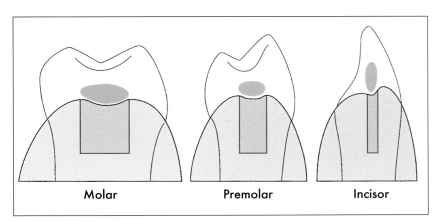

Fig 88 The width and shape of the approximal tooth surfaces and contact surfces of molars, premolars, and incisors.

wooden toothpicks were ineffective on the lingual aspects of proximal surfaces and that the triangular shape was superior. Triangular wooden toothpicks were more effective overall than dental floss. Toothpicks made from wood with low surface hardness and high strength appear to be more efficient (Bergenholtz et al, 1980).

The narrow approximal spaces of the anterior teeth do not require supplementary oral hygiene. In contrast, interdental cleaning is essential for the broad, approximal key-risk surfaces of the molars and premolars, where the buccal and lingual papillae tend to be inflamed and swollen, im-

peding access of the toothbrush bristles (Figs 87 and 88).

On the approximal surfaces of the molars and premolars, correct daily use of triangular-tipped wooden toothpicks and a medium-abrasive fluoride toothpaste should not lead to iatrogenic abrasion. In the adult population, most of these surfaces are restored (see Fig 80). In most of the maintenance patients in the author's practice, EVA reciprocating tips have been used in the EVA-Profin (Dentatus) contra-angle handpiece (see chapter 3) for PMTC one to four times per year, for more than 30 years. The tips are inserted in-

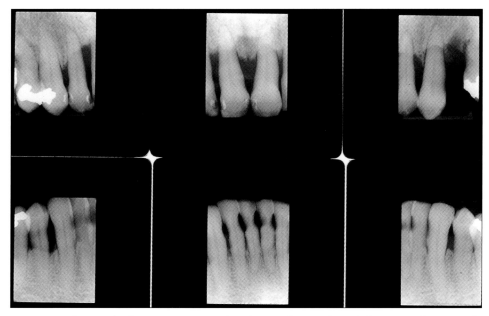

Fig 89 Radiographs from a patient with advanced periodontal disease, whose tooth-cleaning habits were self-taught. "Needs-related" toothcleaning consisted of intensive use of a circular toothpick interdentally in the mandibular incisors, resulting in advanced iatrogenic defects.

terproximally for about 10 seconds, at a speed of about 16,000 reciprocating strokes per minute, ie, approximately 2,500 to 10,000 strokes per year, for more than 30 years. The tips are used in conjunction with a prophylaxis paste that is more abrasive than medium-abrasive toothpastes. In addition to professional cleaning, more than 90% of the patients also use triangular-tipped, wooden toothpicks with fluoride toothpaste daily (Axelsson et al, 2004). No iatrogenic abrasion has been observed on the approximal surfaces of the molars and premolars: The approximal papillae are very healthy, which is confirmed by the average approximal pocket depth of only 2.2 mm (Axelsson et al, 2004) (see chapter 3), and there has been almost no caries development, irrespective of age. There was only one new caries lesion or less per individual during the first 15 years (Axelsson et al, 1991), and the caries incidence remained approximately the same during the following 15 years (Axelsson et al, 2004).

In contrast, the radiographs in Fig 89, from a self-taught subject with advanced periodontal dis-ease, show the effect of constant use of a circular wooden toothpick in the interdental spaces of the delicate mandibular incisor teeth.

The potential benefits of fluoridated toothpicks, dental tape, and floss have largely been overlooked, despite data from as early as 1981 showing high fluoride uptake and release from sodium fluoride (NaF)–impregnated wooden toothpicks (Mörch and Bjorvatn, 1981). Several brands of fluoridated wooden toothpicks, plastic toothpicks, and dental tapes and floss have recently been introduced (see Fig 79).

Studies by Petersson et al (1994) and Kashani et al (1995, 1998a, 1998b) show rapid uptake and release of fluoride from wooden toothpicks: Wood can store NaF crystals, both on the surface and in porosities. The NaF crystals dissolve readily in contact with liquids, such as water or saliva. The toothpick should be moistened in the saliva for a few seconds just before use to accelerate the release of fluoride. In vivo, only the point of the toothpick is used. The in vivo study by Kashani et al (1998b) showed that 4 weeks' daily use of fluoridated wooden toothpicks reduced demineralization of enamel and dentin at approximal sites in situ. In another study, Kashani et al (1998c) used wooden toothpicks impregnated with 4% NaF and 2% chlorhexidine, ie, a combination of mechanical and chemical plaque control and topical fluoride release.

Use of interdental brushes

As an effect of progressive periodontal disease and the associated loss of periodontal support, the interproximal spaces gradually enlarge. The interdental anatomy also changes in response to periodontal treatment. After initial therapy, swelling in the papillae subsides, leaving larger, wide-open spaces at sites previously accessible only to dental floss or tape or at the most slim triangular toothpicks. The establishment of excellent gingival plaque control after treatment is followed by gingival recession and remodeling of the alveolar bone. In such wide-open interdental spaces, interdental brushes are more suitable for cleaning than are dental floss, dental tape, and toothpicks.

From their inception, interdental brushes were available in a variety of sizes to fit varying interdental space widths (Gjermo and Flötra, 1970). The larger brushes are generally held by the wire handle, which is the central filament holding structure. Smaller versions are attached to a metal or plastic handle to facilitate maneuverability and accessibility (Fig 90). Studies have shown that the interdental brush is superior to dental floss in cleaning large interdental spaces (Bergenholtz and Olsson, 1984; Kiger et al, 1991). In a more recent study, in patients with moderate to severe periodontitis, Christou et al (1998) showed the interdental brush to be more efficient than dental floss in removal of approximal plaque and pocket reduction. The subjects stated a preference for the interdental brush over dental floss.

Waerhaug (1976) showed that individuals who habitually used the interdental brush were able to maintain supragingival proximal surfaces

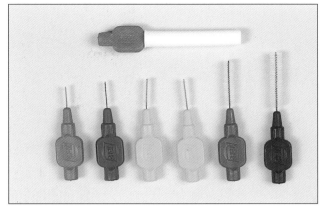

Fig 90 An assortment of different sizes of interdental brushes with plastic handles.

free of plaque and to remove some subgingival plaque below the gingival margin. Although unconfirmed to date, it is believed that the most efficient cleaning is achieved if the largest brush that will fit the embrasures is selected. It is recommended that fluoride toothpaste be applied on interdental brushes to enhance the cleaning and caries-preventive effect on these key-risk surfaces.

To date, the design of all interdental brushes has been circular in cross section. However, the open interproximal space is roughly triangular in cross section and concave in longitudinal section, and the embrasure is wider lingually than it is buccally. A tailored interdental brush, triangular in cross section instead of round, would enhance plaque removal. It should be used in a handle (EVA-H), enabling insertion from the lingual embrasure (Figs 91a and 91b). The length of the bristle in cross section as well as longitudinal section should be tailored to the shape of the interdental space. The flat, smooth base of the T-shaped profile to which the bristles are attached would suppress the papillae apically, optimizing plaque removal even on very broad approximal surfaces 2 to 3 mm subgingivally (Axelsson, 1978).

An electric, rotating interdental brush (Braun/ Oral-B Interclean [ID2], Gillette) has recently been introduced. It resembles a miniature Weed Eater, with a 10-mm Hytrel filament, which ex-

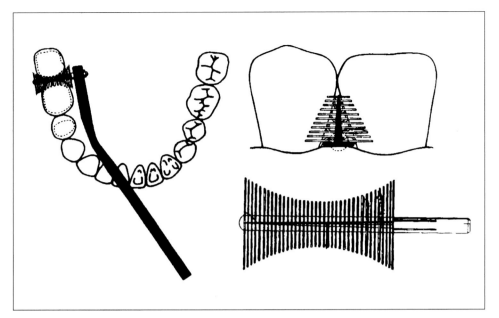

Fig 91a Tailor-made interdental brush as it should look.

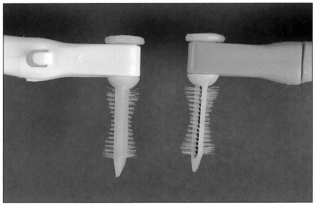

Fig 91b The "tailor-made" interdental brush is shown in long section from the front *(right)* and from the back *(left)*. The flat, smooth back side presses down on the buccal and lingual papillae in the posterior teeth, thereby facilitating 2- to 3-mm subgingival removal of plaque biofilms. Observe the different lengths of the bristles, which optimize the accessibility and cleaning effect of the line angles. The pointed triangular tip facilitates entrance into interproximal spaces.

trudes and rotates on activation. Unlike conventional floss, this device can be used with one hand. The tip is inserted interproximally, and the device is activated. The filament extrudes into the interproximal space, creating an elliptical motion of 6,500 revolutions per minute, removing plaque from the adjacent interproximal tooth surfaces (Figs 92 to 94). The recommended application time is 2 to 3 seconds per interproximal space.

To determine its safety and effectiveness, the device was tested by a group of 52 volunteers with gingivitis in a 30-day study. Half the group used conventional dental floss, and the other half used the Interclean once daily in the evening. They were also instructed to brush twice daily. Both

groups demonstrated significant reductions in plaque, bleeding, and gingivitis. No differences were observed between the groups (Gordon et al, 1996).

Another 6-week study compared the Interclean (ID2) brush and dental floss in a group of 48 volunteers. As in the previous study, plaque, gingivitis, and bleeding scores declined in both groups. The authors concluded that the Interclean (ID2) brush "should benefit patients who find manual floss difficult to use or who have poor manual dexterity" (Cronin and Dembling, 1996). However, the interdental subgingival cleaning effect of a rotating circular brush may be questioned, in comparison with the previously de-

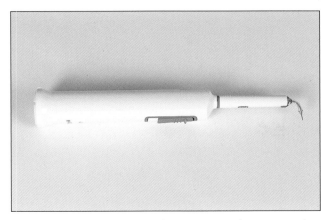

Fig 92 Braun/Oral-B Interclean rotating electric interdental brush.

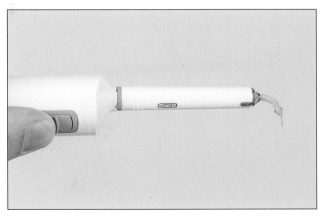

Fig 93 The interdental brush in action.

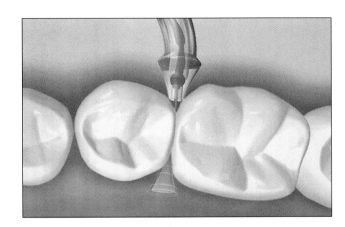

Fig 94 Close-up of Braun Oral-B Interclean brush in situ interdentally.

scribed 2- to 3-mm subgingival cleaning effect of a triangular pointed toothpick and the tailor-designed interdental brush (see Figs 81, 91a, and 91b).

As discussed earlier, the Rota-dent rotating electric toothbrush (see Figs 69 and 70) could also be useful for interdental cleaning in wide-open interdental spaces (for reviews on interdental cleaning materials and methods, see Kinane, 1998 and Warren and Chater, 1996).

Supplemental oral hygiene aids

The single-tufted toothbrush was designed to improve access to tipped, rotated, or displaced teeth and to clean teeth affected by gingival recession or teeth recontoured by odontoplasty to correct grade I furcation involvement. Gjermo and Flötra (1970) demonstrated that the combined use of the single-tufted brush and wooden toothpicks compensated for the lack of effectiveness of wooden toothpicks alone within lingual embrasures. This study emphasizes the need to combine interdental cleaning techniques, given the variation of conditions under which plaque accumulates in the dentition.

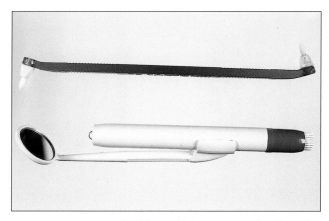

Fig 95 Lactona 27 double-ended interspace brush and a supplementary illuminated mouth mirror for self-diagnosis.

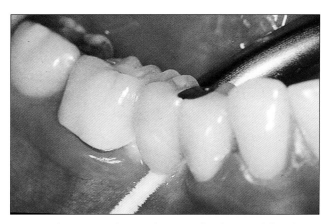

Fig 96 The Super Floss aid, specially designed for cleaning around fixed partial denture pontics.

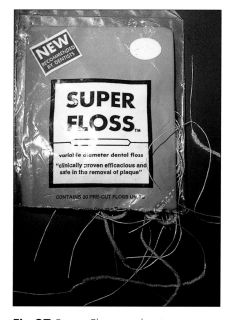

Fig 97 Super Floss packaging.

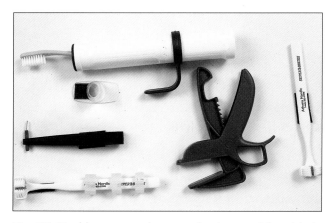

Fig 98 Oral hygiene aids specially designed for elderly or physically disabled persons with poor manual dexterity.

Figure 95 shows the double-ended Lactona 27 single-tufted interspace brush (Voprak Lactona, Bergen op Zoom, The Netherlands). The alternative angulations of the bristles offer excellent accessibility. The supplementary illuminated mouth mirror is essential for self-diagnosis, inspection before oral hygiene procedures, and

evaluation of the outcome. Oral-B Super Floss (Gillette) is specially designed for cleaning around fixed partial denture pontics (Figs 96 and 97). Figure 98 shows a recently introduced range of oral hygiene aids specially designed for elderly or physically disabled persons with poor manual dexterity.

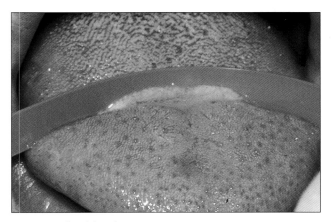

Fig 99 A flexible plastic tongue scraper.

Tongue cleaning

Oral hygiene procedures should include the tongue (Axelsson et al, 1987b; Bollen et al, 1998; Mongardini et al, 1999). This is of particular importance in selected high–caries-risk patients, high–periodontal-risk patients, and patients suffering from halitosis.

In a 30-month longitudinal study, 180 children with more than 1 million mutans streptococci per 1 mL of saliva were selected from a population of more than 600 schoolchildren aged 13 years. The study evaluated the effect on oral mutans streptococci of initial, intensive "complete-mouth" disinfection, comprising PMTC, tongue scraping, mouthrinsing with 0.2% chlorhexidine, interdental application of chlorhexidine-gel, and individualized oral hygiene education four times within a week, followed by one single complete-mouth disinfection treatment every 6 months. The level of mutans streptococci colonization of the approximal surfaces, salivary levels of mutans streptococci, and levels of mutans streptococci on the dorsum of the tongue were assessed before and after tongue scraping.

Two weeks after the initial intensive complete-mouth disinfection, there was a dramatic reduction in mutans streptococci on the approximal surfaces, in the saliva, and on the dorsum of the tongue. However, mutans streptococci was not eliminated from the oral cavity, and a single treatment every 6 months was too infrequent to maintain the initial suppression. The mutans streptococci levels returned to about 50% of the baseline values, but the effect on dental caries incidence was good.

Another interesting finding was that the level of mutans streptococci, sampled with a wooden spatula from the dorsum of the tongue, was consistently lower immediately before, rather than after, use of a flexible plastic tongue scraper (Fig 99). Because the concentration of mutans streptococci in the salivary film on the deposits of the tongue was low before scraping, this indicates that the rough surface of the tongue may serve as a reservoir for mutans streptococci (Axelsson et al, 1987b).

The dorsum of the tongue is the main habitat for Streptococcus salivarius, another very potent cariogenic bacteria, and may also serve as a reservoir for the periopathogens *Porphyromonas gingivalis*, *Prevotella intermedia* (Van Winkelhoff et al, 1988), and *Actinobacillus actinomycetemcomitans* (Asikainen et al, 1995; Müller et al, 1990; Slots et al, 1980). In a longitudinal study, Mongardini et al (1999) showed that, compared with scaling and debridement alone, supplementary complete-mouth disinfection, including daily tongue brushing, significantly reduced probing depths and increased clinical attachment.

Halitosis is caused mainly by sulfur compounds, the end products of protein and amino acid metabolism by the anaerobic gram-negative microflora, which are found in the subgingival plaque biofilms and in the depths of deposits on the dorsum of the tongue. For elimination, debridement of deep, diseased periodontal pockets is recommended, supplemented by daily gingival plaque control and tongue scraping. Figure 100 shows a range of aids for tongue-cleaning. Further reduction in halitosis may be achieved by supplementing mechanical tongue scraping and debridement with zinc oxide lozenges (Fig 101). Zinc oxide neutralizes the malodorous sulfur compounds.

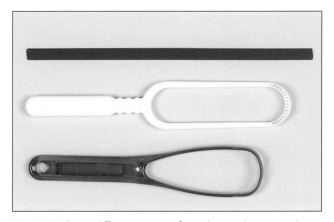

Fig 100 Three different types of mechanical tongue cleaning aids: a flexible plastic scraper (*top*; see Fig 99), a specially designed tongue brush (*middle*), and a specially designed tongue scraper (*bottom*).

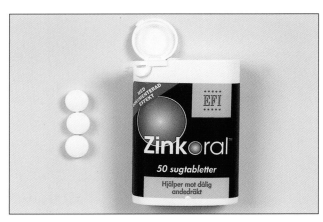

Fig 101 Zinc oxide lozenges for neutralizing malodorous sulfur compounds.

NEEDS-RELATED ORAL HYGIENE AT HOME

A fundamental principle for all preventive action is that the positive effect is greatest where the risk of disease is greatest. Initially, patients are most likely to see positive results of their oral hygiene efforts if key-risk teeth and surfaces are targeted. Subsequently, more stringent oral hygiene standards may be imposed by including the buccal surfaces, where the results are largely indiscernible.

As discussed earlier, although daily manual toothbrushing is a well-established routine in most industrialized countries, interdental cleaning is very uncommon. In the light of current knowledge of normal plaque distribution and the distribution of dental disease in the dentition and on individual tooth surfaces, it cannot be claimed that needs-related toothcleaning is being practiced in toothbrushing populations. The adult patient today cleans mainly the tooth surfaces that are least susceptible to disease. The introduction of needs-related oral hygiene procedures is a potentially powerful preventive measure in this population. Of 8,760 hours per year, the individual patient normally spends no more than 2 hours in the dental clinic. Thus, the patient's own daily oral hygiene efforts may be regarded as potentially greater determinants of oral health than professional services that are provided once or twice a year.

Selection of oral hygiene aids should be based on the individual needs of the patient, and their use should be demonstrated by a dentist or a dental hygienist. Figures 102 to 105 show a range of oral hygiene aids appropriate for preschool children, older children, adult periodontal-risk patients, and adult caries-risk patients, respectively.

Needs-related home care for children and young adults

In children and young adults, the predominant oral disease is dental caries: Fewer than 30% of 12-year-old Europeans are caries free, but only 0.1% to 0.3% have aggressive periodontitis. Recent studies have shown that initiation of caries lesions tends to occur more frequently at specific ages, particularly in children. The key periods seem to be during eruption of the permanent molars and in the early posteruptive period, in which the

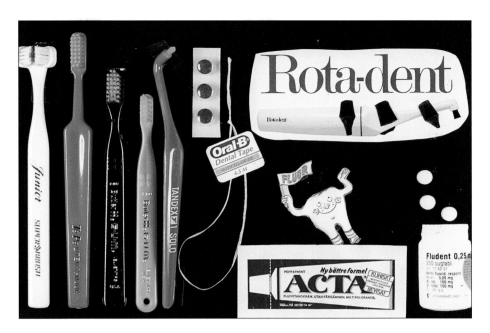

Fig 102 Assortment of oral hygiene aids suitable for preschool children.

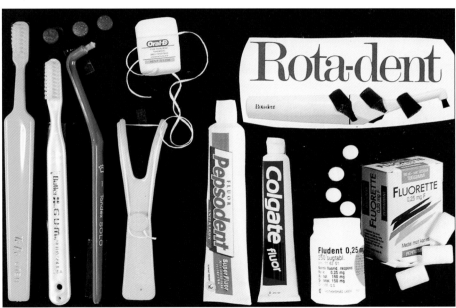

Fig 103 Range of oral hygiene aids suitable for young adults.

enamel undergoes secondary maturation. In this context, it should be noted that, on average, the teeth erupt 6 to 12 months earlier in girls than in boys (Teivens et al, 1996).

Key-risk age group 1: 1 to 2 years

Studies by Köhler and Bratthall (1978) and Köhler et al (1982) showed that mothers with high levels of salivary mutans streptococci frequently transmit mutans streptococci to their offspring as soon as the first primary teeth erupt, leading to increased development of caries. Other stud-

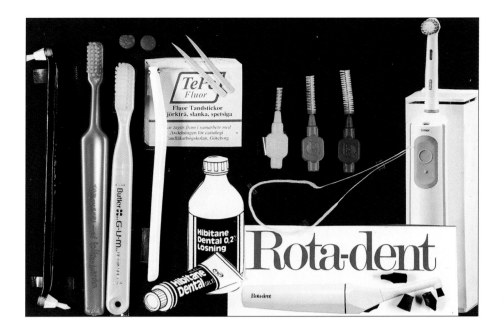

Fig 104 Range of oral hygiene aids suitable for periodontal-risk adult patients.

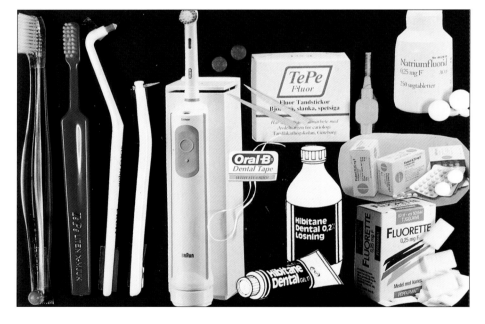

Fig 105 Range of oral hygiene aids suitable for caries-risk adult patients.

ies have shown that 1-year-old children who already have plaque and gingivitis develop several caries lesions during the following years, while infants with clean teeth and healthy gingivae, maintained by regular daily cleaning by their parents, remain caries free (Wendt et al, 1994).

Alaluusua and Malmivirta (1994) reported similar findings in an 18-month study of 92 mothers and toddlers (baseline age of 19 months). The study assessed the potential of four variables, which can be measured at the chairside, to identify those children who would experience caries during the subsequent 1.5-year period: visible plaque on the labial surfaces of the maxillary incisors, the use of a nursing bottle, mother's caries prevalence, and mother's salivary level of mutans streptococci.

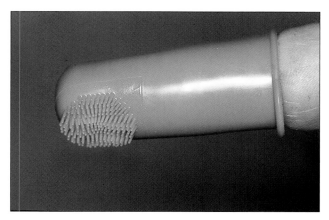

Fig 106 A simple rubber brush to be used on the index finger of parents as the very first oral hygiene aid in their baby's mouth.

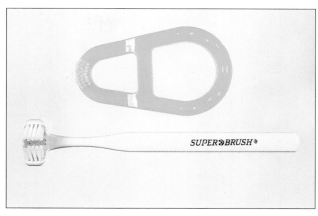

Fig 107 (top) A brush with an extra-wide handle for the primary incisors, making it safe for infant handling. (bottom) A triple-headed toothbrush suitable for parents to use in the cleaning of their infant's teeth.

Of the four variables, visible plaque and the use of a nursing bottle were strongly associated with caries development. The other two variables had weak or no statistically significant associations. The best indicator of risk was visible plaque, with a sensitivity of 83%, specificity of 92%, positive prognostic value of 63%, and negative prognostic value of 97%. On the basis of visible plaque, 91% of the children were correctly classified, compared to 72% to 77% for the other variables. The results suggest that visible plaque on the labial surfaces of maxillary incisors of a young child is a sign of caries risk.

In another investigation, Grindefjord et al (1995) studied the relative risk (odds ratio) that 1-year-old infants would develop caries by the age of 3.5 years. Infants who had poor oral hygiene and unfavorable dietary habits, salivary mutans streptococci, little or no exposure to fluoride, and parents with low educational level or immigrant background were at a 32 times greater risk of developing caries than were other children in the study.

Once established, habits are difficult to change. It is therefore important to establish good habits as early as possible and to postpone or prevent the introduction of bad habits. Moreover, the enamel of erupting and newly erupted primary and permanent teeth is most caries susceptible until the completion of secondary maturation (Kotsanos and Darling, 1991). In 1- to 3-year-old children, the specific immune system, particularly immunoglobulins in saliva, is immature, and poor oral hygiene will therefore favor the establishment of cariogenic microflora such as mutans streptococci.

Regular daily oral hygiene habits should therefore be established as soon as the first primary incisors begin to erupt. This should be done twice a day, while the infant is seated in the parent's lap. A simple rubber brush attached to the index finger could be the very first step (Fig 106), followed by a more efficient brush for cleaning the incisors (Fig 107). The infant can also be allowed to play with the brush in the mouth. As soon as the primary molars start to erupt, the parents should clean the infant's teeth twice a day. At this stage, a more traditional, efficient toothbrush and a pea-sized amount of fluoride toothpaste are recommended. Figure 107 shows a toothbrush specially designed to enable parents to clean the teeth of infants and toddlers comfortably and efficiently.

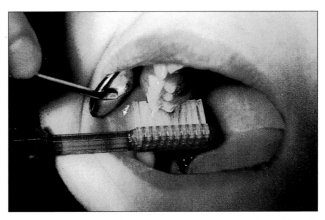

Fig 108 Special technique for using a manual toothbrush to clean the fissures of erupting molars. (From Kuzmina, 1997. Reprinted with permission.)

Key-risk age group 2: 5 to 7 years (eruption of the permanent first molars)

Carvalho et al (1989) studied the pattern and amount of de novo plaque accumulation on the occlusal surfaces of the permanent first molars, 48 hours after PMTC, in relation to eruption stage. Heavy plaque reaccumulation was found on the occlusal surfaces of erupting maxillary and mandibular molars, particularly in the distal and central fossae and related fissures. This is in sharp contrast to the fully erupted molars, which are subject to normal chewing friction (see Fig 54, chapter 1). Abrasion from normal mastication significantly limits plaque formation, which explains why almost all occlusal caries in molars begins in the distal and central fossae, during the extremely long eruption period of 14 to 18 months. By contrast, fissure caries is very rare in premolars, which have a brief eruption period of only 1 to 2 months.

Parents should intensify mechanical plaque control in their children's teeth twice a day using fluoride toothpaste. Special attention should be paid to the erupting first molars. To optimize removal of fissure plaque in these teeth, the toothbrush should be directed at a right angle (90 degrees) along the arch of the posterior teeth. Small, rotating scrubbing movements should be used,

until all the fissure plaque is removed (Fig 108). If the pointed, rotating Rota-dent electric toothbrush is used for this purpose, the parents in effect accomplish semiprofessional cleaning.

In needs-related cleaning for children younger than 8 years, interdental cleaning should target the mesial surfaces of the permanent first molars. When carried out by a parent, this procedure may be facilitated by use of a dental floss holder. Needs-related cleaning of the lingual and buccal surfaces should then commence on the lingual surface of the mandibular posterior teeth, ie, the sites of heaviest plaque reaccumulation. Because of the thickness of the alveolar bone and the lingual inclination of the teeth, the risk of trauma is also smallest in this region.

Key-risk age group 3: 11 to 14 years (eruption of the second molars)

Normally the second molars start to erupt at the age of 11 to 11.5 years in girls and around the age of 12 years in boys (Teivens et al, 1996). The total eruption time is 16 to 18 months. During this period, the approximal surfaces of the newly erupted posterior teeth are undergoing secondary maturation of the enamel and are also at their most caries susceptible (Kotsanos and Darling, 1991). Although 11- to 14-year-old children have by far the highest number of intact tooth surfaces, many are at risk. To protect the intact surfaces and to remineralize any noncavitated lesions, integrated plaque control measures and the use of fluoride agents should be intensified on the approximal surfaces of all the posterior teeth and the occlusal surfaces of the second molars, starting with 11- to 11.5-year-old girls. If this is maintained throughout the secondary maturation period and needs-related self-care habits are established, there is a high probability that the intact tooth surfaces will remain intact for life.

At this age, needs-related cleaning should begin with interproximal cleaning of the lateral segments, up to the distal surfaces of the canines. The most suitable oral hygiene aid is fluoridated

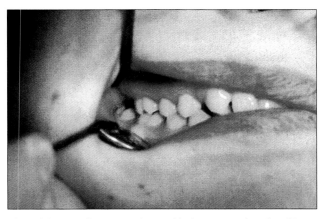

Fig 109 Partially erupted mandibular second molar. (From Kuzmina, 1997. Reprinted with permission.)

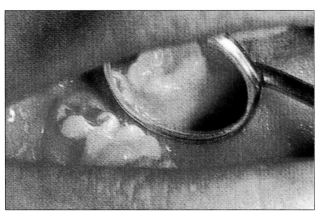

Fig 110 Compare the amount of stained occlusal plaque on the partially erupted second molar with that on the fully erupted first molar with chewing function. (From Kuzmina, 1997. Reprinted with permission.)

dental tape, used with or without a holder. When flat fluoridated tape is used, the rubbing method is most effective. This method is easy to learn and should be combined with application of fluoride toothpaste. If a large number of approximal surfaces are restored, wooden fluoridated toothpicks might also be appropriate. Bitewing radiographs are useful guides for selecting a suitable oral hygiene aid for the interproximal areas.

In the previously discussed in vivo study by Badersten and Egelberg (1972), designed to evaluate the plaque-removing effect of toothpaste during toothbrushing, the toothpaste resulted in a 50% improvement. It is clear that toothpaste should also be used on the key-risk surfaces, ie, in the interproximal molar and premolar regions.

The fissures of erupting second molars without full chewing function accumulate about five times more undisturbed plaque than do molars in full function. This is illustrated in Fig 109, which shows a mandibular second molar that has no chewing function because the opposing tooth is unerupted. Heavy plaque deposits (stained red by disclosing solution) have accumulated undisturbed in the occlusal fissures. No plaque has accumulated in the fissures of the mandibular first molar (Fig 110) because both the mandibular and maxillary first molars are fully erupted and in

chewing function (see also Fig 54 in chapter 1). To clean the fissures of erupting second molars, the child is instructed to use the same toothbrushing technique described earlier for first molars (see Fig 108).

After the approximal surfaces of the posterior teeth and the occlusal surfaces of the second molars, the next surfaces to be cleaned, in right-handed children, are the lingual surfaces of the mandibular right molars and premolars; a toothbrush and fluoride toothpaste should be used. In right-handed subjects, these surfaces are often neglected, as are the corresponding surfaces on the left side in left-handed subjects. The remaining lingual surfaces of the mandibular teeth and subsequently the buccal surfaces of the mandibular and maxillary teeth are then cleaned.

Finally, to optimize the effect of fluoride, the "slurry" of fluoride toothpaste and saliva is swished around and between the teeth and spit out, followed by a similar swish with just a spoonful of water.

Needs-related home care for adults

As discussed earlier, Fig 80 shows the typical dental status of the average Scandinavian adult. It may be assumed that approximal restorations and cavity margins are located subgingivally, and that subgingival plaque is present on the remainder of the restored interproximal surfaces. To clean these surfaces, fluoride toothpaste should be applied interdentally to the molars and premolars with a toothbrush, the index finger, or a syringe. The approximal surfaces of the mandibular molars and premolars are cleaned from the lingual direction with a triangular fluoridated toothpick in a handle, beginning on the right side in a right-handed subject (see Fig 86). The approximal surfaces of the maxillary molars and premolars are cleaned in the same manner. Very large interproximal spaces may require use of an interdental brush instead of a toothpick. Finally, the toothbrush and fluoride toothpaste are used, starting on the lingual surfaces of the mandibular right molars and premolars, as described for 11- to 14-year-old children.

ESTABLISHMENT OF NEEDS-RELATED ORAL HYGIENE HABITS

Motivation and needs

A prerequisite for successfully establishing needs-related oral hygiene habits is a well-motivated, well-informed, and well-instructed patient. *Motivation* is defined as readiness to act, or the driving force behind actions. Greater responsibility has been identified as the most lasting motivating factor. An individual's actions are also governed by perceived needs. According to Abraham Maslow's famous hierarchy of needs, the most primitive needs are physiologic, including the need to breathe, satisfy hunger and thirst, and rest. Until all physiologic needs are satisfied, a person does not give priority to emotional needs; even less attention is paid to social needs and the need for activity.

In 1978, a national survey of the adult Swedish population by Håkansson, indicated that just over 50% of all dentate adults believed that they had good teeth. Only 20% thought they had poor teeth, and 94% felt they could masticate well. A remarkable fact was that as many as 75% of complete denture wearers felt that they could masticate well. The study disclosed the marked discrepancy between the subjective perception by patients and the objective, professional evaluation of actual dental needs based on accurate diagnosis. In a more recent Swedish study by Axelsson and Paulander (1994), in a randomized sample of more than 600 dentate 50- to 55-year-old subjects, very few were dissatisfied with their teeth (color, 8%; shape, 5%; thermal sensitivity, 5%; malocclusion, 3%; temporal joint, 3%; oral mucosal problems, 3%; bruxism, 5%; bleeding gingiva, 5%; and halitosis 4%).

The patient must tactfully be made aware of the true dental conditions. A prerequisite is accurate assessment of the Plaque Formation Rate Index (PFRI), gingivitis, periodontitis, caries, and risk profile. These data allow the dentist to identify dental needs and form a basis for discussion of responsibilities and the potential for improvement in oral health. Of at least equal importance is the personal history of the patient, including attitude toward dental care and toward his or her own teeth, medical conditions that might compromise oral status, and general problems and unsatisfied needs of a higher order than dental problems on the Maslow hierarchy of needs. Other relevant data to be noted include the patient's current oral hygiene routines (if any), dietary habits, medication, and fluoride history.

The diagnosis discloses the current situation, but the history explains why this condition has developed. Of great importance at a patient's first appointment is asking neutral questions during the interview to become acquainted with the individual. Except when the appointment has been made because of an emergency (eg, toothache), a good rule of thumb is that the patient should talk

70% of the time and the clinician only 30%. An important detail that should be elicited during history-taking is the patient's concern about remaining dentate. According to the Swedish national survey (Håkansson, 1978), 84% of all adults definitely wanted to remain dentate, only 1.5% preferring complete dentures.

It is also important to adopt a positive attitude and to regard each new patient as potentially compliant. According to the "Rosenthal" or "Pygmalion" effect, moreover, our expectations of a person are more often than not fulfilled. If we believe that a person is capable of doing something really well, the probability of this actually happening is greater than if we believe the opposite. This is true, irrespective of the actual circumstances and the efforts made in other respects by both ourselves and the patient. Furthermore, rather than be proved wrong, we unconsciously wish to see our own first evaluation borne out. From the very outset, therefore, it is important to show confidence in the patients' ability to comply, for example, with respect to oral hygiene.

Unfortunately, dentists are at a disadvantage. Most adult patients (85%) make dental appointments for restorative treatment, and, according to the national survey, this is mainly what the profession provides. Moreover, 10% of adults regard dental treatment as such an unpleasant experience that they seek care only reluctantly, if at all (Håkansson, 1978). In a more recent study (Axelsson and Paulander, 1994) 50- to 55-year-old adults ranked the importance of positive experiences of dental care as follows:

1. Regular maintenance care
2. Positive, caring attitude by the dental personnel
3. High quality of dental care and in particular information (case presentation) and preventive dentistry
4. Reasonable fees

The following were the most negative experiences of dental care:

1. Pain
2. Negative attitudes by dental personnel
3. Low competence and absence of preventive dentistry
4. High fees

In this context, dental hygienists and preventive dental nurses have an advantage, because they provide solely noninvasive, preventive services, and patients in their care have quite different expectations and wishes.

Information: Choice of vocabulary

At the end of the history taking or patient interview session, attention begins to focus on the patient's own dental health, and the presentation of information becomes relevant. When the patient is motivated to learn more about the etiology and prevention of dental disease, this should be imparted in dialogue form. The informative section designed to broaden the patient's knowledge should be brief, and preferably spread over several visits as the patient's curiosity and interest are stimulated.

It is important that the information be understood, and this necessitates selecting a vocabulary suited to the individual patient. Two extreme cases would be the vocabulary of an immigrant who has not yet managed to learn more than a couple hundred words and the vocabulary of a native speaker, comprising some 5,000 words. Although the dentist may have a vocabulary that includes hundreds of technical dental terms, the dental vocabulary of the immigrant may be confined to five words. Under these circumstances, the language must be extremely simple. However, vocabulary that is too simplified can be uninteresting to the patient, and, in a discussion of this type, the clinician must be prepared for any possible informed questions in the fringe areas of dentistry.

In discussing the etiology and treatment of dental disease, the dentist should bear in mind that

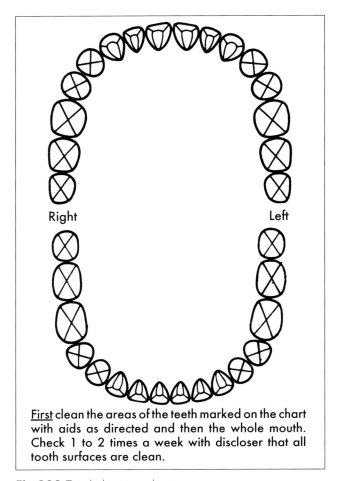

First clean the areas of the teeth marked on the chart with aids as directed and then the whole mouth. Check 1 to 2 times a week with discloser that all tooth surfaces are clean.

Fig 111 Toothcleaning chart.

the patient is only marginally interested in "the average mouth": The purpose of the discussion should be to reinforce the patient's desire to have his or her own dental health documented. Only when the patient asks about his or her own diagnosis, directly or indirectly, should this be presented.

Self-diagnosis

On the basis of individualized knowledge of the etiology, signs and symptoms, and treatment of dental disease, acquired through informative discussion, the patient should be encouraged to take an active role in charting his or her own oral status. It is quite interesting to observe a patient who,

for the first time, searches with curiosity for the signs of dental disease in his or her own mouth. Important aids are a magnifying mirror equipped with a light, an illuminated mouth mirror (see Fig 95), and plaque-disclosing agent. If available, complete-mouth radiographs are an advantage. Other useful equipment includes an intraoral camera (see chapter 8) and computer-aided radiographs (see chapter 6 in volume 3).

Use of a toothcleaning chart

An appropriate beginning is to ask the patient whether he or she has lost any teeth and why they are missing, as well as which teeth he or she thinks are most affected by caries. The patient may be asked to sketch the approximate extent of the damage on a chart (Fig 111) showing all the tooth surfaces of the dentition. Often everything falls into place for the patient even at this early stage. The observations made from the mouth can be compared with the complete-mouth radiographs. Subsequently, the patient should be encouraged to look for signs of disclosed plaque and gingivitis and to draw their general outline on the toothcleaning chart. Later, a detailed professional diagnosis of gingivitis, periodontitis, and dental caries is made and a broad comparison can then be made with the patient's own observations.

Sites with possible loss of alveolar bone support noted from the complete-mouth radiographs can be examined clinically. The patient will probably request a detailed periodontal examination. This should comprise gingival conditions, measurement of probing depth and periodontal attachment loss, and, where indicated, mobility testing and microbiology tests. To allow an evaluation of the individual risk profile, a detailed caries examination should also be carried out.

Use of a risk profile

Risk profiles for tooth loss, dental caries, and periodontal diseases can be presented graphically, by

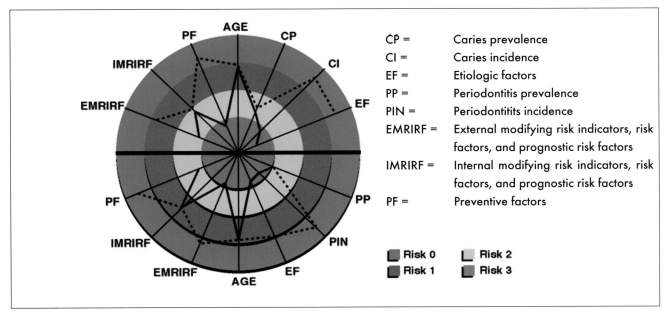

Fig 112 Risk profile for dental caries and periodontal diseases. *(solid line)* C3P3 at baseline; *(dotted line)* C1P1, 2 years later.

combining symptoms (risk markers) of disease (prevalence, incidence, treatment needs, etc); etiologic factors; external modifying risk indicators, risk factors, and prognostic risk factors (EMRIRF); internal modifying risk indicators, risk factors, and prognostic risk factors (IMRIRF); and preventive factors. This can be done manually or by computer. Degrees of risk, 0, 1, 2, and 3, are displayed in green, blue, yellow, and red, respectively (Fig 112). The graphs are very useful tools for communication with the patient during discussions of issues such as oral health status, etiology, modifying factors, prevention, and reevaluations.

Because some patients may suffer from both dental caries and periodontal diseases, risk profiles can be presented separately or combined. (The use of specific risk profiles for high–caries-risk and high–periodontal-risk patients is demonstrated in chapter 8.) Figure 112 illustrates a combined risk profile from a patient, classified after detailed examination and history taking as at high risk for both dental caries and periodontal diseases (C3P3), on the following basis:

1. His prevalence of caries and periodontitis were high.
2. His incidence of caries (CI) and periodontitis (PI) had been very high.
3. He was exposed to many etiologic factors, both nonspecific factors (high PFRI and plaque volume) and specific caries-inducing pathogens and periopathogens (salivary mutans streptococci, salivary lactobacilli, A actinomycetemcomitans, P gingivalis, P intermedia, Bacteroides forsythus, and Treponema denticola.
4. He also exhibited many external and internal modifying risk indicators, risk factors, and prognostic risk factors for both dental caries and periodontal diseases. For dental caries, the most important EMRIRFs were high frequency of intake of sticky, sugar-containing products and a medication with salivary depressive side effects. For periodontal diseases, the most important EMRIRF was regular smoking, 10 to 20 cigarettes per day. Of the IMRIRFs, the most important for dental caries was reduced stimulated salivary secretion rate (0.6

mL/min) and the most important for periodontal diseases was diabetes mellitus.

5. His standard of oral hygiene was very low, and dietary habits were poor. The patient had no preventive dental care habits and was an irregular dental attender.

After presentation of the case findings and a session of self-diagnosis, the dentist and patient discussed a treatment strategy, based on shared responsibility between the patient and the oral health personnel. Two years later, he was classified as a low-risk patient for both dental caries and periodontal diseases (C1P1), on the following basis:

1. The etiologic factors had been dramatically reduced (from red to green) by an initial intensive combination of mechanical and chemical plaque control (self-care and professional) and by maintenance of a high standard of plaque control, ie, a dramatic improvement in the most important preventive factors.

2. Treatment needs (excavation and restoration of open cavities, scaling, root planing, and debridement of diseased periodontal pockets) and plaque-retentive factors were eliminated.

3. Important EMRIRFs were reduced. The patient stopped smoking and reduced the estimated daily sugar clearance time by 80%. In addition there was no further need for medicine with salivary depressive effects. As a consequence of this and regular use of fluoride chewing gum, the salivary secretion rate increased from 0.6 to 1.0 mL/min.

4. The use of fluoride was increased. A new fluoride toothpaste technique was introduced, and fluoride chewing gum was recommended after meals; this was supplemented by professional application of fluoride varnish.

As a consequence of these preventive measures and the healthier lifestyle, the patient developed no new caries lesions and experienced no further loss of periodontal support.

After the dentist and patient jointly locate the signs of dental disease and evaluate the risk profile, they should discuss the probable reasons for the specific location and nature of the signs. It is preferable if the patient broaches the subject. A prerequisite for the appearance of clinical signs of gingivitis, periodontitis, and caries is bacterial colonization of the tooth surfaces. The occurrence of plaque must therefore be an early point of discussion. If the patient's main problems are signs and symptoms of caries, then dietary and salivary factors should also be discussed at this stage.

From a motivational perspective, there are several ways of dealing with the occurrence of plaque. Disclosing plaque by staining is generally regarded as an excellent means of making the location and extent of plaque deposits visible to the naked eye. However, because the papilla occupies most of the interproximal space, the presence of interproximal plaque, particularly in the premolar and molar regions, tends to be underdiagnosed. The use of a disclosing agent should therefore be supplemented by probing of the interproximal surfaces of the premolars and molars. The patient should be encouraged to attempt to locate the disclosed plaque and to transfer his or her observations in broad outline to the toothcleaning chart. This is then supplemented by procedures such as probing.

A "mouth odor test" may also be a useful motivational tool. For human beings to emit unpleasant odors is socially undesirable and, in intimate situations, socially unacceptable. The method combines the use of two senses, sight and smell. The lingual and interproximal regions between the mandibular first and second molars usually harbor malodorous plaque. However, the location of plaque at the first visit shows only where the patient has failed to clean adequately; of greater importance is active involvement in the evaluation of the 24-hour PFRI, described in chapter 1. The patient will then understand (1) how frequently the teeth should be cleaned and (2) which tooth surfaces need particular attention during the toothcleaning procedure.

A normally intelligent patient, made aware of his or her dental needs by actively charting them as outlined, would be unlikely to propose the traditional order of toothbrushing. The patient would see no dental motivation for cleaning maxillary buccal surfaces first, even though this is not only traditionally taught in dental schools but also recommended by the Swedish National Board of Health and Welfare. There is no odontologic justification for cleaning teeth in the same order as they are numbered, and it is also clear that patients will not comply well if they are given no odontologic motivation for cleaning their teeth in the manner recommended by the profession.

In the Karlstad studies on adults, which have been in progress since 1972, patients have been deliberately encouraged in self-diagnosis, primarily of the localization of plaque and gingivitis. As a secondary effect, this has also stimulated their interest in following in detail, from visit to visit, the status of individual gingival pockets and the specific occurrence of caries lesions. On this basis, in their needs-related cleaning, they pay special attention to tooth surfaces in order to arrest noncavitated caries lesions, heal inflamed gingiva, and reduce probing depth (Axelsson and Lindhe, 1978, 1981a; Axelsson et al, 1991; Axelsson et al, 2004).

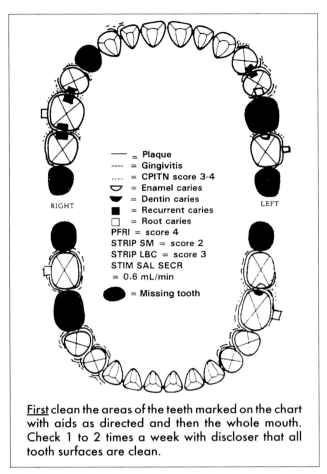

First clean the areas of the teeth marked on the chart with aids as directed and then the whole mouth. Check 1 to 2 times a week with discloser that all tooth surfaces are clean.

Fig 113 Sample toothcleaning chart. (CPITN) Community Periodontal Index of Treatment Needs; (SM) *Streptococcus mutans*; (LBC) *Lactobacillus* count; (STIM SAL SECR) Salivary secretion rate.

Instruction in oral hygiene

On the basis of the joint (patient-professional) observations on dental disease, PFRI, and the location of plaque, the patient should be encouraged to make suggestions as to the choice of oral hygiene aids and, above all, as to the order of priorities for cleaning. The properties and effects of various oral hygiene aids should be discussed at this point.

Needs-related instruction in oral hygiene now commences. The patient is encouraged to practice with the "self-selected" aids and to test them in the mouth. The patient is trained to learn through experience. On the toothcleaning chart, there is ample space to note where the different needs-related, tested aids are to be used, and in

what order. A copy of the toothcleaning chart and the individual risk profile are filed for future reference and the patient is given the toothcleaning chart to use at home as a reminder (Fig 113).

There is great motivational value for the patient in seeing the results of these efforts as quickly as possible. It is equally important not to demand too much or too little in the initial stages. For example, for a high jumper who is used to clearing 2.0 m, 1.5 m is of no interest, clearing 2.2 m can become a reality, and 2.4 m is too much. A fundamental principle for all preventive action is that the effect is greatest where the risk of development of disease is greatest. A consequence of this is that the patient has the greatest chance of being able to discern positive results in oral hygiene efforts by concentrating initially on key-risk teeth and surfaces. Only then should greater demands be made by including areas where results will be barely discernible (for example, on buccal surfaces).

Division of responsibility

When all the joint "diagnoses" have been traced in broad outline on the toothcleaning chart (see Fig 113) and the risk profile (see Fig 112), it is extremely important that the division of responsibility be discussed. The patient's first responsibility is for daily care of the teeth. The necessary measures to be taken should be specified. The responsibility of the dentist, which may involve treating particular signs of dental disease, making regular detailed diagnostic investigations, and planning therapy, should also be specified. The responsibility of the dental hygienist might include providing any necessary information on the subject of dental care, removing calculus, finishing restorations, carrying out PMTC, and providing topical fluoride therapy. If a preventive dental nurse is available, the special areas of responsibility could likewise be specified.

As mentioned earlier, greater responsibility is of considerable importance as a motivating factor. When the division of responsibility is complete, a "contract" might be drawn up and signed by the respective parties involved. New studies in behavioral science substantiate the effectiveness of this method. The responsibility ensuing from an agreement that a person has signed is more binding than a hasty affirmative in a moment of suddenly inspired courage.

Establishment of new habits: The linking method

Most people acknowledge that it is difficult to break established bad habits, as evidenced by all the failed attempts to give up smoking, although most people who try are both well motivated and well informed as to the health hazards associated with smoking. Most people know that overconsumption of sugar is a risk factor for dental caries and can cause obesity. Despite this, if eating sweets has become an established habit, they continue to consume sugar. Relapse is also associated with attempts at weight loss because poor eating habits are so firmly established.

The difficulty of establishing new, good habits can be just as great as the problem of "breaking" already established bad habits. Everyone has experienced the disappointment of not being able to keep a New Year's resolution for long, despite good motives, knowledge, and proficiency at the time the resolution was made. Likewise, in dental practice, ostensibly well-motivated, well-informed, and well-instructed patients fail to comply. Motivation, information, and instruction are no guarantee of the successful establishment of good oral hygiene habits. Moreover, studies (eg, Bratthall, 1966) have demonstrated that one-shot efforts of this nature do not produce long-term effects, ie, ones that last more than a month.

If, however, oral hygiene motivation, information, and instruction are combined with professional toothcleaning, the effect of the one-shot operation, expressed in terms of plaque and gingivitis values, persists after 3 months. In a 30-year longitudinal study in adults, similar effects have been attained with a combination of needs-related oral hygiene motivation, information, and instruction and professional toothcleaning one to four times a year (Axelsson et al, 2004).

A fundamental prerequisite for establishing needs-related toothcleaning habits is a well-motivated, well-informed, and well-instructed patient. Despite these basic prerequisites, behavioral science shows a high risk that these habits will not become firmly established. Therefore, when introduced, new habits should be firmly linked to already established habits. The new habit should always be carried out immediately prior to the established habit, because the risk of overlooking the latter is minimal. This principle of behavioral science is called the linking method and has been described by psychologists in a dental context (Weinstein and Getz, 1978).

In practice, the linking method may be applied as follows: If the patient has irregular oral hygiene habits, an interview should reveal other established habits that, in terms of frequency and point of time during the day, coincide well with the proposed oral hygiene routine. For example, the patient may be in the habit of taking a shower every morning and watching the news on television every evening. According to the linking method, oral hygiene should be "slotted in" the patient's daily routines immediately prior to the morning shower and the evening news.

According to an early Swedish national survey (Håkansson, 1978), although more than 90% of all dentate adults habitually pop the toothbrush into their mouths every day, very few have established daily, interdental toothcleaning habits. The toothpick is nevertheless about six times more common an oral hygiene aid among adults than are dental floss and interdental brushes. For most dentate adults, the toothpick is the most appropriate interdental oral hygiene aid. Although toothpicks are habitually inserted from the buccal direction, the correct toothbrushing technique should effectively clean the buccal aspects of the interproximal space. In the mandibular molar and premolar region, it is preferable to insert the toothpick from the lingual direction (see Fig 86). This is greatly facilitated if the toothpick is fitted with a handle.

According to the linking method, needs-related toothcleaning in adults should commence in the mandibular molar and premolar region, with interdental cleaning from the lingual direction. A toothpick in a handle and an abrasive fluoride toothpaste are used (see Fig 86). Following this, the same aids should be used to clean the approximal surfaces of the maxillary molar and premolar teeth. Because the use of toothbrush and toothpaste on the buccal surfaces is an established, non–needs-related habit, there is no risk involved in leaving this until last. In the most recent analytic epidemiological study in randomized

samples of 35-, 50-, 65- and 75-year-old adults in the county of Värmland, Sweden, it was shown that 50%, 46%, and 36%, respectively, of the 35-, 50-, and 65-year old subjects perform approximal toothcleaning before they use the toothbrush, confirming our significant improvement in the last decade (Axelsson et al, 2000).

When it finally comes to the turn of the toothbrush, it should first be applied to the lingual surfaces of the mandibular posterior teeth; these sites are rarely cleaned efficiently, and plaque usually reaccumulates rapidly and is most adhesive here. Toothbrushing should therefore commence here, while the most toothpaste is on the brush and the bristles are most rigid. Afterward, the mandibular buccal and occlusal surfaces should be cleaned. In the maxilla, toothbrushing can begin on the buccal and occlusal surfaces of the molars, followed by the palatal surfaces. With needs-related toothcleaning in accordance with the linking method, the labial surfaces of the maxillary anterior teeth should be left until last, thus reversing the traditional order.

When the described interdental cleaning method, with toothpick and toothpaste, becomes an established habit, it is expected to have such an effective preventive effect on dental disease that use of the interdental brush will never be necessary. By analogy, establishing the habit of effective interdental toothcleaning with dental tape and toothpaste at an early age can prevent approximal caries and loss of periodontal support in the interproximal regions, so that there will be no reason to change this established habit in adult life.

Without exaggeration, it can be claimed that today, needs-related toothcleaning is practiced by roughly only 15% to 20% of the population in industrialized countries. If the principles of the linking method were successfully applied to establish needs-related toothcleaning habits in all patients, the positive effect on the future oral health of the population and on the demand for dental resources would be inestimable (see the 30-year longitudinal study by Axelsson et al, 2004 presented in chapter 3).

The principles of the linking method should also be applied in clinical dental practice, to emphasize a needs-related modus operandi. For most clinicians, dental prophylaxis with a rubber cup and polishing paste on the buccal and lingual surfaces of the teeth is an established non–needs-related habit. Professional interdental cleaning of the molar and premolar segments, however, does not lend itself in the same obvious fashion to becoming an established habit. According to the linking method, needs-related PMTC should also begin interdentally, from the lingual aspect of the mandibular posterior teeth, with a mechanically powered toothpick and fluoride prophylaxis paste (see chapter 3).

Furthermore, in terms of patient motivation, it is of utmost importance to follow this cleaning sequence, based on analysis of the patient's needs. The clinician loses credibility if PMTC is not carried out in accordance with the needs of a patient who has actively participated in identifying sites of dental disease and plaque deposits. The linking method can also be used in establishing needs-related finishing and scaling procedures. Why finish the occlusal surfaces of restorations to a glossy sheen, but not subgingival approximal restorations? Why remove supragingival plaque and calculus if subgingival plaque and calculus are left undisturbed in the deepest pockets? (For review on needs-related plaque control measures based on risk prediction, see Axelsson, 1998.)

EFFECT OF SELF-CARE ON DENTAL CARIES

Caries-preventive effect of toothbrushes

The claim that daily toothbrushing prevents caries has been challenged on the grounds that it has not been substantiated in clinical studies. In the hundreds of studies based on supervised toothbrushing, the control groups already had an established habit of daily toothbrushing, sometimes more frequent and efficient than the test groups; ie, they were not true control groups (for review, see Bellini et al, 1981). Some studies have, however, shown toothbrushing to provide significant caries prevention, at least on buccal and lingual surfaces (Ainamo, 1971; Finn et al, 1955; Fogels et al, 1982; Fosdick, 1950; Granath et al, 1978; Strålfors et al, 1967). In a large-scale (N = 4,294) double-blind 3-year clinical trial in 12- to 13-year-old children, Chesnutt et al (1995) found that a significantly ($P < .01$) higher percentage of the children who brushed their teeth less than once or once per day developed recurrent caries lesions compared with those who brushed their teeth two times per day—10%, 9.5%, and 6.6%, respectively.

In a more recent 2-year Scottish study in more than 500 selected high-risk children (mean age at baseline, 5.3 years) performed supervised daily toothbrushing in the school. Compared to a randomized control group without intervention, a 36% caries reduction at D3 level in the first permanent molars was achieved (Curnow et al, 2000).

It is regrettable that the relative effects of toothbrushing and fluoride toothpaste have never been properly evaluated in a well-controlled 3-year longitudinal study. Such a study would have to be conducted in a nontoothbrushing population of 12-year-old subjects with high caries incidence and prevalence and the greatest number of newly erupted permanent tooth surfaces at risk. Two test groups and a true negative control group should be randomly selected. The following study design should be used:

- Test group 1: Once a day, at school, a dental assistant should brush the subjects' teeth according to the Bass method, using a placebo toothpaste. Quality control should be monitored by plaque disclosure after regular brushing.
- Test group 2: The subjects would be treated as in test group 1 but would receive fluoride toothpaste.
- Control group: The subjects would be a true negative control group without any intervention.

In test group 1 and in the control group, the caries-preventive effect of toothbrushing should be compared on the surfaces readily accessible to the toothbrush, ie, all surfaces of the anterior teeth and the buccal and lingual surfaces of the posterior teeth. The additional caries-preventive effect of fluoride toothpaste could then be evaluated by comparison of these tooth surfaces in test groups 1 and 2.

Unfortunately, it is probably too late to run such a study, because toothbrushing with fluoride toothpaste is such a widespread, well-established, caries-preventive measure, that it could not ethically be withheld from caries-susceptible subjects for such a long experimental period. However, in a longitudinal clinical study, one test group of 8- to 10-year-old children performed 1-minute supervised toothbrushing per day at school using a nonfluoride toothpaste (for school as well as home use). The quality of the toothcleaning was checked every time. In a matched control group, 99% of the children brushed their teeth with a fluoride toothpaste 1 to 3 times per day, according to their parents. The control group developed significantly more new DSs compared with the test group—2.3 versus 0.40 DSs per individual ($P >$ .001). The authors concluded that this study suggests that children who participate in a 1-minute daily brushing regimen with a nonfluoride toothpaste may show improved dental health. Their increment of dental decay (DSs) over a 1-year period was found to be lower than the increment obtained from a comparable group of children whose oral hygiene regimen was not intentionally altered or influenced and who used their regular fluoride toothpaste at home (Fogels et al, 1982).

Relationship between standard of oral hygiene and dental caries in children

Several studies have indicated that plaque control is superior to fluoride toothpaste as a caries-preventive measure and that comprehensive toothbrushing can control the development of dental caries. A recent meta-analysis of clinical studies on the caries-inhibiting effect of plaque control by chlorhexidine showed an overall caries reduction of 46% (van Rijkom et al, 1996). For fluoride toothpastes and other fluoride agents, similar analysis shows a caries reduction of only about 20% to 25%. Chlorhexidine is twice as effective because it targets the cause of dental caries (the cariogenic plaque) directly, while fluoride is an external factor that modifies enamel demineralization and remineralization. Thus, regular mechanical removal of the cariogenic plaque (biofilms) by toothbrushing should be superior to fluoride toothpaste on all tooth surfaces accessible to the toothbrush.

In a 2-year longitudinal study, Wendt et al (1994) showed that children whose teeth were brushed daily by their parents from the age of 1 year were caries free at the age of 3 years. Most of the children receiving only irregular parental toothbrushing exhibited rampant caries at the age of 3 years. Most children with plaque on the anterior primary teeth and gingivitis at 1 year of age also exhibited rampant caries at the age of 3 years, while children with clean teeth and healthy gingivae at 1 year of age were caries free at the age of 3 years.

In a study of 15 children undergoing orthodontic treatment, Holmen et al (1985a, 1985b) investigated the effect of regular disturbance or removal of dental plaque. Preformed orthodontic bands with two metal posts, 0.3 or 0.5 mm thick, welded to the inner surface were used. The posts created a space between the band and the buccal surface of the tooth. Two homologous premolars were banded for 5 weeks. One tooth in each pair served as a control to which the band remained cemented for the entire test period. The other band was removed weekly, and the buccal surface was cleaned, either by careful pumicing with a nonfluoride toothpaste or simply by wiping with a cotton pellet. No fluoride of any kind was added during the entire test period.

After extraction, the teeth were examined macroscopically, in polarized light, and by scanning electron microscope. The enamel changes in

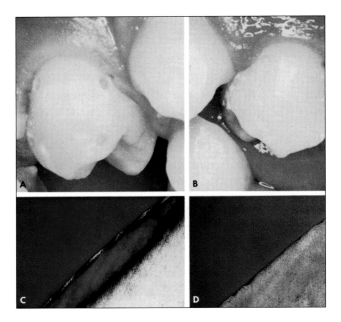

Fig 114 *(A)* Classic subsurface lesions in control tooth banded for 5 weeks of orthodontic treatment. *(B)* Unaffected experimental tooth, debanded for weekly cleaning during 5-week period of orthodontic treatment. No subsurface dissolution is visible. *(C)* Polarized light view of the control tooth. *(D)* Polarized light view of the experimental tooth. (From Holmen et al, 1988. Reprinted with permission.)

the control teeth ranged from slightly accentuated overlapping of the perikymata to pronounced opaque, white lesions. In contrast, all the experimental teeth appeared clinically sound. In polarized light, the control teeth showed classic subsurface lesions of varying severity (Fig 114). No subsurface dissolution could be discerned in any of the experimental teeth, regardless of cleaning procedure.

This study convincingly demonstrated the importance of intraoral mechanical forces for caries initiation and progression. In all subjects, complete elimination of mechanical forces (resulting in undisturbed plaque) led to the development of caries, despite the complex interplay of other individual factors, as indicated by variations noted in the rate at which the lesion progressed (Fig 115). The determining factor is mechanical suppression of bacterial activity, even in the absence of fluoride. The fact that none of the experimental teeth showed any visible evidence of caries offers further support for the principle that "clean teeth do not decay."

Dijkman et al (1990) carried out an experimental crossover study in which teeth with experimentally induced 100-mm-deep enamel le-

sions were placed in situ for 3 months. In the control teeth, which were not brushed, 50% of the lesions progressed. In the test teeth, which were brushed daily with a placebo toothpaste, none of the lesions progressed, and no new lesions developed. In a test in which the teeth were brushed daily with a fluoride dentifrice, 40% of the experimentally induced lesions decreased in depth, ie, there was an additional reduction in caries.

The authors concluded that the remineralizing efficacy of fluoride dentifrice is the result of the cleaning effect of the brushing by the dentifrice (presumably on the pellicle) as well as the fluoride effect on mineral nucleation and growth. As a measure directed against the etiologic factor of caries (dental plaque), cleaning prevents the development of new caries lesions and the progression of enamel lesions.

Granath et al (1978) studied the variation in caries prevalence related to combinations of dietary and oral hygiene habits and chewing fluoride tablets in 4-year-old children. The statistical analysis supported the following order of efficiency of preventive measures:

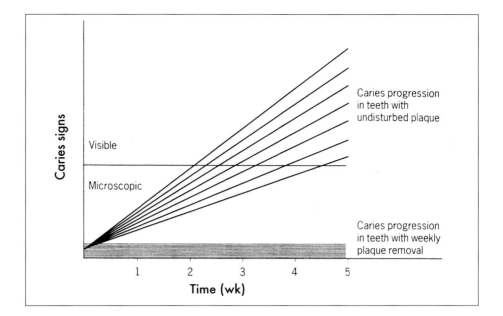

Fig 115 Caries progression (clinical and microscopic signs) in control teeth, with undisturbed plaque, and experimental teeth, with weekly plaque removal. (From Holmen et al, 1988. Reprinted with permission.)

1. Dietary restrictions
2. Oral hygiene
3. Chewing of fluoride tablets

Additionally, it was observed that the effect of one preventive measure was greatest when the others were least favorable. The lower the prevalence of caries, the more limited the effect of each preventive measure; the greatest reduction was for the measure with the weakest potential.

Identification of caries-risk children and prevention of caries in preschool children was studied by Holst et al (1997). The most frequent risk factors identified at 2 years of age were, in descending order of importance, lack of oral hygiene with visible plaque, deep fissures in molars, and frequent intake of sweet drinks.

Similar findings were also reported by Alaluusua and Malmivirta (1994) in their previously discussed study of 92 children and their mothers. The age of the children at baseline was 19 months. Four variables were assessed for their ability to identify young children who would experience caries during the subsequent 1-year period. Visible plaque and the use of a nursing bottle were strongly associated with caries development, while the mother's caries prevalence and the mother's salivary level of mutans streptococci had weak or no statistically significant associations. The best indicator of risk was visible plaque. The results suggest that visible plaque on the labial surfaces of maxillary incisors of a young child is a sign of caries risk.

Similar results were also found by Schröder and Granath (1983). In a study of 3-year-old children, oral hygiene was found to be more effective in preventing caries than good dietary habits. The combination of adverse dietary habits and relatively poor oral hygiene was shown to predict the caries risk in this age group with a predictive value of 0.64. The suggested age for screening was 1.5 to 2.0 years.

In the investigation by Grindefjord et al (1993), the relative risk (odds ratio) that 1-year-old infants would develop caries by the age of 3.5 years was studied. Those who had poor oral hygiene and unfavorable dietary habits, salivary mutans streptococci, little or no exposure to fluoride, and parents with low educational level or immigrant background were at 32 times greater risk of developing caries than were children without the corresponding etiologic and external risk factors.

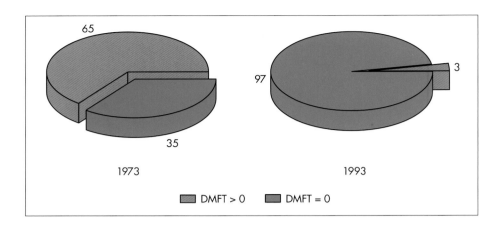

Fig 116 Caries prevalence among 3-year-old children in the county of Värmland, Sweden, in 1973 and 1993. (DMFT) Decayed, missing, or filled teeth.

The importance of establishing good habits as early as possible and postponing or preventing bad habits should not be underestimated. In the county of Värmland, Sweden, large-scale preventive programs at maternal and child welfare centers emphasize the early establishment of good oral hygiene and dietary habits. As a result, from 1973 to 1993, the percentage of caries-free 3-year-olds has increased from 35% to 97% (Fig 116).

In a study involving 4-, 8-, and 13-year-old children, analysis of a number of variables showed that the age at which children are introduced to organized dental care and toothbrushing frequency were more significantly correlated to dental caries than were dietary factors (Stecksen-Blicks et al, 1985).

Sundin et al (1992) analyzed the relationship between caries incidence and a number of caries-related factors in 15- to 18-year-old subjects to estimate the explanatory value of the consumption of sweets and various conditions. Data on oral hygiene, salivary counts of mutans streptococci and lactobacilli, salivary flow rate, and oral sugar clearance times were collected. The results indicated that consumption of sweets is still one of the most important caries-related factors and that it is particularly harmful in combination with poor oral hygiene and low salivary flow rate.

The standard of oral hygiene is usually evaluated by a quantitative plaque index or, in a few re-cent studies, by the subject's gingival health. Frequency of toothbrushing has also been used to investigate the relationship between oral cleanliness and dental caries. Caries prevalence is usually expressed as DMFT and DMFS indices and caries incidence as the number of new decayed surfaces or decayed or filled surfaces per year. Most of the earlier studies evaluating the relationship among toothbrushing, standard of oral hygiene, and dental caries have been conducted on children between the ages of 3 and 15 years. During the past decade, clinical trials have been extended to include older age groups.

Sutcliffe (1977) found a statistically significant association between caries experience and oral hygiene in 3- to 4-year-old children. A greater response to poor oral cleanliness was found on occlusal and approximal surfaces (the key-risk surfaces) of primary teeth. The difference on buccal and lingual surfaces was not significant.

Kleemola-Kujala and Rasanen (1982) analyzed the associations among dental caries, poor oral hygiene, and high sugar consumption. The subjects comprised 543 children, aged 5, 9, and 13 years, from a low-fluoride area. The association between the amount of plaque and dental caries was statistically significant at all levels of sugar consumption. With increasing total sugar consumption, the risk of caries increased significantly only when oral hygiene was also poor.

Beal et al (1979) studied the association between dental cleanliness and the 3-year caries incidence in children, initially aged 11 to 12 years, who received no special fluoride supplements. Children whose oral hygiene was consistently good had lower caries incidence ($P < .05$). The difference was 46.7% on buccal and palatal surfaces, 29.7% on mesial and distal surfaces, and only 7% on occlusal surfaces.

The relative importance of associations between caries occurrence and exposure to sugar-containing products, use of fluorides, medication with drugs that affect salivary secretion, and the standard of oral hygiene was assessed in 9- to 10-year-old children who had mental disabilities and those who did not by Palin-Palokas et al (1987). Based on logistic regression analysis, the standard of hygiene was found to be the most important determinant of caries risk. Compared with the controls, the children with mental disabilities had a very poor level of oral hygiene. Frequent ingestion of sugar-containing snack foods was also found to be an important factor for caries risk in both the children who had mental disabilities and those who did not.

A study by Ogaard et al (1994a) investigated the relationship between oral hygiene level, expressed as the number of nonbleeding papillae, and the development and progression of approximal caries in 14- to 15-year-old subjects. The subjects were assigned to one of two groups, according to their nonbleeding papillae counts. The numbers of carious and restored approximal surfaces were significantly higher ($P < .001$) in the group with fewer nonbleeding papillae. The progression of lesions from 14 to 15 years of age was also significantly higher in the group with fewer nonbleeding papillae. It was concluded that, in populations exposed regularly to fluoride, the standard of oral hygiene may be an important indicator of high caries risk.

In a similar study, Mathiesen et al (1996) investigated the relationship between oral hygiene level and caries experience in 14-year-old subjects who used fluoride dentifrices on a regular basis. Oral hygiene was expressed as gingival bleeding points. Multiple regression analysis indicated that the oral hygiene level was the only factor that accounted for the variation in caries experience. Significantly ($P < .001$) fewer caries lesions and restored approximal surfaces were demonstrated in the group with good oral hygiene than in the group with poor oral hygiene. Additional fluoride in the form of fluoride tablets or mouthrinses resulted in a lower caries experience only in the group with good oral hygiene.

In a large study on the dental health of schoolchildren aged 11 to 12 years, Addy et al (1986) reported highly statistically significant but low Pearson correlation coefficients between plaque scores and DMFT. Although the results supported the etiologic role of plaque, the importance of other factors essential for the development of caries was also highlighted.

As early as 1981, a review by Bellini et al (1981) on the effect of toothbrushing on dental caries in children disclosed that the association between plaque and caries was most apparent in the anterior teeth and on free smooth surfaces. Because these are the areas most readily accessible to toothbrushing, a direct effect of oral hygiene on caries can be assumed. None of these studies provides basic information about the hygiene, fluoride, and caries interrelationship.

Today, there is evidence to suggest that in a population with regular fluoride exposure, poor oral hygiene may promote the development of caries by reducing the efficacy of fluoride. In the study by Stecksen-Blicks and Gustafsson (1986), oral hygiene, among several other factors, had the greatest potential to differentiate between low- and high-caries groups in 13-year-old subjects, even though the use of fluoride toothpaste was more common in the group with high caries activity. Williams and Curzon (1990) also found a clear association between oral hygiene and caries in small children who used fluoride toothpaste. In a recent 2-year prospective study of almost 500 Swedish schoolchildren (10 to 11 years old at baseline), those with very poor oral hygiene (n = 33) developed on average 1.42 compared to only 0.28 new DSs in those with very good oral hygiene (n = 46) (Hänsel Petersson et al, 2002).

In the prefluoride era, even small amounts of plaque had the potential to cause demineralization and caries development, and only "perfect" oral hygiene was able to inhibit caries. In the last two decades, fluoride has become readily available, mostly in the form of fluoride-containing toothpaste, but also as fluoride tablets, mouthrinses, and other topical agents. During a moderate caries challenge, fluoride inhibits demineralization and enhances remineralization. Under these conditions, some plaque probably can be tolerated without the development of caries. However, in the presence of poor oral hygiene, the pH falls so low that plaque liquid is undersaturated with respect to fluorapatite (Larsen, 1990). This may explain why additional fluoride supplementation does not have an enhanced clinical effect (Ogaard et al, 1994b).

Relationship between standard of oral hygiene and dental caries in adults and the elderly

Rajala et al (1980) in a study of male adults, reported consistently higher caries experience in those who brushed only sporadically. Their findings indicated that the positive association between reported daily toothbrushing and low caries experience may be more pronounced in groups with a higher overall risk status, eg, in the strata where education and income are low, frequency of dental visits is irregular, use of sucrose is high, and fluoride exposure is low.

In two studies (Bjertness et al, 1986) of random samples of 35-year-old subjects, in 1973 and 1984, the participants were stratified on the basis of their oral hygiene index scores. In each survey, a statistically significant increase in caries experience was found with increasing oral hygiene scores. In the 1973 study, there was a 17% difference in increase in DMFS between the groups with low and high oral hygiene score. In 1984, the increase was only 3% greater in the group with poor oral hygiene, but still significant ($P < .05$). In another study on patients of the same age by Bjertness (1991), analysis of the relative influence of oral hygiene and other selected independent variables disclosed oral hygiene status to be the most important predictor ($P < .05$) of variation in the number of carious surfaces.

It is well known that neglected oral hygiene may have an influence on recurrent caries. Goldberg et al (1981) studied the associations between oral hygiene, marginal integrity of restorations, and recurrent marginal caries. The probability of recurrent caries increased with decreasing marginal integrity and oral hygiene. Oral hygiene, measured by either a plaque or a gingival index, had a significant effect ($P < .05$).

Hugoson et al (1986) compared the oral health status of individuals aged 3 to 80 years in 1973 and 1983. In the younger age groups, there was an overall reduction in decayed or filled surfaces of about 50%; the reduction for buccal and lingual surfaces was around 90%. Gingival health had also improved over the 10-year period. The number of sites with plaque and calculus had decreased by 1983; buccal surfaces showed the greatest improvement.

The prevalence of caries and caries-risk factors in the elderly has been the subject of several studies in recent years. Kitamura et al (1986) determined the association between overall health status, medication, oral hygiene status, and root caries in nursing home residents with a mean age of 80 years. The best predictors of root caries were the number of remaining teeth, calculus, plaque, and use of pharmaceuticals with adverse side effects on salivary function. The medications were most predictive of maxillary root caries, whereas oral hygiene variables were more closely related to mandibular root caries.

In a prevalence study of root surface caries in 55-, 65-, and 75-year-old subjects, Fure and Zickert (1990) reported that variation in frequency of the lesions was best explained by the salivary levels of mutans streptococci and lactobacilli, the percentage of surfaces harboring plaque, and the frequency of carbohydrate intake. Similarly, Ravald and Birkhed (1991) analyzed a number of different factors in 30- to 78-year-old patients with root caries. Stepwise multiple regression analysis disclosed that lactobacillus count, plaque index, salivary buffering effect, dietary habit index, and number of exposed root surfaces contributed significantly to the coefficient of determination between the groups with no or inactive lesions and the groups with active lesions.

Caries and caries-risk factors in hospital inpatients aged 80 years and older were studied by Budtz-Jorgensen et al (1996). No relationship was found among caries prevalence and degree of dependence, number of drugs, age, or gender. Patients with mental disorders showed increased caries prevalence ($P < .01$). Lower caries prevalence was associated with frequent toothbrushing ($P < .05$). At the tooth level, root caries was associated with high plaque scores ($P < .001$), the degree of gingival recession ($P < .001$), the presence of coronal caries ($P < .001$), and increased probing depth ($P < .01$).

In a 4-year longitudinal study of dental caries and related factors in 88-year-old subjects, Lundgren et al (1997) reported that, from the age of 88 to 92 years, the subjects experienced a significant increase in the proportion of untreated carious root surfaces, plaque score, and the level of lactobacilli. New lesions, predominantly root caries on previously sound surfaces, were found in two thirds of the subjects. Deteriorating dental status was more common in subjects receiving medication for cardiovascular and/or psychiatric disorders than in other subjects. Additionally, a positive correlation was found between plaque score and root caries among the 92-year-old subjects.

Most of the studies cited provide evidence in support of a positive association between good oral hygiene and low caries experience and tend to contradict previous evidence of little or no correlation. However, the results are not in complete agreement. Comparison is difficult, mostly because of wide variations in the protocols. Some studies were of caries prevalence and others of caries incidence, different age groups of subjects were selected, and different variables were used to express oral cleanliness (reported brushing frequency, plaque hygiene index, or assessment of gingival health).

Reported brushing frequency discloses nothing about brushing efficiency. With respect to caries, a plaque index system has certain limitations. Today, few lesions develop on free smooth surfaces where plaque data are easily collected. Most are initiated in contact areas where assessment is more difficult, and in fissures, where plaque is not recorded at all. Nor is the plaque score recorded at a single examination necessarily representative of the subject's average oral hygiene level. A better indicator is gingivitis, which develops in response to a more prolonged period of plaque accumulation arising from unsatisfactory oral hygiene.

Using multiple regression analysis, Mathiesen et al (1996) showed that oral hygiene level was the only factor that could account for the variation of caries experience. This supports the finding by Ogaard et al (1994a) that, in subjects who regularly use fluoride toothpaste, oral hygiene is a good indicator for evaluating caries risk. Most likely the fluoride factor is so strong that it can counteract quite high exposure to sugar, as long as the oral hygiene level is reasonably good. This finding supports an earlier study by Sundin et al (1992), disclosing ingestion of sweets as an important caries-related factor and particularly harmful in combination with poor oral hygiene.

In conclusion, most of the studies of the last two decades provide evidence supporting a positive association between good oral hygiene and lower caries prevalence, contradicting previous evidence of marginal or no association. Since the 1960s and 1970s, caries has steadily declined in most industrialized countries, especially among the younger age groups.

Evidence from longitudinal clinical studies

Studies by Kotsanos and Darling (1991) and Carvalho et al (1992) have shown that the enamel of erupting and newly erupted primary and permanent teeth is most caries susceptible until completion of secondary maturation. For cost effec-

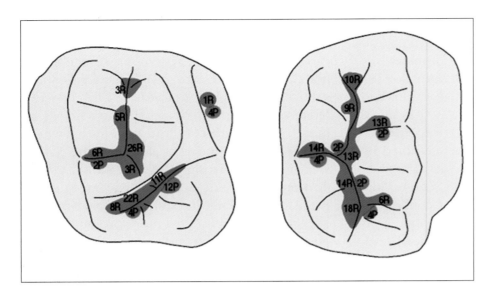

Fig 117 Development of occlusal enamel caries in children in a noninvasive preventive program. (*green* [R]) Arrested caries; (*red* [P]) Enamel caries that has not progressed to cavitation. (Modified from Carvalho et al, 1992 with permission.)

tiveness, plaque control and topical use of fluoride should be intensified during this risk period to maximize the potential for maintaining a caries-free dentition for life.

Carvalho et al (1992) showed that a needs-related, noninvasive preventive program arrested or inactivated about 80% to 90% of active enamel lesions. Figure 117, modified from the studies, illustrates the pattern and number of enamel lesions at baseline that were arrested, and the very few noncavitated enamel lesions that developed in the test group during the 3-year experimental period. After 3 years, only 2% of the occlusal surfaces had to be sealed, and none had to be restored. No progression of lesions into dentin was observed on bitewing radiographs.

The parents were taught to clean the occlusal surfaces of the children's erupting molars with a specific toothbrushing technique and fluoride toothpaste (see Fig 108). In selected children with higher caries risk, home care was supplemented, at needs-related intervals, by PMTC and application of 2% NaF solution. In a matched control group, despite intensive use of topical fluoride and fissure sealing of 70% of the occlusal surfaces, 2% had to be restored (Carvalho et al, 1992).

This low-cost, noninvasive program was subsequently implemented on a large scale in Nexö,

Denmark, and, over a 10-year period, the percentage of caries-free 12-year-old subjects increased from about 30% to almost 90%, without the use of fissure sealants (Thylstrup et al, 1997). Figure 118 compares the mean number of DMFSs by age in Nexö to the national figures for Denmark in 1987 and 1994. The 1994 Nexö data represent children who have followed this program from the age of 6 years. In this context, in the latest epidemiologic data from the World Health Organization's global bank (1997), the lowest caries prevalence in 12-year-old subjects is reported in Denmark, Sweden, Finland, Switzerland, Australia, and some countries in central Africa.

The same concept of intensified plaque control, especially targeting the fissures of erupting molars (see Fig 108), was applied in a longitudinal study in 6- to 8-year-old subjects (first molars) and 11- to 13-year-old subjects (second molars) in Solntsevsky, Moscow, Russia, by Kuzmina (1997). Because there was no school-based preventive program, randomized negative control groups were selected. Figures 119 and 120 show the mean caries incidence in new DMFSs, from 1994 to 1996, in the study and the control groups and in the corresponding age groups in Denmark. The results for the test group compared well with data from Denmark, confirming that a low-cost, needs-

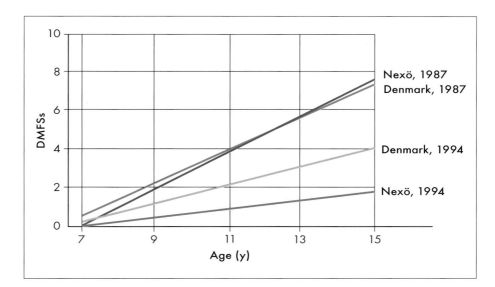

Fig 118 Mean number of DMFSs by age in Nexö and Denmark in 1987 and 1994. (From Thylstrup et al, 1997. Reprinted with permission.)

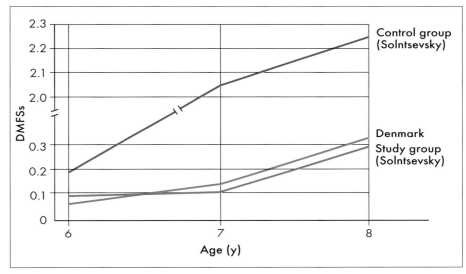

Fig 119 Mean number of DMFSs from 6 to 8 years of age in the study group, compared with the control group and national data of Denmark. (From Kuzmina, 1997. Reprinted with permission.)

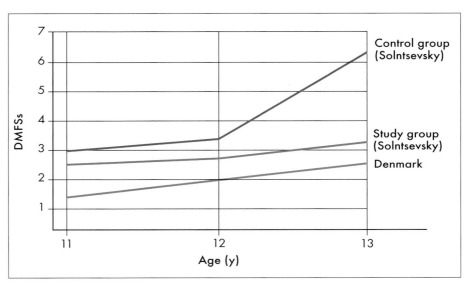

Fig 120 Mean number of DMFSs from 11 to 13 years of age in the study group, compared with the control group and national data of Denmark. (From Kuzmina, 1997. Reprinted with permission.)

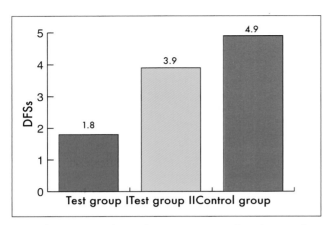

Fig 121 Mean values of new approximal DFSs in molars and premolars per child per 3 years in test group I, test group II, and the control group. (From Axelsson et al, 1994a. Reprinted with permission.)

related plaque control program is efficient even in populations with a very high caries incidence.

These studies show that if the fissures of the molars are maintained free of caries during the risk period, ie, throughout the entire eruption period to the stage of occlusal function, the risk for development of fissure caries is past. In the occlusal surfaces of fully erupted, caries-free molars, general use of fissure sealants represents costly overtreatment, and the procedure should be reserved for very sticky, plaque-retentive fissures in the molars of selected children at high caries risk.

In the Dorchester split-mouth study by Wright et al (1979), the approximal surfaces on the test sides were cleaned daily with quality control. Compared to the contralateral control sides, test sides experienced about a 50% reduction in caries.

The principles described earlier for establishment of needs-related oral hygiene habits, based on self-diagnosis and the linking method, were implemented recently in a 3-year longitudinal study (Axelsson et al, 1994a). Twelve-year-old schoolchildren in São Paulo, Brazil, were randomly assigned to two test groups and one control group. All the children had a well-established habit of daily toothbrushing with a fluoride toothpaste. In addition, all children were exposed to fluoridated drinking water.

Children in test group I were trained and motivated to identify sites with inflamed gingivae, which had the potential to heal, and sites with enamel caries, which had the potential to remineralize. By using the PFRI as a guideline, they deducted how frequently they needed to clean their teeth and which sites required special attention during toothcleaning procedures. In accordance with the linking method (Weinstein and Getz, 1978), they were motivated and trained to apply fluoride toothpaste to the approximal surfaces of molars and premolars first. After these surfaces had been meticulously cleaned with dental tape, used in a rubbing technique, they then carefully cleaned the lingual and buccal surfaces with a toothbrush and fluoride toothpaste. After three initial visits at short intervals, the children were recalled for a monthly checkup during the first 4 months, and then every 3 months, for reevaluation of the results based on self-diagnosis.

The children in test group II were trained, both on models and in their mouths, to clean every tooth surface meticulously, using dental tape, fluoride toothpaste, and toothbrush. They were recalled for reinstruction at the same intervals as test group I.

During the 3-year period, the children in test group I developed 54% and 63% fewer new posterior approximal lesions of dentin, per individual, than did children in test group II and the control group, respectively ($P < .001$) (Fig 121). Conclusions from the study were:

1. In a toothbrushing population using fluoride toothpaste and fluoridated drinking water, a highly significant reduction in the incidence of approximal caries will be achieved with an oral hygiene training program based on self-diagnosis and the linking method.
2. In such a population, frequent repetition of meticulous oral hygiene training is almost redundant (Axelsson et al, 1994a).
3. The effect on plaque and gingivitis was also significantly higher in test group I than in test group II and the control group (Albandar et al, 1994a).

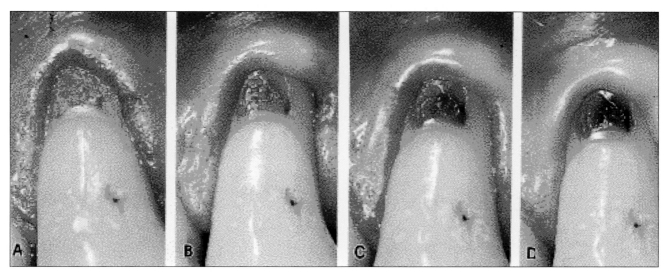

Fig 122 Changes in color and surface structure in an active, plaque-covered root lesion after improvement in oral hygiene. *(A)* Active lesion. *(B)* Lesion after 2 months of improved oral hygiene. *(C)* Lesion at 6 months. *(D)* Lesion at 18 months. (From Nyvad and Fejerskov, 1986. Reprinted with permission.)

Fig 123 *(left)* Active, plaque-covered root caries cavities with a typical yellow color and a soft surface. (From Nyvad and Fejerskov, 1997. Reprinted with permission.)

Fig 124 *(right)* Same cavities shown in Fig 123, 10 years later. With improved plaque control and topical use of fluoride, the cavities are inactive, with a typical dark brown to black color and a semihard surface. (From Nyvad and Fejerskov, 1997. Reprinted with permission.)

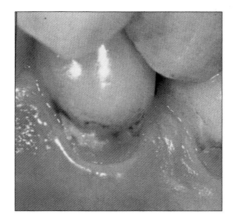
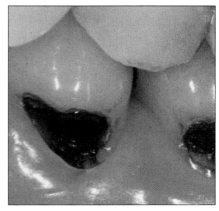

4. Because all three groups had the same exposure to fluoride from toothpaste and drinking water, the greater reduction in caries found on the approximal surfaces in test group I, compared to test group II and the control group, must be the effect of improved interdental plaque control.

The cost effectiveness (cost savings) of large-scale implementation of the system used in this low-cost, low-technology study, based on self-care, is almost inestimable.

Studies by Nyvad and Fejerskov (1986) in elderly patients with active root caries lesions showed that improved oral hygiene converts active lesions to inactive lesions (Fig 122). Even root caries lesions with deep cavities can be successfully converted to inactive lesions by improved plaque control and use of fluoride (Figs 123 and 124).

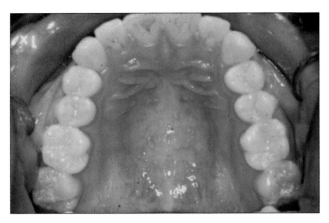

Fig 125 Caries- and gingivitis-free woman, aged 42 years. Meticulous toothcleaning has prevented oral disease, even in the absence of preventive measures such as fissure sealants.

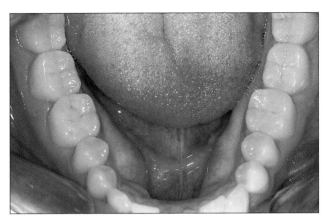

Fig 126 Caries- and gingivitis-free man, aged 40 years.

In a follow-up experimental in situ study, Nyvad et al (1997) showed that daily plaque removal, in combination with fluoride toothpaste, influences the distribution of mineral in sound and carious root surface specimens by increasing the mineral content of the surface. In contrast to plaque-covered specimens, sound specimens developed no caries, and experimentally induced active lesions were arrested.

Finally, most clinicians are aware from personal experience that high-quality mechanical plaque control by self-care prevents not only gingivitis and periodontitis but also dental caries. Up to the age of 20 to 25 years, both the author and his wife developed the same average number of decayed or filled surfaces as the Swedish population. In the 45 years after becoming a dental undergraduate, the author has not developed a single caries lesion nor has he required any scaling. His wife has developed only one caries lesion during the same period, despite the fact that efficient fluoride toothpastes have been available only since 1970. Their two grown children, Eva, 42 years old (Fig 125), and Torbjörn, 40 years old (Fig 126), remain both caries and gingivitis free, despite the genetic makeup they share with

their parents and the fact that they have used no fluoride agents other than a toothpaste, and have received no professional prevention measures such as fissure sealants. This is further anecdotal evidence in support of the claim that clean teeth never decay.

EFFECT OF SELF-CARE ON PERIODONTAL DISEASES

Relationship between standard of oral hygiene and periodontal disease

The effect of mechanical toothcleaning by self-care on gingivitis is generally acknowledged and not as open to question as its effect on dental caries. Supporting evidence is available from several experimental gingivitis studies in humans as well as animals. There is also a strong correlation, at the surface level, between the pattern of plaque reaccumulation and gingivitis (Axelsson, 1987; Ramberg et al, 1994, 1995a; Saxton 1973, 1975).

Recently Hugoson et al (1998) presented data on plaque and gingivitis values from cross-sec-

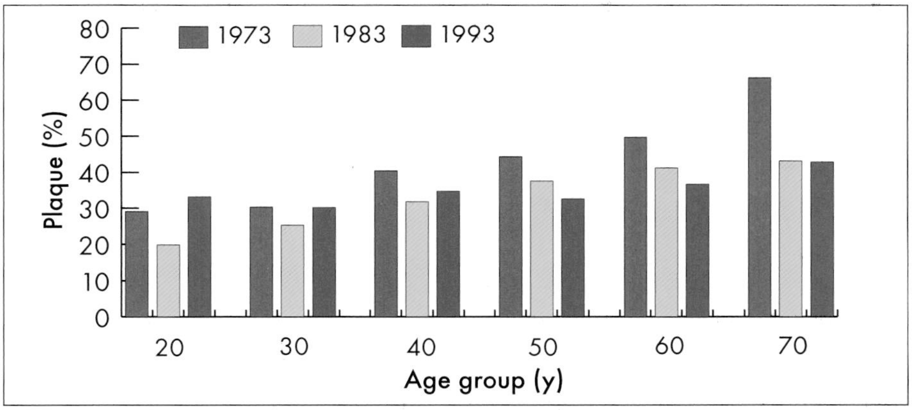

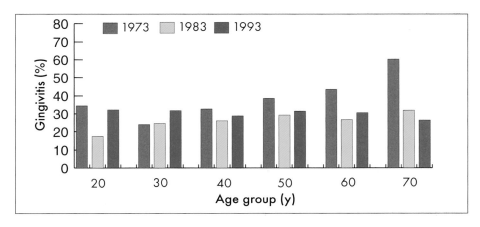

Fig 127 Frequency (%) of tooth surfaces with plaque as a percentage of existing surfaces. There was a significant difference in the mean values in the 20-year-old group between the years 1983 and 1993 (P<.01). (From Hugoson et al, 1998. Reprinted with permission.)

Fig 128 Frequency of sites with gingivitis as a percentage of existing sites. There was a significant difference in the mean values in the 20-year-old group between the years 1983 and 1993 (P < .001). (From Hugoson et al, 1998. Reprinted with permission.)

tional studies in 1973, 1983, and 1993, in randomized samples of adults in the county of Jönköping, Sweden. Population-based dental care programs have been available to all residents of the county for several decades. From 1973 to 1983, the frequency of tooth surfaces with plaque, as a percentage of existing surfaces, decreased in all age groups (Fig 127). However, from 1983 to 1993, it increased in 20-, 30-, and 40-year-old subjects. As an effect of reduced plaque percentage (improved oral hygiene), the frequency of sites with gingivitis, as a percentage of existing sites, decreased in all age groups, from 1973 to 1983. From 1983 to 1993, the frequency of gingivitis increased, particularly in the 20- and 30-year-old subjects (Fig 128), as a consequence of the deterioration in oral hygiene standards (increased plaque percent).

Fewer than 10% of the 20-year-old subjects used toothpicks regularly in 1983 and 1993 (Hugoson et al, 1995). Between 1983 and 1993, there was a decrease in the use of toothpicks among the 30-, 40-, and 50-year-old subjects. As a consequence, most plaque and gingivitis was recorded on approximal sites.

In contrast, a cross-sectional study in a randomized sample of more than 600 subjects aged 50 to 55 years in the county of Värmland, discussed earlier, disclosed that about 65% used toothpicks daily (Axelsson and Paulander, 1994). In the county of Värmland, dental care programs are tailored to the individual patient's needs. As a consequence, in 1998, the percentage of tooth surfaces with disclosed plaque was only 19% in a randomized sample of 60-year-old subjects (Axelsson et al, 1998).

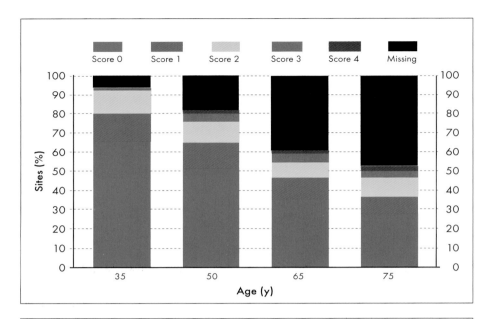

Fig 129 Frequency distribution, at the site level, of CPITN scores 0 to 4 and missing teeth, by age group. (Axelsson et al, 1990. Reprinted with permission.)

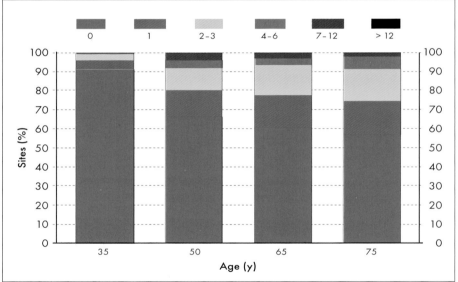

Fig 130 Frequency distribution of sites per individual with CPITN score 4, by age group. (Axelsson et al, 1990. Reprinted with permission.)

Ten years earlier, an analytical epidemiologic study in randomized samples of 35-, 50-, 65-, and 75-year-old subjects in the county of Värmland, Sweden, showed that the percentage of sites with gingivitis (CPITN score 1) was only 12% in 35-year-old subjects and 10% in 50-year-old subjects (Fig 129). The percentage of individuals with sites with probing depths greater than 5 mm (CPITN score 4) was very low in all age groups (Fig 130) (Axelsson et al, 1990).

In the Jönköping cross-sectional epidemiologic studies, the average gingivitis scores were approximately 30% in all age groups in both 1983 and 1993. This is much higher than in a study from the county of Värmland, Sweden, of 225 randomly selected patients, aged 18 to 65 years, examined in 1978 and 1990 (Wennström et al, 1993); in 1990, only 3% to 4% of the gingival units in the various age groups bled on probing, and almost no individuals had gingivitis at more than 30% of the sites.

The subjects of the Värmland study, however, had received regular dental care over a 12-year period at 12 community dental clinics, where particular attention was paid to preventive measures. In contrast, the Jönköping study was based on data from three random samples of individuals in 1973, 1983, and 1993, and the dental care received by the subjects was not uniform. Also, the mean baseline values for plaque and gingivitis were lower in the Värmland study in 1978 than were those in 1973 in the Jönköping study. It is therefore questionable whether a reduction in plaque and gingivitis values can be achieved solely by means of a population-based preventive strategy.

Hugoson et al (1998) discussed possible reasons for the apparent deterioration in the oral hygiene standards of young adults. Professional preventive measures may be less effective today than before. Many individuals have been actively involved for many years, on a daily basis, in oral health education programs for children and adolescents. This may have become a monotonous routine after 20 years or so. Although oral health education and promotion are given high priority, it is possible that the quality has decreased, and people are less aware than they were previously of the importance of good oral hygiene.

Another interesting issue is that, compared to older adults, young adults tend to have a more critical approach to aspects of dental care. In a study by Johansson and Fridlund (1996), participants of both sexes, aged 21 to 30 years, considered fees charged for information and preventive service to be questionable but fees for examination and treatment to be quite acceptable. The patients in the study were informed of the current status of their oral and dental health and given instruction in recommended preventive measures. Instruction was positively received if it was not given as a routine and if the patient considered it to be relevant. If the patient was already well informed on the subject, it was acceptable if the instruction was introduced with the following statement: "Provided that you have been to the dentist every year, the information does not need to be repeated." In other words,

"A medium size suit does not fit everybody. To fit the individual, a suit should be tailor-made."

Using a population-based strategy, it does not seem possible to reduce mean values of plaque and gingivitis below 30%. As stated by Johansson and Fridlund (1996), the preventive message should be revised and presented in a form appropriate for young adults with little experience of oral disease. It is important to prevent exclusion of this age group from dental care, jeopardizing their dental health. In addition to preventive programs aimed at the population as a whole, it is also necessary to have individual programs, targeting areas of risk, to reduce the number of people developing dental disease and to improve the quality of dental care.

Relationship between standard of oral hygiene and gingivitis

In their classic experimental gingivitis study, Löe et al (1965) showed that, in the total absence of oral hygiene, during the 3-week experimental period, there was a gradual increase in thickness of the gingival plaque on the initially clean teeth and healthy gingivae. For the first few days, this plaque was composed of gram-positive cocci and rods, representing the indigenous microflora of the tooth surface. After 4 to 5 days, filamentous organisms and gram-negative cocci and rods "infected" the gingival plaque, nonattaching spirochetes gradually appeared in the gingival sulcus, and the range of microorganisms in the gingival biofilm increased continuously. As a consequence, the first clinical signs of gingivitis developed within about 2 to 3 weeks. When the accumulated plaque was mechanically removed, and daily oral hygiene was reestablished, the gingivae healed within about 1 week.

In their 6-week experimental study, Lang and associates (1973) reported that clinical signs of gingivitis did not occur in subjects carrying out mechanical toothcleaning with a toothbrush and interproximal aids (dental floss and toothpicks)

once every second day. Bosman and Powell (1977) induced experimental gingivitis in a group of students. In those removing plaque only every third or fifth day, gingival inflammation persisted, but healing occurred within 7 to 10 days in the two groups who cleaned their teeth properly once a day or every second day. Proper mechanical toothcleaning every second day, including the key-risk surfaces, is superior to daily toothbrushing concentrated mainly on nonrisk surfaces. The prevention of gingivitis on buccal and lingual surfaces by daily manual toothbrushing has been confirmed in several studies (Ainamo, 1971; Ainamo, 1979; Arno et al, 1958; Lindhe and Koch, 1967; Löe et al, 1965; for review, see Bergenholtz, 1972; Frandsen, 1970, 1986; and Gjermo, 1986).

In an 8-week study, O'Beirne et al (1996) showed that, compared with daily manual toothbrushing according to the Bass method, daily use of an electronically driven sonic toothbrush resulted in improved reduction of gingivitis, gingival crevicular fluid, and interleukin 1b levels in the gingival crevicular fluid.

A 12-month longitudinal study by Boyd et al (1989a) showed that the use of a Rota-dent electric toothbrush resulted in reduction of plaque and gingivitis similar to that achieved by a combination of manual toothbrushing and interproximal cleaning with dental floss or toothpicks. Van der Weijden et al (1994) showed, in an 8-month study, that the Braun Plak Control electric toothbrush was slightly more effective than manual toothbrushing in the reduction of gingival bleeding on probing.

The effect of interproximal oral hygiene aids on plaque and gingivitis has been evaluated in several studies. Gjermo and Flötra (1970) found that the combined use of triangular-tipped toothpicks and a single-tufted brush, or the use of dental floss alone, could reduce plaque on approximal surfaces by an average of 50%. Similar results were reported in the dental floss study by Reitman et al (1980) described earlier and in other studies using triangular-tipped toothpicks or dental floss (Bergenholtz et al, 1974; Bergenholtz and Brithon,

1980; Lobene et al, 1982; Schmid et al, 1976). Improved gingival health, compared to toothbrushing alone, has been reported (Abelson et al, 1981; Kleber and Putt, 1988; Lobene et al, 1982). Daily use of a rotating electric toothbrush (Rota-dent) achieves the same reduction in plaque and gingivitis as the combined effect of manual toothbrushing and interproximal oral hygiene aids (Boyd et al, 1989a; Glavind and Zeuner, 1986).

Caton et al (1993) compared the efficacy of a chlorhexidine mouthrinse (0.12%) plus toothbrushing (mouthrinse group), mechanical interdental cleaning plus toothbrushing (mechanical group), and toothbrushing alone (control group) in reducing and preventing interdental gingival inflammation. Interdental inflammation was recorded with the Eastman interdental bleeding index in 92 male subjects, at baseline, then monthly, for 3 months. The mechanical cleaning group had significant reductions in bleeding sites compared to baseline (56.9%) at 1 month (13.7%), 2 months (6.65%), and 3 months (5.70%). Subjects following the other regimens showed no significant reduction in bleeding at any point in the study.

The effect of location in the mouth on bleeding reduction was also assessed. The percentage of bleeding sites was always higher in the posterior than the anterior region. Analysis of maxillary versus mandibular sites and buccal versus lingual sites revealed no significant differences.

Additional analysis of the data revealed that sites that bled at baseline were more likely to stop bleeding in the mechanical cleaning group. Also, sites that did not bleed at baseline were unlikely to bleed subsequently when mechanical cleaning was used. Neither of these observations applied to the other regimens. These data show that only mechanical interdental plaque removal, combined with toothbrushing, effectively reduces or prevents interdental inflammation, particularly in the posterior teeth, where accessibility for both the toothbrush and mouthrinses is very limited. This highlights the importance of instituting mechanical interdental cleaning to eliminate interdental inflammation.

However, in a recent cross-sectional large-scale study in health care professionals (N = 533), Merchant et al (2002) found no correlation between self-reported oral hygiene habits (frequency of toothbrushing and flossing) and alveolar bone loss (radiographic bitewings), controlling for potential confounders. This could possibly be explained by the following two reasons: *(1)* Health care professionals exhibit less alveolar bone loss than do average individuals, or *(2)* in subjects susceptible to periodontal disease and already exhibiting deep diseased pockets, the frequency of mechanical toothcleaning by self-care is less important than the quality of the cleaning. Therefore, a new supplementary microbrush for subgingival removal of plaque biofilms is a promising approach to improve the effect on periodontal disease by self-performed mechanical plaque control. In 30 periodontally involved teeth (pocket depth, 4 to 10 mm; minimum 30% alveolar bone loss) requiring extraction for periodontal or prosthetic reasons, Carey and Daly (2001) randomly selected one of the two approximal surfaces for subgingival cleaning using a microbrush for 2 minutes prior to extraction. After extraction, the approximal root surfaces were stained. The results showed that 100% of untreated surfaces were completely covered by plaque biofilms. On the brushed surfaces, an average of only 10% of the root-surface area was covered.

CONCLUSIONS

At the end of the recent European Periodontal Workshop on Mechanical Plaque Control (Lang, 1998) the participants (a selected group of 50 European periodontists) were asked to indicate agreement, or otherwise, with the concluding statements from the workshop. The results are presented in Table 2.

Plaque control can be achieved mechanically or chemically by self-care or professionally by dentists or dental hygienists. Plaque control pro-grams based on needs-related combinations of the above are to date the most successful in prevention and control of gingivitis, periodontitis, and dental caries. The effectiveness of mechanical plaque control by self-care depends on the patient's motivation, knowledge, oral hygiene instruction, oral hygiene aids, and manual dexterity. A huge range of oral hygiene aids are available. The dentist and the dental hygienist should assess the individual needs of the patient and recommend appropriate aids.

Materials and methods

Toothbrushing is the most widespread mechanical means of personal plaque control in the world. In Scandinavia, more than 90% of children and dentate adults brush their teeth at least once a day. However, enthusiastic use of the toothbrush is not synonymous with a high standard of oral hygiene.

The design of the toothbrush is much less important than how thoroughly and frequently it is applied to every accessible tooth surface. Generally, the length and width of the handle should be adequate. For example, for preschool children, the handle should be big enough for the parent's grasp and the head of the toothbrush should be small enough for the child's mouth. Soft multi-tufted brushes with high-quality bristles are suitable for most individuals.

The modified Bass method seems to be the most efficient and most frequently recommended toothbrushing technique for removal of gingival plaque.

In most individuals, it should be possible to prevent and control gingivitis and posttreatment periodontitis, as well as dental caries, by meticulous toothbrushing once or twice a day, using an effective fluoride toothpaste, supplemented with needs-related posterior interproximal cleaning. High–caries-risk patients with fast or very fast plaque formation rates (PFRI score of 4 or 5) should clean their teeth just before every meal and use fluoride chewing gum for 20 minutes imme-

Table 2 Professional agreement with concluding statements of the European Periodontal Workshop on Mechanical Plaque Control (n = 50)*

Concluding Statement	Agree (%)	Disagree (%)	No opinion (%)
Periodontitis is caused by subgingival colonization and multiplication of distinct bacterial species	79	13	8
Toothbrushing as commonly performed does not remove interdental plaque.	98	2	0
Interdental cleansing is an essential prerequisite for a successful plaque control program.	94	4	2
Meticulous mechanical dental plaque removal once every 24 hours is adequate to prevent the onset of gingivitis and approximal caries.	92	8	0
Mechanical plaque removal once a day minimizes the risk of gingival trauma.	46	27	27
To prevent recurrence of periodontal disease, optimal plaque removal by personal mechanical plaque control should be supplemented by professional intervention.	71	10	19
Modern designs of automated toothbrushes are sufficiently superior to manual toothbrushes so as to be clinically relevant.	73	21	6
Interdental cleaning should be advised for the whole dentate population, in the absence of methods to identify target groups susceptible to caries and periodontal disease.	77	17	6
Patient compliance is critical for achieving effective plaque control.	100	0	0
Toothbrush design may influence patient compliance.	52	27	21
For the general public, interdental cleaning is needed but not on a daily basis.	51	36	13
Individually tailored oral hygiene measures and supportive care should be based on data of the response to therapy of that particular individual.	92	0	8
Interdental cleaning of the anterior teeth is generally not necessary in individuals who are performing proper toothbrushing.	19	64	17
An adult should participate in the implementation of daily toothbrushing with a fluoridated toothpaste at least once a day in children, beginning with the onset of tooth eruption.	92	0	8
When the aim is periodontal health, the prevention of caries should not be forgotten and vice versa.	96	2	2
In children who have experienced caries and/or periodontal diseases, self-performed oral hygiene procedures are not enough for control of caries and aggressive periodontitis. These individuals must be provided with professional support.	92	4	4
An uncontaminated environment should be ensured after periodontal surgery to achieve undisturbed wound healing.	86	6	8
In the elderly who have experienced caries and/or periodontal diseases, effective, self-performed plaque control and a frequent supportive care program are able to arrest caries and the progression of periodontitis, maintain or improve clinical attachment level, and keep the dentition healthy on a long-term basis in the majority of individuals.	90	4	6

*Adapted from Lang (1998) with permission.

diately after every meal. No-risk or low-risk patients with low or very low plaque formation rates (PFRI 1 or 2) should not need to clean their teeth more than once a day.

Self-taught patients tend to brush with a horizontal scrubbing technique; develop iatrogenic gingival recession and cervical abrasion, particularly on the buccal surfaces of the maxillary left premolars and first molars (if they are right-handed); and fail to remove gingival plaque from the lingual surfaces of the mandibular right posterior teeth.

For disabled people, children, and parents cleaning the teeth of children aged 1 to 8 years, electric (automated) toothbrushes are particularly useful. In the fissures of erupting molars and patients with interproximal spaces widened by periodontal disease, pointed rotating electric toothbrushes (Rota-dent) are particularly suitable. An oscillating-rotating electric toothbrush (Braun/Oral-B) generally seems to be best accepted by patients. Both of these toothbrushes have been proven to be at least as efficient as manual toothbrushing. Abrasion and gingival recession are reduced because less force is applied.

In toothbrushing populations, most DMFSs and the greatest periodontal treatment needs are recorded at posterior approximal sites. There is an enormous potential for prevention and control of periodontal disease and dental caries, simply by supplementing regular daily toothbrushing with gingival plaque removal on these surfaces. In most countries, interproximal toothcleaning is still nonexistent.

For approximal toothcleaning by self-care, dental floss, dental tape, wooden triangular toothpicks, and interdental brushes are available; recently, an electric rotating interdental brush was introduced. In the United States, dental floss is the usual method in children and adults. In Scandinavia, dental tape is recommended in children and young adults, while triangular wooden toothpicks are recommended and preferred in most adults.

It has been shown that a triangular-tipped wooden toothpick can remove gingival plaque biofilms to a depth of 2 to 3 mm subgingivally. Interdental brushes are most suitable for patients with interproximal spaces widened by periodontal disease. Fluoridated dental tape and wooden toothpicks are highly recommended, and, on the approximal surfaces of the posterior teeth (the key-risk surfaces), all interdental cleaning aids should be used in combination with fluoride toothpaste.

To clean tipped, rotated, or displaced teeth, the toothbrush and interdental cleaning aids should be supplemented with a single-tufted toothbrush. Super Floss is recommended for cleaning around fixed partial denture pontics. A specially designed range of oral hygiene aids for the physically disabled patient is also available.

In the so-called complete-mouth disinfection technique, tongue cleaning is included in the oral hygiene program. Tongue cleaning also reduces halitosis.

Needs-related oral hygiene at home

A fundamental principle for all prevention is that the positive effect is greatest where the risk of disease development is greatest. Needs-related mechanical plaque control at home should therefore be intensified on the basis of key risk, at the group, individual, tooth, and surface levels.

Dental caries is the predominant oral disease in children. The key-risk age groups are 1 to 2, 5 to 7, and 11 to 14 years. It is most important that parents establish a routine of cleaning the teeth of infants twice a day as soon as the first primary incisors erupt.

At least up to the age of 8 years, parents should clean all surfaces of their children's teeth, twice a day. Parents should be instructed in intensified mechanical removal of plaque, particularly in the fissures of the erupting permanent first molars, in combination with use of fluoride toothpaste. Up to the age of 5 years only a small amount of fluoride toothpaste, about the size of a pea, should be applied to the brush.

From the start of eruption of the permanent second molars (about 11.5 years for girls and 12 to 12.5 years for boys), daily mechanical plaque control and use of fluoride toothpaste should be intensified in the fissures of the second molars and on the approximal surfaces of the posterior teeth. This focus should continue at least until the so-called secondary maturation of the enamel is completed. Selected caries-risk children should continue with intensified plaque control and use of fluoride toothpaste on all tooth surfaces. Generally, everybody should clean their teeth once or twice a day, using a fluoride toothpaste.

In most adults, cleaning should begin interproximally. The interproximal spaces of the posterior teeth should be filled with toothpaste and cleaned with an appropriate interdental hygiene aid. Right-handed adults should start brushing on the lingual surfaces of the mandibular right teeth; left-handed adults should start on the left side.

Establishment of needs-related oral hygiene habits

The first condition for success in establishment of needs-related toothcleaning habits is a well-motivated, well-informed, and well-instructed patient. *Motivation* is defined as readiness to act or the driving force behind our actions. Greater responsibility has been identified as the most enduring motivating factor. A person's actions are also governed by perceived needs. Therefore, education of the patients in self-diagnosis is of the utmost importance. A toothcleaning chart and risk profile evaluation are very useful tools as supplements to the patient's own mouth for motivation of the patient and education in self-diagnosis.

Establishment of needs-related toothcleaning habits will be facilitated by linking new habits to established habits that, in terms of frequency and time of day, coincide well with the proposed oral hygiene routine. Because established habits are dependable, the new habit should be introduced immediately before the established habit. Indi-

viduals with irregular toothcleaning habits should be advised to clean their teeth immediately before well-established morning and evening habits. Likewise, individuals with well-established daily toothbrushing habits, but irregular or no interproximal cleaning routines, should fill the interproximal spaces with fluoride toothpaste and clean the approximal surfaces of the posterior teeth before beginning to brush.

Effect of self-care on dental caries

To date, the relative caries-preventive effect, on accessible tooth surfaces, of toothbrushing and fluoride toothpaste has not been evaluated in a correctly designed, longitudinal study with a standardized quality of daily toothbrushing, with a placebo or fluoride toothpaste, and compared to a true negative control group, in a high-caries activity, nontoothbrushing population. For ethical reasons, it is now probably too late to conduct such a study because the use of fluoride toothpaste is now so widespread.

Recent meta-analyses of well-controlled longitudinal studies show, however, that the average caries-preventive effect of fluoride toothpaste is about 20% to 25%, while chemical plaque control by daily use of chlorhexidine results in 40% to 50% caries reduction. Plaque control is more effective because it targets the etiology of dental caries, ie, the cariogenic plaque, while fluoride exerts a modifying effect on the demineralization and remineralization processes.

Multivariate analyses of recent longitudinal studies in infants and toddlers show that the standard of oral hygiene at the early age of 1 year is the most important predictive factor for the development of caries.

Other studies have shown that intensified daily plaque control during the eruption of the permanent molars can prevent the development of fissure caries and significantly reduce the need for fissure sealants.

In studies based on the aforementioned principles for establishment of needs-related oral hygiene habits (eg, self-diagnosis, the linking method) in a toothbrushing population, balanced for the use of fluoride toothpaste in an area with fluoridated water, caries reduction of more than 50% was achieved on the approximal surfaces of the posterior teeth, compared to standardized oral hygiene instruction and no instruction at all.

Even in the elderly, it has been shown that active root caries can be arrested by improved oral hygiene.

Effect of self-care on periodontal diseases

In the last 30 to 40 years, following the pioneering studies by the Waerhaug group in the 1950s and the Löe group in the 1960s, the importance of plaque control by self-care for gingival health has not been questioned. (For reviews on mechanical plaque control by self-care, see Axelsson, 1981, 1994, 1998; Echeverria, 1998; Frandsen, 1986; Garmyn et al, 1998; Hotz, 1998; Jepsen, 1998; Kinane, 1998; Lang, 1998; Mayfield et al, 1998; Renvert and Glavind, 1998; Sanz and Herreva, 1998; and Schou, 1998.)

CHAPTER 3

PROFESSIONAL MECHANICAL TOOTHCLEANING

Professional mechanical toothcleaning (PMTC) is a service provided by dental personnel (specially trained dental nurses, dental hygienists, and dentists) and is defined as the selective removal of plaque from *all* tooth surfaces. Gingival plaque biofilms located up to 1 to 3 mm subgingivally are removed with mechanically driven instruments and fluoride prophylaxis paste. Therefore, the procedure is more correctly described as *gingival plaque control* rather than *supragingival plaque control*. If deep subgingival plaque biofilms are also removed, the procedure is referred to as *debridement*, and may be carried out only by dentists and dental hygienists. In new patients with untreated diseased periodontal pockets, debridement may have to be supplemented with removal of calculus (scaling) and root planing. Professional mechanical toothcleaning should not be confused with so-called prophylaxis or polishing, which involves the use of a rotating rubber cup and prophylaxis paste on the buccal, lingual, and occlusal surfaces, ie, the nonrisk surfaces.

MATERIALS AND METHODS

In the so-called Karlstad studies in adults (Axelsson and Lindhe, 1978, 1981a, 1981b; Axelsson et al, 1991, 2004) and children (Axelsson and Lindhe, 1974, 1977, 1981c; Axelsson et al, 1976; for review, see Axelsson, 1981, 1993a, 1994, 1998), standardized procedures were adopted for PMTC. The following materials are required:

1. Plaque-disclosing pellets, eg, Diaplac (Swedish Dental Instruments, Sweden), with erythrosin red or blue (Figs 131 and 132)
2. A contra-angle handpiece made by Dentatus (Hägersten, Sweden), KaVo (Biberach, Germany), or W&H Dentalwerk (Bürmoos, Austria) (a new modified EVA prophylaxis, contra-angle handpiece), and reciprocating tips such as EVA 2000 (medium-universal), EVA 123, EVA 123S, EVA-1, or EVA-7 (Fig 133) (Dentatus)
3. Prophylaxis contra-angle handpiece (W&H or KaVo) and rotating rubber cup such as Young's BS 1800 (Young Dental, St Louis, MO) (see Fig 133)
4. A fluoride-containing prophylaxis paste such as RDA 170 (medium abrasive) (SDI Directa, Upplands-Väsby, Sweden) (Fig 134)
5. A syringe for injecting the paste interproximally (Figs 133 and 135)

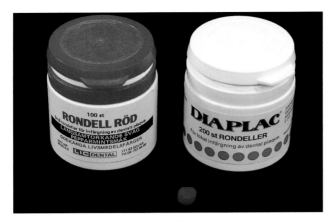

Fig 131 Plaque-disclosing pellets.

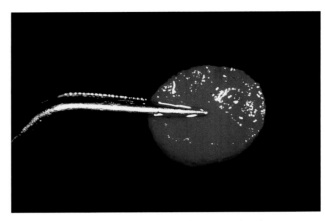

Fig 132 A plaque-disclosing pellet is used with tweezers.

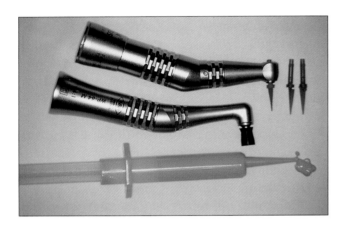

Fig 133 Equipment for PMTC: *(top to bottom)* Profin (Dentatus) contra-angle handpiece and reciprocating tips, prophylaxis contra-angle handpiece and rotating rubber cup, and a syringe for injecting prophylaxis paste interproximally.

Fig 134 A tube of RDA 170 (medium abrasive) fluoride-containing prophylaxis paste.

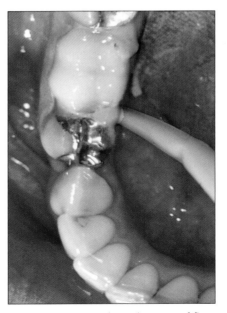

Fig 135 Syringed application of fluoride polishing paste.

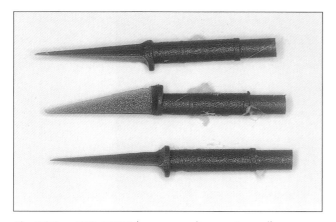

Fig 136a EVA-2000 (green medium, universal).

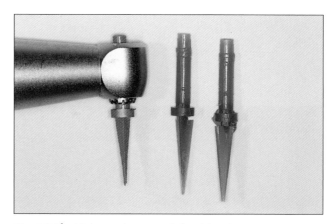

Fig 136b V-shaped, pointed, flexible plastic tips.

Professional mechanical toothcleaning is carried out in the following sequence: (*1*) Plaque is disclosed; (*2*) prophylaxis paste is applied to interproximal areas; (*3*) interproximal areas are cleaned; and (*4*) lingual, buccal, and occlusal surfaces are cleaned.

Plaque disclosure

Because PMTC must target the tooth surfaces normally neglected by the patient, disclosure of plaque is the first step. Application of a disclosing pellet takes less than 1 minute using the following procedure:

1. It is best to start where the plaque deposits are often heaviest, ie, in the mandibular lingual embrasures, where abundant saliva makes disclosure difficult.
2. Plaque in the mandibular buccal embrasures is then disclosed by pressing the pellet lightly into each interproximal space.
3. Finally, the plaque on the maxillary palatal and buccal surfaces is disclosed.

Interproximally, the presence of plaque should be verified by probing; it is invariably present if continuous plaque is visible in the line angles or if the approximal surfaces are carious or restored and the interproximal space is completely occupied by the gingival papilla.

Application of prophylaxis paste to interproximal areas

Use of a disposable syringe facilitates the application of fluoride polishing paste to the interproximal areas (see Figs 133 to 135). A rational procedure is to start from the lingual aspect of the mandibular teeth before the floor of the mouth is filled with saliva. The gingival papillae are pressed down with the point of the syringe before injecting the polishing paste. When paste is already applied to the surfaces requiring most attention, interproximal mechanical cleaning can be carried out very quickly.

Interproximal PMTC

The Profin prophylaxis contra-angle handpiece and the EVA-2000 (green medium, universal) or the V-shaped, flexible EVA-123 (blue large) and EVA-123 (pink small) tips are used for interproximal PMTC (Figs 136a and 136b). An alternative is

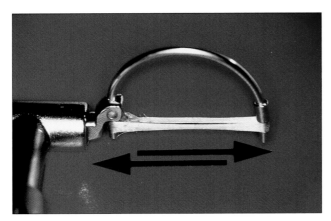

Fig 137 EVA-123 (pink small) and holder with double dental tape.

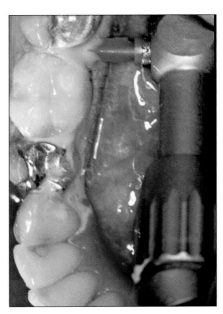

Fig 138 Professional mechanical toothcleaning always commences from the lingual surfaces of the mandibular molars.

the EVA-7 triangular-tipped wooden toothpick, which comes with or without fluoride. The tips are self-steering and reciprocating with 1.0- to 1.5-mm strokes. For children with partially erupted teeth, a special dental tape holder is available for the Profin contra-angle handpiece (Fig 137). The double-tape design will clean the approximal surfaces subgingivally on both sides of the papilla at the same time.

Professional mechanical toothcleaning should start at the linguoapproximal surfaces of the posterior mandible because the average patient usually fails to clean this area properly. This also reinforces the principle that cleaning should always be started where it is most needed to improve motivation for both the patient and the operator (Fig 138). Entering the interproximal space the tip will have a 10-degree coronal angle until the papilla is depressed. Because of the resilience of the papilla, a subgingival cleaning effect can be expected at least 2 to 3 mm subgingivally (see Figs 80 and 81 in chapter 2).

A suitable speed for the contra-angle handpiece is approximately 8,000 rpm (ie, 16,000 strokes per minute or 250 per second). At very low

speeds, (less than 5,000 rpm), vibration will cause discomfort to the patient. The direction of the tip should continually be adjusted in both the vertical and horizontal directions to reach all the approximal surfaces. At the same time, the fluoride polishing paste is applied to all cleaned surfaces.

As described previously, PMTC should always commence from the lingual aspect of the mandibular molars, in accordance with the linking method (see chapter 2). When the approximal surfaces have been carefully cleaned from the fairly easily accessible lingual side, they are then cleaned from the buccal direction. The maxillary interproximal surfaces are cleaned next in the same order: lingual embrasures first and then the buccal embrasures.

Lingual, buccal, and occlusal PMTC

A regular prophylaxis contra-angle handpiece (eg, W&H) and a rotating rubber cup (eg, Young's BS 1800) are recommended for PMTC on the lingual and buccal surfaces (Figs 139 to 141). The

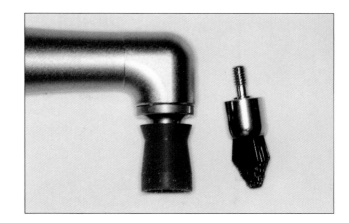

Fig 139 Prophylaxis contra-angle handpiece with Young's rotating rubber cup or pointed brush (for fissures).

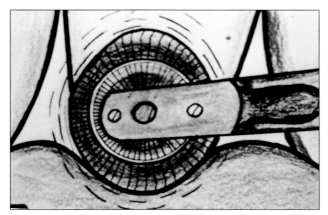

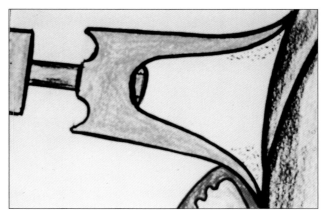

Figs 140 and 141 Illustration of how the rotating rubber cup works. (Illustrations by J. Waerhaug, courtesy of the University of Oslo.)

same medium-abrasive prophylaxis paste used for approximal PMTC is used for these surfaces also. This step should also start on the surfaces most often neglected by the patient, ie, the mandibular lingual surfaces. The rubber cup should clean the line angles and the subgingival surfaces also. It is possible to remove plaque to a depth of at least 1 to 2 mm subgingivally (see Fig 141). To remove plaque from the fissures of erupting molars in particular, a rotating brush (eg, Young's) is used in the prophylaxis contra-angle handpiece (see Fig 139).

After meticulous PMTC, plaque-disclosing pellets are used again to ensure that all tooth surfaces are plaque free.

The pellicle functions as a nonfrictional layer on the tooth surface. For example, in vivo studies by Saxton (1975) established that it takes approx-

imately 5 minutes to remove the pellicle completely from the tooth enamel using a rubber cup and pumice. With an average treatment time of 3 to 7 seconds for each surface, needs-related professional toothcleaning carries minimal risk of abrasive damage to the tooth surface. Resin composite restorations should not be polished with abrasive prophylaxis paste.

During the past 10 years, air polishing devices have been introduced for removal of supragingival plaque and stains (Fig 142). The instruments are efficient, except on the key-risk surfaces, ie, the approximal surfaces of the molars and premolars, because the buccal and lingual papilla impede access of the abrasive powder. There is also a risk of substantial abrasion on exposed root surfaces and resin composite restorations (Boyde, 1984; Horning et al, 1987; Weaks et al, 1984). How-

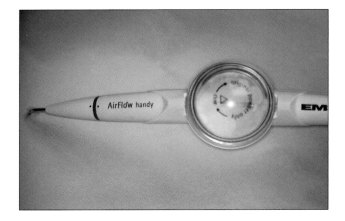

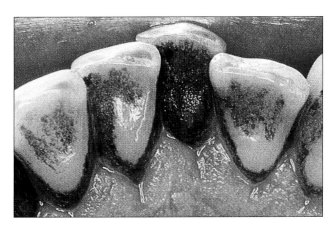

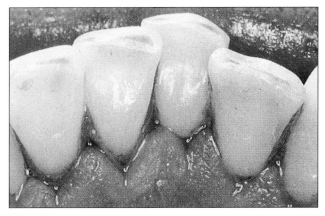

Fig 142 (*above left*) EMS AirFlow air abrasion handpiece (Electro Medical Systems, Nyon, Switzerland).

Fig 143 (*above right*) Staining on the lingual surfaces of the mandibular incisors in a heavy smoker before air abrasion.

Fig 144 Results after air abrasion.

ever, air abrasion is useful for removing extrinsic stains from enamel—particularly smoking- and chlorhexidine-induced stains (Figs 143 and 144).

Hand instruments, such as curettes, may also supplement the reciprocating and rotating instruments for PMTC to remove partly mineralized plaque in the gingival sulcus or more deeply located subgingival plaque biofilms. As earlier discussed, this procedure is also known as *debridement*.

PREVENTIVE EFFECTS OF PMTC

After PMTC, re-formation of perceptible complex plaque in the dentogingival region is normally retarded for several days, compared to about 1 to 2 days after oral hygiene measures carried out by the patient. Professional mechanical toothcleaning should completely remove supragingival plaque from all tooth surfaces and plaque to at least 1 to 3 mm subgingivally; ie, it provides gingival plaque control.

Frequent PMTC also influences the composition of the subgingival microflora and reduces the number of periopathogens (Dahlén et al, 1992; Hellström et al, 1996; Katsanoulas et al, 1992; McNabb et al, 1992; Siegrist and Kornman, 1982; Smulow et al, 1983; Ximénez-Fyvie et al, 2000). After subgingival root surface instrumentation, frequent PMTC can prevent recolonization by subgingival microflora (Magnusson et al, 1984; Mousqués et al, 1980). Some effect could also be expected on caries-inducing pathogens, such as *Streptococcus mutans*, on the approximal surfaces of the molars, which are inaccessible to the toothbrush (Axelsson et al, 1987b).

Experimentally, it has been shown that after a single PMTC, the volume of gingival exudate decreases continuously during the first 24 to 28 hours and does not regain the preexperimental level until 1 week later (Gwinnett et al, 1975). Three sessions of PMTC, at 2-day intervals, will normally induce healing of inflamed gingivae within 1 week. Indirectly, this results in a reduction in plaque formation rate. Studies have shown that the reaccumulation rate of gingival plaque is directly correlated to the degree of gingival inflammation (Axelsson, 1987; Ramberg et al, 1995a) and the quantity of gingival exudate (Saxton, 1975). Therefore, frequent initial PMTC, followed up at needs-related intervals in the maintenance program, enhances the patient's own oral hygiene efforts by removal of mature, partially mineralized plaque and reduces the rate of formation of new plaque.

The fluoride ions in prophylaxis paste gain access to the cleaned approximal surfaces, even subgingivally, increasing the potential for remineralization of enamel caries and root caries on these key-risk surfaces. This reduces the risk for future plaque-retentive factors such as secondary caries, restoration overhangs, and unfinished subgingival margins.

Professional mechanical toothcleaning may also be expected to have a strong patient-motivating effect if it is carried out in a needs-related fashion, similar to the oral hygiene procedures proposed in chapter 2. The patient experiences PMTC as a positive treatment form and attempts to maintain the feeling of cleanliness with his or her own efforts (Glavind, 1977).

Effect of PMTC on caries in children and young adults

Although the preventive effect of plaque control programs on gingivitis and periodontitis is generally accepted, the preventive effect of mechanical plaque control on the development of caries has been questioned. Therefore, a series of longi-tudinal plaque control programs, including PMTC, were initiated to test the effect of mechanical plaque control on gingivitis and dental caries in schoolchildren.

In 1971, a longitudinal clinical trial (Axelsson and Lindhe, 1974, 1977) was initiated to test the hypothesis that gingivitis and dental caries do not develop in schoolchildren maintained on an oral hygiene program that includes PMTC and oral hygiene instruction once every 2 weeks. The subjects were 216 children, aged 7 to 8, 10 to 11, and 13 to 14 years at the beginning of the study. All children were from the same elementary school in the city of Karlstad, Sweden, and all had the same socioeconomic background. All were randomly assigned to a test or a control group.

During a 4-year period, a preventive dentistry assistant carried out PMTC on the test group 16 times a year for the first 2 years and then 4 to 6 times a year for the following 2 years. The control group received ordinary dental treatment once a year, supervised toothbrushing using the Bass method, and fluoride rinsing 10 times a year for the entire trial period. Thus, there were no true negative controls: For ethical reasons, the subjects in the control group were maintained on the regular dental care schedule.

Figure 145 shows the effect of treatment on the amount of plaque according to the Silness and Löe Plaque Index (PI), and the development of gingivitis according to the Löe and Silness Gingival Index (GI). There were no differences between the two groups at baseline. After 1 year, there was a marked reduction in both plaque and gingivitis in the test group, but in the control group, PI and GI remained persistently high. The test group succeeded in maintaining low scores for the following 3 years, but the high plaque and gingivitis scores in the control group did not improve.

More pronounced was the effect on the development of dental caries (Fig 146). During the entire 4-year trial, the test group developed only 62 new decayed or filled surfaces in all, while the control group developed 941 (Axelsson and Lindhe, 1974, 1977).

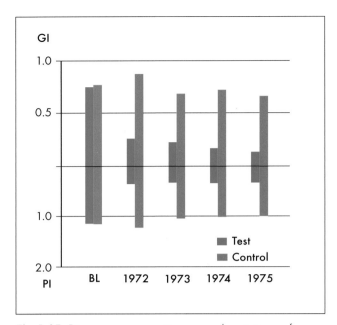

Fig 145 Diagram representing annual variations of mean Gingival Index (GI) and mean Plaque Index (PI) of the test- and control-group children. The average values for groups 1, 2, and 3 have been used. (BL) Baseline. (Modified from Axelsson and Lindhe, 1977 with permission.)

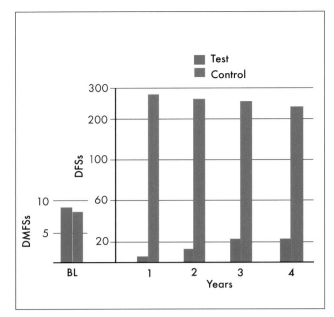

Fig 146 The mean number of new decayed or filled surfaces (DFSs) in the three control groups compared to the three test groups during the 4-year trial. (DMFSs) Decayed, missing, or filled surfaces; (BL) Baseline. (Modified from Axelsson and Lindhe, 1977 with permission.)

The long-term benefits of this early introduction of a program of meticulous mechanical plaque control on dental caries (Figs 147 to 150) and standard of self-performed oral hygiene and gingival health (Figs 151 to 157) should not be underestimated. The long-term caries preventive effect of frequent PMTC in young children was later reconfirmed in another Swedish study (Klock, 1984).

This study was followed up with numerous clinical trials to compare the separate effect of PMTC on caries and gingivitis with other preventive measures such as oral hygiene training, chemical plaque control, and topical use of fluorides. In all these studies, frequent PMTC was superior to the other measures (Axelsson and Lindhe, 1975, 1981c; Axelsson et al, 1976). The first Karlstad study, based on PMTC, aroused interest in other countries such as Norway, Denmark, Great Britain, and Brazil. Researchers in these countries also followed up by carrying out PMTC-based studies, most of which had results similar to those of the original study (Agerbaek et al, 1977; Gisselsson et al, 1983; Hamp et al, 1978, 1982; Kjaerheim et al, 1980; Klimek et al, 1985; Poulsen et al, 1976; Talbott et al, 1977; for review, see Axelsson, 1981, 1993a, 1994; Bellini et al, 1981; Gjermo, 1986; and Hotz, 1998). The only study that failed to reveal a significant effect was that carried out by Ashley and Sainsbury (1981) in 100 schoolchildren with very low caries incidence. For cost effectiveness, it is important that, under field conditions, the frequency of PMTC be based on individual needs (Axelsson, 1998).

Table 3 presents a review of the caries-preventive effect of PMTC in different clinical studies. Direct comparison of the studies is difficult because they differ in frequency of PMTC, concomitant preventive measures, caries prevention in the control groups, and caries risk in the selected populations. All PMTC trials included use of fluoride prophylaxis paste, fluoride rinsing

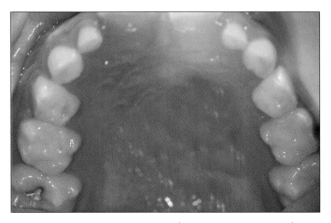

Fig 147 Patient in preventive dentistry test group at baseline, aged 7 years. The patient had one restored maxillary molar.

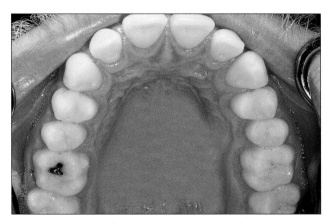

Fig 148 Patient shown in Fig 147, aged 18 years. No additional teeth have developed caries.

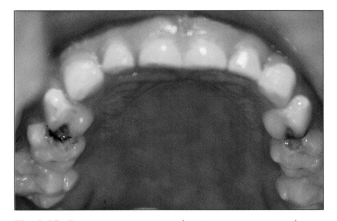

Fig 149 Patient in preventive dentistry test group at baseline, aged 7 years. Several primary teeth are carious.

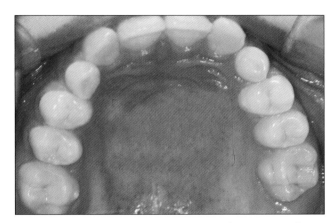

Fig 150 Patient shown in Fig 149, aged 12 years. All permanent teeth are caries free.

Fig 151 Typical patient in control group, aged 18 years. All the permanent premolars and molars have restorations.

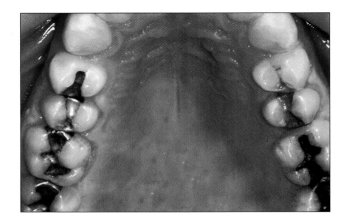

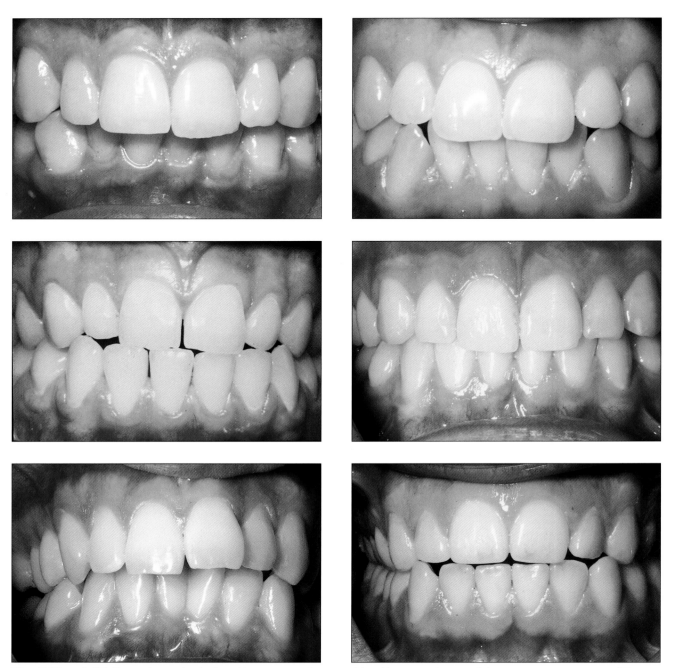

Figs 152 to 157 Typical standard of oral hygiene and gingival health in six subjects of the youngest age group (7 years at baseline) at the age of 18 years (7 years after completion of the study).

Table 3 The caries-preventive effects shown in professional mechanical toothcleaning (PMTC) trials*

Frequency of cleaning[†]	Control[‡]	Annual caries increment (New DS) Test	Control	Caries reduction	Age group (y)	Length of trial (y)	Study
Every 2 wk (F)	F rinse, OHI	2.06	2.38	13%	5–13	2	Westergaard et al (1978)
Every 2 wk (F) (half mouth)	OHI	0.6[§]	2.6[§]	77%	13–14	1.5	Axelsson and Lindhe (1981a)
Every 2 wk (F)	F rinse, OHI, monthly (F)	0.1	3.0	96%	7–12	2	Axelsson and Lindhe (1974)
Every 3–4 mo (F)	F rinse, OHI, monthly (F)	0.15	3.0	95%	9–14	2	Axelsson and Lindhe (1978)
Every 2 wk (F)	F rinse, OHI, monthly (F)	0.18	2.75	90%	13–16	2	Karlsson and Larsson (1976)
Every 2 wk (øF)	0.5% CHX gel every 2 wk (øF)	0.4	4.3	90%	13–14	1	Axelsson et al (1976)
Every 2 wk (F)	F rinse, OHI (F)	0.94	3.40	73%	9–12 (selected high caries risk)	2	Klock and Krasse (1978)
Every 2 wk (F)	OHI (F)	0.43	1.42	70%	7	1	Poulsen et al (1976)
Every 3 wk (F)	F rinse 2 wk	2.10	4.26	51%	7–16	3	Hamp et al (1978)
Every 3 mo (F)	OHI (F)	1.2	3.0	60%	17–19	3	Malmberg (1976)
Every 6 mo (y3)	None	0.33	1.10	70%	16–19	3	Hamp and Johansson (1982)
Monthly (øF)[¶]	Placebo	1.6	2.7	41%	13–14	2	Zickert et al (1982)
Every 3 mo (øF)[¶]	Placebo	1.9	3.5	43%	13–14	2	Zickert et al (1982)
Every 2 mo (F)	F brushing	0.26[#]	0.11[#]	57%	6–8	2	Kaerheim et al (1980)
	F brushing	0.15[#]	0.66[#]	78%	10–12	2	Kaerheim et al (1980)
	F brushing	0.66[#]	1.48[#]	54%	13–15	2	Kaerheim et al (1980)
Every 2 mo (øF)**	None	1.35	2.50	46%	12–14	2	Klimek et al (1985)
Every 2 mo (F)[††]	None	0.57	1.43	60%	7–8	3	Kerebel et al (1985)
Every 2–3 mo (F)	None	0.07	2.47	97%	<35	6	Axelsson and Lindhe (1981b)
	None	0.07	1.80	96%	36–50	6	Axelsson and Lindhe (1981b)
	None	0.07	1.40	95%	>50	6	Axelsson and Lindhe (1981b)
Every 6 mo (F)[‡‡]	F water	0.35	0.89	61%	5–12	5	Bagramian (1982)
Yearly[††]	None	0.32	1.02	69%	11–12	6	Lallo and Solanski (1994)

*Data from Axelsson (1981) and Hotz (1998).

[†](F) Fluoride; (øF) No fluoride.
[‡](OHI) Oral hygiene instruction.
[§]18 months.
[¶]Fluoride rinse after cleaning.

[#] Without occlusal surfaces.
** Fluoride varnish every 6 months.
[††] Daily toothbrushing with fluoride paste (additionally).
[‡‡] Fluoride water, sealants.

after PMTC, or application of fluoride varnish at least twice a year. With one exception (Westergaard et al, 1978), caries reductions in the test groups were always more than 40% (41% to 97%). This remarkable caries-preventive effect achieved by PMTC contrasts with the modest achievements attained by supervised toothbrushing. In several of the trials, subjects in the control groups were also in a fluoride program. Nevertheless, the test groups showed significant caries reduction compared to the "positive" control groups, confirming that meticulous professional mechanical plaque control is an efficient caries-preventive measure.

Effect of combined self-care and PMTC

The experiences gained from early, recent, and ongoing Karlstad studies are continuously implemented in large-scale preventive programs for children and young adults. From a cost effectiveness aspect, the appropriate intervals for education in self-care and PMTC are based on the predicted risk, which may be assessed at a variety of levels, from age group to the individual tooth surfaces.

In 1979, a needs-related preventive program, based on the aforementioned principles, was introduced for all newborn to 19-year-old residents of the county of Värmland, Sweden. The goals for the subjects following the program are:

1. No approximal restorations
2. No occlusal amalgam restorations
3. No approximal loss of periodontal attachment
4. Motivation and encouragement of individuals to assume responsibility for their own oral health

It was hoped that these goals would be attained for 20-year-old residents by 1999. A computer-aided epidemiologic system (Axelsson et al, 1993a, 2001) has evaluated the effect of the program annually since its inception, in almost 100% of all 3- to 19-year-old residents. On average, the number of total and approximal new carious surfaces (DSs) per year was reduced by 85% to 95%. As a consequence, during the same period, caries prevalence has declined dramatically in all age groups. For example, the mean number of DFSs has decreased from 6 to 0.4 in 12-year-old children and from 22 to 2 in 19-year-old young adults from 1979 to 1999. As many as 85% of the 12-year-old children were caries free in 1999. (For details, see chapter 7.)

The Swedish Board of Health and Welfare (Sundberg, 1996) collects annual data on caries prevalence in 5-, 12-, and 19-year-old individuals and the cost per person per year. In comparison to other programs, the needs-related preventive program of the county of Värmland is highly cost effective.

Effect of frequent PMTC on gingival health and subgingival microflora in adults

Gingival plaque control has been considered to have little effect on the subgingival microflora of deep periodontal pockets. However, this may not apply to moderately deep pockets (4 to 6 mm), which may represent an intermediate pathologic state between gingivitis and advanced marginal periodontitis. In a study by Mc-Nabb et al (1992), the subjects had poor oral hygiene and severe gingival inflammation. Four matched sites (one in each quadrant) that harbored at least 20% spirochetes and 15% black-pigmented gram-negative bacilli were selected. During the first 12 weeks (phase 1), supragingival calculus was removed from the right half of the mouth and then the teeth were cleaned by PMTC three times a week. At the beginning of phase 2, supragingival calculus was also removed from the left quadrants, and the entire mouth was subjected to the same protocol used in phase 1. At no time did patients receive oral hygiene instruction. Clinical variables were assessed and microbiologic samples were taken at 3-week intervals.

Significant changes occurred in the composition of the subgingival microflora at cleaned sites. While gram-positive organisms increased proportionally, the number of putative periodontal pathogens, such as *Porphyromonas gingivalis* and spirochetes, decreased. Both PI and GI scores also decreased during the total experimental period of 30 weeks (McNabb et al, 1992).

In another study (Katsanoulas et al, 1992), the effect of mechanical gingival plaque control by PMTC on the composition of the subgingival microflora in untreated 4- to 6-mm-deep periodontal pockets was investigated in subjects with chronic periodontitis. Periodontally diseased sites were subjected to PMTC three times weekly for 3 weeks. Contralateral sites received no prophylaxis and served as controls. No instructions in oral hygiene procedures were issued; the patients continued with their usual oral hygiene routines during the observation period. Clinical examination and darkfield microscopic analysis of bacterial samples were performed every week.

The PI scores decreased markedly for the experimental sites but remained unchanged for the control sites throughout the observation period. The composition of the subgingival microflora at the control sites did not change during the experimental period. At the test sites, the proportion of spirochetes and motile rods decreased continuously. The results indicate that, at periodontally diseased sites with an established subgingival ecosystem, gingival plaque removal may influence the composition of the subgingival microflora (Katsanoulas et al, 1992).

The effect of meticulous gingival plaque control by PMTC on the subgingival microbiota has also been investigated in a long-term study (Dahlén et al, 1992), in which 300 subjects were examined for periodontal disease and monitored, without treatment, for 2 years. After the 2-year examination, 80 subjects were invited to participate in a treatment program intended to improve the standard of plaque control. Of these 80 subjects, 40 had gingivitis and only minor attachment loss, and 40 had moderate periodontitis, 23 with several sites with deep pockets (greater than 4 mm).

After the clinical examination, samples of the subgingival microbiota were harvested. The patients were then recalled for oral hygiene training and PMTC at needs-related intervals. Two years after the year 2 examination, the subjects were reassessed clinically and microbiologically.

The findings demonstrated that meticulous gingival plaque control changes the quantity and the composition of the subgingival microbiota. Two years after initiation of the improved oral hygiene efforts, the total viable counts of bacteria in both deep and shallow pockets were markedly reduced. The number of subjects and sites harboring periopathogens such as *P gingivalis* and *Actinobacillus actinomycetemcomitans* had also decreased markedly between years 2 and 4 (Dahlén et al, 1992).

Hellström et al (1996) investigated the possible influence of careful PMTC on the subgingival microbiota at periodontal sites with suprabony, infrabony, or furcation pockets. None of the participants in the study had undergone periodontal therapy during the previous 12 months or used antibiotics during the 3 months preceding the study. Following a screening examination, six to eight sites per subject were selected, all with probing depths greater than 5 mm: one to three sites with suprabony lesions, one to three with infrabony lesions, and one to three with furcation defects. For each of these sites, a bacterial sample was taken and the following variables were recorded at baseline: plaque, gingivitis, probing depth, and probing attachment level. The case findings were then presented to each subject, followed by thorough supragingival scaling, and instruction in proper plaque control with a toothbrush and a dentifrice. During the subsequent 30 weeks, the subjects were recalled two or three times a week for PMTC by a dental hygienist, about 15 minutes per session. The subjects were reexamined after 30 weeks.

The findings indicated that PMTC, combined with careful self-performed plaque control, had a marked effect on the subgingival microbiota of moderate-to-deep periodontal pockets. Thus, at sites with suprabony and infrabony pockets, as well as at furcation sites, the meticulous, prolonged

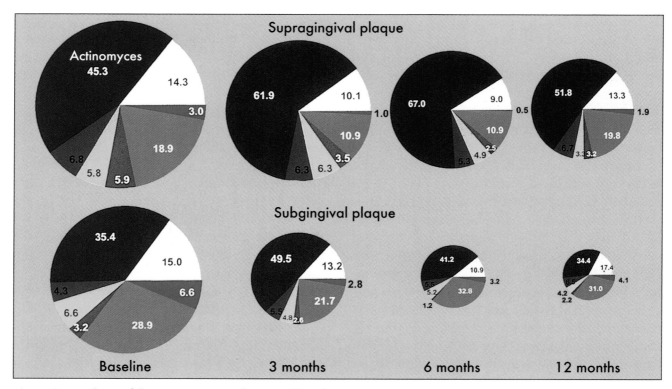

Fig 158 Pie charts of the mean DNA probe count (%) of microbial groups in supra- and subgingival plaque samples at baseline, 3, 6, and 12 months. The species were grouped into microbial groups based on the description by Socransky et al (1998). The areas of the pies were adjusted to reflect the mean total counts at each of the time points relative to the mean count at baseline for the supragingival plaque sample. (From Ximénez-Fyvie et al, 2000. Reprinted with permission.)

gingival plaque removal by frequent PMTC reduced the total number of microorganisms that could be harvested, as well as the percentage of sites with *P gingivalis* (Hellström et al, 1996).

Recently, Ximénez-Fyvie et al (2000) reported a study on the effect of repeated gingival PMTC on the composition of the supra- and subgingival microbiota. Eighteen adult maintenance subjects with periodontitis were clinically and microbiologically examined at baseline and after 3, 6, and 12 months. After the baseline examination, the subjects received scaling and root planing followed by PMTC every week for 3 months. Clinical measures of plaque accumulation—bleeding on probing (BOP), gingival redness, suppuration, pocket depth, and attachment level—were carried out at 6 sites per tooth at each visit. Separate supragingival (n = 1,804) and subgingival (n = 1,804) plaque samples were taken from the mesial aspect of all teeth

in each subject at each time point and evaluated for their content of 40 types of bacteria using a checkerboard DNA-DNA hybridization technique.

The results showed that the mean percent of sites exhibiting plaque, gingival redness, and BOP were significantly reduced during the course of the study. Significant decreases in mean bacterial counts were observed in both supra- and subgingival samples and maintained 9 months after the 3-month treatment period. For example, at baseline, 3, 6, and 12 months, respectively, the subgingival samples contained a mean count ($\times 10^5$) of 2.0, 0.5, 0.6, and .03 of *P gingivalis* ($P < .001$); 2.0, 0.4, 0.4, and 0.1 of *Bacteroides forsythus* ($P < .001$); and 3.4, 0.8, 0.4, and 0.3 of *Treponema denticola* ($P < .01$). Similar reductions were also seen in supragingival samples. While counts were markedly reduced by PMTC, the proportions of the 40 test species were only marginally affected (Fig 158).

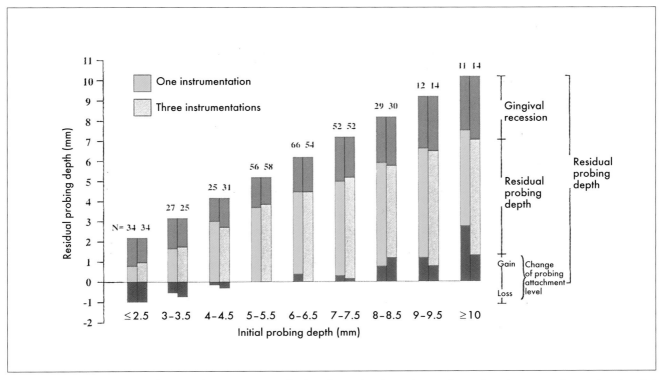

Fig 159 Mean amounts of gingival recession, residual probing depth, and gain/loss of probing attachment level at 24 months in relation to initial probing depth. (Modified from Badersten et al, 1984b with permission.)

In conclusion, the study showed that PMTC profoundly diminished counts of both supra- and subgingival bacterial species, creating a microbial profile comparable to that observed in periodontal health. This profile was maintained at the final examination, 9 months after completion of weekly PMTC.

In a selected group of patients with advanced periodontitis, Badersten et al (1984b) compared the effect of one session of meticulous subgingival scaling and root planing with three sessions, at 1-month intervals, in a 24-month split-mouth study. After the initial instrumentation of single-rooted teeth, the patients were recalled for repeated oral hygiene training and PMTC at intervals based on individual need. The initial probing depths varied from 2.5 to 11.0 mm (Fig 159). Plaque scores, bleeding on probing, probing depths, and probing attachment levels were recorded.

During the initial 9 months of the study, there was a gradual, marked improvement in periodontal status. During the remaining 15 months of the 24-month experimental period, no further changes in the recorded variables were noted. No differences in results were observed between the effects of single and repeated instrumentation. Thus, it appears that deep periodontal pockets in incisors, canines, and premolars may be successfully treated by gingival plaque control and a single, initial session of meticulous instrumentation. The results also suggest that subgingival recolonization by microorganisms during the healing phase may not be a major clinical problem if high-quality plaque control is established (Badersten et al, 1984b) (see Fig 159). These results also should be compared with those of the studies described earlier, based on gingival plaque control (PMTC) without initial subgingival scaling and root planing (Dahlén et al, 1992; Hellström et al, 1996; Katsanoulas et al, 1992; McNabb et al, 1992).

Thus, after a single session of meticulous scaling, root planing, and debridement, without aggressive removal of the root cementum, the health of the periodontal tissues can be maintained by excellent gingival plaque control, comprising self-care supplemented by PMTC at needs-related intervals. The need for repeated subgingival scaling should be regarded as a treatment failure.

Frequent PMTC also has been successfully used in maintenance programs following initial nonsurgical or surgical treatment of marginal periodontitis. In a study by Rosling et al (1976), patients with a high prevalence of infrabony pockets were randomly allotted to a test or a control group. After initial open flap surgery, scaling, and root planing, the test group received PMTC every 2 weeks for 2 years. At reexamination, about 95% of the infrabony pockets had healed, and the condition of the gingiva was excellent. Among the patients in the control group there were some nonresponders with very significant loss of peri- odontal attachment and some lost teeth (Nyman et al, 1977).

In an investigation by Nyman et al (1975) of 20 surgically treated patients with advanced periodontitis, an experimental group subsequently received thorough PMTC and oral hygiene education every 2 weeks; no further clinical loss of attachment could be demonstrated after 2 years. The patients in the control group received the same initial treatment, including surgery, and were recalled for debridement and scaling 6 months postoperatively, but no other attempts were made to maintain gingival plaque control. After 2 years, the control patients exhibited an average clinical loss of attachment of about 2 mm. This very rapid periodontal destruction suggests that, in the absence of proper supportive care based on excellent gingival and plaque control, periodontal surgery may in fact do more harm than good. In the control group, probing depth 2 years postoperatively was approximately the same as before surgery; in the experimental group, the probing depths were maintained at the immediate postoperative level.

In a split-mouth study, Lindhe et al (1982) found excellent healing of marginal periodontitis, irrespective of the initial treatment method—nonsurgical or different types of flap surgery—when this was followed by frequent PMTC for 2 years.

Even in patients with aggressive periodontitis, no recurrence was observed over a 5-year period when initial nonsurgical or surgical treatment was followed by frequent PMTC during the first 2 years (Wennström et al, 1986).

In a split-mouth study of flap surgery, one test quadrant was scaled and all root cementum was removed, exposing root dentin. The root surfaces of the contralateral quadrant were carefully cleaned by PMTC; visible calculus was gently removed with a curette, and removal and planing of cementum were avoided. The flaps were then replaced and sutured. The patients received PMTC every 2 weeks for 24 months. The same gain in clinical probing attachment was observed with both methods, indicating that, when excellent gingival mechanical plaque control is maintained, remaining "diseased" root cementum and possibly some calculus will not interfere with healing (Nyman et al, 1988). Schwarz et al (1993) showed that PMTC removed subgingival plaque biofilms very efficiently during open flap surgery, even without removal of calculus.

Westfelt et al (1983a, 1983b) compared rinsing with 0.2% chlorhexidine digluconate with regular mechanical plaque control during healing after periodontal surgery. Fourteen patients participated: After surgery, seven patients rinsed with chlorhexidine for 2 minutes, twice a day, for 6 months (healing phase). The remaining patients were enrolled in a strict gingival plaque control program that included PMTC once every 2 weeks. After the 6-month reexamination, all patients entered a supportive care program with recalls once every 3 months until the final examination at 24 months. At the end of the healing phase, the chlorhexidine group had a higher frequency of pockets deeper than 4 mm and less gain of attachment in pockets that were initially deeper than 4 mm than did the PMTC group.

Effect of combined self-care, PMTC, and debridement

Effect on periodontal diseases and the outcome of periodontal therapy

The claim that treated pockets with a long junctional epithelium could predispose to repocketing has been refuted by research. The results from one study (Ramfjord et al, 1987) suggest that resistance to progression of periodontal disease is similar for a long epithelial attachment and connective tissue attachment. A higher percentage of shallow pockets (less than 3 mm) lost more than 2 mm of attachment over 5 years compared with deeper pockets, mostly as an iatrogenic effect on buccal surfaces. Thus there appears to be no justification for forceful penetration of the junctional epithelial attachment at recall: If there is no calculus, tightly adapted pocket walls without appreciable subgingival spread of plaque biofilms and inflammation should be treated only by PMTC and cautious debridement.

Using a study design that includes subgingival instrumentation and assessment of subgingival recolonization, several studies have incidentally allowed investigation of the influence of gingival plaque control on subgingival recolonization (Renvert et al, 1990). It is, however, acknowledged that some bacteria may persist at the pocket sites following such instrumentation. Mousqués et al (1980) reported a positive correlation between low supragingival plaque scores and a subgingival flora resembling that associated with health in 14 individuals who had received thorough scaling and root planing but no oral hygiene instruction.

Magnusson et al (1984) observed that, in the presence of supragingival plaque, a subgingival flora with large numbers of motile rods and spirochetes re-formed in 4 to 8 weeks, whereas sustained reductions in motile rods and improved periodontal conditions were achieved following scaling plus supervised oral hygiene and PMTC. Similarly, Sbordone et al (1990) reported that subgingival debridement without gingival plaque control was insufficient to maintain a healthy subgingival microflora.

Axelsson and Lindhe (1981b) also demonstrated the value of carefully designed maintenance program, based on meticulous gingival plaque control by self-care and PMTC, for patients who had been treated for advanced periodontal disease: 77 patients were examined before treatment, 2 months after the last surgical procedure, and after 3 and 6 years. Two thirds (52) of the patients were placed on a supervised maintenance program that included oral hygiene education, meticulous PMTC, and needs-related debridement every 2 months for the first 2 years and every 3 months for the last 4 years of the observation period (recall group). The remaining 25 patients resumed care by the referring dentist, who was informed of the importance of checking their oral hygiene, calculus formation, and gingival and periodontal conditions (nonrecall group).

The data from the second examination showed the positive effect of the initial treatment in both groups. Subsequently, the recall patients were able to maintain excellent gingival health and unaltered attachment levels as an effect of proper oral hygiene and supportive PMTC. In the nonrecall group, plaque scores increased markedly from the baseline values, as did the number of inflamed gingival units (Fig 160). Concomitantly, there were obvious signs of recurrent periodontitis. The mean values for probing depth and attachment levels at the 3- and 6-year examinations were higher than at baseline (Fig 161). Approximately 99% of the tooth surfaces showed improvement, no change, or less than 1 mm of attachment loss in the recall group, compared to 45% in the nonrecall group. In the latter group, 55% of the sites showed 2 to 5 mm of further attachment loss at the 6-year examination, and 20% of the pockets were more than 4 mm deep (Axelsson and Lindhe, 1981b).

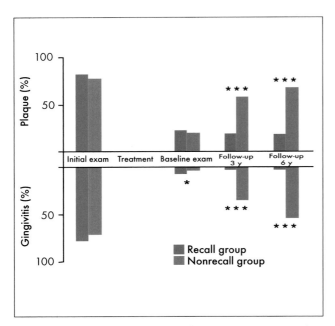

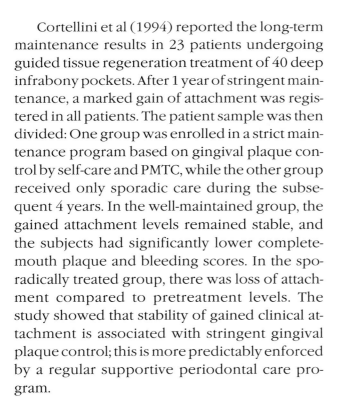

Fig 160 Percentage of tooth surfaces with plaque and inflamed units in the two groups of patients at the initial, baseline, and follow-up (3 and 6 years after the baseline) examinations. $*P<.05$; $***P<.001$. (Modified from Axelsson and Lindhe, 1981b with permission.)

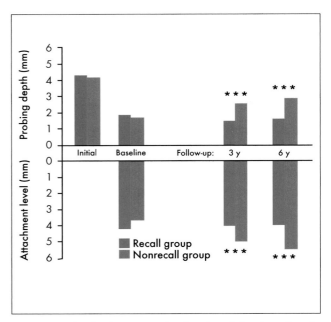

Fig 161 Individual mean pocket (probing) depth and probing attachment level data from the initial, baseline, and follow-up (3 and 6 years after the baseline) examinations. $***P<.001$. (Modified from Axelsson and Lindhe, 1981b with permission.)

Cortellini et al (1994) reported the long-term maintenance results in 23 patients undergoing guided tissue regeneration treatment of 40 deep infrabony pockets. After 1 year of stringent maintenance, a marked gain of attachment was registered in all patients. The patient sample was then divided: One group was enrolled in a strict maintenance program based on gingival plaque control by self-care and PMTC, while the other group received only sporadic care during the subsequent 4 years. In the well-maintained group, the gained attachment levels remained stable, and the subjects had significantly lower complete-mouth plaque and bleeding scores. In the sporadically treated group, there was loss of attachment compared to pretreatment levels. The study showed that stability of gained clinical attachment is associated with stringent gingival plaque control; this is more predictably enforced by a regular supportive periodontal care program.

In another study, Gottlow et al (1992) reported that, with meticulous gingival plaque control by self-care and PMTC, clinical attachment gained by guided tissue regeneration can be maintained for up to 5 years.

Effect on tooth loss, periodontal disease, and dental caries in adults

The overall goals for maintenance programs in randomized samples of adults should be to prevent tooth loss and the recurrence of dental caries and periodontitis after initial active treatment of the diseases, ie, secondary prevention and control. However, for cost effectiveness, such programs must be based strictly on individual needs.

The importance of the patient's oral hygiene efforts, as well as regular recall for PMTC and debridement, to the long-term maintenance of periodontal health was first examined in a study of gingival conditions in 1,428 individuals, 20 to 60 years of age, in an industrial company in Oslo, Norway (Lövdal et al, 1961). Over a 5-year period of regular recall (two to four times a year) for instruction in oral hygiene, PMTC, debridement, and scaling, gingival conditions improved by about 60%. Loss of teeth was reduced to about 50% of what would otherwise have been expected. No surgical treatment of periodontal pockets was undertaken; satisfactory results were very difficult to achieve in pockets more than 5 mm deep. However, there was marked improvement in the periodontal health, even in patients whose oral hygiene remained substandard during the 5 years of the trial.

Suomi et al (1971) measured loss of periodontal support in young individuals with gingivitis or with only minor loss of periodontal attachment. The experimental group received education in oral hygiene measures, PMTC, debridement, and scaling every third month over a 3-year period. Plaque accumulation and gingival inflammation were significantly reduced, and the mean value for clinically measurable loss of attachment was only 0.08 mm per surface during the observation period. Loss of attachment in the control group, in which no special measures were taken to improve oral hygiene, was 0.30 mm (0.10 mm per year).

Söderholm (1979) instituted a program of dental care including oral hygiene education and PMTC every 3 months in 443 shipyard employees who had previously been examined by Björn (1974). After 3 to 4 years, the dental health program had resulted in an improved standard of oral hygiene, and the progressive loss of alveolar bone had been arrested. There was also a decrease in the number of new caries lesions, and the need for dental restorations and the rate of tooth mortality had been reduced.

There are few longitudinal clinical studies in adults on prevention of recurrence of both periodontal disease and dental caries. In 1971 to 1972, a clinical study was initiated in the city of Karlstad, in the county of Värmland, Sweden, to determine whether the development of caries and the progression of periodontitis could be prevented in

adults and whether a high level of mechanical plaque control could be maintained by regularly repeated self-care education based on self-diagnosis, PMTC, and needs-related subgingival debridement. An attempt was also made to study the progression of dental disease in individuals who did not receive special self-care education but who regularly received conventional dental care.

Two groups of subjects from the same region were recruited for the study. Of these, 375 were assigned to test groups and 180 to control groups, stratified by age: 20 to 35 years (group 1), 36 to 50 years (group 2), and 51 to 65 years (group 3). The baseline examination included the following variables:

1. Plaque index (% according to O'Leary)
2. Gingival and pocket bleeding indices (%)
3. Community Periodontal Index of Treatment Needs (CPITN): probing depth (mm), calculus, and restoration overhangs
4. Clinical probing attachment loss (PAL)
5. Alveolar bone loss
6. Numbers of decayed, missing, or filled teeth and surfaces
7. History taking: eg, general diseases, medications, socioeconomic condition, body mass index, lifestyle, dental care and oral hygiene habits, and dietary and smoking habits

During the first 6-year period, the control patients were examined regularly once a year for conventional dental care. After initial scaling and root planing, the test group participants were recalled once every other month during the first 2 years and once every third month during the following 4 years of the study for education in self-care, PMTC, and needs-related subgingival debridement by a dental hygienist.

The oral hygiene training program was based on the principles discussed in chapter 2, with emphasis on self-diagnosis and interproximal cleaning with wooden, triangular-tipped toothpicks and fluoride toothpaste on the approximal surfaces of the posterior teeth. At each visit, a plaque-disclosing pellet was used to detect any residual

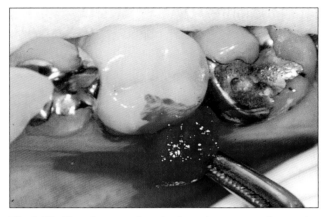

Fig 162 The plaque is first disclosed using a pellet on the linguoapproximal surfaces of the mandibular posterior teeth on the right side.

Fig 163 Result after staining.

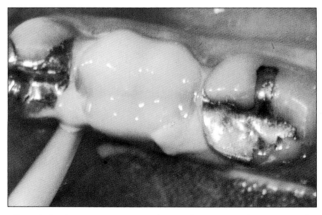

Fig 164 The PMTC procedure is begun on the same side with placement of the fluoride prophylaxis paste interproximally with a syringe.

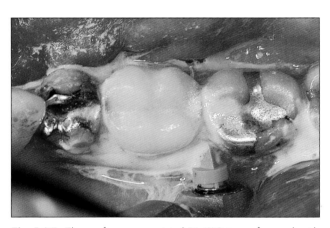

Fig 165 Thereafter approximal PMTC is performed with reciprocating triangular-pointed tips (EVA-2000) in the Profin prophylaxis contra-angle handpiece.

gingival plaque (Figs 162 and 163). Supplementary interdental probing was used for the posterior teeth. After discussion with the patient about any sites with residual plaque, needs-related PMTC was carried out. The first step after plaque disclosure was the application by syringe of fluoride-containing prophylaxis paste to the interdental areas of the mandibular posterior teeth (Fig 164). This was followed by careful approximal PMTC with the Profin prophylaxis contra-angle handpiece and reciprocating flexible pointed tips (Fig 165). The PMTC usually also included use of

a rotating rubber cup and prophylaxis paste, particularly on the lingual surfaces of the mandibular posterior teeth, and was supplemented, as necessary, by nonaggressive subgingival debridement or topical application of fluoride to specific key-risk surfaces.

The subjects were reexamined toward the end of the third and sixth years of the study. Figures 166 and 167 illustrate the mean values of caries incidence (new carious surfaces) and marginal periodontal incidence (vertical loss of attachment), respectively, for the test and control groups over 6

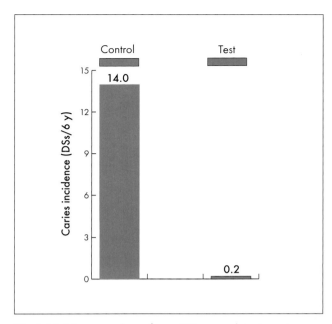

Fig 166 Mean number of new DSs per subject per 6 years in the control groups compared with the test groups. (Modified from Axelsson and Lindhe, 1981a with permission.)

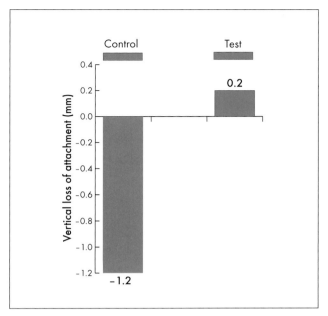

Fig 167 Mean probing attachment loss per subject per 6 years in the control groups and mean probing attachment gain per subject per 6 years in the test groups. (Modified from Axelsson and Lindhe, 1981a with permission.)

years. After 6 years, the number of new carious surfaces per subject was 14.0 in the control groups and only 0.2 in the test groups. On average, patients in the control groups lost 1.2 mm of periodontal attachment during the 6-year period, while those in the test groups had an average gain of 0.2 mm. It is important to note, however, that most of the subjects and sites in the control groups did not lose any or very limited periodontal attachment; the mean data reflect that there were a number of extreme "downhill" cases and sites (Axelsson and Lindhe, 1981a).

After the 6-year reexamination, the control groups were disbanded: For ethical reasons, these subjects were also offered needs-related preventive programs, and most accepted. The few subjects in the test groups who developed new caries lesions and/or lost periodontal attachment during the 6-year period were classified as high-risk or at-risk individuals for caries and/or periodontitis.

During the following years, up to the 15- and 30-year reexaminations, all subjects in the test

groups received a needs-related maintenance program from the same dental hygienist. To ensure maximum cost effectiveness, the recall intervals, as well as the preventive measures used, were based strictly on individual need. Approximately 60% visited the dental hygienist only once a year, 30% twice a year, and 10% (the at-risk and high-risk individuals) three to six times a year; the average annual number of visits for age groups 1, 2, and 3 was 1.2, 1.5, and 1.8, respectively. They were reexamined by a dentist on an average of only once every 3 to 4 years, including the compulsory 15- and 30-year reexaminations.

At the 15- (1987) and 30-year (2002) reexaminations, 317 and 257, respectively, of the original 375 test subjects from the baseline examination in 1972 were available for reexamination. Less than 10 subjects were unwilling to continue in the study, confirming high patient acceptance of the maintenance program. The main reason for dropouts in group 1 (50 to 65 years of age in 2002) and group 2 (66 to 80 years of age in 2002) was

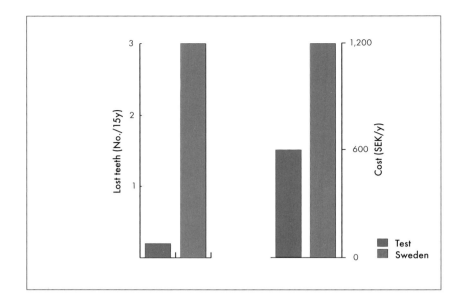

Fig 168 (*left*) Mean number of lost teeth per subject per 15 years (*left*) and the mean yearly cost for dental care (*right*) in the test groups compared to randomized samples of adult Swedes. (Modified from Axelsson et al, 1991 with permission.)

Fig 169 (*below*) Percentage of disclosed plaque on approximal (A), buccal (B), lingual (L), and mean values for all tooth surfaces in the maxilla and mandible in 1972 and 2002 in age groups 1, 2, and 3. (Modified from Axelsson et al, 2004 with permission.)

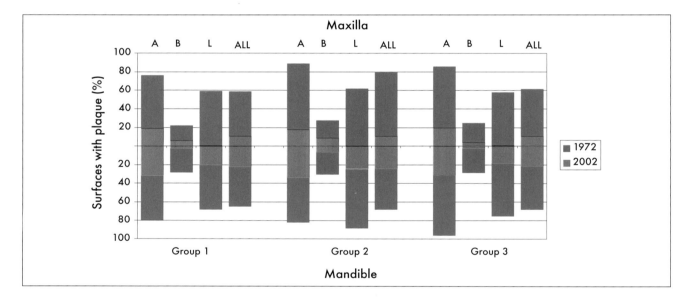

moving to other areas of Sweden. In group 3 (81 to 95 years of age in 2002), most dropouts were a result of death. Only 0.2 teeth per individual were lost over 15 years (70% because of root fractures). During the same period, it was estimated that the randomized samples of the Swedish adult population lost, on average, three teeth per individual (Håkansson, 1991) (Fig 168).

From 1978 until 1987, the total cost per individual per year in the test groups was approximately 50% of the cost per Swedish adult recall patient (see Fig 168). From 1978 until 1987, the mean treatment time by a dentist (including examination) was less than 20% of the average for recall patients in Sweden (Axelsson et al, 1991).

Effect on plaque values

As an effect of self-care education based on self-diagnosis, supplemented with needs-related intervals of PMTC, stained plaque was almost nonexistent at the 30-year reexamination compared to the baseline examination (Fig 169).

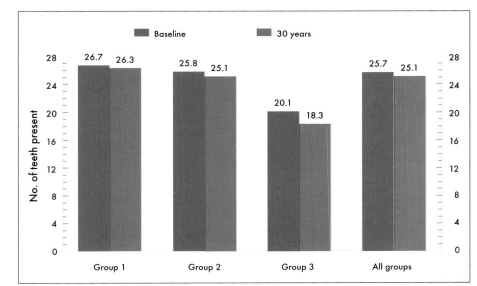

Fig 170 Mean number of teeth (third molars excluded) per individual in age groups 1, 2, and 3 and all groups at the baseline examination and 30-year reexamination. (Modified from Axelsson et al, 2004 with permission.)

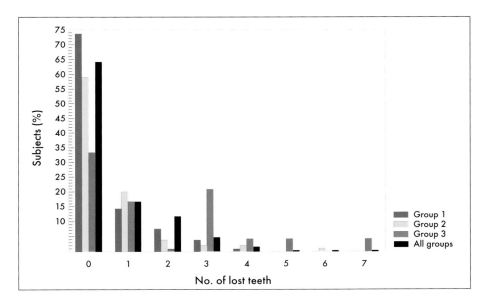

Fig 171 Frequency distribution of lost teeth per 30 years in age groups 1, 2, and 3 and all groups. (Modified from Axelsson et al, 2004 with permission.)

Effect on tooth loss

Figure 170 shows the mean number of teeth per individual in age groups 1, 2, and 3 and in all groups at the baseline examination in 1972 and the 30-year reexamination in 2002. Because the third molars were excluded, the maximum number of teeth is 28. At the 30-year reexamination, groups 1, 2, and 3 were 50 to 65, 66 to 80, and 81 to 95 years old.

As an average, groups 1, 2, and 3 lost only 0.4, 0.7, and 1.8 teeth, respectively. About 70% of these

few lost teeth were lost because of root fractures in teeth with posts. In comparison, randomized samples of 45- to 50-year-old Swedes lost almost 3 teeth per subject per 11 years (ie, almost 10 teeth per 30 years) (Håkansson, 1991). Almost 75% of subjects in group 1 and 60% in group 2 did not lose a single tooth, and 15% to 20% lost only one tooth during the 30-year trial (Fig 171).

In contrast, among a randomized sample of adults in the county of Stockholm, Sweden, during the 20 years from a mean age of 35 years to 55 years, the mean number of lost teeth was 5 (ie, 7.5

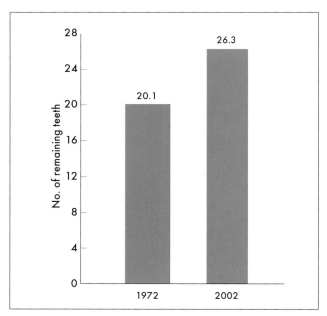

Fig 172 Mean number of remaining teeth in 50- to 65-year-old subjects in 1972 (group 3) compared to 2002 (group 1). (Modified from Axelsson et al, 2004. with permission.)

lost teeth per subject per 30 years). Only 27% did not lose any teeth, 38% lost 1 to 2 teeth, 18% lost 3 to 4 teeth, 7% lost 5 to 6 teeth, 3% lost 7 to 8 teeth, 15% lost 9 to 16 teeth, and 2% lost more than 16 teeth (Jansson et al, 2002).

For ethical reasons, the control groups were offered needs-related preventive dentistry after 6 years; therefore, the baseline data of group 3 (51 to 65 years old in 1972) were used as a cohort control group to group 1 (50 to 65 years old in 2002) at the 30-year examination. Figure 172 shows that 50- to 65-year-old subjects in 2002 (group 1) exhibited an average of 26.3 remaining teeth compared to only 20.1 in 50- to 65-year-old subjects in 1972. The frequency distribution of remaining teeth in 50- to 65-year-old subjects in 1972 compared to 2002 is shown in Fig 173. Only about 30% of the 50- to 65-year-old subjects in 1972 exhibited 24 to 28 remaining teeth compared to about 95% in 50- to 65-year-old subjects in 2002. Figure 174 illustrates the pattern of remaining teeth (%) in 50- to 65-year-old subjects in 1972 compared to 2002.

Effect on dental caries

Figure 175 shows the mean values of new carious surfaces (DS) per subject after 6, 15 and 30 years. In each age group, recurrent caries predominated. Of 257 subjects, 107 developed no new caries lesions, 47 developed only one new caries lesion, and 47 developed two new caries lesions.

The percentage of intact, decayed and filled, and missing surfaces in 50- to 65-year-old subjects in 1972 compared to 2002 is shown in Fig 176. In particular, the percentage of intact surfaces increased from less than 35% to more than 50%, and the percentage of missing surfaces was reduced from almost 30% to only about 5%. Figure 177 shows the frequency distribution of 50- to 65-year-old subjects with specific numbers of intact tooth surfaces in 1972 compared with 2002.

Effect on periodontal disease

Figure 178 shows the changes in mean probing attachment level that occurred at different sites and in different age groups between 1972 and 2002. Except for a mean loss of 0.2 mm on the buccal surfaces in group 1 as an iatrogenic effect, there was no marked change in probing attachment level on the buccal and lingual surfaces, but on the approximal surfaces, gains ranging from 0.3 to 0.4 mm were recorded. Most likely several approximal intrabony pockets at baseline had been "repaired" during the study. Although there was an overall gain of attachment in the sample, further attachment loss of ≥ 2 mm had occurred on a few of the sites (2% to 8%), mainly buccally, as an iatrogenic effect of frequent mechanical cleaning (Fig 179). For comparison, the annual mean loss of periodontal support was 0.1 mm in an 11-year longitudinal study in a randomized sample of Swedish adults (Håkansson, 1991) and two 20-year longitudinal studies in adults in the counties of Jönköping (Hugoson and Laurell, 2000) and Stockholm (Jansson et al, 2002) in Sweden (ie, an average of 3 mm of periodontal support loss over 30 years).

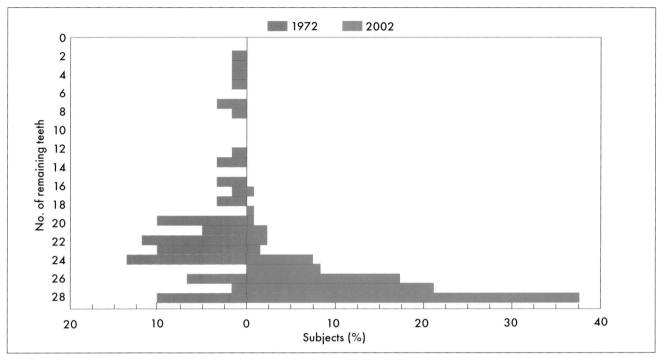

Fig 173 Frequency distribution of remaining teeth in 50- to 65-year-old subjects in 1972 compared to 2002. (Modified from Axelsson et al, 2004 with permission.)

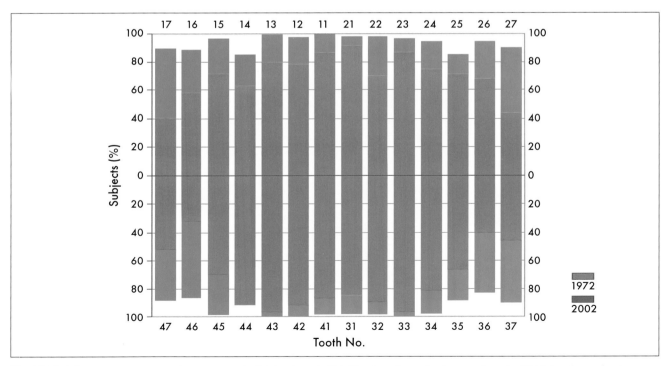

Fig 174 Mean percentage of remaining individual teeth (Fédération Dentaire Internationale [FDI] tooth-numbering system) in 50- to 65-year-old subjects in 1972 compared to 2002. (Modified from Axelsson et al, 2004 with permission.)

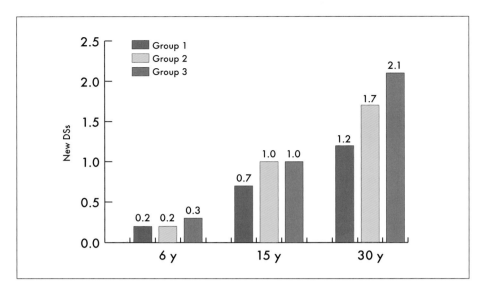

Fig 175 Mean caries incidence after 6, 15 and 30 years, by age group: (Group 1) 50 to 65 years; (Group 2) 66 to 80 years; (Group 3) 81 to 95 years. (DS) Decayed surfaces. (Modified from Axelsson et al 2004 with permission.)

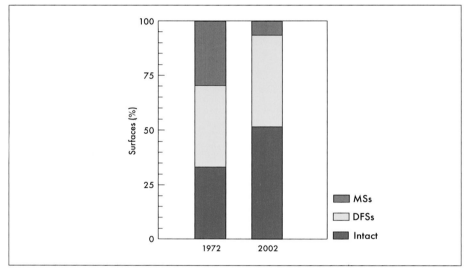

Fig 176 Caries prevalence in 50- to 65-year-old subjects in 1972 compared to 2002. (DFSs) Decayed and filled surfaces; (MSs) Missing surfaces. (Modified from Axelsson et al, 2004 with permission.)

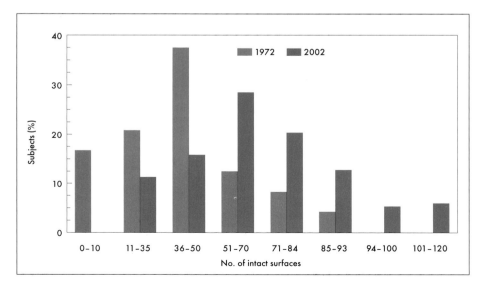

Fig 177 Frequency distribution of 50- to 65-year-old subjects with specific numbers of intact tooth surfaces in 1972 compared to 2002. (Modified from Axelsson et al, 2004 with permission.)

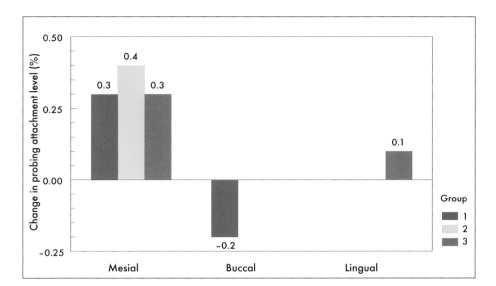

Fig 178 Changes in mean probing attachment level from 1972 to 2002 on mesial, buccal, and lingual surfaces in age groups 1, 2, and 3. (Modified from Axelsson et al, 2004 with permission.)

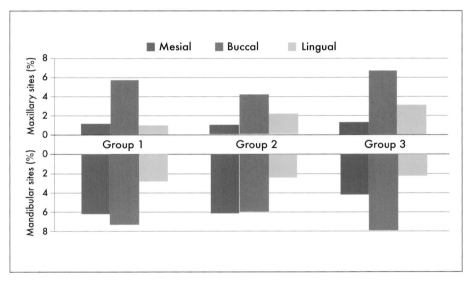

Fig 179 Percentage of mesial, buccal, and lingual sites with ≥ 2 mm probing attachment loss in age groups 1, 2, and 3 from 1972 to 2002. (Modified from Axelsson et al, 2004 with permission.)

Figure 180 shows the mean PAL on mesial, buccal, lingual, and all surfaces in 50- to 65-year-old subjects in 1972 compared to 2002. The buccal surfaces exhibit the most attachment loss, which can be explained by iatrogenic trauma caused by mechanical cleaning, as discussed earlier. Therefore, in order to eliminate the iatrogenic effects on PAL, only approximal PAL should be presented as resulting from periopathogens.

The frequency distribution of 50- to 65-year-old subjects with specific mean mesial PALs in 1972 and 2002 is shown in Fig 181. Figure 182 shows the frequency distribution of mesial sites with different levels of PAL in 50- to 65-year-old subjects in 1972 compared to 2002. Figures 183 to 185 show the pattern of PAL mesially, buccally, and lingually in 50- to 65-year-old subjects in 1972 compared to 2002. Most improvement was achieved mesially and lingually in 2002, particularly in the maxillary teeth.

The frequency distribution of periodontal treatment needs according to the CPITN at the site level in each age group at the baseline examination in 1972 compared to the 30-year reexamination is shown in Fig 186. Irrespective of age, almost 100% of the sites were healthy (score 0) in 2002.

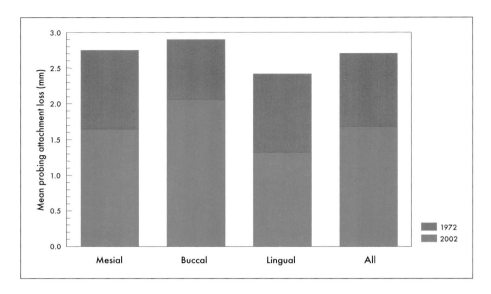

Fig 180 Mean PAL on mesial, buccal, lingual, and all surfaces in 50- to 65-year-old subjects in 1972 and 2002. (Modified from Axelsson et al, 2004 with permission.)

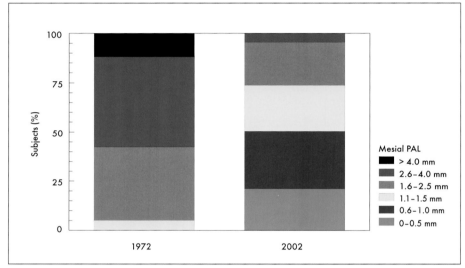

Fig 181 Frequency distribution of 50- to 65-year-old subjects with specific mean PALs mesially in 1972 and 2002. (Modified from Axelsson et al, 2004 with permission.)

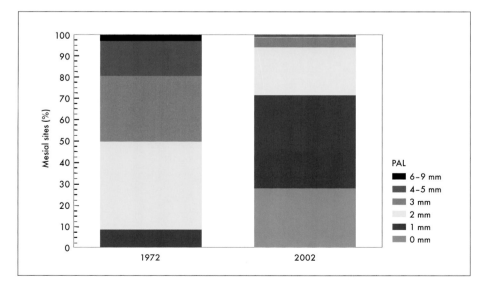

Fig 182 Frequency distribution of PAL on mesial sites in 50- to 65-year-old subjects. (Modified from Axelsson et al, 2004 with permission.)

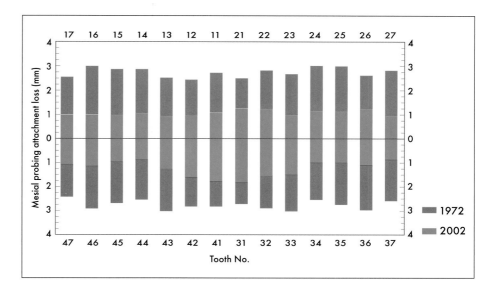

Fig 183 Pattern of mesial PAL in 50- to 65-year-old subjects (FDI tooth-numbering system). (Modified from Axelsson et al, 2004 with permission.)

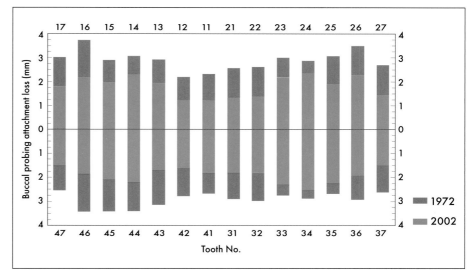

Fig 184 Pattern of buccal PAL in 50- to 65-year-old subjects (FDI tooth-numbering system). (Modified from Axelsson et al, 2004 with permission.)

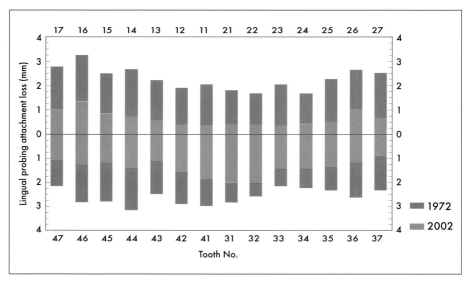

Fig 185 Pattern of lingual PAL in 50- to 65-year-old subjects (FDI tooth-numbering system). (Modified from Axelsson et al, 2004 with permission.)

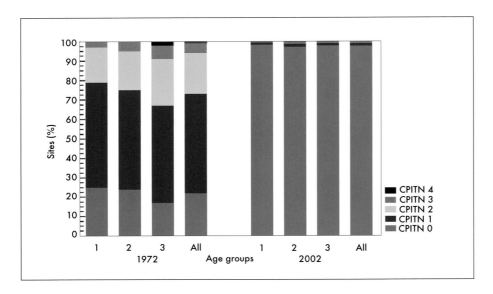

Fig 186 Frequency distribution of sites according to CPITN scores 0 to 4 in age groups 1, 2, and 3 and all groups in 1972 and 2002. (Modified from Axelsson et al, 2004 with permission.)

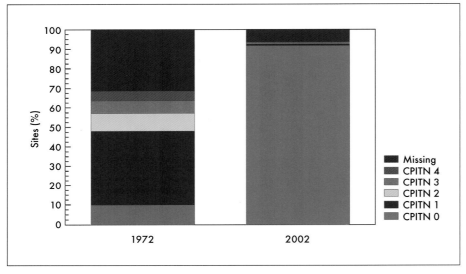

Fig 187 Frequency distribution of missing sites and sites with CPITN scores 0 to 4 in 50- to 65-year-old subjects. (Modified from Axelsson et al, 2004 with permission.)

Missing sites and teeth do not have any periodontal treatment needs. Thus, periodontal treatment needs should take into account missing sites when the "true" treatment need is estimated. Figure 187 shows the frequency distribution of sites according to CPITN score 0 to 4 and missing sites in 50- to 65-year-old subjects in 1972 compared to 2002. In 2002, almost 95% of the sites were healthy and only 5% were missing, compared to only 10% healthy sites and more than 30% missing sites in 1972.

The frequency distribution of 50- to 65-year-old subjects with a specific number of sites exhibiting

CPITN score 3 (4- to 5-mm-deep diseased pockets) in 1972 compared to 2002 is shown in Fig 188. In 2002, more than 75% of the subjects had no CPITN score 3 sites, compared to only 23% in 1972.

Figures 189a to 189h show the pattern of missing sites and sites with CPITN score 0 to 4 mesially, distally, buccally, and lingually in 50- to 65-year-old subjects in 1972 and 2002. Most missing sites were found in the maxillary and mandibular molar regions of the 50- to 65-year-old subjects in 1972. In 1972, most CPITN scores 3 and 4 sites were found mesially and distally in the maxillary posterior teeth, and most CPITN score 2 (calculus)

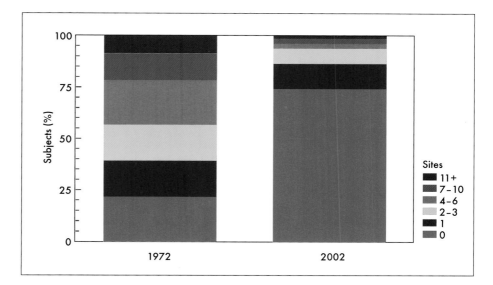

Fig 188 Frequency distribution of 50- to 65-year-old subjects with a specific number of sites exhibiting CPITN score 3 in 1972 compared to 2002. (Modified from Axelsson et al, 2004 with permission.)

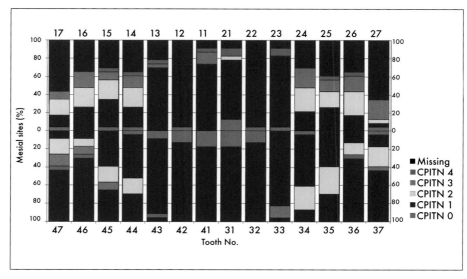

Fig 189a Pattern of missing sites and sites with CPITN score 0 to 4 mesially in 50- to 65-year-old subjects in 1972. (Modified from Axelsson et al, 2004 with permission.)

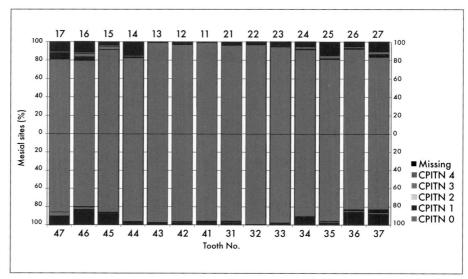

Fig 189b Pattern of missing sites and sites with CPITN score 0 to 4 mesially in 50- to 65-year-old subjects in 2002. (Modified from Axelsson et al, 2004 with permission.)

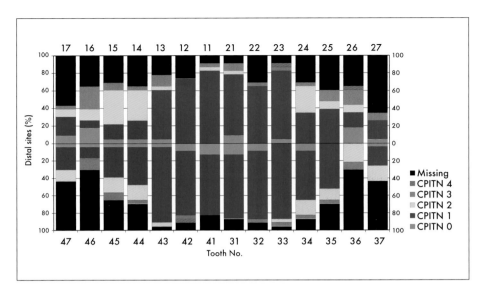

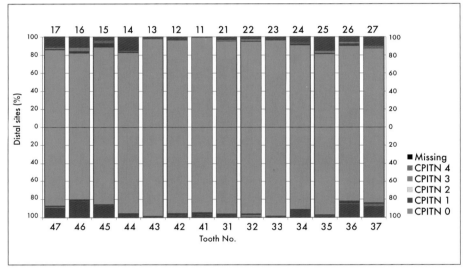

Fig 189c Pattern of missing sites and sites with CPITN score 0 to 4 distally in 50- to 65-year-old subjects in 1972. (Modified from Axelsson et al, 2004 with permission.)

Fig 189d Pattern of missing sites and sites with CPITN score 0 to 4 distally in 50- to 65-year-old subjects in 2002. (Modified from Axelsson et al, 2004 with permission.)

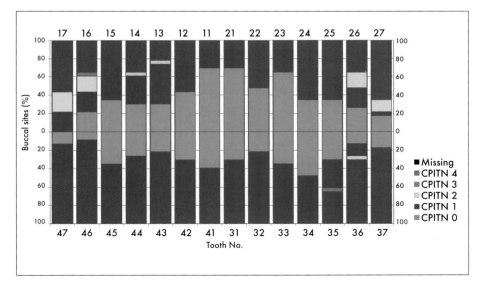

Fig 189e Pattern of missing sites and sites with CPITN score 0 to 4 buccally in 50- to 65-year-old subjects in 1972. (Modified from Axelsson et al, 2004 with permission.)

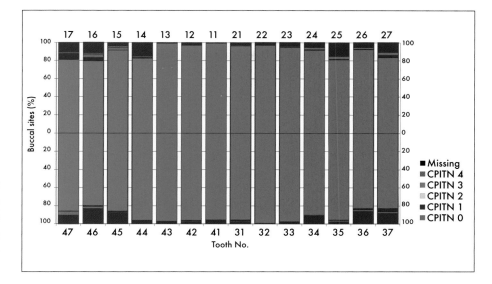

Fig 189f Pattern of missing sites and sites with CPITN score 0 to 4 buccally in 50- to 65-year-old subjects in 2002. (Modified from Axelsson et al, 2004 with permission.)

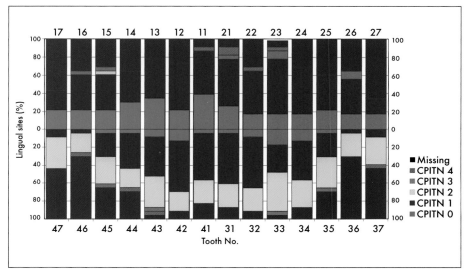

Fig 189g Pattern of missing sites and sites with CPITN score 0 to 4 lingually in 50- to 65-year-old subjects in 1972. (Modified from Axelsson et al, 2004 with permission.)

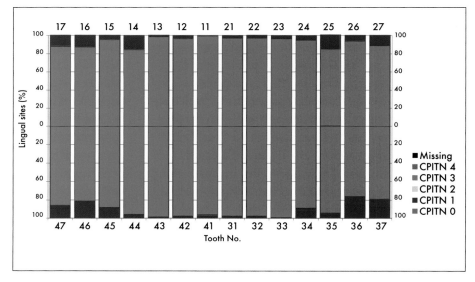

Fig 189h Pattern of missing sites and sites with CPITN score 0 to 4 lingually in 50- to 65-year-old subjects in 2002. (Modified from Axelsson et al, 2004 with permission.)

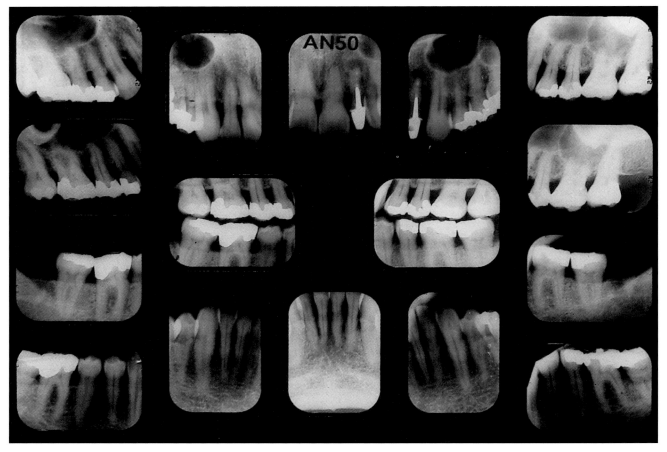

Fig 190a Complete-mouth radiographs of a 50-year-old man in test group 2 at the baseline examination (1972). He has greater-than-average PAL for his age, especially in the maxillary teeth.

sites were found lingually in the mandibular teeth, while almost 100% of all sites were healthy in 2002.

The long-term effect on tooth loss, caries prevalence, loss of periodontal support, gingival health status, and standard of oral hygiene may be illustrated by a subject from age group 2 who exhibited more loss of periodontal support than the average for his age group at the baseline examination in 1972. Figure 190a shows complete-mouth radiographs of this 50-year-old man at baseline in 1972. Figure 190b presents the radiographs taken at the 30-year reexamination in 2002. In 1972, the patient exhibited greater-than-average loss of periodontal attachment for his age, especially in the maxillary

teeth. During the following 30 years, no teeth were lost, there was no further loss of attachment, and no new caries lesions developed. As shown in Fig 190b, the margin of the alveolar crest is well mineralized, indicating periodontal health at the age of 80 years. The buccogingival status and standard of oral hygiene at the 30-year reexamination are shown in Figs 191a and 191b.

This study indicated that a well-trained dental hygienist, supervised by a dentist and using established professional preventive measures (PMTC, etc) to complement improved self-care, was able to prevent further loss of periodontal attachment and reduce the development of new caries lesions to less than 2 carious surfaces per

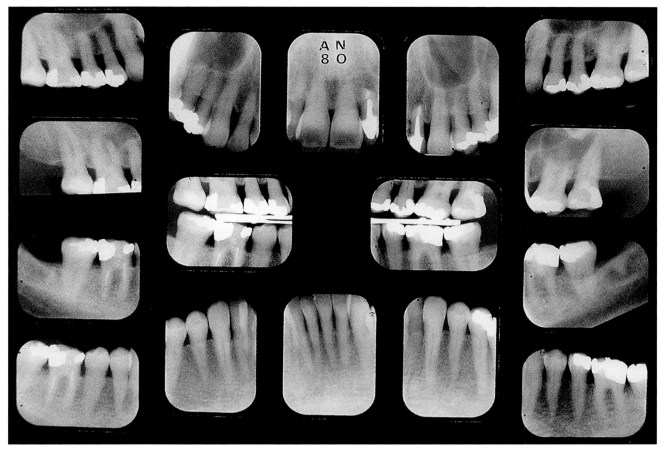

Fig 190b Complete-mouth radiographs of the patient in Fig 190a, aged 80 years, at the 30-year reexamination (2002). No teeth have been lost, there has been no further loss of probing attachment level, and no new caries lesions have developed. Note the well-mineralized margin of the alveolar bone, indicating an absence of active periodontitis.

subject over 30 years, irrespective of age. It was also shown that excellent self-care habits, including self-diagnosis, can be successfully established in adults; age does not matter. At the 30-year reexamination, 92% brushed their teeth ≥ 2 times per day and 8% either in the morning or before going to bed. In addition, 70% used toothpicks daily, 44% used dental tape daily, and 35% used interdental brushes daily. In addition, the percentage of smokers was reduced from 46% in 1972 to only 10% in 2002.

Experience gained from this study is applied in the 2-year training program for dental hygienists at the dental hygiene school in Karlstad and in preventive programs for adults at public dental health clinics and private dental practices in the county of Värmland, which has the highest ratio of dental hygienists per dentist (1:1) in Sweden.

As a consequence, dental health status has improved considerably in the adult population during the last decade, as shown in analytical epidemiologic studies in randomized samples of 35-, 50-, 65-, and 75-year-old adults in 1988 and 1998. (For details, see chapter 9.)

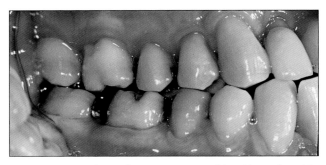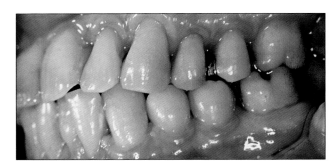

Figs 191a and 191b Buccogingival health and oral hygiene status of the same patient in Figs 190a and 190b at the 30-year reexamination (2002).

Box 3 Subgingival plaque-retentive factors

- Cavitated caries lesions
- Deep, narrow bony pockets
- Furcation involvement
- Root grooves
- Rough, unplaned cementum
- Cementum hypoplasia
- Root resorption
- Calculus
- Iatrogenic effects of subgingival scaling, such as grooves and exposed dentinal tubules on the root surfaces, restoration overhangs, defective and ill-fitting margins of crowns, unpolished restorations, recurrent caries, root caries, etc

Box 4 Supragingival plaque-retentive factors

- Cavitated caries lesions
- Carious restorations
- Restoration overhangs and defective margins
- Ill-fitting margins of crowns and inlays
- Unpolished restorations
- Resin composite restorations
- Supragingival calculus
- Exposed, unplaned root surfaces

Role of plaque-retentive factors

For optimal mechanical plaque control, professional and by the patient, factors conducive to plaque retention must be eliminated or at least minimized. Plaque-retentive factors may be present subgingivally as well as supragingivally, and subgingival plaque control may be complicated by conditions that predispose the site to subgingival plaque retention (Box 3). Several conditions most frequently predispose sites to supragingival plaque retention (Box 4).

CONCLUSIONS

Professional mechanical toothcleaning (PMTC), a service provided by dental personnel (specially trained dental nurses, dental hygienists, and dentists), is defined as the selective removal of plaque from all tooth surfaces, and particularly key-risk surfaces. Gingival plaque biofilms up to 1 to 3 mm subgingivally are removed with mechanically driven instruments and fluoride prophylaxis paste. For cost effectiveness, PMTC should target dental plaque, inflamed gingivae, diseased pockets, incipient caries lesions, and so on. In toothbrushing populations, the approximal surfaces of the posterior teeth are the main focus. Traditional prophylaxis or polishing with a rotating rubber cup and prophylaxis paste on the buccal, lingual, and occlusal surfaces, ie, the nonrisk surfaces, does not constitute PMTC.

Materials and methods

Needs-related PMTC should be carried out in the following sequence:

1. Dental plaque is detected with disclosing agents, supplemented by interdental probing of the posterior teeth.
2. Prophylaxis paste is syringed into the interproximal spaces of the posterior teeth.
3. The approximal surfaces of the mandibular posterior teeth are cleaned from the lingual side using a triangular-pointed tip in the Profin prophylaxis contra-angle handpiece. The procedure continues until all the approximal tooth surfaces are thoroughly cleaned.
4. Lingual, buccal, and occlusal surfaces are cleaned with prophylaxis paste in a rotating rubber cup in a prophylaxis contra-angle handpiece. This procedure starts on the lingual surfaces of the mandibular right posterior teeth in right-handed patients and on the opposite side in left-handed patients.
5. For quality control, plaque is redisclosed after PMTC, and the posterior interdental areas are probed.

Preventive effects of PMTC

This needs-related sequence should not only result in optimal preventive effects, but also motivate the patient to apply the same sequence to self-care. In new-risk and high-risk patients, an initial intensive period, with three to four visits in 7 to 10 days, is recommended (for details, see chapter 8). Apart from comprehensive history taking, diagnosis, elimination of plaque-retentive factors (carious cavities, calculus, unfinished restorations, overhangs, and unplaned root surfaces), and education of the patient in self-care and self-diagnosis, PMTC should be carried out at every visit. These frequent sessions will result in:

1. Healing of inflamed gingivae within 7 to 10 days.
2. Elimination of gingival exudate and a dramatic reduction in plaque formation rate.
3. Reduction in the edema of the periodontal soft tissue and thus the probing depth. As a consequence, the subgingival environment will change from anaerobic conditions favoring prolific growth of the periopathogenic microflora toward an environment that favors the less harmful, facultatively anaerobic microflora.
4. Enhanced potential for arrest of active noncavitated coronal enamel and dentin caries lesions, and arrest of active root caries lesions.

In addition, establishment of needs-related oral hygiene habits is significantly enhanced by frequent PMTC that is consistently performed in the same sequence.

Several well-controlled longitudinal studies in children and young adults have shown that frequent PMTC is more efficient than are topical use of fluoride and supervised toothbrushing programs in prevention and control of dental caries. However, for cost effectiveness, the frequency of PMTC should be strictly related to predicted caries risk. Professional mechanical toothcleaning is usually scheduled more frequently initially and then at gradually prolonged intervals as the patient becomes more efficient in controlling dental caries and periodontal disease by self-care.

The most efficient caries-preventive programs comprise needs-related combinations of PMTC, mechanical and chemical plaque control by self-care, and topical application of fluoride by the patient and by dental personnel.

Well-controlled clinical studies have shown that gingival plaque control by frequent PMTC in patients with periodontal disease will result in reduced probing depth, increased probing attachment, reduced quantities of subgingival microflora, reduced range of subgingival periopathogenic microflora, and improved healing of periodontal soft tissues as well as intrabony pockets after periodontal surgery and regenerative therapy.

Longitudinal clinical studies have shown that oral hygiene education and PMTC, in combination with needs-related debridement, can prevent periodontitis and dental caries in adults because these strategies target the causes of the diseases.

The range of effects reported in different clinical studies is strongly correlated to the materials and methods used, the sample size, and the incidence of the disease in the population. If needs-related plaque control habits for self-care are established and supplemented by needs-related intervals of PMTC, very significant preventive effects will be achieved in high-risk individuals, compared with a matched, truly negative control group (for reviews on professional mechanical plaque control, see Axelsson, 1981, 1993a, 1994; Garmyn et al, 1998; Hotz 1998; and Sanz and Herrera, 1998).

CHAPTER 4

CHEMICAL
PLAQUE CONTROL

Plaque control can be achieved mechanically or chemically through self-care or professional care. By far the most efficient plaque control programs are those combining mechanical and chemical methods: self-care supplemented by needs-related professional mechanical toothcleaning (PMTC) and professional chemical plaque control (PCPC). For example, the toothpaste used for mechanical plaque control by self-care usually contains not only an abrasive agent but also antiplaque or antimicrobial agents, such as sodium lauryl sulfate (SLS), stannous fluoride (SnF_2), triclosan plus zinc citrate, triclosan plus copolymers, triclosan plus pyrophosphate, or chlorhexidine (CHX) digluconate.

Antiplaque and antimicrobial preparations suitable for self-care (excluding antibiotics) are available in a variety of vehicles:

1. Toothpastes
2. Mouthrinses
3. Irrigants
4. Gels
5. Chewing gums

The following groups of antiplaque and antimicrobial preparations are available for PCPC:

1. Irrigants
2. Varnishes
3. Gels
4. Sustained-release agents

Chemical plaque control should always be regarded as a needs-related supplement to, and not a substitute for, mechanical plaque control. Therefore, the choice of agent and frequency of use for self-care or professional care should be related to the individual patient's predicted risk for oral disease (Box 5).

INDICATIONS

Chemical plaque control may be used for a variety of purposes:

1. To prevent plaque formation
2. To reduce the plaque formation rate
3. To control plaque formation
4. To reduce, disrupt, or remove existing plaque
5. To alter the composition of the plaque flora
6. To exert bactericidal or bacteriostatic effects on microflora implicated in caries and periodontal diseases

> **Box 5** Main differences between self-administered and professional chemical plaque control
>
> *Chemical plaque control by self-care*
> - High frequency is required: one to three times per day, regularly or intermittently.
> - Accessibility and efficacy are acceptable supragingivally but very limited subgingivally and interproximally in the molar and premolar regions, particularly for mouthrinsing.
> - Effectiveness is dependent on compliance, and regular daily use is relatively costly unless the agent is incorporated in a toothpaste.
>
> *Professional chemical plaque control*
> - The frequency should be needs related and generally is more frequent during the initial intensive period to arrest enamel (incipient) caries, convert active root caries to inactive status, and heal inflamed periodontal tissue as soon as possible and thereby reduce the plaque formation rate.
> - The accessibility is high because the agent is professionally applied.
> - The duration of effect can be extended by the use of slow-release agents such as CHX-thymol varnish (Cervitec; Vivadent, Liechtenstein) and controlled slow-release agents containing CHX (PerioChip; Astra Zeneca, São Paulo, Brazil).

7. To alter the surface energy of the tooth and thereby reduce plaque adherence
8. To inhibit the release of virulence factors from plaque bacteria

A variety of agents are available, including bisbiguanides (CHX), bispyridines, halogens (iodine and fluorides), heavy metal salts, herbal extracts (sanguinaria extract), oxygenating agents, phenolic compounds (thymol, triclosan, and Listerine [Pfizer, New York, NY]), pyrimidines, quaternary ammonium compounds, amino alcohols (Decapinol, Biosurface Pharma, Sweden), antiplaque enzymes, antiplaque modifying agents, and combinations of different agents.

Although many antimicrobial agents appear to be suitable for plaque control, few have demonstrated clinical efficacy because of inherent problems in the mode of action of agents in the mouth and difficulties in the incorporation of these agents in dental products. Many of these agents exhibit broad-spectrum antimicrobial activity in the laboratory, but they may display valuable selective properties on plaque. The effect of an agent will be concentration dependent. Initially, the inhibitor may briefly attain levels above the minimum inhibitory concentration, but subsequent desorption will reduce the concentrations. At sublethal levels, agents can effectively inhibit bacterial metabolism (eg, acid production and protease activity) and reduce the rate of bacterial growth. To increase antibacterial effectiveness, agents with complementary modes of action can be combined.

Dental products containing antimicrobial agents should be formulated so that long-term use does not *(1)* disrupt the natural balance of the oral microflora, *(2)* lead to colonization by exogenous organisms, or *(3)* lead to the development of microbial resistance. Several products that satisfy these criteria are now available and are clinically effective in helping to control plaque and gingivitis. As new agents and combinations of agents with improved antiplaque and antimicrobial properties are developed, the challenge will be to increase clinical efficacy while preserving microbial homeostasis in the mouth.

Role of the Oral Environment

The oral cavity may be regarded as a single microbial ecosystem, as described in chapter 1. The oral environment is dominated by saliva flow. Total daily secretion has been estimated to be 600 to 700 mL. Around 15% of adults have low stimulated flow rates, which may be related to drugs with anticholinergic action, irradiation, or disease. The resting volume of saliva is approximately 7 mL and is replenished about 10 times per minute during waking hours. At this rate, antibacterial agents administered in the form of a mouthrinse are rapidly removed from the supragingival environment. On the other hand, salivary flow almost ceases during sleep. Chemical plaque control agents should be most effective when the secretion rate is reduced because the rate of removal from the oral cavity is also decreased; therefore, in patients with high caries risk because of xerostomia or hyposalivation, the optimal time for administration should be just before going to sleep.

In addition, it is important to recognize that in the oral cavity there are several major and minor compartments, each constituting a separate microenvironment not easily affected by major events in the oral cavity. These environmental factors significantly influence the access and release of antimicrobial agents from different delivery systems. For example, suspensions of charcoal in water placed in the vestibule of one side of the mouth of an individual who is not talking or chewing will spread within about 5 minutes to the dorsum of the tongue and the hard palate on the same side of the mouth. However, no spread to the other side of the mouth occurs even after prolonged conversation. If placed initially under the tongue, charcoal will coat the entire dorsum of the tongue within 1.5 minutes, but the hard palate on both sides will not be covered for at least 4 minutes (Jenkins and Krebsbach, 1985). Thus, even the major subcompartments within the oral cavity are not in immediate communication. There are also many discrete compartments within the oral cavity, such as the interproximal spaces, the gingival crevice, the gingival pocket, the occlusal fissures, the papillae of the dorsum of the tongue, and the crypts and fissures of the tonsils, each of which creates a special microenvironment within the open system of the oral cavity as a whole.

All the surfaces of the oral cavity are colonized by microorganisms. Facultatively anaerobic streptococci constitute an essential part of the microflora that constantly colonizes the mucous membranes and the teeth. Microorganisms are regularly swallowed with saliva, and the amount within the oral cavity fluctuates, simply because the microbial deposits building up on mucous membranes and in particular on tooth surfaces grow and multiply, and thus provide a reservoir for the oral environment. Fluctuations also occur during sleeping and waking hours and as a result of activities such as eating, drinking, and oral hygiene procedures.

A specific area that supports a bacterial flora is termed a *habitat*. The flora of a habitat develops through a series of stages, collectively called *colonization*. The first important aspect of colonization of a habitat by bacteria is access. The organisms must be able to enter the habitat and consequently they must be able to be transmitted from one habitat to another. In the human mouth, the tongue and tonsils as well as the oral mucosa may serve as a reservoir for bacteria that, given the right conditions, may colonize periodontal pockets.

In defining different habitats in the oral cavity, it is important to recognize that the physical dimensions of a habitat do not fall within specific limits. The whole oral cavity an occlusal tooth surface, or even a defined area on the occlusal surface may be considered a habitat. Very often these habitats, together with their physiologic characteristics, are referred to as *microenvironments*. In oral microbiology, changes in the flora of a habitat such as the mouth may indicate, for example, patients at risk of caries, while changes in tooth surface microenvironments can identify a surface at risk of disease.

CHEMICAL PLAQUE CONTROL AGENTS
Measurement of effectiveness

The mechanisms of action of antimicrobial agents must be restated in the context of current knowledge and treatment goals. For many years, the "gold standard" against which all antiseptic agents have been evaluated is the effect of phenol. Analysis is based on the concentration of the test agent that could kill the bacteria in question. The critical determinant is the effect of the agent on the limitation of bacterial growth over a 24-hour period.

A wide range of concentrations of the test agent are deposited in a series of test tubes, each containing the standard culture of bacteria, and the mixture of bacteria and test agent is incubated for 24 hours at 37°C. The minimum inhibitory concentration is defined as the dilution of the agent that does not permit growth of the target bacteria. To determine the minimal bactericidal concentration, an aliquot is removed from a tube showing no visible growth and reinoculated into a second sterile broth tube. Lack of growth after a second 24-hour incubation period suggests that the agent is *bactericidal*. If, on the other hand, some growth is visible, the agent has not arrested but only retarded growth, and is considered to be *bacteriostatic*.

The results are expressed in units per milliliter or the concentration needed to achieve 100% or 90% kill. The effective dose of the agent is directly related to the concentration and the time it took for the drug to kill the target bacteria. The rate at which the drug kills bacteria depends on two variables: the concentration of the agent and the duration of application. Under the test conditions, the most potent drugs rapidly achieve a maximal kill at low concentrations.

The results are compared with the concentration of phenol needed to achieve a similar result under similar test circumstances and expressed as the *phenol coefficient*. Because chlorhexidine has been shown to be more potent than phenol, the *chlorhexidine coefficient* superseded the phenol coefficient in tests of new chemical plaque control agents (Gjermo et al, 1970, 1974).

However, it has become clear that the interrelationship of the host, drug, and parasite must also be considered. A test tube is an optimal environment in which both the target bacteria and the drug are evenly distributed throughout the liquid, allowing the test drug ready access to the target bacteria. In the host, however, the bacteria may be secluded and under certain circumstances may be inaccessible to the drug. For example, the target bacteria may be enmeshed in a complex plaque matrix interproximally, subgingivally, or in the fissures, and thus be unavailable for direct contact with the drug. Necrotic tissue or pus may adsorb to the drug and prevent its contact with the bacteria. Moreover, in many cases, for a drug to be maximally effective, the target bacteria must be at a high level of biosynthetic activity. In a highly stressed environment such as the plaque biofilm, a low level of bacterial biosynthetic activity may reduce susceptibility of the bacteria to the action of the test drug.

Despite these complications, an antimicrobial effect can be achieved by a potent antiseptic agent if it is retained in the oral cavity and gradually released. The value of substantivity, retention, and slow release is now understood by dental researchers, and in retrospect may have been exaggerated, while many of the other complexities of drug-parasite interactions mentioned may have been overlooked (Fine et al, 1985).

Even in the most favorable circumstances, complete replication of oral conditions for in vitro tests is impossible. However, it is reasonable to screen a product in a logical, systematic manner to determine the ability of the agent to kill a series of putative pathogens, to penetrate intact plaque, or to exert a selective effect on specific segments of the plaque population. To meet these requirements, appropriate test procedures have been devised to measure substantivity, penetrability into plaque biofilms, and the selectivity of the agent to target bacteria.

Substantivity

Substantivity is defined as the ability of an agent to bind to tissue surfaces and be released over time, delivering an adequate dose of the principal active ingredient in the agent. Thus, the agent delivers the sustained activity necessary to confront the bacteria attempting to colonize the tooth surface.

To measure this property, an agent affinity minimal inhibitory concentration, or substantivity assay, has been developed. In the tube dilution assay, varying concentrations of the agent are deposited in a series of tubes, and the effect is visualized as a reduction in growth of the bacterial inoculum. In the drug substantivity assay, tooth analogs are immersed in various concentrations of the test drug, removed, and washed, and the amount of bound drug is assayed by placing the tooth into a broth containing a standard inoculum of bacteria. The efficacy of the experimental agent is determined by analyzing the optical density of the test inocula in the broth after incubation for 20 hours.

Penetrability

For a drug to be effective, it must be able to penetrate deeply into the formed plaque matrix (biofilm). To test penetrability, a bacterial mat or plaque is developed on a nichrome wire by suspending the wire in a broth containing the bacteria to be studied. The bacteria-laden wire is removed and dipped for a brief period in a tube containing the antibacterial agent. The chemically treated plaque is washed and placed in a second broth tube to determine the effect of this treatment on plaque growth as compared with a placebo-treated sample. Results obtained from these experiments have clearly demonstrated that some agents do, in fact, have less success in the preformed plaque assay then they do when placed in solution in a test tube.

Selectivity

The selectivity assay was developed to test chemical agents in a mixed population of bacteria that mimic plaque formation in vivo. A chemostat system has been devised that could house a variety of bacteria that form the backbone of developing plaque. To start, a tooth replica is placed into a continuously mixing nutrient broth, to which different bacterial species are added to develop a mixed flora. The concept of selectivity implies that

the drug has the ability to affect specific bacteria in a mixed population.

In vivo, it is clear that physical removal of plaque only attains plaque reduction in areas where direct contact is made by the mechanical tool (Finkelstein and Grossman, 1979). Plaque hiding in pits and fissures or subgingivally is not reached by mechanical methods of brushing and interproximal toothcleaning and thus remains unaltered. Although chemotherapy appears to be capable of overcoming some of the shortcomings inherent in methods of mechanical intervention, several difficult questions still have to be addressed:

1. What level of suppression of the flora is realistic and desirable in a plaque-laden environment?
2. How will the drug gain entrance to protected domains, such as the tonsillar crypts or the subgingival spaces, potential hiding places for repopulating bacterial species? (At this time, neither site-directed chemotherapy nor lethal doses appears to be possible in the oral cavity.)
3. Is antimicrobial therapy targeted to specific bacteria an attainable or desirable goal for the future? (Perhaps 20 to 30 of the 400 species that reside in the oral cavity are associated with either caries or periodontal disease, but the remainder are either neutral or beneficial to the host. Although few of the overall flora have been implicated in disease, an approach that focuses on specific bacteria, while ecologically correct, must await a better understanding of the pathogenic effects that occur at the molecular level.)

The main advantage of the chemical approach is that the zone of diffusion achieved with a chemical agent is greater than the limited radius of effect of a mechanical agent (Caton et al, 1993). Boxes 6 and 7 list antiseptic and plaque-modifying agents, a number of which have been tested for potency against plaque bacteria. Although the list is extensive, only a few of these agents have

Box 6 Antiseptic agents*

Bisbiguanides
- Chlorhexidine
- Alexidine

Bispyridines
- Octenidine hydrochloride

Halogens
- Iodine
- Iodophores
- Fluorides

Heavy metal salts
- Silver, mercury
- Zinc, copper, tin

Herbal extracts
- Sanguinaria extract

Oxygenating agents
- Peroxides
- Perborate

Phenolic compounds
- Phenol
- Thymol
- Triclosan
- 2-Phenylphenol
- Hexylresorcinol
- Listerine (thymol, eucalyptol, menthol, methyl salicylate)

Pyrimidines
- Hexetidine

Quaternary ammonium compounds
- Cetylpyridinium chloride
- Benzethonium chloride
- Domiphen bromide

*Modified from Newbrun (1985) with permission.

Box 7 Enzymes and other plaque-modifying agents*

Antiplaque enzymes
- Amyloglucosidase, glucose oxidase
- Dextranase
- Fungal enzymes
- Mucinase
- Mutanase
- Pancreatin
- Proteinase-amylase
- Zendium

Plaque-modifying agents
- Ascoxal (ascorbic acid, percarbonate, copper sulfate) (Astra Zeneca)
- Urea peroxide

*Modified from Mandel (1988) with permission.

The most obvious way to affect bacteria is by altering the surface components or metabolic activities of the cell, leading to cell wall permeability and death. Nevertheless, other approaches can also affect bacterial survival (Gjermo et al, 1970; Scheie, 1989, 1994a, 1994b). These approaches impart a more subtle change in the bacterial life cycle but are still effective. What has emerged is awareness that even the most substantive agents work only at sublethal levels in the mouth because of the tremendous diluting effect of saliva and the rapid turnover of epithelial cells. Nevertheless, it appears that sublethal levels of antiseptic agents can affect the attachment of cells to surfaces and to other bacteria; the vulnerability of cells to immune responses and to secondary contact with antimicrobial agents; the growth rate and division time of cells; and the pathogenic potential of the bacterial cells. Studies of antimicrobial agents designed for use in the oral cavity should take these factors into account.

A general requirement for biologic activity of an agent is bioavailability, ie, delivery of the agent to its site of action in a biologically active form and at effective doses. To allow persistence of the effect, the agent must be retained by adsorption or binding at the site of action without loss of bio-

progressed to the stage of testing for human application. Despite advances, even the most stringent in vitro test fails to duplicate the conditions in the oral cavity. Therefeore, it is mandatory that an agent be tested in the oral cavity under conditions that closely resemble the normal mode of use.

logic activity. The clinical efficacy of a chemical plaque control agent depends on its potency and its substantivity, ie, its degree of binding to surfaces and its rate of release from the binding sites. A substantive agent is retained in the oral cavity for protracted periods by adsorption or binding to oral surfaces, including mucosal surfaces, tooth surfaces, pellicle, and gingival plaque, and subsequent release from binding sites. This allows a prolonged association between the agent and the target, ie, plaque bacteria on the tooth surface.

The rate at which a nonsubstantive agent is cleared from the oral cavity is determined by salivary clearance. This allows only a short-term effect of the agent, and bacteria will be able to metabolize and multiply in the interval between applications. Thus, to be clinically effective, a nonsubstantive agent must be applied frequently.

The binding of substantive agents to oral surfaces is usually nonspecific, involving van der Waals forces or ionic, hydrophobic, or covalent interactions. Agent molecules bound to oral surfaces are in equilibrium with agent molecules in saliva. Because the mucosal surfaces represent the largest fraction of the total oral surface, these are the main reservoirs for agent molecules. Molecules are released from their binding sites as the salivary concentration of the agent decreases. The agent will be biologically active as long as the molecules are available at the site in an active form and at effective concentrations.

The amount retained during application depends on the agent's ability to adsorb and bind to oral surfaces, as well as salivary and gingival flow rate, dosage, concentration, contact time, and frequency of application. The release of the agent may be caused by exchange with salivary calcium, and the rate of release depends on pH, the dissociation constant of the agent, and the salivary flow rate. The agent may also be cleared along with desquamating epithelial cells.

Effects on plaque formation

As described earlier, chemical agents may prevent plaque formation by one or more of the following principles:

1. Inhibition of bacterial colonization
2. Inhibition of bacterial growth and metabolism
3. Disruption of mature plaque
4. Modification of plaque biochemistry and ecology

Inhibition of bacterial colonization

Antiplaque effects may be exerted by interference with bacterial adsorption processes rather than by direct antimicrobial activity. Various approaches to modifying the surface characteristics of teeth, pellicle, and bacteria to reduce bacterial adhesion to the tooth surfaces have been explored. Surface-modifying agents include anionic polymers and substituted amino-alcohols. They have high spreading action and may adsorb to enamel, thus lowering its surface free energy. In vitro studies have shown that agents that reduce the surface free energy will reduce bacterial adsorption to the surface (Glantz and Attström, 1986; Quirynen et al, 1990a). However, the effect of in vivo application of such agents has not been clearly demonstrated to date. Polymethyl siloxane has been added to dentifrices, but the main purpose is to polish the teeth.

Agents with an ability to aggregate bacteria in saliva could also reduce bacterial colonization.

Inhibition of bacterial growth and metabolism

Most chemical plaque control agents used today are broad-spectrum antimicrobials that exert direct bactericidal or bacteriostatic effects. They bind to the bacterial membrane and interfere with normal membrane functions, such as transport. This disturbs bacterial metabolism and may kill the bacteria. Adsorption to the bacterial membranes may also lead to alterations in permeability, resulting in leakage of intracellular components, along with protein denaturation and coagulation of cytoplasm contents.

Effective chemical plaque control agents have substantive properties and retain their activity in the oral environment. Antibacterial activity itself is not necessarily reflected in clinical effects. In fact, no direct correlations have been demonstrated between antimicrobial effects in vitro and plaque inhibitory efficacy in vivo (Gjermo et al, 1970).

Disruption of mature plaque

Production of adhesive glucans by cell-bound or cell-free glucosyl transferases is an important feature of plaque accumulation. Attempts to inhibit plaque formation through the hydrolytic action of dextranase or mutanase should be feasible, but experimental results to date have been disappointing. Conditions in vivo are complicated by several factors: The plaque matrix contains several types of polysaccharides, and a further barrier is the diffusion of the agent into the plaque matrix. Frequent application of antimicrobial agents, eg, CHX, may disperse and eliminate existing plaque.

Modification of plaque biochemistry and ecology

In theory, caries may be prevented by chemical agents that reduce or alter the metabolic activity of plaque bacteria. The immediate clinical benefit of such agents is, however, questionable. On the other hand, long-term suppression of microbial metabolism, even by low-potency antimicrobial agents, such as fluoride metal ions, glucan-synthesis inhibitors, and certain sugar substitutes, eg xylitol, may affect plaque biochemistry and ecology because bacteria show varying degrees of susceptibility to the various agents.

Factors affecting delivery to and clearance from the oral cavity

Solubility

Classically, an antiplaque agent must be solubilized in its delivery vehicle to allow rapid release into the oral environment, particularly when the application time is limited. The digluconate salt of CHX, for instance, was selected for the development of CHX mouthrinses on the basis of its high aqueous solubility; triclosan, which has poor aqueous solubility, is solubilized in the flavor and surfactant phase of a dentifrice, which facilitates its release and retention during application.

The bioactivity of an agent is not determined solely by its solubility. However, stabilizing an agent in solution may counteract its tendency to adsorb onto the oral surfaces because its chemical potential is reduced. This may occur for metal ions, for example, by complexation with ligands (Cummins and Watson, 1989). Similarly, this may occur for sparingly soluble organic species by interaction with the surfactant or flavoring oils. The balance between solubility and stability in solution, on the one hand, and bioavailability, on the other, is a very important determinant of clinical benefit.

An alternative delivery is to deposit an antiplaque agent in the form of sparingly soluble particles within the oral cavity so that they deliver low doses of the agent over a long period. This principle has been used for slow-release devices, such as the antibiotic fibers used in the treatment of certain refractory cases of periodontitis, but has yet to be applied to antiplaque agents delivered from dentifrices.

Accessibility

The plaque control agent has to reach the site of action and be maintained at that site long enough to have a sustained effect. A study illustrating this point was conducted by Bouwsma et al (1992): Once-daily use of a wooden triangular interdental cleaner was more effective in reducing interdental bleeding than was twice-daily rinsing with CHX, undoubtedly because the CHX rinse did not reach the interproximal site. These results are especially interesting in light of the supposition that chemical agents are superior to physical agents designed for plaque removal because chemical agents have a greater locus of activity as a result of their ability to diffuse and thus reach inaccessible areas. However, in especially secluded areas of the oral cavity, such as the interproximal and subgingival areas, access is a prerequisite for diffusion to be effective. This suggests that local delivery of antimicrobial agents may have to be targeted to interproximal or subgingival areas. Reports of the efficacy of irrigators, whether or not they are supplemented with antimicrobial agents, have demonstrated the importance of this approach in clinical management (Rosling et al, 2001). Targeted delivery increases the efficacy of a drug, and random, nontargeted delivery of an otherwise potent drug can reduce its clinical benefits.

Ionic interactions (pH and ion binding)

Ionic interactions between agent and receptor sites are important in determining the retention of a number of antiplaque agents because the pH of the delivery vehicle governs the state of ionizaiton of both the agent and the receptor groups (carboxylate, phosphate, and sulfate) (Bonesvoll et al, 1974a). Similarly, the interactions between a positively charged antiplaque agent and its receptor sites can be reduced by the presence of an excess of metal ions, such as calcium or lanthanum, which presumably compete with the antiplaque agent for the receptor sites. Not surprisingly, quaternary ammonium salts, which carry a single positive charge, are displaced at lower calcium concentrations than is chlorhexidine, which has a double-positive charge.

Stability

Chemical breakdown or modification of an antiplaque agent may occur during storage, particularly at elevated temperatures. This may be due to intrinsic instability of the agent or the presence of other ingredients, such as water, abrasives, or surfactants. Modification of an agent may also occur in the oral cavity as a result of metabolic breakdown by salivary or bacterial enzymes. In either case, inactivation of the agent and loss of clinical efficacy ensue. Organic antimicrobials, such as CHX and triclosan, are highly resistant to enzymatic degradation. However, salivary proteases may inactivate the antiplaque enzymes present in some toothpastes, which may explain their lack of clinical efficacy in vivo (Van der Ouderaa and Cummins, 1989).

DELIVERY VEHICLES

Chemical plaque control agents may be delivered to the oral cavity by various vehicles. The vehicle of choice depends, first, on compatibility between the active agent and the constituents of the vehicle. (For instance, the first fluoride toothpastes were ineffective because of the incompatibility of fluoride with the abrasive system.) Second, the vehicle should provide optimal bioavailability of the agent at its site of action. Third, patient compliance is of major importance. Patient compliance is probably reduced with increasing frequency of dose and length and complexity of the treatment. Therefore, chemoprophylaxis is most likely to succeed if the delivery vehicle does not require the establishment of new habits or if the treatment is independent of patient compliance.

Mouthrinses

Mouthrinses are the simplest dosage form for antiplaque agents (Ciancio, 1988; Kornman, 1986); the most common form is a water-alcohol mixture to which flavor, nonionic surfactant, and humectant are added to improve cosmetic properties (Pader, 1988). Most antiplaque agents, such as quaternary ammonium compounds, bisbiguanides, metal salts, essential oils, and plant extracts, are compatible with this vehicle. Stannous fluoride, however, has a short shelf life in a mouthrinse because of loss of stannous ions by precipation.

Gels

Dental gels have been used mainly as delivery vehicles for CHX and stannous fluoride. The most common gel is a simple, thickened aqueous system containing humectant but neither abrasive nor foaming agents. As such, it is compatible with most antiplaque agents. The most appropriate gel for SnF is a low-pH (3 to 4), high-humectant (glycerol), low–water-content system that minimizes stannous ion precipitation. Gels are usually applied in standard or custom-made trays to provide close contact with the agent and its site of action, ie, the plaque-covered tooth surface.

Toothpastes

A typical toothpaste contains an abrasive and a surfactant, which together are intended to remove loosely bound material, including plaque, pellicle, and stains. In addition, flavor is added for mouth freshness and therapeutic agents, particulary fluoride, are added for anticaries efficacy.

To deliver an antiplaque agent effectively, a toothpaste must have further properties without compromising these basic functions. The complex mixture of dentifrice components should be physically and chemically compatible with the antiplaque agent to provide a product that is stable during storage but that allows delivery to the sites of action in the most biologically active form during the time of application.

In practice, relatively few antiplaque agents have been successfully incorporated into clinically effective toothpaste, largely because of the incompatibility of the agents with other ingredients and the nature and complexity of the oral environment. The most striking example of chemical incompatibility is the precipitation of CHX in the presence of long-chain surfactant molecules, such as stearate and SLS. Because precipitation inactivates CHX (Johansen et al, 1975), conventional toothpaste formulations containing CHX have poor clinical efficacy.

Metal ions such as zinc and tin are, in contrast, compatible with anionic surfactants. Provided that care is taken to optimize formulation to avoid undue loss by precipitation or overcomplexation, metal ions are chemically and biologically available from "conventional" toothpaste formulations (Cummins and Watson, 1989).

The nonionic, phenolic antibacterial agents, such as triclosan, are also compatible with conventional toothpaste formulations, provided that the agent is not entrapped in the surfactant phase structure (Van der Ouderaa and Cummins, 1989). Clinical efficacy has been demonstrated for triclosan alone and in combination with zinc citrate or a retentive copolymer of methoxyethylene and maleic acid (PVM/MA) (Gantrez; GAF, New York, NY) (Rosling et al, 1997a, 1997b; Svatun et al, 1989a).

Sustained-release devices and varnishes

Devices and varnishes for sustained release of chemical plaque control agents such as CHX may provide long-term contact between the agent and its site of action. The effect will depend on the degree and rate of release of the agent from the vehicle. Chlorhexidine and CHX-thymol varnishes have been recently introduced and are very effective against plaque formation as well as mutans streptococci (MS) (for review, see Matthijs and Adriaens, 2002). A new delivery system for controlled, slow release of CHX in a biodegradable gelatin chip (PerioChip) has been introduced for treatment of diseased periodonal pockets (Jeffcoat et al, 1998; Soskolne et al, 2003).

Chewing gums and lozenges

The release of various chemical plaque control agents from chewing gums and lozenges has been evaluated. As for sustained-release devices and varnishes, the effect will depend on release of the agent from the gum during chewing or from the lozenges as they dissolve. The contact time will be prolonged, but increased salivation will inevitably increase the clearance rate of the agent from the oral cavity. Nevertheless, further work on chewing gums and lozenges as vehicles is warranted. Administration of chemical plaque control agents via such vehicles may represent effective and acceptable routes, particulary for patients with low toothbrushing compliance. For individuals with reduced salivation, the stimulation of salivary secretion by chewing may relieve discomfort. Recently, a chewing gum that contains CHX has been introduced (Simons et al, 2001).

Irrigants

Vehicles for chemical plaque control agents applied during supragingival and subgingival irrigation are of similar composition to mouthrinses. Special devices are available for high-pressure irrigation through cannulae.

CLASSIFICATION

Different classification systems for chemical plaque control agents, such as grouping them into first- and second-generation antimicrobial agents, have been presented. The first-generation agents appear to be effective in vitro but lack substantivity and thus are not as effective in vivo. The second-generation agents, of which CHX is the principal example, are substantive and effective in vivo (Bonesvoll et al, 1974a, 1974b; Kornman, 1986). Another system for classifying antiseptic agents by group is presented in Boxes 6 and 7.

The most recent trend is to group the chemical plaque control agents as follows:

1. Cationic plaque control agents
2. Anionic plaque control agents
3. Nonionic plaque control agents
4. Other plaque control agents
5. Combination of plaque control agents

(For review, see Addy, 1986, 1997; Addy et al, 1994; Addy and Moran, 1997a; Addy and Renton-Harper, 1997; Adriaens and Gjermo, 1997; Ciancio, 1995; Fine, 1994, 1995; Giertsen, 1990; Gjermo, 1989; Gjermo et al, 1970; Goodson, 1994; Kjaerheim, 1995; Kornman 1986; Löe et al, 1986; Mandel, 1988; Marsh, 1992; Newbrun, 1985; Rölla et al, 1997; Scheie, 1989, 1994a; Wåler, 1989; Wennström, 1997; and Wu and Savitt, 2002.)

Cationic agents

Cationic agents are generally more potent antimicrobials than are anionic or nonionic agents because they bind readily to the negatively charged bacterial surface. Likely binding sites on gram-positive bacteria are free carboxyl groups from teichoic and lipoteichoic acid within the bacterial cell wall. These groups account for the electronegative charge of the gram-positive bacteria. Gram-negative bacteria have less peptidoglycan in the cell wall, and the peptidoglycan molecule

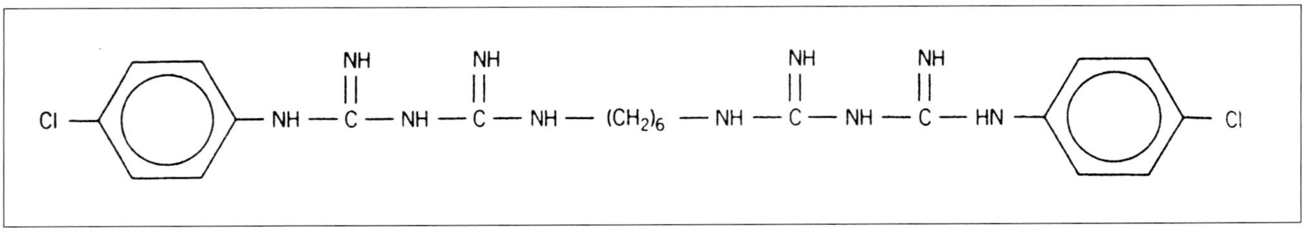

Fig 192 Structural formula of chlorhexidine (1,1'-hexamethylenebis [5-(p-chlorophenyl) biguanide]).

is shielded from the external environment by the outer membrane. On the other hand, lipopolysaccharides of gram-negative bacteria have a strong anionic charge and thus also have high affinity for cations. Gram-negative bacteria have hydrophilic channels that span the bilayer of the outer membrane. Hydrophilic molecules can percolate through these channels, and small ions, such as metal ions, presumably diffuse freely. Thus, cationic agents can interact with both gram-positive and gram-negative bacteria, and, by virtue of their antimicrobial properties, cationic agents may reduce the number of viable bacteria on the tooth surfaces or reduce the pathogenicity of established dental plaque. This has been confirmed in numerous studies.

The following groups of cationic agents have been tested or used as chemical plaque control agents:

1. Bisbiguanide detergents: chlorhexidine and alexidine
2. Quaternary ammonium compounds: cetylpyridinium chloride, benzethonium chloride, and domiphen bromide
3. Heavy metal salts: copper, tin, and zinc
4. Pyrimidines: hexetidine
5. Herbal extracts: sanguinaria extract

Bisbiguanides (detergents)

Chlorhexidine. Chlorhexidine is a cationic detergent that consists of two 4-chlorophenyl rings and two biguanide groups symmetrically connected by a central hexamethylene chain (Fig 192). The phenol rings and the hydrocarbon chain constitute the hydrophobic portion of the molecule, and the chlorine and amino groups are hydrophilic.

Chlorhexidine is the most potent and thoroughly studied antiplaque agent currently available (for review, see Addy et al, 1994; Addy and Moran, 1997b; Ciancio, 1995; Emilson, 1994; Fine, 1995; Giertsen, 1990; Gjermo, 1989; Jones, 1997; Lang and Brecx, 1986; Matthijs and Adriaens, 2000; Rölla et al, 1997; van Rijkom et al, 1996; Wåler, 1989); it has been widely used as an antiplaque agent in Europe for almost 30 years and was recently introduced for this purpose in the United States. In addition, CHX effectively disinfects the oral mucosa and has been extensively used as a topical antiseptic on skin and mucous membranes.

Chlorhexidine has a broad antimicrobial spectrum but is generally more effective against gram-positive than gram-negative bacteria. Of the gram-positive oral bacteria, MS are highly susceptible; the minimum inhibitory concentration values of CHX digluconate for various strains of *Streptococcus mutans* were found to be 1 to 4 µg/mL (Emilson, 1977b). The minimum inhibitory concentration for *Streptococcus sobrinus* was 0.19 µg/mL (Hennessey, 1973), whereas the values for various strains of *Streptococcus sanguis* ranged between 0.39 µg/mL (Hennessey, 1973) and 8 to 128 µg/mL (Emilson, 1977b). However, the minimum inhibitory concentration values are influenced by a variety of factors (eg, growth medium, pH, bacterial strain, inoculum size, and presence of organic material) and thus cannot be directly compared.

Chlorhexidine binds readily to the surface of *S sobrinus* (Davies, 1973). At low concentrations, CHX causes leakage of intracellular potassium and reduction of bacterial acid production (Luoma, 1972). Higher concentrations of CHX may induce ultrastructural alterations, such as disruption of cell walls and coagulation of cytoplasmic constituents.

Numerous clinical studies have confirmed the antiplaque and antigingivitis effects of CHX, which were systematically investigated by Löe and coworkers in the 1970s (for review, see Addy et al, 1994; Addy, 1997; Addy and Moran, 1997b; Addy and Renton-Harper, 1997; Emilson, 1994; Gjermo, 1989; Jones, 1997; and Lang and Brecx, 1986). Twice-daily mouthrinses with 10 mL of 0.2% CHX digluconate inhibit plaque formation almost completely (Löe and Schiött, 1970a, 1970b). In addition, the acidogenicity of established plaque decreases for several hours after a CHX rinse; acid production from sucrose was significantly repressed for 24 hours after a single, 1-minute mouthrinse with 10 mL of 0.2% CHX digluconate (Oppermann, 1979). Long-term studies have confirmed that rinsing with CHX (0.1% to 0.2% CHX digluconate) as an adjunct to mechanical toothcleaning decreases plaque accumulation and gingivitis for a period of 4 to 6 months (Axelsson et al, 1993c; Grossman et al, 1986; Lang et al, 1982) or 2 years (Banting et al, 1989).

Chlorhexidine may cause transient alterations in the composition of the oral microflora, eg, selective suppression of *S mutans* and *Actinomyces* species and an increase in the proportion of *S sanguis* (Axelsson et al, 1993c; Emilson, 1981; Emilson and Fornell, 1976; Mikkelsen et al, 1981; Rindom-Schiött et al, 1976; for review, see Emilson, 1994). Thus, suppression of the susceptible *S mutans* has been recommended for caries prevention in highly infected subjects (Araujo et al, 2002; Axelsson et al, 1987b; Emilson 1977a; Emilson et al, 1982; Gripp and Schlagenhauf, 2002; Ie and Schaeken, 1993; Lindquist et al, 1989a; Sandham et al, 1991; Twetman and Pettersson, 1998; Twetman and Grindefjord, 1999; Wallman and Birkhed, 2002; Zickert et al, 1982b, 1987a, 1987b).

The increased proportion of *S sanguis* may be explained by a selection of less sensitive strains (Emilson and Fornell, 1976; Mikkelsen et al, 1982). It has also been suggested that the increased proportion of *S sanguis* is the result of genetic changes, because susceptible strains of *S sanguis* may develop resistence to CHX.

Although the antibacterial action of CHX is mainly the result of its effects on bacterial membranes, it has been shown that CHX inhibits specific enzymes involved in bacterial adhesion, ie, glucosyl transferase (Ciardi et al, 1978; Scheie and Kjeilen, 1987) or enzymes essential for bacterial metabolism and growth, such as the phosphoenolpyruvate phosphotransferase transport system in oral streptococci. In vivo studies support the hypothesis that CHX is strongly adsorbed to oral surfaces, which act as reservoirs for slow release of the agent (Bonesvoll and Gjermo, 1978; Bonesvoll et al, 1974a, 1974b; Gjermo et al, 1974, 1975). A detailed review of the data from these studies has been published (Gjermo, 1989).

Approximately 30% of the chlorhexidine applied (from both 0.1% and 0.2% mouthrinses and a 1% gel) is retained. The uptake and release is highly pH dependent and is at a maximum between pH 7 and pH 9 (Bonesvoll et al, 1974a). Calcium ions strongly suppress total oral retention, consistent with electrostatic binding of CHX to the carboxyl groups of surface protein and glycoprotein molecules. Elevated levels of CHX in saliva have been reported up to 24 hours after application.

The oral soft tissues have been proposed as the most important structures for the bulk retention of CHX (Bonesvoll and Olsen, 1974). It has been suggested that these tissues release CHX slowly into saliva, whereupon it retards the growth and metabolism of oral bacteria for extended periods. No details of the mechanism have been specified, however. Levels of CHX in plaque and pellicle have been proposed as decisive for its antiplaque effect in vivo (Oppermann, 1980a). Significant transfer from oral tissues to plaque bacteria via the saliva appears unlikely, however, because CHX binds strongly to salivary protein. It is more likely that the concentration of CHX maintained in saliva acts as a buffer, preventing loss of CHX from plaque, with minimal actual mass transfer.

Clinically, CHX has a greater antiplaque effect than do other agents with similar, or even better, antimicrobial efficacy in vitro. This superior effect has been ascribed to the substantive properties of CHX and to the fact that it retains its antimicrobial effect even when adsorbed to tooth surfaces. It is thought that the positively charged CHX molecule binds through electrostatic forces, ie, to phosphate, carboxyl, or sulfate groups on the oral mucosa, on bacteria, and in the pellicle. Persistence of biologic activity and a suitable rate of dissociation of active molecules from these binding sites ensures antimicrobial activity at the site of action for several hours.

Despite widespread clinical use, there are few reports of adverse effects (for review, see Gjermo, 1989). No adverse systemic effects have been reported, but local side effects are common. These include discoloration of teeth, restorations, dentures, and tongue; desquamation and soreness of the oral mucosa; taste disturbances; and a bitter taste (Axelsson and Lindhe, 1987; Eriksen and Gjermo, 1973; Flötra et al, 1971; Löe and Schiött, 1970a, 1970b; Löe et al, 1976; for review, see Axelsson, 1993b, 1994 and Gjermo, 1989). These local adverse effects encouraged the search for alternative antiplaque agents for long-term use.

Chlorhexidine is used as a chemical plaque control agent in mouthrinses, toothpastes, gels, chewing gums, and irrigation solutions for self-care and gels, irrigants, varnishes, and controlled slow-release systems for professional use. Most toothpastes contain anionic compounds, such as SLS and monofluorophosphate, which inactivate the cationic CHX. Therefore toothpastes without anions, such as Zendium (Sara Lee, Chicago, IL), are recommended for use in direct combination with CHX mouthrinse. Chlorhexidine is most effective directly after mechanical removal of plaque, because the large CHX molecule has very limited penetration into plaque. If a toothpaste with anions is used, CHX mouthrinse should not be used until 1 to 2 hours afterward.

Alexidine

Alexidine is also a bisbiguanide. It has structural similarities to CHX and similar clinical efficacy, but alexidine is not widely used.

Quaternary ammonium compounds

Cetylpyridinium chloride, benzalconium chloride, and benzethonium chloride are quaternary ammonium compounds that have been tested experimentally as chemical plaque control agents. Cetylpyridinium chloride has been widely used in mouthrinses, mainly as an antimicrobial agent.

The cetylpyridinium chloride molecule has both hydrophilic and hydrophobic groups, allowing ionic and hydrophobic interactions. Interaction with bacteria is assumed to be similar to that achieved by CHX, ie, via cationic binding.

Although the antimicrobial effect of cetylpyridinium chloride is equal to or better than that of CHX, cetylpyridinium chloride is less effective on plaque (Gjermo et al, 1970), possibly because it loses some antimicrobial activity when adsorbed to surfaces. Notably, the substantive properties are also different: Initial retention of cetylpyridinium chloride is higher than that of CHX, but clearance of the former is more rapid (Bonesvoll and Gjermo, 1978).

Heavy metal salts

Numerous in vitro and in vivo studies have confirmed the antiplaque potential of divalent metal ions, first reported as early as 1940 (Hanke, 1940; for review, see Mandel, 1988; Mandel and Kleinberg, 1986; Rölla et al, 1997; and Scheie, 1989). It has been reported that Cu^{2+}, Sn^{2+}, and Zn^{2+} inhibit in vitro growth or acid production by *S mutans*, *S sobrinus*, *S sanguis*, *Streptococcus salivarius*, and *Actinomyces* species (Bates and Navia, 1979; Maltz and Emilson, 1982). Mouthrinses containing either Cu^{2+} (0.25 to 5.00 mM), Sn^{2+} (0.04% to 0.4% SnF_2), or Zn^{2+} (5.0 to 30.2 mM) inhibit plaque formation in vivo (Ellingsen et al, 1980; Harrap et al, 1983; Skjörland et al, 1978; Tinanoff et al, 1976; Wåler and Rölla, 1982); as well as acid production in established plaque (Afseth, 1983; Ellingsen et al, 1980; Oppermann and Johansen, 1980; Oppermann and Rölla, 1980; Svatun and Attramadal, 1978).

The metal ions exert antimicrobial effects, depending on the ionic concentration as well as the chemistry of the metal ion in the specific system. The bacteriostatic effect of metal ions has long been recognized. As early as 1889, Miller proposed the use of metal ions to treat rampant caries, and in 1940 Hanke reported an antiplaque potential for mouthrinses containing certain metal ions. The antimicrobial effect is proportional to the concentration of free ions, which is the predominant bioactive form. Some positively charged and neutral complexes are also thought to be biologically active (Cummins and Watson, 1989). Thus, hydrolysis of metal ions and complexing of metal ion-ligands reduce the activity. The formulation of the vehicle is therefore crucial.

Metal ions are substantive agents: Several studies have confirmed their retention and slow release in the oral cavity (for review, see Wennström, 1997). The salivary and plaque levels of Cu^{2+}, Sn^{2+}, and Zn^{2+} are elevated for several hours after a mouthrinse; after repeated rinses, metal ions are retained and even accumulated. Metal ions and CHX compete for the same binding sites within the oral cavity. In vivo studies of CHX may thus serve as an explanatory model for the pharmacodynamics of metal ions in the oral cavity.

Metal ions bind strongly to plaque components, possibly through electrostatic forces. The antiplaque effect relates partly to the antimicrobial activity and partly to displacement of Ca^{2+} from pellicle and bacterial surfaces. Binding of metal ions to bacteria alters their surface charge and adherence potential (Olsson and Odham, 1978).

Metal ions interact with both gram-positive and gram-negative bacteria. The antimicrobial effect is nonspecific and may be explained by several mechanisms. Metal ions form metal-salt bridges with anionic groups of enzymes. This in turn may influence substrate interactions because of altered charge or conformational changes of the enzyme. Metal ions have an antiglycolytic effect, shown both in vitro in pure cultures of bacteria and in vivo as reduced acid formation. This effect may be ascribed to interference with the structure and function of glycolytic enzymes. Oppermann and coworkers (1980) suggested that divalent metal ions inhibit glycolysis in dental plaque by oxidative inactivation of SH groups of glycolytic enzymes. Other studies support, indirectly, the concept that the inhibitory effect of metal ions involves oxidation of SH groups of essential enzymes (Scheie et al, 1985). Glycolytic enzymes that are known to contain SH groups, such as enzyme I of the phosphoenolpyruvate phosphotransferase transport system, aldolase, and glyceraldehyde-3-phosphate dehydrogenase, are inhibited by both zinc and tin ions.

Because of their antibacterial properties, Cu^{2+}, Sn^{2+}, and Zn^{2+} may reduce the number of viable bacteria on tooth surfaces (Giertsen et al, 1991; Tinanoff et al, 1980) or reduce the pathogenicity of established plaque by long-term suppression of plaque acidogenicity: Acid production by plaque was significantly repressed for up to 6 or 4 hours, respectively, following a single, 1-minute mouthrinse with 10 mL of solutions containing either Cu^{2+} (1.0 mM) or Zn^{2+} (20.0 mM) (Afseth, 1983; Oppermann and Rölla, 1980). A 1-minute mouthrinse with 10 mL of 0.2% SnF_2 significantly decreased acid production by plaque for up to 24 hours (Ellingsen et al, 1982b).

Whereas Cu^{2+} and Sn^{2+} are known to be more potent plaque inhibitors than Zn^{2+} (Afseth, 1983; Giertsen et al, 1987; Oppermann, 1980b; Skjörland et al, 1978); the antiplaque and antiglycolytic effects of mouthrinses containing divalent metal ions are generally less than those observed with equimolar or lower concentrations of CHX (Helldén et al, 1981; Oppermann and Rölla, 1980; Svatun et al, 1977; Waerhaug et al, 1984; Wåler and Rölla, 1980). Moreover, Cu^{2+}, Sn^{2+}, and Zn^{2+} influence the bacterial composition of the oral flora to a varying degree. The plaque-inhibitory effect of 0.5% zinc citrate, when incorporated into a dentifrice, is maintained after 3 years' use without observable shifts in the oral flora (Jones et al, 1988b). Mouthrinses containing SnF_2 (0.04% or 0.2%) have been found to reduce the relative proportion of *S mutans* and *S sanguis* in plaque, to reduce the populations of *S mutans* in saliva, and to reduce the salivary levels of lactobacilli (Axelsson et al, 1993c; Gross and Tinanoff, 1977; Svanberg and Rölla, 1982; Svanberg and Westergren, 1983). In addition, Cu^{2+} may selectively suppress *S mutans*.

These mechanisms account for the cariostatic effects of Cu^{2+}, Sn^{2+}, and Zn^{2+} in rats. Stannous fluoride has been used as an antimicrobial and caries-preventive agent in humans for many years, applied topically in gels, mouthrinses, or toothpastes. Stannous fluoride has both cariostatic and antiplaque properties. It reduces both the formation of acid by bacteria in dental plaque and the number of MS.

Although there are several studies of zinc in toothpastes (reviewed by Gunbay et al, 1992), since 1988 there have been practically no studies of zinc as the main active ingredient in mouthrinses. Zinc salts have been used mainly in combination with other agents (for review, see Giertsen, 1990).

Zinc ions are considered to act by inhibiting glycolytic enzymes (Scheie et al, 1988) or by displacing magnesium ions and hence inhibiting enzyme systems. There is evidence that zinc ions may inhibit both the adsorption of bacteria to the tooth surface and the growth of existing plaque (Harrap et al, 1984; Saxton, 1986). In addition, zinc citrate has recently been added to dentifrices to inhibit formation of calculus and to increase the substantivity of triclosan. Concerns were raised about a possible interference of Zn^{2+} with the cariostatic effect of fluoride, but this supposition is not supported by available data (Koch et al, 1989; Stephen et al, 1988).

Adverse effects related to the clinical use of metal ions are the unpleasant metallic taste, a tendency to induce dryness in the oral cavity, and the formation of yellowish to brownish dental stain,

particulary from Sn^{2+} (Axelsson et al, 1993c; Hanke, 1940; Skjörland et al, 1978; Svatun et al, 1977; Waerhaug et al, 1984). The staining is probably caused by metal sulfides formed as a result of reactions between the metal ions and sulfhydryl groups of pellicle proteins (Ellingsen et al, 1982a). However, staining from SN^{2+} is generally less than that arising from the use of CHX (Axelsson et al, 1993c; Svatun et al, 1977; Waerhaug et al, 1984). Zinc salts do not stain (Hanke, 1940), probably because zinc sulfide is white to grayish white or yellowish. Because zinc salts have a less pronounced metallic taste than do other metal salts (Hanke, 1940), Zn^{2+} is widely used in mouthwashes and dentifrices.

Copper, tin, and zinc ions are all trace elements essential to life, and as such may be characterized as microminerals or nutrients. Inadequate intake may impair health at the physiologic or cellular level. Zinc is a relatively nontoxic, noncumulative, essential trace element and, after iron, is the second most abundant trace metal in human tissues. A daily intake of 15 mg of Zn^{2+} is recommended for adults and adolescents by the US National Academy of Sciences (National Research Council, 1980). The main source of Zn^{2+} is the diet, where Zn^{2+} generally is bound to proteins. It influences a number of human biologic functions. In the oral cavity, Zn^{2+} is essential for taste perception.

Pyrimidines

Hexetidine is a synthetic hexahydropyridine that has antibacterial and antifungal activity in vitro and in vivo. It is active against gram-positive and gram-negative bacteria, including oral bacteria such as *S mutans*, *S sobrinus*, and *S sanguis*. The in vitro antibacterial activity of hexetidine is reported to be inferior to or essentially similar to that of CHX or cetylpyridinium chloride. Hexetidine-containing mouthrinses are commercially available, but at clinically acceptable concentrations only a very slight antiplaque effect is exerted. Increasing the concentration of hexetidine from 0.10% to 0.4% increases the antiplaque efficacy to approach that of 0.2% CHX, but there is a corresponding increase in the frequency of desquamative lesions.

The exact mechanism of the antiplaque activity is not clear. Hexetidine has been claimed to inhibit glycolysis, but this is not supported by clinical data. The antibacterial effect is reduced in the presence of saliva. Enhanced antiplaque effects observed in combination with divalent metal ions, eg, Zn^{2+29} or Cu^{2+13}, are probably related to increased intracellular uptake of the metal ions. The agent has not been evaluated for its ability to prevent dental caries in humans.

Herbal extracts

Sanguinaria extract is a herbal preparation. It is a mixture of bezophenanthridine alkaloids, obtained by alcohol extraction, from the bloodroot plant *Sanguinaria canadensis*. Sanguinaria extract has been used in homeopathic preparations and in folk medicine for the treatment of topical infections and as an expectorant. Mouthrinses and toothpastes containing sanguinaria extract are commercially available.

Sanguinaria extract is antimicrobial against gram-positive and gram-negative bacteria, including oral bacteria. The exact mode of action is not clear; it seems to interfere with essential steps in the synthesis of the bacterial cell wall and septum (Walker, 1990). Sanguinaria extract reportedly suppresses the activity of several enzymes, possibly through oxidation of SH groups. The antimicrobial activity is thought to be associated with the lipophilic property of the molecules. More important, however, may be that the structure of sanguinaria extract allows the molecule to function as a metal ion ligand.

Commercially available sanguinaria extract preparations contain quite high concentrations of zinc chloride. Zinc ions are chemoprophylactic, and it has been suggested that the effects of sanguinaria extract may be related to the Zn^2 (Southard et al, 1987). Although sanguinaria extract has substantive properties, clinical data on the efficacy of sanguinaria extract mouthrinses are not conclusive (Laster and Lobene, 1990).

Anionic agents

Sodium lauryl sulfate is an anionic detergent that consists of a hydrophilic sulfate group and a 12-carbon hydrophobic chain. Sodium lauryl sulfate is also an effective denaturant. It is the most frequently used detergent in commercial toothpastes; it also has a low innate flavor and low toxicity at the concentrations normally used in oral products (for review, see Pader, 1988). Sodium lauryl sulfate is available as a dry powder and is soluble in water. It inhibits the in vitro growth of *S mutans*, *S sanguis*, *Bacteroides melaninogenicus*, *Actinomyces viscosus*, and *Veillonella alcalescens*.

In a study by Giertsen et al (1989b), twice-daily mouthrinses with SLS were found to significantly inhibit plaque accumulation. The plaque-inhibitory effect of SLS may partially be ascribed to its antibacterial property. Sodium lauryl sulfate inhibits specific enzymes such as glucosyl transferase from *S sobrinus* and *S mutans* as well as glucose-phosphotransferase in membranes of *S sobrinus*.

The high affinity of SLS for Ca^{2+} may contribute to the antiplaque effect: In vitro studies suggest an affinity for Ca^{2+}, hydroxyapatite, and enamel and that the sulfate group of the molecule may bind electrostatically to hydroxyapatite through Ca^{2+} bridges (Barkvoll et al, 1988). Data indicate that negatively charged salivary proteins (eg, phosphoproteins) bind electrostatically to calcium ions in the hydration layer of enamel in the initial stages of pellicle formation. Therefore, SLS may interfere with pellicle formation by binding to calcium ions in the hydration layer and even by displacing adsorbed proteins.

Previous studies have shown that negatively charged compounds with a high affinity for Ca^{2+} (eg, fluoride and phosphate) inhibit adsorption of proteins to hydroxyapatite and desorb adsorbed proteins (Rölla and Melsen, 1975a). Because most oral microorganisms carry a net negative charge, it is probable that calcium ions contribute to both the aggregation and attachment of bacteria to negatively charged proteins of the pellicle by formation of Ca^{2+} bridges. Sodium lauryl sulfate may compete with bacterial surfaces for binding sites of both the pellicle and dental plaque and thus interfere with bacterial adhesion and aggregation. It may also bind to negatively charged groups (eg, carboxyl, phosphate, and sulfate) of glycoproteins on the oral mucosa through Ca^{2+} bridges, and hydrophobic interactions between SLS and the oral mucosa may occur.

As mentioned earlier, in vitro work by Barkvoll et al (1988) has shown that SLS binds to hydroxyapatite and enamel through the hydration layer. This may be a factor in the inhibition of monofluorophosphate protection against caries (Barkvoll, 1991; Melsen and Rölla, 1983) when SLS is incorporated in monofluorophosphate dentifrices.

In a thesis on interactions of SLS and CHX, Barkvoll (1991) drew attention to the neutralization of CHX even when these agents were applied separately, with intervals between applications. The interval between toothbrushing with a dentifrice containing SLS and rinsing with CHX solutions should be "more than 30 minutes, probably nearer 2 hours" (Barkvoll et al, 1989).

Veys et al (1992) studied the effect of SLS on the oral mucosa in patients with allergic stomatitis and suggested that the denaturing effect of SLS on the oral mucin layer increases the exposure of the mucosa to various food proteins, resulting in hypersensitivity.

Nonionic plaque control agents

The most successful and frequently used nonionic plaque control agents (triclosan and Listerine) both belong to the group of noncharged phenolic compounds:

1. Phenol
2. Thymol
3. Listerine (thymol, eucalyptol, menthol, and methyl salicylate)
4. Triclosan
5. 2-Phenylphenol
6. Hexylresorcinol

Listerine

Listerine (named after Lister) was tested for efficacy against oral bacteria as early as 1884, by W. D. Miller. It is a combination of the phenol-related essential oils, thymol and eucalyptol, mixed with menthol and methyl salicylate in a hydroalcoholic solution (Mandel, 1988). Substituted, mixed phenols, such as the "essential oils" of Listerine, are now being used as disinfectants and antiseptics (Scheie, 1989). The mechanisms of action of phenols against bacteria are relatively complex. At high concentrations, there are disruptions of the cell wall and precipitation of cell proteins, and, at lower concentrations, there is inactivation of essential enzymes. Because the nonionized molecule is biologically active, phenols are most active at neutral or slightly acidic pH.

Listerine is one of the most extensively used and researched family health care products on the market today. It has been used by millions of consumers, particulary in the United States, for more than 100 years. In September 1987, Listerine antiseptic mouthrinse was the first nonprescription product to be awarded the American Dental Association Council on Dental Therapeutics' seal of acceptance as an aid in controlling supragingival dental plaque and gingivitis. Short-term clinical data indicated Listerine to be an effective antiplaque agent and provided the basis for long-term studies of its safety and efficacy against gingivitis. These clinical investigations, 6 months or longer in duration, have persuasively demonstrated Listerine's efficacy in helping to prevent and reduce supragingival plaque and gingivitis as an adjunct to normal oral hygiene and regular professional care.

In vitro studies have demonstrated that Listerine kills *S mutans, A viscosus, S sanguis, Bacteroides* species, and *Fusobacterium* in 30 seconds. *Bacteroides* and *Fusobacterium* species are generally recognized as the primary etiologic agents of oral malodor, usually originating on the tongue or in the gingival crevice. Listerine also significantly depresses odorigenic bacteria sampled from the tongue and crevicular spaces.

Like other phenolic compounds, Listerine has anti-inflammatory properties. This may explain in part why Listerine, in one 6-week double-blind mouthrinse study, was as effective as 0.2% and 0.1% CHX in reducing gingivitis (Axelsson and Lindhe, 1987).

Triclosan

Apart from Listerine, the most important oral hygiene product is triclosan, currently incorporated into commercial toothpastes and mouthrinses. Triclosan (2,4,4´-trichloro-2´-hydroxydiphenyl ether) is a nonionic antimicrobial agent with hydrophilic and hydrophobic properties (Fig 193).

Triclosan has been used in consumer products such as deodorants, soaps, and talcum powder for more than 25 years and, more recently, in toothpastes and mouthrinses as a chemoprophylatic agent, with the aim of reducing formation of plaque and the development of gingivitis. One problem in testing the activity of triclosan is its poor aqueous solubility. It is solubilized in the flavor and surfactant phase, and when the antimicrobial effect is tested, possible additive or synergistic effects with the diluent must be considered. In commercial products, triclosan is solubilized in one or more detergents, such as SLS and sodium lauryl sarcosinate, or in propylene glycol or polyethylene glycol.

In vitro studies have shown that, unlike cationic antimicrobials, triclosan is not significantly impaired by the presence of SLS. Because most commercial toothpastes and mouthrinses contain SLS, this is an important advantage. Results from studies showing its positive clinical effects on plaque

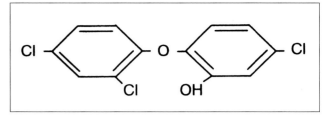

Fig 193 *Structural formula of triclosan (2,4,4´-trichloro-2´-hydroxydiphenyl ether).*

formation and gingival health indicate that triclosan is compatible with sodium fluoride (Nabi et al, 1989) and sodium monofluorophosphate (Svatun et al, 1989b), the two most common fluoride salts in toothpastes. An in situ study showed that triclosan neither enhances nor interferes with remineralization of artificial enamel lesions by fluoride (Mellberg et al, 1991). In a later, similar study, the triclosan- and zinc citrate–containing toothpaste had a slightly better remineralizing effect than did a fluoride control (ten Cate, 1993).

Triclosan acts on the microbial cytoplasmic membrane, inducing leakage of cellular constituents and thereby causing bacteriolysis. Presumably, the hydrophobic portion of the triclosan molecule adsorbs to the lipid portion of the bacterial cell membrane and thus interferes with vital cell membrane functions such as transport. At low concentrations, triclosan is bacteriostatic. Triclosan has a broad-spectrum efficacy on gram-positive and gram-negative bacteria in vivo and in vitro, mycobacteria, and strictly anaerobic bacteria. It also exhibits an effect against fungi, such as several *Penicillium* and *Aspergillus* species as well as *Candida* species and spores.

Since the introduction of triclosan-containing dental products more than a decade ago, investigators have examined the effect on some common oral bacteria. In vitro studies have shown that 0.2% triclosan has an antibacterial effect on early plaque formers, *S mutans*, and most gram-negative species, including periopathogens. Rosling et al (1997b) showed that daily use of a toothpaste containing triclosan and copolymer resulted in significant reduction of subgingival periopathogens compared to a control fluoride toothpaste in a 3-year clinical study.

According to animal experiments, the acute toxicity of triclosan is low. (The median lethal dose is more than 5 g/kg given orally in dogs.) Administered via oral hygiene products, the amount of triclosan in either a standard volume of 10 mL of 0.03% mouthrinse or 1 g of 0.3% toothpaste would be 3 mg.

No adverse systemic effects have been reported in several human studies of up to 3 years' duration on daily use of triclosan-containing products. A slight uptake of triclosan may occur in the oral mucosa. However, surfactants cause a concentration-dependent increase in mucosal permeability to oil-soluble compounds, the anionic and cationic surfactants being especially potent. For example, SLS is shown to damage the permeability barrier, and high concentrations of detergents in dental products might thereby increase penetration of triclosan through the oral epithelium.

Triclosan in oral hygiene products. As mentioned earlier, bioavailability is the ability of an antibacterial agent to be retained in the oral cavity in a biologically active state. To enhance the effects of toothbrushing and mouthrinsing, antiplaque agents must be retained in the oral cavity and must be released, over time, in effective concentrations. This quality, called *substantivity*, depends on release from the delivery vehicle (diffusion-dependent), adsorptive "uptake" at the receptor sites (concentration-driven), and distribution within the oral cavity (receptor-bound or in saliva). Salivary flow rate and composition, binding affinity, pH, and epithelial desquamation will affect oral clearance (Fine, 1995; Scheie, 1989, 1994a). Therefore, the antiplaque effect depends more on the long-term maintenance of a high concentration of the antimicrobial agent than on the initial amount of retention (Bonesvoll and Gjermo, 1978; Gjermo, 1989).

Triclosan shows a positive, linear, dose-response antiplaque effect in vivo. Organic antimicrobials, such as CHX and triclosan, are also thought to be highly resistant to enzymatic degradation in the oral cavity (van der Ouderaa and Cummins, 1989).

Teeth, pellicle, supragingival plaque, saliva, and oral mucosa are the likely binding sites for antiplaque agents. The exact mechanism of retention for triclosan is not known, but several possibilities have been proposed. Two different detection methods revealed similar amounts (38% and 36%) of the administered dose of triclosan to be retained in the oral cavity after a single mouthrinse with a toothpaste slurry. This level of retention is comparable to that reported for CHX (Bonesvoll et al, 1974a, 1974b). It was postulated that triclosan was bound to proteins and, being lipophilic, was also taken up by the oral mucosa. Microreservoirs of triclosan in plaque, mucosa, and tooth surfaces were believed to be sources of triclosan during the postbrushing period.

Saxton and coworkers (1988) have found triclosan to exert its major effect on plaque formation by inhibiting plaque buildup on initially clean tooth surfaces. They deduced that the teeth must be important retention sites for triclosan, as has also been suggested for CHX (Rölla et al, 1971). In vitro, the acquired enamel pellicle has been found to be an additional important binding site for triclosan.

In vivo, salivary concentrations of triclosan decline to about one fifth 2 hours after a triclosan-containing toothpaste is used. Detectable levels of triclosan have been found for up to 8 hours in plaque and saliva and for 3 hours on the oral mucosa. These concentrations of triclosan are considerably higher than the in vitro minimum inhibitory concentration values against various streptococci. The salivary release curve is relatively steep, however, indicating a rapid release from other oral binding sites.

Different concentrations of triclosan in toothpaste, with the addition of a surface-coating copolymer of polyvinylmethyl ether and maleic acid (Gantrez), resulted in statistically significant reductions in bacterial counts in saliva up to 5 hours after use; in contrast, evidence of bacterial count recovery was apparent 30 minutes after use of paste containing triclosan but no copolymer (Addy et al, 1989a). The prolonged effect of the combination product suggests that the copolymer increases retention of triclosan: The copolymer, being lipophilic, is retained on oral surfaces, providing binding sites for triclosan and reducing its rate of diffusion from the surfaces.

The influence of triclosan on mean pH in plaque following a 1-minute sucrose rinse has also been investigated. A significantly greater inhibition of plaque acid production was observed after use of a triclosan- and zinc-containing toothpaste slurry than was found after use of a placebo dentifrice (a 0.70-unit drop in pH for triclosan versus a 1.01-unit drop in pH for the placebo) (Gilbert, 1987). Both the reduction in plaque formation and the drop in pH may explain the additive effect found on fissure caries and smooth-surface caries lesions in rats when triclosan-copolymer was added to a sodium fluoride (NaF) toothpaste and the results were compared to those attained with the NaF toothpaste by itself (Nabi et al, 1989).

Effect on plaque formation. Although triclosan is effective against a wide range of organisms in vitro, clinical studies in subjects rinsing with toothpaste slurries with concentrations of triclosan varying from 0.3% to 0.5% have shown only moderate levels of antiplaque activity (Jenkins et al, 1991b; Saxton, 1986; Saxton et al, 1987, 1988). At low concentrations (0.2%), triclosan brings about no significant reduction in plaque formation, even in combination with 0.5% zinc citrate. However, at higher concentrations of triclosan, the addition of 0.5% zinc citrate greatly enhances the effect on plaque formation (Saxton et al, 1987, 1988; Svatun et al, 1987).

In a double-blind crossover study, Kjaerheim and Waaler (1994) determined the plaque-inhibiting effect of different triclosan- and SLS-containing mouthrinses. An attempt was also made to locate the binding sites of triclosan in the oral cavity. The results indicated that triclosan alone has an antiplaque effect, independent of the ef-

fect of SLS. Furthermore, the results suggest that the SLS monomers may play a role as carriers of triclosan and that the teeth are not the only binding site for triclosan.

Antimicrobial agents should have a broad-spectrum effect against the indigenous plaque flora because the selective reduction of some species could encourage overgrowth of more pathogenic organisms. Quantitative and qualitative changes in the plaque have been reported after the use of triclosan products. Comparing a triclosan–zinc citrate preparation and a placebo, Jones et al (1988a) reported a decrease in the number of anaerobic bacteria as well as the number of actinomycetes in plaque accumulated during a 21-day in vivo trial. In the 3-year clinical study by Rosling et al (1997b), daily use of a triclosan-copolymer toothpaste resulted in significantly greater reduction of subgingival periopathogens than did a control fluoride toothpaste.

Effect of triclosan on calculus formation. Two long-term clinical trials have found significantly lower amounts of calculus in test groups using a combination 0.2% triclosan and 0.5% zinc citrate toothpaste (Svatun et al, 1990). Similar results have been reported after 6-month use of a 0.3% triclosan and 2.0% copolymer dentifrice (Lobene et al, 1991). In a 3-month clinical study Volpe and coworkers (1993) found that a toothpaste containing a combination of triclosan and pyrophosphate was as effective in reducing supragingival calculus formation as the active toothpaste tested by Lobene et al (1991).

Effect on gingivitis. Whereas the antiplaque effect of triclosan alone in toothpastes and mouthrinses has been tested in clinical studies, the effect on gingival health has been investigated only in a 21-day experimental gingivitis model (Saxton, 1986). In addition, it has been tested in combination with zinc salts or the copolymer Gantrez. The combination of zinc salts and triclosan has demonstrated better antiplaque and antigingivitis effects than triclosan or zinc separately (Saxton, 1986; Saxton et al, 1987; Saxton and van der Ouderaa, 1989).

Long-term studies have shown that subjects using toothpastes containing both triclosan and zinc citrate are able to maintain gingival health at baseline levels that had been attained by professional toothcleaning and oral hygiene instruction (Saxton et al, 1987; Svatun et al, 1987 1989a, 1989b). Statistically significant reductions in mean Gingival Index (GI) and Sulcular Bleeding Index (SBI) values have been reported, and it has been noted that the benefits to gingival health were most marked in subjects incapable of adequate toothbrushing (Saxton, 1986; Saxton et al, 1987). Clinical studies with triclosan-copolymer toothpaste have also demonstrated significant reductions in gingivitis (Garcia-Godoy et al, 1990).

There are clearer indications of a positive effect of triclosan on gingivitis than on plaque formation (Saxton and van der Ouderaa, 1989). The antigingivitis effect seems to become more marked both with longer-term use of triclosan and when the initial or sustained oral hygiene of the test subjects is poor (Garcia-Godoy et al, 1990; Saxton et al, 1987) or when there is a high number of bleeding sites.

These findings have led to speculation that the antigingivitis effect may be due to mechanisms other than plaque removal or retarded plaque formation. As mentioned earlier, phenolic compounds are known to have anti-inflammatory properties and exert their activity by inhibiting prostaglandin synthesis. In vitro studies have demonstrated triclosan to be an effective inhibitor of not only cyclo-oxygenase but also lipoxygenase, retarding both prostaglandin and leukotrein production. In a 6-month trial by Lindhe et al (1993), subjects using a triclosan-containing toothpaste exhibited less plaque and gingivitis than did control subjects using placebo dentifrice. Further analysis of the clinical findings also disclosed that "the triclosan-associated microbiota provoked less inflammation than the microflora formed in the control group." This conclusion was validated by findings from a series of studies (Barkvoll and Rölla, 1995; Kjaerheim et al, 1995a) documenting that triclosan, in different human model systems, reduces or inhibits inflammatory

reactions elicited by, eg, SLS, histamine, and dental plaque.

Triclosan may also have an analgesic effect. The pain associated with the use of mouthrinses containing SLS was reduced when triclosan was added to the solutions (Waaler et al, 1993). This biologic effect of triclosan was recently demonstrated by Kjaerheim et al (1995b), who showed that triclosan may have an analgesic effect because of direct interaction with excitable membranes (for review, see Kjaerheim, 1995 and Scheie, 1994a).

Effect on periodontal disease. Rosling et al (1997a) showed, in a 3-year clinical study, that daily use of a toothpaste containing triclosan and copolymer resulted in significantly greater reductions in probing depths and less loss of attachment than did a placebo fluoride toothpaste.

Salifluor

Salifluor, containing 5-n-octanyl-3-trifluoromethylsalicylanilide (A8-F), is a chemical derivative of aspirin with antibacterial and anti-inflammatory properties. It is also substantive, binding to oral surfaces in an active form. Staining of the teeth or the tongue by salifluor is very limited. Salifluor is virtually tasteless and hence may be useful as a mouthrinse or in a toothpaste. Salifluor is an effective antimicrobial agent, with minimum inhibitory concentrations comparable to those of antibiotics against a wide range of oral organisms, including the major periodontal pathogens.

Salifluor has been compared with CHX as an antiplaque agent in mouthrinses, in three short-term, double-blind, randomized, crossover design studies by Furuichi et al (1996). The findings indicated that mouthrinses containing salifluor were significantly more effective than were control rinses and that the salifluor mouthrinses were as effective as 0.12% chlorhexidine mouthrinse in retarding 4-day de novo plaque formation. No significant difference was observed between the 0.12% salifluor and 0.12% CHX mouthrinses in retarding de novo plaque formation and the development of gingivitis during a 14-day period without mechanical plaque control. However, the long-term effects of salifluor have to be compared with those of CHX in double-blind clinical studies, and the side effects of salifluor must be evaluated.

Other agents

Delmopinol

Delmopinol should be regarded as a surface-modifying agent. It belongs to the group of compounds known as the *substituted amino-alcohols*. Investigations of the related compound delmopinol have shown its benefits in control of plaque and gingivitis (Collaert et al, 1992a). Dose-dependent antiplaque and antigingivitis responses to delmopinol were demonstrated in a trial assessing the efficacy of rinsing with solutions of 0.05%, 0.1%, and 0.2%. A statistically significant dose-response effect was found; the 0.05%, 0.1%, and 0.2% groups experienced plaque reductions of 17%, 21%, and 33%, respectively. Gingivitis, assessed by the Gingival Bleeding Index (GBI), was reduced by 20%, 23%, and 33%, respectively, compared to the placebo group (Collaert et al, 1992a).

In another study, Collaert et al (1994) tested a possible dose-response effect of topical application of delmopinol hydrochloride on the salivary microbiology, the healing of experimentally induced gingivitis, plaque development, and supragingival plaque composition. After professional toothcleaning, the test subjects abstained from oral hygiene but used a soft applicator to brush 2 mL of a placebo agent on their teeth twice daily for 2 weeks. At the end of this period, the subjects' teeth were cleaned and they were assigned to one of three treatment groups. They applied 2 mL of 0.1%, 0.5%, or 1.0% delmopinol hydrochloride twice daily for the next 2 weeks and refrained from all other oral hygiene procedures.

At the end of the placebo and delmopinol hydrochloride treatment periods, all groups were subjected to the following tests:

1. Saliva samples were obtained and cultured on a series of media.
2. The degree of gingivitis was evaluated by gingival crevicular fluid and GI.
3. The stainable buccal plaque extension was analyzed planimetrically.
4. The bacterial morphotypes of plaque adjacent to the gingival margin were analyzed.

No changes in the salivary microbiologic counts were detected. All delmopinol groups exhibited less gingivitis (lower gingival crevicular fluid and GI scores) than did the placebo group. Mean plaque extension was reduced by 16% for the 0.1% delmopinol group, 56% for the 0.5% delmopinol group, and 58% for the 1.0% delmopinol group. The results indicated that short-term use of delmopinol promotes the healing of preestablished gingivitis, reduces plaque formation, and delays plaque maturation without detectable changes in the salivary microflora (Figs 194 and 195).

Overall, the effect of delmopinol is similar to that of CHX on plaque and less than that of CHX on gingivitis. Side effects of 0.1% and 0.2% delmopinol are limited to transient anesthesia of the dorsum of the tongue. The exact mode of action is unclear: Antibacterial activity is minimal in vivo, but bacterial matrix formation is disrupted, interfering with bacterial attachment.

Enzymes

Whole saliva contains peroxidase enzymes that oxidize thiocyanate (SCN^-) to hypothiocyanite ($OSCN^-$) in the presence of hydrogen peroxide:

$$H_2O_2 + SCN^- \rightarrow OSCN^- + H_2O$$

Hypothiocyanite is antimicrobial and in vitro inhibits some streptococci and lactobacilli (Lumikari et al, 1991). The activity of the salivary per-

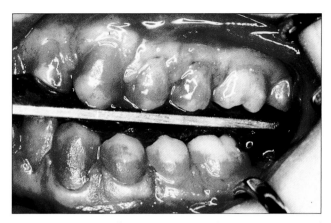

Fig 194 Amount of stained plaque after 2 weeks' use of placebo mouthrinse without mechanical toothcleaning. (From Collaert et al, 1994. Reprinted with permission.)

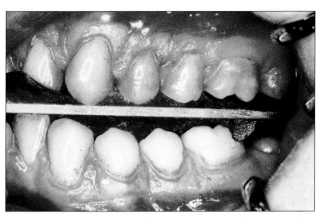

Fig 195 The same person as in Fig 194 after 2 weeks' use of mouthrinse with 0.5% delmopinol HCl without mechanical daily toothcleaning. (From Collaert et al, 1994. Reprinted with permission.)

oxidase system depends on available hydrogen peroxide. Hydrogen peroxide is produced by various bacteria as a metabolic end product but in inadequate quantities for maximum salivary peroxidase activity. The enzyme amyloglucosidase provides glucose, from which glucose oxidase produces hydrogen peroxide. Addition of these enzymes to oral products is intended to ensure the presence of sufficient hydrogen peroxide to control proliferation of bacteria through enhanced peroxidase activity.

Mouthrinses containing enzymes have been tested for their ability to reduce plaque, gingivitis, and dental caries, but the results have been disappointing (Hugoson et al, 1974). Dentifrices containing these enzymes may have slightly improved antiplaque and antigingivitis effects compared to nonenzyme dentifrices, but it is doubtful whether the marginal effect is of clinical relevance.

Plaque removal agents

The idea of the "chemical toothbrush" is perhaps synonymous with the use of detergents for many household cleaning purposes. Indeed, detergents such as SLS have found long and widespread use in dentifrices and some mouthrinses. Theoretically, toothpaste detergents could remove plaque, although there are no data to support this idea. Nevertheless, compared to water, SLS and SLS-containing toothpastes do inhibit plaque regrowth (Jenkins et al, 1991a). This may be due to an antimicrobial action rather than a plaque-removing action. Interest in plaque removal or loosening was renewed following the launch of a prebrushing rinse that contained SLS among other ingredients. There are no published long-term home use studies supporting the claimed benefit.

Combination agents
Heavy metal ions plus detergents

Plaque is a complex aggregation of various bacterial species. It is therefore unlikely that one single agent can be effective against the complex flora. Combinations of two or more agents with complementary inhibitory modes of action may enhance the efficacy and reduce the adverse effects of chemoprophylactic agents, offering promising prospects for new and effective chem-

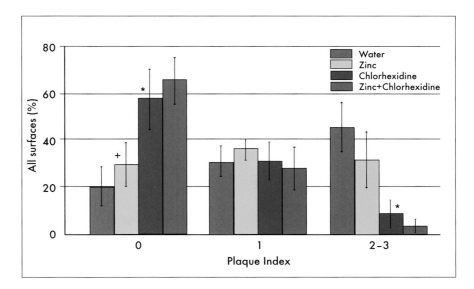

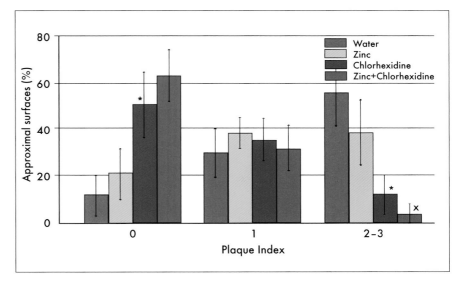

Figs 196 and 197 Effect on total (Fig 196) and approximal (Fig 197) plaque formation of 2 days' twice-daily mouthrinses with 10 mL water or with aqueous solutions of either 10 mmol/L zinc chloride, 0.55 mmol/L chlorhexidine diacetate, or 10 mmol/L zinc acetate and 0.55 mmol/L chlorhexidine diacetate in combination in 10 individuals according to Silness and Löe Plaque Index scores 0 to 3. No mechanical oral hygiene was performed. (+) Significantly better than water; (*) Significantly better than water or zinc chloride; (x) Significantly better than water, zinc chloride, or chlorhexidine diacetate ($P \leq .05$). (Modified from Giertsen et al, 1988, 1989a with permission.)

ical plaque control agents. Several combinations of metal ions with surfactants have been evaluated: For instance, combinations of hexetidine and Cu^{2+} or Zn^{2+}; Zn^{2+} and CHX; CHX, Zn^{2+}, and fluoride; Zn^{2+} and SLS; and zinc and triclosan show enhanced plaque-inhibitory effects. This is probably related to increased intracellular uptake of the metal ions. Giertsen et al (1988) reported that mouthrinses containing $ZnCl_2$ and CHX in combination suppressed dental plaque formation and acid production by established dental plaque

more effectively than did either of the agents individually (Figs 196 and 197).

When used in combination, $ZnCl_2$ and CHX also have a synergistic growth-inhibitory effect on *S sobrinus* and *S sanguis*. In addition, sub-bactericidal concentrations of $ZnCl_2$ and CHX in combination reduce glucose uptake in *S sobrinus* to the same degree as $ZnCl_2$ alone; thus, the synergistic antibacterial effect of $ZnCl_2$ and CHX may be ascribed to mechanisms other than direct inhibition of glucose uptake as a result of the agents in combination. The improved clinical effects of Zn^{2+} and CHX may in part be explained by prolonged antibacterial activity of the combined agents because CHX increases the binding of Zn^{2+} by the bacteria, which results in increased intracellular uptake of the metal ions.

In another short-term mouthrinse study, Giertsen et al (1989b) compared the single and combined plaque-reducing effects of SLS and zinc citrate with the effect of a placebo solution. Compared to the placebo, SLS and zinc citrate increased the frequency of PI score 0 by 52.9% and 98.3%, respectively; use of SLS resulted in 36.9% and use of zinc citrate resulted in 55.7% fewer surfaces with scores 2 or 3 ($P < .05$ in all cases). The combination of zinc citrate and SLS resulted in a threefold increase in the frequency of PI score 0 and a 70.5% reduction in the frequency of scores 2 or 3 ($P < .05$). The enhanced antiplaque efficacy of the combination of zinc citrate and SLS may be ascribed to the increased solubility of zinc citrate in the presence of SLS as well as to the additive antibacterial effect of the two agents.

Triclosan combinations

Triclosan products are the first successfully modified antimicrobial commercial agents to apply available knowledge of adsorption and desorption of chemical agents in the oral environment. Each of the other agents has a different mechanism of action and a different level of substantivity. Unlike CHX, Listerine, and other commercial products, triclosan has been specially modified and designed to optimize antimicrobial potency by being linked to a substantive agent. As a result of this engineered substantivity, the modified drug now has increased efficacy in the oral cavity. To date, the following three modifications of triclosan have been developed and used in dentifrices or mouthrinses:

1. Combination with zinc citrate to take advantage of potential antiplaque and anticalculus properties

2. Incorporation in a copolymer of methoxyethylene and maleic acid to increase retention time

3. Combination with pyrophosphates to enhance calculus-reducing properties

These products also contain fluorides and a silica base to provide an anticaries effect. Combination products containing triclosan at concentrations of 0.2% to 0.5% and zinc citrate at concentrations of 0.5% to 1.0% have significantly reduced plaque, gingivitis, and calculus (Stephen et al, 1990b; Svatun et al, 1990). Significant reductions in plaque, gingivitis, and calculus have also been found with a dentifrice containing 0.3% triclosan and copolymer of methoxyethylene and maleic acid (Cubells et al, 1991; Deasy et al, 1991; Garcia-Godoy et al, 1990; Lindhe et al, 1993). Short-term studies of a mouthrinse and a prebrushing rinse (0.03% triclosan and 0.25% copolymer) have also revealed significant plaque reduction (Abello et al, 1990).

Silicone oil is known to adsorb to hydroxyapatite and to teeth because of its low surface tension. Treatment with silicone oil in vivo will thus change the surface properties of teeth. The thin, resistant layer of silicone oil that adsorbs to teeth renders them hydrophobic, a property that influences both pellicle and plaque formation. An experimental toothpaste containing both triclosan and silicone oil markedly improved gingival health in teenagers with established gingivitis (Rölla et al, 1993). In a follow-up study, Ellingsen and Rölla (1994) showed that teeth treated topically with silicone oil and triclosan showed marked plaque reduction, and those treated with silicone oil alone showed a moderate reduction, compared to those treated with a placebo. The combination of silicone oil and 0.3% triclosan also inhibited plaque formation on proximal surfaces. It appears likely that the thin layer of silicone oil firmly bound to the teeth serves as a reservoir for triclosan, which is then released only slowly, because of its low solubility in saliva.

Chlorhexidine plus fluorides

For caries prevention, commercial products containing efficient antiplaque and *S mutans* inhibitors such as CHX plus zinc and fluorides would be very attractive. Only one commercial toothpaste has recently been produced according to such a formula (discussed later in this chapter). However, some of the most frequently used fluoride compounds (eg, monofluorophosphate) are anionic and inactivate the cationic CHX. This is one reason why no caries-preventive effect was achieved in a 2-year, crossover study in which one test group used 1% chlorhexidine digluconate gel in custom trays, once every 2 weeks, in combination with sodium monofluorphosphate mouthrinse (Axelsson et al, 1976). However, the combination of CHX and sodium fluoride has a synergistic toxic effect on the cytoplasm of bacterial cells and the enzymes for fermentation of carbohydrates. This indicates that acid production by the microbes is also reduced (Luoma, 1972; Meurman, 1988).

The results of a 2-year clinical trial using a combination of CHX (0.05%) and NaF (0.044%) in a mouthrinse were reported by Luoma et al (1978). A total of 164 schoolchildren aged 11 to 15 years rinsed under supervision with 10 mL of CHX-NaF solution, NaF solution (0.044%), or a placebo once daily on 200 school days per year. In addition, they brushed their teeth with a toothpaste of the same composition as the mouthrinse they were using. After 2 years, the group using the combined CHX-NaF agent showed a significant reduction in gingival bleeding as well as the smallest increment of decayed, missing, or filled surfaces. The authors proposed that a synergistic effect takes place when the two agents are used together.

Fluoride is acknowledged as the most important cariostatic agent currently available. It has some limitations and does not prevent lesion formation when the plaque pH is persistently low. This was demonstrated by Ogaard et al (1991), in a study in which shark enamel containing 33,000 ppm F^- (almost pure fluorapatite) was exposed to a high cariogenic challenge. Under these condi-

Table 4 Adjusted values for lesion depth and mineral loss after 4 weeks of rinsing with fluoride (F) or fluoride combined with chlorhexidine (CHX + F)*

Regimen	n	Lesion depth (µm)	Mineral loss (vol% × µm)
CHX + F	15	8 ± 12	545 ± 290
F	14	26 ± 21	706 ± 370
No F[†]	20	82 ± 26	1,175 ± 460

*Data from Ullsfoss et al (1994).
[†]The no-F data are from a previous study (Ogaard et al, 1986), under identical conditions, of five patients, aged 11 to 13 years, with a total of 10 pairs of premolar teeth.

tions, not even additional fluoride in the form of mouthrinses could protect the shark enamel.

One way to increase the effect of fluoride could be the use of antibacterial agents to reduce acid formation in plaque, thereby reducing the acid attack on the mineralized surfaces. Chlorhexidine has been shown to reduce acid formation for up 24 hours in established plaque (Oppermann, 1979). However, the relatively large CHX molecules do not penetrate very thick plaque: For optimal antimicrobial effect, dental plaque should be mechanically removed by self-care and needs-related PMTC before CHX is used.

Ullsfoss et al (1994) evaluated the possible caries-inhibitory effect of combining 2.2-mM CHX mouthrinses twice daily with 11.9-mM NaF rinses and compared it with the effect of only daily NaF rinses. They used an in vivo human caries model in which plaque-retaining bands were placed on premolars scheduled for extraction. In nine subjects, 29 teeth were banded for 4 weeks. Saliva and plaque were sampled for bacterial culture before and after the study period. After the teeth had been carefully extracted, the tooth surfaces were analyzed by microradiography. The combination of CHX and fluoride rinses resulted in enamel mineral loss only slightly higher than that observed in "sound" enamel and clearly less than that observed with fluoride rinses alone (Table 4).

Box 8 0.1% Chlorhexidine digluconate and 0.05% sodium fluoride mouthrinse solution

- Sodium fluoride: 0.05 g
- Saccharin sodium: 0.02 g
- Polysorbatum 80: 0.01 g
- Hibitane solution 20%: 0.50 g
- Menthol pip aetherol: 0.02 g
- Aqua purificata ad: 100.00 g
- Menthol or any other flavoring may be added.

Use in a 10-mL mouthrinse, twice a day for 3 weeks.

Box 9 1% Chlorhexidine digluconate and 0.05% sodium fluoride gel

- Chlorhexidine digluconate: 0.50 wt%
- Sodium fluoride: 0.20 wt%
- Methyl cellulose: 3.00 wt%
- Aetherol menthol pip: 0.02 wt%
- Polysorbatum 20: 0.01 wt%
- Methylium paraoxibenzoas: 0.05 wt%
- Spiritis fortis: 0.50 wt%
- Saccharin sodium: 0.02 wt%
- Aqua purificata ad: 100.00 wt%
- Artificial coloring such as green or orange may be added.

Use as a gel in customized trays, 5 minutes a day for 2 weeks, or as a dentifrice, twice a day for 2 to 3 weeks.

Both total plaque bacteria and *S mutans* were reduced by CHX rinses, confirming the discrete mechanisms of action. As in previous studies, fluoride alone did not totally prevent caries lesions. The combination of CHX and fluoride was significantly more effective in reducing both lesion depth and mineral loss; probably because of both a general inhibition of acid formation and a specific effect on *S mutans* (Ullsfoss et al, 1994).

Although for ethical reasons the cariostatic effect of CHX alone was not investigated in the aforementioned study, Luoma (1972) found that CHX and fluoride in combination had a significantly greater inhibitory effect on acid production by *S mutans* than did CHX alone. The results of the investigation by Ullsfoss et al (1994) supported the hypothesis that combining fluoride with agents that decrease acid formation in plaque may counteract even an extreme cariogenic challenge. On this basis, a strategy for managing high caries risk or caries risk generally would involve mechanical plaque removal combined with the use of a plaque-inhibiting agent and fluoride in a mouthrinse or dentifrice twice daily. Katz (1982) suggested topical applications of a 1% NaF and 1% CHX gel combined with daily rinses with 0.05% NaF and 0.2% CHX as the key elements in a regimen for total caries prevention

after irradiation of the head and neck. The study by Ullsfoss et al (1994) provided the first in vivo human experimental model data to support this concept.

As mentioned earlier, CHX and NaF, used in combination, will have a synergistic caries-preventive and antimicrobial effect in mouthrinses and gel preparations. However, such preparations have a limited shelf life of 6 months and are therefore not commercially available. The pharmaceutical prescriptions listed in Boxes 8 and 9 can be used for mouthrinses and gels, respectively.

According to the studies by Giertsen et al (1988), the prescription preparations would be even more efficient with the addition of 10.0 mM of $ZnCl_2$. A combination of chlorhexidine with a fluoride compound that is more antibacterial than NaF might provide even better results. Fluoride compounds with such antibacterial activity against *S mutans* are stannous fluoride and amine fluoride (AmF).

Ostela and Tenovuo (1990) studied the in vitro susceptibility of *S mutans*, *S sobrinus*, and *Lactobacillus casei* to dental gels containing various combinations of AmF 297, SnF_2, and CHX. The combination of AmF and SnF_2, with a total fluoride content of 1.2%, was the most effective against MS but not against *L casei*. At a much lower total fluoride concentration (0.4%), AmF alone or combined with SnF_2 was significantly less effective against MS than either CHX or a CHX-AmF-SnF_2 combination. The CHX-AmF combination was a slightly more potent inhibitor of streptococcal growth than was CHX-NaF. For *L casei*, only minor differences were observed, but CHX alone seemed to be the most effective agent. Of the agents studied, CHX seemed to be the most potent individual chemotherapeutic compound; its effect on *S mutans* could be enhanced when it was combined with AmF. However, the most effective CHX-AmF and AmF-SnF_2 combinations in these in vitro experiments have yet to be tested for antibacterial efficacy in vivo.

The effect of antimicrobial treatment on the numbers of MS in plaque from the margins of restorations and in saliva was studied by Wallman et al (1994). Nineteen subjects with well-restored dentitions and counts of more than 0.5×10^6 MS per 1 mL of saliva were treated with 1% CHX gel in individually designed applicators for 5 minutes a day for 9 days. Ten of the subjects continued the treatment with 0.4% SnF_2 gel and the remaining nine used a placebo gel for another 14 days. Plaque samples from the margins of selected restorations and stimulated saliva were collected at baseline, after the completion of each gel treatment, and again at regular intervals for up to 24 weeks. The CHX gel treatment suppressed MS at the margins of restorations as well as in saliva. Additional treatment with the SnF_2 gel prolonged this suppression, unlike CHX treatment alone. In the CHX-SnF_2 group, the number of MS at the margins of amalgam and resin composite restorations was still significantly lower at the end of the study than at baseline.

Other fluoride compounds and combinations

Stannous fluoride. Most investigations of the antibacterial mechanism of SnF_2 have focused on

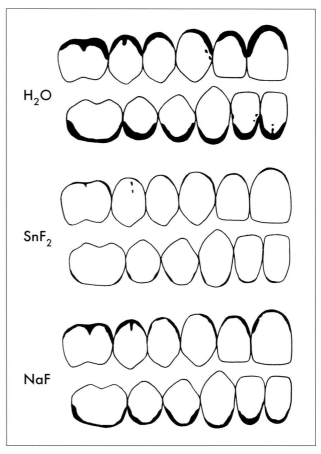

Fig 198 Plaque accumulation on a dentition 4 days after rinses with water (H_2O), 0.3% stannous fluoride (SnF_2), or 0.2% sodium fluoride (NaF). Plaque formation was enhanced with sucrose. (From Bay and Rölla, 1981. Reprinted with permission.)

either alterations in bacterial growth or on bacterial adhesion and cohesion. Investigations have suggested that SnF_2 unbalances bacterial growth, oxidizes thiol groups in enzymes involved in glycolysis and the transport of sugar, reduces the quantity of lipoteichoic acid in plaque, and decreases bacterial growth. Figure 198 illustrates plaque reaccumulation during 3 days' mouthrinsing with 0.3% SnF_2, 0.2% NaF, or water. The effect of SnF_2 on bacterial adhesion could also explain, in part, its antibacterial properties. Early observations suggested that SnF_2 reduces the surface energy of enamel, making adhesion of bacteria less likely, and alters bacterial adhesion and cohesion (Glantz, 1969; Tinanoff et al, 1976).

It has also been shown that, unlike NaF, SnF_2 mouthrinses selectively reduced the numbers of *S mutans* in saliva and in plaque (Svanberg and Rölla, 1982). In concentrations as low as 0.001%, SnF_2 inhibited *S mutans* coherence, whereas NaF showed no effect. Several others have also reported the frequent use of SnF_2 mouthrinses or gels to have significant effects on gingivitis and periodontitis (for review, see Tinanoff, 1990).

Scanning electron microscopic examination of enamel treated by SnF_2 reveals that, in addition to the tin-containing layer, which consists of small granules, larger granules are present. The large particles are probably calcium fluoride (CaF_2) or a CaF_2-like material, and the smaller ones are stannous phosphate (Rölla and Saxegaard, 1990). Enamel treated by stannous fluoride becomes hydrophobic, and this may contribute to the antiplaque effect of stannous fluoride, because hydrophobic surfaces are less easily colonized by bacteria (Rölla et al, 1994).

The cariostatic effect of SnF_2 is clearly dependent on the deposition of fluoride on the tooth surface. The tin-phosphate layer is probably not associated with caries protection. It appears likely that CaF_2 is deposited on the tooth surfaces (as large particles) when teeth are treated with SnF_2 and that CaF_2-containing particles then serve as pH-dependent reservoirs of fluoride, which release fluoride (and calcium) during carious challenges (Rölla, 1988, Rölla and Saxegaard, 1990).

The antiplaque effect of SnF_2 can clearly contribute to its cariostatic activity, particulary when it is applied frequently in toothpastes, gels, or mouthrinses. The selective elimination of *S mutans* and the inhibition of acid formation by dental plaque for several hours after a mouthrinse with aqueous solutions of SnF_2 probably contribute as well. Both the antiplaque effect and the inhibition of acid formation by SnF_2 are most likely caused by the oxidation of thiol groups.

The antigingivitis effect of SnF_2 is presumably a result of its antiplaque effect. It also appears that the inhibition of bacterial metabolism when carbohydrates are used as substrates (the inhibition of acid formation) may also affect other metabolic pathways, where antigens or enzymes associated with gingivitis may be correspondingly inhibited, and that this may contribute to the antigingivitis effect. The selective effect of SnF_2 on *S mutans* may be due to the antiacid effect of SnF_2. A less aciduric environment in plaque will probably be an ecological disadvantage for the aciduric *S mutans*, which consequently is reduced in number.

Stannous fluoride plus amine fluoride. In the last decade, a new fluoride mouthrinse (Meridol, GABA International, Münchenstein, Switzerland) was introduced. It has the following components:

1. Amine fluoride 297 (Olaflur)
2. SnF_2
3. Flavoring agents
4. Sweeteners that are "safe for teeth"
5. Coloring agents
6. Solubilizer

Demineralized, deionized, or distilled water is used as the solvent, and the total ionized fluoride concentration of the solution is 250 ppm. To date it has not been possible to produce ready-to-use stable SnF_2 mouthrinses because aqueous SnF_2 solutions are chemically unstable and become turbid soon after mixing; this is followed by precipitation and oxidation. Olaflur prevents precipitation of SnF_2, making it possible to prepare clear, stable aqueous solutions of SnF_2 for use as mouthrinses.

The mode of action is based on the proven caries-preventive effects of the two active fluorides, AmF and SnF_2. The antiplaque effects of both are well known; the combined antiplaque and antimicrobial effects are synergistic.

The effect of Meridol mouthrinse on dental plaque and gingivitis has been evaluated and

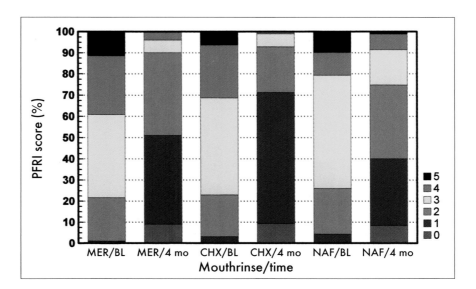

Fig 199 Effect of mouthrinses on Plaque Formation Rate Index (PFRI) as evidenced by frequency distribution of PFRI scores 0 to 5. (MER) Meridol; (CHX) Chlorhexidine; (NAF) Sodium fluoride; (BL) Baseline; (4 mo) 4 Months. (From Axelsson et al, 1993d. Reprinted with permission.)

compared with the effect of placebo solutions in numerous double-blind clinical studies (Axelsson et al, 1994b; Banoczy et al, 1989; Brecx et al, 1993; Zimmerman et al, 1993). The effect on specific pathogens associated with dental disease and the potential for remineralization of incipient caries in vitro has also been tested (Althenhofen et al, 1989; Axelsson et al, 1993d).

In long-term studies, the effect on gingivitis was comparable to that of 0.1% CHX digluconate mouthrinse (Axelsson et al, 1994b; Brecx et al, 1993). There are no known side effects, such as mucosal irritation, and only limited yellowish staining of the teeth occurs with long-term use (Axelsson et al, 1993d; Zimmerman et al, 1993). There are no significant ecological changes in the oral microflora (Axelsson et al, 1993d; Zimmerman et al, 1993).

Studies have also shown that, compared to a placebo, AmF-SnF$_2$ mouthrinse significantly reduces the surface free energy of the enamel, explaining in part why Meridol mouthrinse significantly reduces the Plaque Formation Rate Index (PFRI) (Axelsson et al, 1993d). In a 4-month double-blind study, Meridol, 0.1% CHX, and 0.025% NaF mouthrinses were used twice a day after mechanical toothcleaning. The PFRI and salivary levels of MS and lactobacilli were evaluated.

From more than 1,000 high school students aged 17 to 19 years, 300 subjects with high salivary MS levels and/or PFRI scores were selected and randomly allotted to one of three test groups. Twice a day after toothcleaning, test group I rinsed with Meridol (AmF plus SnF$_2$, equivalent to 0.025% fluoride). Test group II rinsed with 0.1% CHX, and test group III rinsed with 0.025% NaF.

The PFRI, MS, and lactobacilli decreased in all groups after 4 months compared to baseline values. High or very high PFRI (score 4 or 5) decreased by 36% in group I, 29% in group II, and 12% in group III (Fig 199). High MS scores (score 3) decreased by 24% in group I, 52% in group II, and 10% in group III (Fig 200). The percentage of subjects with the highest level of lactobacilli (score 4) decreased by 18%, 8%, and 4%, respectively, in groups I, II, and III (Fig 201). Chlorhexidine (group II) was the most efficient in reducing the mean values of PFRI, followed by Meridol (group I) and NaF (group III). The only side effects observed were heavy brown staining of the teeth in the CHX group and very limited yellow staining in the Meridol group.

The study reconfirmed earlier studies that showed daily rinsing with CHX to be efficient in reducing plaque formation and salivary MS levels. However, for caries prevention, Meridol

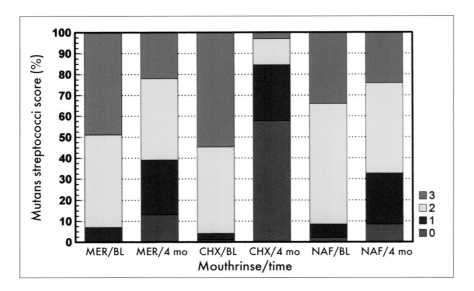

Fig 200 Effect of mouthrinses on MS as evidenced by frequency distribution of Strip Mutans scores 0 to 3. (MER) Meridol; (CHX) Chlorhexidine; (NAF) Sodium fluoride; (BL) Baseline; (4 mo) 4 Months. (From Axelsson et al, 1993d. Reprinted with permission.)

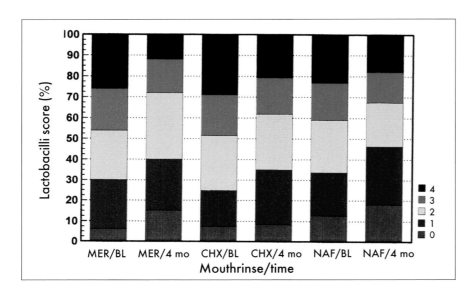

Fig 201 Effect of mouthrinses on lactobacilli as evidenced by frequency distribution of strip lactobacilli scores 0 to 4. (MER) Meridol; (CHX) Chlorhexidine; (NAF) Sodium fluoride; (BL) Baseline; (4 mo) 4 Months. (From Axelsson et al, 1993d. Reprinted with permission.)

could be more advantageous for long-term use in high-risk individuals because of its fluoride content and the absence of side effects, such as brown staining and mucosal lesions. The positive effects of the NaF mouthrinse can partly be explained by the study design, which required all the participants to clean their teeth mechanically before rinsing twice a day, and the so-called Hawthorne (placebo) effect.

CHEMICAL PLAQUE CONTROL BY SELF-CARE

The following vehicles are available for antiplaque and/or antimicrobial agents (excluding antibiotics) for chemical plaque control by self-care:

1. Toothpastes
2. Mouthrinses
3. Irrigants
4. Gels
5. Chewing gums

Arguably, toothbrushing with toothpaste is the most common method of oral hygiene practiced in developed countries (Frandsen, 1986). Chemical plaque control agents can be evaluated from one or both of two perspectives:

1. In the short or long term, are chemicals better and safer alternatives to mechanical tooth-cleaning with a conventional toothpaste?
2. Delivered via toothpaste or as an additional oral hygiene product, do chemicals provide adjunctive benefits to toothbrushing with conventional fluoride toothpaste?

Again, because some plaque control products, notably toothpastes, will be considered for long-term use, convincing evidence of the safety of the active ingredients is essential. Assuming that there is now a consensus that no chemical product can be a long-term replacement for mechanical tooth-cleaning, the role, safety, and cost effectiveness of chemical adjuncts must be debated.

As discussed earlier, plaque forms continuously but at different rates among individuals as well as on different tooth surfaces: If plaque control is not regular and adequate, gingivitis and dental caries may ensue (Löe et al, 1965; Von der Fehr et al, 1970). This raises the question of whether individuals with poor oral hygiene require an effective chemical agent for life. The same question arises for individuals susceptible to chronic periodontitis and high caries risk who are unable to maintain satisfactory supragingival plaque control. For many individuals, the answer logically must be yes.

There is no evidence that the short-term use of chemical adjuncts produces long-term behavioral changes in oral hygiene practices; any improvements are limited to the period of active use of the agent. The long-term adjunctive use of agents in any vehicle other than a toothpaste, however, would have major cost implications. For example, simple arithmetic determines that for an individual, over a period of 1 year, the cost of rinsing twice daily with a proprietary mouthrinse would be several times greater than the outlay for toothbrushes and toothpaste. Therefore, for cost effectiveness, the ideal way forward would be the enhancement of chemical inhibition via the toothpaste vehicle.

For long-term use, however, toothpastes are particulary open to misuse or abuse, and chronic overdosage of agents is always a risk, as observed in the case of fluoride. Safety in long-term use must therefore be a priority if chemical plaque control is to play anything other than its present short-term role in preventive dentistry.

Toothpastes

For many years, most toothpaste manufacturers have shown an interest in producing products containing efficient and safe antimicrobial and antiplaque agents. The anionic detergent SLS has been by far the most common antiplaque agent in toothpastes for decades. Stannous fluoride, amine fluoride, and different enzymes have also been used for years, with varying effects on dental plaque and mostly limited effects on gingivitis.

Toothpastes containing chlorhexidine

Even CHX has been used in toothpastes. As early as the 1970s, a 2-year study was carried out by Johansen et al (1975), but the results compared to a

Fig 202 Newly formulated toothpastes containing 0.4% CHX, 1.3% zinc lactate, and 0.32% NaF: Crest Specialist Care (Procter & Gamble, Cincinnati, OH), Blend-a-med Parosan (Blendax, Mainz, Germany), Blend-a-med Forte (Blendax); and AZ15 (Blendax).

placebo toothpaste were unimpressive, because good mechanical plaque control was achieved by all subjects as a result of their greater awareness of oral hygiene. Owing to the cationic nature of CHX, other (anionic) substances in the toothpaste could have inhibited the efficacy of CHX.

Recent short-term and long-term studies with experimental toothpastes containing CHX, however, have shown significant effects on dental plaque, gingivitis, and oral microflora. Maynard et al (1993) evaluated the effects of 1% CHX and 1% CHX-fluoride toothpastes on supragingival plaque flora in a 6-month home-usage study. In 146 subjects, plaque was collected from six teeth at baseline, 6 weeks, and 24 weeks. Total anaerobic counts were lower at 6 and 24 weeks in both active groups than in the control group, but there were no differences in any other bacterial variables.

In a short-term study, Jenkins et al (1993b) tested the effect of a new experimental CHX toothpaste on the development of plaque, gingivitis, and tooth stains. This study was the first phase in the evaluation of a 1% CHX toothpaste, formulated to ensure high availability of CHX. The study was a 19-day, randomized, double-blind, placebo-controlled, crossover experimental gingivitis trial

in 14 healthy human volunteers. Subjects had 0 plaque and low gingivitis scores at baseline. Plaque, gingivitis, and dental stains were measured on days 12 and 19 of the study. The toothpaste was used twice a day as a slurry rinse with no other form of oral hygiene. The washout period was 21 days. Clinically, there was pronounced reduction of plaque and gingivitis following use of the active product, and, compared to the placebo, the difference was highly statistically significant. However, some staining occured when CHX toothpaste was used. The product warrants further evaluation for potential clinical use.

In another short-term study, Claydon and Addy (1995) showed that after a single brushing with a 1% CHX toothpaste there was significantly less 24-hour plaque regrowth than there was after use of a placebo toothpaste or a fluoride toothpaste.

Sanz et al (1994) evaluated the effect of a toothpaste containing CHX and zinc on plaque, gingivitis, calculus, and staining, in a 6-month randomized, stratified, double-blind parallel study of 208 subjects. The participants used either a 0.12% CHX rinse and a "gum care" toothpaste (positive control), a placebo rinse and the gum care toothpaste (control), or the placebo rinse and an experimental toothpaste containing 0.4% CHX and 0.34% Zn^{2+} (experimental group).

After 6 months' use, all groups had less plaque and less gingivitis and all had developed calculus and tooth stains. The positive control group and the experimental group showed significant reductions in plaque and gingivitis (Gingival Index: number of bleeding sites) compared to the control group. Significantly more calculus developed in the positive control group, but calculus development was similar in the experimental and control groups. Finally, tooth staining was significantly higher in the groups using CHX-containing products than it was in the control group. The experimental toothpaste users had much less staining than did the positive control. It was concluded that, compared to a 0.12% CHX rinse, the experimental toothpaste would

Fig 203 Commercially available antiplaque toothpastes: Colgate Total (silica-based toothpaste containing 0.3% triclosan, 2% Gantrez copolymer, and SLS as plaque-inhibiting agents, plus 0.32% NaF [0.145% F]) (Colgate-Palmolive, New York, NY); Pepsodent Ultra (silica-based toothpaste containing 0.3% triclosan, 0.75% zinc citrate, and SLS as plaque-inhibiting agents, plus 1.1% sodium monofluorophosphate [0.145% F]), recently replaced with Pepsodent Triple (Elida Robert, Helsinki, Finland); Dentosal (silica-based toothpaste containing 0.3% triclosan, 5% soluble pyrophosphates, and SLS as plaque-inhibiting agents, plus 0.32% NaF) (Blendax); Oral-B Tooth and Gum Care (silica-based toothpaste containing 0.4% SnF_2 [0.1% F] as a plaque-inhibiting and caries-preventive agent) (Gillette, Boston, MA); and Elmex (silica-based toothpaste containing AmF as a plaque-inhibiting and caries-preventive agent) (GABA International).

contribute to a significant improvement in oral hygiene, with less staining (Sanz et al, 1994).

Chlorhexidine has a well-known specific effect on MS (for review, see Emilson, 1994), and CHX-NaF combinations seem to have a synergistic caries-preventive effect and suppression of MS (Emilson et al, 1976; Luoma, 1972; Luoma et al, 1978; Meurman, 1988; Zickert et al, 1987b). In addition, zinc has proven to increase the antiplaque effect of CHX (see Figs 196 and 197). Therefore, the aforementioned experimental toothpaste combined with 0.3% NaF also would be very attractive for caries prevention if bioavailabity and stability could be successfully maintained in the formulation. Such toothpastes are now available commercially (Fig 202).

Toothpastes containing triclosan

The most widely available antimicrobial toothpastes contain triclosan, and these have been studied in the greatest detail. Being nonionic, triclosan is compatible with toothpaste formulations and has reasonable substantivity: It is detectable on the oral mucosa and in dental plaque at least 3 and 8 hours, respectively, after use (Gilbert and Williams, 1987). It is a broad-spectrum antimicrobial agent, active against all the major plaque bacteria (Marsh, 1992). The use of triclosan-containing products has been associated with very few adverse side effects. (Gjermo and Saxton, 1991).

Triclosan alone has only moderate antiplaque properties (Jenkins et al, 1989; Saxton, 1986), and attempts to increase its clinical effectiveness have included trying to increase retention and combining it with other toothpaste-compatible antimicrobial agents. To increase its substantivity and plaque-inhibiting effect in toothpastes, triclosan has been combined with three different compounds: Gantrez copolymer, zinc citrate, and pyrophosphate. Figure 203 shows commercially available antiplaque toothpastes.

Triclosan plus copolymer. The effect on plaque and gingivitis of toothpastes containing 0.3% triclosan and 2.0% copolymer has been compared with the effect of a placebo toothpaste in nine long-term studies, in accordance with the 1986 American Dental Association Council on Dental Therapeutics guidelines for acceptance of chemotherapeutic products for the control of supragingival dental plaque and gingivitis (Table 5) (American Dental Association, 1986).

Table 5 Plaque- and gingivitis-reducing efficacy of triclosan-copolymer dentifrice in long-term clinical studies*

Study	Location	No. of subjects[†]	Duration (mo)	Clinical design	Plaque efficacy versus placebo[‡]		Gingivitis efficacy versus placebo[‡]	
					Q-H Index	PS Index	L-S Index	GS Index
Palomo et al (1989)	Guatemala	98	6	Parallel with PMTC at start	-12.7%	-23.1%	-24.1%	-38.4%
Garcia-Godoy et al (1990)	Dominican Republic	108	7	Parallel with PMTC at start	-58.9%	-97.7%	-30.1%	-87.5%
Cubells et al (1991)	Spain	108	6	Parallel with PMTC at start	-24.9%	-50.8%	-19.7%	-57.5%
Deasy et al (1991)	USA	121	6	Parallel with PMTC at start	-32.3%	-73.6%	-25.6%	-57.1%
Mankodi et al (1992)	USA	294	6	Parallel with PMTC at start	-11.9%	-19.3%	-19.7%	-73.6%
Denepitiya et al (1992)	USA	145	6	Parallel with PMTC at start	-18.4%	-29.2%	-31.5%	-57.1%
Bolden et al (1992)	USA	306	6	Parallel with PMTC at start	-17.0%	-18.6%	-29.0%	-47.6%
Triratana et al (1993)	Thailand	120	6	Parallel without PMTC at start	-32.9%	-46.0%	-18.8%	-38.3%
Lindhe et al (1993)	Sweden	110	6	Parallel without PMTC at start	-31.2%	Not reported	-26.6%	Significantly fewer bleeding sites[§]

*From Jackson (1996). Reprinted with permission.

[†]Number of subjects in both the triclosan-copolymer dentifrice group and the placebo dentifrice group who completed the entire study.

[‡]Plaque and gingivitis efficacy results from data obtained at the final clinical examination. All percentages relating to plaque and gingivitis efficacy of the triclosan-copolymer dentifrice were calculated relative to the placebo dentifrice and were statistically significant at the 99% level of confidence: Q-H Index = Quigley-Hein (Turesky modification) Plaque Index; PS Index = Plaque Severity Index of Palomo et al (1994); L-S Index = Löe-Silness (Talbot, Mandel, and Chilton modification) Gingival Index; GS Index = Gingivitis Severity Index of Palomo et al (1994).

[§]At the conclusion of the study, the triclosan-copolymer dentifrice group had significantly fewer bleeding sites (and significantly more gingivitis-free sites) than did the placebo dentifrice group.

In the study by Lindhe et al (1993), the pattern of plaque and gingivitis on the surfaces of the maxillary and mandibular teeth at baseline examination and at the final examination (6 months) was presented for the first time, thereby comparing the effect and accessibility of the triclosan-copolymer toothpaste and the placebo (control) toothpaste on the individual tooth surfaces (Figs 204 and 205).

Two independent studies have reported caries efficacy results from double-blind, long-term (26-month) clinical studies comparing a 0.3% triclosan and 2.0% copolymer toothpaste in a NaF-silica base to a clinically proven, positive control toothpaste containing NaF and silica (Feller et al, 1993; Mann et al, 1993). The studies were conducted in accordance with the 1988 American Dental Association Council on Dental Therapeu-

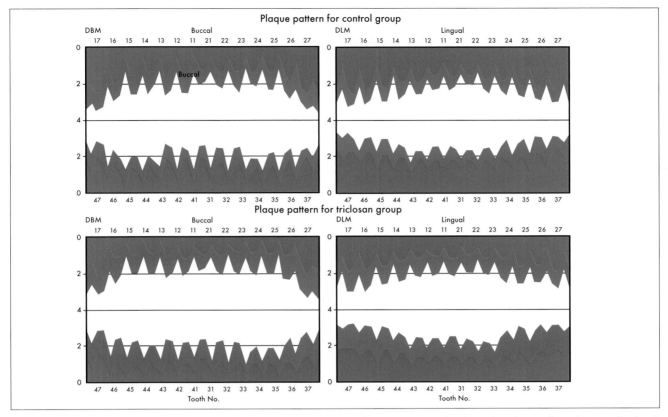

Fig 204 Mean Quigley-Hein Index score for individual tooth surfaces examined at baseline (*green*) and 6 months (*red*) for patients using a placebo toothpaste *(top)* and patients using a triclosan-copolymer toothpaste *(bottom)*. The green area represents the reduction in plaque scores that occurred between the two examinations (Fédération Dentaire Internationale [FDI] tooth-numbering system). (DBM) Distal, buccal, and mesial surfaces; (DLM) Distal, lingual, and mesial surfaces. (Modified from Lindhe et al, 1993 with permission.)

Fig 205 Percentage of sites with bleeding on gentle probing (GI score 2) at the 6-month examination in the triclosan group *(green)* compared to the control group *(red)* (FDI tooth-numbering system). (DBM) Distal, buccal, and mesial surfaces; (DLM) Distal, lingual, and mesial surfaces. (Modified from Lindhe et al, 1993 with permission.)

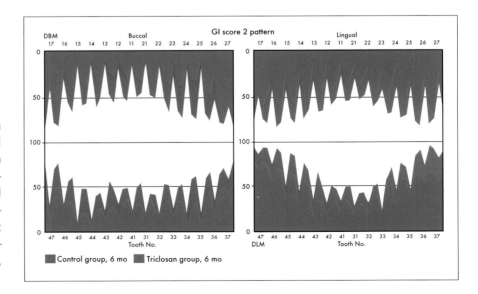

Table 6 Caries-inhibiting efficacy of triclosan-copolymer toothpaste in long-term clinical studies

Study	Location	No. of subjects*	Duration (mo)	Clinical design	Caries efficacy[†] Positive control toothpaste DFS	DFT	Triclosan-copolymer toothpaste DFS	DFT
Feller et al (1993)	USA	1,636	26	Parallel	1.64	0.53	1.55	0.48
Mann et al (1993)	Israel	1,285	26	Parallel	3.84	0.95	3.72	0.90

*Number of subjects in both the triclosan-copolymer toothpaste group and the placebo toothpaste group.
[†]Statistical analysis of the 26-month decayed or filled surface (DFS) and decayed or filled teeth (DFT) caries increments indicated that the triclosan-copolymer toothpaste provided anticaries efficacy equivalent to that of the positive control (clinically proven NaF-silica toothpaste).

tics Guideline for Caries Clinical Trials: Superiority and Equivalency Claims for Anticaries Dentifrices (American Dental Association, 1988c). The results showed equivalent anticaries efficacy for the toothpastes containing 0.3% triclosan and 2.0% copolymer, in either a 0.243% or 0.331% NaF-silica base, and the positive control dentifrice (Table 6).

The study by Feller et al (1993), conducted in the United States, involved 1,636 male and female adult subjects. The test toothpaste contained 0.3% triclosan and 2.0% copolymer in a 0.243% NaF-silica base. After 26 months, caries increments for subjects using the test toothpaste were 1.55 for decayed or filled surfaces and 0.48 for decayed or filled teeth. For the group using the positive control dentifrice, the corresponding values were 1.64 and 0.53, respectively.

The study by Mann et al (1993) was conducted in Israel and involved 1,285 male and female adult subjects. The test toothpaste contained 0.3% triclosan and 2.0% copolymer in a 0.331% NaF-silica base. After 26 months, caries increments for the test group were 3.72 for decayed or filled surfaces and 0.90 for decayed or filled teeth. The corresponding caries values for the control group were 3.84 and 0.95, respectively.

Rosling et al (1997b) evaluated the long-term effects of *(1)* meticulous, self-performed, supragingival plaque control and *(2)* the use of a triclosan-copolymer–containing dentifrice on the subgingival microflora in adult subjects susceptible to destructive periodontitis. Forty individuals were recruited for the trial. They had all been treated, by nonsurgical means, for advanced periodontal disease 3 to 5 years prior to the baseline examination. During the subsequent maintenance phase, all subjects had, at different time intervals, exhibited sites with recurrent periodontitis.

At a baseline examination, six surfaces per tooth were examined for bleeding on probing, probing depth, and probing attachment level. The deepest probing site in each quadrant (ie, four sites per subject) was selected, and samples of the subgingival bacteria were taken. At baseline, all volunteers received detailed information on proper oral hygiene techniques. This information was repeated on an individual needs-related basis during the course of the subsequent 36 months.

No professional subgingival therapy was delivered between the baseline and the 36-month examinations. The subjects were randomly distributed into two equal groups of 20 individuals each; one was a test group and the other a control group. The members of the test group were supplied with a fluoridated dentifrice containing triclosan-copolymer (Colgate Total), while the controls received a similar dentifrice without triclosan-copolymer.

In subjects with advanced and recurrent periodontitis, carefully practiced supragingival plaque control had some effect on the subgingi-

Fig 206 Total viable count (TVC × 10⁶) of microorganisms sampled at baseline and after 36 months of study in a test group (using a fluoridated dentifrice containing triclosan-copolymer) and a control group (using a similar dentifrice without triclosan-copolymer). At baseline, the TVC values varied between 15 × 10⁶ and 18 × 10⁶. At the final examination, the corresponding TVC values were 12 × 10⁶ (control) and 7 × 10⁶ (test). The reduction of the TVC value was statistically significant (*$P < .05$) only in the test group. (NS) Not significant. (From Rosling et al, 1997b. Reprinted with permission.)

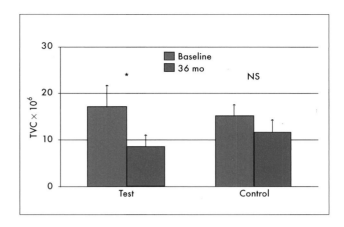

val microbiota, but it was insufficient to prevent progression of disease. In the corresponding group of subjects who used a triclosan-copolymer dentifrice, the subgingival microbiota was reduced in both quantitative and qualitative terms, and recurrent periodontitis was almost entirely prevented (Fig 206). It may be speculated that the gradually reduced pocket depth, because of the earlier discussed combination of antiplaque and anti-inflammatory effects of triclosan, resulted in not only a reduction of the total viable subgingival microorganisms but also a significant shift from anaerobic periopathogens, particularly *Prevotella intermedia* and *Porphyromonas gingivalis*, toward facultatively anaerobic organisms (Rosling et al, 1997b).

Rosling et al (1997a) also presented data from a 3-year clinical study that was performed to determine if triclosan and a copolymer, incorporated in a dentifrice and used by periodontitis-susceptible subjects, could influence clinical symptoms characteristic of recurrent periodontitis. Sixty subjects were recruited for the study. They were randomly selected from a group of patients previously treated for advanced periodontal disease. This treatment had included oral hygiene instruction and subgingival debridement but no surgical therapy. During a 3- to 5-year period following active therapy, the patients had been enrolled in a maintenance care program but had, at various intervals, exhibited signs of recurrent periodontitis.

The patients were stratified into two balanced groups with respect to mean probing depth. The test group included 30 individuals who used a dentifrice containing triclosan-copolymer-fluoride, ie, 0.3% triclosan, 2% copolymer, and 1,100 ppm of fluoride from 0.243% NaF (Colgate Total). The control group included 30 subjects who used a dentifrice that was identical to the one used in the test group but without the triclosan-copolymer content.

Following the baseline examination, including clinical and radiographic assessments, all volunteers received detailed information on how to brush their teeth in a proper way. This information was repeated on an individual needs-related basis during the course of the subsequent 36 months. No professional subgingival therapy was delivered between the baseline and the 36-month examination, but the subjects were recalled every 3 months. Reexaminations were performed after 6, 12, 24, and 36 months of the trial. A second set of radiographs was obtained at the 36-month final examination.

In subjects susceptible to periodontal disease, meticulous, self-performed, supragingival plaque control maintained over a 3-year period failed to prevent recurrent periodontitis. In a similar group of subjects, following the same plaque control program, the daily use of a triclosan-containing dentifrice reduced both the frequency of deep periodontal pockets and the number of sites that

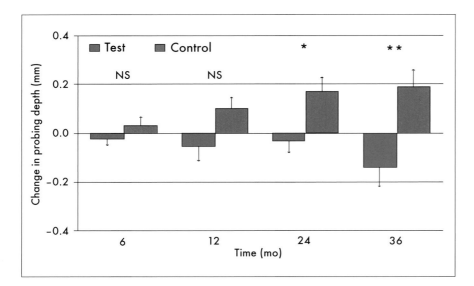

Fig 207 Mean changes in probing depth during a 3-year trial of a triclosan-copolymer toothpaste. In the control group there was a gradual increase in the mean probing depth, while in the test group there was a slight decrease. The difference in changes in probing depth between groups was statistically significant at 24 (*P < .05) and 36 (**P < .01) months. (NS) Not significant. (From Rosling et al, 1997a. Reprinted with permission.)

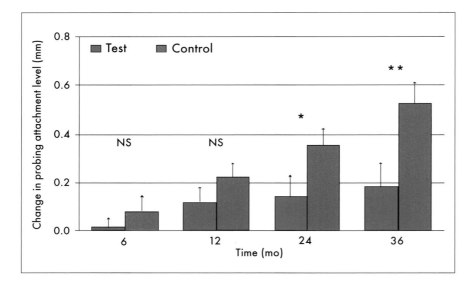

Fig 208 Mean changes in probing attachment level during a 3-year trial of a triclosan-copolymer toothpaste. In the control group, there was a significant further loss of attachment during the 36-month interval. There was also some additional attachment loss in the test group. The difference in additional attachment loss between the test and control groups was statistically significant after 24 (*P < .05) and 36 (**P < .01) months. (NS) Not significant. (From Rosling et al, 1997a. Reprinted with permission.)

exhibited additional probing attachment and bone loss (Figs 207 and 208). The gradually reduced probing depths and thus the retarded attachment loss in the test group most probably are the result of the combined antiplaque and anti-inflammatory effects of triclosan (Rosling et al, 1997a).

In a recent double-blind 5-year study on more than 500 adults, it also was shown that daily use of triclosan copolymer toothpaste resulted in a significant reduction of pocket depths of 3.5 mm or more compared to a placebo toothpaste (Cullinan et al, 2003).

The following conclusions are drawn from a review of the results of more than 15 independent, double-blind, long-term clinical studies, ranging from 3 to 60 months, and involving more than 4,000 male and female subjects in different geographic areas of the world:

1. Compared to a placebo toothpaste, a 0.3% triclosan-copolymer toothpaste in a 0.243% NaF-silica base provides statistically significant (P < .01) and clinically important effects on supragingival plaque, gingivitis, and supragingival calculus.

2. A 0.3% triclosan and 2.0% copolymer toothpaste, in either a 0.243% or 0.331% NaF-silica base, provides statistically significant ($P < .01$) and clinically important anticaries efficacy, equivalent to that of a clinically proven NaF-silica positive control dentifrice.

3. Microbiologic monitoring of the oral microflora, in association with two studies of the effect of triclosan on plaque and gingivitis, indicates that the long-term use (up to 1 year) of a toothpaste containing 0.3% triclosan and 2.0% copolymer in a 0.243% NaF-silica base does not cause the development of pathogenic, opportunistic, or resistant oral microorganisms.

4. Long-term (3-year) daily use of a 0.3% triclosan and 2% copolymer toothpaste in periodontitis-susceptible patients results in reduced probing depths, reduced subgingival microflora, a smaller percentage of periopathogens, and less probing attachment loss than does the use of a placebo fluoride toothpaste.

Triclosan plus zinc citrate. An alternative approach that should improve the antiplaque and antigingivitis effect of triclosan has been to combine zinc and triclosan. This combination is advantageous in toothpaste formulations because of the low toxicity and few reported side effects of the agents when applied in therapeutic doses (Scheie, 1989). In addition, because of its nonionic character, triclosan may be expected to be compatible with other desirable ingredients in toothpastes (Van der Ouderaa and Cummins, 1989).

The potential of the zinc ion as a plaque inhibitor has been demonstrated in several clinical mouthrinse studies using aqueous solutions (Harrap et al, 1984). Zinc is highly substantive and is retained in the mouth for even longer periods than triclosan. Many commercially available toothpastes contain zinc citrate, and the effects have been evaluated in a number of studies (Jones et al, 1988a, 1990; Saxton, 1986, 1989a, 1989b; Saxton et al, 1987, 1988; Stephen et al, 1990b; Svatun et al, 1989a). A general conclusion is that zinc has a moderate inhibitory effect on plaque formation,

most notably in individuals who are heavy plaque formers (Saxton et al, 1987). The effect of zinc ions on gingivitis has not yet been reported.

Additive or synergistic effects have been proposed for a combination of zinc salts and antimicrobial agents and/or surfactants (Giertsen et al, 1987, 1989a, 1989b; Saxer and Mühlemann, 1983), indicating different or complementary modes of action. Studies in human volunteers using the triclosan–zinc citrate toothpaste over a 7-month period have found neither adverse effects on the composition of the normal oral microflora nor any evidence of development of bacterial resistance to triclosan (Jones et al, 1988a). A combination of two antimicrobial agents has the potential advantage of additive or synergistic interactions, resulting in greater clinical benefits than are derived from relatively low concentrations of either agent (Van der Ouderaa and Cummins, 1989).

Successful antimicrobial agents may exert beneficial inhibitory effects at sublethal concentrations, and this may be important for topical delivery of antimicrobial agents from dental products: In the mouth, such agents may be available at high concentrations for relatively short periods. Zinc salts and triclosan at low levels inhibit acid production by *S mutans*, and the trypsinlike protease of the periodontal pathogen *P gingivalis* (Cummins, 1991; Giertsen et al, 1988). An important observation in these studies was the additional inhibitory effect of zinc and triclosan in combination. Additional antiplaque effects and gingival health improvements were achieved by a toothpaste containing both agents compared with similar products formulated with either agent alone (Gjermo and Saxton, 1991; Saxton, 1986).

In an experimental gingivitis model in human volunteers, subjects using a toothpaste slurry containing triclosan and zinc citrate had significantly less plaque and gingivitis than did those using a control slurry without antimicrobial agents. The test paste selectively suppressed the rise in obligately anaerobic bacteria found in the plaque of subjects using the control paste (Jones et al, 1990).

In a 1-year unsupervised clinical trial, more than 100 test subjects used either a toothpaste containing 1% zinc citrate and 0.2% triclosan or a placebo. All subjects had initially good oral hygiene and relatively low levels of gingivitis (Svatun et al, 1987, 1989a). After a prestudy period with PMTC, motivation, and oral hygiene instruction, the plaque and gingival scores at baseline approached 0. Six months later, 93% of the control group but only 40% of the test group had more than 5% bleeding gingival units. During these 6 months, the use of specific interdental cleaning aids such as toothpicks or floss was not allowed.

In the final 6-month period of the study, the participants who had habitually used interdental cleaning devices before the study were allowed to resume their use. The results after 1 year showed essentially no changes from the 6-month values among the test group or among non-flossers in the placebo group. However, the GI values of those in the placebo group who had used interdental cleaning aids regulary during the last 6 months of the study approached those of the experimental group. The authors concluded that, in subjects with good oral hygiene, regular use of the test toothpaste could compensate for lack of mechanical interdental cleaning (Svatun et al, 1987, 1989a).

Several studies have tested reduced concentrations of zinc citrate combined with triclosan. A toothpaste containing 0.5% zinc citrate and 0.2% triclosan has shown promising results in the experimental gingivitis model (Jones et al, 1990; Saxton et al, 1988; Saxton, 1989a; Saxton and van der Ouderaa, 1989). In a 6-month clinical trial in 150 subjects, plaque and gingivitis levels remained approximately 50% below those in the control group (Saxton, 1989b).

These results were confirmed in an unsupervised 6-month study (Svatun et al, 1990) that followed the same protocol and included test subjects similar to those in the previous investigations of the higher zinc citrate toothpaste (Svatun et al, 1987, 1989a). The results of the study by Svatun et al (1990) in relatively young subjects with good oral hygiene encouraged further investigation in which the toothpaste was used by a wider age range of adults with poorer oral hygiene (Stephen et al, 1990b). A similar result was observed; there was approximately a 50% difference between test and control groups, indicating that the toothpaste is also effective in older people with higher initial levels of disease.

Another long-term study was published by Svatun et al (1993a). A double-blind, 7-month parallel clinical study was conducted to compare the effect of toothpaste containing 0.3% triclosan and 0.75% zinc citrate in a 0.8% sodium monofluorophosphate–silica base and toothpaste containing 0.8% monofluorophosphate-silica on supragingival plaque, gingivitis, and supragingival calculus formation. Compared to the control toothpaste, the triclosan–zinc citrate toothpaste resulted in statistically significant reductions of 28% in supragingival plaque formation, 50% in gingival bleeding, and 59% in supragingival calculus formation after the first month of the trial. During the subsequent 6-month period, 72% of the subjects using the triclosan–zinc citrate toothpaste maintained or improved the gingival status achieved in the first month of the trial. In contrast, only 17% of the control group achieved this goal. Thus, compared to a control toothpaste, use of the triclosan–zinc citrate toothpaste over a 7-month period provided statistically significant and clinically relevant benefits in controlling gingivitis. Levels of supragingival plaque and calculus were also significantly reduced.

To date, four medium-term studies of 6 months or more have been carried out according to the American Dental Association's guidelines for the conduct of plaque and gingivitis studies (Stephen et al, 1990b; Svatun et al, 1989a, 1990; Svatun et al, 1993a). The effect on plaque ecology of both short-term (21-day) and long-term (6-month) exposure to the combination of zinc and triclosan in a toothpaste has been monitored microbiologically (Jones et al, 1988a, 1990; Stephen et al, 1990b). There were no indications of gross imbalance of the plaque microflora. In the 21-day study, the proportion of anaerobic organisms increased in the placebo group during the non-brushing period, reflecting the successional changes in the flora during plaque maturation. Complex plaque did not develop in the test group; the plaque remained immature in composition.

In the 6-month study by Stephen et al (1990b), plaque ecology was studied according to the American Dental Association's guidelines for studying plaque and gingivitis. Plaque was sampled at the start of the study, at intervals during the experimental period, and 3 months after completion of the testing. Total aerobes, streptococci, lactobacilli, *Veillonella*, and *Actinomyces* were monitored throughout the study. No significant shift in the ecological balance occurred with respect to time or between treatment groups. None of the participants in the zinc-triclosan group developed detectable levels of triclosan-resistant bacteria. The only side effects observed among the 450 participants using the zinc-triclosan toothpaste for longer than 6 months were two cases of tooth staining. A similar lack of adverse reactions was recorded for the control groups. Furthermore, there have been no reports of subjective side effects in participants in clinical trials who use toothpaste containing combinations of zinc salts and triclosan over prolonged periods of time (Saxton, 1989b; Svatun et al, 1987, 1989a, 1990).

The caries-preventive effect of toothpastes containing triclosan and zinc citrate has been evaluated by ten Cate (1993) in vitro and in situ. A comparison was made among a nonfluoride (negative) control toothpaste, a fluoride (positive) control toothpaste (1,000 ppm of fluoride as sodium monofluorophosphate), and a test toothpaste containing 1,000 ppm of fluoride (as monofluorophosphate), 0.3% triclosan, and 0.75% zinc citrate trihydrate. The two fluoride pastes were superior to the negative control in all tests. No differences were observed between the two fluoride pastes in the in vitro studies. Under conditions of cariogenic challenge, in situ caries progression was inhibited by the fluoride test paste containing triclosan and zinc citrate trihydrate; a significantly better result than that reported for the fluoride control paste, which was associated with an increase in lesion severity.

Triclosan plus pyrophosphate. Pyrophosphate has been added to toothpastes as an anticalculus agent. Studies have shed light on the possible mode of action of this compound (White, 1991). Pyrophosphate diffuses into mineralizing plaque, where it adsorbs to developing crystallites, retarding the development of a discrete calculus layer with structural integrity. Plaque mineralization, or petrification, is delayed, allowing removal of the softer deposits during routine oral hygiene procedures. Thus, the long-term use of an inhibitor of crystal growth reduces the amounts of calculus rather than directly affecting the mineral content of plaque (for review see White, 1992).

Pyrophosphate also has some limited antimicrobial properties. There are few data on the in vitro or in vivo antimicrobial effects of combinations of triclosan and pyrophosphate. Pyrophosphate, however, has low substantivity, and this may restrict its antimicrobial effect in the mouth.

Long-term studies comparing the effects of different triclosan-containing toothpastes. A double-blind, 7-month clinical study was conducted by Svatun et al (1993b) to determine the effects of three triclosan-containing test toothpastes on supragingival plaque, gingivitis, and supragingival calculus formation. A 0.8% monofluorophosphate-silica toothpaste was used as a control. Each test toothpaste contained 0.3% triclosan, which was combined with 0.75% zinc citrate, 2% Gantrez copolymer, or 5% pyrophosphate. Subjects were assigned to one of the three test groups or to the control group by random allocation within nine strata. Subjects were evaluated for supragingival plaque and calculus formation and gingivitis after 1-, 4-, and 7-month use of the toothpaste.

After 7 months, use of the triclosan–Gantrez copolymer and the triclosan-pyrophosphate toothpastes each resulted in a statistically significant reduction of approximately 25% in gingival bleeding compared to the control toothpaste. Neither toothpaste resulted in statistically significant reductions in supragingival plaque or calculus formation. In contrast, the triclosan–zinc citrate toothpaste provided statistically significant reductions of 33% in supragingival plaque, 51% in gingival bleeding, and 67% in supragingival calculus formation. The reductions in gingival bleeding and calculus formation were statistically superior to those of the other two test products. Over a 7-month period, triclosan–zinc citrate toothpaste provided statistically significant and clinically relevant reductions in supragingival plaque and calculus formation as well as control of gingivitis (Svatun et al, 1993b).

Renvert and Birkhed (1995) compared the effects of three triclosan-containing toothpastes in a 6-month, unsupervised toothbrushing study. The effects on plaque, gingival bleeding, and certain salivary microorganisms (MS, lactobacilli, total counts of streptococci, and total counts of microorganisms) were evaluated. A total of 123 subjects were divided into four groups according to the severity of their GBI. Of these, 112 subjects completed the study. Following a 4-week preexperimental period of using a sodium monofluorophosphate toothpaste (placebo), the subjects were issued one of three triclosan-containing toothpastes available in Sweden—Colgate Paradent (0.3% triclosan-copolymer) (n = 26); Pepsodent Gum Health (0.2% triclosan–zinc citrate) (n = 31); Dentosal Friskt Tandkött (0.3% triclosan-pyrophosphate) (n = 28)—or continued with the placebo (n = 27).

Colgate Paradent reduced baseline plaque values by 39% (Quigley-Hein Index) during the 6-month experimental period; the corresponding values for the other products were a reduction of 6% for Pepsodent Gum Health, an increase of 5% for Dentosal Friskt Tandkött, and an increase of 2% for the placebo. There were statistically significant differences in plaque levels ($P < .05$) between the Colgate Paradent and Pepsodent Gum Health groups, and between the Colgate Paradent and placebo groups. The GBI had improved in all four groups. A significant difference ($P < .05$) with respect to bleeding was found between the Colgate Paradent and placebo groups ($P < .05$) at the 3-month registration. A statistically significant increase over time in total number of streptococci and total colony-forming units (CFUs) was noted

in the Dentosal, Pepsodent, and placebo groups but not in the Colgate group. This study seems to verify that a toothpaste containing a combination of triclosan and copolymer effectively reduces supragingival plaque formation and gingival bleeding without causing major shifts in the salivary microflora (Renvert and Birkhed, 1995).

However, both these studies together show that toothpastes containing triclosan plus copolymer as well as triclosan plus zinc citrate result in significantly more plaque and gingivitis reduction compared with placebo toothpastes. However, Nogueira-Fiho et al (2000) recently showed in a partial-mouth experimental model of gingivitis that an experimental triclosan toothpaste containing copolymer (PVM/MA), zinc oxide, and pyrophosphate significantly reduced plaque (PI), gingivitis (GI), and gingival bleeding compared to a placebo toothpaste, in contrast to two commercial triclosan toothpastes containing either copolymer or zinc oxide. In another recent study, Jannesson et al (2002) showed that supplementation of 10% xylitol to a commercial triclosan and copolymer toothpaste significantly reduced salivary MS and dental plaque compared to the original paste and a control toothpaste without triclosan (for review, see Adriaens and Gjermo, 1997 and Volpe et al, 1996).

Mouthrinses

By far the most frequently tested vehicle for chemical plaque control by self-care is the mouthrinse, in use for centuries as breath fresheners, medicaments, and antiseptics. One of the oldest still in use today is Listerine, a combination of essential oils and phenolic compounds. This formulation, derived from Lister's original work with carbolic acid, has been used since the end of the 19th century. Only in more recent times, however, have mouthrinses been given much credence as preventive agents against dental diseases. An increasing awareness of the role of plaque in the initiation of caries and periodontal diseases has changed the attitude of dental professionals and caused an upsurge in the search for antiplaque agents (Mandel, 1988).

A chemical agent that will act as an adjunct or alternative to mechanical cleaning has long been sought. Chlorhexidine was accepted enthusiastically by the profession, and although it has lived up to its early promise, the major disadvantages of staining and unpleasant taste have provided an incentive to continue the search for equally efficient agents that lack side effects. So far, none has been found.

Mouthrinses are the simplest vehicle for chemical plaque control agents and are usually a mixture of the active component in water and alcohol, in addition to a surfactant, a humectant, and flavor. Most chemical plaque control agents are compatible with this vehicle. Figure 209 shows an assortment of commercially available mouthrinses.

Factors influencing the effectiveness of mouthrinsing agents

The effective use of active ingredients in mouthrinses is dependent on several factors. Normally, a therapeutic mouthrinse contains an active ingredient or drug that must be dissolved in the formulation. Mouthrinses currently on the market are aqueous-based formulations but contain numerous other ingredients that must be compatible with the drug. The potential for undesirable interactions among ingredients is a major concern in formulation and manufacture. Some interactions are specially designed, such as the increased solubility of poorly water-soluble drugs (for example, triclosan) by adding surfactants and other ingredients to form a microemulsion. However, incompatible ingredients are sometimes inadvertently included, especially in complex formulations, where the chemistry is not completely understood.

The packaging material can also be a source of incompatibility. Any number of possible interactions can affect, either directly or indirectly, the

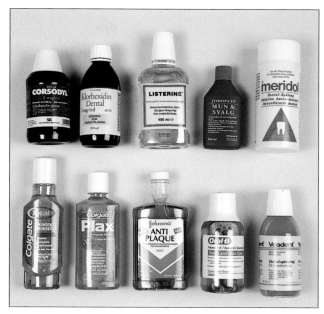

Fig 209 Commercially available mouthrinses: Corsodyl (0.2% CHX) (GlaxoSmithKline, Brentford, Middlesex, UK); Hibitane Dental (0.1% CHX) (ACO, Stockholm, Sweden); Listerine (essential oils: 0.1% eucalyptol, 0.04% menthol, thymol, 0.6% SnF_2, etc); Jodopax (iodine) (Cederroth International, Upplands-Väsby, Sweden); Meridol (AmF, SnF_2 [0.025% F]); Colgate Total (0.03% triclosan, 0.25% copolymer, 0.025% NaF); Colgate Plax (0.03% triclosan, 0.25% copolymer, 0.025% NaF); Johnson & Johnson Anti Plaque (0.05% cetylpyridinium chloride monohydrate, 0.05% NaF) (Johnson & Johnson, New Brunswick, NJ); Oral-B Tooth and Gum Care (0.05% cetylpyridinium chloride monohydrate, 0.05% NaF); Viadent (0.03% *Sanguinaria canadensis*, zinc citrate) (Colgate-Palmolive).

availability of the drug in the formulation. This can usually be evaluated in the laboratory on new and aged samples of the product.

Drugs that are complexed with other materials, although still soluble in the formulation, may exhibit reduced bioavailability in vivo, as discussed earlier in this chapter. The term *bioavailability* is usually used to express a temporal relationship of free drug concentration at the target site. In this case, after mouthrinsing, the bioavailability is the concentration of free drug in the environment of the target site and the rapidity with which it disappears. This can be determined, provided that the site can be sampled and the drug concentration can be measured in the medium contacting the target site (ie, saliva, plaque fluid, or crevicular fluid).

The duration of exposure to the mouthrinse may be important: Because most of the dose in the mouthrinse is spat out, the subject should rinse long enough for optimal retention of the drug. This optimal time period has been determined for some orally used antiseptics, such as CHX and sanguinaria extract. In general, 30 to 45 seconds is sufficient.

Once introduced into the oral environment via a mouthrinse, the residual drug must diffuse in saliva before it can reach its intended sites of action. In saliva, the drug is free to interact with salivary components before reaching oral surfaces. In theory, only the free soluble drug can interact optimally with target sites. Such sites include plaque, enamel, the gingival sulcus, gingival tissue, and the mucous membranes.

The amount of drug retained on oral surfaces after rinsing is also thought to be important, because subsequent desorption of the drug into microenvironment of the target site could provide a sustained effect. This will be determined mainly by the substantivity of the particular drug used. Because of the long interval between doses (mouthrinses are usually used only once or twice a day), highly substantive drugs may have a distinct advantage. Superimposed upon this is the normal clearance process by which materials are removed from oral surfaces by salivary flow (Dawes 1983, 1987; Dawes and Weatherell, 1990). The longer a drug can be retained in the environment of the target site in active form, the greater the potential for a therapeutic effect.

Chemical plaque control by mouthrinsing also depends on patient compliance and several other factors related to the drug and formulation used, including the intrinsic antimicrobial activity of the drug; its concentration and availability in the rinse; and its substantivity, bioavailability, and activity in saliva and plaque.

To be included in a mouthrinse formulation, the drug must be soluble in an aqueous vehicle.

Poorly water-soluble drugs, such as triclosan, can be solubilized with surfactants. However, care must be taken not to affect the availability of free drug in the formulation or its ability to be delivered to oral surfaces. The choice of surfactant can also be critical. Surfactants promote the wetting of surfaces, facilitating penetration of drugs to interproximal areas and into the plaque matrix. Anionic surfactants are generally not compatible with cationic antibacterials such as CHX and sanguinaria extract. Triclosan, in contrast, is not compatible with nonionic surfactants (Furia and Schenkel, 1968).

The pH can also be an important factor. It has been shown that the pH can influence the intraoral retention of cationic antibacterials such as CHX, possibly by affecting the charge of surfaces in the oral cavity (Bonesvoll et al, 1974b). The pH can also affect the charge on some drug molecules themselves (eg, sanguinaria extract and triclosan) and thus influence their water solubility or ability to penetrate cell membranes. It is obvious from these examples that a broad range of formulation factors must be considered during the development and testing of a mouthrinse formulation.

The combination of usage habits and the rapid clearance of drugs from the oral cavity compromises the efficacy of antimicrobial mouthrinses. Mouthrinses are usually used once or twice a day, morning and evening. Thus, the interval between doses may span 12 hours or more, whereas measurements of drug levels in saliva after topical application via mouthrinses and dentifrices show that drugs are cleared rapidly, usually with half-lives of less than 1 hour (Afflitto et al, 1989). Therapeutically effective drug levels in saliva are therefore fleeting.

Early work with CHX indicated that it had considerable affinity for oral surfaces that might explain its efficacy (Hjelfjord et al, 1978; Rölla and Melsen, 1975b; Rölla et al, 1970, 1971). Studies showing an apparent correlation between the degree of intraoral retention of CHX and its clinical effectiveness led to the conclusion that substantivity in the oral cavity was important (Bonesvoll

Table 7 Intraoral retention values and salivary half-lives of antimicrobial mouthrinses*

Agent	Intraoral retention	Salivary half-life
Chlorhexidine (0.12%)	30%	1 h
Triclosan (0.03%)	35%	30 min
Sanguinaria extract (0.03%)	19%	< 15 min

*Data from Bonesvoll et al (1974a); for review, see Gaffar et al (1997).

Table 8 Postrinsing concentration of antimicrobial agents in dental plaque*

Agent	Amount in plaque 1 h postrinsing (μg wet weight ± 50)
Chlorhexidine (0.12%)	467 ± 5.0
Triclosan (0.03%)	121 ± 35
Sanguinaria extract (0.03%)	74 ± 24

*For review, see Gaffar et al (1997).

and Gjermo, 1978; Gjermo, 1974; Gjermo et al, 1974, 1975; Wåler and Rölla, 1985; Wåler, 1989). This experience with CHX has been extrapolated to apply to antimicrobial agents in general: A high degree of substantivity is considered to be an advantage. More drug will be retained on oral surfaces, at concentrations exceeding the minimum inhibitory concentration, for longer periods of time.

Table 7 summarizes intraoral retention data reported for some of the commonly used antiplaque agents. The salivary half-life data, in many cases, may reflect intraoral retention, because it may be assumed that at least some of the retained drug will be desorbed and thus eliminated in saliva. Although other factors can affect this, such as salivary flow rate and loss of retained drug via systemic absorption, it is a property worth documenting. Some differences are apparent among antimicrobial agents. A highly effective antiplaque agent has a high degree of intraoral retention (30%) and a relatively long half-life in saliva. The avid binding of CHX to salivary proteins, however, tends to inactivate it in saliva. These data indicate that CHX is well retained on oral surfaces but is then slowly desorbed, probably in active form, for longer periods than some of the other agents. A combination of various oral reservoirs conceivably could contribute to the prolonged effect of CHX (Wåler, 1989).

Regardless of the mechanism, it is certain that the actual concentration of an antimicrobial agent in plaque following mouthrinsing is important. Table 8 lists the concentrations of antimicrobial agents in plaque 1 hour after the use of three different commercial antiplaque mouthrinses: 0.12% CHX, 0.03% triclosan, and 0.03% sanguinaria extract. All three agents were found to exceed their minimum inhibitory concentration for plaque microorganisms, but most striking was the high concentration of CHX. This was partially a result of the relatively higher concentration of CHX in the mouthrinse formulation. In practice, the concentration of these agents in mouthrinses is limited, either because of adverse side effects (as with CHX) or because of undesirable effects on taste perception or on the flavor of the mouthrinse. Nevertheless, at the concentrations currently used in commercial mouthrinses, all of these agents can inhibit, to varying degrees, the formation of dental plaque.

Another factor that will influence the effect of chemical plaque control by mouthrinses is accessibility (Axelsson et al, 1993d, 1994b; Caton et al, 1993; Jenkins et al, 1993a; Ramberg et al, 1992). A chemical plaque control agent delivered as a mouthrinse will not be equally distributed to all tooth areas. As shown by Jenkins et al (1993a), a significant effect may occur only at certain areas in a dentition with normal height of periodontal support, ie, on the buccal aspects of the mandibu-

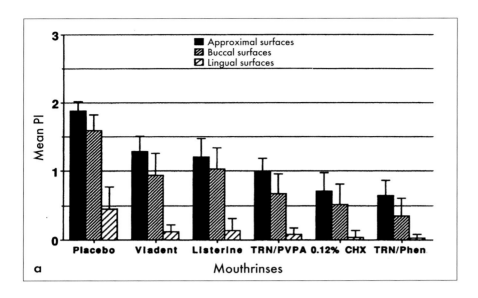

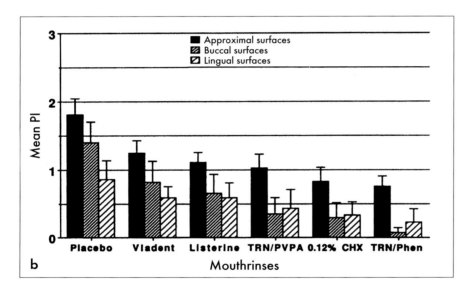

Fig 210 Effects of various mouthrinses on 4-day de novo plaque formation on specific tooth surfaces in the maxilla (a) and the mandible (b). (TRN/PVPA) 0.06% Triclosan plus polyvinyl phosphonic acid; (TRN/Phen) 0.06% Triclosan plus phenolic flavor. (From Ramberg et al, 1992. Reprinted with permission.)

lar teeth and on the posterior lingual surfaces. Mouthrinses may not be particulary effective on interdental gingivitis. Caton et al (1993), studying subjects with gingivitis, compared the effect of 3-month use of a CHX mouthrinse, wooden interdental toothpicks, and a negative control on bleeding interproximal sites. Daily use of the toothpicks reduced the number of interdental bleeding sites by 90%, but gingival conditions in the CHX group and the negative control group showed an improvement of only 15%.

Following surgical periodontal therapy, however, CHX rinsing provides healing conditions almost similar to those obtained through PMTC programs (Westfelt et al, 1983a). This difference in efficacy is probably attributable to the fact that patients with periodontal destruction have open interdental sites, allowing accessibility to the approximal tooth surfaces during mouthrinsing.

Ramberg et al (1992) evaluated the effects of Viadent (sanguinaria extract); Listerine; 0.06% triclosan plus polyvinyl phosphonic acid; 0.06% tri-

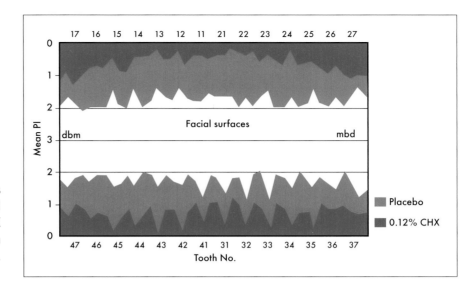

Fig 211 Mean PI of facial surfaces (distal [d], buccal [b], and mesial [m]) on day 4 of using 0.12% CHX and placebo mouthrinses twice a day. (From Ramberg et al, 1992. Reprinted with permission.)

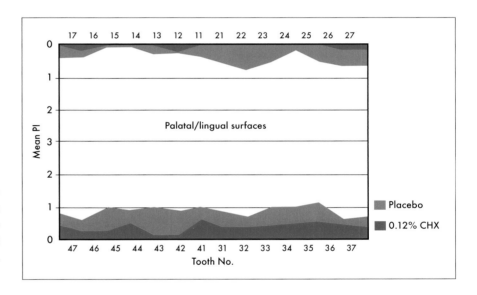

Fig 212 Mean PI of palatal and lingual surfaces on day 4 of using 0.12% CHX and placebo mouthrinses twice a day. (From Ramberg et al, 1992. Reprinted with permission.)

closan plus phenolic flavor; 0.12% CHX (positive control); and placebo mouthrinses twice a day on 4-day de novo plaque formation. All mouthrinse solutions had the greatest inhibiting effect on plaque on the lingual surfaces and a very limited effect on the approximal surfaces (Fig 210).

Figures 211 and 212 compare the effect of 0.12% CHX and a placebo on mean PI. Figures 213 and 214 show the pattern of disclosed plaque at baseline and after 4-month mouthrinsing twice a day with Meridol (AmF plus SnF$_2$) (Axelsson et al, 1994b).

Short- and long-term studies of mouthrinsing agents

Experimental models. In vitro and in vivo studies of mouthrinses containing chemical plaque control agents, especially CHX, are legion, and there are numerous recent reviews (Adams and Addy, 1994; Addy, 1986; Addy and Renton-Harper, 1997; Axelsson, 1993b; Giertsen, 1990; Hull, 1980; Kornman, 1986; Mandel, 1988; Ramberg, 1995a; Rölla et al, 1997; Scheie, 1989, 1994a; Wåler, 1989;

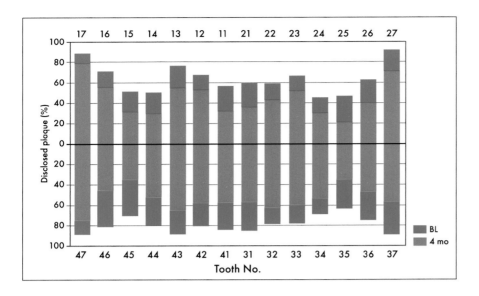

Fig 213 Effect of 4-month mouthrinsing twice a day with Meridol on PI of mesiobuccal surfaces. (BL) Baseline. (From Axelsson et al, 1994b. Reprinted with permission.)

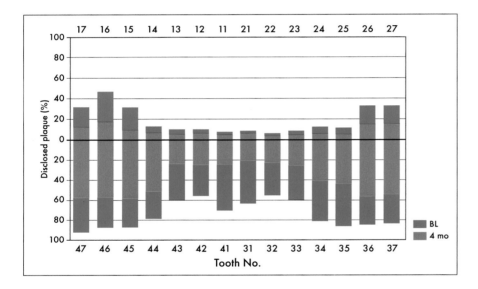

Fig 214 Effect of 4-month mouthrinsing twice a day with Meridol on PI of lingual surfaces. (BL) Baseline. (From Axelsson et al, 1994b. Reprinted with permission.)

Wennström, 1997). Before individual active ingredients are discussed, the plethora of experimental models for testing mouthrinses warrants some clarification. Clinical trials of mouthrinses take several forms, depending on the expected benefit of the active ingredients and, sometimes, the whim of the individual researchers. This makes comparisons among trials of different materials difficult, if not impossible. Trials may be designed to disclose effects on plaque removal or plaque formation, on inhibition or reduction of

gingivitis, and on reduction in calculus formation. Additionally, there are studies on the effects on the oral microflora in vitro and in vivo and on dental caries.

Plaque removal trials are mainly short term, in which plaque is scored before and after a single rinse. They are easily controlled because rinsing is supervised and have been used more recently to assess the plaque-removing effects of several rinses, including a prebrushing rinse (Binney et al, 1992). In plaque regrowth trials, the subjects

Box 10 American Dental Association guidelines for acceptance of chemotherapeutic products for the control of supragingival dental plaque and gingivitis

The 1986 American Dental Association Council on Dental Therapeutics guidelines for clinical studies of efficacy require the following criteria:

- Two independent studies should be conducted.
- The study populations should represent typical product users.
- The test product should be used in a normal regimen and compared to a placebo.
- The study design should be either parallel or crossover.
- Each study should be at least 6 months in duration.
- The plaque and gingivitis scoring procedure should be conducted at baseline, after 6 months, and at an intermediate period of time.
- Microbiologic profile should demonstrate that pathogenic or opportunistic microorganisms do not develop over the course of the study.

Box 11 American Dental Association guidelines for caries clinical trials: Superiority and equivalency claims for anticaries dentifrices

The 1988 American Dental Association Council on Dental Therapeutics guidelines for clinical studies of efficacy require the following criteria:

- Two independent studies should be conducted.
- The study population should represent typical product users.
- Each study should be at least 2 years in duration.
- The studies should have a baseline examination, an intermediate examination, and a final examination.
- The studies should be able to detect a 10% difference with at least 80% power between the clinically positive control product and the test formulation product.

are given a thorough PMTC at baseline, and plaque is allowed to accumulate over the short term, varying from 16 hours to several days, without any other oral hygiene measures. The 4-day model is most frequently used (Ramberg, 1995b). The aim is to assess the plaque-inhibiting ability of the agent on its own.

Studies on reduction of gingivitis are usually of longer duration than are plaque removal or plaque regrowth studies, although in most cases the amount of plaque present is assessed together with the gingivitis. The American Dental Association's guidelines on the use of trials to measure reduction of gingivitis and control of supragingival plaque stipulate a period of at least 6 months (Box 10).

However, shorter studies are still conducted according to the "experimental gingivitis" model; these may last from days to weeks. Trials may be unsupervised; ie, other oral hygiene measures are not controlled and subjects continue with their habitual techniques, using the test rinses as well. In other trials, oral hygiene measures may strictly fol-

low a set regimen. Most of those long-term studies are very dependent on subject compliance. Gingivitis is assessed with a gingival index, used with or without a bleeding on probing score (Hull, 1980).

Trials on the effect of an agent on calculus are of necessity long term, ie, 6 months or longer. Calculus is measured by a calculus index, usually the Volpe-Manhold Index (Volpe et al, 1967), which has gained almost universal acceptance. Most of these trials have concerned the effects of an agent on calculus only, but some have included assessments of plaque, gingivitis, and even staining.

In vitro or in situ experimental models are most frequently used to evaluate the caries-preventive effect of mouthrinses (ten Cate, 1993). For longitudinal clincal caries-preventive trials, however, the American Dental Association's guidelines for evaluation of dentifrices is recommended (Box 11).

Clinical trials use two main methods. One is the parallel study, in which control and test groups of subjects are matched as far as possible for age, sex, and periodontal disease status. Each group uses only the test or the control regimen. The other method is the crossover study, which has the advantage of requiring fewer subjects without loss of power to detect statistically significant differences. Each subject uses either the test or the control rinse, is assessed at the end of the set period, and, after a "washout period," uses the alternative formulation or procedure. Matching of subjects in groups is unnecessary, because every subject uses each procedure and acts as his or her own control.

Crossover studies typically use healthy volunteers and measure the inhibitory effects of agents on the development of plaque and gingivitis, alone or as adjuncts to toothcleaning. Parallel studies may be used similarly but are also used to study the action of agents in reducing plaque and gingivitis.

Another variant relates to the bias of the clinician or subject. Trials can be single blind, in which the clinician assessing the subject is unaware of which regimen the subject has used, or double blind, in which neither the subject nor the clinician is aware of the "treatment" until the end of the trial. Finally, studies are designed to test the action of a specific ingredient or ingredients or designed to evaluate a complete formulation or product. Controls will vary according to the type of study.

In this review, the main thrust will be to present studies in which each agent has been evaluated as the active ingredient in a mouthrinse, as well as studies of specific mouthrinse formulations commercially available as oral hygiene products. However, it is difficult to organize mouthrinse studies on the basis of classification of the agents as cationic, anionic, nonionic, or combined chemical plaque control agents. Because the outstanding antiplaque and antigingivitis effect of CHX is so well documented, it is regarded as the gold standard of mouthrinse efficacy and serves as a positive control in most recent studies of mouthrinses. Cationic, nonionic, anionic, or combination agents may be evaluated in the same study. Chemical plaque control agents that are commercially available as mouthrinse products with proven long-term effects on plaque and gingivitis will be considered more comprehensively.

Listerine. The mouthrinse with the longest history of use is the nonionic Listerine, a hydroalcohol solution of thymol, menthol, eucalyptol, and methyl salicylate. Since 1890, when Miller demonstrated that caries is caused by microorganisms (Miller, 1973), Listerine has been used as a mouthwash to try to prevent dental diseases. It has gained much popular acceptance, probably because of its "disinfectant" smell and taste and its ability to dispel odors and create a clean sensation in the mouth. Antiplaque activity has been demonstrated in several studies (reviewed by Ross et al, 1989), and Listerine is of sufficient standing to have recieved the approval of the American Dental Association (1988a).

There are several antimicrobial agents in the hydroalcohol base, and the mode of action of the active ingredient or ingredients in the mouthrinse has not been established. Phenolic compounds act by disrupting the cell wall and inhibiting bacterial enzymes. There is also some evidence that lipopolysaccharide-derived endotoxin may be extracted from gram-negative bacteria. In both short- and long-term studies, efficacy has been independent of the standard of oral hygiene.

Microbiologic studies have demonstrated no emergence of opportunistic, potential, or presumptive pathogens: The only adverse effects have been an initial burning sensation and bitter taste.

Short-term studies. Menaker et al (1979) evaluated the effect of Listerine mouthrinse on dental plaque in a 21-day double-blind, controlled clinical trial in adult subjects. Use of Listerine twice daily as a supplement to regular toothbrushing over a 21-day period significantly retarded the accumulation of dental plaque by approximately 43% compared to toothbrushing and rinsing with the control mouthwash.

A double-blind mouthrinse study evaluated the effects of 0.2% and 0.1% Hibitane (chlorhexidine digluconate) mouthrinse, Listerine, and a

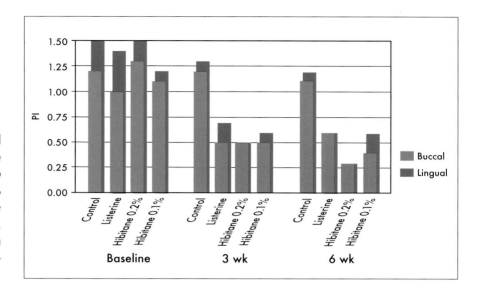

Fig 215 Mean PI scores for buccal and lingual surfaces at the baseline examination and after 3 and 6 weeks of twice-daily use of 0.2% and 0.1% Hibitane (chlorhexidine digluconate) mouthrinse, Listerine, and a placebo mouthrinse. (From Axelsson and Lindhe, 1987. Reprinted with permission.)

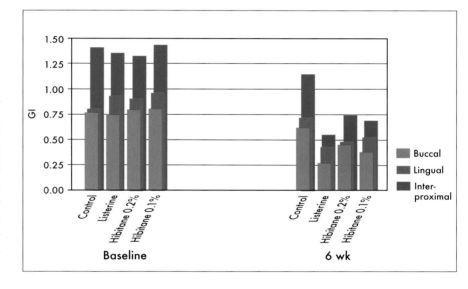

Fig 216 Mean GI scores for interproximal, buccal, and lingual units, calculated from measurements made at the baseline examination and after 6 weeks of twice-daily use of 0.2% and 0.1% Hibitane (chlorhexidine digluconate) mouthrinse, Listerine, and a placebo mouthrinse. (From Axelsson and Lindhe, 1987. Reprinted with permission.)

placebo mouthrinse twice a day for 6 weeks on dental plaque and gingivitis in young adults (Axelsson and Lindhe, 1987). The study showed that 0.2% Hibitane was the most efficient in plaque reduction (Fig 215), but Listerine and 0.1% Hibitane were just as efficient in the reduction of gingivitis as was 0.2% Hibitane (Fig 216) indicating that Listerine may have a specific anti-inflammatory effect. A higher extrinsic stain index and some cases of oral mucosal irritation were recorded in the 0.2% Hibitane group.

Brecx et al (1990) used a double-blind experimental gingivitis model to compare the antiplaque, antigingivitis, and antimicrobial efficacies of Listerine, Meridol, and CHX. Meridol is a mixture of AmF and SnF_2. After PMTC, 36 subjects rinsed twice daily for 21 days with one of the three rinses or a placebo control as their only oral hygiene measure. At the end of 3 weeks, plaque indices were lowest in the CHX group and much higher in the Listerine and Meridol groups, although not as high as in the placebo group. Listerine and Meridol were not significantly better than the control at inhibiting gingivitis, whereas CHX reduced gingivitis scores by 50% compared with the placebo. The authors used a vitality dye test on plaque bacteria to demonstrate that CHX had the greatest effect on the proportion of viable organisms, followed by Meridol; Listerine and the placebo had no significant effect compared to the baseline values.

In a later study, Brecx et al (1992) reported the effect of similar rinse formulations on plaque and gingivitis when used as adjuncts to normal tooth-cleaning. Two Meridol preparations were compared with Listerine, CHX, and a placebo. Plaque indices were lowest with CHX; one of the Meridol formulations and Listerine provided significant reductions in plaque compared with the placebo rinse. The gingivitis scores with the Listerine group were little better than those in the placebo group. The authors concluded that there were some advantages to combining Listerine and mechanical oral hygiene measures, which had not been used in the earlier study.

In a double-blind, controlled clinical study in 94 subjects, Ross et al (1993) evaluated the effect of rinsing time on the antiplaque and antigingivitis efficacy of Listerine mouthrinse. The effectiveness of 30- and 60-second Listerine rinses in both inhibiting the development of and reducing existing supragingival plaque and gingivitis was tested in an experimental gingivitis model. Following the baseline examinations, subjects recieved half-mouth PMTC and began twice-daily supervised rinsing either with Listerine for 30 or 60 seconds or with a control mouthrinse for 30 seconds as their sole oral hygiene measure.

Statistical analysis revealed that both the 30- and 60-second Listerine rinses were significantly ($P < .01$) more effective than the control in inhibiting and reducing plaque, gingivitis, and gingival bleeding. Although 60-second rinses with Listerine were significantly more effective ($P < .01$) than 30-second rinses in controlling plaque, the two rinse durations were similarly effective in controlling interdental bleeding and gingivitis. The results of this study confirmed that the rec-

ommended twice-daily rinsing with Listerine for 30 seconds is an effective regimen for gingivitis control (Ross et al, 1993).

Long-term studies. Lamster et al (1983) investigated the effect of Listerine mouthrinse on reduction of existing plaque and gingivitis. All 145 subjects entering the study had plaque scores greater than or equal to 1.8 on the Turesky modification of the Quigley-Hein Index and scores greater than or equal to 2.0 on the Modified Gingival Index. The clinical trial was double blind. No prophylaxis was performed. Subjects rinsed twice a day with either 20 mL of Listerine or its vehicle control. Plaque, gingivitis, and the condition of the soft tissues were evaluated at baseline and at 1, 3, and 6 months. The Listerine group showed a 20.8% reduction in plaque scores and a 27.7% reduction in gingivitis scores compared to subjects using the vehicle control.

In another 6-month double-blind study, De-Paola et al (1989) evaluated the effect of Listerine mouthrinse on inhibition of supragingival plaque and gingivitis: After a complete PMTC, 108 adult subjects rinsed for 30 seconds twice daily for 6 months with 20 mL of either Listerine or a hydroalcoholic control rinse. At 3 and 6 months, subjects were evaluated for soft tissue condition, extrinsic tooth stain, plaque surface area, and gingivitis. Listerine mouthrinse, as an adjunct to normal oral hygiene, significantly inhibited the development of both plaque and gingivitis by 34% compared to the hydroalcoholic control rinse.

Gordon et al (1985) evaluated the efficacy of Listerine mouthrinse in inhibiting the development of plaque and gingivitis. This 9-month double-blind, controlled clinical trial in 85 subjects further demonstrated the effectiveness of twice-daily rinsing with Listerine as an adjunct to professional care and normal oral hygiene. To maximize initial gingival health and reduce plaque scores to 0, all subjects received three complete PMTCs at weekly intervals and an additional PMTC at the outset of the study. The effectiveness of this regimen was further enhanced in the Listerine group.

Plaque development in the Listerine group was significantly lower, 18.6% and 13.8%, respectively, at 6 and 9 months than in the vehicle control group. There was virtually no change in gingivitis in the control group over the 9-month period following the multiple PMTCs and normal oral hygiene. In the Listerine group, at the conclusion of the study, there was 16.9% less gingivitis than at baseline and significantly (22.1%) less than in the control group (Gordon et al, 1985).

Fine et al (1985) evaluated the effect of rinsing twice a day for 9 months with Listerine on the properties of developing plaque. In a double-blind, controlled clinical experiment, plaque was collected from a group of volunteers who used Listerine antiseptic, its vehicle control, or a water control twice daily in addition to their normal toothbrushing. At the end of the study, plaque collected from the supragingival surfaces of 20 teeth from each of the 78 subjects was weighed wet, freeze dried, reweighed, resuspended, sonicated, and estimated for protein. In addition, endotoxin activity was evaluated by means of the limulus lysate assay. The results demonstrated that Listerine has a dramatic effect on the toxic activity of plaque, expressed as a decrease in limulus lysate assay, as well as on its biomass.

In a 6-month study in 83 subjects of the effect on plaque flora of rinsing twice daily with Listerine, no significant increases in numbers of presumptive oral pathogens, spirochetes, black-pigmented *Bacteroides*, *S mutans*, or *Candida albicans* were found. It was concluded that long-term use of Listerine did not induce antimicrobial-resistant strains in plaque or encourage undesirable oral pathogens (Minah et al, 1989).

In a 6-month double-blind trial, the same research group tested the effect of Listerine as a supplement to regular oral hygiene measures on the development of gingivitis (DePaolo et al, 1989). Both plaque and gingivitis were inhibited by 34% compared with the hydroalcohol control. The results were highly statistically significant.

Kato et al (1990) reported that the bactericidal effect of Listerine on the oral microflora is not as great as that of CHX, further supporting the theory that the reduction in gingivitis achieved by Listerine in long-term studies is due not solely to the reduction in plaque volume but also to the anti-inflammatory effect.

In two recent 6-month studies, twice-daily rinsing with Listerine was significantly more efficient in controlling gingivitis compared with placebo rinsings and almost equivalent to daily flossing in the interproximal areas (Charles et al, 2002; Bauroth et al, 2002).

Listerine also has an effect on recurrent aphthous ulcers (Meiller et al, 1991). Compared with either a baseline period of observation or a control 5% hydroalcoholic mouthrinse, Listerine was found to reduce the duration of ulcers by 2 days, on average, as well as the severity as perceived by the patient. The number and frequency of ulcers were reduced by both the Listerine and control mouthrinses.

Listerine has few disadvantages. The taste is unpleasant, with a burning sensation on the mucosa, although this does not deter its use; patients seem to adapt to it. It causes little if any staining and no enhancement of pathogenic organisms.

Chlorhexidine digluconate. The cationic-detergent CHX is still regarded as the gold standard among chemical plaque control agents used in commercial mouthrinses. The first report of its antiplaque activity was made by Löe and Schiott (1970b). Interestingly, this report of the meeting in Dundee, Scotland, in 1969 states that, "The primary objective of this experiment is not to introduce CHX as a possible antimicrobial agent for the clinical prevention and control of plaque." By 1974, there were already more than 70 articles on the oral use of CHX, and it has been proven many times over as the most effective agent not only against plaque but also in long-term studies on gingivitis.

The initial study on CHX mouthrinses (Löe and Schiott, 1970b) demonstrated that, in the absence of normal mechanical toothcleaning, 0.2% CHX solutions used as 10-mL rinses for 1 minute twice per day prevented plaque formation and the development of gingivitis. Numerous studies

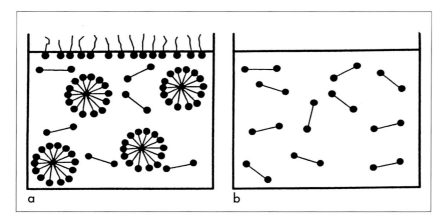

Fig 217 Molecular configuration of chlorhexidine (b) compared with other common antiseptics (a) when dissolved in water. (From Rölla et al, 1997. Reprinted with permission.)

have confirmed the antiplaque activity of CHX mouthrinses used alone, and these have been extensively and repeatedly reviewed (Adams and Addy, 1994; Addy, 1986; Addy et al, 1994; Addy and Moran, 1997b; Addy and Renton-Harper, 1997; Albandar et al, 1994b; Gjermo et al, 1974; Hull, 1980; Jones, 1997; Kornman, 1986; Lang and Brecx, 1986; Mandel, 1988; Wåler, 1989). Therefore, 0.1% CHX mouthrinse is now used as a positive control in many clinical trials of new mouthrinse formulations. The reduction in plaque indices is usually found to be about 60%, while the gingival indices are reduced by about one third (Grossman et al, 1986). The American Dental Association (1988b) has approved its use in 0.1% formulations.

The major advantage of CHX over most other compounds lies in its substantivity: Binding to soft and hard tissues prolongs the effect for hours after rinsing (Wåler, 1989). Bacterial counts in saliva consistently drop to between 10% and 20% of baseline after single rinses and remain at this level for at least 7 hours (Addy and Wright, 1978) and probably more than 12 hours (Schiött et al, 1970).

In addition to the high substantivity, a new finding by Kjaerheim et al (1994) may explain the superior clinical antiplaque effect of CHX. During a research project unrelated to CHX, a positive correlation was observed between the antiplaque effect of antiseptics and their critical micelle concentration: The higher the critical micelle concentration, the greater the antiplaque effect. This

is in general accordance with the observation that nonionic detergents (all with low critical micelle concentration) have poor antiplaque effects. Kjaerheim et al (1994) reduced the critical micelle concentration of SLS by addition of salt and demonstrated that this impaired the plaque-inhibiting effect. It is known that CHX does not form micelles (ie, it has an extremely high critical micelle concentration), and this was proposed as an explanation for the superior clinical effect of CHX. The reason for the lack of micelle formation is that the CHX molecules have two "heads," one in each end of the molecule (Fig 217). The detergents that easily form micelles have a head-and-tail configuration.

Mouthrinses with CHX contain only monomers (and presumably some dimers and trimers), whereas all other relevant anionic, nonionic, or cationic antiseptic agents used in mouthrinses are in the form of micelles, each of which usually contains 100 or more molecules. Unlike other antiseptic agents, present mainly as micelles, with a much lower density of monomers, CHX exposes the teeth and the oral mucosa to a very dense solution of molecules present as monomers (Rölla et al, 1997).

Chlorhexidine has not only a broad antimicrobial spectrum but also a specific effect on MS (Axelsson et al, 1987b, 1993d; Emilson, 1977a; 1981; Emilson et al, 1976, 1982, 1987; Emilson and Fornell, 1976; Krasse and Emilson, 1986; Löe et al, 1976; for review, see Emilson, 1994). The large

CHX molecule does not readily penetrate thick plaque: To facilitate accessibility, heavy plaque deposits should be mechanically removed by self-care and PMTC.

The plaque-inhibitory action of CHX is dose dependent, a function of concentration and volume of the rinse. Although considerable plaque inhibition is attained with doses as low as 1 mg twice a day, the effect increases with dose, leveling off at around 10 to 20 mg twice per day, with minor improvements at even higher doses. This dose response means that concentrations of solutions can be varied; doses are maintained by increasing or decreasing volumes of the rinse. Nevertheless, tooth staining is also largely dose dependent, and if the dose is maintained, there is no evidence that staining is reduced by using low concentrations of CHX. Side effects such as poor taste and mucosal erosions are also concentration dependent, and it is probable that concentrations above 0.2% would generally not be tolerated.

Most of the early commercially available rinses contained 0.2% CHX. For all oral hygiene products containing antiseptics, the inclusion of an agent in the formulation does not guarantee the availability and efficacy of the product. During the last decade, other effective mouthrinses containing 0.12% CHX have been introduced: The recommended volume for rinsing is 15 mL twice per day, providing an 18-mg dose on each occasion. Rinsing once a day with CHX is less effective (Löe and Schiott, 1970a, 1970b), presumably because the substantivity is between 12 and 14 hours (Schiött et al, 1970). Studies by Lang et al (1982) have shown that the effect of CHX mouthrinse on gingivitis is strongly correlated to the concentration and frequency of use.

The main disadvantage of CHX is its taste, and a barrier to compliance is its affinity for dietary compounds, some of which cause staining (Addy et al, 1985). In some cases, staining of the teeth is severe, and removal requires PMTC. Additionally, tongue brushing may be required to remove discoloration of the soft tissue. To overcome the problem of staining, lower concentrations have been formulated, eg, Peridex (0.12%) (Zila, Phoenix, AZ) and Hibitane Dental (0.1%). A comparison in vitro of the 0.1% and 0.2% formulations (Addy et al, 1989b) showed that 0.1% CHX produced no greater staining of acrylic resin specimens than did the control, whereas the 0.2% formulation caused heavy staining.

A 6-week double-blind clinical study also confirmed that, in contrast to heavy staining and some oral mucosal lesions caused by twice-daily use of 0.2% CHX mouthrinse, little or no staining was caused by 0.1% CHX rinse, but similar reductions in gingivitis were achieved (see Fig 216) (Axelsson and Lindhe, 1987). Figures 218 to 221 show typical staining of the teeth and the tongue after 2- and 4-week rinsing with 0.2% CHX twice a day. Typical oral mucosal lesions developed after twice-daily rinsing with 0.2% CHX are shown in Figs 222 and 223. However, when 0.1% CHX mouthrinse was used twice daily after mechanical toothcleaning by self-care in a long-term (4-month) double-blind study, it resulted in heavy staining (Fig 224), in contrast to Meridol (AmF plus SnF_2) and NaF mouthrinses (Axelsson et al, 1993d, 1994b).

As emphasized earlier, long-term unsupervised use of dental products containing antimicrobial agents should not result in adverse alterations to the ecology of dental plaque. Several studies have monitored the composition of the oral microflora during routine use of such products by human volunteers: Such studies are prerequisites for acceptance of a product by the American Dental Association. Undesirable alterations to the composition of plaque would include the suppression of the predominant organisms, leading to overgrowth by either exogenous species (eg, yeasts or enteric bacteria) or organisms associated with oral diseases. Likewise, the development of resistance or tolerance to these agents would be undesirable.

During 6 months' use of CHX mouthrinse, the total numbers of bacteria in plaque were reduced, but there were no detectable shifts in the balance of individual microorganisms (Briner et al, 1986a). At the end of the study, the overall sensi-

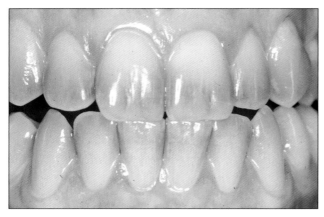

Fig 218 Typical staining of the teeth after 2 weeks of twice-daily rinsing with 0.2% CHX. (Courtesy of P. Ramberg.)

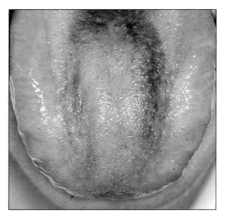

Fig 219 Typical staining of the dorsum of the tongue after 2 weeks of twice-daily rinsing with 0.2% CHX. (Courtesy of P. Ramberg.)

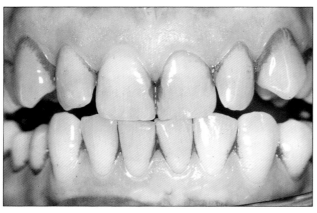

Fig 220 Staining of the teeth after 4 weeks of twice-daily rinsing with 0.2% CHX. (Courtesy of P. Ramberg.)

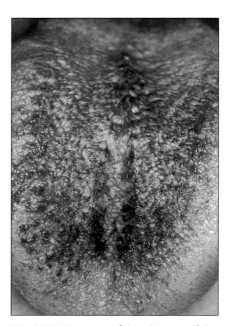

Fig 221 Staining of the dorsum of the tongue after 4 weeks of twice-daily rinsing with 0.2% CHX. (Courtesy of P. Ramberg.)

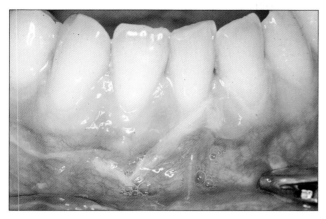
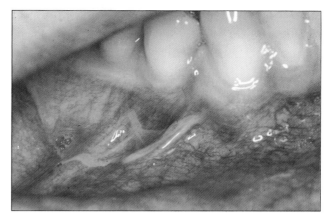

Figs 222 and 223 Oral mucosal lesions developed in sensitive patients after twice-daily rinsing with 0.2% CHX. (From Axelsson and Lindhe, 1987. Reprinted with permission.)

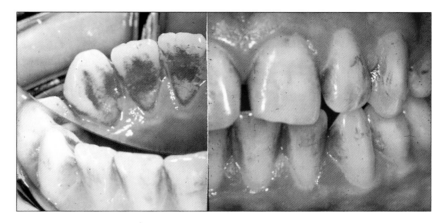

Fig 224 Staining after 4 months of twice-daily rinsing with 0.1% CHX. (From Axelsson et al, 1994. Reprinted with permission.)

tivity of the plaque microflora to CHX was unchanged, although occasionally less sensitive streptococci and *Actinomyces* strains were isolated (Briner et al, 1986b). In a long-term study of 0.1% CHX, Meridol, and NaF mouthrinses, used twice daily for 4 months, none caused any significant effect on the total counts of salivary microflora, apart from a limited reduction of salivary streptococci levels in the CHX group. However, the reduction of salivary *S mutans* levels was significant in all three groups and significantly greater in the 0.1% CHX and Meridol groups than in the NaF group (Axelsson et al, 1993d).

In a recent review on antiplaque biocides and possible bacterial resistance, Sreenivasan and Gaffar (2002) concluded that the results of studies on the real-life use of oral care formulations with antiplaque biocides show no emergence of resistant microflora or alterations of the oral microbiota. Moreover, such formulations have been found to provide the benefits of reducing plaque and gingivitis.

No allergies to CHX have been reported from the Western countries. However, in a previously discussed 6-week double-blind mouthrinse study, some subjects who rinsed with 0.2% CHX twice a day exhibited limited oral mucosal lesions and irritations. None of the subjects in the 0.1% CHX and Listerine groups experienced such problems (Axelsson and Lindhe, 1987).

In vitro growth of epithelial cells is inhibited to some extent by 2 hours' exposure to a 250-fold dilution of 0.2% CHX gluconate. Although in vitro studies should not be directly extrapolated to clinical conditions, the results suggest that, when the oral mucosa has been breached, CHX may delay healing. This must be weighed against the beneficial effects of bacterial inhibition: On balance, CHX enhances healing of periodontal surgical wounds.

Cline and Layman (1992) evaluated the effects of CHX on attachment and growth of human fibroblasts and periodontal ligament cells. Treatment of root surfaces with up to 0.12% CHX had no effect on attachment, but both morphology and attachment were affected by 0.2% to 2% solutions. Direct exposure of cells to 0.4% CHX caused a reduction of 90% in H-thymidine uptake: This may contraindicate the use of concentrations greater than 0.1% CHX as a subgingival irrigant.

Relevant to the evaluation of CHX and other antiseptic rinses is the selection of subjects for studies of antiplaque activity. Most studies are conducted in healthy volunteers, in whom very effective inhibition can be demonstrated. Relatively few studies have investigated the effects of CHX, alone or in combination with toothbrushing, on established disease; in those that have, the gingivitis was mild to moderate. No studies have been conducted in subjects with severe gingivitis. Thus, it is clear that CHX is a major plaque inhibitor. Provided that supragingival and subgingival plaque deposits are mechanically removed, it is extremely effective in promoting or maintaining gingival health. In the absence of mechanical plaque control, plaque removal by CHX mouthrinse and consequently the resolution of gingivitis appear to be poor.

As suggested earlier, the size of the CHX molecule may limit access interproximally and into heavy deposits of plaque, supporting the concept that CHX and other antiseptics have greater preventive than therapeutic potential. Moreover, CHX, when delivered in rinse form, has no effect on established chronic periodontitis because the vehicle does not facilitate penetration of pockets

(Flötra et al, 1971). The original study by Löe and Schiott (1970b) indicated that twice-daily rinsing for 60 seconds with 10 mL of 0.2% CHX completely prevents plaque and gingivitis; however, removal of established plaque deposits requires multiple rinsing at greater frequencies than does prevention of plaque formation.

Short-term studies. Caton et al (1993) compared the efficacy of a CHX mouthrinse (0.12%) plus toothbrushing (mouthrinse group); mechanical interdental cleaning plus toothbrushing (mechanical group); and toothbrushing alone (control group) in reducing and preventing interdental gingival inflammation. Ninety-two male subjects were examined for interdental inflammation using the Eastman Interdental Bleeding Index at baseline and then monthly for 3 months while they followed one of the oral hygiene regimens. At 1 month, the mechanical cleaning group had significant reductions in bleeding sites compared to baseline (56.9% versus 13.7%); this result persisted throughout the study (2 months: 6.65%; 3 months: 5.70%). The other regimens resulted in no significant reduction in bleeding at any point in the study. The mechanical interdental cleaning group showed improvement over baseline at 1 month, and the full benefit was apparent after 2 months.

The effect of location in the mouth on bleeding was also assessed. The percentage of posterior sites that bled was always higher than the percentage of anterior sites. That is because the mouthrinse has limited accessibility to the wide approximal surfaces of the molars and premolars, which also have buccal and lingual papillae. Analysis of maxillary versus mandibular sites and buccal versus lingual sites revealed no statistically significant differences.

Additional analysis of the data revealed that sites that bled at baseline were more likely to stop bleeding in the mechanical cleaning group. Sites that did not bleed at baseline were unlikely to bleed subsequently when mechanical cleaning was used. Neither of these observations applied to the other regimens. These data show that only mechanical interdental plaque removal combined with toothbrushing effectively reduces or prevents interdental inflammation, particularly in the posterior region with buccal and lingual papillae, which limit accessibility for the mouthrinse. This underscores the importance of instituting mechanical interdental cleaning to eliminate interdental inflammation (Caton et al, 1993).

These findings were partly confirmed in a study by Asikainen et al (1984). Thirty subjects brushed their teeth only on one side, chosen randomly, without toothpaste twice a day and rinsed directly afterward with 0.2% CHX for 7 days. After 7 days, gingival bleeding had decreased on test sides with mechanical cleaning and CHX but had increased on sides treated only with CHX (Fig 225).

The effects of 0.12% CHX, 0.06% triclosan plus 1.6% copolymer, and placebo mouthrinses on de novo plaque formation at sites with healthy and inflamed gingivae were recently evaluated by Ramberg et al (1996). On day 0, gingival crevicular fluid was obtained at predetermined sites, and gingivitis (GI) was assessed. After meticulous PMTC, the subjects then refrained from all mechanical plaque control measures for 18 days. During the first 14 days (rinse phase I), they rinsed with 0.12% CHX, 0.06% triclosan, or the placebo solution. Clinical examinations (gingival crevicular fluid and GI) were repeated, and the amount of plaque formed was determined on days 4, 7, and 14.

On day 14, after a new PMTC, rinse phase II was initiated: The participants now rinsed for 4 days with the same preparation and in the same manner as before. The examinations were repeated on day 18. Each participant received a comprehensive PMTC and was instructed to perform meticulous mechanical plaque control during the following 4 weeks. A second experimental period was then initiated. A total of three experimental periods were repeated, until all subjects had rinsed with the three different mouthrinse preparations.

Significantly more plaque formed at sites with gingivitis than at surfaces adjacent to healthy gingival units. Preexisting gingivitis significantly in-

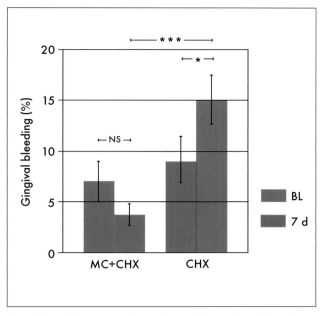

Fig 225 Changes in GBI after 7 days of combined mechanical cleaning (MC) and CHX rinsing or CHX rinsing alone (*P < .05; ***P < .0001). (BL) Baseline; (NS) Not significant. (From Asikainen et al, 1984. Reprinted with permission.)

creased the amount of de novo plaque in subjects rinsing with preparations containing CHX and triclosan. However, the overall plaque scores for the placebo mouthrinse group were higher than those in the CHX and triclosan groups, and plaque scores were significantly higher in the triclosan group than in the CHX group (Ramberg et al, 1996).

Short-term use during active treatment. During active treatment (nonsurgical, surgical, and particularly regenerative therapy), there are periods when mechanical plaque control is not recommended. Numerous antiseptic mouthrinses have been tested, but to date CHX seems to be the most effective (for review, see Addy and Renton-Harper, 1997; Gjermo, 1974; Hull, 1980). Despite its side effects, CHX mouthrinse is well accepted by most patients for short-term use.

In a recent 2-week study, Francetti et al (2000) compared the effect of mouthrinse with 0.12% CHX spray twice daily for 1 week after initial nonsurgical periodontal treatment followed by periodontal surgery. After another week with mechanical self-performed toothcleaning, there was no difference in plaque scores between the groups, but significantly more staining was observed in the mouthrinse group compared with the spray group.

Westfelt et al (1983a) compared mouthrinsing with 0.2% CHX with regular plaque control during postoperative healing of 14 patients who had undergone periodontal surgery. Seven rinsed twice a day for 2 minutes with CHX for 6 months (healing phase), and the remainder were enrolled in a strict plaque control program with PMTC once every 2 weeks. Following reexamination 6 months postoperatively, all patients were enrolled in a maintenance care program once every 3 months until the final examination at 24 months.

At the end of the healing phase, there was a higher frequency of pockets deeper than 4 mm and less gain of attachment in pockets initially deeper than 4 mm in the CHX group. However, the differences were small and tended to diminish during the maintenance phase. There was no difference in gingival scores between the groups. It was concluded that CHX is an acceptable alternative to mechanical plaque control during healing after periodontal surgery (Westfelt et al, 1983a). In such patients, the interproximal spaces normally are open because of attachment loss, permitting access for the mouthrinse.

For patients undergoing regenerative therapy, rinsing with 0.1% CHX is recommended for 1 week preoperatively and about 6 weeks postoperatively. During this period, the patient should also receive regular complementary PMTC of areas inaccessible to CHX. Maintenance (supportive) care starts when the flaps are sutured in position. After regenerative procedures it is important to keep the surgical area as clean as possible and not to disturb the healing by mechanical brushing. Patients who feel discomfort during rinsing can wipe the wound with gauze soaked in CHX and applied with tweezers. The patient should be recalled for healing control and PMTC once a week during the first 6 weeks

and can thereafter resume conventional home care. This very strict postsurgical program presumes a stringent selection of suitable patients prepared to comply with the postoperative program.

Long-term studies. The excellent clinical results achieved in the initial short-term trials with CHX by Löe and Schiött (1970a) and others have also been achieved in several long-term studies. In the first of these (Löe et al, 1976), 150 medical students rinsed daily for 2 years with a solution of CHX at a nominal level of 0.2%. Although interaction with a flavoring agent reduced the effective concentration of CHX, there were significant reductions in plaque, gingivitis, total numbers of facultative and anaerobic bacteria, and salivary *S mutans*. In this same clinical trial, several medical tests, such as blood count, urinalysis, and sedimentation rate were carried out at regular intervals. No significant differences were observed between test and control subjects for any variables evaluated. These data indicate that CHX can safely be used long term, under professional supervision.

A potential danger of prolonged application of an antimicrobial agent is the possibility that drug-resistant mutants will develop. In an animal study in which CHX was applied three times a day for 4 years, no reduced sensitivity of any bacteria to CHX was found. In a previously discussed human study, Löe et al (1976) applied 0.2% CHX solution for 2 years and found only a slightly reduced sensitivity of microorganisms to CHX (see also Axelsson et al, 1993d, 1994b; Banting et al, 1989; Briner et al, 1986b; for review, see Sreenivasan and Gaffar, 2002).

In a previously mentioned 6-month study, Lang et al (1982) tested the effects of supervised rinsing with CHX in 158 schoolchildren, aged 10 to 12 years, in three experimental groups and a control group. The first group rinsed with a 0.2% solution of CHX six times a week. The second rinsed with the same concentration but only twice a week. The third group rinsed six times a week with a 0.1% solution, and the control group rinsed

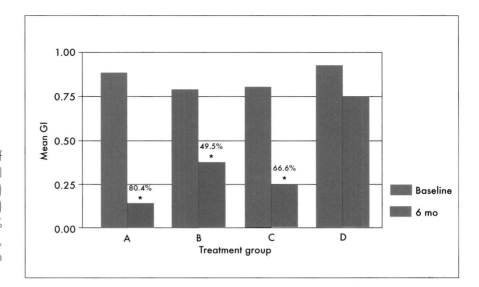

Fig 226 Mean GI in four groups of children in a 6-month supervised trial of CHX mouthrinse (*P < .05). (A) 0.2% CHX, six times weekly; (B) 0.2% CHX, twice weekly; (C) 0.1% CHX, six times weekly; (D) Placebo, six times weekly. (Modified from Lang et al, 1982 with permission.)

six times a week with a placebo. All rinsing was supervised. No attempt was made to change the children's oral hygiene habits.

After 6 months, the group rinsing six times a week with a 0.2% solution of CHX had an 80.4% reduction in mean GI (reduced to 0.13). The group rinsing only twice a week with an identical solution had a 50.0% reduction, and the group rinsing six times a week with the 0.1% solution had a reduction of 66.6%. All reductions were statistically significant compared to the change in the placebo group (Fig 226). It was concluded that gingivitis can indeed be controlled by regular rinsing with 0.1% or 0.2% CHX mouthrinses over an extended period (Lang et al, 1982).

In another 6-month clinical study, Grossman et al (1986) studied the effects of CHX mouthrinse on plaque and gingivitis in 430 adults. The subjects were divided into two comparable treatment groups, matched for age, sex, and initial severity of gingivitis. Following thorough PMTC, they were instructed to rinse twice a day for 30 seconds with 15 mL of a 0.12% CHX mouthrinse or a placebo mouthrinse.

After 3 and 6 months, the CHX group had significantly less gingivitis, gingival bleeding (35% to 40%), and plaque accumulation (61%) than did the placebo group. As expected, accumulation of

dental calculus and extrinsic dental stain increased in the CHX group. No significant differences in adverse oral soft tissue effects were observed between the two groups. It was concluded that a 0.12% CHX mouthrinse is a valuable adjunct to regular personal oral hygiene procedures and professional care for the prevention and control of gingivitis (Grossman et al, 1986).

Briner et al (1986a, 1986b) evaluated the antimicrobial effect of a 0.12% CHX mouthrinse on plaque bacteria in a controlled trial in 80 subjects: Forty subjects rinsed twice daily with 15 mL of CHX or placebo mouthwash for 6 months, while continuing their usual oral hygiene practices. Supragingival plaque was collected from each subject and assayed for eight representative microbial populations at baseline, after 3 and 6 months of treatment, and after a 3-month posttreatment period.

During the trial, significant reductions in the numbers of total aerobes, total anaerobes, streptococci, and actinomycetes were observed in plaque samples taken from subjects using the CHX. The effect of CHX on accumulation of actinomycetes was especially pronounced, with reductions ranging from 85% to 97%. This was considered to be particularly important because *Actinomyces* is one of the bacteria associated with the pathogenesis of gingivitis in both experimental animals and in humans. Although CHX treatment reduced the number of microbes on teeth, it did not produce a detectable shift in microbial populations. At the end of the 3-month posttreatment period, the microbial profile was similar to the pretreatment profile for both groups. Therefore, no residual effects of CHX on plaque bacteria were observed after cessation of rinsing (Briner et al, 1986a, 1986b).

Banting et al (1989) evaluated the long-term clinical effects of twice-daily use of a bioavailable 0.12% CHX mouthrinse in a large group of adult subjects (n = 476 at baseline), representative of the normal population. All subjects received initial scaling, root planing, debridement, and PMTC. The PMTC and debridement were repeated twice each year during the 2-year longitudinal trial.

The authors reported that, compared with a placebo control, twice-daily use of 0.12% CHX as a 30-second rinse for 2 years:

1. Reduced gingivitis (mean GI scores) by 40%
2. Reduced gingival bleeding (the proportion of sites with GI scores of 2 or 3) by 50%
3. Reduced supragingival dental plaque by 35%
4. Reduced the total gingival probing depth by an average of 9 mm per subject
5. Increased formation of supragingival calculus
6. Resulted in a higher proportion of subjects who were free of subgingival calculus

They concluded that subjects who rinsed with 0.12% CHX twice daily for 2 years had significantly better periodontal health than did those in the control group (Banting et al, 1989).

During the same study, plaque was sampled at baseline and recalls to evaluate the long-term effect on plaque bacteria of rinsing twice daily with 0.12% CHX. It was found to be highly effective in reducing the quantity of several potentially periopathic plaque bacteria (*Actinomyces*, *Veillonella*, and *Fusobacterium* species) throughout 2 years of continuous use. These data were in accordance with clinical observations of reductions in plaque, gingivitis, and probing depths

(Banting et al, 1989). There was no emergence of resistant strains; this fact, along with the continued efficacy, underscored the safety of the CHX formulation.

These findings were also confirmed in a previously mentioned, 4-month, double-blind trial of rinsing with 0.1% CHX, Meridol, or NaF twice daily after toothcleaning (Axelsson et al, 1993d, 1994b).

Sanguinaria extract. Sanguinaria extract, present in the proprietary mouthrinse Viadent in the United States, is an alkaloid from the plant *Sanguinarina canadensis*. It has been incorporated into dentifrices and mouthrinses, but reports on its efficacy are conflicting.

Wennström and Lindhe (1985) conducted a short-term, double-blind, crossover study of plaque regrowth and gingivitis. For each 2-week experimental period, the subjects used a mouthrinse containing 0.03% sanguinaria extract or the control, a placebo rinse similar in taste and color. Plaque indices were 40% lower, and gingivitis scores were 25% lower, when the sanguinaria extract rinses were used. Normal toothbrushing was suspended during the trial period, and, in some subjects, slight discoloration of the teeth and dorsum of the tongue was noted.

In a single-blind crossover, experimental gingivitis study, Moran et al (1988) compared the effects of the sanguinaria extract–zinc mouthwash Viadent with those of 0.2% CHX in the absence of all other oral hygiene. The positive control, CHX, was significantly more effective than Viadent in inhibiting both plaque and gingivitis. That study did not have a placebo control, but, in a later study, Moran et al (1992a) used a 4-day plaque regrowth model to compare several rinses, including sanguinaria extract with and without zinc chloride. A saline rinse served as the placebo control. The use of sanguinaria extract alone produced results that differed little from those produced by the use of saline. The combination of zinc chloride and sanguinaria extract produced a "modest" reduction in plaque growth, possibly attributable to the zinc chloride.

In a single-blind crossover study of plaque regrowth, Quirynen et al (1990b) have shown zinc chloride on its own to be almost as effective as the combination of sanguinaria extract and zinc chloride. During the 18-day experimental period, rinsing with Viadent, Viadent without sanguinaria extract, or CHX was the sole oral hygiene measure. Although plaque growth was assessed on only four teeth in each of 12 subjects, a limited number of sites, the differences between the areas of the buccal surfaces of the teeth covered by plaque were highly significant in a comparison of CHX and Viadent. Viadent was only slightly better than the control, and at only one period in the trial (3.5 days) did Viadent provide a statistically significant improvement in plaque inhibition. The authors concluded that Viadent mouthrinse would have at most a limited role as a plaque inhibitor.

The possible long-term benefit of sanguinaria extract mouthrinse is difficult to assess because most studies have evaluated rinsing in combination with use of a sanguinaria extract dentifrice. For example, Harper et al (1990a), in a 6-month double-blind, parallel study, assessed gingivitis and plaque in subjects using Viadent dentifrice together with a mouthrinse containing sanguinaria extract and zinc chloride. The negative controls used the same products but without the sanguinaria extract or zinc chloride. Plaque and gingivitis scores were 21% and 25% lower, respectively, in the group using sanguinaria extract than in the control group at the end of 6 months. Bleeding on probing was also reduced in the test group compared with controls.

Harper et al (1990b) also reported on changes in the microflora of the buccal mucosa and the supragingival and subgingival plaque in the same clinical trial. No opportunistic overgrowth of pathogens was found, but there were reductions in the numbers of organisms associated with gingivitis, which may have accounted for the reduction in gingivitis.

Similarly beneficial effects were found by Hannah et al (1989) in a 6-month trial of gingivitis prevention. They studied a group of orthodontic patients using both toothpaste and rinses

containing sanguinaria extract and zinc chloride. However, it seems likely that if sanguinaria extract is to have any benefit, it must be used in combination with zinc chloride and as both a toothpaste and mouthrinse. The necessity of such combinations of products to attain efficacy would have cost-benefit implications.

Delmopinol. An alternative approach to antimicrobial control of dental plaque is to prevent the attachment or retention of bacteria on the tooth by affecting its surface characteristics. Relatively little work has been carried out in this field, but Collaert et al (1992a, 1992b) investigated the surface-active agent, delmopinol hydrochloride, a substituted amino-alcohol with little if any effect on salivary microflora. They reported a significant dose-response effect for plaque regrowth after 2 weeks, during which rinsing was the only oral hygiene procedure. There was no significant difference in gingival bleeding indices between subjects who rinsed with 0.2% CHX and those who used 0.2% delmopinol solution for 2 weeks.

At the end of a preliminary 2-week placebo period and after the use of the active rinses, saliva samples were taken, gingival crevicular fluid flow was measured, gingivitis was scored using the GI, plaque was measured planimetrically, and the PI was scored. Although there were significant reductions in salivary counts of anaerobes, aerobes, and *S mutans* in subjects rinsing with CHX, no such changes were detected for those using delmopinol. There was no difference in effect on crevicular fluid flow and gingivitis, but CHX had a greater inhibiting effect on mean extent of plaque (Collaert et al, 1992a, 1992b).

Moran et al (1992b) studied plaque regrowth in a 4-day crossover study of delmopinol, CHX (as a positive control), and a placebo rinse. No mechanical oral hygiene was allowed. Although the plaque scores were higher with delmopinol rinsing than they were with CHX rinsing, the difference between the scores obtained with placebo rinsing and those obtained with delmopinol was significant, suggesting that delmopinol would be of benefit in reducing gingivitis. In a separate investigation in the same report, it was shown that delmopinol had very little effect on salivary bacteria after one rinse, and the effect was lost in 1 hour. Some subjects using delmonipol complained of the taste and a burning sensation in the mucosa.

Elworthy et al (1995), in a 6-month home-use trial, evaluated the effect of 0.1% and 0.2% delmopinol mouthrinses, used twice daily, on the plaque microflora in 141 subjects. Plaque was collected at baseline and at 12, 24, and 36 weeks. Overall, there were no consistent effects on specific microflora or total microflora counts. However, throughout the trial, the active groups experienced a significantly greater reduction in the proportion of dextran-producing streptococci than did the control group. There was no colonization by *Candida* species or gram-negative aerobic bacilli in the active groups, nor was there any decrease in susceptibility to delmopinol. The authors concluded that delmopinol appears to mediate its antiplaque effect without causing a major shift in bacterial populations, although dextran-producing bacteria appear to be affected, which may be relevant to the agent's mode of action.

In these recent studies, delmopinol has shown promise as an antiplaque agent, warranting further investigation.

Cetylpyridinium chloride. Short-term studies on mouthrinses containing cetylpyridinium chloride have not shown them to provide significantly better plaque inhibition than placebo mouthrinses (Binney et al, 1992).

Triclosan. In a clinical crossover trial, 20 subjects used triclosan-copolymer rinses for 7 days without brushing (Abello et al, 1990). The effects were compared with those of rinses containing alcohol and water placebos and a prebrushing rinse containing sodium benzoate (referred to generically as PLAX). The triclosan-copolymer reduced plaque by approximately one half, compared with the water placebo or the PLAX rinse. Compared with the alcohol placebo, the triclosan-copolymer rinse reduced plaque by 31%.

A similar study compared a 0.03% triclosan-copolymer combination, used as a prebrushing rinse for 6 days, with a water-based placebo rinse similar in color and flavor (Hunter et al, 1994). Twice-daily rinses with the test agent followed by toothbrushing with toothpaste resulted in a 31% reduction in plaque compared with the placebo. The reduction obtained with water rinsing alone in the previously mentioned study seemed greater than the reduction found when rinsing was followed by toothbrushing in the present trial.

Using virtually the same protocol, 60 subjects rinsed with water, PLAX, or triclosan-copolymer (Rustogi et al, 1990). Again, the triclosan-copolymer agent reduced plaque indices to 40% of the values obtained with the placebo mouthrinse. The effect of PLAX was not significantly different from that of the water placebo. Recently PLAX has been marketed with a revised formulation containing 0.03% triclosan, 0.25% copolymer, and 0.025% NaF. Colgate Total mouthrinse also has the same formulation (see Fig 209). (For review of the effects of prebrushing mouthrinses containing triclosan-copolymer, see Angelillo et al, 2002.)

As noted previously, triclosan in combination with either zinc citrate or a copolymer appears to achieve better results than the agents do alone. Kjaerheim and Waaler (1994) conducted a double-blind, clinical, crossover study in eight volunteers to determine the plaque-inhibiting effect of different triclosan-SLS mouthrinses. An attempt was also made to locate the oral binding sites of triclosan. After the volunteers had rinsed for 4 days with solutions of various concentrations of triclosan and/or SLS, plaque deposits were scored according to the Silness and Löe PI.

The 0.15% and 0.30% concentrations of triclosan yielded comparable plaque-inhibiting effects in vivo. Furthermore, the 0.10% triclosan with 1.5% SLS exhibited a greater (though not significantly greater) effect than did 0.10% triclosan with 0.75% SLS. The mouthrinse containing 0.05% triclosan and 0.25% SLS was as effective as the two formulations containing 0.10% triclosan. Collectively, the results indicate that triclosan alone has an antiplaque effect, independent of the effect of

SLS. Furthermore, topical application of 0.3% triclosan on the teeth failed to produce a clinically discernible effect, suggesting that the SLS monomers may play a role as carriers of triclosan and that the teeth are not the only binding site (Kjaerheim and Waaler, 1994).

Amine fluoride plus stannous fluoride
Short-term studies. The ability of Listerine, CHX, and an $AmF-SnF_2$ mouthrinse (Meridol) to control dental plaque and gingival inflammation was compared during a 3-week experimental gingivitis trial (Brecx et al, 1990). The study demonstrated that CHX was superior to Listerine and Meridol in its ability to maintain low plaque scores and gingival health when no mechanical oral hygiene was performed. Meridol was as effective as Listerine in reducing plaque accumulation, and both solutions were effective in maintaining healthy gingiva. Furthermore, in contrast to Listerine, Meridol showed a pronounced but transient antibacterial effect in vivo.

The antiplaque, antigingivitis, and antimicrobial efficacies of Listerine and two different $AmF-SnF_2$ mouthwashes (Meridol I and Meridol II) were evaluated as supplements to routine mechanical cleaning (Brecx et al, 1992). In this double-blind study, a placebo preparation served as a negative control, and a CHX solution was used as a positive control. After PMTC, 49 volunteers continued habitual, unsupervised oral hygiene for a period of 2 weeks to develop more realistic baseline values. At day 0, subjects began to rinse twice daily with one of the five mouthwashes.

After 3 weeks of rinsing, plaque indices were lowest in the CHX and the Meridol I groups. Subjects using Listerine or Meridol II had indices similar to each other, significantly lower than that of individuals rinsing with the placebo solution. Through this period, the GI scores were similar in the Meridol, Listerine, and CHX groups. At day 21, the mean GI score in the CHX group was significantly lower than that in the placebo group. The plaque vitality scores showed that CHX, and to a lesser extent the Meridol solutions, had an antibacterial effect in vivo, but Listerine did not.

This study demonstrated that, when mouthrinses are used to supplement habitual mechanical oral hygiene, CHX remains the most powerful solution. Furthermore, it was shown that a combination of routine, self-performed, unsupervised mechanical oral hygiene with Meridol or Listerine is more beneficial for plaque control than mechanical oral hygiene alone (Brecx et al, 1992).

Horwitz et al (2002) evaluated the effect of 3 weeks of twice-daily use of Meridol mouthrinse and 0.1% CHX mouthrinse after surgical flap debridement. Both mouthrinses resulted in similar significant improvements in probing depth and clinical attachment level after 12 weeks. However, CHX mouthrinse resulted in a significantly higher staining index after 3 weeks compared with Meridol.

Netuschil et al (1995) evaluated counts and vitality of bacteria in plaque during CHX, Meridol, and Listerine rinsing in a double-blind study. After PMTC, 40 students refrained from all oral hygiene measures for 3 days, during which they rinsed with Listerine, Meridol, 0.2% CHX, or a control solution. The PI was recorded at the start and the end of the investigation. Total bacterial counts and CFUs of 1-, 2-, and 3-day-old dentogingival plaque were determined.

The PI, total bacterial counts, and CFUs of the CHX and Meridol groups differed significantly from those of the control group. Because of the strong antibacterial action of CHX and Meridol, dying or nonproliferating bacteria were found on the tooth surfaces. Thus only a thin plaque could develop. As a clinical consequence, both substances retarded plaque development, reflected in significantly reduced plaque indices (Netuschil et al, 1995).

Long-term studies. The effect on plaque growth of a dentifrice and mouthrinse containing a combination of AmF and SnF_2 or NaF was compared in a 5-month trial (Nemes et al, 1991). The mouthrinse was used after the subjects brushed with the dentifrice. Relative to baseline values, the plaque indices of the group using the $AmF-SnF_2$ combination decreased by 64% and those of the group using NaF decreased by 40%. Counts of *S mutans* and *Lactobacillus* decreased in both groups, but there were no significant differences between the groups.

Brecx et al (1993) compared the long-term effect of Meridol mouthrinse on plaque growth and gingival response during a 3-month investigation. A placebo preparation served as a negative control, and a 0.2% CHX solution served as a positive control, in a double-blind design. After PMTC, the 36 volunteers continued their usual oral hygiene for a period of 2 weeks. After another PMTC (month 0), they rinsed twice daily (morning and evening) with one of the three mouthrinses in addition to performing their habitual mechanical toothcleaning.

After 3 months of rinsing, PI scores remained lowest in the CHX group, although the subjects using Meridol had PI scores significantly lower than those rinsing with the placebo solution. Throughout the experiment, the GI scores in the Meridol group were higher than those in the CHX group and lower than those in the placebo group. The plaque vitality scores showed CHX and Meridol to have a bactericidal effect. Meridol caused more tooth staining than the placebo but significantly less staining than CHX. The study demonstrated that Meridol reduced accumulation of plaque, retarded development of gingivitis, exerted a definite bactericidal action on plaque, and caused only slight staining (Brecx et al, 1993).

Zimmerman et al (1993) investigated whether Meridol could be recommended as safe for long-term use with respect to the control of plaque and gingivitis. In a double-blind clinical study, Meridol rinse was tested over a period of 7 months in 102 subjects with chronic gingivitis. Gingival indices (GI and SBI) and plaque indices (PI and approximal PI) were recorded at baseline and after 3.5 and 7 months. The composition of the supragingival plaque was evaluated by darkfield microscopy.

During the 7 months, the GI in the test group decreased from 1.36 to 0.95, and the SBI decreased from 52.0% to 29.3%. The PI fell from 1.17 to 0.68, and the approximal PI went from 61.3% to 50.6%. No significant changes were recorded in the control group. In the test group, the proportion of cocci in the plaque increased while the proportion of rods and other plaque bacteria decreased significantly ($P < .001$). In the control group, the composition of the plaque microflora was stable throughout the study. No side effects were reported (Zimmerman et al, 1993).

In a 4-month double-blind study, the effect of Meridol, 0.1% CHX, and 0.025% NaF mouthrinses on gingivitis (GI), plaque (PI), and plaque formation rate (PFRI) was evaluated (Axelsson et al, 1994b). From more than 1,000 high school students aged 17 to 19 years, 300 subjects with relatively high gingivitis and plaque scores were selected and randomly assigned to one of three test groups. Twice a day after mechanical toothcleaning, test group I rinsed with Meridol, test group II rinsed with 0.1% CHX, and test group III rinsed with 0.025% NaF.

The reduction in GI in test groups I, II, and III was 34%, 44%, and 28%, respectively. For PI, the reductions were 38%, 57%, and 26%, in groups I, II, and III, respectively. The PFRI was reduced by 60%, 72%, and 29% in groups I, II, and III, respectively (Figs 227 to 229; see also Figs 199 to 201). The only side effects observed were heavy brown staining of the teeth in the CHX group (see Fig 224) and some cases of light yellow staining in the Meridol group. The positive effects of the NaF mouthrinse can partly be explained by the study design, which required all the participants to clean their teeth twice a day before rinsing (Axelsson et al, 1994b).

Experimental chlorhexidine combinations. In a recent study by Giertsen and Scheie (1995), the relative bacteriostatic and/or bactericidal effects of mouthrinses containing CHX and zinc ions combined with fluoride was assessed by documenting changes in the viability and glycolytic activity of dental plaque. Following 2 days of plaque accumulation, four groups of 10 students rinsed with 12 mM of NaF, 0.55 mM of CHX diacetate plus NaF (NaF-CHX), 10 mM of zinc acetate plus NaF (NaF-Zn), or with the three agents in combination (NaF-CHX-Zn). Plaque samples were collected before

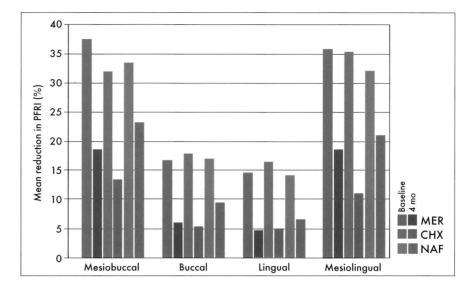

Fig 227 Effect of mouthrinses on PFRI. (MER) Meridol; (CHX) Chlorhexidine; (NAF) Sodium fluoride; (B) Baseline; (4 mo) 4 months. (From Axelsson et al, 1994b. Reprinted with permission.)

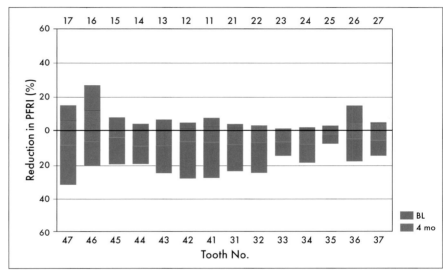

Fig 228 Effect of Meridol on PFRI at lingual surfaces. (BL) Baseline; (4 mo) 4 months. (From Axelsson et al, 1994b. Reprinted with permission.)

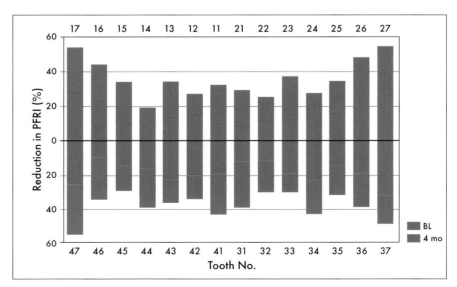

Fig 229 Effect of Meridol on PFRI at mesiobuccal surfaces. (BL) Baseline; (4 mo) 4 months. (From Axelsson et al, 1994b. Reprinted with permission.)

and 90 minutes after rinsing. Thereafter, the pH response to sucrose was monitored in each student.

Both NaF-CHX and NaF-CHX-Zn significantly inhibited the drop in pH, whereas the effect of NaF-Zn was not significant. Bacterial viability was assessed by counting CFUs. All mouthrinses except NaF reduced glucose consumption and acid formation, and thus the fall in pH. The NaF-CHX rinse reduced the CFUs equal to the reduction of glucose consumption, indicating that inhibition of plaque acidogenicity was a bactericidal rather than a bacteriostatic effect. Neither NaF nor NaF-Zn reduced the number of CFUs; thus, NaF-Zn decreased glucose metabolism without affecting plaque viability. The NaF-CHX-Zn rinse reduced both the number of CFUs and glucose metabolism of surviving plaque microorganisms (Giertsen and Scheie, 1995).

As mentioned earlier, Ullsfoss et al (1994) evaluated the potential caries-inhibiting effect of combining 2.2-mM CHX mouthrinses used twice daily with 11.9-mM NaF rinses. They compared the results with the effect of daily rinsing with NaF in a caries model that involved the use of plaque-retaining bands on premolar teeth scheduled for extraction. Nine subjects (a total of 28 teeth) were fitted with the bands for 4 weeks. Saliva and plaque samples were collected before and after the study period for bacterial cultures. After the teeth had been carefully extracted, specimens of the tooth surfaces were analyzed by microradiography. The combination of CHX and NaF rinses resulted in enamel mineral loss that was only slightly higher than that observed in "sound" enamel and clearly less than that observed with NaF rinses alone. Both total plaque bacteria and *S mutans* were reduced by CHX rinses, confirming the discrete mechanisms of action.

At present, no CHX-NaF mouthrinse is commercially available because of problems in formulating products with long-term stability and bioavailability, although a combined CHX-NaF rinse is available by prescription from pharmacies (see Box 8). However, efficient commercial toothpastes containing CHX and NaF separately are available.

Irrigants

As discussed earlier, chemical plaque control agents in mouthrinses have limited access to interproximal and subgingival areas. To overcome this disadvantage, powered oral irrigation methods have been introduced, not only for professional use but also for self-administration.

Supragingival irrigation

The first powered oral irrigation device for self-care accepted by the American Dental Association was the Waterpik oral irrigator (Waterpik Technologies, Fort Collins, CO) in 1968. Acceptance was based on studies documenting its safety and ability to flush food particles from tooth surfaces, particularly those difficult to access by other methods. Subsequently, a number of studies evaluated the effect of this device not only on debris but also on bacteria. In one study, patients with varying degrees of gingival health were studied to determine the effect on the removal of oral debris (Tempel et al, 1975). Using a scoring system for particulate matter, the researchers found that powered irrigation removed three times as much debris and bacterial products as did rinsing, a statistically significant difference. Also, water irrigation was more effective than rinsing, regardless of whether the comparison was before or after brushing and flossing.

The reported effects of water irrigation on plaque removal have varied among studies, although effects on gingival health have been consistently reported as beneficial. The explanation for this apparent lack of correlation may be found by evaluating the effect of irrigation on plaque composition and metabolism. Although studies of plaque metabolism are lacking, electron microscopic studies have provided information documenting disruption of bacterial cells following water irrigation. For example, in an animal study, Brady et al (1973) found that the irrigation device not only removed plaque but also produced changes in the adherent plaque left on irrigated tooth surfaces. These changes were of three types: *(1)* ruptured bacterial cell walls; *(2)* production of bacterial ghosts, with cell walls intact but empty of contents; and *(3)* bending and invagination of bacterial cell walls, with the contents apparently unaffected. These results indicate that residual dental plaque may be irreparably damaged and rendered less pathogenic by irrigation.

In an electron microscopic study, Cobb et al (1988) directed a powered pulsating irrigation stream of normal saline at the gingival crevice from a distance of 3 mm. Control patients had similar periodontal statuses as subjects but received no treatment. A comparison of calibrated periodontal pocket specimens (16 control and 16 test specimens) from these patients revealed qualitative differences in microbial morphotypes at various probing depths. Control specimens, at all probing depths examined (0 to 6 mm), exhibited a mixed microbial flora consisting of cocci and short rods at 0 to 4 mm and spirochetes and fusiform and branching organisms at 5 to 6 mm. In contrast, test specimens exhibited a few cocci and short rods at 0 to 4 mm and a mixed flora at 5 to 6 mm. This study showed that supragingival irrigation had a maximum effect on subgingival plaque at probing depths up to 3 mm; the effect gradually decreased as probing depth increased. These studies confirm the findings of Eakle et al (1986), who found that supragingival use of the Waterpik oral irrigator delivered an irrigant to 44% to 71% of the probing depths.

White et al (1988a) evaluated the effect of supervised oral irrigation over 6 weeks on motile bacteria at 3- and 6-mm probing depths in the gingival pockets of 20 patients. *P intermedia* was measured with a fluorescent antibody method. Water irrigation reduced total counts by 50% compared with baseline in the experimental group. Counts in the untreated controls increased by 27% compared with baseline. In deeper pockets, a significant reduction in the numbers of *Bacteroides*, motile rods, and spirochete groups was found ($P < .01$).

Flemmig et al (1990) assessed the efficacy of supragingival irrigation with 0.06% CHX on naturally occuring gingivitis. The relative benefits of CHX irrigation in comparison with CHX rinsing, water irrigation, and normal oral hygiene were evaluated. In a blind, placebo-controlled, 6-month study, 222 patients were assigned to one of four groups: once-daily irrigation with 200 mL of 0.06% CHX gluconate (experimental); twice-daily rinsing with 15 mL of 0.12% CHX (positive control); once daily irrigation with 500 mL of water (irrigation control); and use of NaF dentifrice for normal oral hygiene only (negative conrol). All groups used the same NaF dentifrice for toothbrushing. At baseline, 3 months, and 6 months, patients were examined for GI, bleeding on probing, PI, probing depth, Calculus Index, and stain. After the baseline visit, all patients received a supragingival and subgingival PMTC.

At 6 months, GI and bleeding on probing were significantly reduced ($P < .05$) by adjunctive CHX irrigation (42.5% and 35.4%, respectively), CHX rinsing (24.1% and 15.0%, respectively), and water irrigation (23.1% and 24%, respectively) compared to toothbrushing alone. Plaque was reduced only by CHX irrigation (53.2%) and CHX rinsing (43.3%), while calculus and staining increased in the two CHX groups. The reduction in probing depth was minimal after CHX irrigation (4.6%).

The study demonstrated that the effect of chlorhexidine on gingivitis is enhanced when CHX is used with this oral irrigator and treatment regimen. Irrigation with water had less effect than did chlorhexidine irrigation but significant clinical benefits. Irrigation may increase the access of solutions beneath the gingival margin, resulting in improved tissue health (Flemmig et al, 1990).

In another part of the same study, Newman et al (1990) showed that, compared to toothbrushing only, daily irrigation with 0.06% CHX resulted in significant reduction of gram-negative, anaerobic rods. Irrigation with water had only a very limited effect on the composition of the supragingival and subgingival microflora.

Brownstein et al (1990) compared oral irrigation and rinsing with CHX and placebos in the treatment of naturally occurring chronic gingivitis. Forty-four subjects with at least six interproximal sites that bled on probing were randomly distributed on a double-blind basis into four treatment groups: placebo rinse, CHX rinse (0.12%), placebo irrigation, and CHX irrigation (0.06%). One half of the mouth was scaled 2 weeks prior to therapy in all groups. Rinses were performed twice daily, and irrigation once a day by means of an oral irrigator with the tip directed at a right angle to the tooth. The subjects continued with routine oral hygiene without instruction. The active treatment period was 2 months. Variables were recorded at baseline and at 60 days.

Both the CHX rinse and CHX irrigation significantly reduced plaque. Gingival bleeding decreased by 26% at both scaled and unscaled sites following CHX rinses and by 40% at both types of sites following CHX irrigation. There was greater reduction in bleeding in the CHX irrigation group than there was in the placebo irrigation group. These data indicate that delivery of 0.06% CHX by an oral irrigator is an effective means of treating naturally occurring gingivitis (Brownstein et al, 1990).

In a recent 12-week study in patients with type 1 and type 2 diabetes, the test group supplemented regular oral hygiene with CHX irrigation twice daily. Compared with a control group using only regular oral hygiene, the test group exhibited lower scores for bleeding on probing, gingivitis, plaque, probing depth, and subgingival bacterial counts. Inflammatory cytokines IL-1 and prostaglandin E_2 were also reduced in the test group (Al-Mubarak et al, 2001).

A number of other studies have found beneficial reductions in bacteria in subgingival plaque when antimicrobial agents such as Listerine, iodine, or SnF_2 are used as irrigants. Ciancio et al (1989) evaluated the effect of Listerine delivered by an oral irrigation device (Waterpik) on plaque, gingivitis, and subgingival microflora. Sixty-one adults completed a 6-week, double-blind, controlled clinical study with 5% alcohol (control) and with Listerine used in a powered pulsating oral irrigator. All subjects irrigated twice daily, once under supervision and once at home, with the irrigator tip directed gingivally at a 45-degree angle to the tooth surface. All subjects received half-mouth PMTC 7 days prior to study entry.

Both irrigants reduced gingivitis significantly at both the 3- and 6-week examinations. A 54% to 62% decrease in gingival inflammation was found at 6 weeks on both sides of the mouth. Between groups, there were no statistically significant differences in gingivitis. Crevicular fluid volumes correlated well with reduction in gingivitis; both groups showed significant reductions from baseline. However, the magnitude of change was greater when Listerine was the irrigant. In addition, the subgingival bacterial cell counts were significantly lower in the Listerine group than they were in the control group (Ciancio et al, 1989).

Subgingival irrigation

Soh et al (1982) investigated the effect of direct application of CHX to periodontal pockets and the practicality of patient self-therapy using a technique of subgingival irrigation. After initial assessment of clinical variables, patients received scaling and PMTC and then were instructed in the irrigation of designated pockets with CHX or a placebo using a disposable syringe and blunt needle. Patients received no other oral hygiene instruction.

During the 28-day irrigation period with CHX, there was a highly significant reduction in periodontal inflammation, which was maintained at levels significantly below the baseline values for a further 28-day period without irrigation. There was a deterioration in the periodontal state of those patients who used the placebo. The irrigation technique itself caused no discernible injury in this group of typical periodontal patients. Also, staining in the CHX group was minimal. It was concluded that subgingival irrigation with CHX by self-care is effective in reducing periodontal inflammation and in controlling subgingival plaque

(Soh et al, 1982). Intermittent treatment of this kind by the patient at home might reduce to more manageable levels the frequency of hygiene visits and the need for rigorous interdental oral hygiene.

Braun and Ciancio (1992) evaluated the probing depth–related delivery effect of a specially designed irrigation tip called Pik Pocket (Waterpik) (Fig 230), used with the Waterpik powered oral irrigator. Ninety posterior sites on teeth untreatable because of advanced periodontal disease were irrigated by placing the irrigator tip approximately 2 mm subgingivally. All pockets were irrigated at mesiobuccal, midbuccal, and distobuccal locations with an erythrosin solution placed in the irrigation reservoir. Thirty similar sites on untreatable teeth in five patients served as controls and were rinsed with erythrosin solution. Following rinsing or irrigating, the teeth were extracted, and the level of dye disclosed along the root surfaces was measured from a fixed point made with a bur at the gingival margin prior to extraction.

No adverse tissue reactions were observed, and patients reported no discomfort associated with irrigation. The average probing depth was 7.5 mm. The results showed that subgingival irrigation with this specialized tip safely delivered an aqueous solution to approximately 80% of the depth of a pocket, while rinsing reached only 0.2 to 0.5 mm subgingivally (Braun and Ciancio, 1992).

Harper et al (1991) examined the effect of subgingival irrigation with Listerine antiseptic on the microbial flora of periodontal pockets in a 6-week double-blind, controlled clinical trial. A total of 50 subjects with at least four 4- to 6-mm pockets were randomly assigned to a Listerine or placebo group. They received a half-mouth scaling and professionally applied complete-mouth subgingival irrigation at baseline and then irrigated subgingivally once daily at home.

At 42 days, Listerine irrigation produced a significant reduction in black-pigmented *Bacteroides* species versus placebo in both scaled and unscaled sites, as well as significant reductions in total anaerobic flora and *Capnocytophaga* species versus placebo in unscaled sites. The numbers of

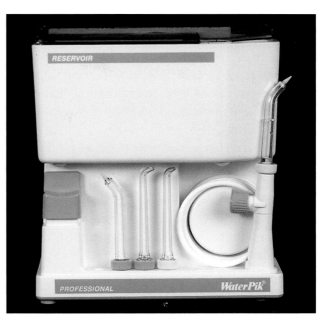

Fig 230 The Pik Pocket pulsating irrigator for subgingival irrigation by self-care.

motile rods were significantly lower and the numbers of coccoid cells were significantly higher in Listerine-treated sites than they were in control sites. Changes from baseline were also significant in Listerine-treated sites. The results indicate that subgingival irrigation with Listerine promotes significantly greater and more persistent decreases in the periopathic flora than does irrigation with water alone and may thus be a useful adjunct to conventional periodontal therapy (Harper et al, 1991).

Combined self-administered and professional irrigation

Wolff et al (1989) examined the effect of combining a professionally applied subgingival antimicrobial agent (SnF_2), delivered at debridement and PMTC, with home personal subgingival delivery of an antimicrobial agent (iodine) in patients with gingivitis and early periodontitis. A control group of 32 individuals was instructed to use dental floss and a sulcular toothbrushing method as a regular oral hygiene regimen. The test group of 42 subjects received professional subgingival irrigation with a 1.64% SnF_2 solution delivered after debridement and PMTC. These subjects were also instructed in the home subgingival delivery of an iodine solution along with the use of dental floss and sulcular toothbrushing. Subjects were clinically evaluated at baseline and after 8 weeks.

Compared to the control group, individuals in the test group had significantly lower GI scores at 8 weeks. In addition, the test group had significantly lower bleeding and GI scores at 8 weeks than at baseline. The authors concluded that professional subgingival irrigation with SnF_2, when combined with home subgingival delivery of an iodine solution, is an effective means of improving gingival health for patients with gingivitis and early periodontitis (Wolff et al, 1989).

The clinical and microbiologic effects of 0.04% CHX delivered daily by home-applied marginal irrigation, in combination with a single professional irrigation of 0.12% CHX, was tested over a 3-month period by Jolkovsky et al (1990). Sixty periodontal maintenance patients, each with at least two pockets deeper than 4 mm and bleeding on probing, were assigned to the following groups:

1. One professional subgingival 0.12% CHX (Peridex) irrigation followed by adjunctive daily home marginal 0.04% CHX irrigation (Pik Pocket)
2. One professional subgingival 0.12% CHX irrigation followed by adjunctive daily home marginal water irrigation
3. One professional subgingival water irrigation followed by adjunctive daily home marginal water irrigation
4. Control

At baseline and after 3 months, subgingival plaque samples were taken from two sites per patient. Probing depths, PI, GI, and gingival recession were assessed. Scaling and root planing (supportive periodontal treatment) was provided for each patient followed by subgingival irrigation as outlined.

At 3 months, both the GI and probing depths were significantly reduced in all irrigation groups compared to baseline. There were no significant changes in clinical variables in the control group from baseline to 3 months. At 3 months, the GI was significantly lower in group 1 than it was in group 4. The numbers of *Wolinella recta* and black-pigmented *Bacteroides* species were significantly reduced in group 1 at 3 months compared to baseline. The findings suggest that it is possible to achieve beneficial clinical and microbiologic effects from adjunctive professional 0.12% CHX and home 0.04% CHX subgingival irrigation in periodontal maintenance patients who are receiving supportive periodontal treatment (Jolkovsky et al, 1990).

Based on these studies, reductions in gingivitis are not always correlated with similar reductions in plaque. A possible explanation is that, although the mass of plaque may not be changing, its composition or metabolism may be changing, and toxic products may be leaching. Also, because plaque scores only measure supragingival plaque, the Waterpik and Pik Pocket powered pulsating irrigators may also be producing changes subgingivally that are not reflected in traditional plaque indices. Furthermore, changes in plaque thickness may not be reflected in area measurements.

Oral irrigators have been modified so that they can be used with either a large chamber to hold water or a smaller chamber to hold chemical irrigating agents. In addition, modified irrigator tips that can be safely placed subgingivally are available. These deliver the irrigant deeper into pockets than is possible with tips for supragingival irrigation (see Fig 230) (for review, see Ciancio, 1995 and Wennström, 1997).

Gels

A gel used as a vehicle for chemical plaque control agents is a thickened aqueous system that contains a humectant but neither abrasive material nor foaming agents. As such, gels are generally compatible with the chemoprophylactic agents. Gels are colored and flavored to encourage compliance. Gels are usually applied in standard or customized trays to provide close contact between the agent and its site of action, ie, not only on buccal, lingual, and occlusal tooth surfaces but also on the approximal surfaces. Such trays can be used for both daily self-administration and professional administration after PMTC.

Gels containing chemical plaque control agents are generally used for daily home care (for so-called brush on) in combination with mechanical toothcleaning. Gels containing CHX have been commercially available for several years, but the lack of abrasives and detergents in such formulations make them poor alternatives to conventional toothpastes for most people. Therefore the new, previously described toothpastes containing CHX, zinc, and NaF or triclosan plus copolymer or zinc oxide will be very much appreciated. The most commonly used commercial CHX gel is Corsodyl (1% CHX). Earlier in this chapter, a pharmaceutical prescription for a gel containing 1% CHX and 0.05% NaF was presented (see Box 9). Figure 231 shows three commercially available gels. For caries prevention, Elmex, in particular, and Gel-Tin offer the additional benefit of relatively high fluoride concentrations.

Chlorhexidine

Brushing with CHX-containing gel or toothpaste has been less extensively studied than rinsing. There are reports, however, demonstrating significant effects with regard to both plaque and gingival inflammation scores following brushing with a CHX-containing gel.

An intraindividual, double-blind crossover study was designed by Lie and Enersen (1986) to determine whether a 1% CHX gel used twice a day could substitute for ordinary toothpaste in a group of maintenance care patients with poor oral hygiene. A 4-week test period with CHX gel or placebo gel was followed by an intermediate

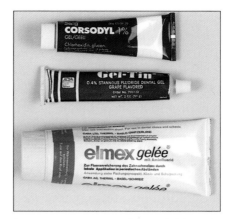

Fig 231 Commercially available gels: Corsodyl (1% CHX), Gel-Tin (0.4% SnF$_2$) (Young Dental, St Louis, MO), and Elmex (3.3% AmF, 2.2% NaF [1%F]).

period with ordinary toothpaste, and then by another 4-week test period. Plaque-covered surfaces, bleeding sites, and extra staining were recorded at each sitting.

The results demonstrated that brushing with CHX gel significantly improved the oral hygiene level and reduced gingival inflammation. The only adverse effect observed was a marked increase in the amount of brown staining, especially on the facial surfaces of nonsmokers. In this short-term study, the staining was not considered cosmetically unacceptable by the patients. The total amount of staining was always greater in smokers than in nonsmokers (Lie and Enersen, 1986).

Treatment with CHX gel in humans has been found to reduce MS to low levels in saliva and dental plaque, although after treatment there is a gradual return to pretreatment levels (Axelsson et al, 1987b; Emilson et al, 1976; Emilson, 1981; Zickert et al, 1982b). The observation that CHX could suppress MS to low levels led, in the 1980s, to clinical trials for the prevention of caries, in which MS were specifically targeted by antimicrobial measures. The subjects were mainly schoolchildren with high caries risk, determined clinically or bacteriologically by assessment of MS. These criteria for caries risk were based on studies demonstrating: (1) that children with high salivary counts of MS have a higher incidence of dental caries than do children with low counts and (2) that the approximal surfaces of premolar and molar teeth are more heavily colonized by MS and have a higher prevalence of caries than do the corresponding surfaces of anterior teeth (Axelsson et al, 1987b; Kristoffersson et al, 1986).

Studies of this treatment strategy showed clearly that, when CHX was used in subjects at high risk for caries, significant reductions in dental caries were obtained. Zickert et al (1982a, 1982b) provided directed and controlled antimicrobial therapy with a 1% CHX gel. The gel was used in custom-fitted vinyl applicators 5 minutes a day at home for 14 days only in children with more than 2.5×10^6 MS per 1 mL of saliva. The effect of treatment was examined every 4 months, and, on each occasion, only the children with high numbers of salivary MS in the test group were treated again. After 3 years, the children in the control group had developed 9.6 new caries lesions and the treated group only 4.2, a difference of 56%. A more impressive reduction of carious activity was obtained in children with greater than 1 million MS per 1 mL of saliva at the start of the study in the test group, in whom a caries reduction of 81% compared with that in the control group (3.9 versus 20.8 lesions) was found.

In a 12-month self-care program, Joyston-Bechal (1992) evaluated the effect of a combination of CHX and NaF gel on caries incidence, MS, and lactobacilli in patients who had received radiotherapy near the salivary glands. Such therapy almost invariably results in a reduced salivary secretion rate and increased caries risk. The regimen consisted of twice-daily rinsing with 10 mL of 0.2% CHX, diluted 1:1 with water, for 1 week before, during, and for 4 weeks after radiotherapy. Chlorhexidine was then substituted with a 0.05% NaF rinse daily. A saliva substitute containing 2 ppm of fluoride was used as required.

Scaling was carried out before radiotherapy, and patients received dietary advice and oral hy-

giene instruction. Radiographs were taken at baseline and after 6 and 12 months. Bacterial samples were taken from the tongue halfway through and at the end of radiotherapy, and at 6, 8, 12, 24, 40, and 52 weeks postirradiation. Whenever levels of MS exceeded 2×10^6 CFUs per 1 mL of sample, 1% CHX gel in custom-made applicator trays was applied at home for 5 minutes daily for 14 days.

In 25 subjects completing the program, there was a total of only three new caries lesions after 12 months. Thirteen preexisting enamel lesions were arrested. There were significant reductions in levels of MS from baseline values during and 4 weeks after radiotherapy. There were no significant increases in levels of MS throughout the study. Nineteen of the 25 subjects required at least one course of CHX gel to maintain this low level of MS. Levels of lactobacilli rose steadily after radiotherapy and remained high throughout the study. There was significant improvement in gingival health at 6 and 12 months. The mean stimulated whole salivary flow rate was significantly reduced after radiotherapy; this reduction persisted for 12 weeks and had not returned to baseline values at the end of 12 months. It was concluded that the CHX-NaF regimen used in this study can be recommended for the control of caries in this group of highly susceptible individuals (Joyston-Bechal, 1992).

Stannous fluoride

Stannous fluoride gels for self-care are mostly used according to the brush-on method, ie, in combination with toothbrushing and other oral hygiene aids for mechanical toothcleaning at home. The gel tested most extensively in brush-on studies is Gel-Tin, containing 0.4% SnF_2.

Boyd et al (1988) showed that a 0.4% SnF_2 gel used twice daily was significantly (65%) more effective in controlling gingivitis in adolescent orthodontic patients for a 9-month study period than was a SnF_2 mouthrinse, whether used once or twice daily, or toothbrushing alone. Tinanoff et al (1989) achieved a 47% reduction of gingivitis in a group of adults who used a 0.4% SnF_2 brush-on gel twice daily for 6 months.

Several studies of brush-on 0.4% SnF_2, used twice daily, have found substantial reductions in MS (Table 9). Wallman et al (1994) evaluated the effect of initial CHX treatment followed by SnF_2 gel application on MS in the margins of restorations. Nineteen persons with well-restored dentitions and more than 0.5×10^6 MS per 1 mL of saliva were treated with 1% CHX gel in individually designed applicators 5 minutes a day for 9 days. Ten of the subjects continued the treatment with 0.4% SnF_2 gel, and the remaining nine with a placebo gel for another 14 days. Plaque samples from margins of selected restorations and stimulated saliva were collected at baseline, after the completion of each gel treatment, and again at regular intervals for up to 24 weeks.

The CHX gel treatment suppressed MS in the margins of restorations as well as in saliva. Additional treatment with the SnF_2 gel prolonged this suppression compared with CHX treatment alone. In the CHX-SnF_2 group, the number of MS in margins of amalgam and resin composite restorations was still significantly lower at the end of the study than it was at baseline. In the CHX-placebo group, the margins of amalgam restorations, mainly placed in premolars and molars, were recolonized somewhat faster than were the margins of resin composite restorations in anterior teeth (Wallman et al, 1994).

Chewing gums and lozenges

The release of various chemoprophylactic agents from chewing gums and lozenges has been evaluated. As with sustained-release devices and varnishes, the effect will depend on release of the agent from the gum during chewing or from the lozenges during dissolution. The contact time will increase, but increased salivation will inevitably increase the clearance rate of the agent from the oral cavity. Nevertheless, further work on chewing gums and lozenges as vehicles is justified. Administration of chemoprophylactic agents via such vehicles may represent effective and acceptable routes, particularly in patients with low toothbrushing compliance. For individuals with reduced salivation, the stimulation of salivary secretion by chewing may also be beneficial and may relieve discomfort.

To date, CHX-containing chewing gum seems to be the only commercial product with proven antiplaque and antigingivitis effects. As discussed earlier in this chapter, it has been believed that the plaque-inhibitory action of CHX is mainly dependent on a slow release of the antiseptic from an oral reservoir. However, a considerable amount of data also indicate that plaque inhibition is dependent on the CHX that adsorbs to the tooth surface (for review, see Giertsen, 1990; Jenkins et al, 1988; Rölla et al, 1997; and Wåler, 1989).

Chewing gum has the potential to deliver CHX throughout the mouth; however, by the very nature of the delivery system, intimate contact with certain teeth and certain tooth surfaces is assured. It would therefore appear that, for plaque inhibition, there is a need only to saturate the receptor sites on the tooth and adjacent gingiva with CHX. This may explain why low doses (approximately 3 mg) delivered to the teeth by sprays are effective in inhibiting plaque (Kalaga et al, 1989a) and equivalent in effect to twice-daily 20-mg rinses (Kalaga et al, 1989b). The findings in a recent study by Smith et al (1996), that 20-mg daily doses of

Table 9 Reduction in MS with twice-daily brush-on application of SnF_2 (0.4%)

Study	No. of subjects	Length of study	Reduction in MS (%)
Potter et al (1984)	16	14 d	99
Tinanoff and Zameck (1985)	10	21 d	47
Vierrou et al (1986)	10	6 wk	90
Keene and Fleming (1987)	17	4 mo–3 y	31*
Tinanoff et al (1989)	31	6 mo	60*

*Compared to sodium fluoride control.

CHX from chewing gum were as effective as 40 mg per day of CHX from rinses, are consistent with previous observations.

As early as in 1987, the use of CHX-containing chewing gum has been reported to inhibit plaque growth (Ainamo and Etemadzadeh, 1987); therefore, the combined effect of CHX in the chewing gum vehicle may provide benefits to gingival health.

In the aforementioned study by Smith et al (1996), the efficacy of a CHX chewing gum on plaque and gingivitis was evaluated. Subjects (151) were screened for baseline plaque and gingival indices before receiving PMTC and being randomly assigned to one of three treatment groups: Group 1 chewed two pieces of CHX gum (Fertin A/S) for 10 minutes twice a day (total daily CHX of 20 mg); group 2 chewed two pieces of placebo gum for 10 minutes twice a day; and group 3 rinsed with 10 mL of 0.2% CHX mouthwash for 1 minute twice a day (total daily CHX of 40 mg). Plaque, gingivitis, and stain evaluations were made at 4 and 8 weeks.

Plaque and bleeding scores were significantly lower at 4 and 8 weeks in the CHX chewing gum group than in the placebo gum group and similar at 8 weeks to those in the rinse group. The intensity of staining at week 8 was significantly less in the CHX gum group than in the CHX rinse group. At week 8, the extent of staining was also significantly smaller in the CHX gum group than in the CHX rinse group. The results demonstrated that this CHX chewing gum, used with normal toothcleaning, provides adjunctive benefits to oral hygiene and gingival health that are similar to the benefits provided by a 0.2% CHX rinse (Smith et al, 1996).

The antigingivitis effect of CHX chewing gum was also evaluated in periodontal maintenance patients by van Moer et al (1996) in a 3-month clinical study. The use of a CHX chewing gum provided a significantly greater reduction in gingivitis than did the use of placebo chewing gum.

The effects of 14-day use of either a chlorhexidine-xylitol or a xylitol chewing gum on salivary levels of MS, lactobacilli, and yeasts were determined for 53 subjects (mean age: 79.49 ± 7.7 years) participating in a randomized, double-blind, placebo-controlled trial (Simons et al, 1997). Salivary flow rates and enamel staining were measured, and the attitudes of the subjects to gum chewing were evaluated by the use of structured questionnaires. The chlorhexidine-xylitol gum significantly reduced the salivary levels of MS ($P < .001$), lactobacilli ($P < .05$), and yeasts ($P < .05$), while the xylitol placebo gum produced significant reductions in MS ($P < .01$) only.

The study population found chewing gum twice a day for 10 minutes an acceptable method of receiving medication and improving oral health. Prior to gum use, participants recognized persistent symptoms of dry mouth, and their subjective evaluation of dry mouth was significantly ($P < .001$) related to the number of prescribed medications with xerostomic side effects. The desire to continue gum use was significantly related to the subjects' evaluation of their oral dryness at baseline and to their perceived oral health gain from gum usage. Only those participants with stained enamel at baseline exhibited increased enamel staining following CHX-xylitol gum usage (Simons et al, 1997).

In a 5-day, double-blind, randomized crossover study, Simons et al (1999) also showed that CHX-xylitol chewing gum used for 15 minutes twice daily resulted in significant plaque and gingivitis reduction compared with chewing gum containing xylitol alone. Subsequently, in a 1-year randomized, controlled study, Simons et al (2001) evaluated the long-term effect of chewing CHX-xylitol gum twice daily for 15 minutes on plaque and gingivitis in 111 elderly occupants of residential homes. Compared with a group using gum containing only xylitol, plaque and gingivitis scores were significantly reduced. The acceptance of both chewing gums was high.

Recently, in a short-term double-blind study, it also was shown that use of a chewing gum containing mastic (a resinous exudate obtained from the mastic tree) resulted in significant reduction of salivary bacterial counts, plaque, and GI compared with the use of a placebo gum (Takahashi et al, 2003).

PROFESSIONAL CHEMICAL PLAQUE CONTROL

Various delivery systems are available for professional chemical plaque control (antibiotics are not included):

1. Subgingival irrigants
2. Varnishes
3. Gels
4. Controlled slow-release delivery systems

In addition, some prophylaxis pastes for PMTC may contain chemical plaque control agents.

Subgingival irrigants

In endodontic treatment, root canals are routinely irrigated with antiseptics such as iodine, diluted sodium hypochloride (Dakin) solution, and ethylenediamine tetraacetic acid solution to supplement mechanical preparation by flushing away mechanically dislodged infected debris from the root canal walls, killing as many bacteria as possible.

Similar principles may be applied to nonsurgical treatment of diseased periodontal pockets. Initial healing would be enhanced if motile subgingival microflora, bacterial plaque biofilms, and plaque-covered calculus, dislodged by subgingival scaling, root planing, and debridement, could be flushed out of the pocket by subgingival irrigation with an antimicrobial irrigant that kills most of the residual microflora.

The importance of prevention of bacteremia and spread of endotoxins and lipopolysaccharides from the subgingival gram-negative microflora into the vascular systems has recently been highlighted, because lipopolysaccharides will increase the risk of preterm, low–birth weight deliveries and development of systemic inflammatory diseases such as cardiovascular diseases (heart infarcts, strokes, etc). Therefore, it is desirable to supplement subgingival scaling, root planing, and debridement with subgingival irrigation. Various direct irrigation systems are available to deliver the antimicrobial irrigant solution to the depth of the pocket, including syringes and pulsating irrigation devices. Antimicrobial agents may also function as coolants during ultrasonic subgingival scaling and debridement.

The most frequently used solutions for subgingival irrigation are CHX, SnF_2, iodine, and hydrogen peroxide (H_2O_2). To minimize bacteremia, the pocket should be irrigated and filled with a bactericidal solution (preferably iodine) before and during subgingival intrumentation. In this way, nonattaching subgingival microflora (spirochetes, etc) are flushed out and as many bacteria as possible are killed. The irrigant solution is released very rapidly from the pocket; therefore, bacteriostatic agents have very limited effect.

Accessibility

Use of a syringe with a blunt-ended needle for irrigation has been tested by many investigators. Hardy et al (1982) reported that a solution from a syringe with a blunt-ended needle would penetrate to the apex of the pocket if the needle was inserted 3 mm subgingivally. Others have found that pocket irrigation with a blunt-ended cannula attached to a Waterpik oral irrigator resulted in penetration of 71.5% in 3.5- to 6.0-mm pockets. Other professional irrigation systems have also been introduced (Periodontal Pik handpiece [Waterpik Technologies] and PerioSelect [Sultan Chemists, Englewood, NJ]), as have coolant systems for ultrasonics.

Professional devices

The Waterpik Professional Dental System (WP-32E) includes the earlier mentioned Pik Pocket tip and Pik Pocket reservoir, as well as the standard jet tip and standard reservoir. The Pik Pocket tip facilitates access to the subgingival space and

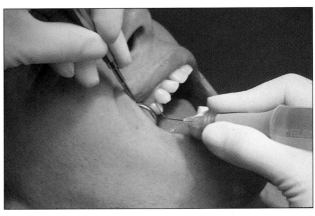

Fig 233 The cannula of the Periodontal Pik handpiece is placed into the pocket.

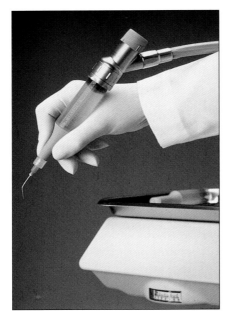

Fig 232 The Periodontal Pik handpiece for professional subgingival pulsating irrigation with antiseptic solutions (iodine solution).

more targeted delivery of antimicrobial or therapeutic solutions (see Fig 230). The tip is designed to ensure low-pressure, low-flow delivery, regardless of setting. It is made of flexible rubber for greater comfort, enhancing patient compliance. The Pik Pocket reservoir is designed for ease of measurement and economical use of all the solution. The Periodontal Pik subgingival handpiece is designed for attachment to the air turbine tube during subgingival irrigation (Figs 232 and 233).

The PerioSelect periodontal irrigation system features a dial-selected delivery mechanism that allows the user to efficiently dispense any of five individual irrigation solutions. Among the solutions employed in the system are 1.7% H_2O_2, which can be used to debride loosened plaque, calculus, and necrotic tissue following scaling, to inhibit anaerobic bacteria, and to mechanically flush and cleanse the site; a 1% $ZnCl_2$ astringent designed to reduce inflammation and to promote healing; and a 0.12% CHX compound formulated to kill pathogenic organisms, inhibit amino acid metabolism, retard acid production, and inhibit sugar-transport synthesis. These three irrigating solutions are formulated for sequential subgingival application to provide a "flush-neutralize-kill" effect in the pocket. The other two solutions include a 2% NaF compound for treating subgingival caries and controlling sensitivity and Perio Rinse (Biotrol, Louisville, CO), a specially developed bacteriostatic compound that can be used as a mouthrinse before and after therapy and should remain in the delivery lines after the completion of therapy to minimize bacterial accumulation in the tubing.

The solutions are mint flavored and ready to use. Each is contained in 1,000-mL, color-coded packaging that matches the color of the individual solution. The rack holding the packaged solutions can be mounted in the operatory or kept on a mobile cart (Fig 234).

The system's delivery unit uses a quiet peristaltic pump that permits delivery under constant

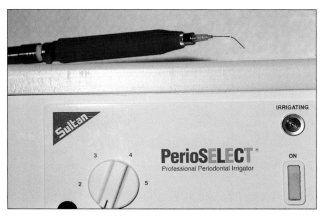

Fig 235 The handpiece of the PerioSelect unit.

Fig 234 The PerioSelect unit for professional subgingival pulsating irrigation with antimicrobial solutions.

volume and pressure, and the system warms the solutions to a comfortable temperature. The system also includes an autoclavable handpiece designed for comfortable operation (Fig 235), clear tubing to facilitate identification of the solution being dispensed, and a specially designed probe with a sideport to allow precise solution delivery. The system is offered with the handpiece, sideport probes, self-contained racks, and an optional mobile cart.

The following procedure is recommended by the manufacturer of PerioSelect for subgingival irrigation after mechanical debridement:

1. Bend the 24-gauge sideport "exact" cannula into a half-moon shape. Do not crimp the cannula.
2. Establish a firm but comfortable fulcrum grip on the PerioSelect handpiece. Finger-to-finger fulcrums work well.
3. Check the periodontal chart to ascertain the probing depth of the site to be treated.

4. Position the cannula parallel to the long axis of the tooth.
5. Place the cannula at the base of the pocket and retract 0.5 mm to establish a safe, even flow of solution and to allow the infected site to be properly irrigated at a low, 25-mL-per-minute constant pressure setting.
6. Using minimal pressure, "walk" the cannula as you would a probe and irrigate at the deepest point of each mesial-midline-distal surface. Do not drag the cannula on the floor of the pocket, which can injure inflamed tissues.
7. Irrigate the diseased pockets with hydrogen peroxide for 5 to 6 minutes.
8. To use the next agent, zinc chloride, turn the selector valve to the appropriate position and irrigate the same pockets with zinc chloride for 5 to 6 minutes. Follow the same procedure to irrigate with the CHX solution to kill bacteria, and, for dentinal sensitivity, apply NaF in the same manner.

9. When solutions are depleted, snap out the empty container and replace it with a fresh one.

10. Following treatment of an individual patient, remove and dispose of the covered cannula and remove the handpiece. Clean the handpiece ultrasonically and then sterilize it in an autoclave or chemical vapor sterilizer.

Some ultrasonic devices deliver antimicrobial agents as coolants through the tip during subgingival scaling and debridement, for example, the magnetostrictive Cavi Med 200 (Cavitron/Dentsply, York, PA), the piezoelectric Piezon Master 400 (Electro Medical Systems, Nyon, Switzerland), and Vector (Dürr Dental, Brietigheim-Bissingen, Germany).

Nosal et al (1991) evaluated the penetration depth of the water coolant for medicament lavage of an ultrasonic device into the periodontal pocket. Teeth previously planned for extraction and exhibiting probing depths of 3 mm or greater were used in this study. A reference notch was placed on the tooth at the level of the gingival margin, and the probing attachment level was measured from the base of the notch to the base of the pocket. The magnetostrictive ultrasonic device, equipped with a special tip and a reservoir of erythrocin dye–colored coolant, was activated and moved in a vertical direction from the gingival margin to the apical extent of the pocket. The tooth was extracted and the penetration depth of the dye-colored water spray was measured from the reference notch to the apical limit of the stained subgingival plaque. The tooth was counterstained with methylene blue to determine the coronal extent of the connective tissue attachment.

Dye-stained root surface was observed along the full extent of the probe tip's penetration path. The dispersion of the dye-colored stain was localized to the area of the ultrasonic probe with very little lateral dispersion. The authors concluded that the instrument may be an effective system for simultaneous mechanical removal of plaque and calculus and delivery of a chemotherapeutic agent. The limited dispersion of the liquid dye would indicate that chemical plaque control with this delivery system is dependent on thorough debridement with the instrument so that all affected surfaces are instrumented. The liquid coolant component of the ultrasonic unit was observed to extend the probe tip's penetration into the periodontal pocket. Regardless of location, probing depth, or arch, the dye was observed to penetrate to the full extent of the probing depth (Nosal et al, 1991). The water coolant of the ultrasonic unit does extend apically as far as the probe tip, thereby providing coolant at the tip of the instrument.

The combination of an ultrasonic unit and chemotherapeutic agent may be effective in light of the probe's ability to remove plaque biofilms and calculus, disrupting the subgingival microbial environment and possibly exposing more organisms to the antimicrobial effect of the subgingival irrigant. However, the minimum dispersion of any medicament lateral to the ultrasonic tip requires a comprehensive technique with overlapping strokes to ensure thorough dispersion of medicament throughout the subgingival environment.

Effects

Subgingival irrigation has been evaluated as a monotherapy, in combination with scaling and root planing with and without the use of antimicrobial agents, and with different delivery devices. When the many published studies on the effect of pocket irrigation on the subgingival microflora are evaluated, the standard of gingival plaque control by self-care must be considered. Although it is unlikely that gingival plaque control at home significantly affects the composition of the subgingival microflora in deep periodontal pockets (Kho et al, 1985; McNabb, 1992; Smulow et al, 1983), the rate of subgingival recolonization after scaling and root planing has been shown to be related to the standard of oral hygiene (Magnusson et al, 1984). Furthermore, where gingival plaque control is inadequate, the effect of irrigation may be difficult to interpret: Any improvements in periodontal conditions after pocket irrigation may be attributed to the effect of the antimicrobial agent on gingival plaque located in the orifice of the pocket only or on the plaque present in the deeper portion of the pocket.

In addition, the ability of subgingival irrigation to establish effective concentrations of a chemotherapeutic agent in the pocket over a prolonged period of time can be questioned. The relatively short contact time achieved during pocket irrigation may not be sufficient to obtain a therapeutic effect of antiseptic solutions. Some studies have indicated that chemical plaque control agents with high substantivity in the oral cavity, such as CHX, have a relatively limited effect in periodontal pockets (for review, see Wennström, 1997). Therefore, more bactericidal agents, such as iodine solutions, are recommended.

Stabholz et al (1993) evaluated the maintenance of antimicrobial activity on human root surfaces after in situ subgingival irrigation with tetracycline or CHX. The substantivity of tetracycline hydrochloride and CHX was assessed in extracted teeth. Fifty periodontally compromised teeth scheduled for extraction, with probing depths ranging between 6 and 12 mm, were root planed and then irrigated in situ with one of four solutions: tetracycline hydrochloride at concentrations of 10 or 50 mg/mL, 0.12% CHX, or 0.9% sterile saline. Each tooth was exposed to 150 mL of the respective irrigation solution. Following extractions, the teeth were incubated at room temperature for 22 days. Incubation solutions were replaced at 24-hour intervals. The solutions removed were examined for desorbed antimicrobial activity.

Tetracycline, 50 mg/mL, exhibited significantly greater antimicrobial activity than CHX for 12 days and saline for 16 days. Tetracycline, 10 mg/mL, exhibited significantly greater antimicrobial activity than both CHX and saline for 4 days. Chlorhexidine did not exhibit any significant antimicrobial activity at any time. The findings demonstrate long-lasting substantivity of tetracycline hydrochloride, but not CHX digluconate, on

teeth exposed to a single episode of pocket irrigation of their periodontally exposed roots. The amount of antimicrobial activity retained was proportional to the concentration of tetracycline hydrochloride used for irrigation (Stabholz et al, 1993).

Although in vitro studies have shown that the majority of isolated subgingival bacteria are inhibited by CHX at concentrations that would be attainable by subgingival application (62 to 125 μg/mL), its in vivo effect when used for subgingival irrigation is limited. One reason is most likely the relatively short contact time achieved during pocket irrigation. However, even when application was prolonged by incorporation of the antimicrobial agent in a gel (Oosterwaal et al, 1991b) or a sustained-release device (Addy et al, 1988), no significant prolonged effect was achieved. The inability of the large CHX molecules to penetrate the plaque biofilm, the presence of blocking protein in the crevicular fluid, the fast turnover of the fluid, and the limited surface area available for retention of the drug may also be factors that limit the effect of CHX in the subgingival area. There are studies indicating that other (and more) bactericidal agents, such as iodine solutions (Jodopax), would be more efficient for subgingival irrigation than CHX (Rosling et al, 1986, 1998, 2001; for review, see Greenstein, 1999). Iodine is able to penetrate the cell walls of the microorganisms quickly. Iodine's bactericidal effects probably result from a disruption of protein and nucleic acid structure and synthesis.

In studies by Wennström et al (1987a, 1987b), a design was used to allow the evaluation of the antimicrobial agent itself. The design included an initial 3-month period of intensive gingival plaque control to minimize the effect of improved oral hygiene on the variables used for describing the effect of the subsequent subgingival treatment. In the first part of the study, professional irrigation was performed every 2 or 3 days during a 6-week period without concomitant mechanical debridement. The pockets (deeper than 6 mm) were irrigated with a 0.2% solution of CHX, 3% hydrogen peroxide, or saline solution. Also, in each patient, some pockets were left without any subgingival treatment. The treatment effect was evaluated up to 26 weeks following the termination of irrigation.

The results revealed that subgingival irrigation, irrespective of the type of solution used, had a transient effect on clinical and microbiologic parameters. Similar short-term effects of irrigation of unscaled periodontal pockets have been reported in several other studies. Furthermore, because CHX- and H_2O_2-irrigated pockets did not show any greater improvements than saline-irrigated sites, the ability of subgingival antimicrobial irrigation to compensate for a less comprehensive mechanical debridement must be questioned. Despite the fact that each pocket was irrigated for 2 minutes every 2 or 3 days, a minimum contact time with the subgingival plaque bacteria required to kill bacteria was not reached. Hence, the transient effect observed could be attributed to a washing effect, since the effect was similar for both antimicrobial and saline solutions (Wennström et al, 1987a, 1987b).

In a second part of the same study, the pockets were treated by subgingival scaling and root planing in combination with irrigation. As in many other studies, a marked resolution of the clinical symptoms of periodontal disease and an improvement in the microbiologic variables were noted following subgingival debridement (Badersten et al, 1984a; Lindhe et al, 1982; Listgarten et al, 1978, 1986; MacAlpine et al, 1985). The use of adjunctive subgingival irrigation with CHX or H_2O_2, however, did not improve the healing results above the level obtained after mechanical debridement alone or in combination with saline irrigation.

However, the supplementary long-term effect of subgingival irrigation may be related to how bactericidal the irrigant solution is and the quality of the daily gingival plaque control. Schlagenhauf et al (1990) compared the effect of subgingival scaling and subgingival pocket irrigation with 0.1% CHX or saline controls on the repopulation of subgingival periodontal sites with disease-associated microorganisms following a single procedure of scaling and root planing. Pertinent clinical variables (attachment level, plaque index, bleeding on probing) were also recorded. In 30 individuals with previously untreated periodontal disease, 375 sites were thoroughly scaled and subsequently either rescaled, irrigated, or not treated at all for the following 6 months at 1-month intervals.

The initial scaling and root planing procedure led to significant clinical and microbiologic improvements in all experimental groups. These improvements were maintained in all but the untreated sites. Based on the observed clinical and microbiologic changes, subgingival irrigation of periodontal pockets at 1-month intervals was as effective as a similar regimen of scaling and root planing. However, 0.1% CHX, used as test irrigant, was no more effective than saline controls (Schlagenhauf et al, 1990).

Schlagenhauf et al (1994) also assessed the effect of repeated subgingival oxygen irrigations in previously untreated deep periodontal pockets. They selected 112 pockets that were 4 mm or deeper in 14 subjects. Probing attachment level and bleeding on probing were recorded, as was the presence of disease-associated microorganisms within the pockets. Subsequently, the pockets were irrigated with gaseous oxygen once a week during a continuous 8-week period. Irrigation with nitrogen served as control. Reevaluation of all clinical and microbiologic variables at the end of the study revealed that repeated oxygen insufflations resulted in significantly greater clinical improvement in the periodontal baseline conditions than did the control (nitrogen).

Wikesjö et al (1989) evaluated the effect of subgingival irrigation for 1 minute with approximately 20 mL of a 3% aqueous H_2O_2 solution. This treatment was repeated twice weekly for 6 months or until *Actinobacillus actinomycetemcomitans* could not be detected at two consecutive appointments. A total of 24 periodontal pockets in seven patients with aggressive periodontitis were treated. Three of the patients were adolescents and the remaining four were adults. At completion of the irrigation protocol, 46% of sites were found to be free from detectable levels of *A actinomycetemcomitans*. All sites were reexamined after a period of 5 months among the sites where *A actinomycetemcomitans* had been eliminated, only two were found to contain the organism. The authors concluded that this treatment regimen has some potential to suppress *A actinomycetemcomitans* and that generally the effect of the treatment lasts for at least 5 months.

Rosling et al (1983) evaluated the microbiologic and clinical effects of subgingival application of a H_2O_2-NaCl and sodium bicarbonate mixture, followed by subgingival irrigation with 1% povidone-iodine solution (Betadine; Purdue Frederich, Norwalk, CT), in the treatment of periodontal disease. Twenty adults with moderate to severe periodontal disease were included in a split-mouth design study. All patients were given oral hygiene instructions and were subjected to supragingival scaling in all four quadrants and subgingival scaling and root planing in half the dentition. Ten patients were instructed to use the chemical antimicrobial mixture twice a day instead of dentifrice and received professional application of the mixture once every 14 days for 3 months in connection with reinstruction in oral hygiene procedures. The remaining 10 patients received oral hygiene instructions combined with PMTC without the use of chemicals once every 14 days for 3 months. The effect of treatment was evaluated by monitoring the subgingival microflora, clinical periodontal parameters, and by computer-assisted subtraction analysis of serial standardized radiographs to determine changes in the mass of the supporting alveolar bone.

Subgingival debridement combined with mechanical plaque control resulted in decreased numbers of subgingival microorganisms, including spirochetes and motile rods, and arrested the progressive breakdown of the periodontal tissues. Topical antimicrobial agents used in combination with subgingival scaling further reduced the subgingival microflora and substantially improved early periodontal healing, including gain of probing attachment level and gain in radiographic alveolar bone mass during the 12 months of observation. No clinical improvement but rather a tendency to further periodontal breakdown was found in the unscaled quadrants, even in those patients who were subjected to a personal application of the topical antimicrobial mixture.

The study indicated that professional and personal subgingival application of a mixture of H_2O_2-NaCl and sodium bicarbonate will significantly enhance the microbiologic and clinical effects of periodontal scaling and root planing. These agents and the topical mode of antimicrobial therapy seemed to be efficient in the management of periodontal diseases (Rosling et al, 1983).

The three previously discussed studies and other studies seem to indicate that H_2O_2 solution is an efficient antimicrobial agent for subgingival irrigation. However, because H_2O_2 releases toxic free oxygen radicals, its suitability for frequent use is questionable.

In a study by Southard et al (1989), eight patients with moderate periodontitis volunteered to participate in a study to assess the effect of subgingival 2% CHX irrigation, with and without scaling and root planing, on clinical variables and the level of *P gingivalis* in periodontal pockets. Each quadrant was required to have at least one site exhibiting a probing depth of 6 mm or greater and bleeding on probing. The patients were treated in a randomized four-quadrant design: one quadrant received no treatment; a second quadrant received scaling and root planing only; a third quadrant received CHX irrigation only; and the fourth quadrant received scaling and root planing plus CHX irrigation. Sites to receive CHX were irrigated at 0, 1, 2, and 3 weeks.

Clinical and microbiologic variables were measured and recorded at 0, 5, 7, 11, and 15 weeks. The clinical variables measured included PI, GI, probing depth, bleeding tendency, and attachment level. The level of *P gingivalis* was also evaluated.

All variables at most time periods were significantly reduced in the root planing group. However, the PI was significantly reduced only at 5 weeks, and the level of *P gingivalis* was significantly reduced only through 7 weeks. Chlorhexidine irrigation alone significantly reduced all variables from baseline. The reduction in levels of *P gingivalis* extended through the 11-week period. Scaling and root planing plus irrigation reduced the levels of *P gingivalis* significantly more than did irrigation alone. This reduction occurred through the 11-week period but did not persist to the 15th week. The combined therapies resulted in significantly greater attachment gain at 5 and 7 weeks than did root planing alone.

The study showed that meticulous root planing was an effective treatment in deep pockets through the 7-week period. Irrigation with a high concentration of CHX (2%) alone was nearly as effective as root planing. Attachment levels were further enhanced and the level of *P gingivalis* was further reduced by the addition of 2% CHX to scaling and root planing (Southard et al, 1989). On the other hand, lower-concentration CHX subgingival irrigants (0.02% to 0.2%) have provided limited or no additional effect, as discussed earlier. Furthermore, in most studies when subgingival irrigation with antimicrobial agents has been used alone, initial reductions in gingival plaque, gingivitis, and subgingival microflora have been achieved, but treatment has failed to eliminate all signs of inflammation.

Only one study has directly compared the efficacy of a syringe with that of a pulsating jet irrigator with a cannula (Itic and Serfaty, 1992). The investigators reported similar improvements in clinical indices, but gingival crevicular fluid was decreased when a jet irrigator was used.

Use as coolants during ultrasonic subgingival scaling

Reynolds et al (1992) compared the effects of 0.12% CHX and water as coolants during subgingival debridement with an ultrasonic scaler (Cavi Med 200). The results showed greater reductions in probing depth for CHX- than water-irrigated sites. The percentage of subgingival spirochetes also tended to be lower in CHX- than water-irrigated sites.

Chapple et al (1992) also compared the effects of irrigation with CHX and water during subgingival debridement with ultrasonic instruments (Cavitron). They used a conventional tip or a new, modified tip that allows the irrigant solution to pass through the inside of the tip. After a 2-week initial oral hygiene education, 17 patients with periodontitis were treated in a split-mouth design. All four quadrants were randomly treated with conventional tip plus water irrigants, conventional tip plus 0.2% CHX irrigant, modified tip plus water irrigant, or modified tip plus 0.2% CHX irrigant.

As an effect of the initial oral hygiene education, plaque scores as well as gingival bleeding index were reduced during the first 2 weeks. Twenty-two weeks after subgingival debridement, some additional reduction in plaque scores was achieved in the two groups of quadrants irrigated with 0.2% CHX. Similar improvements in bleeding scores and probing attachment levels were observed following all four modes of treatment (Chapple et al, 1992).

Recently a 12-year maintenance study in 223 subjects with advanced periodontal disease was reported (Rosling et al, 2001). Following an initial examination comprising ordinary complete-mouth radiographs and measurements of probing attachment levels and probing depths, the patients were randomly distributed among four treatment groups. Two of the groups were then assigned as test and the other two as control. All patients were subjected to basic periodontal therapy consisting of supragingival and subgingival scaling with an ultrasonic device (Odontoson; Flex Dental, Hørsholm, Denmark), which delivered either 0.9% saline or 0.1% povidone-iodine water solution (Jodopax) during the scaling procedure. One group each of test and control patients received traditional scaling performed by specially trained dental hygienists, while the remaining test and control groups were subjected to scaling combined with surgery performed by a periodontist (modified Widman flap). Sites in the two test groups were flushed with iodine solution, and the control sites were flushed with saline during the entire scaling procedure. All patients were reexamined after 3, 5, and 12 years and received supportive periodontal treatment according to their personal needs during the study.

Nonsurgical scaling combined with subgingival flushing with 0.1% iodine solution performed by dental hygienists resulted in a mean attachment loss of only 0.28 mm per subject per 12 years compared with 0.87 mm in the control group with saline flushing ($P < .001$). The corresponding mean pocket depths at the 12-year reexamination were 2.9 mm in the test group and 3.3 mm in the control group. The mean number of lost teeth was 1.3 compared with 2.4 in the test and control groups, respectively (Rosling et al, 2001).

Forabosco et al (1996) compared the outcome of a surgical access flap (modified Widman flap) and conventional scaling and root planing versus nonsurgical irrigation with an ultrasonic instrument connected to a container housing iodine solution in a 5-year longitudinal study. At the 5-year reexamination, the authors concluded that the use of an ultrasonic instrument and an iodine irrigant solution achieved a statistically comparable outcome to that of surgical treatment, even in pockets initially up to 7 mm in depth.

In an earlier study, Rosling et al (1986) achieved a significantly greater gain in probing attachment 12 months after a single subgingival scaling and debridement with an ultrasonic scaler and 0.05% iodine solution as a coolant than was obtained with ultrasonic instrumentation and water cooling (for review, see Rams and Slots, 1996; Greenstein, 1999; and Wennström, 1997).

Varnishes

Effects on oral microflora

Varnishes for sustained release of CHX have been developed to increase the substantivity and the effectiveness of delivery of CHX to sites colonized by MS (Balanyk and Sandham, 1985; Sandham et al, 1988; Schaeken and de Haan, 1989). These stud-

ies differ with respect to varnish composition, CHX concentration, number of treatments, and full or partial treatment of dentition. Despite these differences, the results show, so far, the most persistent effect on the population of MS, ranging from strong suppression to undetectable levels of the organisms for extended periods of time. The enhanced inhibitory effect is most likely due to the prolonged contact of the antimicrobial with the dentition.

The best effects have been obtained when an extra layer of polyurethane sealant (Sandham et al, 1988, 1991) or bonding resin (Fure and Emilson, 1990) has been applied over the CHX varnish, which increases the retention of the CHX varnish on the teeth for up to approximately 5 days (Sandham et al, 1991). Without an extra layer, the long-term effect seems to be highly dependent on the concentration of CHX in the varnish. A dose-response effect has been observed in plaque studies in which a varnish containing 40% CHX was found to be more effective against MS than a varnish containing either 10% or 20% CHX (Schaeken et al, 1989) or 25% or 33% CHX (Schaeken et al, 1991b).

In the study by Sandham et al (1988), MS were not detected in saliva for a mean period of 35 weeks (range, 4 to 89 weeks) in 21 of 33 persons treated with up to four weekly applications of a 20% CHX varnish. In the remaining 12 subjects, however, repeated applications of varnish did not eliminate MS for at least 4 successive weeks, which was the criterion for a successful treatment. In this group, the past caries experience was twice that of the subjects who were treated successfully, again indicating the importance of retentive sites in the dentition in relation to efforts to eliminate MS.

An abbreviated treatment with a CHX-containing varnish (20% CHX) was compared to a similar treatment with a placebo varnish and to a prophylaxis alone for its effects on the numbers of detectable salivary MS in 51 adults (Sandham et al, 1991). The varnishes, applied once weekly for 4 weeks, were held in place with a covering layer of either of two polyurethane sealants (Fluor Protector; Vivadent, or Adhesit; Vivadent). At the first appointment, the varnish-sealant combination was applied to all tooth surfaces, but at succeeding appointments only the occlusal and approximal surfaces were covered.

The CHX varnish, covered with either sealant, reduced the salivary MS by an average of 99.9% in all 20 subjects treated, and below detectable levels for at least 4 weeks in nine patients. In the groups receiving the placebo varnish-sealant combination, the number of MS was reduced only by approximately 32%, and none of the subjects experienced absence of detectable MS for 4 weeks, although one subject did so for 3 weeks. No significant difference between the effects of the two polyurethane sealants was observed, indicating that Fluor Protector, which contains silane-fluoride, will not reduce the effect of CHX. Treatment with a single PMTC had no effect on MS levels.

Subjects treated with CHX varnish also experienced an increase in *S sanguis* and a small decrease in yeasts. Loss of detectable MS did not cause changes in the numbers of other microorganisms examined beyond those observed with CHX varnish treatment alone (Sandham et al, 1991). It has also been shown that this complete-coverage method is effective in suppressing MS levels for long periods, even in children with fixed orthodontic appliances (Sandham et al, 1992). Furthermore, the period of undetected and reduced salivary levels of MS is significantly prolonged when the varnish treatment is preceded by a gel treatment in subjects with high caries experience and exposed root surfaces (Fure and Emilson, 1990).

Ie and Schaeken (1993) evaluated the effects of one and two applications of 40% CHX varnish on the numbers of MS in human dental fissure plaque from molars and premolars. Twenty-nine subjects (aged 20 to 30 years) participated in the study and were randomly assigned to one of three groups. In each subject, two fissures with high levels of MS were selected. The fissures in group 1 (control group) were treated with a placebo varnish containing no CHX. Fissures in group 2 received a single application of 40% CHX varnish. Fissures in

group 3 received an additional application of CHX varnish 1 week after the first. Fissure plaque samples were taken prior to the first application of CHX varnish and 1, 2, and 4 months thereafter.

Compared with the results in the control group, the suppression of MS was significantly greater in plaque from group 2 for up to 2 months and in plaque from group 3 for up to 4 months after application. Mutans streptococci were suppressed more strongly in premolars than in molars and more strongly and for a longer period of time in the fissures of premolars treated twice than in the fissures of premolars treated once (Ie and Schaeken, 1993).

Schaeken et al (1996) also showed that, while 40% CHX varnish significantly reduced MS in molar fissure plaque after a single 15-minute application, the percentage of *Actinomyces naeslundii* was increased. Schaeken et al (1991a) have also shown that treatment with 40% CHX varnish every 3 months in a group of patients resulted in a decreased number of carious and restored root surfaces, comparable with that obtained after fluoride varnish treatment.

Pienihäkkinen et al (1995) compared the efficacy of a 40% CHX varnish (EC40; Certichem, Nijmegen, The Netherlands) with that of a 1% CHX–0.2% NaF gel for decreasing the level of salivary MS. The subjects were screened for a high level of MS with a Dentocult-SM strip method (Orion Diagnostica, Espoo, Finland). In varnish groups with fluoride (V_{CHXF}, n = 20) and without fluoride (V_{CHX}, n = 19), the varnish was applied to dry teeth from an ampoule and an anesthetic syringe with a blunt needle and removed after 15 minutes. In group V_{CHXF}, an additional 2.26% fluoride varnish (Duraphat; Rorer, Cologne, Germany) was applied. The CHX-NaF gel treatment included the application of the gel with rubber cups and dental tape for 5 minutes on three occasions during 1 week (G_{CHXF}, n = 21).

In group G_{CHXF}, a significant decrease ($P =$ .001) in MS was observed after 4 weeks only. In groups V_{CHX} and V_{CHXF}, the strip values for MS were still reduced after 12 weeks (Fig 236). In groups V_{CHX} and G_{CHXF}, a small, although statisti-

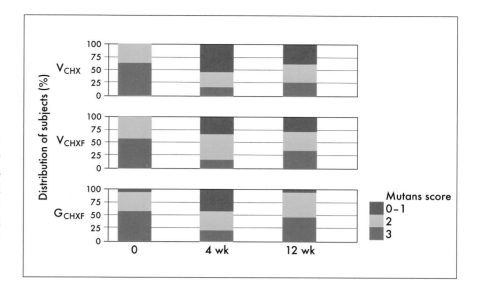

Fig 236 Distribution of subjects in Dentocult-SM categories before treatment (0) and 4 and 12 weeks after treatments. (V_{CHX}) 40% CHX varnish; (V_{CHXF}) 40% CHX and 2.26% NaF varnishes; (G_{CHXF}) 1% CHX-0.2% NaF gel. (Modified from Pienihäkkinen et al, 1995 with permission.)

cally significant, increase was observed in the total number of microorganisms after 4 and 12 weeks. Opinions on taste sensations associated with the treatments were generally negative but were least negative in the V_{CHXF} group; fewer side effects were also reported in the V_{CHXF} group. The results indicated that one treatment with CHX varnish provides suppression of salivary MS equal to or longer than three treatments with the gel form. Although the concentration of CHX was high, the varnish was associated with fewer side effects and complaints of bad taste than the 1% CHX–NaF gel, especially when coated by a fluoride varnish. This is probably because it allowed targeted application to teeth and minimal contact with the oral mucosa (Pienihäkkinen et al, 1995).

In a 30-month study on 13- to 14-year-old children in Surinam, no effect on caries incidence, MS, or lactobacilli was achieved by the use of 40% CHX varnish every 6 months compared to a placebo gel. (De Soet et al, 2002). However, in a group of mothers with high salivary MS, PMTC followed by the use of 40% CHX varnish prevented transmission of MS to their infants at 2 years of age compared to an untreated control group of mothers (Gripp and Schlagenhauf, 2002).

In a 3-year, randomized controlled trial in 1,240 selected high–caries-risk Scottish 11- to 13-year-old schoolchildren, needs-related use of a 10% CHX varnish (Chlorzoin; Imperial Chem, Macclesfield, UK) (4 to 6 times the first year and 1 to 3 times the following 2 years) resulted in initial reduction of MS compared to a placebo varnish. However, no significant caries reduction was achieved (Forgie et al, 2000).

Chlorhexidine plus thymol

Almost a decade ago, a new commercial varnish (Cervitec) was introduced (Fig 237). This varnish contains two antimicrobial agents as active components, CHX and thymol, both in concentrations of 1 wt%, in a varnish system. The varnish consists of polyvinylbutyrol, ethanol, and ethylacetate. The antimicrobial concentration has been chosen in such a way that an optimum effect can be realized without the disadvantages mentioned previously.

The use of two antimicrobial agents is important. As discussed earlier in this chapter, CHX attacks a large number of bacterial strains and specifically MS. Thymol is less specific but has broad-banded antimicrobial influences on other bacteria not, or less, influenced by CHX. The two antimicrobials in this varnish have been investigated and tested for decades. Both these agents

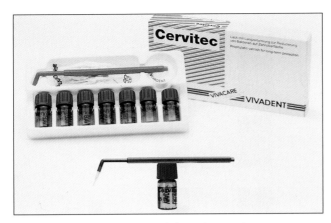

Fig 237 Cervitec CHX and thymol varnish.

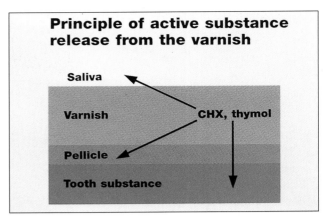

Fig 238 Illustration showing how CHX and thymol are released from the varnish to the pellicle, tooth surface, and saliva.

are accepted for the control of plaque and gingivitis by the American Dental Association Council of Dental Therapeutics.

Cervitec should not be considered a sealant. It is an antimicrobial-releasing varnish system that delivers the active antimicrobial ingredient to the hard tissues, dentin, and enamel (as well as to pellicle, plaque, and soft tissues). Unlike sealants, which function by forming a strongly adherent, impermeable polymer, the Cervitec system protects by means of an antimicrobial depot mainly in and partly on the hard tissues. The varnish interacts directly with the hard tissue surfaces as well as with the pellicle (Fig 238). The varnish penetrates initial lesions to at least 25 μm and releases antimicrobial agents for at least 3 months. The adsorption and subsequent desorption of CHX from dentin, enamel hydroxyapatite, and pellicle is the most important local antimicrobial action.

Because the varnish remains at the sites for a period of several days to weeks, the active ingredients are strongly bound to the tissues and diffuse into the dentin and enamel. This provides excellent site protection because the antimicrobials' presence prevents bacterial action and the antimicrobials are released very slowly and gradually. Thus, Cervitec builds in a few days an antimicrobial depot on the application site without

interfering with the natural protective mechanisms of remineralization by saliva and fluorides. The depot of the antimicrobials is still active after several months, but the varnish stays on the teeth for a much shorter period.

Indications. Cervitec is recommended for the following situations:

1. Prevention and control of enamel caries by application on risk surfaces, such as molar fissures, approximal surfaces of molars and premolars, and around orthodontic brackets
2. Arrest of non-cavitated enamel caries lesions
3. Prevention and control of recurrent caries along margins of crowns and restorations in selected caries-risk patients
4. Prevention and control of root caries in selected caries-risk patients
5. Primary prevention of enamel caries in erupting primary and permanent teeth in selected caries-risk children

Application. Accurate application of the varnish ensures minimal contact with the oral mucosa, resulting in fewer side effects and better patient compliance. Also, it is possible to treat preschool children, in whom the liquid and gel forms cannot be used, with the varnish.

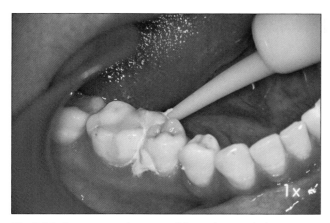

Fig 239 Syringe application of fluoride prophylaxis paste interproximally before PMTC of the selected key-risk surfaces.

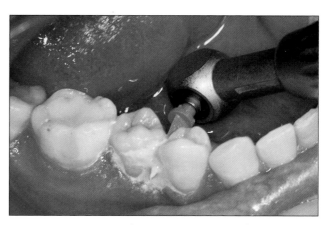

Fig 240 Interproximal PMTC using the Profin prophylaxis contra-angle handpiece (Dentatus, Hägersten, Sweden) and EVA-2000 (Dentatus) reciprocating tips.

Fig 241 After comprehensive removal of plaque from selected key-risk surfaces by PMTC, Cervitec varnish is taken from the bottle using a special disposable pointed brush.

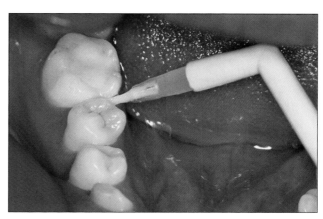

Fig 242 Application of the slow-release chemical plaque control agent (Cervitec) in order to kill remaining cariogenic microorganisms (particularly MS) and protect the selected tooth surface from these organisms for more than 3 months.

Before application of the varnish, dental plaque is mechanically removed by PMTC in particular from surfaces at risk for caries, such as fissures and approximal surfaces (Figs 239 and 240). After a thorough rinse, the working field (normally a quadrant) is isolated with cotton rolls and dried by air.

The varnish is applied with a small, disposable pointed brush (Figs 241 and 242), microbrush, syringe, or disposable micropipette, only where required, ie, spot application (Fig 243). Such spots include surfaces at risk for primary enamel caries (molar fissures, cervical areas, areas around orthodontic brackets, and approximal surfaces); recurrent caries (along crown margins, particularly in patients with xerostomia or salivary dysfunction); and root caries (exposed root surfaces in caries-risk patients). The varnish will be used selectively and not for complete-coverage application. About 10 seconds after application, the varnish is air dried.

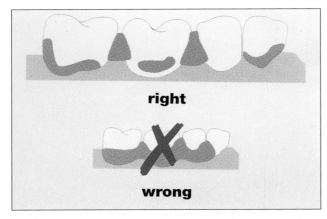

Fig 243 The Cervitec varnish is used selectively on key-risk surfaces to prevent initiation of caries and to arrest non-cavitated caries lesions.

The patient may rinse immediately after completion of treatment. A medicinal taste subsides after a short time. To allow diffusion of the active components, patients should wait approximately 3 hours before eating and about 20 hours before the resumption of normal oral hygiene. Varnish fragments can be swallowed without risk.

Cervitec is stored at normal room temperature (not exceeding 21°C, or 70°F). It should be discarded after the expiration date. A rapidly evaporating solvent is employed in Cervitec. Therefore, only the required quantity for the application to the isolated teeth should be removed from the bottle, particularly if the ambient temperature is high. For the same reason, the bottle should be closed immediately after dosage. Opened bottles should be used within 3 months.

Frequency of application. The optimal frequency of varnish application for caries inhibition probably varies by individual and has not yet been fully evaluated. Normally, the varnish is applied once every 2 or 3 months, but an intensive mode, three times within 2 weeks, has been suggested for patients with high caries risk. To promote remineralization of the tooth structure, subsequent treatment with a fluoride varnish is recommended.

It is recommended that the use of antibacterial varnishes be based on microbial diagnosis and estimated caries risk. The effect of the treatment should be monitored by follow-up samplings.

Effects on oral microflora and dental caries. Several initial experiments on Cervitec were carried out by Huizinga et al (1991). Three important questions must be answered to determine if a varnish is caries preventive:

1. Do both active ingredients initially come out of the varnish fast and are they subsequently released over prolonged periods?
2. Does the varnish have an effect on root surface caries?
3. Does the varnish have an effect on enamel caries?

The release of CHX and of thymol from Cervitec was studied in detail in vitro by Huizinga et al (1991), who arrived at the following conclusions:

1. Both CHX and thymol are released, initially quickly and subsequently more slowly.
2. The release takes place over periods much longer than 3 months.
3. After a release period of 3 months, about 80% of the available antimicrobials is still left in the varnish.
4. The combination of CHX and thymol is important for the release of both antimicrobials. When both agents are present in the varnish, the released amounts of CHX and of thymol are much larger than are the amounts released from varnishes containing CHX or thymol alone.

The preventive effects of Cervitec on root caries in situ were studied by Huizinga and Arends (1991) and Huizinga et al (1990), who concluded that a single varnish treatment with Cervitec on roots in situ reduced root caries by about 80% with respect to the control (Fig 244). In this study, the roots were under an extremely high and exaggerated continuous plaque attack for 2 weeks. Under these extreme conditions, some varnish effects were still noticeable after 4 weeks. Because in vivo roots are seldom covered continuously by plaque for more than 2 days, one application of Cervitec can be expected to be effective for periods of about 3 months.

The preventive effects of Cervitec on enamel caries were first shown by Huizinga et al (1991). Under the extreme conditions in this study—2-week continuous plaque presence on the enamel—one Cervitec treatment reduced enamel caries by at least 45% (see Fig 244). A reduction in enamel caries was still noticeable after 1 month. In vivo, Cervitec can be expected to have a preventive effect on enamel caries for periods of 3 to 6 months.

An important early in vivo experiment on Cervitec was carried out by Petersson et al (1991b). They determined the number of MS present in interdental spaces and in the saliva in 40 schoolchildren after 3 months if, according to split-mouth design, Cervitec or a control varnish was applied. There was an intermediate reduction of the number of interdental MS as well as a slower recolonization in interdental spaces after application of Cervitec. The long-term effect of the varnish was also strongly indicated by the fact that, 3 months after application, the levels of MS in saliva were significantly reduced.

In an in vitro study, Petersson et al (1992) also compared the antimicrobial effect of Cervitec on gram-positive and gram-negative oral bacterial strains and yeast to that of a placebo varnish. Cervitec varnish showed antimicrobial activity against all gram-positive and gram-negative microorganisms tested, including a candidal strain. Toothpicks and dental floss treated with Cervitec showed antimicrobial effect on MS even after storage at room temperature for up to 12 months.

In vitro experiments indicated that dentin specimens that had been treated with Cervitec released CHX for prolonged periods of time (Arends and Ruben, 1993). These authors suggested that, after application of Cevitec, root dentin acts as a CHX-releasing reservoir for a period of 6 months. Ten Cate et al (1993) also showed

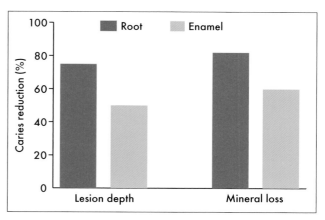

Fig 244 Summary of the in situ preventive effect of Cervitec varnish on root and enamel caries. (Modified From Huizinga and Arends, 1991 with permission.)

that Cervitec prevents the colonization of bacteria on root surfaces and disturbs the bacterial metabolism. The lactic acid production is also reduced.

Bratthall et al (1995) evaluated the possibility of reducing development of fissure caries by using Cervitec. Children aged 7 to 8 years and 12 to 13 years, 251 in each age group, were selected. Each child had at least two sound contralateral permanent molars. A split-mouth method was used with one test and one control tooth within the same arch. At baseline and after 2 years, all children were examined for decayed, missing, or filled surfaces and teeth. In addition, the size of any cavities was estimated. From 200 children, plaque samples of test and control occlusal surfaces were collected at baseline and after 1 year and processed to estimate the number of MS in saliva using the Dentocult-SM method. Cervitec varnish was applied at baseline, after 3 to 4 months, and after 8 to 9 months.

Cervitec varnish reduced fissure caries development significantly; the levels of salivary MS at baseline were significantly correlated with caries status at baseline and with total caries increment over the 2-year period. Caries development in a fissure was significantly correlated to the level of MS in plaque at the same site. Three months after the last application of varnish, a certain reduction in the numbers of MS in plaque was seen in the test teeth. A greater number of large caries lesions was found in the untreated teeth. It was concluded that varnish should be considered as a further option for the prevention of fissure caries, possibly in more individualized programs or in combination with already established methods (Bratthall et al, 1995).

In a recent study, Araujo et al (2002) evaluated the effect of Cervitec on MS in plaque and on caries development on occlusal surfaces of erupting permanent molars in a 2-year split-mouth study. Cervitec varnish was used on the test teeth at baseline and after 3 and 6 months. Thereafter, the teeth had reached full eruption. After 3 and 6 months there was a significant reduction of MS in plaque, and after 2 years all test teeth were caries free, while 50% of the untreated control teeth were decayed.

Twetman and Petersson (1997a) compared the effects of an intensive and a monthly mode of antibacterial varnish application on the levels of MS in interdental plaque and whole saliva. Eighty-eight healthy schoolchildren (aged 11 to 13 years) with high scores of salivary MS were selected by a screening procedure and randomized into two groups. Mutans streptococci were enumerated at all mesial interdental sites of the permanent first molars with the aid of a modified chairside technique. The method disclosed a total of 161 sites with moderate or high colonization levels. The subjects were treated with the 1% CHX-thymol–containing varnish (Cervitec), either in an intensive mode with three appliations within a 2-week period or in a monthly mode over a 3-month period. The varnish was applied with a miniball burnisher after the teeth had been cleaned interdentally with dental floss and dried with air. Follow-up samples of saliva and plaque from the interdental areas were collected after 1, 3, and 6 months.

Both groups exhibited a statistically significant ($P < .05$) reduction in interdental MS after 1 month compared with baseline. Absence of MS growth resulted more frequently following the in-

tensive mode than after the monthly application. After 3 months, a significant reduction compared with baseline was still found in the intensive mode group but not in the monthly mode group. No reduction was found in either group after 6 months. Mutans streptococci levels in saliva were mainly unaffected at the follow-up samplings, with the exception of a slight reduction in the intensive mode group after 1 month. The results suggested that an intensive mode of CHX-thymol varnish application is more effective against interdental MS than is the monthly mode of application. Bacterial growth should be monitored in a site-specific way because interdental reductions were not adequately reflected in whole-saliva samples (Twetman and Petersson, 1997a).

Recently, Wallman and Birkhed (2002) compared the effect of Cervitec varnish, placebo varnish, and a 1% CHX-containing gel on MS in margins of restorations in adults. The varnish was applied twice with a 3- to 4-day interval on all tooth surfaces. The gel was applied in three individual 5-minute applications on 2 consecutive days. Plaque samples from eight to nine margins of premolars and molars, as well as salivary samples, were collected from every subject after 1, 4, 8, and 12 weeks. The general finding was that the CHX gel provided the most pronounced reduction of MS, followed by the Cervitec varnish. However, no significant differences could be demonstrated either in MS counts in margins or in saliva among the three groups, but within the groups, the reduction of MS over time was significant. This indicates that PMTC followed by varnish application on the tooth surface has a preventive effect.

Twetman and Petersson (1998) also compared the effects of three different CHX-containing preparations on MS levels in interdental plaque and whole saliva. Ninety-three healthy schoolchildren (8 to 10 years old) with high scores of salivary MS were selected by a screening procedure and randomized into three equally sized groups. Mutans streptococci were enumerated at all mesial interdental sites of the permanent first molars with the aid of a modified chairside technique. The patients were then treated three times within 2 weeks with either a 1% CHX-thymol–containing varnish (group A) or a 1% CHX gel (group B), or subjected to daily supervised toothbrushing with a 0.4% CHX dentifrice for 1 month (group C). Follow-up samples of saliva and plaque from the interdental sites were collected 1 and 3 months after termination of treatment.

A statistically significant reduction in MS levels was found in the saliva and interdental plaque in all groups after 1 month. The CHX-containing dentifrice (group C) was the most effective method in reducing MS levels in saliva, and it provided a significantly stronger ($P < .05$) suppression after 1 and 3 months than did either the gel (group B) or the varnish (group A). The reduction of MS was less marked in interdental plaque than in the saliva, and the three groups exhibited MS reductions of similar magnitude (20%) and duration, persisting up to 3 months. However, a high proportion (approximately 50%) of all interdental sites were relatively unaffected by the treatments. Interdental MS colonization was difficult to combat, irrespective of the CHX preparation and method used, while the salivary levels were more easily affected. Daily toothbrushing with a CHX-containing dentifrice was more effective in reducing MS in saliva than were gel or varnish applications (Twetman and Petersson, 1998).

Twetman and Petersson (1997b) also reported that combined treatment with equal amounts of Cervitec varnish and a fluoride varnish (Fluor Protector) resulted in a longer effect on interdental plaque samples of MS than Cervitec alone.

In another study, Banoczy et al (1995) evaluated the effect of the simultaneous application of Cervitec and an AmF-SnF$_2$–containing toothpaste (Meridol) on *S mutans* counts in saliva and dental plaque of schoolchildren (12 to 14 years of age) during a 6-week period. The children were separated into group 1 (Cervitec varnish plus fluoride-containing toothpaste), group 2 (Cervitec varnish plus Meridol toothpaste), and group 3 (Meridol toothpaste alone). Over the 6 weeks, the greatest improvement in salivary *S mutans* count occurred in group 2. Overall, a statistically significant decrease in total microbiologic count and *S mutans* was found in all three groups.

In a recent 3-year study in selected caries-susceptible teenagers, Pettersson et al (2000) evaluated the effect of quarterly treatments with Cervitec varnish and fluoride varnish (Fluor Protector) on approximal caries. The mean caries incidence per 3 years on the approximal surfaces were 2.7 ± 3.1 and 3.1 ± 3.5 in the fluoride varnish and the Cervitec groups, respectively; however, the difference was not significant. It must, however, be observed that the caries-preventive effects of CHX and fluoride varnishes also represent the sum effect of initial removal of cariogenic plaque biofilms by PMTC. Thereafter, the cleaned tooth surface is protected from direct contact with reaccumulated plaque for at least 1 week by the varnish in addition to the effect of the CHX and/or fluoride. Therefore, placebo varnish and a negative control group without use of varnish should be compared with the test varnish in longitudinal, randomized controlled varnish studies on caries prevention. (For review on CHX varnishes, see Matthijs and Adriaens, 2002.)

Gels

The three most common types of commercial gels available for chemical plaque control are gels containing CHX (Corsodyl, 1% CHX), gels containing SnF$_2$ (Gel-Tin, 0.4% SnF$_2$), and gels containing AmF plus NaF or SnF$_2$ (Elmex, 3.3% AmF and 2.2% NaF). For professional use, the gels are applied in custom-made mouth trays or with syringes supragingivally (Fig 245), subgingivally, or in combination.

Effects on mutans streptococci

In a 2-year crossover study, 0.5% CHX gel applied professionally in customized trays once every 2 weeks and PMTC once every 2 weeks were compared for their effects on dental caries, gingivitis,

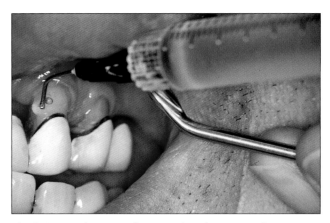

Fig 245 Application of SnF$_2$ gel on the root surfaces using a syringe.

and oral microflora in highly caries-active 13- to 14-year-old patients. Compared with PMTC, application of CHX gel had very limited effect on caries and gingivitis (Axelsson et al, 1976). However, in contrast to PMTC, treatment with CHX gel once every 2 weeks resulted in a significant reduction of MS (Emilson et al, 1982).

As discussed earlier in the chapter, Zickert et al (1982a) achieved an impressive reduction in caries activity with a 1% CHX gel. These results were confirmed in another study of schoolchildren with greater than 1 million MS per 1 mL of saliva. Chlorhexidine gel (1%) was applied every third month if the children had more than 2.5×10^5 MS per 1 mL of saliva (Lindquist et al, 1989a). After 2 years, this resulted in a 52% reduction in caries. In comparison, topical application of fluoride varnish did not reduce caries incidence in children highly colonized by MS.

The data from these two studies show that controlled antimicrobial treatment was the main reason for the reduction in caries incidence. That the treatment effect was related to the level of colonization of MS was supported by the observation that the period of time MS in saliva were found to be above the threshold level during the study period correlated significantly with the number of new lesions. This also indicated that salivary numbers of MS that were occasionally greater than the threshold level, as seen before the CHX treatment, were not as harmful as was the constantly high population of MS in the control group. A further important factor contributing to the low caries activity in the CHX-treated children could have been the increased potential for remineralization during the periods in which fewer acid-producing bacteria were present.

Another interesting observation in these studies in children was that, after termination of antimicrobial treatment, caries reduction was maintained during the 2-year follow-up period. The subjects in the test group with more than 2.5×10^5 MS per 1 mL saliva at the start of the study showed a slight tendency to have a continued lower caries increment than did the control children.

Gisselsson et al (1988) applied a 1% CHX gel four times a year in approximal spaces with a syringe. This was followed by dental flossing. This treatment resulted, after 3 years, in a caries reduction of 52% compared with a control group. The somewhat higher reductions in approximal caries, 68% and 60%, obtained by Zickert et al (1982a) and Lindquist et al (1989a), respectively, may be due to the fact that in these studies all surfaces on all teeth were exposed to the 1% CHX gel, thereby preventing recolonization of interproximal spaces by MS from other tooth surfaces. In addition, these two studies were carried out in selected groups with high salivary MS levels.

Gisselsson et al (1994) also evaluated the effect of CHX gel treatment on the incidence of approximal caries in preschool children. One hundred seventeen 4-year-old children, divided into two groups, a CHX gel group (n = 59), and a placebo gel group (n = 58), participated. Group 1 was treated four times a year with a 1% CHX gel, and group 2 was treated with a placebo gel. Approximately 0.7 mL of gel was applied interdentally by means of flat dental floss. A control group (group 3), which did not receive any flossing or gel treatment, was also included in the study (n = 116).

After 3 years, when the children were 7 years old, the mean incidence of caries on approximal surfaces, including both enamel and dentin lesions, was 2.59 in the CHX gel group, 4.53 in the placebo gel group, and 4.20 in the control group. The mean number of approximal restorations at the end of the study was 0.33 in the CHX gel group, 1.04 in the placebo gel group, and 0.80 in the control group. The progression of approximal caries lesions, diagnosed on bitewing radiographs from the age of 5 to 7 years, was slower in the CHX group than in the placebo gel group (the control group was not evaluated in this respect). A cost analysis, based on the total treatment time in minutes, showed a small increase in cost for the flossing program. Thus, professional application four times a year of CHX gel in combination with dental flossing has a caries-reducing effect on approximal caries in primary teeth (Gisselsson et al, 1994).

A possible explanation for the superior caries-preventive effect in this study may be a repeated effect on the recolonization of MS on the approximal tooth surfaces, which may have favored the fluoride-induced remineralization. A similar synergistic effect, using a CHX-NaF gel, has been reported in 1- to 4-year-old children by Tenovuo et al (1992).

In a double-blind study, 45 young adults were divided into three groups of equal size and given a PMTC (three times during 1 week) with dental gels containing either CHX (1%), or a SnF_2-AmF combination (fluoride content: 1.20%). The control group received PMTC with a placebo gel. The number of salivary S $mutans$ was monitored for 11 weeks after the gel treatment, by means of both mitis-salivarius-bacitracin agar plates and the chairside method based on the adhesion of S $mutans$ on plastic strips (Dentocult-SM).

Professional mechanical toothcleaning with a CHX gel was clearly the most effective treatment, but the baseline levels of streptococci were returned in 11 weeks. The SnF_2-AmF gel also reduced S $mutans$ significantly ($P < .001$), but recolonization occurred within 7 weeks. The placebo gel, ie, PMTC by itself, did not show any statistically significant effect on the numbers of salivary S $mutans$. None of the treatments affected the levels of lactobacilli or the total aerobic flora

in saliva samples. For individuals with high levels of carious activity, PMTC with a CHX gel or a SnF_2-AmF gel can be regarded as an alternative to the commonly used application of gel in trays at home, which requires greater patient compliance.

The combination of SnF_2 and AmF is stable, but according to these results this gel should be regarded only as an alternative for those patients who, for some reason (eg, allergy), are unable to use CHX for targeted antistreptococcal treatment. However, in future studies, SnF_2-AmF gel should be studied as an alternative to fluoride varnishes: Both have a high fluoride content and have to be professionally applied, but, unlike varnishes, the SnF_2-AmF combination reduces the numbers of oral S $mutans$.

Effects of subgingival use

The short-term bactericidal effect of 2% CHX gel, 4% SnF_2 gel, or 1.25% AmF gel on the subgingival microflora was determined in 40 periodontal pockets in 10 patients (Oosterwaal et al, 1991b). The antimicrobial gels or a placebo gel was applied in 5- to 9-mm-deep periodontal pockets three times within 10 minutes. Before and 30 minutes after the applications, samples were taken of the subgingival microflora to determine the total number of bacteria as well as the number of black-pigmented $Bacteriodes$ species.

Reductions in the total number of bacteria were found in all test groups. The reductions in the pockets treated with CHX gel or SnF_2 gel were significantly greater than reductions in the pockets treated with the placebo gel. A significant reduction in black-pigmented $Bacteroides$ was found after treatment with CHX gel or AmF gel. It was concluded that, within 30 minutes of application, 2% CHX gel or 4% SnF_2 gel reduces the microflora of periodontal pockets by more than 99% (Oosterwaal et al, 1991b).

The long-term clinical and microbiologic effects of locally applied 2% CHX gel, 1.25% AmF gel, 4% SnF_2 gel, or placebo gel in 40 periodontal pockets of 10 patients were also studied by Oosterwaal et al (1991a). The gels were applied three times within 10 minutes of mechanical debridement of the pockets. The treatment effect on the subgingival microflora was evaluated by microscopic and cultural studies of subgingival plaque samples. In addition, supragingival plaque, bleeding after probing, and probing depth were scored. Examinations were carried out before and during a period of 36 weeks after treatment.

At the start, the cultured microflora consisted mainly of anaerobic gram-negative bacteria. Following treatment, the clinical variables were significantly improved. There was a significant decrease in proportions of spirochetes, motile rods, and nonmotile rods. A significant decrease was also found in the total anaerobic count; the facultative counts remained at the pretreatment levels. This suggested that the treatment resulted in a mainly facultative subgingival microflora. The percentage of gram-negative rods showed a significant reduction after treatment, but returned to baseline at week 12.

Statistical analysis of the bacteriologic and clinical examinations failed to demonstrate any significant differences among the four treatment groups. Thus, like the placebo gel, subgingival application of 2% CHX gel, 1.25% AmF gel, or 4% SnF_2 gel did not augment the effect of mechanical debridement on bacteriologic and clinical parameters during the experimental period. However, the treatments resulted in a facultative subgingival microflora that is compatible with the host (Oosterwaal et al, 1991a).

Sustained-release devices

To be effective in vivo, a pharmacologic agent must reach its site of action and be maintained there at a sufficient concentration for the effect to occur. These three criteria—accessability, concentration, and substantivity—are strictly correlated to the effect of an antimicrobial agent.

Accessibility

The targets of the pharmacologic agents delivered locally for the treatment of periodontitis include the bacteria residing in the periodontal pocket and possibly the bacteria invading both the soft and hard tissue walls of the pocket—the pocket epithelium, the exposed cementum, or radicular dentin (Adriaens et al, 1988a, 1988b; Saglie et al, 1982a, 1982b, 1985).

In well-organized subgingival plaque biofilms, the microorganisms are inaccessible to locally delivered antimicrobial agents (as discussed earlier). Extracellular components of the bacterial biofilm may impair diffusion and/or scavenge or inactivate a significant proportion of the applied active agent, and thus protect biofilm bacteria from the action of the antimicrobial agent. These problems are well recognized in infections of implanted devices, such as catheters or prostheses, when sessile bacteria form extensive, well-organized biofilms that are thought to be associated with the observed lack of antibiotic efficacy in such infections.

Concentration and substantivity

To be effective in vivo, a pharmacologic agent should reach the site of action at a concentration higher than its minimal efficacious concentration. Definition of the desired concentration range is a key aspect in maximizing therapeutic efficacy and minimizing expected side effects.

A first approximation of the desired in situ concentration of antimicrobial agents comes from in vitro experiments looking at the susceptibility of the target microorganisms to different concentrations of the drug in terms of growth inhibition (minimum inhibitory concentration) or bacterial killing (minimum bactericidal concentration). The in vitro concentration inhibiting or killing 90% of tested isolates is frequently selected as the target in vivo concentration.

Inhibitory drug concentrations are based on in vitro experiments in which bacteria are grown under planktonic conditions. However, when bacteria are part of highly organized biofilms, such as those present in periodontal pockets and on the root surface in particular, significantly higher concentrations are needed. No direct evidence is available to estimate the minimum inhibitory or bactericidal concentrations of periodontal pathogens growing in organized biofilms; estimations from other biofilm infections, however, indicate that the necessary concentrations are at least 50 times higher than those needed to inhibit or kill bacteria growing under planktonic conditions.

Once a drug reaches the site of action in an effective concentration, it must remain at the site long enough for its pharmacologic effect to occur. Different classes of antimicrobial agents inhibit and/or kill infecting microorganisms by specific mechanisms, eg, inhibition of protein synthesis, interference with cell wall growth, or DNA synthesis. Bacterial inhibition and killing by these different specific mechanisms have been shown to require different durations of exposure to effective concentrations of antimicrobials in vitro. Thus, the effect of some antimicrobials is exquisitely time dependent, while others may require significantly shorter exposure times at effective concentrations to damage the microorganisms. The duration of antibacterial levels of the drug at the site of infection is therefore considered to be of critical importance for the effectiveness of the so-called time-dependent antimicrobials.

The gingival crevicular fluid flow into periodontal pockets averages 20 mL per hour and markedly increases with gingival tissue inflammation. Total pocket fluid volume thus may turn over about 40 times an hour in a moderate-sized periodontal pocket (0.5-µL volume), which is more frequent than the oral cavity salivary turnover rate of about 28 times an hour. The rapid clearance of substances from the periodontal pocket limits the efficacy of locally applied, nonbinding antimicrobial agents in periodontitis treatment.

This explains why antimicrobial agents that exert bactericidal effects within a 5-minute time period are preferable for subgingival irrigation.

Studies have been performed to determine the in vitro antimicrobial concentrations needed to kill *A actinomycetemcomitans*, *P gingivalis*, *P intermedia*, and *Fusobacterium nucleatum* within a 5-minute period (Kunisada et al, 1997). Iodine showed 5-minute bactericidal action against the test organisms at concentrations therapeutically attainable in subgingival sites (0.25% to 0.50%). Clinically applicable solutions of povidone-iodine, a water-soluble combination of molecular iodine and the hydrophilic polymer polyvinyl-pyrroliidone, also exhibits rapid bactericidal effects against several putative periodontal pathogens, including *A actinomycetemcomitans*, *P gingivalis*, *P intermedia*, *F nucleatum*, *Eikenella corrodens*, and *Streptococcus intermedius*, which may explain (Anderson et al, 1990; Nakagawa et al, 1990) the positive effects of iodine solution (Jodopax) as a coolant during subgingival debridement with an ultrasonic scaler in the previously discussed study by Rosling et al (2001). In comparison, the 5-minute bactericidal concentrations required of CHX (0.5% to 2.0%) and SnF_2 (0.5% to 20.0%) generally exceeded concentrations of the agents present in most commercial products. Metronidazole and amoxicillin required microbial contact times of 60 minutes or more for bactericidal activity. Interestingly, the bactericidal effects of povidone-iodine and sodium bicarbonate are potentiated in vitro by hydrogen peroxide. (For review, see Addy and Moran, 1997a, 1997b; Adriaens, 2000; Goodson, 1989; and Wilson, 1993.)

In addition, environmental conditions in periodontal pockets may alter effective in vivo concentrations of antimicrobial agents. For example, CHX may be inactivated in periodontal pockets via binding to serum proteins, which are markedly elevated in gingival crevicular fluid.

Based on these considerations, controlled, slow-release delivery systems for subgingival use of antimicrobials are more attractive than irrigants. Therefore, a new approach that uses local delivery systems containing antibiotic or antiseptic drugs has been introduced for the treatment of periodontal disease. These systems allow the ther-

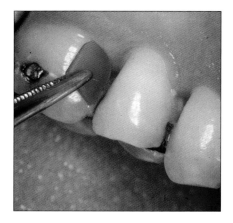

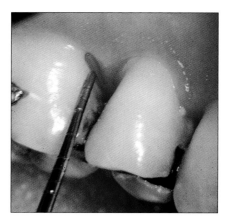

Figs 246 and 247 Placement of the PerioChip in the periodontal pocket using suitable forceps.

    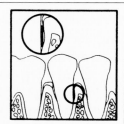

1. Open individual foil packet.

2. Grasp PerioChip at flat end with suitable forceps.

3. Insert PerioChip, curved end first, into the periodontal pocket.

4. Press PerioChip apically to the base of the pocket.

5. After proper insertion, PerioChip should rest subgingivally at the base of the pocket.

Fig 248 Stepwise instructions for using the PerioChip.

apeutic agents to be delivered to the diseased site with minimal side effects. The new approach also addresses the critical concern of unnecessary exposure of the patients to large amounts of systemic antibiotics.

Sustained-release chlorhexidine chip

A biodegradable chip for sustained and direct delivery of chlorhexidine to the periodontal pocket has been developed (Steinberg et al, 1990). This sustained, direct CHX delivery system (PerioChip) has been developed into a small CHX chip and studied in several multicenter, randomized clinical trials. PerioChip has been approved for use in the United States and is available in most European countries.

The CHX chip is an orange-brown rectangular chip, rounded on one end. It measures $4.00 \times 5.00 \times 0.35$ mm. It weighs 7.4 mg and contains 2.5 mg of CHX gluconate in a gelatin matrix (Fig 246). The chip is biodegradable and indicated for use in periodontal pockets that are 5 mm or deeper.

The chip, which is very user friendly, is inserted in the pocket by means of suitable forceps. By grasping the flat end of the chip, the clinician can place the rounded end into the periodontal pocket (Figs 247 and 248). It is recommended that the area to be treated be dry, because a wet chip may become soft and more difficult to insert. The entire procedure takes less than 1 minute. No retention system is required.

The patient should refrain from mechanical interdental cleaning for 10 days to avoid dislodging the chip. The chip biodegrades in 7 to 10 days and does not require an additional appointment for removal.

Recent in vitro studies demonstrated two phases of CHX release from the cross-linked protein matrix of the PerioChip. An initial burst effect occurred in the first 24 hours, whereby 40% of the CHX was released, probably because of diffusion. This was followed by a constant, slower release over about 7 days, occurring partially in parallel with enzymatic degradation of the chip (Lerner et al, 1996).

Two hours after installation in the periodontal pocket, the CHX concentration in the gingival crevicular fluid reaches approximately 2,000 mg/mL (Soskolne et al, 1998). During the next 4 days, the concentration fluctuates between 1,300 and 1,900 μg/mL, and from the 5th day onward the concentration gradually decreases to reach a value of 128 mg/mL at day 8. This means that during this entire period of time, the concentration in the gingival crevicular fluid remains above the 125 mg/mL required to inhibit the growth of 99% of the subgingival microflora (Stanley et al, 1989). During the first 4 days, the release kinetics are indicative of a zero-order delivery system (MacNeil et al, 1999). From day 5, the concentration decreases rapidly, most likely because of the biodegradation of the gelatin matrix.

The first multicenter study with the CHX chip was conducted in Europe with 118 subjects (Soskolne et al, 1997). It was a randomized, single-blind, split-mouth design of 6 months' duration, providing a test and a control group. Only maxillary quadrants were used. Each of the maxillary quadrants was required to have at least one pocket 5 to 8 mm in depth that bled on probing. The study groups consisted of scaling and root planing plus a CHX chip or scaling and root planing alone. Both groups received a subgingival scaling and root planing at baseline and a supragingival scaling at 3 months. In the test group, at baseline following scaling and root planing and at 3 months after scaling, chips were placed in sites where probing depths were 5 mm or more. All patients received a complete-mouth supragingival prophylaxis at 3 months according to clinical

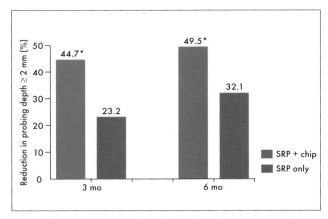

Fig 249 Percentage of probing depths improving 2 mm or more among sites that were 7 mm or deeper at baseline. (SRP) Scaling and root planing; (Chip) PerioChip. *Statistically significant at $P \leq .0001$. (Data from Soskolne et al, 1997.)

need. Evaluations were performed at baseline and at 3 and 6 months. The primary analysis was probing depth. Secondary analysis was made of probing attachment level and bleeding on probing.

In the scaling and root planing plus chip group, there was a significantly greater reduction in probing depth and improvement in attachment level than there was in the group receiving scaling and root planing alone. In sites that measured 7 mm or more at baseline, there was an even greater improvement with statistical significance at all intervals. Bleeding on probing improved in both groups, with a significant difference favoring the chip-treated group at the 3-month interval ($P \leq .05$). When analyzed on a per-patient basis, there was a significant difference in the percentage of pockets that improved 2 mm or more from baseline between the chip-treated group and the scaling-only group. In sites that were 7 mm or more at baseline, there was an even more dramatic shift (Fig 249) (Soskolne et al, 1997).

Later, two multicenter studies were conducted in the United States to compare the PerioChip plus scaling and root planing, scaling and root planing alone, and the use of a placebo chip (Jeffcoat et al, 1998). The findings from these two stud-

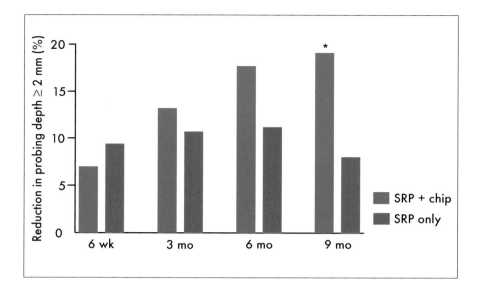

Fig 250 Percentage of probing depths improving 2 mm or more. (SRP) Scaling and root planing; (Chip) PerioChip. *Statistically significant at $P \leq .0001$, analyzed on a per-patient basis. (Data from Jeffcoat et al, 1998.)

ies were combined. A total of 447 patients participated, each with at least four teeth with probing depths between 5 and 8 mm. If supragingival calculus was present, it was removed before baseline scores were taken. All patients received a 1-hour scaling and root planing appointment, which included all teeth. Half the test group received Perio-Chip in two test sites and only scaling and root planing in two test sites. The other half of the patients received placebo chips in two test sites and scaling and root planing in the other two sites. Clinical data were collected at 7 days, 6 weeks, and 3, 6, and 9 months. If any of the chip sites measured 5 mm or deeper at 3 or 6 months, another chip was placed.

At the 9-month evaluation, the attachment level was statistically greater when scaling and root planing were combined with a CHX chip than when scaling and root planing were used alone. Bleeding on probing was significantly reduced in the CHX chip group compared with the scaling-only group at 9 months ($P = .012$). The percentage of sites with a reduction of 2 mm or more in probing depth was significantly greater in the CHX chip group than in sites treated with scaling alone (Fig 250).

In summary, these multicenter trials indicated the following (Jeffcoat et al, 1998):

1. The adjunctive use of a CHX chip resulted in a significant reduction in probing depths from baseline compared to controls at 6 and 9 months.
2. The adjunctive use of a CHX chip resulted in significant increase in probing attachment level from baseline compared to controls at 9 months.
3. The adjunctive use of a CHX chip resulted in a clinically and statistically significant outcome, as demonstrated by a greater proportion of probing depth reductions of 2 mm or more from baseline compared to controls at 9 months.
4. The CHX chip is most effective when it is placed every 3 months in pockets that remain 5 mm or more in depth.
5. No serious adverse events were reported for the CHX chip.
6. The CHX chip is a safe and effective adjunctive form of chemotherapy for the treatment of chronic periodontitis.

Later, Jeffcoat et al (2000) followed up on these multicenter studies by evaluating the effect of PerioChip on alveolar bone loss as a supplement to scaling and root planing in 45 subjects with at least four 5- to 8-mm periodontal pockets in a double-blind, controlled trial over 9 months. Control groups received either a placebo chip with scaling and root planing or scaling and root planing alone. Test groups received active Perio-Chip with scaling and root planing or scaling and root planing alone. Standardized radiographs for quantitative digital subtraction radiography were taken at baseline and after 9 months. At 9 months, 15% of subjects treated with scaling and root planing alone experienced loss of alveolar bone in one or more sites, but no subjects treated with active PerioChip and scaling and root planing lost bone ($P < .01$).

In a randomized, split-mouth, single-blind study in 26 nonsmokers over 6 months, Heasman et al (2001) showed that use of PerioChip with scaling and root planing reduced the bleeding index significantly ($P > .05$) compared with scaling and root planing alone. However, no differences in plaque score, pocket depth, or clinical attachment level were shown.

In a 6-month, randomized split-mouth study in 20 subjects with chronic periodontitis, Azmak et al (2002) recently showed that use of PerioChip with scaling and root planing significantly reduced gingival crevicular fluid and matrix metalloproteinase 8 (MMP-8) levels compared to scaling and root planing alone. However, they also showed no significant differences in pocket bleeding, pocket depth, and clinical attachment level between the two treatment procedures, which could be explained partially by the small number of sites in the study.

Recently, the effect of PerioChip on the outcome of periodontal regeneration was evaluated by Reddy et al (2003). Forty-four subjects with one or more sites with probing pocket depth and clinical attachment loss of 5 mm or more following initial therapy (scaling and root planing with oral hygiene education) and radiographic evidence of bone loss were selected for a blind, two-arm parallel study. By random, a PerioChip or placebo chip was placed in the selected pocket 1 week prior to regenerative therapy, which included graft placement and site coverage with guided tissue membranes. After 9 months, the mean gain in bone height based on quantitative subtraction digital radiography was 3.5 ± 0.45 mm in the Perio-Chip sites compared with only 1.5 ± 0.55 mm ($P < .001$) in the placebo chip sites.

The first long-term evaluation of PerioChip in periodontal maintenance therapy was recently published by Soskolne et al (2003). At baseline, a PerioChip was placed in all pockets 5 mm or more in depth in 835 patients. The patients were scheduled to receive routine periodontal maintenance therapy at 3-month intervals with repeated PerioChip placement at sites where the pocket depth remained at 5 mm or more. After 2 years, 23.2% of patients had at least 2 pockets showing a reduction of 2 mm; 59% of the sites had been reduced to a pocket depth of less than 5 mm. No or only very mild adverse reactions were observed.

COMPLETE-MOUTH DISINFECTION CONCEPT

All the surfaces of the oral cavity are colonized by microorganisms. Thus, not only the tooth surfaces but also other specific environments, such as the tongue, tonsils, oral mucosa, and the periodontal crevice, may serve as a reservoir for bacteria, which, given the right conditions, may colonize tooth surfaces and periodontal pockets. Therefore, oral hygiene procedures in highly infected, at-risk individuals should not just include toothcleaning. In such individuals, an initial intensive complete-mouth disinfection should be the method of choice.

Control of dental caries

It is well known that the dorsum of the tongue is the main reservoir for *S salivarius*, which is a very potent cariogenic (acidogenic) bacteria. In one study, however, higher numbers of MS were repeatedly found on the dorsum of the tongue after five thorough scrapings with a tongue scraper than prior to scraping, indicating that the dorsum of the tongue is an important reservoir for MS (Axelsson et al, 1987b). Lindquist et al (1989b) found a significant correlation between the prevalence of MS in saliva and its prevalence on the dorsum of the tongue. These data support the inclusion of the dorsum of the tongue in oral hygiene procedures, at least in patients highly infected by periopathogens and/or cariogenic bacteria, such as *S mutans*.

Such a tongue-cleaning procedure should include initial thorough mechanical removal of deposits from the dorsum of the tongue, followed by mouthrinsing with an efficient antimicrobial agent (CHX), because the dorsum of the tongue is normally covered by thick deposits, which prevent penetration of the antimicrobial solution. Studies using DNA fingerprints have shown that there are reservoirs, or fugitive habitats, where MS can survive CHX treatment (Kozai et al, 1991).

The whole oral cavity, an occlusal tooth surface, or even a defined area on the occlusal surface may be considered a habitat. These habitats, together with their physiologic characteristics, are often referred to as *microenvironments*. In oral microbiology, changes in the flora of a habitat such as the mouth may indicate, for example, patients at risk of caries, while changes in the microenvironments of tooth surfaces can identify a surface at risk of disease.

A study in 14-year-old children with more than 1 million MS showed the surfaces most heavily colonized with MS to be the approximal surfaces of the molars and the second premolars (see Fig 51 in chapter 1; Kristoffersson et al, 1984).

A 30-month longitudinal follow-up study evaluated the effects of PMTC as well as chemical plaque control, including CHX gels, on salivary and interproximal MS and approximal dental caries (Axelsson et al, 1987b). One hundred eighty-seven 13-year-old children with more than 1 million MS/mL saliva were selected and randomly distributed into three groups. Group I initially received PMTC; tongue scraping; CHX treatment, including mouthrinse plus 0.5% CHX gel applied interproximally by a syringe; and oral hygiene instruction with an emphasis on the approximal surfaces most colonized by MS. The treatment was given four times at intervals of 2 days, followed by one single treatment every 6 months throughout the experimental period. The initial treatment period for group II, also consisting of four visits, included the same oral hygiene instructions as for group I, but without the CHX treatment. The instructions were repeated every 6 months. Group III was maintained in the preventive program provided by the local dental health office and based on mechanical plaque control and topical use of fluorides and CHX at individualized intervals.

Group I experienced a significant, immediate reduction of MS in saliva as well as on approximal tooth surfaces. After 6 months, there were no differences among the three groups regarding these variables. Compared with baseline, there was a significant reduction of MS in all groups (Figs 251 and 252). There was no significant difference in caries progression among the three groups. However, the selected high-risk individuals in group I developed only 0.25 new manifest approximal carious lesions per year, compared with 0.27 for all children of the same age group in the area. Seventeen individuals had approximal surfaces with consistently high or consistently low MS levels. It was found that 46% of the surfaces with high values developed new or progressive caries compared with 2% of the surfaces with low values (see Figs 49 and 50 in chapter 1).

The study showed that an initial intensive combined professional mechanical and chemical plaque control treatment, including tongue scraping, may dramatically reduce salivary as well as approximal levels of MS but total elimination was not possible in any of the selected highly infected subjects. However, only one professional treatment every 6 months plus daily, needs-related oral hygiene by self-care could not maintain the low levels of MS after initial treatment (Axelsson et al, 1987b). From a caries-preventive aspect, more frequent and needs-related intervals of treatment have to be evaluated in such selected caries-risk individuals.

Fig 251 Percent of subjects with different levels of salivary *S mutans* on the different sampling occasions in groups I (PMTC, tongue scraping, CHX treatment, and oral hygiene instruction four times at intervals of 2 days, followed by one single treatment every 6 months), II (same oral hygiene instructions as for group I repeated every 6 months), and III (maintained in the preventive program provided by the local dental health office). (CFU) Colony-forming units; (BL) Baseline. (Modified from Axelsson et al, 1987 with permission.)

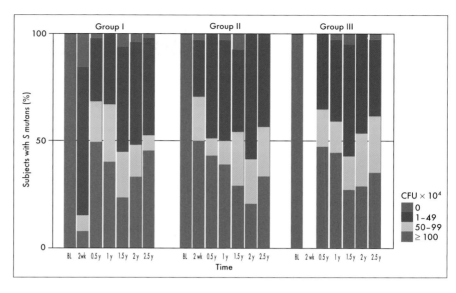

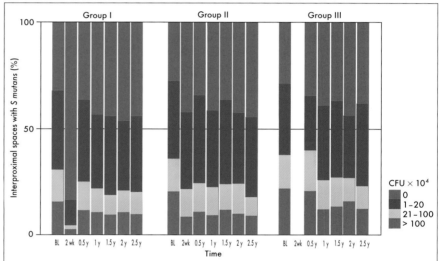

Fig 252 Percent of interproximal spaces with different approximal *S mutans* scores on the different sampling occasions in the three groups (I, II, and III) in Fig 251. (CFU) Colony-forming units; (BL) baseline. (Modified from Axelsson et al, 1987 with permission.)

The main clinical problem encountered in these studies was the difficulty in suppressing or eliminating MS for extended periods. After treatment, these organisms proliferated and recolonized the dentition. This underlines the importance of an effective disinfection of the tooth surfaces as well as other reservoirs (tongue, etc) for a long-lasting effect. In studies where complete removal of the MS from the dentition had not been achieved, treatment was followed by a gradual reappearance of the organisms.

Recolonization time varies among subjects. Bacteria returned more slowly in persons in whom cultivable MS had been reduced by 99.9% after treatment with CHX gel than they did in persons with a smaller reduction. Also, in cases where the MS had been decreased to low or undetectable levels by the gel, they generally reached pretreatment levels after 2 or 6 months (Emilson, 1981; Maltz et al, 1981; Ostela et al, 1991). In irradiated patients at high risk for caries, a very rapid recurrence of MS has been reported after CHX gel treatment (Epstein et al, 1991).

Recolonization of MS after CHX treatment follows a specific pattern. The organisms reappear first on molar surfaces and then in a posteroanterior direction on premolars and anterior teeth (Emilson et al, 1987). Heavily colonized surfaces are found more frequently on molars before treatment, and these teeth are also the most difficult to disinfect (Emilson and Lindquist, 1988). This observation indicates that these surfaces need more intensive treatment with antimicrobial agents than do other surfaces less colonized with the organisms. Another relevant observation in this study was that tooth surfaces with high levels of MS before treatment are more rapidly recolonized even if the MS has been suppressed to undetectable levels by treatment.

The most likely explanation for the reappearance of the MS is regrowth (Kozai et al, 1991). This suggests that there must be reservoirs in the dentition and other microenvironments in the oral cavity that are hardly affected or not affected at all by chemotherapy and from which the MS recolonize the dentition after the antimicrobial pressure is removed. The molar teeth, the most difficult to disinfect, are also in general more subject to caries and tend to have more restorations, thus providing more sites for bacterial retention. This is supported by studies showing that teeth of subjects with more restorations are more rapidly recolonized with MS after CHX than are those of subjects with few restorations (Maltz et al, 1981; Sandham et al, 1988, 1991). Other retentive sites include occlusal fissures, enamel cracks, active incipient lesions with rough surfaces, or under relatively intact enamel surfaces from which MS could slowly recolonize the external tooth surface. This may also explain why a more long-lasting effect is obtained in the anterior region, which has fewer retention sites (Emilson et al, 1987).

Patients with orthodontic appliances generally have many retention sites and increased populations of MS and lactobacilli (Lundström and Krasse, 1987a; Scheie et al, 1984) and increased caries. Treatment of these subjects with CHX gel has been associated with rapid recolonization by MS, thus illustrating the problem of reaching and killing the bacteria in the newly concealed areas in these subjects (Lundström and Krasse, 1987a).

This, together with lack of effect on the lactobacillus population, meant that there was no significant reduction in caries during orthodontic treatment (Lundström and Krasse, 1987b)

However, a recent meta-analysis of well-controlled clinical studies on the caries-inhibiting effect of plaque control by CHX showed an overall caries reduction of 46% (Van Rijkom et al, 1996). For fluoride toothpastes and other fluoride agents, similar analysis shows a caries reduction of only about 25%. Chemical plaque control by CHX is twice as effective because it targets the cause of dental caries (the cariogenic plaque) directly, whereas fluoride is an external factor that modifies enamel demineralization and remineralization.

These findings indicate that combinations of meticulous mechanical cleaning and use of chemical antimicrobial agents according to the so-called complete-mouth disinfection concept should be recommended initially for control of dental caries in high-risk individuals.

Control of periodontal diseases

Several studies have shown that the important periopathogens *A actinomycetemcomitans*, *P gingivalis*, *Bacteroides forsythus*, and *P intermedia* colonize not only the periodontal pocket and sulcus but also other niches and environments in the oral cavity, such as the dorsum of the tongue, the tonsils, and the oral mucosa. Via the saliva, microorganisms may be transmitted from such reservoirs and reinfect single periodontal pockets after subgingival instrumentation, irrigation, and use of local controlled slow-release delivery systems for antimicrobial agents.

Traditionally, the initial treatment of chronic periodontitis consists of four to six consecutive sessions of comprehensive scaling, debridement, and root planing at 1- to 2-week intervals. Such quadrant or sextant therapy might result in a reinfection of a previously disinfected area by bacteria from an untreated region. Within 1 week, a mature biofilm can be formed. In addition, not all the subgingival microflora and subgingival biofilms are eliminated by nonsurgical mechanical instrumentation alone.

Thus, an initial intensive combination of mechanical cleaning and use of chemical antimicrobial agents according to the complete-mouth disinfection concept is recommended for optimal elimination of periopathogens from all the reservoirs of the oral cavity in untreated periodontitis-susceptible patients (particularly those with aggressive periodontitis), in patients with so-called refractory periodontitis, before regenerative therapy, and in diabetics and patients with cardiovascular disease. To prevent reinfection, this initial complete-mouth disinfection might be prolonged, based on the individual's needs, and followed up with excellent gingival plaque control by self-care and needs-related intervals of supportive care, including PMTC, debridement, and use of chemical plaque control agents.

Some studies of the complete-mouth disinfection concept have been carried out in patients with periodontitis in Leuven, Belgium. Vandekerckhove et al (1996) investigated, over an 8-month period, the clinical benefits of complete-mouth disinfection within a 24-hour period in the control of chronic periodontitis. Ten adult patients with advanced chronic periodontitis were randomly assigned to test and control groups. The control group received the standard regimen of initial periodontal therapy, consisting of scaling and root planing in quadrants at 2-week intervals. In the complete-mouth disinfection group, scaling and root planing of the four quadrants were performed within 24 hours and immediately followed by a thorough supragingival and subgingival application of CHX to limit any transfer of bacteria. The latter involved tongue brushing with a 1% CHX gel for 60 seconds, mouth rinsing with a 0.2% CHX solution twice for 60 seconds, repeated subgingival irrigation of all pockets with a 1% CHX gel (three times within 10 minutes), and mouth rinsing twice daily with a 0.2% CHX solution for 2 weeks. In addition, both groups received thorough oral hygiene instruction. The PI, GI, probing depth, gingival recession, and bleeding

on probing were recorded prior to the professional cleaning and at 1, 2, 4, and 8 months afterward.

When the GI-PI ratio was considered, the latter was lower in the test group at all follow-up visits. For pockets of 7 mm or more, complete-mouth disinfection resulted in a significantly (P = .01) higher reduction in probing depth at each follow-up visit; at month 8, there was a reduction of 4 mm (from 8 to 4 mm) in comparison to a reduction of 3 mm (from 8 to 5 mm) for the classic therapy. The gain in clinical attachment level was 3.7 mm for the test group versus 1.9 mm for the control group. A radiographic examination also indicated that improvement in the test group was superior to that in the control group. This pilot study suggested that carrying out a complete-mouth disinfection in 1 day results in a better clinical outcome in patients with chronic periodontitis than do scalings by quadrant at 2-week intervals over several weeks.

Oral microbiologic evaluation with phase-contrast microscopy and culturing techniques showed a significantly greater reduction in periopathogens, lasting up to 2 months in the single-rooted and up to 8 months in the multirooted teeth, in the test group compared to the control group (Vandekerckhove et al, 1996).

This pilot study was followed up by Mongardini et al (1999) in 40 patients with generalized aggressive periodontitis (n = 16) or severe chronic periodontitis (n = 24) in a randomized, single-blind, parallel, 8-month study. The participants were randomized and matched for age, smoking habits, and degree of periodontal destruction in a test group (n = 20) and a control group (n = 20).

Within 24 hours, the 20 patients in the test group underwent a complete-mouth scaling and root planing in combination with application of different CHX agents for complete-mouth disinfection. The 20 control patients underwent scaling and root planing quadrant by quadrant, resulting in a total of four sessions at 2-week intervals. Except for repeated oral hygiene instructions the control group received no adjunctive therapy. During the 8-month follow-up, no additional pocket instrumentation was allowed in either the test or the control group to compare the ability of the treatments to maintain results obtained by a single session of scaling and root planing.

Scalings and root planings were performed by the same investigator in all patients with an assortment of periodontal curettes. The time spent per quadrant was on average 1 hour. In the control group, this was always started in the maxillary right quadrant, to be continued clockwise. For the test patients, the scaling and root planing was completed during two sessions within 24 hours, starting with the mandible to prevent an early contamination of the maxilla. Prior to the start of the scaling and root planing in the test group, the pockets in the respective quadrants were irrigated subgingivally (one time) with 1% CHX gel (Corsodyl) to prevent bacteremia and to detect eventual allergic reactions toward the antiseptic.

All patients received standard oral hygiene instructions immediately after the first session of scaling and root planing. These instructions included interdental plaque control (dental floss, toothpicks, and/or interdental brushes) and toothbrushing and tongue (dorsum) brushing twice daily. All patients were provided with toothbrushes and fluoride-gel toothpaste. Plaque control and oral hygiene instruction were repeated at several occasions (months 1, 2, and 4). Moreover, at the 2-month recall visit, all teeth were professionally cleaned by PMTC. The elimination of all CHX staining in the test group allowed blind clinical measurements.

Immediately after each instrumentation (thus twice within 24 hours), patients in the test group received an additional mouth disinfection by (in chronologic order):

1. Brushing of the tongue dorsum (by the patient) for 60 seconds with 1% CHX gel (Corsodyl gel)
2. Rinsing of the mouth with 0.2% CHX solution (Corsodyl mouthrinse) twice during 1 minute
3. Spraying of the pharynx with a 0.2% CHX spray (Corsodyl spray)

4. Repeated subgingival irrigation of all pockets by means of a syringe with a blunt needle (three times within 10 minutes) with 1% CHX gel (subgingival CHX application was repeated a third time at day 8)

The test group was also instructed to rinse twice daily for 1 minute with a 0.2% solution of CHX and to spray the tonsils twice daily with 0.2% CHX during the following 2 months. This study differs from the previous pilot study in two aspects: the prolonged use of CHX mouthrinse (2 months instead of 2 weeks) and the application of the 0.2% CHX spray to the tonsils.

Figures 253 to 256 show the effects over time of the test and control programs on probing depth reduction in single- and multi-rooted teeth. Generally, the complete-mouth disinfection (test group) resulted in significantly greater reductions than did the standard therapy (control group), irrespective of the disease category.

The changes in clinical attachment level over time are shown in Figs 257 to 260. In comparison to baseline, significant gains in attachment were always recorded for the test group but were only recorded for the initially deep pockets (≥ 7 mm) in the control group. In addition, mouth malodor was significantly reduced in the test group compared to the control group as an effect of healed periodontal pockets and tongue brushing with CHX gel. This should further motivate the patients to maintain a clean and healthy oral cavity (Mongardini et al, 1999)

Quirynen et al (1999) evaluated the effect of full- versus partial-mouth disinfection on the microflora in the treatment of chronic or generalized aggressive periodontitis. Sixteen patients with generalized aggressive periodontitis and 24 patients with chronic periodontitis were randomly selected to test and control groups. The control groups were scaled and root planed per quadrant at 2-week intervals and educated in oral hygiene. The test group received one-stage full-mouth disinfection treatment according to the principles described in the studies by Vandekerckhove et al (1996) and Mongardini et al (1999). At baseline

Fig 253 Changes in probing depth over time for deep (7-mm) and medium (4.5- to 6.5-mm) pockets in the maxillary right quadrant of patients with severe chronic periodontitis: Single-rooted teeth. Statistically significant differences between control and test groups at each follow-up visit are indicated (**P ≤ .005). (Modified fom Mongardini et al, 1999 with permission.)

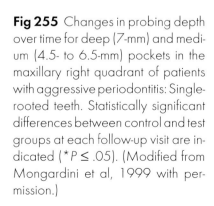

Fig 254 Changes in probing depth over time for deep (7-mm) and medium (4.5- to 6.5-mm) pockets in the maxillary right quadrant of patients with severe chronic periodontitis: Multi-rooted teeth. Statistically significant differences between control and test groups at each follow-up visit are indicated (*P ≤ .05; **P ≤ .005). (Modified from Mongardini et al, 1999 with permission.)

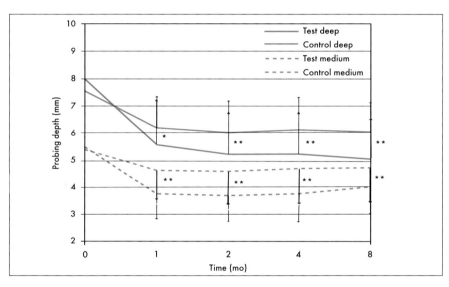

Fig 255 Changes in probing depth over time for deep (7-mm) and medium (4.5- to 6.5-mm) pockets in the maxillary right quadrant of patients with aggressive periodontitis: Single-rooted teeth. Statistically significant differences between control and test groups at each follow-up visit are indicated (*P ≤ .05). (Modified from Mongardini et al, 1999 with permission.)

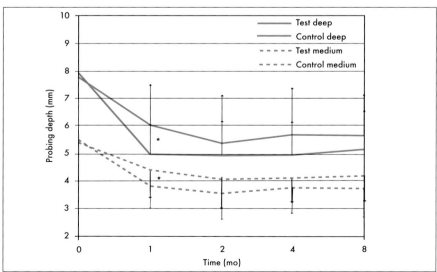

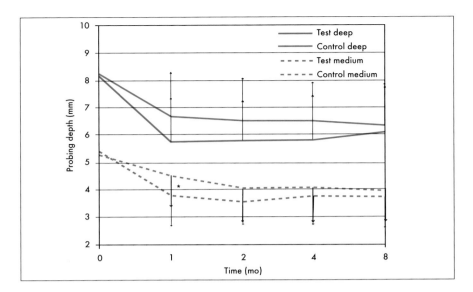

Fig 256 Changes in probing depth over time for deep (7-mm) and medium (4.5- to 6.5-mm) pockets in the maxillary right quadrant of patients with aggressive periodontitis: Multi-rooted teeth. Statistically significant differences between control and test groups at each follow-up visit are indicated (*P ≤ .05). (Modified from Mongardini et al, 1999 with permission.)

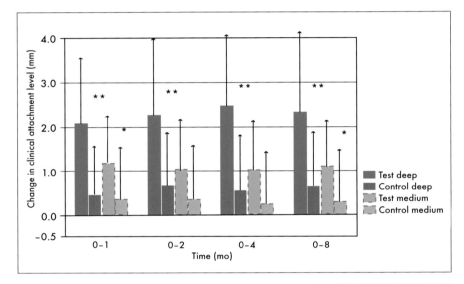

Fig 257 Changes in clinical attachment level over time for deep (7-mm) and medium (4.5- to 6.5-mm) pockets in the maxillary right quadrant of patients with severe chronic periodontitis: Single-rooted teeth. Statistically significant differences between control and test groups at each follow-up visit are indicated (*P ≤ .05; **P ≤ .005). (Modified from Mongardini et al, 1999 with permission.)

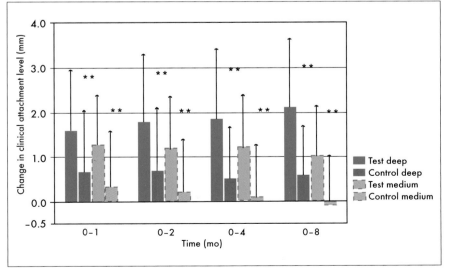

Fig 258 Changes in clinical attachment level over time for deep (7-mm) and medium (4.5- to 6.5-mm) pockets in the maxillary right quadrant of patients with severe chronic periodontitis: Multi-rooted teeth. Statistically significant differences between control and test groups at each follow-up visit are indicated (**P ≤ .005). (Modified from Mongardini et al, 1999 with permission.)

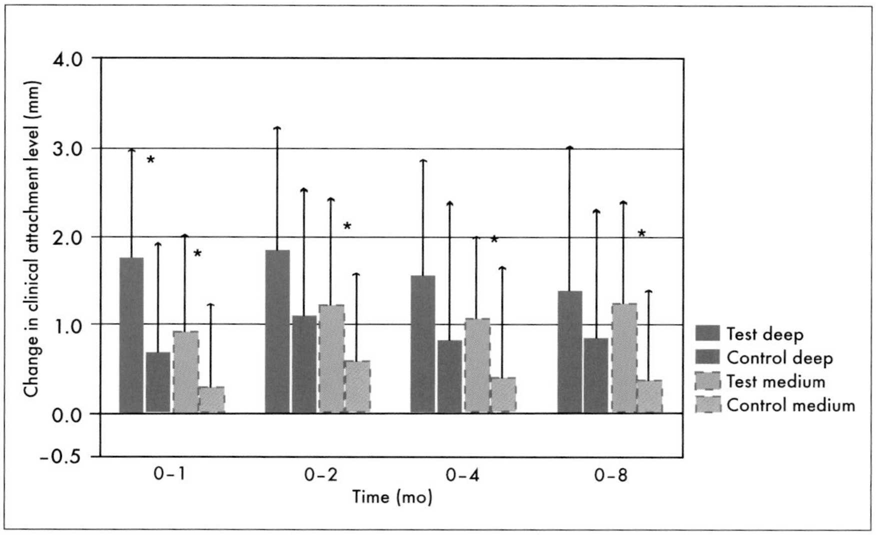

Fig 259 Changes in clinical attachment level over time for deep (7-mm) and medium (4.5- to 6.5-mm) pockets in the maxillary right quadrant of patients with aggressive periodontitis: Single-rooted teeth. Statistically significant differences between control and test groups at each follow-up visit are indicated (*$P \leq .05$). (Modified from Mongardini et al, 1999 with permission.)

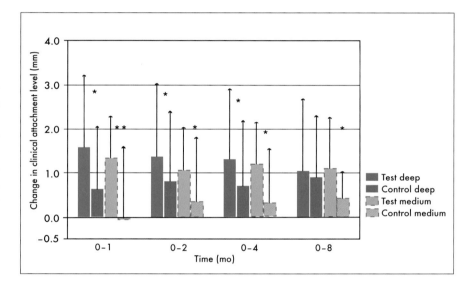

Fig 260 Changes in clinical attachment level over time for deep (7-mm) and medium (4.5- to 6.5-mm) pockets in the maxillary right quadrant of patients with aggressive periodontitis: Multi-rooted teeth. Statistically significant differences between control and test groups at each follow-up visit are indicated (*$P \leq .05$; **$P \leq .005$). (Modified from Mongardini et al, 1999 with permission.)

and after 1, 2, 4, and 8 months, microbiologic samples were taken from all oral cavity niches (tongue, oral mucosa, saliva, and pooled samples from single- and multi-rooted teeth). In comparison to the standard scaling and root planing, the one-stage, full-mouth disinfection resulted in significant additional improvements. The test group showed larger reductions in the proportions of spirochetes and motile organisms in the subgingival flora and more significant reductions in the

density of key periopathogens, with eradication of *P gingivalis*. The beneficial effects in the other niches were primarily restricted to the number of CFUs/mL of black-pigmented bacteria, especially on the mucosa and in the saliva, and to a lesser extent on the tongue. Both treatments were well tolerated by the patients.

In an earlier study, Bollen et al (1998) evaluated the effect of one-stage full-mouth disinfection in patients with severe periodontitis. Sixteen patients were selected by random to a control group (scaling and root planing treatment) and a test group (scaling and root planing with one-stage full-mouth disinfection). Plaque samples were taken from pockets, tongue, mucosa, and saliva at baseline and after 2 and 4 months. In addition, changes in pocket depth, attachment level, and bleeding on probing were recorded. The full-mouth disinfection resulted in a statistically significant additional reduction/elimination of periopathogens, especially in the subgingival pocket, but also in the other niches. These microbiologic improvements were reflected in a statistically significant higher probing depth reduction and attachment gain in the test patients.

Quirynen et al (2000) recently evaluated the relative importance of the use of CHX in the one-stage, full-mouth disinfection protocol. Three groups of 12 patients each with advanced periodontitis were followed, both from a clinical and microbiologic point of view, over a period of 8 months. The patients from the control group received scaling and root planing quadrant per quadrant at 2-week intervals. The other two groups underwent a one-stage full-mouth scaling and root planing with or without the adjunctive use of CHX. Plaque and gingivitis indices, probing pocket depth, bleeding on probing, and clinical attachment level were recorded at baseline and after 1, 2, 4, and 8 months. Microbiologic samples were taken from different intraoral niches (tongues, mucosa, saliva, and pooled samples from pockets of single- and multi-rooted teeth).

All treatment strategies resulted in significant improvements for all clinical variables, but patients in the test groups consistently reacted significantly more favorably than the control group in pocket depth reduction and attachment gain. Also, from a microbial point of view, both test groups showed greater improvement when compared to the control group, especially when periopathogens in the pockets were considered. However, the differences between the two test groups were negligible.

In a supplementary study, the long-term additional microbiologic effect of one-stage, full-mouth disinfection was evaluated in comparison with scaling and root planing alone using the checkerboard DNA-DNA hybridization technique (de Soete et al, 2001). Nineteen patients with advanced chronic periodontitis and 12 patients with aggressive periodontitis were randomly assigned to test and control groups. The control group (9 chronic and 6 aggressive periodontitis patients) received scaling and root planing per quadrant at 2-week intervals. The test group (10 chronic and 6 aggressive periodontitis patients) received one-stage, full-mouth disinfection according to the earlier described principles. At baseline and after 2, 4, and 8 months, pooled subgingival plaque samples were taken from single- and multi-rooted teeth. The presence and levels of 30 subgingival taxa were determined using whole genomic DNA probes and checkerboard DNA-DNA hybridization. Both treatments resulted in important reductions of the pathogenic species up to 8 months after therapy, both for their detection level and frequency. In the test groups, an additional improvement was achieved, especially in the chronic periodontitis group, where *P gingivalis* and *B forsythus* were reduced below detection level. The number of beneficial species remained nearly unchanged.

SCHEDULES FOR NEEDS-RELATED CHEMICAL PLAQUE CONTROL

An attempt to summarize and systematize the jungle of chemical plaque control agents and products is presented in Tables 10 and 11.

Nonrisk or low risk for caries and periodontitis

- Daily use of toothpastes containing efficient fluoride components and detergents

Table 10 Self-care delivery systems for chemical plaque control agents

| Agent | | Toothpaste | | Effect† | | | | | Mouthrinse | | Effect† | | | | | Delivery system* Irrigation | | Effect† | | | | | Gel | | Effect† | | | | | Chewing gum | | Effect† | | | |
Main group	Subgroup	CA	PR	PI	GI	C	MS	P	CA	PR	PI	GI	C	MS	P	CA	PR	PI	GI	C	MS	P	CA	PR	PI	GI	C	MS	P	CA	PR	PI	GI	C	MS
Cationic agents	Bisbiguanides CHX								+		3	2	2	3	2	+		3	2	2	3	2	+		3	2	3	3	2	+		3	2	2	3
	Alexidines																																		
	Cetylpyridinium chloride		+						+		1	1	–	–	–																				
	Heavy metal ions																																		
	Sn²⁺	+		2	2	2	2	–	+		2	2	2	2	–		+	2	2	2	2	–		+	2	2	2	2	–						
	Zn²⁺		+	1	1	1	–	–		+	1	1	–	–	–		+	1	1	–	–	–													
	Cu²⁺		+					–		+	2	–	–	–	–		+	1	1	–	–	–													
	Pyrimidines (hexetidine)									+	2	–	–	2	–																				
	Herbal extracts		+							+																									
	Sanguinaria extracts	+		1	–	–	–	–	+		1	–	–	–	–	+		1	–	–	–	–													
Anionic agents	Sodium lauryl sulfate	+		1–2	–	–	–	–	+		1	–	–	–	–																				
Nonionic agents	Listerine		+	1	2	–	–	–	+		2	2	–	1	1	+		2	2	–	1	1													
	Triclosan		+	1	2	–	–	–		+	1	2	–	–	–		+	1	2	–	–	–													
	Salifluor								+		2	2	–	2	2			2	2	–	2	–													
Other agents	Delmopinol								+		2	2	–	–	–																				
	Enzymes	+		1	1	1	–	–	+		1	–	–	–	–																				
Combination agents	CHX–zinc chloride	+		3	2	3	3	–	+		3	3	2	3	–	+		3	2	3	3	–													
	CHX–NaF								+		3	2	3	3	–																				
	CHX–zinc lactate–NaF	+		2	2	2	2	2								+		3	2	3	3	–	+		3	2	3	3	–						
	CHX-thymols																																		
	Triclosan-copolymer-NaF-SLS	+		2	2	2	2	–	+		2	2	1	–	–																				
	Triclosan–zinc citrate-NaF-SLS	+		2	2	2	2	–																											
	Triclosan–zinc citrate-monofluorophosphate-SLS																																		
	Triclosan-pyrophosphate-NaF-SLS	+		1	1	1	–	–																											
	SnF₂-SLS	+		2	1	2	2	–	+		2	1	2	2	1	+		2	1	2	2	–	+	+	2	1	2	2	–						
	AmF-SnF₂-SLS	+		2	1	2	2	–															+	+	2	1	2	2	1						
	AmF-NaF-SLS																						+		2	1	2	2	1						
	Iodine compound								+		2	1	1	2	2	+		2	1	1	2	2													

*(CA) Commercially available product; (PR) Experimental or individual preparation; (–) No available data or no effect.

†Author's estimation of effect on plaque (PI), gingivitis (GI), caries (C), mutans streptococci (MS), and periopathogens (P), on a scale of 0 to 3, based on available data.

Table 11 Professional delivery systems for chemical plaque control agents

Delivery system*

Agent Main group	Subgroup	Subgingival irrigation CA	PR	Effect† P	PD	AL	Varnish CA	PR	Effect† PI	GI	C	MS	Gel CA	PR	Effect† PI	GI	C	MS	P	Subgingival controlled slow-release chips CA	PR	Effect† P	PD	AL
Cationic agents	Bisbiguanides																							
	CHX	+	+	2	–	–	+		3	–	3	3	+	+	3	2	2	3	–	+		3	2	2
	Alexidines																							
	Heavy metal ions																							
	Sn²⁺	+		2	–	–								+	2	2	2	2	–					
	Zn²⁺																							
	Cu²⁺																							
Other agents	Delmopinol																							
	Enzymes																							
Combination agents	CHX–zinc chloride																							
	CHX–NaF											+	+	3	2	3	3	–						
	CHX–zinc lactate–NaF																							
	CHX–thymols						+		3	–	3	3												
	Triclosan–copolymer–NaF																							
	Triclosan–zinc citrate–monofluorophosphate																							
	Triclosan–pyrophosphate–NaF																							
	SnF₂		+	2	1	–						+	+	2	1	2	2	–					
	AmF–NaF											+	+	2	1	2	2	1					
	Iodine compound	+	+	3	1	2																	

*(CA) Commercially available product; (PR) Experimental or individual preparation; (–) No available data or no effect.

†Author's estimation of effect on periopathogens (P), probing depth (PD), attachment level (AL), plaque (PI), gingivitis (GI), caries (C), and mutans streptococci (MS), on a scale of 0 to 3, based on available data.

Caries risk and high caries risk

Self-care

- Alternative 1: Use of toothpaste containing triclosan plus copolymer, zinc citrate, or silicon oil plus NaF (> 0.3%), two or three times per day
- Alternative 2: Use of toothpaste containing CHX plus zinc lactate plus NaF (> 0.3%), two or three times per day

After comprehensive mechanical removal of plaque from all tooth surfaces, the slurry of remaining toothpaste is rinsed in between the teeth for 30 seconds.

To supplement CHX-containing toothpaste, intermittent use of CHX plus NaF gel in customized trays could be recommended daily for 3 weeks, three or four times per year. To supplement toothpastes containing triclosan, NaF, and SLS, a mouthrinse containing amine fluoride and SnF_2 could be recommended for long-term use after every toothcleaning.

For patients with reduced salivary secretion rate, the use of chewing gum containing CHX and NaF directly after every meal would be the first alternative.

Professional care

- Alternative 1: After three initial intensive treatments in 1 week, selective use of varnish containing CHX-thymol on risk surfaces three or four times per year after PMTC
- Alternative 2: Use of CHX plus NaF or AmF plus SnF_2 gels in customized trays three times for 5 minutes three or four times per year after PMTC

Periodontal risk and high periodontal risk

Self-care

- Alternative 1: Use of toothpaste containing triclosan plus copolymer, zinc citrate, or silicon oil plus NaF twice daily
- Alternative 2: Use of toothpaste containing CHX–zinc lactate and NaF twice daily

After mechanical removal of plaque from all tooth surfaces, the slurry of remaining toothpaste is rinsed between the teeth for 30 seconds. To supplement toothcleaning with these toothpastes, selective interproximal irrigation with CHX or SnF_2 and subgingival irrigation with iodine solution could be recommended. An alternative to interproximal irrigation could be intermittent use of a CHX mouthrinse twice daily for 3 weeks, three or four times per year. For long-term use, Meridol, Listerine, or triclosan-copolymer mouthrinses also could be recommended.

Professional care

- Irrigation with iodine (or SnF_2) solutions during scaling, root planing, and debridement
- Application of a controlled slow-release system for subgingival delivery of chlorhexidine (Perio-Chip)

The above recommendations are summarized in Table 12.

CONCLUSIONS

Chemical plaque control mainly means control of plaque formation by use of chemical antimicrobial agents (antibiotics not included). Chemical plaque control can be performed by self-care as well as by professionals. By far the most efficient plaque control programs are those combining mechanical and chemical methods. For chemical plaque control by self-care, the following delivery systems are available: toothpastes, mouthrinses, irrigants, gels, and chewing gums. For professional chemical plaque control, pocket irrigants, varnishes, gels, and controlled slow-release agents are used. Chemical plaque control should be regarded as a needs-related supplement and not a substitute for mechanical plaque control.

Table 12 Recommended use of chemical plaque control agents

	Self-care†										Professional use†‡							
	Toothpaste		Mouthrinse		Gel		Chewing gum		Irrigant		Gel		Varnish		Irrigant		Controlled slow-release chips	
Risk*	Agent	Int	Agent	Int	Agent	Int	Agent	Int	Agent	Int	Agent	Int	Agent	Int	Agent	Int	Agent	Int
C0	NaF-SLS	2×d																
C1	Tri-cop-NaF	2×d																
C2	CHX-Z-NaF or tri-cop-NaF	2×d 2×d	CHX-NAF IM3W	2×d									CHX-thym	3-4×y				
C3	CHX-Z-NaF or BM tri-cop-NaF		CHX-NaF IM3W or AmF-SnF2 LT	2×d	CHX-F-NaF	1×d	CHX-NaF	AM			CHX-NAF or SnF2-AmF	3×5 min 4×y	CHX-thym	4-6×y				
P0	NaF-SLS	2×d																
P1	Tri-cop-NaF	2×d																
P2	Tri-cop-NaF or CHX-Z-NaF	2×d	CHX-NaF IM3W	2×d					CHX or SnF2	1×d					Iod	During SC, RP, deb	CHX	3×y after deb
P3	Tri-cop-NaF or CHX-Z-NaF	2-3×d	CHX-Z-NaF IM3W or List LT	2×d					CHX or SnF2 or Iod	1×d	CHX	3×5 min 4×y			Iod	During SC, RP, deb	CHX	3×y after deb
C2P2	Tri-cop-NaF or CHX-Z-NaF	2-3×d	CHX-NaF IM3W or AmF-SnF2 LT	2×d	CHX-NaF	1×d	CHX-NaF	Am	CHX-NaF or SnF2	1×d	CHX-NaF	3×5 min 4×y			SnF2 or Iod	During SC, RP, deb	CHX	3×y after deb
C3P2	CHX-Z-NaF or BM Tri-cop-NaF	BM	CHXcopNaF IM3W or AmF-SnF2 LT	2×d	CHX-NaF	1×d			CHX-NaF	1×d	CHX-NaF	3×5 min 4×y	CHX-thym	4-6×y	SnF2	During SC, RP, deb	CHX	3×y after deb
C2P3	Tri-cop-NaF or CHX-Z-NaF	2-3×d	CHXcopNaF IM3W	2×d	CHX-NaF	1×d			CHX-NaF	1×d	CHX-NaF	3×5 min 4×y			Iod or SnF2	During SC, RP, deb	CHX	4×y after deb
C3P3	Tri-cop-NaF or BM CHX-Z-NaF	BM	CHXcopNaF IM3W	2×d	CHX-NaF	1×d	CHX-NaF	AM	CHX-NaF	1×d	CHX-NaF	3×5 min 4×y	CHX-thym	4-6×y	SnF2 or Iod	During SC, RP, deb	CHX	4×y after deb

*Predicted risk: (C0) Non-caries risk; (C1) Low caries risk; (C2) Caries risk; (C3) High caries risk; (P0) Non-periodontal risk; (P1) Low periodontal risk; (P2) Periodontal risk; (P3) High periodontal risk.

†(Int) Interval; (Tri) Triclosan; (cop) Copolymer; (Z) Zinc salts; (thym) Thymol; (List) Listerine; (Iod) Iodine solutions; (BM) Before meals; (AM) After meals; (IM3W) Intermittent for 3 weeks; (LT) Long term; (d) Day; (y) Year.

‡(SC) Scaling; (RP) Root planing; (deb) Debridement.

Chemical plaque control by self-care is characterized by high frequency (one to three times a day). It is used regularly or intermittently, has good accessibility supragingivally except interproximally in the posterior teeth, and is compliance dependent unless incorporated in toothpastes. Professional chemical plaque control should be carried out at needs-related intervals directly after mechanical removal of the plaque biofilms by PMTC or PMTC plus subgingival debridement to optimize the accessibility of the antimicrobial agent. The duration of effect by PCPC can be extended by the use of slow-release agents such as CHX-thymol varnish and controlled, slow-release agents containing CHX.

Indications

Different chemical plaque control agents may have a variety of effects, such as prevention or control of plaque formation, reducing the plaque formation rate (PFRI), reducing or disrupting existing plaque, altering the composition of the plaque flora, or altering the surface energy of the tooth and thereby the plaque adherence.

Chemical plaque control agents

Several factors influence the effect of different chemical plaque control agents:

1. *Substantivity.* This is the ability of an agent to bind to tissue surfaces and be released over time, delivering an adequate dose of the active principal ingredient in the agent.
2. *Penetrability.* For an agent to be effective, it must be able to penetrate deeply into the formed plaque biofilms.
3. *Selectivity.* The agent has the ability to affect specific pathogenic bacteria in a mixed population, eg, the specific effect of CHX on MS.
4. *Solubility.* The antiplaque agent must be solubilized in its delivery vehicle to allow rapid release into the oral environment, particularly

when the application time is limited, eg, in toothpastes, mouthrinses, and irrigants.

5. *Accessibility.* The ability of the chemical plaque control agent to reach the site of action and be maintained at that site for long enough to have sustained effect is crucial. For example, mouthrinses have very limited accessibility to the approximal surfaces of the posterior teeth, unlike gels and varnishes, which can be actively placed interproximally by a syringe and a pointed brush, respectively.
6. *Concentration.* The antiplaque effects as well as side effects of chemical plaque control agents are strongly related to the concentration. Therefore, the optimal concentration for plaque control without unacceptable side effects has to be evaluated for every plaque control agent.
7. *Ionic interactions (pH and ion binding).* Ionic interactions between agent and reception sites are important in the determination of the retention of a number of antiplaque agents because the pH of the delivery vehicle governs the state of ionization of both the agent and the receptor groups. Similarly, the interactions between a positively charged antiplaque agent and its receptor sites can be reduced by the presence of an excess of metal ions, such as calcium.
8. *Stability.* Chemical breakdown or modification of an antiplaque agent may occur during storage, particularly at elevated temperature. Commercial products should have stability for at least 2 years during normal conditions, ie, the expected effect of the antimicrobial agent should still be valid.

Delivery vehicles

Chemical plaque control agents may be delivered to the oral cavity and the individual tooth surfaces by various vehicles. The vehicle of choice depends, first, on compatibility between the active agent and the constituents of the vehicle. Second, the vehicle should provide optimal bioavailability of the agent at its action. Third, patient compliance is of major importance for self-care products. From this point of view, toothpastes should be the most suitable delivery system for self-care. Other delivery systems for self-care are mouthrinses, gels, chewing gums, and irrigants. For professional use, the recently introduced varnishes for slow-release of CHX and thymol and degradable chips for subgingival controlled, slow release of CHX seem to be the most efficient delivery systems.

Classification

The most recent trend is to group chemical plaque control agents as follows: cationic, anionic, and nonionic plaque control agents and combination plaque control agents. Cationic chemical plaque control agents are generally more potent antimicrobials than are anionic or nonionic agents because they bind readily to the negatively charged bacterial surface. The following groups of cationic agents have been tested or used as chemical plaque control agents: bisbiquanide detergents (CHX and alexidine), quaternary ammonium compounds, heavy metal salts (copper, tin, and zinc), pyrimidines (hexetidine), and herbal extracts (sanguinaria extract). Of these, CHX is clearly the most efficient and frequently used, followed by SnF_2. Chlorhexidine is regarded as the gold standard among the chemical plaque control agents and is frequently used in positive control groups for evaluation of the effect of other agents. However, there are limitations to long-term daily use of CHX, particularly in mouthrinses but also in toothpastes, because of staining.

Among anionic chemical plaque control agents, SLS is the most commonly used. It is an effective detergent and used in most toothpastes.

Triclosan and Listerine are the most successful and frequently used nonionic chemical plaque control agents. Both belong to the group of noncharged phenolic compounds and, like other phenolic compounds, they have not only docu-

mented antiplaque effects but also anti-inflammatory properties. Listerine has been used as mouthrinse for more than 100 years. Triclosan has successfully been used in toothpastes during the last decade. To increase its substantivity, triclosan is combined with either copolymer or zinc citrate. Toothpastes containing triclosan have been used in well-controlled long-term studies (up to 3 years) without any side effects.

Among other chemical plaque control agents to date, delmopinol is the most promising. It should be regarded as a surface-modifying agent and belongs to the group of substituted amino-alcohol compounds. The antiplaque effect has been documented in mouthrinses.

Plaque is a complex aggregation of various bacterial species. It is therefore unlikely that one single agent can be effective against the complete flora. Combinations of two or more agents with complementary inhibiting modes of action may enhance the efficacy and reduce adverse effects of chemical plaque control agents. Combining such agents, therefore, may offer promising prospects for new and more effective chemical plaque control agents. Examples of improvement by combination of agents are heavy metal ions (Zn^{++}) plus CHX or SLS, triclosan plus copolymer or zinc citrate, and SnF_2 plus AmF.

Chemical plaque control by self-care

Fluoride toothpastes containing CHX plus zinc lactate, triclosan plus copolymer, or triclosan plus zinc citrate seem to be the most efficient commercially available products for chemical plaque control or caries prevention. Other plaque-inhibiting toothpastes containing SnF_2 (Oral-B), AmF (Elmex), or enzymes plus 0.3% NaF plus xylitol (Zendium) may also be useful. The last is recommended in combination with CHX mouthrinse because, in contrast to the other toothpastes, it contains no anions such as SLS or monofluorophosphate that inhibit the effect of the cationic CHX.

A recent 3-year longitudinal study has shown that daily use of a triclosan-copolymer toothpaste in a selected group of periodontal susceptible patients resulted in significant reduction of probing depth, attachment loss, and subgingival periopathogens compared to a placebo fluoride toothpaste.

At present, the most efficient formula for a toothpaste for chemical plaque control and prevention of both gingivitis and periodontitis and caries is:

- Alternative 1: A silica-based toothpaste with SLS, triclosan ($\geq 0.3\%$), > 2% Gantrez copolymer or zinc citrate ($\geq 0.7\%$), NaF ($\geq 0.3\%$), and xylitol.
- Alternative 2: A silica-based toothpaste with CHX ($\geq 0.4\%$ CHX), zinc ($\geq 1.3\%$ zinc lactate), NaF ($\geq 0.3\%$), and xylitol.

The following technique is recommended to ensure that key-risk surfaces benefit from optimal accessibility and substantivity of chemical plaque control agents and fluorides in the toothpaste. The approximal spaces of the molars and premolars are filled with toothpaste from a toothbrush, a cotton roll, the forefinger, or a syringe before mechanical approximal toothcleaning with fluoride dental tape, a fluoride-impregnated wooden toothpick, or interdental toothbrush. Thereafter, the remaining tooth surfaces are cleaned with the toothbrush and toothpaste. The remaining toothpaste slurry is expectorated without subsequent rinsing with water. After regular needs-related toothbrushing procedures without special interproximal toothcleaning, the slurry of remaining toothpaste is rinsed between the teeth for 30 seconds and expectorated, followed by only one very brief rinse with water.

From compliance and cost-effectiveness aspects, the strategy for chemical plaque control by self-care should be to optimize the use of safe, efficient toothpastes.

Mouthrinses containing 0.1% or 0.2% CHX are the most efficient chemical plaque control agents for rinsing. Chlorhexidine also has a specific effect on MS. As an antigingivitis agent, 0.1% CHX is approximately as efficient as 0.2% CHX. Because of heavy staining problems even with 0.1% CHX, long-term use is questionable. In high-risk patients, short-term or intermittent rinsing, for 2 to 6 weeks, three or four times per year, is recommended. The optimal effect is achieved by rinsing twice a day directly after mechanical toothcleaning. However, the cationic CHX mouthrinse is incompatible with toothpastes containing anions such as SLS and monofluorophosphate. Therefore, a nonionic toothpaste such as Zendium (0.32% NaF) is recommended for use in direct conjunction with CHX mouthrinses as previously discussed.

In some clinical studies, Listerine mouthrinse has resulted in an antigingivitis effect similar to CHX. Because Listerine is also accepted for long-term use according to the American Dental Association's regulations, it is an alternative to CHX for long-term use as an antigingivitis agent.

For caries prevention, antiplaque mouthrinses should contain fluorides. Therefore, products such as Colgate Total and Plax (both containing triclosan-copolymer-NaF) and Meridol (AmF-SnF$_2$) are recommended for long-term or intermittent use in individuals with high caries activity.

For short-term and intermittent use, a stable and bioavailable mouthrinse containing CHX, zinc, and NaF would be attractive; a mouthrinse containing the antiplaque agent delmopinol and NaF would be appropriate for long-term use in high-risk individuals.

Studies of powered, pulsating irrigation devices such as the Waterpik oral irrigator and Pik Pocket for subgingival irrigation have shown that irrigation safely reduces not only food particles but also bacteria from areas that are inaccessible to mouthrinses. The plaque biofilms will not be completely removed, but their composition may be changed. Therefore, chemical plaque control by irrigation should only be used as a supplement to mechanical plaque removal.

The effect of water irrigation on bacteria and the enhanced effect of chemical irrigation on reducing bacteria explain the clinically significant improvements in gingival health found when powered oral irrigators are used as adjuncts to preventive dental therapy, especially for plaque control problems. It has also been shown that daily use of effective antimicrobial agents as irrigants safely reduces subgingival bacterial counts. Irrigation, especially with chemotherapeutic agents, offers a site-specific approach to dental therapy.

Irrigants containing chemical plaque control agents such as CHX, SnF_2, Listerine, and iodine have proven to be more efficient than placebo irrigants on gingivitis, plaque, and subgingival microflora when applied daily as self-care irrigants. However, recommendations for long-term use of daily irrigation with well-known and safe chemical plaque control agents should be restricted to selected high-risk individuals.

Of the commercially available gels, Corsodyl (containing 1% CHX) has proved to be the most efficient on plaque and gingivitis. For caries prevention, SnF_2 gels (Gel-Tin) or AmF plus NaF gels (Elmex) may be at least as effective as CHX gels. A commercially available CHX plus NaF gel with ensured stability and bioavailability would find wide application.

Chlorhexidine and SnF_2 gels have proven to provide a specific and substantial reduction of MS in controlled clinical trials.

The effect of chemical plaque control agents in gels for self-care is related to the frequency of use, exposure time, accessibility, and the concentration of the agents. Therefore, application of gels in individually made trays for 5 minutes should be more efficient than so-called brush-on gels.

Chewing gums containing CHX seem to be as efficient as CHX mouthrinses against plaque, gingivitis, and MS and cause much less staining. In high–caries-risk patients, particularly those with xerostomia or salivary dysfunction, use of CHX chewing gum immediately after meals would be preferable to a CHX mouthrinse because of the additional benefit of salivary stimulation by the chewing gum. In such patients, use of a chewing gum containing both CHX and NaF after every meal would be highly appropriate.

Professional chemical plaque control

Mechanical removal of subgingival microflora by scaling, root planing, debridement, and PMTC results in significantly greater improvement in clinical and microbial variables related to periodontal diseases than does subgingival irrigation with chemical plaque control agents. Several clinical studies have shown that professional subgingival irrigation with chemical plaque control agents will have a supplementary effect to subgingival scaling and debridement. Pulsating jet devices will probably result in more efficient accessibility for the irrigation solution than will syringes. Subgingival irrigation with antimicrobials (iodine) as coolants through the inside of ultrasonic tips during meticulous scaling and debridment seems promising.

The effect of CHX in subgingival irrigation solution is very limited, in contrast to the superior effect of CHX mouthrinses in the oral cavity. More bactericidal agents, such as iodine solutions, seem to be more efficient for subgingival irrigation than does CHX.

The long-term outcome of professional subgingival irrigation with antimicrobials as a supplement to subgingival scaling and debridement is related to the quality of the daily gingival plaque control by self-care. Subgingival irrigation with efficient antimicrobial solutions immediately before, during, and after subgingival scaling and debridement will reduce the risk for bacteremia in selected risk patients. This is of great importance because endotoxins from the subgingival microflora may increase the risk for cardiovascular and other systemic diseases. Subgingival pulsating jet irrigation by professionals supplemented with subgingival pulsating jet irrigation by self-care is more efficient than either of the two methods used separately.

At present, two different varnishes are available for clinical use: EC40, containing 40% CHX, and Cervitec, containing 1% CHX and 1% thymol. These two varnishes should be regarded mainly as chemical plaque control agents for professional use to selectively prevent and control root caries as well as enamel caries and recurrent caries.

Several in vitro and in vivo studies have recently documented the effect of these varnishes on MS, root caries, and enamel caries, including fissure caries. Compared to baseline data, MS levels are decreased up to 3 to 6 months after repeated application. In selected caries-risk patients, two or three applications in 7 to 10 days are recommeded during the initial intensive treatment. During the maintenance program, about four applications a year are recommended in selected caries-risk patients.

A varnish containing CHX, thymol, and fluoride with ensured stability, substantivity, and bioavailability would be very attractive for caries prevention.

Supragingival use of CHX gels has a well-documented caries-preventive effect as well as a specific effect on MS. Both SnF_2 gels and AmF gels have a caries-preventive effect similar to that of CHX but less effect on dental plaque and MS. The effect of professionally used gels is correlated to accessibility, exposure time, and the concentration of the chemical plaque control agent. To optimize accessbility, the dental plaque has to be mechanically removed by comprehensive PMTC. Thereafter, gel is syringed into the interproximal areas. Finally, gel is applied in customized trays. To prolong exposure time, new gel is reapplied three times in 5 minutes during the same visit. Subgingivally, gels have some initial antimicrobial effects, but the long-term effect is very limited. From a caries-preventive aspect, a CHX-NaF gel with ensured stability and bioavailability would be the agent of choice (see prescription for individual preparation of 1% CHX plus 0.05% NaF gel in Box 9.) Alternatives are 1% SnF_2 gel, AmF plus NaF gels, or AmF plus SnF_2 gels.

Recently a biodegradable chip (PerioChip) for controlled, sustained, and direct delivery of CHX to the periodontal pocket has been developed. The CHX chip is an orange-brown rectangular chip rounded on one end. It contains 2.5 mg of CHX gluconate in a gelatin matrix. The chip is biodegradable and indicated for use in periodontal pockets that are 5 mm or more. The chip is inserted into the pocket by means of forceps after comprehensive mechanical removal of subgingival biofilms by debridement. The entire procedure takes less than 1 minute. No retention system is required. The patient should refrain from mechanical interdental cleaning for 10 days to avoid dislodging the chip. The chip biodegrades in 7 to 10 days and does not require an additional appointment for removal.

Recent multicenter trials have indicated the following:

1. The adjunctive use of the CHX chip resulted in significant reduction in probing depths from baseline compared to controls at 6 and 9 months.
2. The adjunctive use of the CHX chip resulted in significant gain in attachment level from baseline compared to controls at 9 months.
3. The CHX chip is most effective when it is placed every 3 months in pockets that remain 5 mm or greater in depth.
4. No serious adverse events were reported for the CHX chip.
5. The CHX chip is a safe and effective adjunctive chemotherapy for the treatment of chronic periodontitis.

Complete-mouth disinfection concept

Results from studies based on the so-called complete-mouth disinfection concept open new possibilities for successful prevention and control of dental caries as well as periodontitis. This concept includes initial, intensified, and comprehensive mechanical cleaning as well as use of antimicrobial agents to eliminate plaque biofilms and microorganisms not only from every tooth surface (supragingivally and subgingivally) but also from other environments and reservoirs in the oral cavity, such as the dorsum of the tongue, the tonsils, and the oral mucosa. Such complete-mouth disinfection should be prolonged and repeated, based on the individual need, as a supplement to improved daily plaque control by self-care.

Schedules for needs-related chemical plaque control

Schedules for expected effects and the self-care and professional use of chemical plaque control agents based on predicted caries and periodontitis risk are shown in Tables 10, 11, and 12.

CHAPTER 5

USE OF FLUORIDES

Fluorides have unique external modifying effects on the initiation, progression, and arrest of caries. However, a prerequisite for fluoride to have an optimal effect is a combination of excellent mechanical and chemical plaque control directed toward the cause of dental caries: the cariogenic plaque.

The caries-inhibiting effect of fluoride (F) has been known for about 60 years. For 30 to 40 years following pioneering work by Dean and coworkers (1942), it was generally believed that the most significant caries-preventive effect of fluoride was preeruptive. Recommendations for use were based on the assumption that incorporation of fluoride in the enamel apatite lattice would confer on the enamel a resistance to acid dissolution; ie, a high intake of fluoride during tooth formation and mineralization would result in enamel that was rich in fluoride, with enduring resistance to dental caries.

Preventive measures based on this assumption included fluoridation of public water supplies to the 1-mg/L level or, alternatively, supplying fluoride in salt or milk or in tablet form to children. The above approach, ie, systemic use of fluorides as an important caries preventive method, is no longer accepted because the preeruptive caries preventive effect is almost nonexistent. Epidemiologic studies (Driscoll et al, 1982) have shown that, even where the water supply is optimally fluoridated, the topical effect of fluoride in the tooth environment is important. In addition, children with erupted permanent teeth who moved to a region with fluoridated water exhibited a reduction in caries incidence, ultimately demonstrating an incidence similar to that experienced by children born in the region. At the same time, there was an increasing recognition that caries is a disease resulting from an imbalance between processes of mineral loss and gain rather than an irreversible process of demineralization.

An increasing number of inconsistencies gradually emerged between the concept of enamel resistance and actual clinical and experimental observations. It became clear that high fluoride content in the dental hard tissues was of less importance than a moderate increase in fluoride concentration in oral fluids. Modern concepts of the mechanism of action of fluoride emphasize the importance of a daily supply to establish and maintain a significant concentration of fluoride in saliva and plaque fluid to control enamel dissolution.

There is general agreement today among scientists in the field of fluoride research that the caries-preventive and caries-controlling effects of fluorides are almost exclusively posteruptive, ie, topical. The vehicle may be drinking water, slow-release tablets, or specific topical agents such as toothpastes, gels, or varnishes. Much current fluoride research is concerned with improving the efficacy of topical treatments, based on an understanding of the mechanisms underlying the cariostatic action of fluoride (for review, see ten Cate and Featherstone, 1996). Laboratory studies have shown that fluoride not only reduces the equilibrium solubility of enamel (more or less apatite) but also exerts a wide range of effects on calcium phosphate chemistry, including the kinetics of dissolution and precipitation (ten Cate, 1994; ten Cate et al, 1995). Fluoride also affects bacterial metabolism, particularly acid production and acidurance. It has also been shown that the formation of fluoride reservoirs, in the form of calcium fluoride (CaF_2) on the tooth surfaces and in the tooth environment, is of great importance (Ogaard et al, 1992).

Natural Occurrence of Fluoride

In biology, fluoride is usually classified as a trace element and belongs to the halogen group (fluorine, chlorine, iodine, and bromine). In biologic materials, the concentration of fluoride is generally as low as a few parts per million (ppm). However, fluorides occur in the environment at far higher concentrations than do so-called trace elements.

Fluoride enters the atmosphere by volcanic action. It is returned to the earth's surface by deposition as dust, rain, snow, or fog. Fluoride enters the hydrosphere by leaching from soil and minerals into groundwater and by entry with surface water. Fluoride enters vegetation by processes such as uptake from soil and water and absorption of gaseous fluorides from air. It returns to the soil by plant waste or may enter the food chain and be returned as animal or human waste. Directly or indirectly, fluoride will also enter these pathways via different industrial processes and products.

Because of the small radius of the fluorine atom, its effective surface change is greater than that of any other element. As a consequence, fluorine is the most electronegative and reactive of all the elements. Because it reacts promptly with its environment, it rarely occurs in the free or elemental state in nature and occurs most frequently in the form of inorganic fluoride compounds. Fluorides reach their highest concentration in siliceous rocks, alkaline rocks, geothermal waters, hot springs, and volcanic fumaroles and gases.

There are about 150 known fluoride-containing minerals, of which fluorspar (fluorite [CaF_2]; 49% F), fluorapatite ($Ca_{10}F_2 [PO_4]6$; 6.3% F), and cryolite (N_3AlF_6; 54% F) are the most important. Fluorspar and fluorapatite (FA) are widespread in many countries.

Concentrations of fluoride in groundwater are influenced by such factors as availability and solubility of fluoride-containing minerals, porosity of the rocks or soils through which water passes, pH, temperature, and the presence of other elements such as calcium, aluminum, and iron that may complex with fluoride. Normally the fluoride concentration in groundwater is limited to 0.2 to 2.0 ppm, but in the United States, for example, fluoride concentrations greater than 60.0 ppm have been reported.

By contrast, most surface water contains less than 0.1 ppm of fluoride. In rivers, the concentration may range from 0.1 to 1.0 ppm. Seawater contains 1.2 to 1.4 ppm of fluoride, depending on the chlorinity. Concentrations may be altered locally by undersea volcanic activity. Thermal streams associated with volcanic activity may exhibit extremely high fluoride concentrations, ranging from 10 to 6,000 ppm in acidic spring waters. Most fluoride in water exists as free fluoride ions, but complexed fluoride increases with increasing salinity, reaching 50% to 60% in seawater.

INTAKE OF FLUORIDE

Intake of fluoride is mainly derived from drinking water and beverages. It is estimated that about 60% to 65% comes from such sources in regions with fluoride levels of less than 0.3 mg/L in the drinking water and about 75% to 80% in regions with higher fluoride concentrations. Researchers have estimated the average fluoride intake by adults from the following dietary sources:

1. Water and nondairy beverages: 60% to 80%
2. Grain and cereal products: 6% to 8%
3. Meat, fish, and poultry: 5% to 7%
4. All other foods: 10% to 14%

Mineral water may contain 1.8 to 5.8 mg of fluoride per liter. Tea leaves are a particularly rich source of fluoride, most of which is rapidly released into tea infusions, within 5 to 10 minutes. Fluoride concentrations of brewed tea commonly range from 0.5 to 4.0 ppm (Duckworth and Duckworth, 1978). As would be anticipated, fluoride concentrations in tea made with fluoridated water are somewhat greater than are those found in tea brewed from water with low fluoride content.

Although the fluoride content of most meat, fish, and poultry products is quite modest, it is extremely high in a few items; in canned sardines, the content may be as high as 16 mg/kg. The fluoride content of mechanically deboned meat products is high because of the presence of bone particles.

Use of fluoridated water in commercial food preparation slightly increases the fluoride content of canned fruits, vegetables, soups, and stews, but overall such foods do not contribute large amounts of fluoride to the diet.

Fluoride intake from diet (including drinking water and beverages with fluoride levels of less than 6 mg/L) and recommended use of fluoride-containing dental products such as toothpastes, mouthrinses, lozenges, and chewing gums will normally have no adverse effect on general health in young adults and adults. However, in children up to the age of 6 years, a high intake of fluoride will result in visible fluorosis of the teeth. The maturation phase of the maxillary incisors, when susceptibility to fluorosis is greatest, occurs when an individual is 22 to 26 months of age (Evans and Stam, 1991). To prevent the development of visible and esthetically disturbing fluorosis, fluoride intake in infants and preschool children should be limited and controlled.

The intake of fluoride associated with development of enamel fluorosis of the permanent teeth has been estimated to range from 40 to 100 μg/kg per day. Infants consuming formulas made from concentrated liquids, or powders diluted with water providing fluoride at 1,000 μg/L, are at risk of dental fluorosis. A major effort should be made to avoid use of fluoridated water for dilution of formula powders. For young infants, formulas prepared from concentrated liquids should be made with nonfluoridated water if it is economically feasible. For older infants, who consume appreciable amounts of solid or semi-solid baby food, the daily intake of formula falls substantially below 0.15 L a day, and the risk of using fluoridated water to dilute concentrated liquid formulas decreases.

Ophaug et al (1985) determined the dietary intake of fluoride by 6-month-old formula-fed infants living in three communities with different water fluoride concentrations: less than 0.3 ppm, 0.3 to 0.7 ppm, and more than 0.7 ppm. Fluoride intakes averaged 230, 320, and 490 μg/d, respectively, or approximately 29, 40, and 61 μg/kg of body weight for an 8-kg infant. The authors pointed out that concentrated liquid formula diluted with fluoridated water (1 ppm) may provide a fluoride intake greater than 100 μg/kg per day.

The vessel used for cooking may influence fluoride content of the food. Teflon-coated vessels release fluoride, whereas aluminum vessels absorb fluoride during cooking. When water containing 1 mg of fluoride per 1 kg of water was boiled for 15 minutes in a stainless steel or Pyrex vessel, there was little change in fluoride concentration. In a Teflon-coated vessel, the fluoride concentration increased to about 3 mg/kg; in an aluminum vessel, the fluoride concentration decreased to about 0.3 mg/kg (Full and Parkins, 1975). As mentioned earlier, tea may contain from 0.5 to 4.0 ppm of fluoride. In some subtropical and tropical countries, even young (1- to 6-year-old) children drink a lot of tea every day, which may explain the relatively high prevalence of fluorosis in such regions.

The predominant diet of infants is human or cow milk, which has a very low fluoride concentration compared to plasma, ranging in human milk from only 0.005 to 0.01 mg/L, even when the mother's intake of fluoride is high. The promotion of breastfeeding of infants has been increasingly successful in the United States and most European countries in recent decades, whereas commercially prepared formulas were more common during the 1960s and 1970s.

When infant formulas are manufactured in regions with fluoridated water supplies, most of the fluoride is removed from the water before it is incorporated into the infant formula. Thus the fluoride content of infant formulas is at present influenced only to a small extent by water used in commercial preparation; the major determinant is the fluoride content of the mineral mix. Concentrations of fluoride are generally somewhat greater in isolated soy protein–based formulas and in protein hydrolysate–based formulas than in milk-based formulas, primarily because the sources of calcium that must be added to non-milk-based formulas contain appreciable amounts of fluoride. In the case of isolated soy protein–based formulas, additional fluoride is included as part of the isolated soy protein.

Among commercially prepared baby food items, chicken is also notable for its fluoride content. The chicken meat comes from aged laying hens with high fluoride concentrations in the bones. The meat used in commercial preparation of strained chicken for infants is obtained by mechanically deboning the neck and spine. The fluoride content of the resultant product may be as high as 5,000 µg/kg.

For children 1 to 6 years of age, as for older individuals, fluoride intake is dervied predominantly from beverages rather than from food. Fluoride intake is likely to be quite low when the predominant beverage is cow milk but will be considerably greater when consumption of fruit juices, fruit-flavored drinks, and carbonated beverages is high.

Fluoride ingestion also occurs with topical use of dental products containing fluoride, such as dentifrices, mouthrinses, varnishes, and gels (Table 13). Intake from these sources should not be overlooked in establishing recommendations for use of fluoride supplements. It is estimated that about 10% of the world population uses fluoride toothpaste, predominantly in the industrialized countries. The conventional fluoride concentration in toothpastes was 0.10% but has recently been increased in most brands to 0.15%.

Studies have shown that 20% to 60% of toothpaste is swallowed and almost completely absorbed in children younger than 6 years of age. If a 3-year-old child uses 1 g of 0.15% fluoride toothpaste per day and swallows 50%, the total fluoride absorption should be about 0.75 mg per day, which, in combination with other sources of fluoride, such as drinking water, will increase the risk of visible fluorosis in the permanent incisors. Therefore, for children under the age of 6 years, only small amounts of fluoride toothpaste (0.5 g, the size of a pea) should be used, and brushing should be followed by thorough rinsing with water.

Because the major caries-inhibiting effect of fluorides is almost exclusively posteruptive, there is no scientific basis for recommending fluoride supplements for preschool children. It is surprising that in some countries supplementary fluoride tablets (0.25 to 0.50 mg per day) are still recommended from the age of 6 months. The intake of fluoride by 5-year-old children is often more than 50 µg/kg a day. A daily fluoride supplement of 0.50 mg given to an 18-kg, 5-year-old child results in an additional 28 µg/kg of fluoride each day. Thus, for many preschool children who are given supplements, total fluoride intake will exceed 100 µg/kg per day.

Table 13 Fluoride content and dosage of toothpastes and other common topical fluoride preparations*

Route	Type of preparation[†]	Fluoride concentration (ppm)	Amount used	Total fluoride dose (mg)
Topical application by dental personnel	2% NaF solution	9,100	2–3 mL	18–27
	10% SnF$_2$ solution	24,250	2–3 mL	49–73
	5% NaF varnish	22,600	0.5–1 g	11–23
	1.23% APF gel	12,300	3–4 g	37–49
	9% SnF-ZnSiO$_4$ paste	22,500	2 g	45
Home treatment	0.05% NaF solution	226	10 mL/d	2.3
	0.2% NaF solution	910	10 mL/wk	9.1
	0.5% APF gel	5,000	3–4 g/tray	15–20
	0.4% SnF$_2$	970	2 g/brush	2
Toothpaste	0.22% NaF	1,000	2 g[‡]	2
	0.76% MFP	1,000	2 g[‡]	2
	0.145% F (NaF + MFP)	1,450	2 g[‡]	2.9

*From Fomon and Ekstrand (1996). Reprinted with permission.

[†](NaF) Sodium fluoride; (SnF$_2$) Stannous fluoride; (APF) Acidulated phosphate fluoride; (ZnSiO$_4$) Zinc silicate; (MFP) Monofluorophosphate.

[‡]1 g twice daily.

TOXICOLOGY OF FLUORIDE

Topical fluoride agents are safe and harmless if used strictly as directed. However, systemic intake of fluoride must be limited because fluoride is a toxic substance. Based on the very few known cases of accidental death attributed to ingestion of fluoride by children, it has been concluded that if a child ingests a fluoride dose in excess of 15 mg/kg, then death is likely to occur. A fluoride dose as low as 5 mg/kg may be fatal for some children. Therefore, the probable toxic dose (PTD), defined as the threshold dose that could cause serious or life-threatening systemic signs and symptoms necessitating immediate emergency treatment and hospitalization, is 5 mg/kg of fluoride.

It is essential that the fluoride concentrations in dental products be known to the persons who use them. It is even more important to know the amounts of fluoride that are contained in the unit packages (bottles or tubes) as well as the amounts involved during routine usage and how these amounts relate to the PTD. The fluoride concentrations in toothpastes are approximately 1,000 to 1,500 ppm (see Table 13). In some European countries, toothpastes for young children contain lower fluoride levels (because of concerns about dental fluorosis); in others, fluoride levels in toothpaste range up to 2,500 ppm.

Most fluoride-containing mouthrinses contain either stannous fluoride (SnF$_2$) or sodium fluoride (NaF). The concentrations in mouthrinses are usually expressed as percentages, ie, the number of grams of SnF$_2$ or NaF per 100 mL of rinse solution. Most over-the-counter mouthrinses contain 0.05% NaF or 0.0226% F. Thus, the fluoride concentration is 22.6 mg/100 mL or 226 mg/L (mg/L or mg/kg = ppm). Some mouthrinses, such as the 0.2% NaF rinses, are available only by prescription and have a fluoride concentration of 910 mg/L. Those containing 0.4% SnF$_2$ (24.3% F by weight), have a fluoride concentration of 970 mg/L.

Table 14 Fluoride in dental products*

Product		Compound (%)	Concentration of fluoride		Amount usually used		Amount containing PTD‡	
			%	ppm	Product	Fluoride	10-kg child§	20-kg child‖
Rinse	NaF	0.05	0.023	230	10 mL	2.3 mg	215 mL	430 mL
	NaF	0.20	0.091	910	10 mL	9.1 mg	55 mL	110 mL
	SnF_2	0.40	0.097	970	10 mL	9.7 mg	50 mL	100 mL
Dentifrice	NaF	0.22	0.10	1,000	1 g	1.0 mg	50 g	100 g
	MFP†	0.76	0.10	1,000	1 g	1.0 mg	50 g	100 g
	MFP†	1.14	0.15	1,500	1 g	1.5 mg	33 g	66 g
Topical gel or solution	NaF (APF)	2.72	1.23	12,300	5 mL	61.5 mg	4 mL	8 mL
	SnF_2	0.40	0.097	970	1 mL	0.97 mg	50 mL	100 mL
	SnF_2	8.0	1.94	19,400	1 mL	19.4 mg	2.5 mL	5 mL
Varnish	NaF	5.0	2.26	22,600	0.5–1 g	11–23 mg	2.2 mL	4.5 mL
Tablet	0.25 mg F	–	–	–	1/d	0.25 mg	200 tab	400 tab
	0.50 mg F	–	–	–	1/d	0.50 mg	100 tab	200 tab
	1.00 mg F	–	–	–	1/d	1.00 mg	50 tab	100 tab

*Modified from Whitford (1996) with permission.
†(MFP) Monofluorophosphate.
‡(PTD) Probable toxic dose: 5 mg F/kg. If this amount or more is ingested, the individual should receive emergency treatment and be hospitalized.
§Approximate average body weight of a 1-year-old child.
‖Approximate average body weight of a 5- to 6-year-old child.

One of the standard textbooks on dental drugs, *Accepted Dental Therapeutics*, published by the American Dental Association, states that the preferred SnF_2 solution for topical application to the teeth contains 8% SnF_2. Such a solution has an extremely high fluoride level (19,400 mg/L). The acidulated phosphate fluoride (APF) gels generally contain 1.23% F, added as NaF, and 1% phosphoric acid (about pH 3.5). Thus, they have a fluoride concentration of 12,300 mg/L.

Among fluoride products for professional use, some fluoride varnishes for slow-release purposes may contain fluoride concentrations of greater than 2% (Duraphat [Rorer, Cologne, Germany]: 5% NaF; Bifluorid 12 [Voco, Cuxhaven, Germany]: 6% NaF + 6% CaF_2).

Table 14 shows the fluoride concentrations of a variety of dental products, the quantities of product and fluoride involved in normal use, and the quantities of the products that would contain enough fluoride to reach the PTD for a 10-kg child (about 1 year old) or a 20-kg child (5 to 6 years old).

If dental products are used according to the recommendations, they will not result in any life-threatening acute fluoride toxicity. Young children, however, do not always use dental products according to the recommendations of manufacturers or dental professionals. Some children occasionally drink mouthrinses, eat dentifrices, or swallow too many fluoride tablets. Mouthrinses are available in bottles containing 0.5 L. This quantity is 2.5 times the PTD for a 10-kg child and

1.2 times the PTD for a 20-kg child. Fluoride toothpastes usually are available in 100-g (75-mL) tubes. For toothpastes containing 0.10% to 0.15% F, this amount contains enough fluoride to exceed the PTD. The 1.23% gels, the 8% SnF_2 solutions, and the 5% NaF varnish are clearly hazardous to the health of young children or even adults. A 100-count bottle of 0.25-mg tablets of fluoride would not comprise a lethal dose, whereas there is a potentially lethal dose in 100-count bottles of 0.5- or 1.0-mg fluoride tablets.

The following recommendations should be followed for mouthrinses, toothpastes, 0.4% SnF_2 gels, and tablets:

1. They should not be used by young children without the supervision and presence of an adult.
2. They should be kept out of the reach of young children.

For 1.23% APF gels, high-concentration SnF_2 solutions, and other products with similarly high fluoride concentrations, the following recommendations should be followed:

1. They should be applied only by dental professionals.
2. The patient should not be left unattended.
3. The quantities used and, after application, retained in the mouth should be minimized.

Fluoride compounds differ widely with respect to fluoride bioavailability (absorbability) and, therefore, in their acute toxic potentials. Thus, SnF_2 is slightly more toxic than NaF, apparently because high doses of the stannous ion adversely affect the kidneys and other organs. A variety of other factors may also influence the toxicity of fluoride compounds, eg, route of administration, age of the individual, rate of absorption, and acid-base status. In animal experiments (Mornstad, 1975), the same dose of fluoride per kilogram of body weight administered intravenously at a constant rate for 20 minutes resulted in a level of fluoride that was 4.5 times higher and a plasma concentration curve that was 10 times higher in adult animals than in growing animals. Thus, fluoride was removed from the plasma and soft tissues much more rapidly in the younger animals, mainly because of the remarkable ability of the developing skeleton to rapidly take up the fluoride ions.

There are indications that one important determinant of the clinical outcome in fluoride poisoning is the magnitude of the peak plasma fluoride concentration achieved after ingestion. Fluoride is concentrated in the renal tubular fluid, so that when the entire organ is analyzed, it has a higher concentration than that of plasma. Because the kidneys are the only significant route for removal of fluoride from the body, the adequacy of kidney function is also an important variable in cases of fluoride toxicity.

The blood-brain barrier is effective in restricting the passage of fluoride into the central nervous system, where the fluoride concentration, like that of fat, is only about 20% of that in plasma. The fluoride concentration is higher in plasma and extracellular fluid than in the intracellular fluids. Low pH will significantly increase the intracellular fluoride concentration.

Characteristics of acute fluoride toxicity

Toxic signs are alarmingly rapid after ingestion of large amounts of fluoride. In nearly all cases of fluoride poisoning, the victims experience nausea, vomiting, and abdominal pain within minutes of ingestion. There may or may not be a variety of nonspecific symptoms, such as excessive salivation; tearing; mucous discharges from the nose and mouth; diarrhea; headache; cold, wet skin; or convulsions. As the episode progresses, generalized weakness, carpopedal spasms, or spasm of the extremities and tetany often develop.

These myopathologic signs are accompanied by declining plasma calcium concentrations, which may fall to extraordinarily low values and rising plasma potassium levels, which indicate a generalized toxic effect on cell membrane function. The pulse may be thready or not detectable. Blood pressure often falls precipitously to dangerously low levels. Respiratory acidosis, which diminishes the pH gradient across most cell membranes and results in a net migration of fluoride from extracellular fluid into the intracellular fluids, develops as the respiratory center is depressed. Cardiac arrhythmias may develop in association with the hypocalcemia and hyperkalemia. Extreme disorientation or coma usually precedes death, which may occur within the first few hours of fluoride ingestion.

Treatment of acute fluoride toxicity

The immediate treatment of acute fluoride toxicity should be aimed at reducing the amount of fluoride available for absorption from the gastrointestinal tract: Vomiting should be induced by administration of an emetic, such as ipecac. This should be followed by the oral administration of 1% calcium chloride or calcium gluconate. If these solutions are not available, then as much milk as can be ingested should be given.

The hospital emergency department should be informed that a case of fluoride poisoning is in progress while these procedures are being carried out. The patient should be transported to the hospital at the earliest possible time. Vomiting should not be induced if the victim has no gag reflex or while the patient is unconscious or experiencing convulsions because of the danger of aspiration. In these cases, a cuffed endotracheal tube should be inserted and gastric lavage should be performed with a solution containing calcium or activated charcoal. At the hospital, specific routines should be available for medical treatment, depending on the severity of the signs and the symptoms (for details on fluoride toxicology, see Whitford, 1996).

PREERUPTIVE EFFECTS OF FLUORIDE

Positive effects

As discussed earlier, the caries-inhibiting effect of fluoride is predominantly (almost totally) posteruptive. Preeruptive effects are very limited. However, some positive effects can be described. These would be most beneficial during eruption of the molars, a critical period for the initiation of fissure caries (Carvalho et al, 1989).

In teeth, as in all the mineralized tissues, fluoride levels tend to be greatest at the surface because this region is closest to the tissue fluid that supplies the fluoride. Therefore, preeruptive fluoride accumulation is highest in the pulpal aspect of the dentin and the outer surface of the enamel (Fig 261). A much higher total fluoride concentration is found in the dentin because of an endogenous fluoride supply obtained from the vessels of the pulp. The outer surface of the enamel will receive a "topical" supply of fluoride from the surrounding follicular fluid, explaining why fluoride concentrations decrease from the inner surface of the dentin and the outer surface of the enamel.

The concentration of fluoride is also higher in those parts of the enamel that are the first to develop and mature, ie, the incisal edges and the occlusal surfaces of the molars and premolars. These preeruptive effects of fluoride may reduce susceptibility to the initiation of caries in the molar fissures during eruption and possibly around the approximal contact surfaces before the second maturation is completed after eruption.

In the enamel, the dentin, and the root cementum, some fluoride is incorporated within the interior of the mineral crystallites as an integral part of the crystal lattice. However, fluoride may also be more superficially located, perhaps absorbed on crystal surfaces or loosely entrapped in the hydration shells of the mineral crystallites. Most of the fluoride ions that enter the apatite lattice probably replace an OH^- ion or at least occupy an equivalent space. It has also been suggested

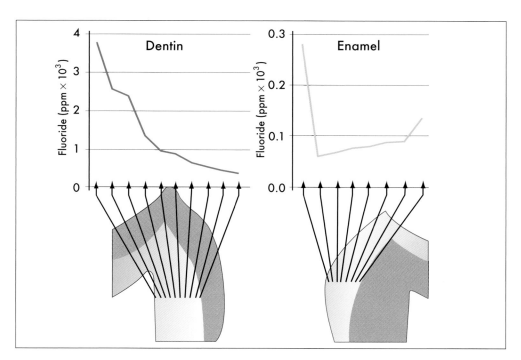

Fig 261 *(left)* Fluoride distribution across section of molar dentin. The fluoride concentration falls steeply from the pulpal surface to the dentinoenamel junction. *(right)* Fluoride distribution across section of molar enamel. The fluoride concentration is high in the surface region and falls to a plateau in the enamel interior. (Modified from Robinson et al, 1996 with permission.)

that fluoride may replace the larger PO_4^{3-} ions and even substitute for CO_3^{2-} or HCO_3^- ions present in the mineral. Most of the fluoride buried within the crystallites will have been acquired during the period of crystal growth, a process sometimes known as *accretion*.

Being built into the crystal as it forms, such fluoride is quickly buried and will remain locked in the lattice interior for as long as the crystal exists. The incorporation of fluoride can significantly alter the properties of mineralized tissues because the inclusion of any extraneous element in a crystalline lattice will alter its reactivity. If it occupies a position normally occupied by a hydroxyl group, fluoride can greatly increase stability of the lattice. Fluoridated apatite lattices are more crystalline, more stable, and, therefore, less soluble in acid. A decrease in carbonate content is generally found in highly fluoridated tooth mineral. This is probably due to direct substitution of carbonate by fluoride: The enamel will be more acid resistant because enamel with relatively high carbonate content is much more soluble. Fluoride will also

reduce the citrate content of the enamel, which may increase initial acid resistance.

On the other hand, more superficially located fluoride may have relatively little effect on the behavior of the crystalline lattice. However, it can dramatically affect fluid-crystal equilibrium, which involves the interaction between ions at crystal surfaces and those in solution.

Negative effects: Fluorosis

By far the best-known preeruptive effect of excessive fluoride intake is fluorosis, first described by Black and McKay (1916) as "mottled enamel." They suggested that it could be related to the water supply in the endemic areas. When it was subsequently shown in humans as well as in experimental animals that mottled enamel was an effect of fluoride on enamel formation and maturation, the condition was termed *enamel fluorosis*.

271

As early as the early 1940s, Dean et al (1941, 1942) demonstrated a positive correlation between the fluoride concentration in the drinking water and the prevalence and severity of fluorosis. Numerous studies have subsequently confirmed that the risk of developing fluorosis is strongly correlated to the regular intake of fluoride during tooth mineralization and particularly during the maturation phase of the enamel.

Classification

Dean and Elvove (1936) suggested a classification of each person into one of seven categories according to the degree of enamel changes (fluorosis): 0 = normal enamel; 1 = questionable; 2 = very mild fluorosis; 3 = mild fluorosis; 4 = moderate fluorosis; 5 = moderately severe fluorosis; and 6 = severe fluorosis. Later, Dean (1942) combined the categories *moderately severe* and *severe* into one score (severe) to include all enamel surfaces with any type of surface destruction, irrespective of degree. At the time, Dean did not know that histopathologically the entire tooth surface is affected, even in the mildest forms of fluorosis, and the distinction between very mild, mild, and moderate fluorosis was based on the area of tooth surface involved.

From studies in which the histopathology of fluorosis in human teeth was related to the clinical severity, Thylstrup and Fejerskov (1978) developed the so-called Thylstrup-Fejerskov (TF) Dental Fluorosis Index. The enamel changes observed on a single tooth surface can be arranged into 1 of 10 classes according to the severity of the fluorosis (Fig 262 and Box 12), scored from 0 (normal) to 9.

According to the TF Index, the first signs of dental fluorosis (score 1) are thin, white striae across the enamel surface. The fine, opaque lines follow the perikymata pattern and can best be distinguished after the surface of the tooth is dried (Fig 263). If the surface is covered with plaque, it must be cleaned (eg, with a cotton wool roll). Even at this stage of dental fluorosis, the cusp tips, incisal edge, or marginal ridges may appear opaque with the "snow cap" phenomenon (Fig 264).

In slightly more affected teeth (score 2), the fine, white lines are broader and more pronounced. Occasional merging of several lines occurs to produce smaller, irregular, cloudy, or paper-white areas scattered over the surface (Fig 265). These changes may be recorded without drying the teeth but are more evident after the teeth are wiped and dried.

In these mild forms of dental fluorosis, the enamel changes may vary somewhat along the surface, reflecting the structure of the enamel, the variation in enamel thickness, and the presence and variation in thickness of the underlying dentin. Fluorosis will be more readily apparent along incisal edges, cusp tips, and marginal ridges because the arrangement of enamel rods is very irregular and there is no underlying dentin. Moreover, crystal and prism arrangement in the outermost enamel varies among individuals and within the single tooth, so a slight increase in tissue porosity (opacities) of the same degree may manifest itself differently on different parts of the tooth surface.

With increasing severity (score 3), the entire tooth surface exhibits distinct, irregular, opaque, or cloudy white areas. Between these irregular opacities, the perikymata are often accentuated (Fig 266). Certain variations may occur at this stage of severity as a result of the above-mentioned variations in tooth structure. The cervical enamel frequently appears more homogenously opaque, and the mesioincisal part of the maxillary incisors may exhibit various degrees of brownish discoloration (Fig 267), the result of posteruptive staining. In rare cases, the patchy, cloudy areas may exhibit small surface enamel defects because of damage to the surface layer that covers particularly pronounced subsurface porosities.

The next degree of severity (score 4) is manifested as irregular opaque areas that merge. Some of the tooth surface appears chalky white (Figs 268 and 269). At the time of eruption, this stage may vary clinically from a white opaque tooth that, on probing, is relatively hard to a totally chalky tooth that exhibits surface damage

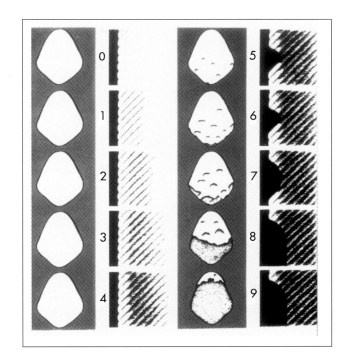

Fig 262 Thylstrup-Fejerskov Dental Fluorosis Index. Scores range from 0 (normal) to 9 (most advanced form of fluorosis). The classification correlates clinical appearance and histopathologic changes. (From Thylstrup and Fejerskov, 1978. Reprinted with permission.)

Box 12 Thylstrup-Fejerskov (TF) Dental Fluoride Index	
0	The normal translucency of the glossy, creamy white enamel remains after the surface is wiped and dried.
1	Thin, white, opaque lines are visible across the tooth surface. Such lines are found on all parts of the surface. The lines correspond to the position of the perikymata. In some cases, a slight "snowcapping" of cusps and incisal edges may also be seen.
2	The opaque white lines are more pronounced and frequently merge to form small, cloudy areas scattered over the whole surface. Snowcapping of incisal edges and cusp tips is common.
3	The white lines are merged, and cloudy areas of opacity occur, spread over many parts of the surface. Between the cloudy areas, white lines can also be seen.
4	The entire surface exhibits a marked opacity or appears chalky white. Parts of the surface exposed to attrition or wear may appear to be less affected.
5	The entire surface is opaque, and there are round pits (focal loss of outermost enamel) less than 2 mm in diameter.
6	The small pits may have merged in the opaque enamel to form bands of less than 2 mm in vertical height. Surfaces where the cuspal rim of facial enamel has been chipped off are also included, and the vertical dimension of the resulting damage is less than 2 mm.
7	There is loss of the outermost enamel in irregular areas, but less than half the surface is so involved. The remaining intact enamel is opaque.
8	The loss of the outermost enamel involves more than half the enamel. The remaining intact enamel is opaque.
9	The loss of the major part of the outer enamel results in a change of the anatomic shape of the surface or tooth. A cervical rim of opaque enamel is often present.

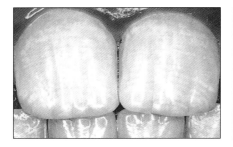

Fig 263 TF Index score 1. The earliest clinical sign of dental fluorosis appears as thin, white opaque lines running across the tooth surface corresponding to the position of the perikymata. (From Fejerskov et al, 1996a. Reprinted with permission.)

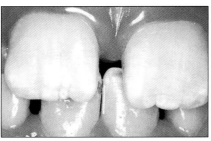

Fig 264 In addition to the thin, white opaque lines, the earliest signs of dental fluorosis may include small, white opaque areas along cusp tips, incisal edges, or marginal ridges. (From Fejerskov et al, 1996a. Reprinted with permission.)

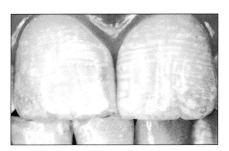

Fig 265 TF Index score 2. The white opaque lines are more pronounced and frequently merge to form wider bands. (From a Danish child born and raised in an area with 1.4 ppm F in the drinking water.) (From Fejerskov et al, 1996a. Reprinted with permission.)

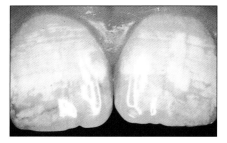

Fig 266 TF Index score 3. The entire tooth surface exhibits cloudy, white opaque areas between which accentuated perikymata lines are evident. (From a child born and raised with 2.1 ppm F in the water supply.) (From Fejerskov et al, 1996a. Reprinted with permission.)

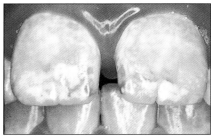

Fig 267 A case classified as TF Index score 3, exhibiting a mesio-incisal brownish discoloration on the maxillary incisors, in addition to an overall cloudy opaque appearance of the tooth surface. (From Fejerskov et al, 1996a. Reprinted with permission.)

immediately following eruption. When such surfaces are probed vigorously, part of the surface enamel may flake off.

In even more severe stages (score 5), the tooth surface is entirely opaque with focal loss of the outermost enamel. Such small enamel defects are usually designated *pits* (Figs 268 and 270). The pits may vary in diameter and may be scattered over the surface, although most frequently they occur along the incisal or occlusal half of the tooth. With increasing severity (score 6), these pits merge to form horizontal bands. In more se-

verely affected teeth (scores 7 and 8), confluence of the pitted areas produces larger, "corroded" areas (Figs 270 to 272). Along the incisal edges and cusps, the surface enamel often flakes off. The pits and other damaged areas frequently appear discolored.

Ultimately, the most severely fluorotic teeth (score 9) exhibit almost total loss of surface enamel, severely affecting tooth morphology (Fig 273). The loss of surface enamel may be so extensive that only a cervical rim of intact, markedly opaque enamel is left. The remaining part of the tooth

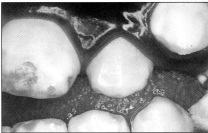

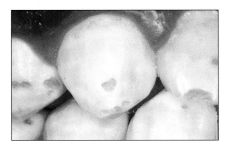

Fig 268 TF Index score 4 (maxillary right incisor) and score 5 with distinct pitting (maxillary left incisor). The entire tooth surface appears white opaque. (From an African adolescent born and raised with 2.8 to 3.0 ppm F in the water supply.) (From Fejerskov et al, 1996a. Reprinted with permission.)

Fig 269 TF Index score 4 (maxillary left lateral incisor and canine and mandibular left canine and first premolar) and score 7 (maxillary left central incisor). (From an African child with more than 5.0 ppm F in the water supply.) (From Fejerskov et al, 1996a. Reprinted with permission.)

Fig 270 TF Index score 7 (maxillary right premolar), score 5 (maxillary right canine), and score 6 (maxillary right lateral incisor). Note that in addition to the posteruptive damage, the entire tooth surfaces appear chalky white. (From Fejerskov et al, 1996a. Reprinted with permission.)

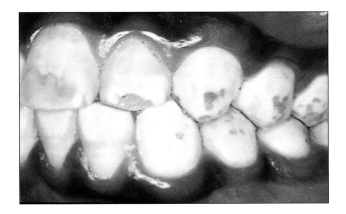

Fig 271 Posteruptive damage classified as TF Index score 7. (From a child born and raised in eastern Africa with about 6 ppm F in the water supply.) (From Fejerskov et al, 1996a. Reprinted with permission.)

often exhibits a dark brownish discoloration. The discoloration is entirely dependent on posteruptive environmental conditions such as dietary habits; the degree of discoloration should therefore not be used as an indication of severity of fluorosis as such.

When the teeth are highly opaque at the time of eruption, they are very susceptible to attrition, and extensive occlusal abrasion is often observed in high-fluoride areas, even in young individuals (Fig 274). Nevertheless, the loss of enamel in dental fluorosis, whether focal or extensive, involves only the surface enamel and not the full thickness.

To date, the TF Index is the only classification of fluorosis to correlate clinical appearance and histopathologic changes. Score 1 is a result of increased porosity along the striae of Retzius in the outer enamel surface; this effect is more pronounced in score 2. The volume of the striae of Retzius involved is about 5%. In score 3, there is subsurface porosity, 80 to 100 µm thick, with a pore volume greater than 5%. Score 4 represents extensive porous enamel lesions with greater than 10% pore volume beneath a 50-µm well-mineralized surface layer. The porous areas are severely hypomineralized because of increased intercrystalline spaces in both rod and interrod enamel. However, the thickness and cross-sectional shape of the individual crystals are normal. The intercrystalline spaces are occupied by water and proteins.

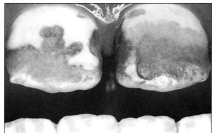

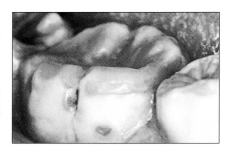

Fig 272 TF Index score 7. (Another case from the same area as the child in Fig 271.) (From Fejerskov et al, 1996a. Reprinted with permission.)

Fig 273 TF Index score 8 (maxillary right central incisor) and score 9 (maxillary left central incisor). (From a child living in eastern Africa with 6.4 ppm F in the water supply.) (From Fejerskov et al, 1996a. Reprinted with permission.)

Fig 274 In areas where the teeth are highly porous and opaque at the time of eruption, extensive occlusal abrasion occurs shortly after eruption. Note pitting of the white opaque buccal surface. (From Fejerskov et al, 1996a. Reprinted with permission.)

The more extensive the zone of hypomineralization beneath a mineralized thin surface layer, the greater the susceptibility of the enamel to posteruptive mechanical damage. For this reason, the index includes scores 5 and 6. Scores 7 to 9 represent fluorosis in which almost all porous enamel is exposed to the oral environment as a result of collapse of the thin outer enamel surface soon after eruption. In these teeth, the pulpal part of the dentin is also hypomineralized. The porous enamel will tend to stain posteruptively, and the fluoride content will increase because of uptake of topical fluoride.

On eruption, the surface is intact, but the clinical appearance will be determined by the severity of subsurface porosity: If the porosity is severe (scores 6 to 9), the thin, well-mineralized surface layer is vulnerable to mechanical trauma and will be destroyed shortly after eruption. Even in cases of mild fluorosis, ie, scores 1 to 3, the occlusal surfaces and cusps will be much more susceptible to attrition and wear than the normal enamel, as discussed earlier.

Patterns

The risk and severity of fluorosis are closely correlated with the plasma fluoride level during enamel maturation, as will be discussed in more detail later. However, even with a constant daily fluoride intake per kilogram of body weight, the plasma fluoride level is age related. Plasma fluoride concentrations in 1- to 3-year-old children are lower than those in 5- to 7-year-old children and much lower than those in adults because, in contrast to enamel, the mineralizing bone tissue will not only gradually take up and accumulate fluoride but also release fluoride during remodeling. Therefore, the earlier in life the individual is exposed to elevated systemic levels of fluoride from sources such as drinking water, infant formulas, fluoride tablets, and fluoridated salt, the greater the risk of fluorosis.

This would also imply that, the later in life enamel mineralization occurs, the more severe enamel fluorosis might be, even assuming a constant dose of fluoride from birth. In individuals exposed from birth to a constant daily fluoride intake per kilogram of body weight, the pattern of fluorosis in the permanent dentition is strongly correlated to the time of maturation of the tooth enamel for the homologous pairs of teeth.

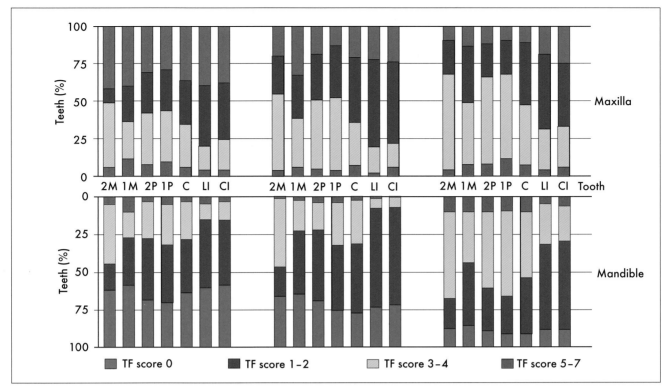

Fig 275 Severity and pattern of fluorosis, according to the TF Index, in three very low–fluoride areas in East Africa. (2M) Second molar; (1M) First molar; (2P) Second premolar; (1P) First premolar; (C) Canine; (LI) Lateral incisor; (CI) Central incisor. (Modified from Manji et al, 1986 with permission.)

A further important implication is that the same daily fluoride ingestion by, for example, children aged 4 to 6 years and of the same weight, may mean widely differing risks of dental fluorosis, depending on prior (and not necessarily current) fluoride exposure since birth, eg, from sources such as fluoridated water and infant fluoride tablet programs.

In severely affected populations, the posterior teeth are more affected than the anterior teeth, in both the maxilla and mandible. In low-fluoride areas, however, this pattern is slightly different. Manji et al (1986) classified the severity of fluorosis in each tooth type from subjects living in three very low–fluoride areas in East Africa (Fig 275). In subjects in all three areas, the incisors and first molars are the least affected, and the second molars and the premolars are most affected. It is apparent that the teeth that mineralize early in life

develop less dental fluorosis; ie, the later any given tooth undergoes mineralization, the greater the prevalence and severity of dental fluorosis of that particular tooth type. These data support observations made as early as 1936 by Dean and Elvove that premolars and second molars, even in low-fluoride areas, show manifestations of mild dental fluorosis.

Given this association between the period of mineralization of the individual teeth and the severity of fluorosis, and assuming a constant exposure to fluoride, it is not surprising that dental fluorosis has seldom been reported in primary teeth. It was previously believed that fluorosis reflected excessive fluoride ingestion during the secretory phase of amelogenesis, but, as mentioned previously, it is now known that the tooth continues to be vulnerable throughout the period of maturation.

Studies by Evans and Stam (1991) determined the relative risk of fluorosis of the central incisors since discoloration associated with fluorosis is a serious esthetic problem in these teeth. They found that the maxillary central incisors are most susceptible to fluorosis at 22 to 26 months of age; for the incisors, fluoride exposure prior to this period carried less risk than exposure for up to 36 months subsequently, a period corresponding to the maturation phase of the incisors.

The clinical implications are that central and lateral incisors are susceptible to fluorosis as a result of excessive fluoride intake up to the age of 5 years; peak susceptibility is at around the age of 2 years. Previously it was believed that once the incisor crowns were complete radiographically, at about 1 year, there was little or no further risk of fluorosis. In the context of this new knowledge about the development of fluorosis and the fact that the caries-inhibiting effects of fluoride are almost exclusively posteruptive, there is no justification for use of systemic fluoride supplements (tablets). However, in caries-risk children, slow-release fluoride lozenges may be recommended for posteruptive (topical) use, once the permanent first molars begin to erupt.

Differential diagnosis

The generalized nature of dental fluorosis within the dentition and over the entire tooth surface makes it easy to distinguish these lesions from symmetric defects of nonfluoride origin. The issue has been confused by claims that it is extremely difficult to decide just how many cases of enamel defects in endemic areas are the result of etiologic factors other than fluoride. However, to date, no other single etiologic factor has been associated with diffuse symmetric dental opacities in humans. Differential diagnosis of defects arising from systemic or local infection or trauma is not difficult (Table 15).

Mechanisms of development

The fact that the dentin of fluorotic teeth is also hypomineralized indicates that fluoride exerts its effect on very basic processes involved in biomineralization in general, irrespective of whether crystal formation and growth occur in mesenchymally or ectodermally derived mineralized tissues. However, relatively little work has been done to identify the mechanisms by which low serum levels of fluoride, which result in dental fluorosis, affect the development of mineralizing tissues.

Mechanisms that have been proposed to explain the formation of fluorotic enamel include a systemic effect of fluoride on calcium homeostasis, altered matrix biosynthesis (protein secretion, synthesis, or mineral composition), a direct or indirect effect on matrix proteinases affecting protein removal, and specific effects on cellular metabolism and function. Most available evidence indicates that fluoride has an effect on cellular function, either directly through interactions with the developing ameloblasts or more indirectly through interactions with the extracellular matrix.

Although the protein composition of secretory enamel is not altered, several studies have shown that high levels of fluoride exposure do inhibit protein synthesis and reduce the total amount of secretory enamel present. Several investigators have suggested that the secretory enamel matrix proteins bind fluoride: This may either alter enamel formation in the secretory stage, or as proposed by Crenshaw and Bawden (1981), act as a reservoir for fluoride, which is later released when proteins are hydrolyzed at the maturation stage. An increase in the influx of fluoride at this early maturation stage may be partially responsible for the susceptibility of the early maturation enamel to the effects of fluoride. The high concentrations of fluoride in the enamel matrix at the transition and early maturation stages of enamel formation could reduce the available ionic calcium, resulting in reduced proteolytic activity at this critical stage.

Table 15 Differential diagnosis: Milder forms of dental fluorosis (TF scores 1–3) and enamel opacities of nonfluoride origin*

Characteristic	Dental fluorosis	Enamel opacities
Area affected	Entire tooth surfaces (all surfaces); often enhanced on or near tips of cusps/incisal edges.	Usually centered in smooth surface; of limited extent.
Lesion shape	Resembles line shading in pencil sketch that follows incremental lines in enamel (perikymata). Lines merging, and, in score 3, a cloudy appearance. At cusps/incisal edges, formation of irregular white caps ("snow cap").	Round or oval.
Demarcation	Diffuse distribution over the surface; of varying intensity.	Clearly differentiated from adjacent normal enamel.
Color	Opaque white lines or clouds; even a chalky appearance. Snow caps at cusps/incisal edges. Score 3 may become brownish discolored at mesioincisal part of maxillary central incisors after eruption.	White opaque or creamy yellow to dark reddish orange at time of eruption.
Teeth affected	Always on homologous teeth. Early-erupting teeth (incisors/first molars) least affected. Premolars and second molars (and third molars) most severely affected.	Most common on labial surfaces of single or occasionally homologous teeth. Any tooth may be affected, but mostly incisors.

*From Fejerskov et al (1996b). Reprinted with permission.

Fluorotic enamel has been shown to have an increased magnesium concentration and decreased levels of carbonate, citrate, and zinc. These changes in the mineral chemistry could affect mineral-matrix interactions and enzyme activity. For example, it has been suggested that enamel proteins produced in the presence of fluoride may be more tightly bound to fluorapatite, thereby making them less accessible to degradation by enamel proteinases (Eastoe and Fejerskov, 1984).

Studies of human fluorotic enamel by Fejerskov et al (1974, 1975) suggested that fluoride interferes with the complex process involved in protein removal and subsequent mineral acquisition during enamel maturation. Amelogenin (the major secretory enamel protein) is hydrolyzed and removed from the matrix begin-ning in the secretory stage and at a more rapid rate during the transition and early maturation stages of enamel formation.

Fluoride has been shown to cause a dose-dependent delay in the hydrolysis and removal of amelogenin protein at the maturation stage of enamel development (Dajean and Menanteau, 1989; den Besten and Crenshaw, 1984; den Besten, 1986): This is most likely due to changes in the function or secretion of enamel matrix proteinases during enamel maturation. A number of enamel matrix proteinases have been identified, including both metalloproteinases and serine proteinases. Beginning at the early maturation stage of enamel development, a serine proteinase (proteinases that contain a serine at the active site), which is highly active against amelogenin, is present in the enamel matrix. It has been suggested that fluoride may specially alter the quantity and/or activity of the extracellular proteinases needed to degrade enamel proteins during the maturation stage of amelogenesis.

It appears that removal of protein during maturation is a critical step for final enamel mineralization. Both amelogenins and nonamelogenins have been shown to inhibit crystal growth. Studies that used proteinases to hydrolyze enamel matrix protein resulted in increased crystal growth in the maturation enamel (Robinson et al, 1990). Therefore, a delayed withdrawal of amelogenin may be a critical mechanism in the formation of enamel fluorosis by delaying the growth of enamel crystals so that, when the tooth erupts, the enamel remains incompletely mineralized. This is currently believed to be the most important mechanism behind the development of fluorosis.

Controlled studies in pigs by Richards et al (1985) clearly showed that the start of fluoride exposure after completion of enamel caused a subsurface type of enamel hypomineralization similar to that observed in human fluorosis. The most likely explanation for the pathogenesis of enamel fluorosis as it appears clinically in humans may therefore lie in an effect on processes occurring predominantly during enamel maturation.

Severity

Drinking water is the major source of fluoride intake. Early epidemiologic studies throughout the world showed a positive relationship between the concentration of fluoride in water and the prevalence of fluorosis (Dean et al, 1941, 1942; Larsen et al, 1989; Manji et al, 1986). However, climate influences the daily intake of fluids and thereby the intake of fluoride (Galagan et al, 1957; Richards et al, 1967). In addition, dietary habits have changed during the last decades: More people consume soft drinks instead of drinking water. Together with increasing exposure to fluoride from topical agents for home care and professional use and other systemic fluoride sources, such as salt and tablets, it is difficult to evaluate the total intake of fluoride, and thereby the dose response that will result in fluorosis.

The highest prevalence of fluorosis is found in the premolars and second molars, as earlier discussed. Because these teeth do not erupt until the age of 12 to 14 years, there is a considerable time lapse (about 7 to 8 years) between exposure to fluoride during preeruptive maturation of the enamel and the eruption of the teeth.

Studies of both water fluoride and tablets show that a daily fluoride intake of only 0.02 mg/kg of body weight during enamel maturation may result in fluorosis (fluorosis prevalence of about 50%) (Butler et al, 1985; Richards et al, 1967). The data also show that, even with very low intake from water, a certain level of fluorosis will be found. In addition, the dose-response relationship is clearly linear (Butler et al, 1985; Richards et al, 1967); ie, there is no "critical" value for fluoride intake, below which the effect on the enamel will not be manifest. Therefore, the conclusions by Hodge (1950) that dental fluorosis will not occur at a water fluoride concentration below 1 ppm are no longer valid.

In this context, administration of additional fluoride in tablet form to infants and small children in areas with water fluoride concentrations of less than 0.7 ppm/L will result in an increase in the prevalence and severity of fluorosis. It is surprising that such recommendations persist in some countries, particularly because of the extremely limited benefit of preeruptive administration. It also should be noted that the additive effect of ingestion from daily use of fluoride toothpaste (0.10% F) may increase the prevalence of fluorosis in areas with fluoride levels of more than 0.2 to 0.3 ppm/L of drinking water (Leverett et al, 1988). It is estimated that 2- to 3-year-old children using fluoride toothpaste (0.10% F) twice a day will ingest fluoride at a rate of about 0.04 mg/kg of body weight per day (for review, see Fejerskov et al, 1996a, 1996b and Clarkson, 2000).

POSTERUPTIVE EFFECTS OF FLUORIDE

For many years, the most important mode of action of fluoride was thought to be its incorporation into the apatite-like enamel crystals during development, resulting in crystals that were highly resistant to subsequent posteruptive acid attack. However, compared to the posteruptive effects, the preeruptive mode of action is now considered to be very minor.

Fluoride interferes posteruptively with the carious process in various ways, such as the inhibition of demineralization, enhancement of remineralization, reduction of acid production in the plaque, and reduction of plaque adhesiveness. The most important cariostatic role of fluoride is its action in the aqueous phase on the tooth surface and between the enamel crystals during demineralization and remineralization. Figure 276 (left) shows an active noncavitated enamel lesion on the mesiolingual surface of a mandibular second molar. Fluoride accumulates in the plaque fluid and as CaF_2 on the enamel surface. During the acid challenge, CaF_2 is dissolved. The surface acts as a so-called micropore filter, and F^- and H^+ ions (HF) diffuse into the subsurface lesion, increasing the amounts of fluoride in the active lesion compared to the surrounding intact enamel. Within the lesion, the F^- ions retard demineralization of the enamel crystals during acid challenge and enhance remineralization by crystal growth and accumulation of fluorapatite on the crystal surfaces when the pH rises.

Such a lesion can be arrested successfully if the patient maintains a high standard of approximal plaque control and uses fluoride toothpaste (Fig 276, right). Remineralization of the lesion is usually incomplete. "Continuous" access to a low concentration of fluoride results in more complete remineralization than does a high concentration of fluoride, which induces more rapid remineralization of the outer surface of the lesion (ie, sealing the micropore filter). As a result, the remineralized enamel surface will be less caries prone than the original intact surface. The total amount of fluoride is increased in the arrested lesion. These and other cariostatic effects of fluoride will be considered in more detail later in this chapter.

On a subclinical, microscopic level, repeated cycles of acid challenge followed by a rise in pH, combined with frequent (daily) access to low concentrations of fluoride from sources such as water and toothpaste, will result in so-called secondary maturation, and the tooth enamel will gradually become more caries resistant. In vitro studies on the development of experimental carieslike lesions have shown that the depth of the induced lesions in extracted unerupted teeth with mature enamel is almost 1.5, 2.0, and 3.0 times greater than the depth in extracted teeth that had been exposed to the oral environment for 0 to 3 years, 4 to 10 years, and more than 30 years, respectively, before extraction (Kotsanos and Darling, 1991).

In vitro studies by Pearce et al (1995) have reconfirmed the results by Weatherell et al (1977), charting fluoride concentrations in enamel caries lesions. On microradiographs of thin sections through the lesions, measurements were made of the surface zone (more mineralized and thereby radiodense) and the subsurface body of the lesion (less mineralized and radiolucent). Each section was then scanned with a proton probe to chart the fluoride concentration at different areas of the lesion (Fig 277).

The results showed extreme variations (from 1,700 to 22,700 ppm) in the maximum fluoride concentrations in the surface zone, and these rarely corresponded with the deepest part of the lesion. Fluoride levels were also elevated in the subsurface lesion, but not markedly. The highest fluoride concentration in the subsurface lesion was associated with a thin surface layer of relatively low mineralization, possibly a result of frequent topical application of fluoride from water or toothpaste. On the other hand, relatively low fluoride concentration in the subsurface lesion was often associated with a thick, well-mineralized surface layer with markedly increased fluo-

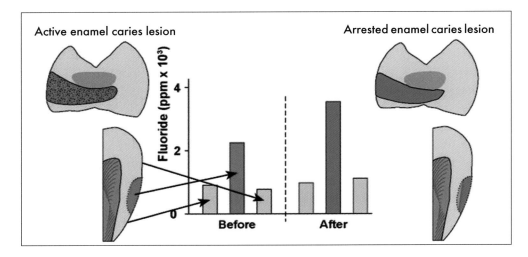

Fig 276 *(left)* Active, noncavitated enamel caries lesion on the mesiolingual surface of a mandibular right second molar. *(right)* Lesion arrested by plaque control and administration of topical fluorides. (Modified from Weatherell et al, 1977 with permission.)

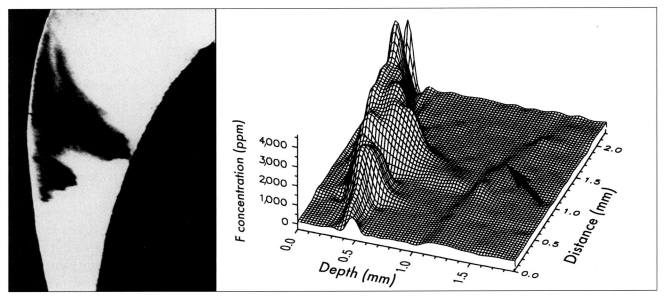

Fig 277 *(left)* Microradiograph of a noncavitated enamel lesion. *(right)* F concentration scan representing the surface. (From Pearce et al, 1995. Reprinted with permission.)

ride level (more than 10,000 ppm), attributable to topical application of agents with high fluoride concentrations, such as acidulated gels and varnishes (1% to 2% F). Some lesions were also associated with a low increase in fluoride in the subsurface lesion and in a thin surface layer, indicating an active or arrested enamel lesion with very limited exposure to fluoride.

Assuming that fluoride is a "marker" of past remineralization events, this study shows that a daily low concentration of fluoride can induce remineralization in the subsurface area of natural enamel caries lesions, especially when the surface layer is thin. Topical application of a high fluoride concentration induces rapid formation of an extremely thick, dense surface layer that may obstruct natural remineralization of the subsurface lesion. The long-term advantages and disadvantages of these two principles for managing enamel caries have yet to be evaluated.

Composition and structure of tooth enamel

The physicochemical interactions between fluoride and enamel during caries development and arrest are closely related to the structural and chemical composition of the enamel. Although macroscopically dental enamel may seem very solid and is the most highly mineralized tissue in the body, at high magnification it is relatively porous. Compared to enamel exposed to the oral environment for years, ie, enamel that has undergone primary and secondary maturation, the enamel of erupting teeth, which has undergone only primary maturation, is porous.

Although the mineral phase is about 96% of the total weight, it is only 85% by volume. The remaining 15% by volume—11% water and 4% protein and lipid (fatty material)—present in approximately equal amounts, constitutes the diffusion channels between crystals and prisms, allowing acid, minerals, and fluoride to pass in or out of the enamel during demineralization or remineralization.

The enamel mineral is hydroxyapatite (HA), which has the following unit cell: $Ca_{10}(PO_4)_6(OH)_2$. The chemical composition may, in various contexts, be shortened: $Ca_5(PO_4)_3OH$. By weight, 37% is calcium, 52% is phosphate (18% is phosphorus), and 3% is hydroxyl. These are the major components, because they are the elements of the apatite proper. Their relative amounts vary only marginally, depending on the origin of the enamel. These percentages do not alter very much when enamel is demineralized by a carious process because the density of the apatite is large relative to the density of the organic components, and the high water content in the lesion is usually disregarded.

The composition of dental enamel is not homogenous. All constituents are present in different concentrations at the anatomic surface and at the dentinoenamel junction. The density and content of organic material and water also vary. Such variations are found not only in transverse sections but also between different areas of teeth; for instance, cervical enamel has a lower mineral density than does occlusal enamel.

In contrast to the major components (HA), the so-called minor elements are seldom evenly distributed through the enamel. The concentration of some elements (fluoride, zinc, and lead) is higher in the surface layers and considerably lower internally. The opposite trend, a low concentration in the external layers and a higher internal concentration, applies to other elements (sodium, carbonate, and magnesium).

The minor components, mostly carbonate and sodium, amount to about 3% to 5% of enamel. This incorporation of carbonate in the HA crystal lattice means that it should be regarded as carbonated HA. Most of these ions are foreign to the HA lattice and may be considered as impurities, included in the mineral by chance during its formation. It is characteristic that they are released quite readily from the enamel during dissolution, which is why the carious process affects the presence of sodium, carbonate, and magnesium. Other elements are incorporated into the apatite lattice as substitutes for very similar apatitic ions (fluoride with hydroxyl; strontium and lead with calcium). These minor components behave as parts of the crystal and are released only when the crystal is dissolved.

The unit cell is repeated in all directions, to form the single enamel crystal or crystals, roughly hexagonal in cross section ($0.03 \times 0.04 \times 0.20$ μm). Each crystal is surrounded by a layer of firmly bound water. The presence of this hydration shell indicates that the enamel crystal is electrically charged. The hydration shell is believed to contain at least some of the above-mentioned foreign ions.

The crystals are densely packed and arranged in rods extending from the dentin to the enamel surface. The rods are held together and support each other through their complicated cross-sectional keyhole shape, the irregularity of their surface, and the wavy pattern in which bands or rods are intermingled. The rods run from the dentinoenamel junction to the enamel surface at an angle of 90 to 120 degrees. The width is about 4 μm at the dentinoenamel junction and 6 μm at the enamel surface.

By volume, a few percent of the fully matured enamel is protein, so-called enamelin. Principally, enamelins are located in the irregularities between the rods and have a delicate framework extending into the rod itself. During carious decalcification in vivo, some of the dissolved apatite is replaced by organic material, presumably precipitated from saliva, plaque, or food.

In the intact enamel, up to 4% by weight and 11% by volume is water, partly located in the hydration shell around each apatite crystal and the remainder filling the pores between the rods and the minute pores within the rods. These pores form the main diffusion pathways into the enamel.

As a result of precipitation and exchange reactions with the external fluids, the composition of enamel changes, both during the preeruptive formation phase and after eruption. One result of this is that both preeruptive and posteruptive maturation, described earlier, yields a fluoride gradient in the enamel, with typical fluoride values of 3,000 ppm or more at the anatomic surface and 100 ppm at the dentinoenamel junction. The fluoride content of the outer enamel surface depends on many variables, such as preeruptive administration of fluoride, posteruptive fluoride regimens, exposure to dietary acids (erosion), the presence of plaque, and the existence of abrasion and attrition.

The variability in composition and the presence of impurities, particularly carbonate, in the crystal lattice explain why dental enamel apatite is more soluble than pure HA and why solubility varies throughout the enamel.

Dissolution of enamel

The physicochemical integrity of dental enamel in the oral environment is entirely dependent on the composition and chemical behavior of the surrounding fluids: saliva and plaque fluids.

The main factors governing the stability of enamel apatite are pH and the free active concentrations of calcium, phosphate, and fluoride in solution.

Whether the apatite crystals will dissolve or remain intact under any given circumstance depends on the product of the free active concentrations of the ions in the liquid phase, the ion activity product. At equilibrium, the solution is saturated and the ion activity product is referred to as the *solubility product*.

The development of a clinical caries lesion involves a complicated interplay among a number of factors in the oral environment and the dental hard tissues (Fig 278). The carious process is initiated by bacterial fermentation of carbohydrates, leading to the formation of a variety of organic acids and a fall in pH. Initially, H^+ will be taken up by buffers in plaque and saliva, but when the pH continues to fall (H^+ increases), the fluid medium will be depleted of OH^- and PO_4^{3-}, which react with H^+ to form H_2O and HPO_4^{2-}. Once the depletion is complete, the pH can then fall below the critical value of 5.5, where the aqueous phase becomes undersaturated with respect to hydroxyapatite. Therefore, whenever surface enamel is covered by a microbial deposit, the ongoing metabolic processes within this biomass result in pH fluctuations and occasional steep falls in pH, which may result in dissolution of the mineralized surface.

Dissolution of enamel can result in the development of either a caries lesion or erosion. *Caries* is defined as the result of chemical dissolution of the dental hard tissues caused by bacterial degradation products, ie, acids from low–molecular sugar consumption. The *erosive lesion* is defined as chemical dissolution of tooth substance caused by any other acid-containing agent. Mixed lesions may well exist, particularly when erosion penetrates to the dentin, causing hypersensitivity, which may lead to inadequate plaque control and caries. This condition occurs frequently on exposed root surfaces.

The appearance of the two lesions differs: the caries lesion is characterized by a subsurface

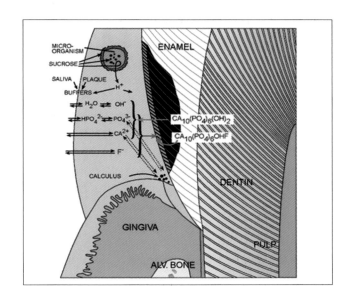

Fig 278 Development of noncavitated enamel caries. (Modified from Larsen and Bruun, 1994 with permission.)

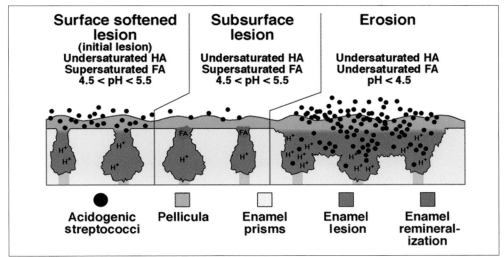

Fig 279 Mechanisms for enamel dissolution at different pH challenge and saturation of minerals.

demineralized lesion body, covered by a rather well-mineralized surface layer (see Figs 276 to 278). In erosion, the surface has been etched away layer by layer. No subsurface demineralization can be seen in the erosive lesion.

In principle, dental enamel can be dissolved under two different chemical conditions (Fig 279). When the surrounding aqueous phase is undersaturated with respect to HA and supersaturated with respect to FA, HA dissolves and FA is formed. The resulting lesion is a caries lesion in which the dissolving HA originates from subsurface enamel and FA is formed in the surface enamel layers. The higher the supersaturation with respect to FA, the more fluoride is taken up in the enamel surface, the better mineralized becomes the surface enamel layer, and the less demineralized becomes the subsurface body of the lesion.

At pH levels above 6.0, all oral fluids are supersaturated with respect to both HA and FA, and apatite formation is to be expected (calculus formation and redeposition of minerals in surface enamel). In the pH range 5.5 to 4.5, saliva becomes undersaturated with respect to HA but remains supersaturated with respect to FA; carious demineralization occurs. Even very low fluoride concentrations in the oral fluids (less than 0.10 ppm F) establish this supersaturation with respect to FA. Fluoride is always present in the oral fluids as a physiologic component of the secreted saliva at the 0.01-ppm level. Importantly, saliva remains supersaturated with respect to FA at low pH, which establishes an undersaturation with respect to HA. These conditions imply that during drops in the plaque fluid pH to the range 5.5 to 4.5, at which HA may dissolve, FA will simultaneously be formed in the demineralized area, thereby reducing the loss of dental mineral. This type of remineralization during drops in pH is therefore a natural part of any carious process.

Evidence for this mechanism is found in the increased fluoride levels observed in the outer layers of early caries lesions (see Figs 276 and 277). This mechanism implies that the effect will increase with increasing supply of fluoride. This is the crux of the explanation of the most important cariostatic action of fluoride: At concentrations below 1.00 ppm in the oral fluids, even modest elevations in fluoride concentration will result in significant reductions in the mineral loss caused by plaque acids.

On the other hand, if undersaturation exists with respect to both HA and FA, both apatites dissolve concurrently, and layer after layer is removed. This will result in an erosion-type lesion. The above mechanisms for enamel dissolution are illustrated in Fig 279. Fresh acidic fruit, fruit juices, acidic carbonated soft drinks, and some champagnes are all undersaturated with respect to both apatites and are able to cause erosive demineralization of the teeth.

The early signs of carious destruction appear as dissolution of enamel crystals at the very outer surface, whereby the intercrystalline spaces become widened. The porosity of the surface increases significantly, allowing more extensive diffusion of acid into the enamel and resulting in diffusion of calcium and phosphate ions out of the enamel to a depth of several hundred microns. In addition to these surface reactions, after initial dissolution, the outer 20 to 50 µm of enamel remains relatively well mineralized compared with the increasingly porous underlying enamel, as long as the intermittent dissolution-redeposition phenomena can occur within the pH ranges at which FA is available.

Progression of caries from ultrastructural changes to visible decay should be regarded as the cumulative effect of a long-alternating series of dissolutions at low pH and partial reprecipitations when pH rises. Eventually, after months or years, depending on the cariogenic challenge of the plaque, a clinically detectable white-spot lesion appears in the enamel.

Role of fluoride in enamel demineralization and remineralization

Influence on enamel solubility

Fluoride has a strong affinity for apatite because of its small ionic size and strongly electronegative character. Two kinds of fluoride-apatite interaction occur: incorporation into the crystal lattice and binding to crystal surfaces. Both interactions have important consequences for the solubility and dissolution properties of apatite. The rate at which caries lesions progress is clearly heavily dependent on the rate at which the apatite crystals dissolve. The dissolution rate can be reduced by fluoride even without any reduction in solubility of the bulk mineral. This effect is the basis for topical fluoride treatments.

Thus, the presence of dissolved fluoride at concentrations as low as 0.5 mg/L in acidic solutions causes a reduction in the dissolution rate of initially fluoride-free apatite (Christoffersen et al, 1984). Furthermore, pretreatment of apatite with fluoride solutions significantly reduces its susceptibility to acid dissolution (Cutress, 1966). These two effects are both due to uptake of fluoride by the surfaces of the apatite crystals.

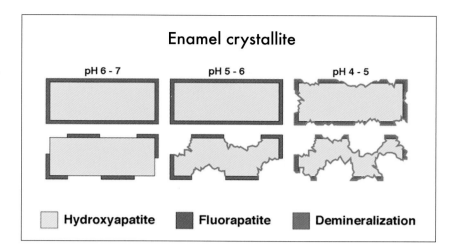

Fig 280 Dissolution of enamel crystallite.

Two types of fluoride uptake have been considered (Arends et al, 1984). In nonspecific binding, F⁻ ions are adsorbed to the crystal surface without reacting chemically with it. In specific binding, ion exchange is involved, and as a result F⁻ ions become incorporated into the crystal surface. Nonspecifically adsorbed F⁻ ions might inhibit dissolution by restricting diffusion of ions from the solid to the solution, and this has been proposed as the exclusive mechanism (Hoppenbrouwers et al, 1987a, 1987b). However, there is abundant evidence that F⁻ ions react with the crystal surface by exchanging with other anions, especially OH⁻ (Lin et al, 1981; White, 1988; White et al, 1988b, 1988c; White and Nacollas, 1990). Such ion-exchange processes are increasingly favored as the pH falls.

Fluoride adsorption with subsequent ion exchange effectively converts part or all of the surface to FA, which will show reduced solubility and a lower rate of dissolution in acidic buffers. The effect of surface fluoridation on dissolution depends on the proportion of the surface converted, because it appears that the nonfluoridated and fluoridated surfaces dissolve independently, in accordance with the solubility properties of HA (or defective apatite in the case of dental mineral) and FA, respectively (Christoffersen et al, 1984; Crommelin et al, 1983). The effect of different pH levels on enamel crystallites completely or partly

covered by FA is illustrated in Fig 280. In support of these concepts, the addition of even low concentrations of fluoride to demineralizing systems significantly reduces in vitro lesion progression (Crommelin et al, 1983; ten Cate and Duijsters, 1983a, 1983b), and concentrations on the order of 0.5 to 2.0 mg/L can prevent lesion formation under partially saturated conditions, such as those found in plaque fluid.

Figure 281 shows the amount of mineral dissolved during an acid challenge at various pH and fluoride concentrations (ten Cate and Duijsters, 1983a). The data pertain to given calcium and phosphate concentrations in solution; the specific concentration values at which demineralization inhibition occurred cannot therefore be extrapolated to physiologic conditions. When lower initial Ca and PO₄ ion concentrations were used, the same degree of inhibition was found for higher fluoride concentrations in solution. In principle, a similar graph showing mineral ion concentrations and ionic strength as variables in saliva or plaque fluid would indicate the fluoride concentrations at which demineralization would be inhibited. This, however, is not possible except in general terms.

When fluoride-deficient enamel, such as subsurface enamel, is subjected to an in vitro caries attack, the resultant defect does not have the characteristic pattern of a subsurface lesion. Instead, a

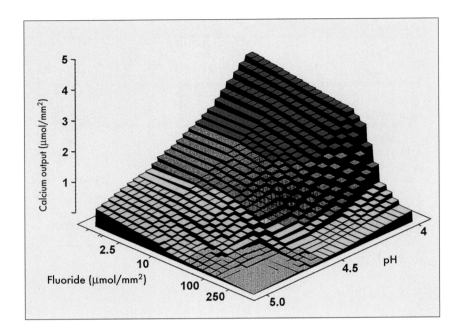

Fig 281 Mineral loss related to pH and fluoride concentration. (Modified from ten Cate and Duijsters, 1983a with permission.)

demineralization defect without a surface layer, ie, an erosion-like lesion, with a rough, microcavitated surface is formed. However, when fluoride is added to the demineralizing solution, the resultant lesion has a completely different histologic appearance: a well-formed surface layer, with a mineral content considerably higher than that of the underlying lesion body.

The mineral content and thickness of the surface layer are correlated with the fluoride concentration in solution. For example, microradiograms of lesions formed in undersaturated calcium phosphate solution at pH 4.5 and different fluoride concentrations show that an increase in fluoride ion concentration results not only in a thicker surface layer but also in a shallower lesion, in which less mineral is lost from the lesion body (Fig 282).

The surface layer in white-spot lesions does not develop solely by this mechanism. When fluoride-rich (surface) enamel is subjected to caries-promoting solutions, a surface layer is formed as a result of the lower local acid solubility as well as the favorable precipitation conditions. In this approach, less mineral is lost from this region and, in addition, mineral released from deeper regions may precipitate in the surface zone instead of diffusing out to the external solution.

At decreased pH, the most loosely attached fluoride on the crystal surface is released. Eventually, even the firmly bound, acid-resistant fluoridated hydroxyapatite (FHA), $Ca_{10}(PO_4)_6OHF$, and FA, $CA_{10}(PO_4)_6F_2$, on the surface of the enamel crystallites is dissolved (see Fig 280), releasing fluoride into the external fluid environment. The fluoride in solution will subsequently retard an acid attack. In this way, structurally incorporated fluoride can also affect the carious process, provided that the acid challenge is sufficient to dissolve some apatite with fluoride but not so strong that only dissolution occurs.

When acid penetrates between the crystals during the carious process, it is important that fluoride be present to retard or inhibit the acid attack. Preferably, the fluoride should be available in the external environment or from readily soluble materials at the crystal surface. This will occur if fluoride is applied frequently to the tooth surface at concentrations low enough to diffuse into the enamel and if there is a reasonably soluble fluoride form present either at the surface (CaF_2) or in the subsurface enamel.

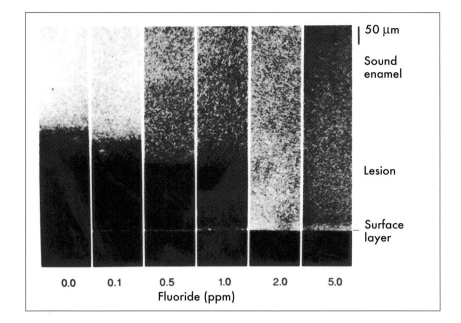

50 μm

Sound enamel

Lesion

Surface layer

0.0   0.1   0.5   1.0   2.0   5.0

Fluoride (ppm)

Fig 282 Microradiographs of representative specimens showing the effect of fluoride concentration in solution on the formation of enamel lesions. Demineralization based on data in Fig 281. (From ten Cate and Duijsters, 1983b. Reprinted with permission.)

Subsequent to an acid challenge, as the pH rises externally and within the tooth structure, if fluoride is present at low concentrations in the intercrystalline fluid, it will enhance either the precipitation of new mineral, incorporating calcium, phosphate, and fluoride, or the growth of crystals on existing, partly dissolved, enamel crystals. Either process will be dramatically enhanced by the fluoride present in the intercrystalline fluid. The crystalline surface so formed will then be more resistant to subsequent acid attack, as discussed earlier.

Thus, during a cariogenic challenge, both the concentration of fluoride at the crystal surfaces and the concentration of fluoride in the liquid phase are important. To reduce the dissolution rate, fluoridation of the crystal surfaces, by whatever means, is necessary, but the surface fluoridation will only be maintained if the solution bathing the crystals contains enough fluoride; otherwise all parts of the crystal surfaces will dissolve (see Fig 280). After all, lesions can be induced even in shark enameloid, which is almost pure FA (Ogaard et al, 1991). The interplay between surface fluoride and dissolved fluoride is a feature of any fluoride treatment and is also an integral part of lesion formation.

During lesion formation, the fluoride content of enamel mineral increases, even in the subsurface region, where the fluoride content of intact enamel is low (Clarkson et al, 1986), probably because subsurface demineralization is diffusion regulated and the solution within the lesion tends to be approximately saturated with respect to HA (Vogel et al, 1988). Consequently, the solution can be supersaturated with respect to FA at low fluoride concentrations (a few μmol/L), which would promote fluoridation of crystal surfaces. The necessary fluoride could be derived from plaque fluid and from limited dissolution of the naturally fluoride-rich surface enamel, as mentioned earlier. It appears that this process of fluoride uptake helps to maintain mineral levels in the surface layer of the caries lesion (Margolis et al, 1986).

Role of calcium fluoride formation

As mentioned earlier, the conventional aim of fluoride application has been to increase the content of firmly bound fluoride as FA or FHA in the enamel in order to improve resistance to acid. The relatively high solubility of CaF_2 (10 to

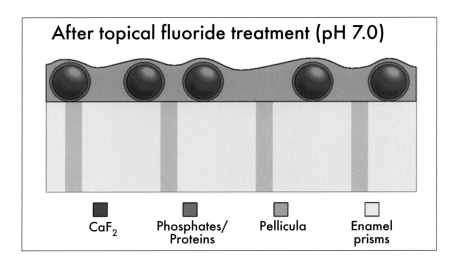

Fig 283 Tooth surface after topical fluoride treatment (pH 7.0) with formation of CaF_2 crystals coated with phosphates/proteins.

12 times greater than that of HA and FA) precludes its formation in vitro. Experiments in which CaF_2 deposits on enamel were leached in tap water showed rapid removal of any loosely bound fluoride (Kalter et al, 1980). It was recently demonstrated that conditions in the oral fluids are different: In vivo, CaF_2 is retained for much longer periods, because of the high calcium and fluoride content of saliva compared to tap water and, more importantly, because of the formation of a phosphate- and protein-rich protective layer on the CaF_2 globules (Ogaard et al, 1992). After a single fluoride mouthrinse, a considerable portion of fluoride was found to be present as calcium fluoride, even after 7 days' leaching in vivo. Additionally, under conditions of cariogenic challenge, CaF_2 was found to be converted to FA (Ogaard et al, 1983).

At neutral pH, precipitation of CaF_2 at apatite crystals and on the surface of the enamel (Fig 283) can be induced by application of dissolved fluoride at high concentration. At low pH, because the concentration of Ca in equilibrium with HA increases, the concentration of fluoride required for CaF_2 precipitation falls according to the formula:

$$Ca_{10}(PO_4)_6(OH)_2 + 20F^- + 11H^+ \rightarrow 10CaF_2 + 3H_2PO_4^- + 3HPO_4^{2-} + 2H_2O$$

By slow dissolution and hence prolonged retention, the solid CaF_2 is assumed to act as a reservoir for fluoride to be released into the liquid environment of the teeth (Fig 284). When all CaF_2 is dissolved its effect is lost, and the CaF_2 depot has to be replenished by repeated application of fluoride.

Formation of CaF_2 is possible at fluoride concentrations greater than 100 ppm, and the amount increases with increasing fluoride activity, prolonged exposure, and lower pH in the solution (Bruun and Thylstrup, 1984; Bruun and Givskov, 1991, 1993; Saxegaard and Rölla, 1988). Thus, significantly greater amounts of CaF_2 are precipitated after a 4-minute application of a 2.0% NaF solution than after a 2-minute mouthrinse with a 0.2% solution. Calcium fluoride formation is considerably increased after use of an APF solution because of the enhanced availability of calcium ions dissolved from the dental apatite. Modest amounts of CaF_2 may be formed from the use of NaF dentifrices. However, because of the low concentration of ionic fluoride in the dentifrice-saliva slurry during toothbrushing with monofluorophosphate (MFP) dentifrice, only negligible amounts of CaF_2 are produced (Bruun and Givskov, 1993). In contrast, extremely high amounts of CaF_2 may be expected from slow-release fluoride var-

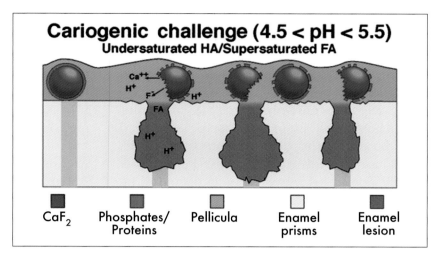

Fig 284 Tooth surface during a cariogenic challenge (4.5 < pH < 5.5): HA undersaturation; FA supersaturation.

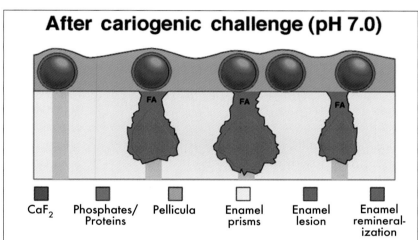

Fig 285 Tooth surface after cariogenic challenge (pH 7.0). The micropores of the enamel surface have been sealed and remineralized.

nishes with high fluoride concentration; for example, one varnish contains 6% NaF and 6% CaF$_2$ (Bifluorid 12).

The dissolution of the CaF$_2$ seems to be the key to its caries-preventing effect, because only the free fluoride ion has an effect on enamel solubility, as discussed earlier. On the other hand, when all CaF$_2$ is dissolved, the effect has come to an end as well. As mentioned already, the dissolution of the CaF$_2$ "globules" covered with phosphate/proteins increases with decreased pH (see Fig 284) so that more ionic fluoride can be expected to be released during cariogenic challenge. Fluoride ions will react with hydrogen ions as well as diffuse into the caries lesion, thereby slowing the rate of progression. However, this mechanism also implies that the CaF$_2$ reservoir will be depleted more rapidly in patients with high caries activity. Therefore, efficient plaque control has to be introduced to increase the pH to 7.0, and meticulous oral hygiene must be combined with frequent reapplication of fluoride. Thereby, the enamel caries lesion will be arrested and the surface of the lesion will be remineralized with FA (Fig 285).

After topical application of concentrated fluorides, considerably more CaF$_2$ is formed in the micropores of caries lesions than on sound enamel surfaces (see Figs 276 and 277). The modest amounts produced on sound enamel surfaces can be expected to be lost within a relatively short period, ie, a few weeks, because of continuous exposure to physical forces in the oral cavity, including toothbrushing, and chemical attacks from food and beverages. In the more inaccessible microporous environment of the early caries lesion, however, CaF_2 may persist for prolonged periods, ie, months, strategically providing elevated fluoride levels at just the sites where the risk of caries progression would be expected to be greatest (Rölla and Saxegaard, 1990). The precipitation of CaF_2 in early caries lesions with subsequent release of fluoride is therefore believed to be a key mechanism for the caries-reducing effect of concentrated topical fluorides such as APF gels, and varnishes (1% to > 2% F). (For review on the role of CaF_2, see Ogaard, 2001.)

Role in remineralization

In most individuals, plaque fluid and saliva are supersaturated with respect to FA but not necessarily with respect to CaF_2. However, precipitation of mineral does not occur automatically from supersaturated solutions. A substrate on which the deposit can form is necessary. Mineral deposition in enamel defects such as the caries lesion may result in complete or partial replacement of the lost mineral and is therefore called *remineralization*. White-spot caries lesion remineralization is a widely documented phenomenon in epidemiologic, in vivo, and in vitro research.

An epidemiological survey revealed that about 50% of the white spots diagnosed in 9-year-old children disappeared during the following 5 years, as a result of the remineralizing potential of saliva (Backer-Dirks, 1966). At the same time, it was demonstrated that saliva and artificial mineralizing fluid produce an increased hardness in demineralized enamel (Koulourides et al, 1961, 1965). Since then, many investigators have studied the mechanism of this phenomenon. Remineralization of enamel lesions was found to occur by deposition of crys-

talline hydroxyapatite, which in vitro can give rise to complete repair of the defect. However, in vivo, the process is considerably slower and there is seldom full recovery. Moreover, although remineralization in vivo is crystalline in form, the crystallites never acquire the large dimensions recorded in vitro (Arends and Jongebloed, 1979; Silverstone, 1983; ten Cate and Arends, 1977).

The role of fluoride in the remineralization process was found to be quite complex, although in theory it is relatively simple: In the presence of fluoride, apatite may precipitate as FA or FHA. Morphologically, precipitation of fluoridated apatites onto an HA crystallite matrix is unrestricted, because the minerals in this group all have closely related crystalline structures. Because FA and FHA are less soluble than HA, the thermodynamic driving force for their precipitation is greater. The general conclusion is that fluoride stimulates apatite precipitation, as shown in many in vitro experiments. Precipitation of FA (or FHA) on an apatite matrix is accelerated by fluoride (Amjad and Nancollas, 1979). After a few molecular layers have been deposited, the substrate material reacts like a bulk FA or FHA mineral, eg, when it is subjected to a dissolution experiment. In vitro, the initial remineralization of white-spot lesions or softened enamel is enhanced by the addition of fluoride to the remineralizing medium, and this results in an increase in the hardness of the surface of the demineralized region (Koulourides et al, 1961).

When the impact of fluoride on lesion remineralization is studied, a distinction should be made between the effects in the remineralizing fluid of high doses of fluoride of short duration and the effects of a continuous low concentration of fluoride, the former simulating topical application of APF gels and varnishes (1% to > 2% F) or even fluoride dentifrices, and the latter simulating fluoridated water, the intervals between toothbrushing, or fluoride mouthrinsing. During and after short-term fluoride treatment, large amounts of fluoride are adsorbed in the lesion.

Caries lesions in enamel contain considerably higher amounts of fluoride than does the surrounding intact enamel; ie, the F^- ions have great affinity for demineralized regions where free Ca^{2+} and PO_4^{3-} ions are available in abundance for "marriage" (see Figs 276 and 277). Frequently the F^- ion "marries" both the Ca^{2+} and PO_4^{3-} ions, forming FA, but the favorite "wife" is the Ca^{2+} ion, resulting in CaF_2 (see Fig 283). The marriage of F^-, Ca^{2+}, and PO_4^{3-} (FA) is more stable than the marriage between the F^- ion and the Ca^{2+} ion (CaF_2), because they divorce easily during cariogenic challenge ($4.5 < pH < 5.5$) (see Fig 284). After cariogenic challenge, F^- and Ca^{2+} marry again at pH 7.0. These marriages (precipitation of FA and CaF_2), which are very important for remineralization, will be accelerated in the outermost region of the lesion (see Fig 285), drawing away many of the free mineral ions (wives) from the inner parts of the lesion (subsurface lesion or lesion body) and effectively slowing diffusion toward the interior of the lesion. A high fluoride concentration results in delayed and incomplete mineralization in the lesion body compared to the remineralization that occurs at a very low fluoride concentration. Excess deposition of FA and CaF_2 may block the lesion pores, resulting in even more pronounced inhibition of diffusion.

These mechanisms were first shown by ten Cate et al (1981) using radioactively labeled solutions and enamel caries lesions treated topically with fluoride. Parallel studies by Silverstone et al (1981) on remineralization of enamel caries showed similar results. Thus, low supersaturation of Ca^{2+} and PO_4^{3-}, combined with low fluoride concentration, at pH 7, resulted in almost complete remineralization throughout the lesion (Figs 286 and 287). With threefold concentrations, remineralization was limited to the surface layer (Fig 288). Apparently, both topical fluoride concentrations and solutions with high mineral concentrations preferentially cause thick surface remineralization. The two mechanisms for remineralization of enamel caries are shown in Fig 289.

Conditions are different in the continuous presence of fluoride at low concentrations, eg, from fluoridated water (1 ppm F). Fluoride is now available concurrently with calcium and phosphate and can diffuse into the lesion and precipitate as FA or FHA. In vitro investigations have shown that low concentrations of fluoride do indeed accelerate the initial mineral deposition in lesions or softened enamel. At a constant concentration of 1 ppm F in the remineralizing solution, a twofold to threefold increase in the rate of precipitation was found (ten Cate and Arends, 1977).

When fluoride was added to remineralizing solutions or saliva, and teeth with carieslike lesions were immersed in these solutions, an increase in hardness was observed both in vitro and in vivo (Gelhard and Arends, 1984; Koulourides et al, 1961, 1965). In these experiments, hardness was measured on the enamel surface exposed to the solution, which could mean that the actual increase in hardness, reflecting mineralization, was restricted to the outermost lesion layers. In situ remineralization of experimentally developed caries lesions with low-concentration fluoride (1 ppm) resulted in significantly higher microhardness on the enamel surface than did remineralization with 10 to 100 ppm of fluoride (Koulourides et al, 1974).

These findings are supported by epidemiologic data showing that natural enamel caries (white-spot) lesions remineralize to a glossy enamel surface in individuals living in an area with fluoridated drinking water. In contrast, many individuals living in a nonfluoridated area exhibit white-spot lesions with a chalky surface, indicating active lesions (Backer-Dirks, 1966; Groeneveld, 1976). Figure 290 illustrates active enamel caries lesions developed at different pH challenges and then arrested by plaque control and topical fluoride or by plaque control without fluoride.

 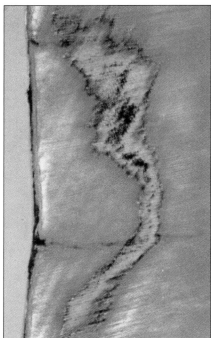 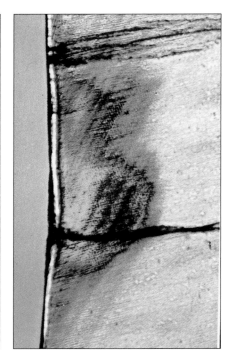

Fig 286 Cross section of an in vitro experimentally developed noncavitated enamel lesion in polarized light. Note the outer surface zone (the micropore filter) and the inner lesion body, with more extensive mineral loss. (From Silverstone, 1973. Reprinted with permission.)

Fig 287 Successful arrest of an experimentally developed enamel lesion in vitro through application of low concentrations of fluoride. Only at the base of the lesion body is there some residual net loss of minerals. (From Silverstone, 1973. Reprinted with permission.)

Fig 288 Remineralization of noncavitated enamel lesions similar to that shown in Fig 286 and with three times the F concentration as that shown in Fig 287. The result was a faster remineralization (sealing of the micropores) of the enamel surface but less complete remineralization of the body lesion in comparison with that seen in Fig 287. (From Silverstone, 1973. Reprinted with permission.)

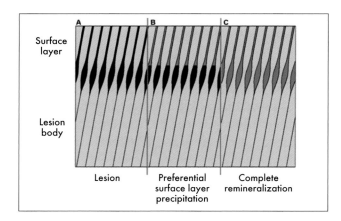

Fig 289 Development and arrest of an enamel lesion. (A) Active noncavitated enamel caries lesion. Observe the intraprismatic perforation of the surface "micropore filter" and the body lesion (black) beneath the surface; (B) Short-term use of high-concentration fluoride; (C) Long-term use of low-concentration fluoride. (Modified from ten Cate and Featherstone, 1996 with permission.)

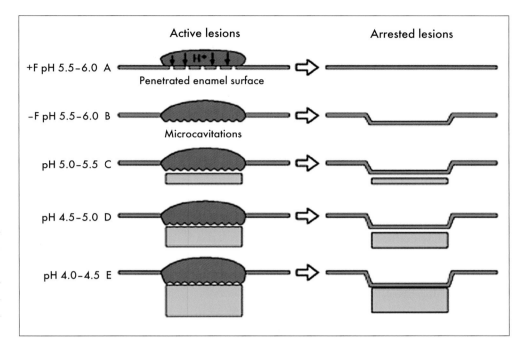

Fig 290 Arrest of caries lesions in enamel related to fluoride and pH during the development of the lesions. (Modified from Thylstrup et al, 1994 with permission.)

In the oral environment there is continual change or cycling of pH, from acid challenge to neutralization by saliva and other factors such as mineral saturation and fluoride concentration. The so-called recycling technique has shown in vitro that fluoride concentrations as low as 0.01 to 0.10 ppm in the mineralizing solution can enhance remineralization (Featherstone and Zero, 1992). As a consequence, regular use of fluoride products such as fluoride toothpaste or slow release of fluoride from glass-ionomer restorations, resulting in prolonged salivary fluoride concentrations of 0.04 to 0.10 ppm, may enhance remineralization of enamel lesions and probably root lesions. Based on these in vitro, in vivo, and epidemiologic studies, it must be concluded that fluoride, particularly when constantly present, will actively enhance remineralization and arrest of caries lesions (for a review of the physicochemical role of fluoride on enamel demineralization and remineralization, see ten Cate and Featherstone, 1996).

Sources of fluoride in the oral cavity

The posteruptive cariostatic effects of fluoride are correlated with fluoride concentration as well as with total exposure time. The total exposure time is also influenced by the substantivity of fluoride in the oral cavity.

For topical fluorides such as dentifrices and mouthwashes, delivery of the fluoride to the site of action and the subsequent retention of fluoride are as important for overall treatment efficacy as the chemical and biochemical interactions discussed earlier. This is partly because the treatments are applied for only a short period of time, often less than 1 minute, and partly because of subsequent washout by saliva.

Analysis of saliva after a single application of a fluoride dentifrice or mouthwash show that much of the retained fluoride is cleared from the mouth within 1 hour. However, studies by Duckworth et al (1991, 1994) also indicate a secondary clearance phase of 2 hours or more during which the salivary fluoride concentration decreases more slowly. These authors postulated that the initial rapid clearance phase is the result of salivary washout and the second phase is the result of the release of fluoride initially retained in oral reservoirs into saliva. This explanation is consistent with a theoretical model that assumes three processes with different rates: fluoride uptake by a reservoir, fluoride release from a reservoir, and fluoride clearance from the mouth. The clearance data of Zero et al (1992a, 1992b) are consistent with this model.

Those results are a measure of average clearance rates, based on analyses of mixed saliva. By sampling at specific intraoral sites, Weatherell et al (1988) were able to show that fluoride clearance varied around the mouth. Fluoride uptake within the first 15 minutes after a NaF rinse was measured on small pieces of dentin mounted on buccal tooth surfaces. Clearance rates, which were inversely related to fluoride uptake, were high on maxillary incisors and low on mandibular incisors compared to rates at molar sites. A qualitatively similar pattern in plaque fluid fluoride concentration following a NaF rinse was observed by Vogel et al (1992).

These variations are linked to the pattern of salivary flow; clearance is fastest in regions "downstream" of the main saliva ducts. Vogel et al (1992) also showed that the clearance rate was slower from plaque fluid than from whole saliva, to the extent that plaque fluid fluoride concentration remained at levels capable of preventing lesion formation for at least 1 hour following a fluoride rinse.

Slow fluoride clearance at a particular site should not be equated with low caries risk. The clearance of other soluble materials from the mouth, such as potentially cariogenic sugars, might follow a pattern similar to that of fluoride (Dawes and Weatherell, 1990). Thus, there may be regions where both the cariogenic challenge and the means for its neutralization are high and others where both characteristics are low. On the other hand, it must be stressed that clean teeth never decay.

Other work suggests that long-term oral accumulation of fluoride is possible (Duckworth and Morgan, 1991; Duckworth et al, 1992; Geddes and McNee, 1982). During regular daily use of fluoride dentifrices or mouthwashes, the baseline fluoride concentration in both saliva and plaque has been found to increase over about 2 weeks. A corresponding decrease occurs when such treatments are discontinued. These findings may indicate the predominance of a different oral reservoir for fluoride from that implicated in the slow secondary phase of salivary clearance discussed earlier.

Among possible reservoir sites are the teeth; plaque; the soft tissues of the gingiva, tongue, and cheeks; and stagnation zones between the teeth, under the tongue, and in the buccal sulcus. The relative importance of the different sites is currently unclear. Plaque is important because of its proximity to the teeth. However, the soft tissues could be a major reservoir because of the relatively large surface area available (Zero et al, 1992a). Jacobsen et al (1992) reported elevated levels of fluoride in oral soft tissue after use of a NaF rinse. Moreover, the fluoride concentrations in the tissue appeared to be higher than values recorded in whole-mouth saliva at similar times, suggesting a slower clearance rate. Duckworth (1993) showed that the aforementioned findings were qualitatively consistent with predictions based on the salivary fluoride clearance model of Duckworth and Morgan (1991).

The importance of the different clearance phases for the anticaries efficacy of topically applied fluoride is also unclear. Few researchers would doubt the importance of the initially rapid clearance phase, when fluoride concentration is relatively high. However, with frequent use of fluoride toothpaste and fluoride mouthrinses, the later phase may also be very important for reduced demineralization and enhanced remineralization. Although the fluoride concentration is low, it may be sustained for a long period, eg, from one topical application to the next. In addition, glass-ionomer materials may act as slow-release reservoirs for fluoride (Hatibovic-Kofman and Koch, 1991).

The chemical nature of intraoral reservoirs may vary depending on the ecology (approximal molar surfaces versus lingual surfaces of mandibular incisors) and topical fluoride agent used (fluoride toothpaste [0.1% F] versus fluoride varnish [> 2% F]). From a cariostatic aspect, the most important fluoride reservoirs are CaF_2 and fluoride bound to plaque bacteria. Application of dissolved fluoride at a high concentration, for example from APF gels and fluoride varnishes (1% to > 2% F), causes precipitation of CaF_2 at neutral pH on apatite crystal surfaces and on the surface of enamel (see Fig 283; Arends and Christofferson, 1990; Ogaard et al, 1992; Rölla and Saxegaard, 1990). Because of the relatively high solubility of the phosphate-and-protein–coated CaF_2 particles, an elevated fluoride concentration will be maintained during cariogenic challenge in the tooth environment until all the particles are dissolved (see Fig 284).

With regard to plaque, binding of fluoride to organic substances seems to be more important than precipitation of CaF_2. Given its high calcium content, plaque fluid could provide a suitable milieu for CaF_2. Oral streptococci acquire fluoride in a tightly bound intracellular form and possibly a loosely bound, buffer-extractable form. Plaque itself contains fairly large quantities of fluoride, but only a small proportion is in the ionic state. However, most of the plaque fluoride is ionizable (Duckworth et al, 1994). The free fluoride concentration of plaque increases when the pH falls and thus becomes available to counteract demineralization.

To create a long-term reservoir close to the site of lesion formation, Pearce (1981) devised a urea-monofluorophosphate treatment intended to promote formation of FA within the plaque. Ammonia liberated by plaque urease raises the pH and thus promotes apatite precipitation, while hydrolysis of monofluorophosphate by phosphatases provides fluoride ions that are incorporated into the apatite. In vivo tests have shown that this system can prevent lesion formation and promote remineralization (Pearce, 1982; Pearce and Moore, 1985).

Effect of fluoride on dental plaque

The most important effects of posteruptive (topical) use of fluoride are the inhibition of demineralization and enhancement of remineralization. Fluoride exerts physicochemical effects not only in the oral fluids, such as the interrod and intercrystalline fluid, pellicle fluid, plaque fluid, and saliva, but also bound in CaF_2, FA, and FHA, as earlier discussed.

Fluoride also reduces acid formation in the dental plaque. It may reduce plaque formation rate and adhesion and may change the ecology of the plaque microflora. Of these effects, the most important is the reduction of acid formation. The fall in plaque pH following exposure to sucrose is smaller when the plaque fluoride content has been enhanced by repeated topical treatment (Geddes and McNee, 1982). However, it must again be emphasized that the most efficient way to prevent acid formation on the tooth surfaces is by frequent mechanical removal of cariogenic plaque supplemented by chemical plaque control to reduce reaccumulation of plaque.

Fluoride alone is inadequate because its cariostatic effect is limited. If the pH of plaque falls below about 4.5, the plaque fluid becomes undersaturated with respect to FA, and demineralization will occur, regardless of the presence of fluoride (Fig 291; Rölla et al, 1991). Human in situ experiments have shown that shark enamel, which consists of FA (33,000 ppm F), can be demineralized in the human oral environment: This occurred within 1 month when the specimens were covered by a 1-mm-thick layer of plaque and also occurred in the presence of excess fluoride in the liquid phase above the enamel. Under these conditions, rinsing with fluoride did not prevent demineralization of shark enamel (Ogaard et al, 1991).

The oral hygiene of many patients at high risk for developing caries is often inadequate (Ogaard et al, 1994a). This probably causes the pH to fall to such a low level that even fluoride is unable to inhibit caries development completely. In such cases, oral hygiene has to be improved, or fluo-

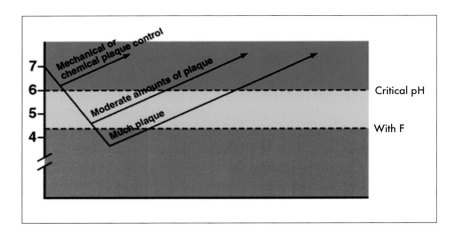

Fig 291 Influence of plaque control and fluoride on the critical pH of enamel. (Modified from Kjaerheim, 1995 with permission.)

ride administration must be supplemented by antibacterial agents, the role of the latter being to prevent the pH from falling to a level at which FA dissolves (see Fig 291).

The pH of plaque fluid is obviously dependent on the bacterial composition of the plaque, particularly the presence of acidogenic and aciduric strains. The number of bacteria at a specific site is also important because more bacteria produce more acid (Figs 42 to 45 and 49 to 51 in chapter 1; Axelsson et al, 1987b; Kristoffersson et al, 1984). This explains why studies in toothbrushing populations of 13- to 18-year-old subjects show that caries development is limited mainly to the approximal surfaces of the molars and premolars: Toothbrushing leaves the interproximal cariogenic plaque undisturbed (see Fig 52 in chapter 1). Therefore, for caries prevention, plaque control and topical fluorides should be directed toward these key-risk surfaces (see Figs 56 and 57 in chapter 1).

The thickness and age of the plaque biofilm have been related to its pathogenicity. It has also been shown that the volume of extracellular polysaccharides in plaque is important in this respect (Zero et al, 1986). One explanation is that fermentable carbohydrates penetrate deeper in the loose, polysaccharide-containing plaque, and acid formation occurs closer to the enamel surface than it does in denser plaque biofilms with less polysaccharides, in which the fermentable

carbohydrate is catabolized close to the plaque surface and further from the enamel surface (Figs 292 and 293).

Fluoride in plaque fluid and plaque

The plaque fluid, the aqueous phase within the plaque, is in contact with the enamel surface, the plaque bacteria (which produce the organic acids), saliva, and gingival fluid. Plaque fluid transports organic acids as well as fluoride, calcium, phosphate, and other ions to the enamel surface. These factors, of which fluoride and pH are the most important, determine whether or not the tooth mineral will dissolve. It is thus the fluoride present in plaque fluid that is so important in demineralization or remineralization of enamel. The plaque fluid is the site at which fluoride most significantly influences the demineralization and remineralization process (Ekstrand et al, 1990).

The fluoride in the plaque fluid may originate from many sources: CaF_2 on the enamel beneath the plaque, CaF_2 in plaque, and fluoride in saliva and gingival fluid. Although there is a correlation between the levels of fluoride in saliva and in plaque fluid (Dawes and Weatherell, 1990; Vogel et al, 1992), the concentration is much higher in plaque fluid. This may be attributed to slower elimination of the ion from the plaque, the thick-

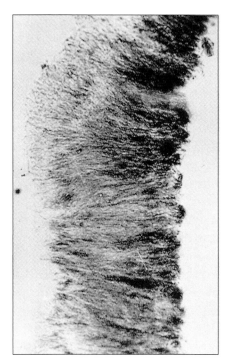

Fig 292 Acid accumulation *(black)* down to the enamel surface in plaque biofilm with a high percentage of acidogenic bacteria after a sucrose rinse.

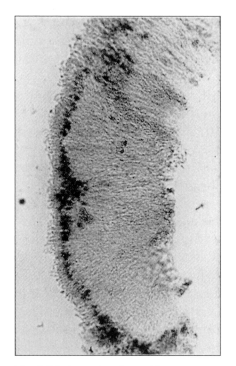

Fig 293 Superficial acid formation in a noncariogenic plaque biofilm after a sucrose rinse.

ness of the salivary film, or the release of fluoride from sources of fluoride in the plaque, such as calcium fluoride. The concentration of fluoride in plaque fluid seems to vary markedly at various sites in the mouth. Plaque fluid collected in the region of the maxillary incisors has a much higher concentration of fluoride than does plaque from other sites (Vogel et al, 1992). The pH of the plaque appears to be an important factor; low pH is associated with low fluoride concentrations.

It is well known that dental plaque is normally richer in fluoride than are the fluids to which the plaque is exposed, ie, saliva, plaque fluid, and gingival fluid (Ekstrand et al, 1990). Plaque thus appears to be able to retain and concentrate fluoride (Tatevossian, 1990). This is probably an important aspect of the cariostatic effect of fluoride because an inverse relationship has been demonstrated between high fluoride content in plaque and low carious activity.

Fluoride in plaque exists in ionic and bound forms: As in saliva, the concentration is determined by the frequency of fluoride exposure and the fluoride concentration of the source. Sources of plaque fluoride include the diet, saliva, and crevicular fluid. Dental plaque (wet weight) contains fluoride at 5 to 10 ppm (Tatevossian, 1990). Only a small part is present as free ions (less than 5%), and the amount is dependent on exposure to fluoride.

The nature of fluoride in plaque is uncertain. It has been reported that strong mineral acid is needed to dissolve all the fluoride in plaque (Ophaug et al, 1987). Although it has been suggested that fluoride may be stored inside the plaque bacteria (Jenkins et al, 1969), this would account for only a very small amount.

Another possibility is that CaF_2 forms in the plaque during mouthrinsing or toothbrushing with fluoride. During and after mouthrinsing (or toothbrushing), the plaque fluid is supersaturated with respect to CaF_2 because plaque fluid contains considerable amounts of calcium. This fluoride will be mobilized when the pH falls below 6 (Rölla and Saxegaard, 1990), accounting not only for the fluoride in plaque, and the release of fluoride and calcium from plaque during carious challenges, but also for the need for strong mineral acid to dissolve plaque fluoride.

A third possible mechanism would be retention of fluoride on the surfaces of the plaque bacteria (Rölla and Saxegaard, 1990), a formidable total surface area with a net negative charge and abundant phosphate and carboxyl groups (Marquis et al, 1993). In the oral environment, which is rich in calcium, the acidic groups on the surfaces of bacteria will acquire counter-ions, mainly calcium. Fluoride can thus be associated with calcium counter-ions (Rölla, 1977; Rölla and Saxegaard, 1990). This fluoride would also be released (with calcium) when the pH approaches the pK of the acidic groups: Both F^- and Ca^{++} are released at low pH. Because the total surface area is so large, this mechanism could also account quantitatively for the fluoride retained in plaque.

Effects on acid formation

As mentioned earlier, the most important effect of fluoride on plaque bacteria is to reduce production of acid, which occurs at relatively low concentrations of fluoride (Kashket et al, 1977). The inhibition of acid production increases, not only with increasing fluoride concentration, but also with decreasing pH. The overall result is reduced acid tolerance (Bender et al, 1985), as the fall in pH is reduced in the presence of fluoride.

The sensitivity of acid production to fluoride varies considerably both between and within species. For example, mutans streptococci are much more sensitive than is *Lactobacillus casei*.

It has long been known that fluoride inhibits the glycolytic enzyme, enolase. Such inhibition could directly reduce acid production and exert indirect effects, because the resulting depletion of phospho*enol*pyruvate would reduce sugar transport into the cell via the phospho*enol*pyruvate-dependent phosphotransferase system, in turn reducing both acid production and glycogen synthesis. Reduced glycogen formation would adversely affect the ability of the bacteria to survive during periods when exogenous fermentable carbohydrate is unavailable; ie, the acidogenic bacteria of the plaque would be "too hungry," and thereby the plaque ecology could be changed.

The cell membrane is impermeable to the F^- ion; therefore, fluoride must enter the cell as the un-ionized acid (hydrogen fluoride). Given the low pH of hydrogen fluoride (3.2), it seems that enolase inhibition would require much higher fluoride concentrations and lower pH levels than those at which fluoride actually inhibits acid production. Fluoride reduces production of bacterial acid principally by interfering with the control of intracellular pH. When external conditions are acidic, bacteria maintain the near-neutral cytoplasmic pH required for optimal enzyme function by pumping H^+ ions out of the cell. There is considerable evidence that fluoride affects permeability of the cell membrane to H^+ (Kashket and Kashket, 1985). Inward H^+ transport by diffusion of hydrogen fluoride would probably not be significant, except at unusually low pH and high fluoride concentrations. Instead, fluoride appears to act on the H^+ export system by inhibiting the adenosine triphosphatase enzyme responsible. The adenosine triphosphatase is located within the cell membrane, and it appears that extracellular fluoride acts directly on the enzyme; transport of fluoride into the cell is not required (Hayes and Roden, 1990). The inhibition of acid production would be a consequence of cytoplasmic acidification, rather than the primary effect of fluoride.

By inhibiting acid production, and thus reducing the extent to which plaque bacteria can lower the pH of their environment, fluoride will directly affect the rate of dissolution of tooth mineral by reducing the undersaturation, which is the main driving force for dissolution (Wahab et al, 1993).

In addition, because the buffering capacity of plaque increases as the pH falls (Shellis and Dibdin, 1988), a higher terminal pH implies that there will be a more rapid return to resting pH following the exhaustion of available carbohydrate.

Effects on ecology, amount, and formation rate of plaque

Dental plaque is a bacterial biofilm community comprising a wide variety of genera, species, and varieties of organisms (Bowden et al, 1979; Bowden, 1990; Bowden and Edwardsson, 1994; Darveau et al, 1997). As a biofilm, it follows a series of stages during its development to a point where its composition is relatively stable and in balance with the local environment (see chapter 1). The major phases of development are:

1. Adsorption and adherence of bacterial cells to the tooth pellicle.
2. Growth of the adherent cells and competition among different bacterial populations.
3. Stability of the biofilm community.

Throughout its development, dental plaque is subject to a variety of factors governing its composition, including nutrients, immune mechanisms, environmental pH, and ecosystem-generated antibacterial compounds. The changes in plaque that result from the influences of such factors may be very subtle and difficult to detect. The composition of plaque microflora and pattern of plaque formation rate may vary significantly from site to site in the same mouth.

Fluoride is one of many factors that influence the ecology, amount, and formation rate of plaque. Clearly with respect to the physiology of oral bacteria and their role in the various oral ecosystems, the plaque pH is a very important variable (Bowden, 1990; Marsh, 1992). As mentioned earlier, there is convincing evidence from several in vitro and in vivo studies (Bowden, 1990; Marsh and Bradshaw, 1990) that fluoride reduces the production of acid by plaque, although the duration of the effect varies, depending on the type of fluoride application (concentration and clearance time). As a consequence of reduced acid formation, a higher environmental plaque pH is achieved, which can influence competition among plaque bacteria. Thus, overgrowth of acidogenic and aciduric bacteria such as mutans streptococci is prevented. However, L casei seems to be less sensitive than mutans streptococci to the effect of fluoride on acid formation.

The long-term effects on plaque ecology by relatively low concentrations of fluoride cannot be predicted because of the possibility of phenotypic or genetic adaption to fluoride. This could explain why, for example, Kilian et al (1979) did not find significant differences in the bacterial composition of plaque samples taken from subjects living in areas with relatively high (3 to 21 ppm F) or low (< 0.3 ppm F) levels of naturally occurring fluoride in the water supply.

However, other animal and human studies have shown that higher concentrations of fluoride (> 250 mg/L) can influence the bacterial composition of plaque (Bowden, 1990). A review of studies (Hamilton and Bowden, 1996) on this effect of fluoride shows that, rather than causing complete elimination of bacterial populations from plaque, fluoride has the potential to affect the proportions of different bacteria in plaque ecosystems. Other studies have shown that a high concentration of fluoride may influence the synthesis of intracellular and extracellular carbohydrate polymers (glucan and fructan) (Hamilton, 1977, 1987, 1990; for review, see Carlsson and Hamilton, 1994).

Reduced synthesis of carbohydrate polymers may result in less survival of "hungry" acidogenic bacteria, less plaque volume, and less adhesive plaque (Carlsson and Egelberg, 1965). It is also possible that fluoride may affect early adhesion in vivo by modifying the selective deposition of salivary macromolecules on enamel (Rölla, 1977). Changed surface energy and wettability could be some of these effects (Glantz, 1969).

After adherence, cells on the surface begin to divide and extend the biofilm. There is little doubt that cellular division accounts for a major increase in biofilms in vitro and plaque in humans (Li and Bowden, 1994). However, it is difficult to determine the rate of cellular division in the plaque community in humans. Although some studies have been done in experimental animals, there are no human studies differentiating between an effect of fluoride on the rate of cellular division and effects on other aspects of accumulation, such as continuous adherence and coaggregation. Therefore, most of the data from studies of plaque development in vivo and in vitro record overall accumulation of plaque rather than cellular growth of the biofilm.

When the effects of fluoride on plaque in vivo are considered, it is important to differentiate between the amount of environmental fluoride that can occur naturally in drinking water and higher levels that might be introduced by application of different topical fluoride agents, such as APF gels and fluoride varnishes (1% to 2% F). Sufficiently high concentrations of NaF (> 0.2%) will affect the numbers of organisms in human plaque (Bowden, 1990). The antimicrobial effect of fluoride is enhanced in combination with cations such as tin and amine (for details see chapter 3). Two recent long-term in vivo studies have shown that regular use of different topical fluoride agents more than once a day may influence salivary mutans streptococci levels as well as the amount of plaque (Plaque Index [PI] and Plaque Formation Rate Index [PFRI]).

In a previously discussed 4-month double-blind study (Axelsson et al, 1994b; see chapter 3), 300 mutans streptococci–positive subjects with relatively high plaque and gingivitis scores were selected from among 1,000 subjects aged 17 to 19 years and randomly assigned to one of three test groups. Twice a day, after mechanical toothcleaning, test group I rinsed with a mouthrinse containing amine fluoride and SnF_2 (equivalent to 0.025% F), test group II rinsed with 0.1% chlorhexidine (CHX) as a positive control, and test group III rinsed with 0.025 NaF (0.012% F).

Figure 294 shows the effect on PI. The CHX mouthrinse resulted in the most significant reduction, followed by AmF plus SnF_2 (Meridol; GABA International, Müchenstein, Switzerland). However, even the NaF mouthrinse resulted in about 26% reduction. The effect on 24 hours of de novo plaque reaccumulation (PFRI) is illustrated in Fig 295. Again the CHX mouthrinse was the most efficient agent, followed by AmF plus SnF_2, but the NaF mouthrinse also reduced plaque reaccumulation (29%). Some of the effect can be explained by the study design, which required that all the participants clean their teeth twice a day before rinsing.

Chlorhexidine was significantly more efficient in reducing salivary mutans streptococci than were the other two mouthrinses. Rinsing with NaF did not achieve any significant reduction. None of the three mouthrinses resulted in significant reduction of *Lactobacillus*. None of the rinsing programs reduced the total amount of cultivable salivary microflora or the total number of cultivable salivary streptococci (Axelsson et al, 1993d; see chapter 3).

In the second study, 52 subjects with stimulated salivary secretion rates of less than 0.7 mL/min (mean of 0.4 mL/min) were selected for a 6-month longitudinal fluoride chewing gum study (Axelsson et al, 1997a). The subjects were instructed to chew a piece of fluoride chewing gum containing 0.25 mg F (Fluorette; Fertin Pharma, Vejle, Denmark) for 15 minutes directly after every meal or snack, ie, four to six times per day. The effects on stimulated salivary secretion rate, plaque (PI), gingivitis (Gingival Index [GI]), plaque formation rate (PFRI), and salivary mutans streptococci scores were evaluated.

The 6-month results showed an increase in the mean value of stimulated salivary secretion rate from 0.4 to 0.6 mL/min. The effects on PI, GI, and PFRI are shown in Fig 296. An average reduction of about 30% was achieved for all three variables. The percentage of subjects with salivary mutans streptococci scores 0 to 3 at baseline and after 6 months are shown in Fig 297. There was a pronounced shift from high scores to low, in particular a decline in score 2 and an increase in score 1 (Axelsson et al, 1997a).

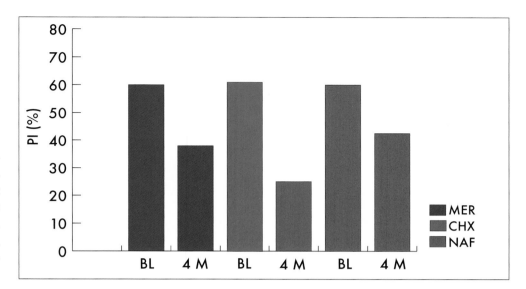

Fig 294 Effect of mouthrinses on PI. (MER) Meridol; (CHX) Chlorhexidine; (NAF) Sodium fluoride; (BL) Baseline; (4 M) 4 Months. (From Axelsson et al, 1993c. Reprinted with permission.)

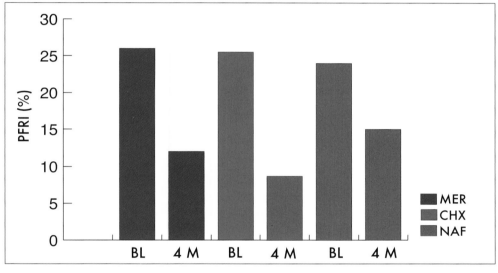

Fig 295 Effect of mouthrinses on PFRI. (MER) Meridol; (CHX) Chlorhexidine; (NAF) Sodium fluoride; (BL) Baseline; (4 M) 4 Months. (From Axelsson et al, 1993c. Reprinted with permission.)

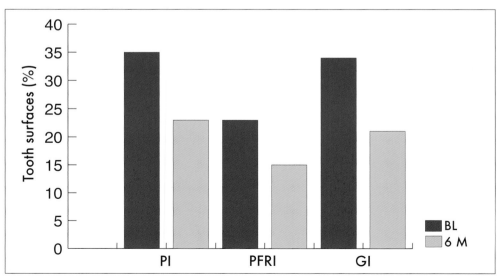

Fig 296 Effect of fluoride chewing gum: Percentage of tooth surfaces with disclosed plaque (PI), de novo reaccumulated plaque 24 hours after PMTC (PFRI), and inflamed gingiva (GI) at baseline (BL) and after 6 months. (From Axelsson et al, 1997a. Reprinted with permission.)

303

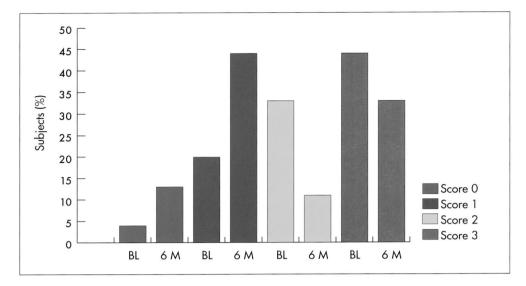

Fig 297 Effect of fluoride chewing gum: Percentage of subjects with strip mutans scores 0, 1, 2, and 3 at baseline (BL) and after 6 months (6 M). (From Axelsson et al, 1997a. Reprinted with permission.)

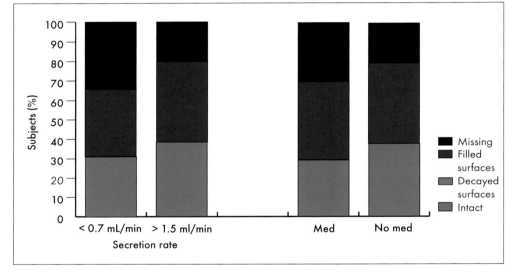

Fig 298 Frequency distribution of missing, filled, decayed, and intact surfaces in individuals with low (< 0.7 mL/min) and high (> 1.5 mL/min) stimulated salivary secretion rates and in individuals using drugs that impair salivary function (Med) and those not taking any medication (No med). (From Axelsson and Paulander, 1994. Reprinted with permission.)

Sjögren et al (1995) found that, after use of fluoride chewing gum and slow-release fluoride tablets, salivary fluoride concentration is greater, and fluoride clearance time is longer in subjects with reduced salivary secretion rates than in subjects with normal secretion rates. These results, together with the aforementioned data and the physiochemical effects of fluoride previously described, indicate that regular long-term use of fluoride chewing gum would have a significant effect on caries reduction in individuals with a reduced salivary secretion rate. Data from analytical epidemiologic studies in a randomized sample of more than 600 adults aged 50 to 55 years show a significant correlation between stimulated salivary secretion rate and caries prevalence (Fig 298; Axelsson and Paulander, 1994). Individuals with stimulated saliva rates of 0.7 mL/min or less should be regarded as risk subjects for dental caries. About 15%, 20%, and 25% of individuals in randomized samples of 50-, 65-, and 75-year-old subjects, respectively, have this type of stimulated salivary secretion rate (Axelsson et al, 1990; for reviews on the effect of fluoride on dental plaque, see Bowden, 1990; Hamilton and Bowden, 1996; Rölla and Ekstrand, 1996; and van Loveren, 2001).

Posteruptive effects of systemically administered fluoride

Fluoridated water

As long ago as the 1920s, McKay concluded that the drinking water in some areas of the United States caused so-called enamel mottling (fluorosis) and that subjects with mottled enamel were less susceptible to caries. A few years later it was reported independently from three different regions that mottled enamel was caused by fluoride in the water, and the condition was therefore called *fluorosis* (Churchill, 1931; Smith et al, 1932; Velu and Balozet, 1931).

Some years later, Dean observed an association between mild fluorosis and unexpectedly low caries experience in regions of the United States, and after the mid 1930s directed his research to the fluoride-caries relationship (Dean, 1934; Dean and Elvove, 1935). He demonstrated that caries experience in 12- to 14-year-old children in Galesburg, Illinois, and Monmouth, Illinois, which had a water fluoride concentration of 1.8 mg/L, was less than half that in nearby Quincy, Illinois, which had a water fluoride concentration of 0.2 mg/L (Dean et al, 1939). He also noted that the low caries experience in Galesburg and Monmouth was accompanied by an unacceptably high level of dental fluorosis and set out to define the water fluoride levels that represented the best compromise between low caries experience and a level of fluorosis that could be considered acceptable. This was done through a series of investigations that have become known collectively as *The 21 Cities Study*. The first part consisted of clinical data from children aged 12 to 14 years with lifetime residence in eight suburban Chicago communities with various, but stable, mean water fluoride levels (Dean et al, 1941). Thirteen communities in four other American states were later included (Dean et al, 1942).

The 21 Cities Study was landmark epidemiology and led to the adoption of 1.0 to 1.2 mg/L as the appropriate concentration of fluoride in drinking water for temperate climates in the United States. Although the results were very impressive, they were nevertheless based on cross-sectional data, ie, indicating a caries-preventive effect of fluoride in drinking water. Therefore, in 1943, a prospective test of the fluoride-caries hypothesis was proposed (Ast, 1943). In January 1945, Grand Rapids, Michigan, became the first community to add fluoride to the municipal water supply.

Because Dean's 21 Cities Study was the basis on which a fluoride level of 1.0 mg/L was selected as the optimum concentration for water fluoridation, some aspects of this study warrant closer scrutiny. Figure 299 shows the relationship between caries experience and dental fluorosis reported by Dean in 12- to 14-year-old subjects in 9 of the 21 cities, which had water fluoride levels ranging from 0.2 to 1.2 mg/L. Caries experience drops sharply toward fluoride concentrations of 1.0 to 1.2 mg/L and then tends to level out. Figure 299 shows the prevalence of fluorosis designated by Dean as very mild or severe; the intermediate, "questionable," cases have been excluded. These data indicate that a 1.0 mg/L water fluoride level leads to substantial caries prevention, accompanied by about a 12% prevalence of fluorosis, all of it in the mild to very mild categories. If the questionable cases are included, however, a fluoride level as low as 0.4 mg/L resulted in about a 40% prevalence of fluorosis. Dean et al (1941) stated that the fluorosis was observed "almost exclusively" on the later-erupting teeth, second molars and second premolars, and that "mottled enamel as an esthetic problem" was not encountered. Dean et al (1941) also claimed that because their index was based on the two most severely affected teeth in the mouth, the prevalence data overestimated the number of teeth affected.

The aforementioned data originate from areas of the United States with a temperate climate. Because fluorosis is dose related, in tropical or subtropical climates a much lower water fluoride concentration may result in fluorosis because daily consumption of water is higher: For this reason, after water fluoridation was introduced in

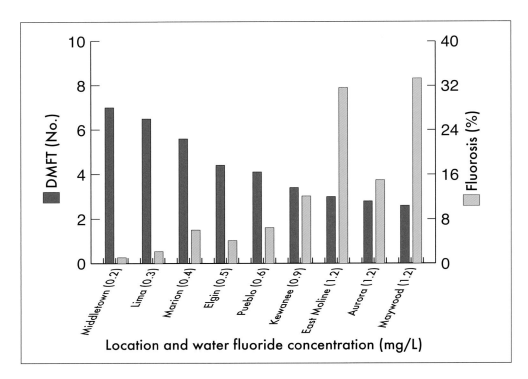

Fig 299 Relation between mean dental caries prevalence and the prevalence of dental fluorosis among children aged 12 to 14 years in 9 of the 21 cities with water fluoride concentration from 0.2 to 1.3 mg/L. (DMFT) Decayed, missing, or filled teeth. (Data from Dean et al, 1941.)

Hong Kong, it was necessary to reduce the water fluoride level successively to 0.5 mg/L. As discussed in the beginning of this chapter, the cariostatic effect of fluoride is almost 100% posteruptive, ie, topical. Therefore, it is questionable whether, for example, drinking water with a 0.5-mg/L fluoride content ingested 10 times per day in a tropical climate will have the same posteruptive cariostatic effect as water with a 1.0-mg/L fluoride content ingested five times per day in a temperate climate.

In this context, a recent large-scale study in Alexandria, Egypt, which has a very hot climate, is of interest: A total of 1,354 schoolchildren aged 12 years from the same district of the city of Alexandria, which has a natural water fluoride concentration of about 0.5 to 1.0 mg/L, were examined for caries prevalence, fluorosis, salivary mutans streptococci levels, sugar clearance score, oral hygiene habits, and tea consumption. In Alexandria, many children start drinking tea as early as 2 years of age. As discussed earlier in this chapter, tea may contain fluoride levels of up to 4 mg/L. More than 90% of the children did

not brush their teeth regularly, and there was therefore no regular exposure to fluoride toothpaste (Axelsson and el Tabakk, 1999a, 1999b).

Figure 300 shows the frequency distribution of children with different fluorosis scores. An established habit of tea consumption was more frequent among the children with fluorosis. Children without fluorosis had significantly more decayed, missing, or filled surfaces (DMFSs) than did children with fluorosis (Fig 301; Axelsson and el Tabakk, 1999a). Children without fluorosis developed significantly more new caries lesions than did children with fluorosis (Fig 302). The latter also consumed more tea, which resulted in a greater posteruptive (topical) cariostatic effect (Axelsson and el Tabakk, 1999b).

With respect to general health, the following conclusions were drawn from the early research on people in temperate climates living in areas with different natural fluoride concentrations in the drinking water (Ast et al, 1950; Blayney and Tucker, 1948; Dean et al, 1950; Hutton et al, 1951):

Fig 300 Frequency distribution of fluorosis scores 0 (none), 1 (mild), 2 (moderate), and 3 (severe) among 12-year-old schoolchildren in Alexandria, Egypt, in 1989 (N = 1,354). (From Axelsson and el Tabakk, 1999a. Reprinted with permission.)

Fig 301 Relationship between fluorosis and caries prevalence among 12-year-old schoolchildren in Alexandria, Egypt. Fluorosis scores: 0 = none; 1 = mild; 2 = moderate; 3 = severe. (DMFSs) Decayed, missing, or filled surfaces. (From Axelsson and el Tabakk, 1999a. Reprinted with permission.)

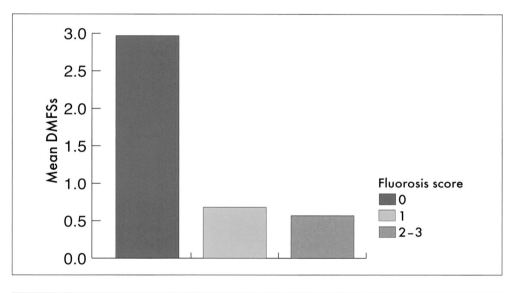

Fig 302 Relationship between fluorosis and caries incidence among 12-year-old schoolchildren in Alexandria, Egypt, from 1989 to 1991. Fluorosis scores: 0 = none; 1 = mild; 2 = moderate; 3 = severe. (DMFSs) Decayed, missing, or filled surfaces. (From Axelsson and el Tabakk, 1999b. Reprinted with permission.)

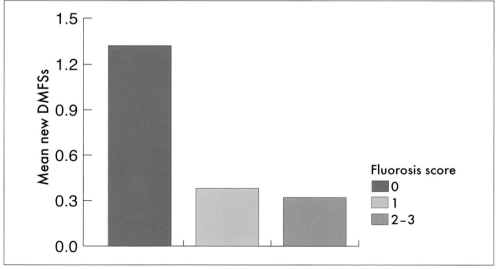

307

1. Any potential health hazards from long-term accumulation of fluoride in areas where the fluoride level in drinking waters is low are minimized by the normal renal excretory mechanism.
2. Ingested fluoride is partly deposited in bone and partly excreted.
3. Although skeletal fluoride concentrations increase with age, skeletal damage cannot be demonstrated in people ingesting water that contains natural fluoride.
4. Apart from dental fluorosis, no impairment of general health has been found among people drinking water containing fluoride levels of up to 8.0 mg/L for long periods of time.
5. People who drink water that contains fluoride at levels around 1.0 mg/L have substantially less caries than do those who drink water with negligible concentrations of fluoride. (For review, see Burt and Fejerskov, 1996.)

However, very high concentrations of natural fluoride may occur in tropical and subtropical climates. The high concentration in combination with high daily consumption of drinking water may result in changes in both the mineralized and the soft tissues, such as the kidney and the thyroid gland.

Cariostatic effects. Four independent studies on controlled fluoridation were begun in 1945 and 1946. The fluoride concentrations in the water supplies of the test communities were increased from negligible to 1.0 to 1.2 mg/L. The purpose of these field trials was to test the hypothesis that the addition of fluoride to low-concentration drinking waters would reduce the caries experience. These original four studies were in the following locations:

1. Grand Rapids, Michigan. Began in January 1945, with nearby Muskegon as the control city. This study was directed by Dean and his colleagues (Dean et al, 1950).
2. Newburgh, New York. Began fluoridation in May 1945, with Kingston, New York, as the control city (Ast et al, 1950).
3. Evanston, Illinois. Began in 1946, with Oak Park, Illinois, as the control city (Blayney and Tucker, 1948).
4. Brantford, Ontario, Canada. Began in 1946 with Sarnia, Ontario, as the control city. Naturally fluoridated Stratford, Ontario, was also included in this study (Hutton et al, 1951).

At the end of terms up to 15 years, a marked reduction in caries experience was shown among children in each of the study populations, despite some differences in study design and examination criteria. The reductions in decayed, missing, or filled teeth (DMFT) among 12- to 14-year-old children in the Grand Rapids, Evanston, Brantford, and Newburgh studies were 55%, 48%, 57%, and 70%, respectively (Arnold et al, 1962; Ast and Fitzgerald, 1962; Blayney and Hill, 1967; Hutton et al, 1956).

These studies also showed that the occurrence and severity of dental fluorosis were similar to earlier observations by Dean (1934) in areas of natural fluoridation, namely that approximately 12% of children growing up with water fluoridated at around 1.0 mg/L developed mild to very mild fluorosis. The results of these pioneer studies were based on cross-sectional data, analyzed with very simple statistical methods and without any adjustment for background factors, such as socioeconomic conditions, oral hygiene, and dietary habits. However, these early studies were followed by additional studies in many countries, usually on children.

Perhaps the most striking feature of these studies, which vary in the quality of their design and operation, has been the general uniformity of the results. A review of fluoridation studies among children prior to 1980, about the time that the caries decline first became fully appreciated, showed that 55 of 72 published studies reported 40% to 70% reductions in caries in children's permanent teeth (Murray et al, 1992). Children in fluoridated Dublin, Ireland, still have DMFT scores that are 45% lower than those among children of the same age in nonfluoridated Glasgow, Scotland (Blinkhorn et al, 1992).

In those countries where fluoridation is widespread, ie, reaching more than half the population, the differences in caries experience between children in fluoridated and nonfluoridated communities are now more commonly in the range of 0% to 30%. This is clearly less than the approximately 50% difference reported in early studies, probably resulting from the influence of fluoride from other sources and improved oral hygiene.

Recently, Seppä et al (2002) presented cohort data in 10-year intervals showing that caries prevalence was significantly lower in a fluoridated area compared to a nonfluoridated area at the first cohort, but 10 years later this difference was eliminated. Six years after the second cohort (1992), water fluoridation was stopped in the area (Kuopia, Finland). In spite of that, reexamination in 1992, 1995, and 1998 in randomized samples of 3-, 6-, 9-, 12-, and 15-year-old children in Kuopia and the nonfluoridated control city, Tyväskylä, showed similar continued decrease of caries prevalence. In addition, the mean number of fluoride varnish and sealant applications had markedly decreased in 1993 through 1998 compared to 1990 through 1992. The fact that no increase in caries was found in Kuopia despite discontinuation of water fluoridation and decrease in use of fluoride varnish and sealants suggests that not all of these measures were necessary for each child. But it also indicates that improved daily self-care is most important (Seppä et al, 2000).

The use of fluoride toothpaste is almost universal in the industrialized countries. In the highly fluoridated countries there is also considerable indirect exposure to fluoride (Brunelle and Carlos, 1990; Burt, 1992; Ismail et al, 1993; Newbrun 1989). In addition, the standard of oral hygiene and other preventive measures targeting the etiology of dental caries has improved significantly during the last two decades, lessening the overall effect of any fluoride method. Meanwhile, caries prevalence has been dramatically reduced in countries without water fluoridation. For example, the Scandinavian countries exhibit the world's lowest caries prevalence among children. (See chapter 6.)

According to 1994 World Health Organization (WHO) statistics, about 210 million people used fluoridated water, with unknown millions drinking water with natural fluoride (≥ 1.0 mg/L). According to the most recent estimation, about 300 million people use fluoridated water (Clarkson, 2000), while about 500 million people use fluoride toothpaste regularly (predominantly in the industrialized countries). However, more than 90% of the world's population, eg, in most developing countries in Asia, Africa, and Latin America, use neither fluoridated drinking water nor fluoride toothpaste regularly.

The posteruptive cariostatic effectiveness of fluoridated water is shown in those communities where fluoridation was introduced and subsequently discontinued. The outcomes have been described from three such communities in Scotland, one in Germany, and one in the United States (Attwood and Blinkhorn, 1991; Künzel, 1980; Lemke et al, 1970). In each instance, a decline in caries was observed in young children after fluoridation began, but this decline did not continue after fluoridation was discontinued. At the time of the studies, drinking water was the only substantial exposure to fluoride.

The most striking example of what happens when fluoridation is discontinued comes from the Tiel-Culemburg study in the Netherlands, a major longitudinal study in which Culemburg was used as a control city while the water supply in Tiel was fluoridated at 1.0 mg/L from 1953 to 1973. In 1968 to 1969, after 15 years of fluoridation in Tiel, there was a major difference in the caries prevalence (DMFSs) among 15-year-old subjects in the two communities (Fig 303): This gradually leveled out over subsequent years, until 1987 to 1988. The caries decline is evident in both communities, although far more pronounced in Culemburg (Kalsbeek et al, 1993).

Regarding the effect of fluoridation on adults, constant exposure to fluoridated water produces the positive effects that would be expected in light of the posteruptive physiochemical effects of a constant, low fluoride concentration in the oral fluids on enamel dissolution and remineral-

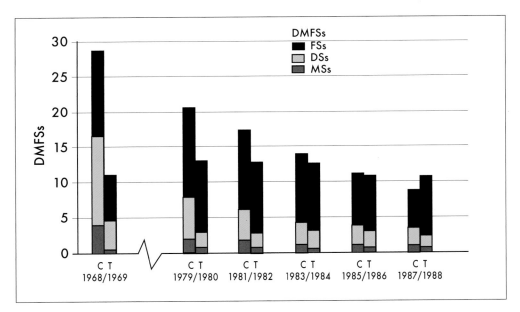

Fig 303 Caries prevalence among 15-year-old subjects in Culemburg (C) and Tiel (T), the Netherlands. The water supply of Tiel was fluoridated at 1.0 mg/L from 1953 to 1973. (FSs) Filled surfaces; (DSs) Decayed surfaces; (MSs) Missing surfaces; (DMFSs) Decayed, missing, or filled surfaces. (Modified from Kalsbeek et al, 1993 with permission.)

ization, as described earlier. Such effects are no longer as pronounced, because of several decades of regular use of fluoride toothpaste in many industrialized countries. However, years ago, before the development of efficient fluoride toothpastes, McKay (1948) noted that the oral health of 45-year-old adults who drank fluoridated water was better than the oral health of those who did not.

Russel and Elvove (1951) found that adults born and raised in naturally fluoridated Colorado Springs, Colorado, had 60% lower mean DMFS scores than did their counterparts in low-fluoride Boulder, Colorado. Residents of Colorado Springs also had far fewer missing teeth and lower caries experience in all tooth types. In particular, the inhabitants of Colorado Springs were far less susceptible to approximal caries (Russel, 1953). A similar profile was observed in adults from Aurora, Illinois, where the drinking water had naturally occurring fluoride at 1.2 mg/L (Englander and Wallace, 1962), and in a British study of adults in Hartlepool (1.5 to 2.0 mg/L at the time) and York (0.2 mg/L) (Murray, 1971a, 1971b).

In order to assess the lifetime effects of fluoridated water, the subjects of both the Colorado Springs and the Aurora studies were limited to permanent residents. In light of current knowledge of the posteruptive cariostatic effects of fluoride, this choice is regrettable, because data on the purely posteruptive effects of fluoridated water were not collected.

The effect on prevalence of root caries has also been correlated with the fluoride concentration of the drinking water (Burt et al, 1986b; Stamm et al, 1990). This is of particular importance because of the increasing numbers of caries-prone elderly people with poor oral hygiene and dietary habits, increasing numbers of natural teeth with exposed root surfaces, and reduced salivary secretion rate because of general ill health and regular medication.

In the aforementioned early studies, no attempts were made to apply multivariate analyses to evaluate the role of other modifying factors, such as differences in socioeconomic conditions, gender, oral hygiene, and dietary habits. In some Swedish studies, Wiktorsson et al (1991a, 1991b, 1992a, 1992b, 1995) evaluated oral health status in selected 30- to 40-year-old subjects from two neighboring cities. One third of the residents who were born and had lived permanently in the city of Uppsala, which had naturally occurring fluoride at a level of 1.0 mg/L in the drinking water, were randomly selected for the study, as were all subjects in the same age group from the city of Enköping, which had a fluoride level of 0.3 mg/L in the drinking water.

The prevalence and incidence of caries were significantly higher in the subjects from Enköping than they were in the residents of Uppsala, although there were no differences in socioeconomic status, gender, salivary secretion rate, salivary buffering capacity, and salivary levels of lactobacilli. However, about 10% of the subjects from Uppsala exhibited mild fluorosis, while none of the subjects from Enköping did.

Approximately 25 to 30 years ago, children in Uppsala had by far the lowest caries prevalence in Sweden. At that time, caries prevalence and incidence in Sweden were more than 10 times higher than they are today because oral hygiene habits were poor and there were no efficient fluoride toothpastes available. Today, the caries prevalence in children in some counties in Sweden is similar to or less than that in the county of Uppsala. For example, in the county of Värmland (fewer than 3% of the population have ≥ 1.0 mg/L of natural fluoride in the drinking water), the mean DMFT in 19-year-old individuals is only 2, compared to 3 in Uppsala, largely as a result of highly significant improvement in plaque control directed toward the cause of dental caries (the cariogenic plaque) combined with fluoride toothpaste and needs-related use of other topical fluoride agents. From 1974 to 1999 the caries prevalence in 12-year-old children in the county of Värmland has decreased from 25 to only 0.4 DFS, which might be the lowest value in the world (for details, see chapter 6.)

In countries with extensive fluoridation, many people spend part of their lives in a fluoridated area and part in a nonfluoridated area. The evidence shows that partial exposure to fluoridation reduces caries experience proportional to the length of exposure (Ast and Fitzgerald, 1962; Burt et al, 1986a). A 4-year British study found that children who were 12 years old when fluoridation began in their community had a 27% lower caries incidence than did age-matched

controls in nonfluoridated areas (Hardwick et al, 1982). This well-conducted study clearly demonstrated the posteruptive effects of fluoride in the drinking water.

Further evidence that the preeruptive effects of fluoridated water are weak has come from Okinawa, Japan. In Okinawa, there was no difference in caries status among nursing students aged 18 to 22 years when those who had received fluoridated water only until 5 to 8 years of age (and none thereafter) were compared with those who had never received fluoridated water (Kobayashi et al, 1992). In Perth, Australia, the main factors associated with freedom from caries were early use of toothpaste and residence in a fluoridated area from 4 to 12 years of age, as opposed to residence from birth to 4 years of age (Riordan, 1991).

Cost effectiveness. Both the WHO and Fédération Dentaire Internationale (FDI) have endorsed water fluoridation as the most cost-effective cariostatic public health measure for countries with relatively high caries prevalence. The average annual cost of water fluoridation in the United States has been estimated at $0.51 per person, although in any one community the range was from $0.12 to $5.41 per person. Factors that influenced the per capita cost were determined to be:

1. The size of the community: The bigger the population, the lower the per capita cost.
2. The number of fluoride injection points required.
3. The amount and type of equipment used.
4. The amount and type of fluoride chemical used, its price, and the cost of transportation and storage.
5. The expertise of the water plant personnel.

The figures on fluoridation costs in the United States were based on actual costs from water treatment plants, although of course these figures will soon become dated. Perhaps more important than the actual costs of fluoridation were the estimates of its savings in treatment costs. Health economists at the 1989 Michigan workshop concluded that water fluoridation was one of the very few public health measures to demonstrate true cost savings: The measure actually saved more money than it cost to operate (Results of the Michigan workshop, 1989). Estimates from the workshop were that fluoridation cost $3.35 per carious surface saved, far less than the fee for any restoration.

Data on the cost savings of fluoridation go back to the Newburgh-Kingston study, where initial dental care for 6-year-old children cost 58% less in fluoridated Newburgh than in nonfluoridated Kingston (Ast et al, 1970). These savings came from fewer extractions, fewer restorations, and a smaller proportion of complex restorations. There were similar findings in British studies: A savings of 49% for children aged 4 to 5 years and 54% for children aged 11 to 12 years (Downer et al, 1981). These savings were maintained even after the caries decline was recognized (Attwood and Blinkhorn, 1989, 1991).

These figures are impressive, but they must be seen in the context of very limited effects in populations with low caries prevalence because of high standards in oral hygiene and regular use of fluoride toothpaste, conditions that now apply in most industrialized countries. While it is clear, for example, that fluoridation reduces the costs of restorative care for children, its effect on the costs of adult dental care is less clear. On the other hand, as discussed earlier, fluoride is an efficient cariostatic measure against root caries, an increasing problem as a consequence of increasing numbers of dentate elderly.

There is no doubt that water fluoridation is a highly effective means of reducing caries experience in a community with relatively high caries prevalence and insufficient preventive programs, and it is equally clear that it is safe and presents no great technical problems in existing municipal water supplies. While it is only one of several means of supplying fluoride to a community, no other method is as economical or as comprehensive when infrastructure conditions are favorable (ie, an extensive water treatment system, people who drink the tap water, trained technicians and engineers, and availability of fluoride compounds). Although socioeconomic status is a major determinant of caries experience, the gradient in caries experience between socioeconomic groups is smaller in fluoridated areas than it is in nonfluoridated areas. Fluoridation reduces, but does not eliminate, the impact of social class on caries, and it does this more cost effectively than other fluoride methods.

There is no central agency that stores global information on water fluoridation projects. However, in the 1980s more than 30 countries reported fluoridation projects covering some 250 million people, not including naturally occurring fluoride in drinking waters. Although fluoridated water is consumed by more than half of the populations of the United States, Canada, Ireland, Australia, and New Zealand, and by virtually everyone in the city-states of Singapore and Hong Kong, some of the more than 30 nations have only one community fluoridating, with little prospect of more. Ireland is the only nation to have a mandatory fluoridation law.

Fluoridation is not technically feasible for much of Asia and Africa, where reticulated municipal water systems are uncommon. There are very few water fluoridation plants in Latin America. In addition, most of the countries in Asia, Africa, and Latin America have tropical or subtropical climates, which result in a high daily consumption of drinking water. As a consequence, 1 mg F/L water would result in a high prevalence of fluorosis. Thus, water fluoridation should not be the method of choice. In Europe, where it is technically feasible, fluoridation is widespread only in Britain. There is no fluoridation at all in Germany, Austria, France, Belgium, Italy, Denmark, Sweden, or Norway. (For reviews on water fluoridation, see Burt and Fejerskov, 1996.)

Fluoride tablets and lozenges

As discussed earlier in this chapter, the preeruptive and systemic effect of fluoride was originally believed to be the most important for caries prevention. Daily ingestion of small amounts of fluoride was assumed to be necessary for optimal caries prevention. The impressive caries reductions observed in fluoridated areas during the 1950s, and the recognition that water fluoridation is not always feasible, stimulated investigation into fluoride supplements for children in nonfluoridated areas. The supplements were designed to provide a daily dose equivalent to the estimated daily ingestion by children in fluoridated areas. Fluoride tablets became the most widely accepted alternative to water fluoridation.

On the basis of elaborate estimates of daily fluoride ingestion in water-fluoridated areas in relation to age, Arnold et al (1960) conducted one of the first successful clinical studies on the cariostatic effect of daily ingestion of fluoride tablets. Children younger than 3 years of age received 1 mg of F^- every second day, and children older than 3 years received 1 mg of F^- per day. This dose remained the basis of dietary fluoride recommendations for about 30 years. The caries reduction reported in this study corresponds with that observed with water fluoridation at 1 ppm of fluoride.

However, scrutiny of the study design and outcome shows that the subjects were highly selected, ie, the children of physicians, dentists, and other employees of the US Public Health Service in Washington. The study was originally planned as an experiment without a control group, and the final analysis comprised those children who continued to the end of the planned study period, only half of the original experimental group. Nearly two thirds of the children began to take fluoride tablets before the age of 3 years and almost all before the age of 6 years.

In this study, as well as in many other studies, the fluoride tablets were not simply swallowed but dissolved in water throughout the study period; ie, all erupting teeth were exposed daily to small amounts of fluoride in effectively the same manner as teeth erupting in children who consume fluoridated water daily. Compliance by parents in regular daily administration of fluoride tablets to small children is very low. Therefore, the positive influence of compliance in this selected group of well-motivated, dentally conscious parents, ie, parents encouraging oral hygiene, and the daily posteruptive exposure to fluoride will fully explain the caries reduction, rather than any preeruptive effect. These aspects of studies are not always apparent in the large tables of data on fluoride tablets and caries prevention that, over the years, have been compiled in textbooks and review papers.

A double-blind study examined the effects of administration of fluoride tablets to pregnant women. Compared to children who received a placebo prenatally, children who received fluoride prenatally experienced no reduction in caries in their primary teeth at the age of 5 years (Whitford, 1992). Thylstrup et al (1979) observed, in agreement with Fanning et al (1980), no decrease in caries levels in children who had been given tablets daily from birth but who discontinued their use at around 2 years of age.

To date, no properly designed prospective fluoride tablet study has shown any significant preeruptive caries-preventive effect. Therefore, it is surprising that many dentists around the world still firmly believe in the value of fluoride supplements given from birth. As examples, the American Dental Association (ADA) recommendations that supplements be taken from birth persisted until 1994; if the drinking water contains 0.30 mg/L or less of fluoride, a 0.25-mg fluoride supplement is still recommended from the age of 6 months by the ADA and from birth in France, Germany, and Austria.

If it is assumed that there is almost no positive preeruptive effect, a reasonable recommendation is to start no earlier than the eruption of the primary molars, or during the third year of life and only in high-risk children, as proposed by the Scientific Committee of the European Union (Clarkson, 1992). However, because of the risk of

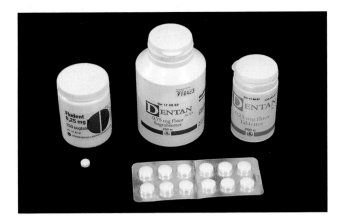

Fig 304 Three alternatives for supplementary topical use of fluorides after every meal in high–caries-risk individuals: chewing gum (Fluorette), 0.75-mg-F lozenges for adult xerostomia patients (Dentan; ACO Läkemedel, Solna, Sweden), and 0.25-mg-F lozenges for children (Dentan and Fludent; Fertin Pharma).

fluorosis during the maturation phase of the permanent teeth, fluoride tablets should not be recommended before the age of 5 years, and use should be based on individual needs rather than general public health programs. Studies by Carvalho et al (1989) showed that almost 100% of fissure caries in the molar teeth starts during eruption: In a 3-year study, intensified plaque control and topical use of fluoride during this risk period almost completely prevented the development of manifest fissure caries, without application of fissure sealants (Carvalho et al, 1992). It is also well known that, on average, teeth erupt 6 to 12 months earlier in girls than boys. Tooth surfaces are most susceptible to caries during eruption, before secondary maturation. To optimize the posteruptive effect of fluoride on the permanent teeth, administration of fluoride tablets should not start before the age of 5.5 years in girls and about 6 years in boys. Administration should be restricted to selected risk individuals. Slow-release fluoride lozenges (0.25 mg F) are recommended for "dessert," ie, immediately after meals, prolonging fluoride exposure directly after the fall in plaque pH.

Well-conducted clinical trials have shown that fluoride tablets provide posteruptive cariostasis in school-aged children. American studies in which the tablets were chewed, swished, and swallowed by schoolchildren under supervision have reported caries reductions of 20% to 28% over 3 to 6 years (DePaola and Lax, 1968; Driscoll

et al, 1978). These studies were designed with controls, placebos, and double-blind conditions. Caries reductions were higher for those teeth erupting during the study period, and, in one study, beneficial effects were still discernible 4 years later (Driscoll et al, 1981). More spectacular results, an 81.3% reduction in caries incidence, were reported from a study in Glasgow, Scotland; children from lower socioeconomic groups, initially aged 5.5 years, sucked a 1.0-mg fluoride tablet or a placebo under supervision at school, every school day, for 3 years. The major effect was in the erupting permanent first molars (Stephen and Campbell, 1978).

There have been well-conducted school-based clinical trials of fluoride tablets in combination with other fluoride therapies. Driscoll et al (1992) found that tablets used with the swish-and-swallow procedure over 8 years gave slightly better results than did fluoride mouthrinsing, although the incidence of caries in all study groups was small. In Scotland, no difference in caries incidence was found over 6 years in three groups of children using fluoride tablets, mouthrinses, or combinations of both and children using placebos (Stephen et al, 1990a). A Swedish study comparing fluoride tablets, fluoride toothpaste, and fluoride varnish found no difference between the groups (Petersson et al, 1985).

Evidence for the efficacy of any preventive procedure should come, as much as possible, from clinical trials that meet specific quality criteria

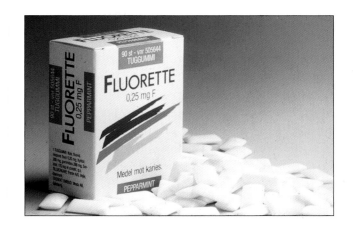

Fig 305 Fluoride chewing gums containing 0.25 mg F per piece.

(randomized controlled trials). From the few trials of fluoride tablets that meet these standards, the evidence is that tablets are effective in school-aged children and when used topically (chewed or slowly dissolved in the mouth). As discussed earlier, there is no incontrovertible evidence for the efficacy of fluoride tablets used from birth or early infancy.

To prolong fluoride exposure and clearance as much as possible, only lozenges, which release fluoride slowly, should be recommended directly after meals. This recommendation should be restricted to selected caries-risk subjects: not only children older than 5 years but also adults, particularly elderly patients with reduced salivary secretion rates. This achieves needs-related intervals of self-administered topical fluoride.

Fluoride lozenges containing 0.25, 0.50, 0.75, and 1.00 mg of fluoride are commercially available (Fig 304). The 0.25-mg fluoride lozenges are recommended for children more than 5 years old, and 0.25- to 1.00-mg fluoride lozenges are recommended for selected young adults and adults, according to individual need. However, compliance in administration of fluoride tablets to children is low. According to WHO statistics, it is estimated that only about 20 million people in the world use fluoride tablets.

Fluoride chewing gum

Recently fluoride chewing gum has become commercially available (Fluorette and Fludent). It is sugarless; xylitol and sorbitol are used as sweeteners. Each piece contains 0.25 mg of fluoride (Fig 305).

It is well known that patients with decreased salivary secretion rate (hyposalivation: ≤ 0.7 mL/min; xerostomia: ≤ 0.1 mL/min of stimulated saliva) are highly susceptible to caries (Axelsson et al, 1990; Axelsson and Paulander, 1994; Dreizen and Brown, 1976). Such patients are predominantly elderly; 20% and 25% of 65- and 75-year-old individuals, respectively, have decreased salivary secretion rates (Axelsson et al, 1990). In addition, most elderly people have exposed root surfaces and therefore a relatively high prevalence of root caries.

As mentioned earlier in this chapter, Axelsson et al (1997a) conducted a study in subjects with a mean stimulated salivary secretion rate of only 0.4 mL/min. For 6 months, fluoride chewing gum (Fluorette) was chewed regularly for 15 minutes, directly after every meal (four to six per day). Plaque quantity (Pl), plaque formation rate (PFRI), and gingival inflammation (GI) decreased by about 30%. The salivary secretion rate increased by almost 50%, and there was a very significant reduction in high salivary mutans streptococci scores (see Figs 296 and 297).

Sugarless chewing gum is unique because in normal use it is chewed for a prolonged period, usually more than 30 minutes, but at the same time it is noncaloric: Both characteristics are important in the context of salivary stimulation. Chewing gum has been shown to elicit a continued flow of saliva during prolonged mastication (Dawes and Macpherson, 1992). The level of stimulus declines, however, during mastication. The gustatory stimulatory component is rapidly depleted as the flavoring agents are released and swallowed, and, because of the softening of the gum, the intensity of the masticatory stimulus falls. Studies of the effect of varying not only the duration of chewing but also the interval between the prior caries challenge (pH drop) and the subsequent gum chewing showed that, for maximum neutralization of the pH, the gum must be chewed for at least 15 minutes, immediately following the caries challenge (Park et al, 1993).

Studies in patients with low salivary flow rates (Abelson et al, 1990; Markovic et al, 1988) have shown that use of a sorbitol-sweetened gum raises the pH of plaque on both enamel and root surfaces. Together, these studies provide powerful evidence that, following consumption of fermentable carbohydrates, chewing of sugarless gum rapidly elevates plaque pH toward resting levels, where it persists for the duration of the experiment. A similar, but less rapid and persistent effect occurs with a sugared gum. Dodds et al (1991) also observed significant increases in salivary flow rate, pH, and buffer capacity, together with raised resting and minimum pH values in plaque in a 2-week intensive regimen with a nonfluoride chewing gum. In situ experiments have shown that the effects of salivary stimulation with sugarless chewing gum will enhance remineralization of experimentally developed caries lesions in enamel (Creanor et al, 1992; Leach et al, 1986, 1989; Manning and Edgar, 1992).

The environment of the enamel lesion during remineralization in these studies is the fluid phase of plaque, which differs from saliva in pH and concentration of calcium and phosphate ions. Although an increase in the supersaturation of saliva arising from stimulation would be expected to have a direct effect on the supersaturation of plaque fluid, to date no such effect has been demonstrated. Steinberg et al (1992) found, however, that chewing of sorbitol and xylitol gum, besides reducing plaque accumulation and gingival inflammation, increased the concentration of acid-extractable calcium in plaque by more than one third. This would be expected to increase the remineralizing potential of the plaque. It was suggested that the effect was due to complexing of calcium by both xylitol and sorbitol, leading to their retention in plaque. Equally, it may be that the elevation of plaque pH that would have followed gum chewing resulted in increased retention of calcium in plaque in the form of insoluble calcium phosphate deposits.

The potential clinical effect of salivary stimulation per se has not been tested in a clinical trial, although it is possible to interpret results from certain clinical studies as effects of salivary stimulation. Thus, the caries-preventive effects of xylitol—including apparent reversals of caries (remineralization) shown in the Turku, Finland, chewing gum trial, in which sucrose- or xylitol-sweetened gum was chewed ad libitum over a 12-month period (Scheinin et al, 1975)—could be the result of the enhanced remineralizing potential, although inhibition of plaque acidogenicity and other effects of xylitol cannot be excluded.

More conclusive indications of a beneficial effect of salivary stimulation are disclosed in studies by (1) Möller and Poulsen (1973), in which chewing of sorbitol gum was associated with a small but significant reduction in caries incidence; (2) Isokangas et al (1989), in which significant long-term benefits were provided by use of xylitol gum, two or three times daily; and (3) Kandelman and Gagnon (1990), in which a decrease of 65% in caries progression was reported in children who chewed xylitol gum (15% or 65%) as part of a preventive program. The study designs did not stipulate routine postprandial chewing for 20 minutes, which would have maximized the influence on saliva. The long-term effect of chewing xylitol gum during the high–caries-risk period (12 to 15 years of age)—should not be underestimated. In the above study by Isokangas et al, 12-year-old children in the test group used xylitol chewing gum to the age of 15, when the second maturation of the enamel should be achieved. At the age of 20, the test group still exhibited lower caries prevalence than the control group (Isokangas et al, 1993).

In the remineralization studies discussed previously, a therapeutic fluoride environment was provided by the use of a fluoride dentifrice. The essential role of fluoride in the remineralizing potential of sugared gum was shown in preliminary data, in which subjects chewed sucrose gum for 20 minutes after meals and snacks for successive 21-day periods, during which the fluoride content of the dentifrice was varied between 0 and 1,000 ppm. In the presence of fluoride, the observed demineralization was statistically nonsignificant; with the nonfluoridated dentifrice, significant demineralization occurred (Manning and Edgar, 1993). Therefore, use of the new sugarless fluoride chewing gum directly after meals offers a very attractive alternative to fluoride tablets, particularly in caries-susceptible patients with impaired salivary secretion.

Compared to subjects with normal salivary flow, patients with reduced salivary flow have prolonged fluoride clearance time in saliva and other oral fluids. In a clinical study by Sjögren et al (1993), two fluoride lozenges (Fludent and Dentan) and two fluoride chewing gums (Fluorette and Fluomin; Ferrosan, Copenhagen, Denmark) were compared. All four products contained 0.25 mg of fluoride. After a single intake of each product, fluoride concentration was determined in whole saliva in three groups of subjects: (1) children, 10 to 12 years of age; (2) adults; and (3) patients with dry mouth. The tablets were sucked until they had completely dissolved in the mouth, and the gum was chewed for 15 minutes. Nine separate saliva samples were collected by having the subjects spit out 0.3 to 0.5 mL of saliva up to 45 minutes after intake. The highest salivary fluoride concentrations were found in the patients with dry mouth. The fluoride tablets and chewing gums investigated had similar salivary clearance patterns. However, there were great intersubject variations.

In a later study, Sjögren et al (1997) showed that the fluoride concentration was two to three times higher in saliva on the chewing side compared to the nonchewing side after use of fluoride chewing gum in subjects who had refrained from oral hygiene for 3 days and thereafter rinsed with 10% sucrose solution for 1 minute. In addition, plaque pH was lower on the nonchewing side, which indicates the importance of continuously changing the chewing side when fluoride chewing gum is used. About 50% of the fluoride was released from the chewing gum after 5 minutes and 80% after 20 minutes. However, the salivary fluoride concentration was still higher on the chewing side than on the nonchewing side after 60 minutes of chewing.

In an early study, Bruun et al (1982) showed that use of a stick of chewing gum containing 0.5 mg of fluoride elevated salivary fluoride levels above baseline values for at least 60 minutes. In an experimental in situ study Lamb et al (1993) showed that only subjects chewing fluoride gum three times per day for 21 days had significantly increased acid resistance and remineralization of enamel lesions compared with nonchewing controls. In two other studies, the remineralizing effect of fluoride chewing gum (0.1 mg F), used five times per day for 21 days, was compared with the effect of an in situ slow-release fluoride device releasing 0.5 mg of fluoride per day on experimentally induced enamel and root caries. A nonfluoride toothpaste was used for oral hygiene three times a day during the test period.

The degree of remineralization of enamel lesions was 35.5% for fluoride chewing gum and 34.0% for the slow-release fluoride device (Wang et al, 1993). Similar results for remineralization (36.0% and 35.8%) were also achieved on root caries (De los Santos et al, 1994). However, fluoride chewing gum resulted in a higher stimulated salivary flow rate than did the slow-release device and the control (2.1, 1.8, and 1.7 mL/min, respectively) and a higher mean salivary fluoride concentration during stimulation (3.00, 0.20, and < 0.02 ppm F, respectively).

Recently, in an intraoral experimental caries model, Sjögren et al (2002) compared the effect of 4-week use of placebo chewing gum (positive control), fluoride chewing gum containing urea, and nonuse of chewing gum (negative control) on demineralized enamel and dentin blocks in situ. The results showed that use of all sugar-free chewing gums (with or without fluoride or urea) inhibited further demineralization, in contrast to the nonuse of chewing gum. However, fluoride chewing gum resulted in the most significant inhibitory effect on the chewing side, compared to the other chewing gums.

Recently, a chewing gum containing CHX (10 mg per stick) has shown significant antiplaque effects (Van Moer et al, 1996; Simons et al 1997, 1999, 2001; Smith et al, 1996; Tellefsen et al, 1996; for details, see chapter 4) similar to those of 0.2% CHX mouthrinse. Both the CHX and the fluoride chewing gums (Fluorette and Fludent) are made by the same company (Fertin Pharma). Therefore, it seems likely that in the near future a combined CHX-fluoride chewing gum, ie, a two-in-one product, will become available. Until then, the clinician could recommend that one stick each of the fluoride and the CHX chewing gums be chewed simultaneously, changing chewing sides continuously, for 15 minutes directly after every meal by high–caries-risk (C3) patients with reduced salivary secretion rate, high or very high plaque formation rate (PFRI score 4 or 5), and high salivary mutans streptococci levels.

Fluoridated salt

The proposal of salt as a vehicle for fluoride in caries prevention is attributed to Wespi (1948, 1950). In the mid-1950s, domestic salt supplemented by potassium fluoride, up to 90 mg/kg, became available in various cantons of Switzerland.

The first 5-year results following consumption of fluoride-rich domestic salt were published by Marthaler and Schenardi (1962). The documented caries reduction of 32% fewer DMFSs in the permanent teeth of 7- to 9-year-old children was not statistically significant. Only with the subsequent caries data that became available from studies in Colombia (250 mg F/kg as NaF), Spain (200 mg F/kg as NaF), and Hungary (250 mg F/kg as NaF) was it shown that fluoride-induced caries reductions could reach 50%.

A prerequisite was the availability of domestic salt with a high fluoride concentration. The state of knowledge on the subject, up to the mid-1970s, was summarized by Marthaler et al (1978). The conclusions were that fluoride ingested via salt prevents dental caries in man, the cariostatic effect being similar to water fluoridation: The fluoride content of salt is adjusted so that urinary fluoride excretion levels are similar to those in areas with optimal water fluoride content.

There has been little progress in the practical application of salt fluoridation. Information is available for Colombia, Jamaica, Costa Rica, and the Federal District of Mexico. Recently, France commenced the use of domestic salt with a fluoride level of 250 mg/kg, which has been available since 1986. Evaluation of the statistical data from these surveys has to take into account variations in objectives, planning, design, duration, supervision, and interference with other fluoride sources, such as fluoride toothpaste. The best surveys are the Colombian trials, begun in 1965, and organized in collaboration with the WHO and the National Institute of Dental Research in the United States.

In these trials, three Colombian towns adopted a fluoride regimen; two introduced salt fluoridated at 200 mg/kg and the third fluoridated the drinking water. A fourth town with no fluoride regimen was included as a control. After 7 years, DMFT scores in children fell in all three towns using the fluoride regimens, while they remained unchanged in the control community. The reduction was similar (about 50%) for water fluoridation (1 mg/L), and fluoridated salt (Gillespie and Roviralta, 1985).

Interesting observations were that the caries reduction in 12- to 14-year-old children was about 50% for water fluoridation and 40% for salt fluoridation, although these subjects had been 5 to 7 years old when the fluoride program started; ie, in these trials, the overwhelming caries-reducing effect was posteruptive (Gillespie and Roviralta, 1985). After 9 years of salt fluoridation in Mexico, caries prevalence in 12-year-olds was reduced from 4.4 DMFT in 1988 to 2.5 DMFT in 1997 (44% reduction). However, the supplementary effects of improved oral hygiene, more frequent use of fluoride toothpaste, and other caries-preventive measures were not evaluated because it was not a randomized controlled trial study (Irigoyen and Sanchez-Hinojosa, 2000).

Recently, Meyer-Lueckel et al (2002) presented caries prevalence in 6- to 16-year-old schoolchildren 12 years after the introduction of salt fluoridation in Jamaica. The results indicated caries reduction in all age groups from the introduction of salt fluoridation in 1987 to 1999. However, the authors concluded that the observed reduction of caries prevalence based on cross-sectional examinations might have been the result of improvements in other caries-preventive measures during the same period. It should also be observed that almost 50% of the children in 1999 exhibited fluorosis. With the current understanding that the effects of any systemically administered fluorides are almost 100% posteruptive, the arguments in support of salt fluoridation are very weak. The only exception might be as a public health measure in tropical countries with high caries prevalence, undeveloped water supplies, and inadequate dental resources.

Fluoridated milk

The rationale for adding fluoride to milk is that the procedure "targets" fluoride directly to children aged 6 years or older and thus would be less expensive than fluoridating the drinking water and exclude the risk of fluorosis. The availability of both fluoridated and nonfluoridated milk also maintains consumer choice. However, despite considerable interest in many quarters, the selection of milk as a vehicle for fluoride is questionable. The first issue is absorption because it was shown many years ago that while fluoride is absorbed almost as completely from milk as it is from water, the process takes considerably longer, therefore calling into question the posteruptive effect of fluoridated milk (Ericsson and Andersson, 1983). If fluoride acted preeruptively to prevent caries this would not be a disadvantage. There are also practical concerns, such as the considerable numbers of children in most countries who do not drink milk.

However, in a recent randomized crossover study, Petersson et al (2002) evaluated the concentration of fluoride in saliva and dental plaque after the following 3-day drinking regimen: *(1)* 200 mL fluoride-free tap water, *(2)* 200 mL tap water with 1 mg F, *(3)* 200 mL standard milk, and *(4)* 200 mL standard milk with 1 mg F. Fluoride-free toothpaste was used prior to and during the study. A significant increase of fluoride was disclosed in saliva 15 minutes after drinking fluoridated water as well as fluoridated milk (0.06 mg/L and 0.05 mg/L, respectively). In plaque, a twofold significant increase in fluoride concentration was achieved 2 hours after drinking fluoridated water as well as fluoridated milk (Petersson et al, 2002).

Although there are few studies on the cariostatic efficacy of fluoridated milk, and some have faults in design or operation, the 5-year double-blind fluoridated milk study conducted in Scotland by Stephen et al (1984) meets the highest methodological standards. Caries was registered both clinically and radiographically. The test group, selected randomly, had 200 mL of milk each day containing 1.5 mg of fluoride (7.5 ppm F). The difference in DMFSs between test and control groups was 48%. As a result of these data and the investigations of Banoczy et al (1985), it was concluded that the daily consumption of fluoridated milk from birth might be beneficial.

However, whether fluoridated milk will be widely accepted as a community public health measure cannot be said with certainty, given the logistic problems of handling and delivery (Stamm, 1972). Stamm summarized the following four weaknesses of milk fluoridation as a public health measure:

1. Because children from the lower socioeconomic groups tend to consume the least fresh milk, they would benefit the least.
2. Any benefit ceases as an individual matures and drinks less milk.
3. Because fluoride in milk is absorbed slowly, there is a very limited topical effect. (However, this must be questioned according to the recent study by Petersson et al, 2002.)
4. Figures show that the procedure can be relatively costly, despite proponents' claims to the contrary.

In part, some of this criticism still may be valid. But school-based fluoridated milk programs in developing countries with high caries prevalence could be an alternative to water fluoridation, since most caries lesions develop during and soon after eruption. A definitive evaluation of fluoridated milk will be based on the results of ongoing large-scale experiments in Bulgaria, Chile, New Zealand, Italy, and Alaska.

FLUORIDE AGENTS FOR TOPICAL USE

Topical fluoride application is without a doubt the most important means of fluoride administration for caries prevention and control. In particular, the widespread use of fluoride-containing toothpaste is thought to be a major factor in the worldwide decrease in dental caries. As discussed earlier in this chapter, there is overwhelming scientific evidence that fluoride exerts its major and almost exclusive cariostatic effect at the plaque-saliva-tooth interface during periods of caries dissolution and arrest. This also implies that the concentration of fluoride in the dental hard tissues is not in itself an important determinant of cariostatic effect.

These findings have shed new light on the understanding and interpretation of results from different modes of topical fluoride applications. Innumerable clinical studies on the caries-preventive effect of topical fluorides have been conducted over the past few decades. The subjects, usually children and young adults, were often highly caries active and living in areas without water fluoridation. The earlier studies were conducted before effective fluoride toothpastes were readily available, ie, the conditions under which the studies were conducted were different from the conditions today. The duration of some studies was several years, but the results of other studies were presented after less than 1 year. The control groups were placebo groups or positive control groups exposed to another well-established, caries-preventive fluoride measure. The diagnostic level at which caries was recorded, clinically and in some studies radiographically, varied considerably among studies. Moreover, the design of the studies was all too often inadequate; very few were randomized, controlled trials. In addition, caries-preventive studies should be 3 years in duration and start with 12-year-old children, who have the highest numbers of permanent tooth surfaces at risk.

Despite these confounding factors, the studies have shown overall the caries-preventive effect of most topical fluoride measures to range between 20% and 40%. Topical fluoride agents are available for self-care or for professional application (eg, by dentists, dental hygienists, or dental assistants). For self-care, the following fluoride agents can be used: toothpastes, toothpicks, dental floss, mouthrinses, gels, lozenges, and chewing gum. Professionally applied fluoride agents are paints, gels, prophylaxis pastes, varnishes, glass-ionomer cements, and other slow-release agents.

Figure 306, based on data extracted from Murray et al (1992), shows that fluoride toothpaste is by far the most frequently used topical fluoride agent (used by more than 450 million people). Only 20 million people use mouthrins-

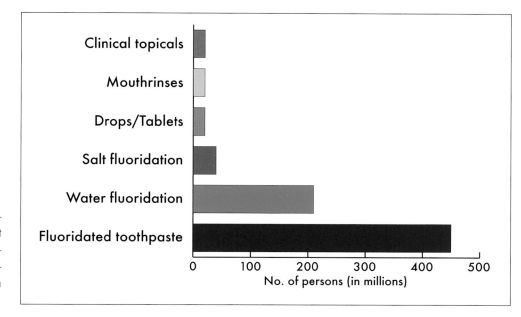

Fig 306 Estimated number of persons throughout the world who use different types of fluoride therapy. (Based on data from Murray et al, 1992.)

es or tablets, while 20 million receive professional applications of fluoride. However, although more than 450 million people use fluoride toothpaste regularly, more than 90% of the world's population does not have access to topical fluoride agents.

There is a great diversity of fluoride compounds used in fluoride agents available to the public and to professionals. The three main categories are:

1. Inorganic compounds, including NaF, SnF$_2$, and ammonium fluoride (NH$_4$F): The salts are readily soluble, providing free fluoride.
2. Monofluorophosphate-containing compounds, such as sodium monofluorophosphate (Na$_2$FPO$_3$): The fluoride is covalently bound in the FPO$_3^{2-}$ ion and apparently requires hydrolysis to free the F$^-$.
3. Organic fluorides, such as AmF and silane fluorides.

Sometimes two or more of these compounds are combined in the same topical fluoride agent. The fluoride concentration in topical fluoride agents for self-care, such as toothpaste and mouthrinses, may vary from 0.012% to 0.150% F, while up to 1.000% F is used in gels. The fluoride concentration in topical fluoride agents for professional application is usually much higher, ranging from about 0.7% to 6.0%, which, as discussed earlier, will promote precipitation of calcium fluoride reservoirs.

Sodium fluoride

Sodium fluoride is by far the most widely used fluoride compound in topical fluoride agents for self-care (toothpastes, mouthrinses, gels, lozenges, and chewing gums) as well as for professional use (aqueous solutions, gels, varnishes, prophylaxis pastes, and slow-release agents). In lozenges and chewing gum, NaF is used exclusively. Compared to other fluoride compounds, its use in toothpastes and varnishes is still increasing. Sodium fluoride will immediately provide free F$^-$ ions at low concentration in the saliva and in plaque, pellicle, and intercrystalline fluids and thereby interact with the carious process during dissolution and remineralization. Higher concentrations of NaF yield greater reservoirs of CaF$_2$.

Stannous fluoride

Stannous fluoride is used in toothpastes, mouthrinses, and gels for self-care, and in aqueous paint-on solutions, gels, and prophylaxis pastes for professional application. Stannous fluoride provides free fluoride (F^-) ions and stannous (Sn^{2+}) ions, which also have an antimicrobial effect (see chapter 4 on chemical plaque control). Additionally, SnF_2 can produce stannous phosphate fluoride precipitates that should retard the carious process but tend to stain demineralized enamel. Unfortunately, this is a particular problem in caries-susceptible patients with poor oral hygiene. In aqueous solution, SnF_2 is rapidly hydrolyzed and oxidized, reducing or eliminating its effectiveness: A fresh solution is therefore required for each treatment. However, in special gels, toothpastes, and prophylaxis pastes, SnF_2 is stable if stored in plastic tubes or cans.

Ammonium fluoride and titanium tetrafluoride

Although in vitro tests have shown that ammonium fluoride and titanium tetrafluoride (TiF_4) deposit fluoride in enamel, the few clinical studies reported to date do not indicate that they are superior to conventional fluoride agents (Büyükyilmaz et al 1994, 1997). The titanium (Ti^{4+}) ions may help to bind F^- ions to apatite surfaces. The unpleasant taste of ammonium fluoride is a disadvantage. So far, no commercial products have been successfully introduced.

Neutral sodium monofluorophosphate

Sodium monofluorophosphate (SMFP) is used mainly in toothpastes and gels for self-care and in gels and prophylaxis pastes for professional application. The SMFP-containing products provide the MFP ion together with some free F^- ions, both of which can diffuse through the plaque and into the enamel; this provides a source of the F^- ion subsequent to hydrolysis, to interact as described earlier. Monofluorophosphate can be hydrolyzed in plaque and under acidic conditions, is also hydrolyzed at the surface of apatite crystals, providing phosphate and fluoride ions. Because the fluoride is covalently bound in Na_2FPO_3, FPO_3^{2-} is more compatible with dentifrice abrasives that react with F^-. From 1965 to 1985, when abrasives such as chalk and pumice comprised the bulk of toothpaste and prophylaxis paste formulations, SMFP was the major fluoride compound in these products.

Although both the interactive role of the F^- ion in calcium phosphate chemistry and its influence on bacterial metabolism are well understood, the role of the intact FPO_3^{2-} ion in caries prevention is unclear. It appears that FPO_3^{2-} ions can bind to apatite crystal surfaces, especially at reduced pH, by substitution for acid phosphate ions (HPO_4^{2-}). Although the FPO_3^{2-} ion may interact as such with apatite in vitro, the cariostatic effect of Na_2FPO_3 in vivo is probably exerted partly or entirely by the F^- ion, derived from FPO_3^{2-} ions through hydrolysis by salivary and plaque phosphatases. In thin plaque, most of the FPO_3^{2-} would reach the tooth surface intact, but fluoride that reached the tooth surface after penetrating thick plaque would be mostly in the form of F^-. The total amount of fluoride reaching a plaque-covered tooth surface from a given dose of FPO_3^{2-} would be somewhat less than that supplied by an equivalent dose of fluoride as F^- as a result of interacting reaction and diffusion processes (Pearce and Dibdin, 1992).

The rate of production of F^- from FPO_3^{2-} obviously will depend on the level of phosphatase activity. It has been suggested that sodium lauryl sulfate (SLS), the detergent most commonly used in toothpastes, may inhibit oral phosphatases and thereby reduce the cariostatic effect of MFP. The MFP ion is taken up by oral bacteria but, unlike F^-, is not bound within the cells: Following toothbrushing, intracellular MFP could be released, providing a substrate for phosphatases, thus acting as a fluoride reservoir. Monofluorophosphate seems to have no effect on acid production.

Acidulated sodium monofluorophosphate

Acidulated SMFP is most commonly used in gels, both for self-care and professional application. Acidulated SMFP is also used in toothpaste and prophylaxis paste formulations with low pH. Acidulated SMFP readily etches the enamel surface, providing calcium ions that can interact with the fluoride and precipitate large amounts of calcium fluoride. Further, the hydrogen ions present complex with the fluoride, producing hydrogen fluoride, which readily diffuses deep into the enamel. There is therefore a chemical rationale for the observed clinical effectiveness of these topical agents.

Sodium fluoride and sodium monofluorophosphate

Duckworth et al (1994) compared fluoride retention in the mouth after use of NaF and Na_2FPO_3 (SMFP) toothpastes. In a large study linked to a clinical trial, plaque was collected from 474 subjects who had been brushing with one of six silica-based test toothpastes at least twice daily for 2 years. Significantly more fluoride was found in plaque from subjects using NaF toothpaste than in plaque from subjects using Na_2FPO_3 dentifrices of equivalent fluoride content.

Similar findings have been reported from other oral fluoride retention studies comparing NaF with Na_2FPO_3. Stookey et al (1993) found that fluoride uptake to demineralized enamel in situ was significantly greater from NaF-silica toothpaste than from Na_2FPO_3–calcium carbonate toothpaste in studies lasting 4 and 9 weeks, as did Reintsema et al (1985) in a 2-week study. However, Mellberg and Chomicki (1983) reported little difference in enamel fluoride uptake in similar studies involving up to 8-week use of the same toothpaste. Bruun et al (1984) found no significant difference between salivary clearance profiles monitored for 2 hours after use of silica toothpaste containing either fluoride compound.

There are several possible explanations for the fact that oral fluoride retention is greater from F^- than it is from FPO_3^{2-}. First, F^- ions diffuse more rapidly than do FPO_3^{2-} ions, by a factor of 1.4 in water and by a factor of 1,000 in dental enamel. These differences could be important under the dynamic conditions prevailing during the limited duration of exposure of the enamel to the toothpaste or mouthrinse. Second, there is no monofluorophosphate analog of CaF_2. Third, FPO_3^{2-} ions do not bind as well to tooth mineral and plaque bacteria as F^- ions do, as discussed earlier.

These differences are likely to be greater in experimental models than in vivo, because in vitro the major determinant of the clinical efficacy of MFP, namely the hydrolysis of FPO_3^{2-} ions to F^- ions, is absent. However, a meta-analysis of clinical studies of the efficacy of fluoride toothpastes revealed that NaF toothpaste is slightly but significantly (6%) better than SMFP toothpaste (Johnson, 1993).

This tendency for NaF to be clinically superior to Na_2FPO_3 (SMFP) may be related to the similar superiority of NaF with respect to oral fluoride retention. There is some evidence that SLS may impair the clinical effect of SMFP. Data from salivary fluoride clearance experiments suggest that SLS reduces oral fluoride retention from SMFP rinses, possibly through competitive adsorption to binding sites and/or via enzyme inhibition (Duckworth and Jones, 1994). Most of the fluoride toothpastes on the market contain SLS.

Amine fluoride

Amine fluoride is an organic fluoride compound—*N*-octadecyltrimethylenediamine-N,N,N-tris (2-ethanol)dihydrofluoride—used in toothpastes, gels, and mouthrinses for self-care and gels and prophylaxis pastes for professional use. This is essentially an ionic fluoride compound that readily provides free fluoride. Its enhanced reactivity has been attributed to the greater affinity of hydrophilic counter-ions to the enamel, which will reduce the surface energy and thereby plaque adhesiveness (see chapter 4). In addition, AmF provides a complexed store of fluoride ions and may enhance diffusion through carious enamel, releasing fluoride at appropriate times and sites.

Silane fluoride

Silane fluoride is also an organic fluoride compound, used in fluoride varnishes for professional application. Like AmF, silane fluoride provides a complexed store of fluoride ions and may enhance diffusion through carious enamel, releasing fluoride at the appropriate time and site. In contrast to other fluoride compounds, such as NaF, SMFP, and SnF_2, which dissolve in water and release fluoride ions, silane fluoride is insoluble. It reacts on contact with saliva, releasing small amounts of hydrogen fluoride. Because of its considerably greater diffusion coefficient, hydrogen fluoride penetrates enamel more readily and rapidly than do fluoride ions. The hydrogen fluoride molecules that have penetrated the intercrystalline fluid of the porous enamel lesion react with water and provide free fluoride ions, which will further retard dissolution and enhance remineralization according to the mechanisms described earlier.

Combination fluoride compounds

Sodium fluoride and SMFP are combined in toothpastes and prophylaxis pastes. Stannous fluoride and AmF are combined in mouthrinses and toothpastes, and NaF and AmF are combined in prophylaxis pastes.

Chemical plaque control agents

Most current fluoride toothpaste formulations contain some chemical plaque control agent or at least a detergent, such as SLS. Fluoride mouthrinses, gels, and prophylaxis pastes may also contain chemical plaque control agents (see chapter 4). In fluoride toothpastes, the following combinations are used:

1. NaF, SMPF, or SnF_2 + SLS
2. NaF or SMPF + triclosan + SLS + agents to enhance the substantivity of triclosan
3. NaF + CHX + zinc lactate

The following combinations are used in mouthrinses:

1. NaF + triclosan + copolymer + SLS
2. NaF or SnF_2 + SLS
3. SnF_2 + AmF
4. NaF + CHX (limited stability)

In gels, NaF, SMFP, or SnF_2 is combined with SLS, and NaF is combined with CHX. As with the mouthrinse, the NaF-CHX gel combination has to be freshly prepared on prescription because of its limited stability (see Box 9 in chapter 4). It should also be noted that SnF_2, solely or in combination with AmF, has documented plaque-reducing effects (Tinanoff et al, 1980; Axelsson et al, 1994b). From a cariostatic aspect, NaF and CHX seem to be the most efficient combination to date, followed by the combination of NaF, triclosan, SLS, and zinc citrate or copolymer and the combination of SnF_2 and AmF. Further research is necessary to formulate the optimal cariostatic combination of fluoride compounds, chemical plaque control agents, and delivery systems.

Topical fluorides for self-care

The following topical fluoride agents are available for self-care: toothpastes, toothpicks, floss, mouthrinses, gels, artificial saliva, tablets, lozenges, and chewing gum. The posteruptive cariostatic effects of tablets, lozenges, and chewing gum have been addressed in detail earlier in this chapter.

Fluoride toothpastes

The cariostatic effect of fluoride toothpastes was recognized more than 30 years ago, and, in conjunction with improved daily oral hygiene, fluoride toothpaste is regarded as the major factor responsible for the dramatic caries reduction in children and young adults in most industrialized countries during the last two decades. At present, more than 90% of toothpastes in the industrialized countries contain fluoride (almost 100% in Scandinavia but less than 50% in Japan). Although it is estimated that more than 450 million people use fluoride toothpaste regularly (see Fig 306), this represents less than 10% of the world's population. In other words, there is a huge untapped market for the introduction of a public dental health program based on inexpensive, effective fluoride toothpastes combined with improved oral hygiene habits.

Following disclosure in the 1930s of the caries-inhibiting effect of fluoride in the drinking water, attempts were made to incorporate fluoride into other vehicles. There was considerable urgency, because tooth decay in schoolchildren had reached epidemic proportions in virtually all the industrialized countries. The first clinical trial of a fluoride toothpaste was conducted in 1942. The active agent was NaF, added to a conventional dentifrice that contained dicalcium phosphate as the abrasive. The results showed no statistically significant reduction in caries because of the formulation of the test dentifrice. Additional studies were undertaken, but NaF in conventional pastes failed to achieve any clinical effect. It was later shown that NaF was incompatible with the abrasive: The fluoride from the NaF dissociated readily but was quickly bound to the calcium-containing abrasive particles in the pastes.

In 1955, Muhler et al reported a clinical trial of SnF_2 in a paste with a new calcium pyrophosphate abrasive system. This study was the first to report a statistically significant, clinically important caries-preventive effect. This major breakthrough created tremendous excitement in the dental community, which aggressively promoted the use of the fluoride toothpaste. A further important development, 11 years later, was the report by Torell and Ericsson (1965) that SMFP was compatible with a toothpaste with the conventional abrasive dicalcium phosphate and would be significantly effective in reducing caries.

In a clinical trial in 1967, Koch showed that a new combination of NaF with acrylic abrasive particles in a toothpaste resulted in a statistically significant caries reduction. By this time, three fluoride compounds dominated as toothpaste additives, although NaF was the least favored. In the early 1980s, however, it was reported that NaF in combination with a silicone dioxide (SiO_2) abrasive was clinically effective, and since then the NaF-SiO_2 formulation has been gaining in popularity.

The ultimate test of the effect of fluoride toothpaste is in human clinical studies. More than 100 such studies over the past 30 years have demonstrated the caries-inhibiting effect of fluoride toothpaste. The pre-1980 clinical trials tested the effects of 1,000-ppm fluoride toothpastes by comparing them with a placebo paste of the same formulation without fluoride. Most of these trials were conducted in the United States, the United Kingdom, and Sweden. The effect was generally evaluated in 2- to 3-year longitudinal trials on 11- to 14-year-old children. Compared to subjects using placebo toothpastes, subjects using different fluoridated toothpastes experienced an average reduction in caries incidence on the order of 20% to 35%.

The effect of fluoride is cumulative: Like periodontal disease, caries is a chronic condition, with periods of exacerbation and quiescence related to the amount and quality of the dental plaque as well as external and internal modifying risk indicators and risk factors. In this context, the caries reduction of 20% to 35% over a period of 2 to 3 years is comparable with the 50% reduction achieved by water fluoridation at that time, when the incidence and prevalence of caries were much higher in industrialized countries than they are today.

After these early toothpaste studies, it was considered unethical to use nonfluoride toothpaste in clinical trials because this would deny trial participants their customary fluoride toothpaste, the benefits of which were considered proven.

Formulations. The main functions of toothpastes are to facilitate mechanical plaque removal by brushing and to serve as vehicles for active agents (eg, fluorides, chemical plaque control agents, and anticalculus agents). Toothpaste formulations may contain the following ingredients:

1. Active agents:
   a. One fluoride compound or two in combination
   b. Agents for enhancement of the fluoride effect
   c. Chemical plaque control agents
   d. Anticalculus agents
   e. Antimercury agents
   f. Buffer systems
2. Abrasive particles
3. Detergents
4. Flavoring agents, preservatives, and coloring agents
5. Thickeners, agents to regulate viscosity
6. Water

The following fluoride compounds are used as active agents in toothpastes:

1. Inorganic fluorides:
   a. NaF
   b. SMFP (Na_2PO_3F)
   c. SnF_2
   d. Potassium fluoride (KF)
   e. Aluminum fluoride (AlF_3)
2. Organic fluorides:
   a. AmF (Olafluor; GABA International)
3. Combinations of fluorides:
   a. NaF and SMFP
   b. AmF and SnF_2
   c. AmF and NaF

Sodium fluoride and SMFP are by far the most commonly used compounds, followed by SnF_2 and AmF. Almost all the NaF, SnF_2, and AmF in toothpastes will be dissolved in the mouth during brushing, releasing optimal amounts of free F^- ions. On the other hand, SMFP initially releases fewer free F^- ions, but also supplies FPO_3^{2-} ions, which within about 1 hour are broken down by phosphate enzymes in the mouth, releasing F^- ions.

From 1955 to 1985 the standard fluoride concentration in toothpastes was about 1,000 ppm (0.1% F = 1 mg of fluoride per 1 g of toothpaste), supplied as 0.20% NaF, 0.76% SMFP, and 0.40% SnF_2. A wide range of fluoride concentrations, from 0.025% to 0.28% F, has been used.

Recent clinical studies have shown that the cariostatic effect of fluoride compounds in toothpastes is significantly dose related. The increased benefit is estimated to be about 6% additional caries reduction for each 500 ppm (0.05%) above 1,000 ppm (0.10%). However, the cariostatic effect below 500 ppm (0.05%) seems to be very limited.

In 1997, the European Community recommended an upper limit of 1,500 ppm (0.15% F) in toothpastes sold without prescription. Therefore toothpastes sold in Europe today usually contain 0.15% F, mostly as 0.30% NaF. Figure 307 shows the range of fluoride toothpastes containing 0.15% F available in Scandinavia in the late 1990s. It is estimated that about 25% to 50% of the toothpaste is ingested; to minimize the risk of fluorosis, in children younger than 6 years only a limited amount of toothpaste, the size of a pea (5 mm), should be used under adult supervision. In areas with a fluoride level of 1 mg/L or more in drinking water, or where fluoridated salt is used, toothpastes with 0.05% F could be recommended for children up to the age of 6 years.

Because a maximum concentration of 0.15% F has been recommended, a recent approach to improve the cariostatic effect of fluoride toothpastes is to add chemicals that increase the efficacy of the fluoride. At low pH, "impurities," such as highly soluble carbonates, will be removed from the outer surface of the enamel and be replaced by more acid-resistant fluorapatite and hydroxyapatite. A more important effect is that the hydrogen (H^+) ions, together with free F^- ions, will form hydrogen fluoride, which readily penetrates enamel caries and the intercrystalline fluid. Hydrogen fluoride then dissolves and delivers free F^-, which at increased pH will further retard enamel dissolution and enhance remineralization.

Salivary constituents known to enhance remineralization, mainly phosphorus and calcium compounds, eg, simple molecules such as calcium carbonate ($CaCO_3$) or trimetaphosphate, as well as a combination product, calcium glycerophosphate (CaGP), have been tested. Toothpastes containing SMFP-NaF-$CaCO_3$-CaGP as well as SMFP–dicalcium phosphate dihydrate are now commercially available.

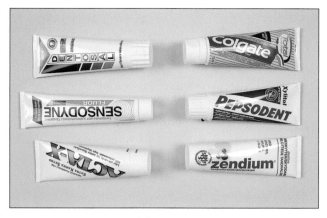

Fig 307 Assortment of fluoride toothpastes containing 0.15% F available in Scandinavia: Dentosal (0.3% NaF) (Blendax), Sensodyne (0.3% NaF) (Stafford-Miller, Stockholm, Sweden), Acta-X (0.3% NaF) (Acta, Falun, Sweden), Colgate Total (0.3% NaF) (Colgate-Palmolive, New York, NY), Pepsodent Xylitol (0.3% NaF) (Elida Robert, Helsinki, Finland), and Zendium (0.3% NaF) (Sara Lee, Chicago, IL).

Box 13 Toothpastes with antiplaque effects

- 0.30% NaF (0.15% F) + 0.40% CHX digluconate + 1.30% zinc lactate (eg, Blend-a-Med Parosan [Blendax, Mainz, Germany] and Crest [Procter & Gamble, Cincinnati, OH])

- 0.30% NaF (0.15% F) + 0.30% triclosan + 2.00% polyvinylmethyl ether–maelic acid copolymer + SLS (Colgate Total)

- 1.10% SMFP (0.15% F) + 0.30% triclosan + 0.75% zinc citrate + SLS (Pepsodent Ultra; Elida Robert)

- 0.30% NaF (0.15% F) + 0.30% triclosan + pyrophosphate + sodium alkyl sulfate (Dentosal)

- 0.15% AmF + 0.40% SnF_2 (0.10% F) (Elmex; GABA International)

- 0.40% SnF_2 (0.10% F) (Oral-B Tooth and Gum Care; Gillette, Boston, MA)

Intensive research is in progress to improve the oral retention (substantivity) of fluoride by precipitation of fluoridated calcium phosphate, crystals of fluoride salt, and fluoride-polymer complex in new toothpaste formulations.

The cariostatic effect of fluorides is limited (see Fig 291). If plaque pH falls below about 4.5, the plaque fluid becomes undersaturated with respect to fluorapatite, and demineralization will occur, even in the presence of fluoride. Therefore, toothpastes containing a combination of fluoride, chemical plaque control agents, and abrasives are currently of interest. A number of chemical plaque control agents have a known antiplaque and thereby cariostatic effect (Box 13; also see Figs 202 and 203 in chapter 4). Even the anionic SLS, which is used as a detergent in most toothpastes, has an antiplaque effect.

Toothpastes specially formulated for patients with rapid calculus formation are available. The main anticalculus agent is tetrasodium phosphate, used alone or in combination with tetrapotassium pyrophosphate. The most commonly used buffering systems in toothpastes are $CaCO_3$ and baking soda.

Almost all toothpastes contain abrasive particles to improve the mechanical removal of plaque during brushing and to reduce staining of the pellicle. The abrasive agent represents 20% to 55% of the total weight of toothpaste. The abrasiveness is expressed as radioactive dentin abrasivity (RDA) values according to the ADA's method of assessment, or dentin abrasive value (DAV) according to the British Standards Institution. The abrasiveness of most toothpastes ranges from 40 to 80 RDA, or 35 to 85 DAV, but toothpastes specially formulated for stain removal range from 100 to 130 RDA.

The following are the most commonly used abrasives in toothpastes:

1. Calcium pyrophosphate ($Ca_2P_2O_7$)
2. Hydrated silica (SiO_2)
3. Sodium bicarbonate ($NaHCO_3$)
4. Insoluble sodium metaphosphate ($NaPO_3$)
5. Acrylic polymer
6. Alumina trihydrate ($Al_2O_3 \cdot 3H_2O$)

of toothpaste trials on ethical grounds. In 1988, the ADA set up guidelines for evaluation of fluoride toothpastes in caries clinical trials (see Box 11 in chapter 4).

In two parallel, large-scale studies, silica-based toothpastes containing NaF and SMFP have been compared. In a US study by Marks et al (1994) in Florida, more than 8,000 schoolchildren, aged 6 to 14 years, were randomly allotted to one of five test groups in a 3-year, double-blind study. All five toothpastes were based on a silica gel and contained the following test substances: SMFP (0.10%, 0.15%, 0.20%, or 0.25% F) and NaF (0.20% F). The incidence of caries in the NaF (0.20% F) group was significantly less ($P < .005$) than that in the 0.20%-F SMFP group. Compared to the 0.15%-F SMFP group, the caries incidence was 18%, 15%, and 5% less in the 0.20%-F NaF, 0.25%-F SMFP, and 0.20%-F SMFP groups, respectively.

In the second study, Stephen et al (1994) compared the effect of four silica-based toothpastes containing NaF or SMFP, at concentrations of 0.10% or 0.15% F, in a 3-year, double-blind study. More than 4,000 11- to 12-year-old Scottish schoolchildren were randomly assigned to test groups. The caries incidence was 6.2% ($P < .05$) lower in the NaF groups than in the SMFP groups. However, for neither of the two fluoride compounds was there any significant difference in effect between concentrations of 0.10% and 0.15%.

In a recent systematic review, Ammari et al (2003) compared the effect of fluoride toothpastes with 600 ppm F or less with fluoride toothpastes containing 1,000 ppm F on caries incidence in randomized controlled trials. On average, 250 ppm F resulted in 0.6 to 0.7 more new DFSs per subject compared with 1,000 ppm F in studies with a mean caries incidence ranging from 3 to 7.5 new DFSs. Thus it can be concluded that the caries-preventive effect of fluoride toothpastes is strongly correlated to the fluoride concentration.

The clinical advantage of NaF over SMFP in silica-based toothpastes may be eliminated in the future because SMFP can be combined with other compounds that synergistically enhance the caries-inhibiting effect: These compounds may not be compatible with NaF. In situ studies have shown that dicalcium phosphate dihydrate significantly enhances the caries-reducing effect of SMFP in toothpastes compared to the caries-reducing effect of SMFP in silica-based toothpastes. Dicalcium phosphate dihydrate is incompatible with NaF.

The caries-inhibiting effect of one fluoride toothpaste that incorporates chemical plaque control agents has been evaluated according to the ADA's guidelines. Two independent, double-blind, 26-month studies showed that a toothpaste containing 0.30% triclosan and 2.00% copolymer in either a 0.24% or 0.33% NaF–silica base has a statistically significant ($P < .01$) and clinically important anticaries effect similar to that of a comparable clinically proven, positive control, NaF-silica toothpaste (Feller et al, 1993; Mann et al, 1993). Because many patients suffer from both caries and periodontal disease, these results are encouraging. Data from a 3-year, double-blind study in a selected group of patients treated for advanced periodontal disease revealed significantly more gain of probing attachment and alveolar bone in subjects using a silica-based toothpaste containing 0.3% NaF, 0.3% triclosan, 2.0% copolymer, and SLS than was found in subjects using a silica-based positive control toothpaste containing 0.3% NaF and SLS (Rosling et al, 1997a).

To date, no other fluoride toothpastes with specific chemical plaque control formulation have been evaluated according to the ADA's guidelines. However, in an in situ study by ten Cate (1993), a silica-based toothpaste containing SMFP (0.10% F), 0.30% triclosan, and 0.75% zinc citrate inhibited caries progression under conditions of cariogenic challenge; results were significantly better in the test group than in the group using a silica-based, positive control toothpaste containing SMFP (0.10% F), which was associated with an increase in lesion severity. Also awaited with interest is a study fulfilling the ADA's guidelines that compares the long-term anticariogenic effect of a newly formulated toothpaste containing

0.3% NaF, 0.4% CHX, and 1.3% zinc lactate to that of a positive control fluoride toothpaste. Also of importance is a comparison, according to the ADA's guidelines, of the anticaries efficacy of the most common fluoride toothpastes containing chemical plaque control agents and a positive control fluoride toothpaste, carried out in a double-blind study on randomized test groups of 12- to 14-year-old subjects with high caries incidence.

Influence of oral care habits. The effect of both fluorides and chemical plaque control agents is modified by many factors, such as the delivery system, stability, concentration, bioavailability, substantivity (clearance time), pH, frequency of application, penetrability, and accessibility.

To date, most research on fluoride toothpastes has concerned the effect of different toothpaste formulations; little attention has been paid as to how the paste is used. However, in recent years, there have been a number of studies on the influence of oral care habits on the efficacy of fluoride toothpastes. As anticipated, it has been shown that the benefit is directly related to the frequency of toothbrushing. As early as 1982, a review of controlled clinical trials disclosed that the greatest caries reductions tended to occur when toothbrushing was supervised, suggesting a higher frequency of toothpaste use in the supervised trials (Bruun and Thylstrup, 1988). The reliability of data on toothbrushing frequency depends on accurate reporting by children, who may not always be willing to disclose how seldom they brush their teeth. In some studies with large numbers of subjects, statistically significantly lower caries increments have been found in trial participants claiming to brush twice or more daily than in those brushing once daily or less (Chesters et al, 1992; Tucker et al, 1976).

In the study in Scottish schoolchildren by Chesters et al (1992) of toothpastes containing 0.10%, 0.15%, or 0.25% fluoride, the subjects were also asked about frequency of toothbrushing and how they rinsed their mouths afterward. The results showed that the caries-inhibiting effect was significantly reduced in children who routinely rinsed thoroughly with a glass of water afterwards than it was in those who either did not rinse at all or used a less efficient means of diluting the toothpaste, such as simply wetting the toothbrush, using handfuls of water, or placing the open mouth under running tap water.

The data indicate that frequent toothbrushing, two times a day or more, with 0.10% fluoride toothpaste without rinsing with water may be more effective than infrequent brushing with a toothpaste containing more than 0.10% fluoride followed by thorough rinsing. These clinical data confirm the results from earlier laboratory-based intraoral fluoride clearance studies by Duckworth et al (1991).

More recent intraoral fluoride clearance studies by Sjögren et al (1994) have shown that fluoride clearance will be significantly prolonged if, instead of rinsing with water, the subjects use the remaining slurry of fluoride toothpaste for rinsing after toothbrushing. In a 3-year clinical study in preschool children, Sjögren et al (1995) also showed significantly greater caries reduction in the test group, who rinsed only briefly with water, than in the control group, who rinsed thoroughly with water.

Delivery of fluoride to the approximal surfaces of the molars and premolars is very important because these are more caries susceptible than are surfaces accessible to toothbrushing. The index finger, a toothbrush, or a syringe should be used to fill the posterior interproximal spaces with fluoride toothpaste before mechanical plaque removal with toothpicks, dental tape, or interdental brushes. This is particularly important before bedtime. In very caries-susceptible patients, the posterior interproximal spaces could be refilled with fluoride toothpaste after the oral hygiene procedures, as a reservoir during the night: Because salivary secretion is minimal during sleep and fluoride clearance in the oral fluids is prolonged, the cariostatic effect of the fluoride toothpaste will be optimized at the key-risk surfaces.

Fig 308 Assortment of over-the-counter fluoride mouthrinses containing 0.025% F for daily use: Colgate Total (NaF), Meridol (AmF + SnF$_2$), Anti Plaque (NaF) (Johnson & Johnson), Oral–B (NaF), Dentan (NaF), ListerFluor (NaF) (Pfizer, New York, NY), Pepsodent (NaF), Thera-med (NaF) (Henkel-Barnagen), and Vademecum (NaF) (Henkel-Barnagen, Ekero, Sweden).

0.20% NaF solution, well below any potentially toxic level. After a corresponding rinse with 0.05% NaF solution, on average approximately 0.35 mg of fluoride per day may be inadvertently ingested. Intentional or accidental swallowing of the entire 10-mL dose of 0.20% NaF solution will result in ingestion of 9.00 mg of fluoride, and the corresponding ingestion from 0.05% NaF solution would be about 2.30 mg of fluoride.

The dose necessitating medical intervention and hospitalization is estimated to be about 5 mg of fluoride per 1 kg of body weight, approximately 120 mg of fluoride in a 5-year-old child (Whitford, 1989). Therefore, although rinsing with fluoride solution is considered safe if the directions are followed, it is not recommended as a general measure for preschool children because of the high risk of their swallowing the entire rinse volume.

Most of the aforementioned fluoride mouthrinses can be used long term by caries-susceptible adults. The combined CHX-NaF solution, used twice a day, will result in severe staining of the teeth and the back of the tongue after some months. To overcome this problem and improve patient compliance, intermittent use for periods of 3 weeks is recommended (see also chapter 4).

From a caries-preventive and toxicologic point of view, recent experimental studies by Chow et al (2000, 2002) show that a 250-ppm-F mouthrinse containing NaF + hexafluorosilicate (Na$_2$SiF$_2$) resulted in significantly more remin-

eralization than a 250-ppm-NaF mouthrinse and had results similar to a 1,000-ppm-NaF mouthrinse. Thus a more efficient low-concentration fluoride mouthrinse could be used daily.

Accessibility and effects on dental caries. The purpose of fluoride mouthrinsing is to provide frequent, relatively low concentrations of fluoride to sites with a predilection for caries. Many studies have evaluated the intraoral distribution of fluoride following mouthrinsing. Three such studies will be briefly reviewed.

Duckworth et al (1987) had volunteers rinse with a placebo and with solutions containing 0.01%, 0.025%, and 0.10% fluoride. Although rinsing with the 0.01% solution did not produce a significant increase in plaque fluoride concentration compared to the placebo, higher plaque fluoride concentrations were attained after rinsing with solutions containing 0.025% and 0.10% F. These concentrations are similar to the low- and high-potency rinses used in home-based and school-based programs, respectively. As caries develops beneath plaque, the fluoride must be able to penetrate the plaque. This study clearly demonstrated that fluoride concentrates in plaque after an individual swishes with an aqueous fluoride solution.

Weatherell et al (1988) used volunteers who rinsed for 15 seconds with a neutral NaF solution containing 0.1% fluoride. Saliva was sampled for 15 minutes afterward from different mucosal sites throughout the mouth. Fluoride was not evenly distributed; the highest concentrations were recorded in the maxillary incisor region, the lowest in the mandibular incisor and molar regions, and intermediate levels in the premolar regions. Assuming that an even fluoride distribution is preferable, these findings have important clinical implications: The solution should be swished as vigorously as possible to distribute the fluoride ions uniformly, and rinsing must be frequent so that oral clearance factors do not selectively reduce fluoride levels at different intraoral sites. For comparison, see also the pattern of PFRI and remaining plaque and the influence of limited accessibility at specific sites (for example, approximal molar surfaces) when mouthrinses containing chemical plaque control agents are used (see Figs 227 and 228 in chapter 4).

Zero et al (1988) instructed volunteers to brush with a conventional fluoride dentifrice containing 0.1% fluoride, to rinse with a neutral NaF mouthrinse containing 0.025% fluoride, or to do both. The procedures were performed twice a day, in the morning and before retiring, and unstimulated whole saliva was collected and analyzed for fluoride concentration. There were five different brushing and rinsing groups. The combination of brushing with a fluoride toothpaste followed by swishing with a fluoride rinse produced three times higher salivary fluoride concentrations than did brushing alone.

The reason that a mouthrinse with a low fluoride concentration supplements the caries reduction already achieved by the use of a toothpaste with a higher concentration of fluoride may be that an uncomplicated aqueous rinse is a better vehicle for delivering fluoride than a more complex dentifrice formulation. Zero et al (1988) also reported that retention of fluoride was better after evening oral hygiene procedures than it was in the morning, suggesting that the best time to use a fluoride rinse at home once a day is immediately before retiring.

Numerous studies have been carried out to assess different fluoride compounds, fluoride ion concentrations, and rinsing frequency of various mouthrinses. The two regimens regarded as standard for individual programs of patient care or for school-based programs are 0.05% NaF (230 ppm F) used daily and 0.20% NaF (900 ppm F) used weekly or biweekly: These are also referred to as the low-potency/high-frequency technique and the high-potency/low-frequency technique, respectively.

In establishing guidelines for selection of fluoride concentration and frequency, the study by Torell and Ericsson (1965) was most important. Two different mouthrinsing regimens were tested on children residing in a fluoride-deficient community. One group rinsed daily at home, unsupervised, with a neutral NaF solution containing 0.025% fluoride. A second group rinsed every 2 weeks under supervision with a neutral NaF solution containing 0.090% fluoride. After 2 years, the caries reduction in the group that rinsed daily with the lower fluoride concentration was 50%, while that in the group rinsing with the higher concentration every 2 weeks was 21%.

In the United States, the 0.09% fluoride rinse was tested in schoolchildren once a week, rather than once every 2 weeks, and two studies compared the effectiveness of weekly mouthrinsing at school with 0.20% NaF and daily rinsing with 0.05% NaF; one study was conducted in a nonfluoridated community and the other in a fluoridated community. In each location, both regimens were shown to be effective in caries prevention; although the daily regimen tended to be slightly better, the difference in effectiveness was not statistically significant (Driscoll et al, 1992).

Because a daily mouthrinsing program costs about four times as much as a weekly program, the daily regimen was not considered to be sufficiently more cost effective. Consequently, weekly fluoride mouthrinsing programs have become standard for organized school-based programs in the United States.

Weekly school-based programs. As a consequence of the results from the early Swedish studies, the Swedish Board of Health and Welfare in

significant differences in root caries incidence among 99 periodontally treated patients in three different fluoride programs: professional application of fluoride varnish (Duraphat) three or four times a year, 0.4% SnF_2 gel three or four times a year, and daily mouthrinsing with a 0.05% sodium fluoride solution.

The 1986 ADA guidelines for fluoride use in adults with active caries recommended the daily use of fluoridated mouthrinse and fluoride toothpaste as well as semiannual professionally applied fluoride treatments. For patients with rampant caries, quarterly visits to the dentist or dental hygienist for topical fluoride treatments and a minimum 4-week course of self-applied fluoride gel were recommended.

However, fluoride chewing gum, used directly after every meal, is probably a more efficient self-care fluoride delivery system to supplement fluoride toothpaste in caries-susceptible adult patients, particularly those with reduced salivary function.

Daily rinsing with combination products: Fluoride mouthrinses with chemical plaque control agents. Fluoride has a very limited potential to reduce demineralization and enhance remineralization when the pH of dental plaque drops to very low values (pH < 5). Therefore, the addition of chemical plaque control agents such as CHX, triclosan, SLS, and certain metal ions (Sn^{2+}, Cu^{2+}, Zn^{2+}, and Sr^{2+}) to fluoride mouthrinses for daily use would be an advantage.

For caries prevention, mouthrinses containing both fluorides and efficient inhibitors of plaque and *Streptococcus mutans*, such as CHX plus zinc, would be very attractive. In combination with sodium fluoride (NaF), CHX has a synergistic toxic effect on the cytoplasm of bacterial cells and the enzymes for fermentation of carbohydrates, indicating that acid production by the bacteria is also reduced (Luoma, 1972; Meurman, 1988).

The results of a 2-year clinical trial using a combination of CHX (0.05%) and NaF (0.044%) in a mouthrinse were reported by Luoma et al (1978). A total of 164 schoolchildren aged 11 to 15 years rinsed with 10 mL of CHX-NaF solution, NaF solution (0.044%), or a placebo under supervision once daily, 200 school days per year. In addition, they brushed their teeth with a toothpaste of the same composition as their mouthrinse. After 2 years, the group using the combined CHX-NaF agent showed significant reduction in gingival bleeding as well as the smallest DMFS increment. The authors proposed that a synergistic effect occurs when the two agents are used together. These findings were subsequently confirmed in a short-term mouthrinse study with CHX plus NaF. The Gingival Bleeding Index (GBI) was reduced by 50%, and the salivary *S mutans* counts were significantly reduced.

Although for ethical reasons the cariostatic effect of CHX alone was not investigated in the aforementioned study, Luoma (1972) found that CHX and fluoride in combination had a significantly greater inhibitory effect on acid production by *S mutans* than did CHX alone. Their investigation supports the hypothesis that combining fluoride with agents that decrease the formation of acid in plaque may counteract even an extreme cariogenic challenge. On this basis, a strategy for managing high caries risk or caries risk generally would be mechanical plaque removal combined with the use of plaque-inhibiting agents and fluoride in a mouthrinse or dentifrice twice daily.

Ullsfoss et al (1994) evaluated the possible caries-inhibitory effect of combining 2.2-mmol/L CHX mouthrinses twice daily with 11.9-mmol/L NaF rinses and compared the results with those of daily NaF rinses alone. The design was an in vivo human caries model in which plaque-retaining bands were placed on premolars scheduled for extraction. In nine subjects, 29 teeth were banded for 4 weeks. Saliva and plaque were sampled for bacterial culture before and after the study period. After the teeth had been carefully extracted, the tooth surfaces were analyzed by microradiography. The combination of CHX and fluoride rinses resulted in enamel mineral loss that was only slightly greater than that observed in "sound" enamel and clearly less than that observed with fluoride rinses alone. Both total plaque bacteria and *S mutans* were reduced by CHX rinses, confirming the discrete mechanisms of action. As in previous studies, fluoride alone did not totally prevent caries lesions. The combination of CHX and fluoride was significantly more effective in reducing both lesion depth and mineral loss, probably because of both a general inhibition of acid formation and a specific effect on *S mutans*.

Topical applications of a 1% NaF and 1% CHX gel, combined with daily rinses with 0.05% NaF and 0.20% CHX, have been recommended as the key elements in a regimen for total caries prevention after irradiation of the head and neck. The study by Ullsfoss et al (1994) provided the first data from an in vivo human experimental model to support this concept. The combination of CHX and NaF will have a synergistic caries-preventive and antimicrobial effect in mouthrinses as well as gel preparations, but, because such preparations have a limited shelf life of 6 months, they are not commercially available. They must be prepared at the pharmacy by prescription (see chapter 4). To date, there are very few commercially available fluoride mouthrinses with documented plaque-reducing effects (see chapter 4). Of these, the following combinations seem to be the most successful:

1. 0.05% NaF + 0.03% triclosan + copolymer + SLS (Colgate Total)
2. AmF (0.0125% F) + SnF_2 (0.0125% F) (Meridol)

(For reviews on fluoride mouthrinses, see Horowitz and Ismail, 1996.)

Fluoride gels for self-care

Gels were introduced as fluoride delivery systems more than 30 years ago. In the United States, fluoride gels have long been the predominat supplement to fluoride toothpastes. Fluoride gels are also widely used in Europe, particularly in Sweden, Germany, Switzerland, and the Netherlands. The current trend, however, is preference for newer products, such as fluoride lozenges or chewing gum for self-care and varnishes for professional use.

Fluoride gels may be used daily for self-care or applied professionally at needs-related intervals. The fluoride concentration is usually lower in gels for self-care (0.1% to 0.5% F) than in gels for professional use (1% to 2%). The active ingredients may be compounds with fluoride effects (neutral NaF and APF) or combinations of fluoride and agents with antiplaque and antigingivitis effects (SnF_2, AmF, AmF and NaF, and CHX and NaF).

For self-care, fluoride gels are usually applied in prefabricated or customized trays, but some, eg, SnF_2 gels, are applied with a toothbrush. Professionally, the gel is also applied in prefabricated or customized trays, but a syringe may be used for supplementary application in the posterior interproximal spaces. Customized trays are more efficient than the prefabricated trays, ensuring closer contact with the approximal surfaces and reducing the total amount of gel required.

Customized trays are more expensive for short-term use but are very cost effective for long-term self-care in adult high–caries-risk (C3) patients, eg, those with xerostomia. As with other delivery systems, the effect of fluoride gels is strongly correlated to the concentration of the active agents, accessibility, substantivity, and total exposure time. Accessibility is enhanced by prior removal of the dental plaque. The total exposure time is related to the frequency and duration of applications and to the thoroughness of removal of the remaining gel by rinsing after application.

Fluoride gels were developed because their viscosity would make them easier to manipulate and would readily permit their application in trays so that the entire dentition could be treated at one time. A gel used as a vehicle for active agents such as fluoride compounds and chemical plaque control agents is a thickened aqueous system containing cellulose compounds for viscosity but neither abrasives nor foaming agents. As such, gels are generally compatible with the active agents. To encourage compliance, gels are usually colored and flavored (eg, orange flavor).

Some gels are formulated to be thixotropic; that is, they tend to flow when under pressure but remain viscous when not. Although in theory,

which is cheaper and much simpler. In populations with low or relatively low caries prevalence, using fluoride toothpaste or fluoridated drinking water daily, the supplementary cariostatic effect of fluoride gels is very limited.

In a meta-analysis of clinical studies on the caries-inhibiting effect of fluoride gel treatment, van Rijkom et al (1998) concluded that the cost effectiveness of the additional effect of fluoride gel in child populations with low or moderate caries incidence must be questioned. For example, 18 subjects were needed to prevent 1.0 DMFS in a population with 0.25 new DMFSs per subject per year.

Self-applied fluoride gels have, however, been used in two groups of patients who are highly susceptible to caries attack: *(1)* those undergoing orthodontic treatment and *(2)* those with xerostomia related to radiation therapy of the head and neck. A reduction in the incidence of demineralized areas was found in orthodontic patients who added daily use of a glycerin-based 0.4% SnF_2 gel (1,000 ppm F) to their usual oral hygiene procedures (Shannon and West, 1979). A reduction in salivary levels of mutans streptococci in orthodontic patients who brushed twice daily with a 0.4% SnF_2 gel has been reported (Vierrou et al, 1986).

Several investigators have reported self-applied topical fluoride to be successful in preventing or controlling rampant caries in patients with radiation-induced xerostomia (Dreizen et al, 1977; Joburi et al, 1991). Daily use of a 0.4% SnF_2 gel has been shown to be significantly more effective in preventing root caries in patients with head and neck cancer than the 1.1% NaF gel (Wescott et al, 1975). Recently it was shown that a 1% neutral NaF gel (Top Gel), applied in customized trays for 5 minutes once daily, was sufficient to inhibit caries almost completely during a 1-year trial in patients with severely impaired salivary function (unstimulated flow ≤ 0.1 mL/min) following radiotherapy for tumors of the head and neck. Carbamide peroxide gels are increasingly used for bleaching vital teeth, which has been shown to decrease

calcium, phosphate, and fluoride content of the enamel. Recent experimental studies by Attin et al (2003) have shown that such demineralization of the enamel can be prevented by using a fluoridated carbamide peroxide gel.

Fluoridated artificial saliva

By far the most caries-susceptible patients are those with xerostomia (< 0.1 mL unstimulated saliva per minute). Xerostomia, or dry mouth, is most commonly related to radiotherapy for tumors of the head and neck or to Sjögren syndrome. Various artificial saliva products have been formulated as gels or sprays to relieve the extremely distressing subjective problems of dry mouth. The sprays are well accepted and applied 20 to 30 times a day. Because of the extremely high caries risk (C3) in patients with xerostomia, all artificial saliva products should contain fluoride (Fig 310). Because of the extremely low salivary flow or the absence of saliva, 20 to 30 spray applications of artificial saliva containing fluoride will markedly prolong fluoride clearance time. In addition, the following self-care program should be recommended for patients with xerostomia:

1. Mechanical removal of plaque should be performed just before main meals with oral hygiene aids and a fluoride toothpaste that contains chemical plaque control agents to minimize acid production from plaque during meals. Cleaning before instead of directly after meals also reduces the risk of erosion and abrasion in patients with xerostomia.
2. Directly after every meal or snack, one piece each of fluoride gum and CHX gum should be chewed for 15 to 20 minutes. This increases salivary secretion and pH in the oral fluids, enhancing remineralization and suppressing the remaining cariogenic microflora.

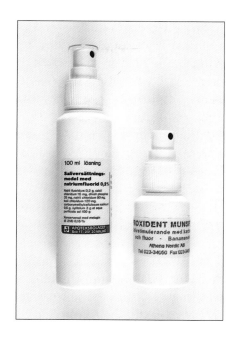

Fig 310 Fluoride-containing artificial saliva sprays: *(left)* ACO, *(right)* Broxident (Athena Nordic, Falun, Sweden).

Topical fluorides for professional application

The following fluoride agents are used for professional application: aqueous solutions for painting; gels; prophylaxis pastes; and slow-release agents, such as varnishes, intraoral slow-release devices, and glass-ionomer cements. The concentration, usually 1% to 2% F, is much higher than it is in agents for self-care. The most common fluoride compounds for professional use are neutral NaF, APF, and SnF_2. Amine fluoride and silane fluoride are used in some commercial products.

Professional applications of fluoride are no longer generally recommended in public health programs, except in regions where there is relatively uniformly high caries prevalence, absence of fluoride toothpaste, fluoride-deficient drinking water, and personnel resources for school-based preventive programs. On the basis of cost effectiveness, professionally applied fluoride agents are justified for special risk groups such as the mentally handicapped; elderly people with dementia who have reduced saliva flow, exposed root surfaces, and multiple restorations; and in children at specific ages (5 to 7 and 11 to 13 years), ie, during eruption of the first and second molars.

In populations with low caries prevalence, daily use of fluoride toothpaste, and possibly fluoridated drinking water, supplementary professional fluoride application is recommended strictly on the basis of individual needs. For details on needs-related preventive programs, see chapters 7 and 8.

Aqueous fluoride solutions

The most common fluoride solutions for painting are neutral 2% NaF (1% F), 8% SnF_2 (2% F), and APF (1.23% F). Amine fluoride solutions are also used. Introduced in the 1940s, this was the first method of professional fluoride application (Knutson and Armstrong, 1943; Knutson, 1948a).

Neutral 2% sodium fluoride solutions. The first studies on painting with neutral 2% NaF were conducted in children in the early 1940s. After thorough PMTC, the teeth were isolated with rubber dam and dried. Thereafter, all tooth surfaces were thoroughly painted with 2% neutral NaF solution, which was allowed to dry for at least 3 minutes. This technique, modified and described by Knutson (1948a), including PMTC, placement of rubber dam, and application and drying of the NaF solution, took an average of 15 to 20 minutes per treatment.

In 1948, the US Public Health Service (USPHS) recommended a series of four weekly applications of 2% NaF at age 3, 7, 11, and 13 years, to coincide with the eruption of new groups of teeth, because the tooth surfaces are more caries susceptible during and just after eruption, before the so-called secondary maturation phase. This recommended regimen was designed for public health programs in which a dentist or dental hygienist could treat several children concurrently, provided that PMTC was restricted to the first of the four weekly visits. Because the NaF solution was allowed to dry on the teeth for 3 minutes, it was theoretically possible to apply fluoride to other children during the drying phase. The series of four treatments was provided only four times between the ages of 3 and 13, allowing different groups of children to be treated each year.

Many studies conducted throughout the world have confirmed the effectiveness of NaF solutions in preventing caries (for review, see Horowitz and Ismail, 1996). Reductions in caries increments in permanent teeth, averaging about 30%, have been reported among children living in communities with fluoride-deficient water supplies. Although there have been fewer studies of the primary teeth, most have found reductions in the incidence of dental caries, but not as great as for the permanent teeth.

Nowadays, the USPHS program cannot be regarded as a cost-effective public health program for children using fluoride toothpaste daily, but topical application of 2% NaF solution is still being used with success, for example, in selected risk groups of Danish schoolchildren. In a 3-year clinical trial, Carvalho et al (1992) showed that, in selected high–caries-risk children, fissure caries can be almost totally prevented and controlled without fissure sealants if daily toothcleaning by the parents using fluoride toothpaste is supple-

mented at needs-related intervals by PMTC and painting with 2% NaF solution during eruption of the molars. These principles are being successfully implemented on a large scale in different regions of Denmark. Slow-release fluoride agents such as fluoride varnish or glass-ionomer cements could have been successful substitutes for the 2% NaF.

Stannous fluoride solutions. In the early 1960s, a method for applying an SnF_2 solution to the teeth for caries prevention was first described. The important differences from the technique for NaF were that a thorough PMTC was to precede every SnF_2 application, the teeth were kept wet with the solution for 4 minutes, making a saliva ejector essential, and applications at 6-month intervals were recommended. Both 8% and 10% SnF_2 solutions have been tested, with little difference in effectiveness between the two concentrations. Overall, studies have shown a reduction in dental caries of just more than 30% with the use of 8% SnF_2 in nonfluoridated communities (Dudding and Muhler, 1962; Ripa, 1990).

Stannous fluoride is relatively unstable: An aqueous solution undergoes rapid hydrolysis and oxidation, which reduces or eliminates its effectiveness. Consequently, a fresh solution is required for each treatment. Highly concentrated SnF_2 solutions are astringent and have a disagreeable taste, but because SnF_2 is so reactive, flavoring to mask the taste is contraindicated. Patients with poor gingival health may experience a reversible gingival tissue irritation. Staining of active caries lesions, probably by the stannous component, has frequently been reported after topical application of SnF_2, particularly in patients with poor oral hygiene. Topical application of highly concentrated (8%) SnF_2 solutions is therefore no longer recommended as a public health measure, nor would it be cost effective in populations using fluoride toothpaste daily. On the other hand, it is considered a highly efficient measure for caries control in adults who are susceptible to root caries.

Acidulated phosphate fluoride solutions. In 1963, researchers at the Forsyth Dental Center in the United States introduced an acidified NaF solution based on the premise that greater fluoride uptake by enamel occurs under acidic conditions. Acidulated phosphate fluoride has a pH of approximately 3.0, is buffered with 0.1 M phosphoric acid, and contains a fluoride ion concentration of 1.23% (Wellock and Brudevold, 1963). The application technique is the same as that for SnF_2.

Although initial studies of solutions of APF indicated that it might be superior to neutral NaF and SnF_2, the entire body of literature on the agent indicates comparable effectiveness. Benefits have averaged about 28% in nonfluoridated communities (Ripa, 1990).

Because of its low pH, APF has an acidic taste; flavoring, however, is not a problem. Acidulated phosphate fluoride is stable when stored in a plastic container and should not be stored in a glass container because it may etch the glass. Repeated or prolonged exposure of porcelain or composite restorations to APF may result in loss of contour, surface roughening, and possibly cosmetic changes. The American Dental Association (1988c) suggested, therefore, that it may be prudent not to use an acidic topical fluoride agent in patients with these types of restorations. In addition, the caries-reducing effects (about 30%) in public health programs were achieved 25 to 40 years ago in fluoride-deficient areas, before the advent of efficient fluoride toothpastes.

Fluoride gels for professional use

Gels for professional use are similar to those for self-care but have a higher fluoride concentration. The most commonly used compounds are APF (1.2% F), neutral 2% NaF (0.9% F), and 2% SnF_2 (0.5% F). The gel may be applied in prefabricated (see Fig 309) or customized trays, but syringes are also used for professional application, most frequently for the approximal surfaces in elderly adults with exposed root surfaces (see Fig 245 in chapter 4).

Because of their ease of handling and clinically proven anticaries effectiveness, APF gels have been used professionally for decades in the United States; the 2% neutral NaF gels (9.04 ppm F) were introduced only recently. Developed in response to concerns that APF gel may etch glass filler particles in composite restorations and the glaze on porcelain crowns, the neutral NaF gels are more commonly used in Europe, particularly in Scandinavia. The concentration is the same as that in the NaF solution. However, 2% NaF gel has not been adequately tested in a 6-month application interval; the anticaries effectiveness requires further clinical validation.

In the United States, APF gels (1.23% F) have been professionally applied with positive results in studies of schoolchildren in fluoride-deficient communities. Ripa (1989) concluded that, with an application frequency of either once or twice a year, the pooled average caries reduction in these studies was 21.9%. When studies with an application frequency of only once a year were excluded, the average caries reduction was 26.3% for fluoride applied twice annually, the recommended frequency. Furthermore, two studies in which children were reexamined 1 and 2 years after discontinuation of APF gel applications reported an apparent persistence of the caries-protective effect, probably attributable to a combined effect of low pH and high fluoride concentration of the gel, which was applied before secondary enamel maturation was completed (Horowitz and Kau, 1974; Ingraham and Williams, 1970).

There are very few studies of the effect of professionally applied topical fluoride in optimally fluoridated communities. Based on the limited evidence available, routine professional topical fluoride applications are not recommended for children who are lifelong residents of optimally fluoridated communities and have very low caries prevalence. It should also be noted that the studies carried out in the United States were conducted 20 to 30 years ago; today such programs would not be cost effective.

A recent fluoride workshop recommended that, because of the high fluoride concentrations, the gels be used selectively in patients with moderate or high caries risk (Clarkson, 2000). As with gels for self-care, professional gel application is most cost effective in high–caries-risk patients, particularly those with xerostomia. Gels with combined fluoride and antiplaque effects, such as NaF with CHX or SnF_2, are preferable. To improve the effectiveness of the gels in these extremely high-risk patients, the teeth should be professionally cleaned (PMTC) and dried before application of the agent. Gel is syringed into the posterior interproximal spaces and then applied in a customized tray three times for 5 minutes each per visit, at needs-related intervals, four to six times per year. Because thorough rinsing immediately after one single application will reduce the effect of accumulated fluoride by about 50%, rinsing is not recommended for 30 minutes after treatment.

Fluoride prophylaxis pastes

The fluoride concentrations of prophylaxis pastes for PMTC and polishing restorative materials range from 0.1% to 1.0%. Sodium fluoride is the most commonly used agent, but SnF_2, AmF, and SMFP are also used. Fluoride prophylaxis pastes remove plaque biofilms, polish restorative materials, and concurrently deliver fluoride. Therefore, the prophylaxis pastes contain abrasives as well as fluoride.

As with fluoride toothpastes, the abrasiveness of the commercial products may be expressed as RDA, according to the ADA's assessment, or as DAV, according to the British Standards Institute. The abrasiveness of most toothpastes ranges from 40 to 80 RDA (35 to 85 DAV); the abrasiveness of prophylaxis pastes may range from 25 to 300 RDA. For PMTC, an RDA of 100 to 200 is most appropriate; an RDA of 40 to 90 is most appropriate for polishing restorations and root surfaces; and an RDA of more than 200 is best for finishing dental materials. Silica (SiO_2) and alumina trihydrate are used as abrasives in pastes with an RDA of 25 to 120 and pumice is used in pastes with an RDA of more than 150. Most commercial products also contain thickeners to modify viscosity. Figures 311 and 312 show an assortment of commercially available prophylaxis pastes.

Concurrent delivery of fluoride during PMTC and finishing and polishing of the cervical margins of restorations should have several advantages: Accessibility is optimal once the plaque biofilms are removed and the posterior interproximal papilla is depressed. However, the effect of the fluoride delivered by the prophylaxis paste should not be overestimated, and in caries-susceptible patients, PMTC should be supplemented by application to key-risk surfaces of more efficient fluoride agents such as varnishes, which release fluoride slowly.

To date there are very few reports of the effect of fluoride in prophylaxis pastes. Clinical trials based on fortnightly PMTC resulted in almost total control of dental caries, irrespective of whether the prophylaxis paste contained fluoride (Axelsson and Lindhe, 1974; Axelsson et al, 1976); ie, in preventive programs based on very high-quality control of gingival plaque, the supplementary effect of fluoride is marginal.

Fluoride varnishes

Fluoride varnishes have been available in Europe for more than 30 years and are widely used for professional application in caries-susceptible patients. Duraphat, the first commercial varnish, was introduced in Europe in 1964; it was followed in 1975 by Fluor Protector (Vivadent, Liechtenstein). More recently, a third varnish has been introduced (Bifluorid 12). Although their use is increasing worldwide, fluoride varnishes have not been accepted, for example, by the ADA or the Canadian Dental Association, although they have been available in Canada for several years. In 1994, the US Food and Drug Administration approved the marketing of Duraphat, under the brand name DuraFlor (Colgate-Palmolive, New York, NY). To date Fluor Protector is approved for use in the United States only as a cavity varnish.

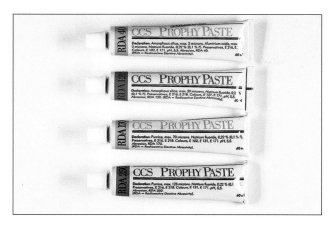

Fig 311 CCS prophylaxis pastes (NaF 0.22%) with four different abrasive values (RDA): *(top to bottom)* yellow (RDA 40), red (RDA 120), green (RDA 170), and blue (RDA 250) (DAB, Stockholm, Sweden).

Fig 312 Two other prophylaxis pastes containing fluoride: Profylan (0.22 % NaF; red/low abrasive and blue/abrasive) (Scania Dental, Knivsta, Sweden) and Hawe Cleanic (CaF$_2$ 0.3%; RDA 27) (Hawe Dental, Lugano, Switzerland).

Fluoride varnishes may be regarded as slow-release or semi-slow-release agents: Prolonged exposure time and high fluoride concentrations (except for Fluor Protector) result in the formation of large CaF$_2$ reservoirs, which are available during demineralization and remineralization (see Figs 283 to 285). To improve the retention of the varnish and thereby prolong the total fluoride exposure time, the tooth surfaces should be professionally cleaned (PMTC) and dried before application.

The total caries-preventive and caries-control effects of professionally applied fluoride varnish represent the sum of the following procedures:

1. Professional mechanical removal of plaque biofilms from the most caries-susceptible tooth surfaces.
2. Protection of these tooth surfaces from direct contact with reaccumulated cariogenic plaque, as long as the varnish is retained on the teeth.
3. The direct release of ions from the fluoride compound in the varnish and the indirect formation of reservoirs of CaF$_2$ for subsequent slow release.

The separate effect of these three procedures has yet to be evaluated for the different fluoride varnishes. Such an evaluation should be designed as a randomized controlled study in a high–caries-risk population and conducted in boys from the time they are age 12 until they are age 15, with three test groups and one control group for every tested varnish. Test group 1 should receive PMTC strictly according to the method described in chapter 3. Test group 2 should have PMTC and placebo varnish application. Test group 3 should have PMTC and the tested fluoride varnish. All test groups should be treated three times within 1 week initially and thereafter four times per year for 3 years. The effect on the posterior approximal surfaces would be evaluated by the use of computer-aided subtraction radiography and clinically by temporary tooth separation for diagnosis of cavitated versus noncavitated lesions.

Fluoride varnishes are recommended for selective application at individual intervals (two to four times per year) to key-risk tooth surfaces in caries-susceptible patients to supplement daily use of other fluoride agents by self-care, such as toothpastes, lozenges and chewing gum. In caries-susceptible children, fluoride varnishes are particularly appropriate for the occlusal and posterior approximal surfaces in erupting and newly erupted teeth and around orthodontic brackets. In caries-susceptible adults (particularly xerostomic patients), key areas to be protected are the margins of crowns and exposed root surfaces.

A wide range of caries-inhibiting effects of fluoride varnishes has been reported, from 30% to 70% (for review, see Petersson, 1993; Petersson et al, 1997; and Horowitz and Ismail, 1996). In a systematic review of caries-preventive methods, Bader et al (2001) concluded that the use of fluoride varnish resulted in the most significant caries reduction among topical fluoride methods.

Duraphat. Duraphat contains 5 wt% NaF (2.26% F), an extremely high fluoride concentration compared to other fluoride agents for professional use. It is a viscous, resinous varnish that should be applied with pointed bristles or syringes to the dried tooth surfaces after PMTC. In contact with saliva, Duraphat hardens into a yellowish brown coating. Patients are instructed not to eat within 2 hours and not to clean their teeth within 4 to 6 hours to improve the effect and retention of the agent. Duraphat usually remains on the tooth surfaces for about 24 hours. It is supplied in tubes (10 mL or 5 × 30 mL) for application with pointed brushes or in glass ampoules (5 × 1.6 mL) for application with a syringe (Figs 313 to 315).

Despite the high fluoride concentration, when applied by syringe only to key-risk surfaces two to four times per year, the total exposure to fluoride is very limited compared to that resulting from other fluoride agents. Ekstrand et al (1980) found no toxic effects with respect to fluoride plasma levels or renal function in preschool children and schoolchildren treated with Duraphat. This is attributable to the fast-setting varnish base, the slow release of fluoride over time, and the comparatively small amounts of the varnish required for selective application to the key-risk surfaces.

No frequent or serious side effects have been reported from Duraphat. However, the varnish should not be applied in contact with bleeding gingival tissues because of the risk of contact allergy to the colophonium base. Duraphat, particularly when packaged in a tube, must be kept safely from the access of small children.

Fluor Protector. Fluor Protector is a polyurethane-based varnish containing 0.9 wt% silane fluoride (0.1% F). The standard package consists of 50 glass ampoules of Fluor Protector (1 mL each), one rubber foot, one brush holder, 50 disposable brushes, and three ampoule breakers (Fig 316). After plaque is removed by PMTC and the tooth surfaces are dried, the colorless varnish is selectively applied with the disposable brush or a minipipette. The varnish is acidic and hardens in air to a colorless, transparent film within 2 to 3 minutes. It is retained on the teeth as a slow-release fluoride agent for 1 to 2 weeks and is particularly well retained on the caries-susceptible posterior approximal surfaces, which are not subjected to abrasion from chewing. Compared to Duraphat and Bifluorid 12, the exposure time is markedly prolonged, and the tooth surface is protected from direct contact with cariogenic plaque as long as the varnish film remains.

In contrast to other fluoride compounds, the silane fluoride in Fluor Protector varnish is insoluble in water but reacts on contact with saliva, releasing small amounts of hydrogen fluoride. Because of its considerably higher diffusion coefficient, hydrogen fluoride penetrates enamel more rapidly and readily than do fluoride ions. This process occurs at the contact area between the varnish and the enamel, given the necessary traces of moisture there. The molecules of hydrogen fluoride that penetrate the enamel or enamel lesion again react with water in the intercrystalline fluid and yield fluoride ions, which will influence demineralization and remineralization at a greater depth in the enamel than can be achieved by other fluoride agents with higher fluoride concentrations.

No severe side effects have been reported from the use of Fluor Protector. However, the varnish should not be applied in contact with bleeding gingival tissue to prevent the development of contact allergy to the polyurethane vehicle.

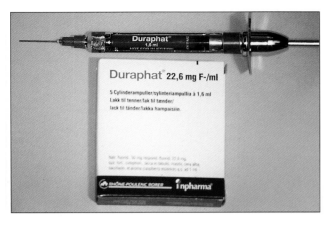

Fig 313 Duraphat fluoride varnish (5% NaF): 30-mL and 10-mL tubes or 1.6-mL glass ampoules for syringe.

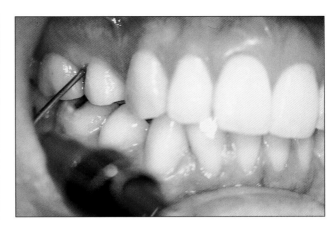

Fig 314 Application of Duraphat varnish interproximally using a syringe.

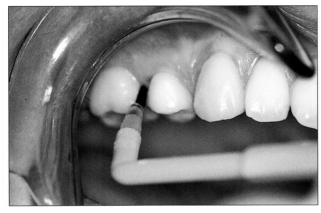

Fig 315 Application of Duraphat varnish from a tube using a disposable pointed brush on selected key-risk surfaces.

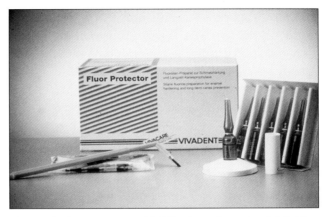

Fig 316 Fluor Protector (0.9% silane fluoride; 0.1% F) in glass ampoules and a disposable pointed brush for application.

Bifluorid 12. Bifluorid 12 is a clear varnish containing 6% NaF and 6% CaF$_2$, which corresponds to 60 mg of each fluoride compound per 1 g of the preparation. Only 1 g of the preparation contains 56.3 mg of fluoride. However, CaF$_2$ is not water soluble and thus is nontoxic. The varnish base consists of collodion and an organic solvent. Toxicologic tests confirm the safety compared to the most common varnish bases. Bifluorid 12 varnish is not only a cariostatic agent but also an agent for treatment of hypersensitivity of exposed root dentin. To ensure an optimum distribution of particle sizes, it contains a homogenous

mixture of highly dispersed NaF and CaF$_2$ particles, between 0.5 and 2.0 μm in size, which allows them to penetrate and block the dentinal tubules and even fill small gaps and spaces around the margins of restorations, ensuring the formation of CaF$_2$ depots.

Bifluorid 12 is supplied as a bottle containing 4- or 10-g fluoride varnish, solvent (bottle), and special foam pellets for application (Fig 317). A large pack containing three 10-g bottles of Bifluorid 12 is also available.

Bifluorid 12 is applied with the following technique:

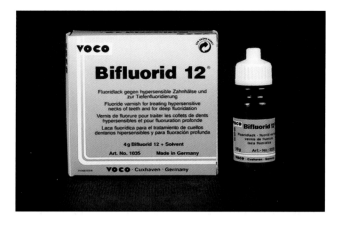

Fig 317 Bifluorid 12 fluoride varnish (6% NaF + 6% CaF$_2$).

1. Prior to application of the varnish, dental plaque is thoroughly removed by PMTC and the tooth surfaces are dried.
2. The bottle is shaken thoroughly to mix the sediment of NaF and CaF$_2$ completely with the solvent.
3. The varnish is applied with a foam pellet to the selected tooth surfaces. Because of the low viscosity, only about 0.02 mL is used per tooth. After 10 to 20 seconds of absorption, the tooth surfaces are dried with an air syringe.
4. Abrasive chewing of food and mechanical toothcleaning should be avoided for 4 and 24 hours, respectively, after treatment.

With proper application and good patient compliance, the varnish is retained on the tooth surfaces for several days.

Cariostatic effects. As mentioned earlier in this chapter, the cariostatic effects of fluoride agents are related to *(1)* the total exposure time, which means frequency of application and clearance time (eg, substantivity and retention of the agent); *(2)* accessibility; *(3)* fluoride concentration and the fluoride compound used; and *(4)* bioavailability. Professional application of fluoride varnish at needs-related intervals in caries-susceptible patients should, therefore, be a very cost-effective supplement to frequent daily home use of fluoride agents. However, as earlier discussed, the total cario-static effect of professional application of fluoride varnish is the end result of the various steps in the method:

1. Mechanical removal of cariogenic plaque by PMTC from caries-susceptible tooth surfaces.
2. Protection of these surfaces from direct contact with reaccumulated cariogenic plaque as long as the varnish remains on the surfaces.
3. Slow release of fluoride from the high fluoride concentration of the varnish, resulting in rapid remineralization of the outer surface of enamel caries lesions, sealing the micropore filter (see Fig 289).
4. Optimal accessibility, ensured by professional, targeted application of the varnish with brushes, syringes, or foam pellets to coat the plaque-free, key-risk tooth surfaces.

These factors may explain some inconsistencies in results from different experimental and clinical trials with fluoride varnishes. However, variability among test and control groups with respect to caries prevalence and incidence, age, number of participants, standard of oral hygiene, and daily use of fluoride toothpaste, would also significantly influence the outcome of fluoride varnish treatment. Factors such as fluoride concentration in the drinking water and the degree of compliance with instructions for use might also have influenced the results.

Over the last 30 years, fluoride uptake in vitro and in vivo, acid resistance, and the caries-preventing effect of fluoride varnish have been investigated in laboratory, animal, and human experimental studies. Laboratory investigations and in vivo experimental studies have shown that varnishes supply fluoride more efficiently than do other topical agents. Fluoride varnish treatments effectively inhibit demineralization, resulting in highly significant reductions in caries; these reductions are 30% to 50% in fissures and sometimes even higher on the proximal surfaces (de Bruyn and Arends, 1987).

Most controlled clinical trials using Duraphat, generally in children of school age, have revealed significant reductions in caries incidence (20% to 50%) in the permanent dentition. Petersson et al (1991a) used Duraphat in an intensified program with three consecutive applications within a 10-day period, once a year, for 3 years. This mode of application was significantly better than two applications a year. In this study, the three PMTCs within 10 days might have healed gingivitis, significantly reducing the plaque formation rate: This may have enhanced the effect of the applications of Duraphat, which contains 5% NaF, in remineralizing the outer surface of existing enamel caries lesions. In a similar 4-year study, Sköld et al (1994) showed that three applications of Duraphat in 1 week every year resulted in significant caries reduction compared with one single application per year.

Based on a similar rationale, an initial intensive program based on frequent PMTC and fluoride varnish is recommended for high–caries-risk patients (see chapter 8). In a study by Koch et al (1979), application of Duraphat twice a year resulted in a 30% caries reduction compared to supervised weekly NaF mouthrinsing. Similar results were also achieved by Seppä and Pöllänen (1987). In a 3-year clinical study, Seppä et al (1994) showed that, compared to application of APF gel twice per year, use of Duraphat was somewhat more effective against approximal caries. Seppä et al (1994) also showed, in a 3-year trial, that there was no difference in caries incidence between groups of subjects treated with annual applications of Duraphat with 5.0% or 2.5% NaF.

In a recent experimental study, Fontana et al (2002) evaluated the effect of Duraphat fluoride varnish on secondary caries progression compared to a Duraphat placebo varnish. The fluoride varnish significantly slowed down the progression of active secondary caries lesions. However, the placebo varnish, which remains in place for a maximum of 24 hours, also showed a trend of slowing down lesion progression, suggesting that the effect of fluoride varnish is not only due to its fluoride release, as earlier discussed. For comparison, Fluor Protector remains as long as 1 to 2 weeks after PMTC.

In another recent experimental in vitro study on enamel blocks with early lesions, Maia et al (2003) showed that a frequently used fluoride toothpaste (1,000 ppm F as NaF) resulted in a higher percentage of surface microhardness recovery compared with Duraphat fluoride varnish and had a similar effect on fluoride uptake when compared with its combination with a single application of the fluoride varnish.

Fluor Protector has not been clinically tested as extensively as Duraphat but has in general shown superior laboratory results, especially, for example, for fluoride uptake in enamel and inhibition of artificial lesion formation (Petersson, 1976; de Bruyn, 1987). In clinical trials with Fluor Protector, some bias from other fluoride sources has been discovered. Some studies have been conducted in areas with an optimal water fluoride concentration or designed as split-mouth studies (de Bruyn and Arends, 1987; Seppä et al, 1982a). Results of studies with a split-mouth design must be interpreted with caution: A crossover effect cannot be discounted (Seppä, 1982; Seppä et al, 1982b) because fluoride varnishes cause a marked elevation in fluoride concentration in mixed saliva. However, in a 3-year longitudinal study in 12- to 15-year-old schoolchildren who used fluoride toothpaste twice daily, application of Fluor Protector four times a year resulted in a 30% ($P = .05$) reduction of approximal caries compared with supervised weekly NaF rinsing or daily use of NaF toothpaste alone (Axelsson et al, 1987a).

Twetman et al (1996) evaluated the effect of Fluor Protector used twice a year in 4- to 5-year-old preschool children in an area with 0.1 ppm of fluoride in the drinking water (A; n = 448) and an area with 1.2 ppm of fluoride in the drinking water (C; n = 206) and compared the results to a control group living in an area with 0.1 ppm of fluoride (B; n = 374). The caries incidence in groups A, B, and C was 0.65, 1.10, and 0.50 new carious surfaces per individual per 2 years, respectively. The difference between group B and groups A and C was statistically significant ($P <$ .05). However, there was not a significant difference in effect between children living in areas with 0.1 ppm and 1.2 ppm of fluoride in the drinking water.

There are few clinical trials testing Duraphat and Fluor Protector under relevant comparable clinical conditions. Therefore, randomized controlled trials including placebo varnishes and running from 12 to 15 years of age should be performed, as earlier discussed. Clark et al (1985), however, reported no difference between the two fluoride varnishes and a nonsignificant reduction in caries for both. Because caries experience and incidence were rather low in these studies, no definite conclusions should be drawn about the relative cariostatic effects of the two varnishes.

In a clinical caries model, de Bruyn et al (1988a, 1988b) compared mineral loss of enamel under undisturbed plaque after one single application of Fluor Protector (0.70% F, 0.10% F, or 0.05% F), Duraphat, or a placebo varnish and found no difference in effect among the three Fluor Protector preparations at 4 or 6 months. At 4 months, all three Fluor Protector varnishes showed better preventive effects than did Duraphat. No difference was observed between Duraphat and the placebo varnish.

In a recent 3-year clinical study, Petersson et al (2000) evaluated the effect of Fluor Protector on approximal caries compared to Cervitec (CHX + thymol) varnish (Vivadent). The mean values of approximal caries incidence per 3 years was 2.7 ± 3.1 and 3.1 ± 3.5 for Fluor Protector and Cervitec, respectively. However, the differences between the two varnishes when used every 3 months was not significant.

To date, data on the cariostatic effect of the recently introduced Bifluorid 12 varnish are available from only one clinical trial. In a 2-year, double-blind study in 12- to 14-year-old schoolchildren, Borutta et al (1991) reported that, compared to a placebo varnish, two and four applications a year of Bifluorid resulted in caries reductions of 25% and 35%, respectively. On the approximal surfaces, the reductions were 36% and 40%, respectively.

To date, Bifluorid 12 has not been compared with Duraphat and Fluor Protector in clinical trials. However, experimental studies have shown that the uptake of fluoride in dentin is 50% higher with Bifluorid 12, which contains 6% NaF and 6% CaF_2, than it is with varnish containing only 6% NaF. In addition, one application of Bifluorid 12 resulted in three times more structurally bound fluoride (CaF_2) after 5 days than did Fluor Protector (Hellwig and Attin, 1994). Recently, Twetman et al (1999) showed that the fluoride concentration in whole saliva and separate gland secretions was higher up to 6 hours after one single application of Bifluorid 12 compared with Duraphat and Fluor Protector. Duraphat resulted in higher concentration of fluoride than did Fluor Protector. These experimental data indicate that the cariostatic effect of Bifluorid 12 should at least be equivalent to that of Duraphat and Fluor Protector.

Sorvari et al (1994) evaluated the efficacy of a CHX solution with fluoride varnish in preventing enamel softening by mutans streptococci in an artificial mouth: The labial surfaces of pieces of bovine incisors were treated with 0.2% CHX solution, with Duraphat fluoride varnish, or with both, while one group was treated with distilled water, and one was left as an untreated control. A placebo varnish was used in the CHX- and distilled water–treated groups. All the varnishes were removed after 24 hours. The enamel slabs were mounted pairwise in an artificial mouth to form approximal contacts. The teeth were continuously rinsed with a common pool of artificial saliva containing 3% sucrose and infected on the first day with *S mutans*, strain Ingbritt. The saliva was renewed daily, and the specimens were incubated at 37°C for 10 days. The appreciable softening found in the distilled water– and placebo varnish–treated group tended to be prevented by the CHX and even more by the fluoride treatment, while the CHX-fluoride treatment completely prevented enamel softening.

In an experimental in vitro demineralization model, van Loveren et al (1996) examined the protective effect of two CHX varnishes, Cervitec and EC40 (Certichem, Nijmegen, The Netherlands) (40% CHX), and Fluor Protector fluoride varnish. Enamel specimens were best protected by Fluor Protector (Fluor Protector > Cervitec = EC40 > no treatment), while the dentin specimens were best protected by the CHX treatments (Cervitec = EC40 > Fluor Protector > no treatment). A 1:1 mixture of Cervitec and Fluor Protector was as effective as the most effective component alone. It was concluded that a varnish containing both fluoride and CHX may be useful, since it could give optimal protection to both enamel and dentin.

In a recent 3-year clinical study, Petersson et al (1998) evaluated the effect of semiannual applications of a varnish containing Fluor Protector and Cervitec (1:1 mixture) in 12-year-old children. A positive control group received semiannual treatment with Fluor Protector alone. After 3 years, the mean approximal caries incidence including enamel caries was 3.0 ± 3.7 and 3.8 ± 4.3 for the test and control groups, respectively.

These in vitro and clinical results indicate that a varnish containing fluoride as well as CHX with high stability and bioavailability would be a highly appropriate cariostatic agent for professional application (see also chapter 4).

Effect on hypersensitivity. Although hypersensitive necks of teeth and exposed root dentin are a common phenomenon from the age of 30 years onward, management of the condition is unsatisfactory for two main reasons: *(1)* Although there is transient pain during activities such as eating cold food and brushing, the patient may forget to mention it in the anxiety of a dental ap-

pointment; and *(2)* agents offering long-term relief have been unavailable, so that treatment measures were effective only temporarily. To date, specially formulated toothpastes, SnF_2 solutions, fluoride varnishes, and dentin bonding materials as well as diluted glass-ionomer cements have been used. Bifluorid 12 seems to offer the most long-lasting effect; about 92% of patients in a multicenter study reported significant improvement or relief of pain after a single application (Schroers, 1994).

Thus, fluoride varnishes seem to be the most efficient caries-preventive agents for professional application. Indications for use are selective, targeted application to the most caries-susceptible surfaces, such as posterior approximal surfaces, fissures, exposed root surfaces, and around orthodontic brackets and margins of restorations. The main advantage of varnishes is their prolonged retention at these risk sites, allowing not only slow release of fluoride but also protection of the professionally cleaned tooth surfaces from direct reaccumulation of plaque as long as the varnish is retained.

In caries-risk (C2) and high–caries-risk (C3) patients, an initial intensive period with two to four treatments within 7 to 10 days is recommended: The combination of frequent PMTC, isolation of key-risk surfaces from direct contact with cariogenic plaque, and continuous exposure to relatively high concentrations of fluoride should result in:

1. Rapid remineralization of the outer surfaces of enamel caries, sealing the micropore filter
2. Transformation of active root caries to inactive caries
3. Accumulation of CaF_2 reservoirs on the tooth surfaces
4. Healing of inflamed gingival tissues and thereby a reduction in the plaque formation rate (PFRI)

In maintenance programs, fluoride varnish should also be applied at needs-related intervals, eg, two to three times per year in C2 patients and four to six times per year in C3 patients (particularly in patients with xerostomia). Finally, a varnish with high stability and bioavailability that contains high concentrations of fluoride as well as CHX and is retained on the teeth for more than 1 week would be highly appropriate for high–caries-risk patients (for reviews on fluoride varnishes, see Clark, 1982; de Bruyn and Arends, 1987; Petersson, 1993; Petersson et al, 1997; and Primosch, 1985).

Slow-release fluoride agents

Fluoride varnishes, with retention of about 1 week, may be regarded as semi–slow-release agents. However, as early as the late 1960s, Epoxylite 9070 (Lee Pharmaceuticals, South el Monte, CA), a polyurethane-based semipermanent fissure sealant material with a high fluoride concentration was introduced. Applied after PMTC and chemical "cleaning" of the fissures with citric acid, it was retained in the fissures of erupting or newly erupted molars for some months, providing a continuous slow release of fluoride. In vivo measurements by electron microprobe x-ray spectrometry showed very high accumulation of fluoride reservoirs in these extremely caries-susceptible fissures (Lee and Schwartz, 1971; Lee et al, 1972).

Consistent levels of fluoride in the mouth have been related to remineralization of early caries lesions. A number of studies have shown that individuals with a higher salivary fluoride content have lower caries levels than individuals with lower salivary fluoride levels (Bruun and Thylstrup, 1984; Duggal et al, 1991; Leverett et al, 1987; Sjögren and Birkhed, 1993). It is the activity of the fluoride ion in the oral fluid that is of most importance in reducing the solubility of the enamel rather than a high content of fluoride in enamel (Fejerskov et al, 1981). Thus, a constant elevation of intraoral fluoride levels may be beneficial to priority groups at risk of developing caries, eg, irregular dental attendees, disabled and high–caries-risk individuals (C3), such as patients with xerostomia and patients with very limited compliance.

The ideal method of providing fluoride to such high–caries-risk individuals would have the following properties:

- Safe to administer
- Cheap and cost effective
- Act topically at the tooth surface
- Prevent caries clinically
- Easily manufactured
- Easy and quick to apply
- Provide long-term fluoride release of at least 1 year
- Robust
- Provide a continuous low concentration of intraoral fluoride
- Effective regardless of patient compliance and motivation

Most of the currently available methods do not satisfy all the criteria, most commonly because they do not exhibit long-term fluoride release or because they rely on patient compliance.

Intraoral slow fluoride release is provided by a device that can deliver a constant supply of fluoride over a period of at least 2 years (Toumba and Curzon, 1993) or by dental materials that release fluoride slowly, and can be repeatedly replenished with fluoride from topical agents used by the patient at home or applied professionally.

Intraoral slow-release devices. The intraoral devices currently available are of two types: the copolymer membrane device and the fluoride glass device.

The copolymer membrane device, developed by Cowsar et al (1976), is used in the United States. It is a membrane-controlled reservoir that consists of an inner core of hydroxyethyl methacrylate–methyl methacrylate copolymer (50-50 mixture) containing a precise amount of NaF. This core is surrounded by a 30-70 hydroxyethyl methacrylate–methyl methacrylate copolymer membrane that controls the rate of fluoride release: Depending on the amount of fluoride in the inner core, this can be between 0.02 and 1.00 mg of fluoride per day. The duration of release ranges from 30 to 180 days, and salivary fluoride levels have been shown to be elevated throughout a 100-day test period (Corpron et al, 1986, 1991; Cowsar et al, 1976; Kula et al, 1987; Mirth et al, 1982).

A fluoride glass device was developed in Leeds, England, based on a similar glass device used to treat trace element deficiencies of cobalt, selenium, and copper in cattle. The device releases trace elements over a period of at least 1 year. It is 4 mm in diameter and is attached to the buccal aspect of maxillary permanent molar teeth with an acid-etch resin composite. It dissolves slowly when moist with saliva, releasing fluoride into the oral environment.

A number of glass devices have been used (ranging from 13.3% to 21.9% F), all of which raised salivary fluoride levels, in some mouths for up to 2 years (Toumba and Arizos, 2000). As mentioned earlier, Leverett et al (1987) showed that zero-caries subjects had higher salivary fluoride than did high-caries subjects. Shields et al (1987) showed that zero-caries subjects from both fluoridated and nonfluoridated communities had salivary fluoride levels of 0.04 µm/mL or greater, whereas high-caries subjects from both fluoridated and nonfluoridated communities had salivary fluoride levels of 0.02 µm/mL or less.

All the intraoral device studies to date have resulted in elevation of salivary fluoride levels. The salivary fluoride levels vary erratically within a 24-hour period, but remain elevated and stable on a daily basis.

Although the location of fluoride devices in the mouth is likely to be of considerable importance for optimal caries protection, to date the distribution of fluoride from slow-release devices, either the copolymer or the glass, has not been studied. It is important to determine whether fluoride moves from one side of the mouth to the other or remains localized.

In a human study (Toumba and Curzon, 1994), the fluoride glass devices were found to be completely safe regarding the possibility of fluoride toxicity following their ingestion, when compared with the plasma levels achieved from swallowing one 2.2-mg NaF supplement tablet. The results showed that the baseline plasma fluoride levels of 0.01 to 0.02 mg/L did not change when the fluoride glass slow-release devices were ingested. The findings of this study suggest that, when swallowed, the devices either pass through the stomach and small intestine very quickly or remain insoluble.

All studies of slow-release devices have found an uptake of fluoride, as demonstrated by increased microhardness or increased fluoride levels in caries lesions (Abrahams et al, 1980; Bashir, 1988; Corpron et al, 1986, 1991; Toumba and Arizos, 2000; Toumba et al, 1996). The cariostatic effect of intraoral fluoride slow-release devices was first evaluated in an experimental animal study: A 63% reduction in caries incidence in rats was achieved by Mirth et al (1983), in a 1-month test of a copolymer device releasing 0.15 mg of fluoride a day.

The only human caries study using the glass slow-release device presented to date is a 2-year, double-blind study involving 174 children aged 8 years in Leeds, United Kingdom (Toumba and Curzon, 1996). The results showed that the test group (fluoride device) developed 67% fewer new carious teeth ($P < .01$) and 76% fewer new carious surfaces ($P < .001$) than the control group (nonfluoride device). The mean salivary fluoride concentration for the test group was 0.17 mg/L after attachment of the device and 0.11 mg/L at the completion of the study, compared with 0.03 mg/L ($P < .001$) in the control group at the end of the study. There were 55% fewer new occlusal fissure carious cavities in the test group than in the control group after 2 years, showing that occlusal surfaces were also protected by the fluoride released from the devices (Toumba and Curzon, 1998). The fluoride glass slow-release devices have recently been patented and commercial products will, after further development, soon be available.

In a recent short-term study, Marini et al (2000) showed in a double-blind study that the copolymer fluoride membrane device resulted in significant reduction of dentinal sensitivity after

4 weeks in a selected group of patients with painful dentinal sensitivity after periodontal therapy. (For review of fluoride slow-release devices, see Toumba, 2001.)

Fluoride-releasing dental materials. Silicate restorative materials and glass-ionomer cements contain large amounts of fluoride (about 15% to 20%), explaining the documented higher concentration of fluoride in the tooth structures surrounding such restorations and the lower incidence of recurrent caries (Forss and Seppä, 1990; Seppä, 1994; Svanberg, 1992; ten Cate and van Duinen, 1995; Weerheijm et al, 1993). In an attempt to achieve a similar advantage, fluoride has been incorporated into other restorative materials, such as resin composites and amalgam. Although the clinical performance of the fluoridated resin composites is still inconclusive, an in situ study by Dijkman and Arends (1992) suggested a possible benefit in prevention of recurrent caries.

Fluoridated amalgam has been shown to increase the fluoride concentration in surrounding enamel and dentin, although this should not be extrapolated to imply a reduced risk of recurrent caries. Skartveit et al (1994) in a 4-year, split-mouth study, showed that, compared to traditional amalgam, fluoridated amalgam resulted in 43% reduction of recurrent occlusal caries.

With respect to release and reaccumulation of fluoride, however, the glass-ionomer cements are unique among dental materials. All the original pure glass-ionomer cements contain a liquid and powder. The liquid is based on an ionic polymer (polyacrylic acid). The powder is a glass containing silicon oxide, aluminum oxide, and CaF_2 (for details, see chapter 6).

In the early 1970s, Wilson and Kent (1972) showed that glass-ionomer materials could form strong bonds to enamel apatites and adhered to collagen in dentin. This physicochemical bond to enamel and dentin provided a watertight seal at the margin, and the coefficient of thermal expansion of the material was similar to that of natural tooth structure. The other important property was the slow release of fluoride. However, the original material has several disadvantages. The setting reaction occurs in two phases, and the material is very sensitive to moisture during the first phase and to dehydration during the second. Further disadvantages are its poor color stability and wear resistance.

To overcome some of these problems, resin-modified glass-ionomer cements were introduced, in which the glass-ionomer material is supplemented by different types of resins (light cured as well as chemically cured), eg, Photac-Fil (Espe, Seefeld, Germany), Vitremer (3M, St Paul, MN), and Fuji II LC (GC Corp, Tokyo, Japan). Resin composites have also been supplemented with glass-ionomer materials; these hybrids are known as *polyacid-modified composite resins* or *compomers*, eg, Compoglass (Vivadent) and Dyract (Dentsply, Milford, DE). In general, the higher the resin supplement, the lower the fluoride content and the fluoride release. Thus, pure glass-ionomer cements release the most fluoride, while compomers and fluoridated resin composites release very limited amounts of fluoride. On the other hand, compomers are much more wear resistant than glass-ionomer cements.

Studies by Hatibovic-Kofman and Koch (1991) showed that, even 1 year after application of glass-ionomer restorations in primary teeth, the fluoride concentration in unstimulated saliva is about six times higher than normal. This continuous release of fluoride is attributable not only to the fluoride in the product but also to the capacity of glass-ionomer cement to take up fluoride from toothpastes and then slowly release it. If the fluoride reservoir is regularly replenished by topical application, glass-ionomer restorations may be regarded as intraoral slow-release fluoride devices. Recharging and release of fluoride from glass-ionomer materials after topical application of fluoride agents such as toothpastes, solutions, and gels are strongly dose related (Creanor et al, 1995; Seppä et al, 1993). Plaque adhering to glass-ionomer restorations contains more fluoride and fewer mutans streptococci than does plaque on composite restorations (Benelli et al, 1993; Forss et al, 1991b).

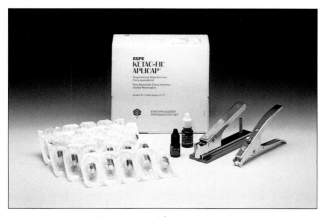

Fig 318 Ketac-Fil resin-modified glass-ionomer cement in capsules (Espe, Seefeld, Germany).

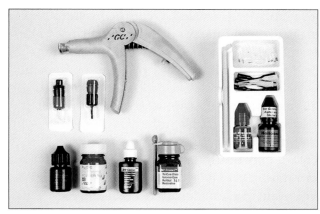

Fig 319 Four different fluoride slow-release restorative and fissure sealant materials: Fuji II LC light-cured resin-modified glass-ionomer restorative material in capsules, Fuji III resin-modified glass-ionomer cement for fissure sealant, Vitremer resin-modified glass-ionomer cement for fissure sealant, and Helioseal F fissure sealant (Vivadent).

Because glass-ionomer materials are very sticky, for ease of handling they are generally supplied in capsules and inserted into the cavity with a syringe (Figs 318 and 319).

Also available as fissure sealants are specially formulated low-viscosity glass-ionomer materials (Fuji III), resin-modified glass-ionomer materials (Vitrebond [3M] and Fuji III LC), and fluoridated resins (Helioseal F, Fissurit F [Voco], and Ultra-seal XT [Ultradent, South Jordan, UT]). In a 5-year longitudinal study, Mejáre and Mjör (1990) compared the effect of a glass-ionomer cement (Fuji III) with two resin-based sealants (Delton [Johnson & Johnson] and Concise [3M]). No caries developed in the occlusal surfaces sealed with glass-ionomer cement, but 5% of the surfaces sealed with resin developed caries. Because more than 60% of the glass-ionomer sealants were totally or partially lost after 5 years, compared to only 10% of the resin sealants, the results indicated that slow release of fluoride from the glass-ionomer cement is more important than the sealing effect. Similar caries-preventive results were achieved in a large-scale, split-mouth, 3- to 5-year study by Arrow and Riordan (1995). Other studies have shown that the retention of glass-ionomer

sealants can be improved by about 50% if, after PMTC, the fissures are conditioned with poly-acrylic acid before application of the sealant (Rasmusson and Rasmusson, 1994). (For review on the caries-preventive effects of fluoride-releasing glass ionomers and resin-modified glass ionomers, see Gao et al, 2000. For details on the effect of glass ionomers and resin-modified glass ionomers as fissure sealants, see chapter 6.)

SCHEDULES FOR NEEDS-RELATED USE OF FLUORIDES

To optimize cost effectiveness, the use of fluorides should always be related to the level of caries prevalence, caries incidence, socioeconomic conditions, and dental care resources in the population, as well as the predicted caries risk in the individual subject.

In countries with high prevalence and incidence of caries and little or no daily brushing with fluoride toothpaste, water fluoridation or the availability of fluoridated salt for the population and daily supervised cleaning with fluoride toothpaste in the schools would be very cost effective, at least in children and young adults. In regions with high caries prevalence and incidence and low socioeconomic levels, school-based, supervised toothcleaning with fluoride toothpaste or fluoride mouthrinse is also recommended. However, a major obstacle is that in developing countries, most of the population lives in rural areas that lack reticulated water supplies or where the hot climate is unsuitable for water fluoridation and fluoridated toothpaste is too expensive.

On the other hand, in most industrialized countries with well-developed dental care resources, relatively high socioeconomic standards, and moderate to low caries prevalence and incidence, home and professional use of fluoride, based strictly on predicted caries risk, would be most cost effective. Even in industrialized countries, there are some individuals and groups within the population with irregular dental attendance habits and poor oral hygiene. In most countries, water fluoridation would therefore be highly beneficial, and use of fluoride toothpaste should be recommended to everyone, irrespective of predicted caries risk. The only exceptions are children younger than 6 years old in areas with more than 1.5 mg of fluoride in the drinking water, because of their risk of developing fluorosis.

Tooth enamel is about 1.5 to 2.0 times more susceptible to caries during eruption and the following 2 years before so-called secondary maturation is completed (Kotsanos and Darling, 1991). In addition, because the plaque formation rate is significantly higher on erupting teeth than on fully erupted teeth (Carvalho et al, 1989), use of fluoride should be intensified during this period. Results from several studies in which different fluoride agents were evaluated on teeth that were already erupted when the study was initiated and those that erupted during the investigation show about a 50% greater reduction in caries among the latter than among the former (Horowitz and Ismail, 1996).

Because the cariostatic effect of fluorides is significantly related to exposure time, concentration, and accessibility, combinations of fluoride agents are more efficient than separate use of the agents. It is estimated that only 3% to 4% of the world's population has access to artificially or naturally fluoridated drinking water. Outside the United States and United Kingdom, water fluoridation is not widespread in the industrialized countries. Therefore, in the following recommendations for rational use of fluorides related to predicted caries risk and age, the fluoride concentration of the drinking water need not be considered, except in the case of children younger than 6 years.

Tables 16 and 17 present recommendations for use of fluorides by self-care and professionals, respectively. Because of the risk for fluorosis in young children and the eruption time of the permanent teeth, the following age groups were selected: 1 to 4 years, 5 to 7 years, 8 to 11 years, 12 to 14 years, 15 to 19 years, 20 to 69 years, and 70 years or older.

Table 16 Recommendations for needs-related use of fluorides by self-care * (continued on next page)

Age group	Caries risk	FTP Alt	FTP Int	FCTP Alt	FCTP Int	OHFDT Alt	OHFDT Int	OHFTP Alt	OHFTP Int	FM Alt	FM Int	FCM Alt	FCM Int	FG Alt	FG Int	FCG Alt	FCG Int	FT Alt	FT Int	FCHG Alt	FCHG Int	FCCG Alt	FCCG Int	FAS Alt	FAS Int
1–4 y	C0	A1	1–2 x d																						
	C1	A1	2 x d																						
	C2	A1	2 x d																						
	C3	A2	2 x d	A1	2 x d																				
5–7 y	C0	A1	2 x d																						
	C1	A1	2 x d																						
	C2	A1	2 x d															A1	AM						
	C3	A2	2 x d	A1	2 x d													A2	AM	A2	AM	A1	AM		
8–11 y	C0	A1	2 x d																						
	C1	A1	2 x d																						
	C2	A1	2 x d															A1	AM						
	C3	A2	2 x d	A1	2 x d							A4	2 x d			A4	2 x d	A2	AM	A2	AM	A1	AM		
12–14 y	C0	A1	2 x d																						
	C1	A1	2 x d			A1	1 x d																		
	C2	A1	2 x d			A1	2 x d											A2	AM	A1	AM				
	C3	A2	BM	A1	BM	A1	2 x d					A4	2 x d			A4	2 x d	A3	AM	A2	AM	A1	AM		
15–19 y	C0	A1	2 x d																						
	C1	A1	2 x d			A1	1 x d																		
	C2	A1	2 x d			A1	2 x d											A2	AM	A1	AM				
	C3	A2	2 x d	A1	2 x d	A1	2 x d											A3	AM	A2	AM	A1	AM		

Table 16 Recommendations for needs-related use of fluorides by self-care* (continued)

Age group	Caries risk	FTP Alt	FTP Int	FCTP Alt	FCTP Int	OHFDT Alt	OHFDT Int	OHFTP Alt	OHFTP Int	FM Alt	FM Int	FCM Alt	FCM Int	FG Alt	FG Int	FCG Alt	FCG Int	FT Alt	FT Int	FCHG Alt	FCHG Int	FCCG Alt	FCCG Int	FAS Alt	FAS Int
20-69 y	CO	A1	2 × d																						
	C1	A1	2 × d					A1	2 × d																
	C2	A2	2 × d	A1	2 × d			A1	2 × d									A3	AM	A2	AM	A1	AM		
	C3	A2	BM	A1	BM			A1	BM			A4	2 × d			A4	2 × d	A3	AM	A2	AM	A1	AM		
70+ y	CO	A1	2 × d																						
	C1	A1	2 × d					A1	2 × d																
	C2	A2	2 × d	A1	2 × d			A1	2 × d			A4	2 × d			A4	2 × d	A3	AM	A2	AM	A1	AM		
	C3	A2	BM	A1	BM			A1	BM			A4	2 × d			A4	2 × d	A3	AM	A2	AM	A1	AM		
Xerostomia	C3	A2		A1	BM			A1	BM			A4	2 × d			A4	2 × d	A3	AM	A2	AM	A1	AM	A1	20-30 × d

*(FTP) Fluoride toothpaste; (FCTP) Fluoride toothpaste with chemical plaque control agent; (OHFDT) Fluoridated dental tape; (OHFTP) Fluoridated toothpick; (FM) Fluoride mouthrinse; (FCM) Fluoride-chlorhexidine mouthrinse; (FG) Fluoride gel; (FCG) Fluoride-chlorhexidine gel; (FT) Fluoride tablet; (FCHG) Fluoride chewing gum; (FCCG) Fluoride-chlorhexidine chewing gum; (FAS) Fluoridated artificial saliva; (Alt) Alternative; (A1) Alternative 1; (A2) Alternative 2; (A3) Alternative 3; (A4) Alternative 4; (Int) Interval; (CO) No caries risk; (C1) Low caries risk; (C2) Caries risk; (C3) High caries risk; (BM) Before meals; (AM) After meals.

Table 17 Recommendations for needs-related use of professionally applied fluorides*

Age group	Caries risk	PMTC+FPP Alt	PMTC+FPP Int	AFS Alt	AFS Int	FG Alt	FG Int	FCG Alt	FCG Int	FV Alt	FV Int	FCV Alt	FCV Int	FS Alt	FS Int
1–4 y	C0														
	C1	A1	1 x y							A1	1 x y				
	C2	A1	2 x y							A1	2 x y				
	C3	A1	3 x y							A1	3 x y				
5–7 y	C0	A1	1 x y							A1	1 x y				
	C1	A1	2 x y							A1	2 x y				
	C2	A1	3 x y					A2	3 x y	A2	3 x y	A1	3 x y	A1	Once
	C3	A1	4 x y					A3	4 x y	A2	4 x y	A1	4 x y	A1	Once
8–11 y	C0														
	C1		1 x y							A	1 x y				
	C2	A1	2 x y							A1	2 x y				
	C3	A1	3 x y									A1	3 x y		
12–14 y	C0	A1	1 x y							A1	1 x y				
	C1	A1	2 x y							A1	2 x y				
	C2	A1	3 x y					A3	3 x y	A2	3 x y	A1	3 x y	A1	Once
	C3	A1	4 x y					A3	4 x y	A2	4 x y	A1	4 x y	A1	Once
15–19 y	C0														
	C1		1 x y								1 x y				
	C2	A1	2 x y					A3	2 x y	A2	2 x y	A1	2 x y		
	C3	A1	3 x y					A3	3 x y	A2	3 x y	A1	3 x y		
20–69 y	C0														
	C1	A1	1 x y												
	C2	A1	2–3 x y							A2	2–3 x y	A1	2–3 x y		
	C3	A1	4 x y							A2	4 x y	A1	4 x y		
70+ y	C0														
	C1	A1	2 x y							A1	2 x y				
	C2	A1	3–4 x y	A4	3–4 x y			A3	3–4 x y	A2	3–4 x y	A1	3–4 x y		
	C3	A1	4–6 x y	A4	4–6 x y			A3	4–6 x y	A2	4–6 x y	A1	4–6 x y		
Xerostomia	C3	A1	6 x y	A4	6 x y			A3	6 x y	A2	6 x y	A1	6 x y		

*(PMTC + FPP) Professional mechanical toothcleaning plus fluoride prophylaxis paste; (AFS) Aqueous fluoride solution; (FG) Fluoride gel; (FCG) Fluoride-chlorhexidine gel; (FV) Fluoride varnish; (FCV) Fluoride-chlorhexidine varnish; (FS) Fluoride-containing sealant (glass-ionomer cement or resin-modified glass-ionomer cement); (Alt) Alternative; (A1) Alternative 1; (A2) Alternative 2; (A3) Alternative 3; (A4) Alternative 4; (Int) Interval; (C0) No caries risk; (C1) Low caries risk; (C2) Caries risk; (C3) High caries risk.

Conclusions

Negative preeruptive effects of fluoride: Fluorosis

Excessive fluoride intake during tooth development will result in fluorosis of varying severity, which is strictly dose related. Ingestion of as little as 0.02 mg of fluoride per 1 kg of body weight per day will result in mild fluorosis. The critical period for development of fluorosis is the enamel maturation phase. The period from 20 to 26 months of age, the maturation phase of the maxillary central incisors, is important for esthetics. The greatest prevalence of fluorosis is found in the premolars and second molars and always on homologous teeth.

Dean's Index (scores 0 to 7) and Thylstrup-Fejerskov (TF) Dental Fluorosis Index (scores 0 to 9) are the indices most frequently used for scoring the severity of fluorosis. The TF index correlates histopathologic characteristics with clinical severity.

For decades, it was claimed that water fluoride concentrations of 1 ppm or less posed no risk of fluorosis. However, no such threshold value exists because of the influence of other contributing factors, such as the age of the individual, total exposure time to fluorides, climate, and exposure to other fluoride sources such as ingested fluoride toothpaste and slow-release fluoride tablets.

Posteruptive effects of fluoride

The caries-preventive effects of fluoride are almost 100% posteruptive; eg, people born and reared in regions with water containing fluoride levels of 1 mg/L who move to an area of low water fluoride concentration after tooth eruption are as susceptible to caries as others in the low fluoride area (assuming they are exposed to a comparable cariogenic challenge). Topical use of fluorides achieves levels of caries reduction unrelated to how much fluoride is incorporated into the enamel preeruptively.

Enamel is a highly porous material consisting of apatite-like mineral crystals surrounded by water and organic components. These crystals contain carbonate and other impurities that make them more reactive to acid than pure HA or FA.

Dental caries is a dynamic process, characterized by periods of acid challenge (acid from bacterial metabolism in plaque) to the tooth mineral, alternating with increases in pH because of the neutralizing effect on the acid of saliva and other factors, such as toothcleaning procedures. The earliest clinically detectable enamel lesion (white-spot lesion) normally consists of an outer surface lesion (a more resistant micropore filter) and a lesion body that can be successfully remineralized or sealed by plaque control and topical use of fluoride.

Fluoride is present in the saliva, plaque, and intercrystalline fluid, incorporated in the mineral crystals, and more or less bound to its surface and on the soft tissues in the oral cavity. In addition, CaF_2 in the pellicle and on the enamel surface may act as a source of fluoride.

The total posteruptive cariostatic effect of fluoride is a combination of several single effects:

1. Reduction of mineral solubility through formation of FA on the crystal surfaces
2. Inhibition of mineral dissolution, primarily through its actions in the aqueous phase on the tooth surface and among the crystals in the tooth
3. Promotion of remineralization by fluoride available on the tooth surface as well as in the intercrystalline fluid
4. Reduction of acid formation by the acidogenic bacteria (primarily mutans streptococci) of the plaque
5. Reduction of plaque adhesiveness and formation by reduction of surface energy, wettability, and extracellular polyglucan and polyfructan formation
6. Influence on cellular metabolism and growth

The first three effects are by far the most important, followed by the fourth. The reduction of caries in humans is *not* dependent on *(1)* systemic administration of fluoride; *(2)* high levels of fluoride in sound enamel; or *(3)* the effect of fluoride on bacterial metabolism or bacterial colonization of enamel. The inhibition of caries is, however, dependent on the presence of fluoride at the plaque-enamel interface during active development of caries: Fluoride will directly alter the dynamics of mineral dissolution and reprecipitation. This does not imply that fluoride incorporated into enamel apatite does not reduce the rate of dissolution to some extent but merely that this factor alone cannot explain the caries reductions observed. These considerations should govern the use of fluoride in caries prevention and treatment:

1. Demineralization (loss of mineral resulting from partial dissolution of enamel crystals) during the acid attack stage of caries is markedly inhibited if fluoride is present in solution at the time of the acid challenge. Fluoride diffuses with the acid from plaque into the enamel pores and acts at the crystal surface to reduce mineral loss. Fluoride present on crystal surfaces, for example, as FA deposited during a previous cycle, will be highly resistant to subsequent acid attack.

2. Fluoride present in solution at the crystal surface during a rise in pH following demineralization can combine with dissolved calcium and phosphate ions to precipitate or grow FA-like crystalline material within the tooth. Fluoride enhances this mineral gain (remineralization) and provides a material that is more resistant to subsequent acid attack. In the case of high fluoride concentrations, CaF_2 can be precipitated: This material slowly redissolves, providing a source of F^- ions to inhibit demineralization or promote remineralization.

3. Fluoride regimens for the prevention of dental caries should be designed and used on the basis of the aforementioned principles. High-concentration topical preparations, used at intervals of months or years by the dentist, work not only initially on both demineralization and remineralization but also by providing fluoride as CaF_2 or adsorbed in early lesions and subsequently available over prolonged periods.

4. The most effective caries-preventive fluoride regimen, both from a theoretical perspective and as proved clinically, is the frequent (daily) low-concentration application of fluoride toothpaste and/or mouthrinses. Very low levels of fluoride (1.0 mg/L and less) in the oral fluids affect the rate of mineral dissolution and enhance redeposition of calcium and phosphate, forming acid-resistant FA and fluoridated HA. Frequent, long-term topical use of low-concentration fluoride will result in almost complete remineralization of subsurface enamel lesions. On the other hand, use of topical fluoride agents with high fluoride concentration, such as APF gels and varnishes (1% to 2% F), will result in a faster remineralization and blocking of the "micropores" of the surface lesion, while the lesion body is less remineralized. In addition, more phosphate/protein–coated CaF_2 particles will be formed on the enamel surface and in the pellicle. To date, there are no long-term data documenting the separate cariostatic effect of these two principles. However, frequent use (more than once a day) of low-concentration fluoride agents, supplemented by application of high-concentration topical fluoride agents ($\geq$ 1% F) at needs-related intervals, should be more effective than either of the two methods used separately.

5. Finally, it must be stressed that "clean teeth never decay," because plaque control is directed toward the etiology of dental caries: the cariogenic plaque. If the pH in the plaque fluid on the tooth surface drops frequently or nonstop to less than pH 5, then fluoride alone, regardless of concentration or frequency of application, cannot prevent formation of caries lesions.

Posteruptive effects of systemically administered fluoride

There are overwhelming data from in vitro, in situ, in vivo, and well-controlled clinical studies and field trials that the predominant cariostatic effects of systemically administered fluorides are posteruptive:

1. There is no relationship between in vitro acid solubility of enamel and clinical reductions in caries.
2. There is no relationship between the fluoride content of enamel and the caries experience of the individual in either the primary or the permanent dentition.
3. Topical fluoride agents achieve levels of caries reduction unrelated to how much fluoride is incorporated in the enamel.
4. Children born and reared in areas with drinking water fluoridated at 1 mg/L develop caries to about the same degree as do those in a low-fluoride area if they move into these low-fluoride areas after tooth eruption and are exposed to a comparable caries challenge.
5. The opposite is also true: Individuals born and reared in a low-fluoride area exhibit about the same low-caries incidence as individuals in a fluoridated area (1 mg/L) if they move to such an area just at the time of eruption of the first permanent teeth.
6. The prevalence and incidence of root caries are correlated with the fluoride concentration of the drinking water.
7. At relatively low concentrations, fluoride in plaque fluid and saliva can inhibit the progression of a caries lesion.

Fluoride can be systemically administered via drinking water, drops, tablets, salt, milk, or chewing gum. Of these vehicles, fluoridated drinking water is to date by far the most common method (about 300 million people worldwide), followed by fluoridated salt (40 million) and fluoride tablets (20 million). However, the current trend is toward increasing the use of fluoride lozenges and chewing gums by caries-susceptible adults with reduced salivary secretion rates and a reduction in use of fluoridated milk and salt.

Fluoridated drinking water is by far the most cost-effective public health measure for prevention and control of caries and reversal of subclinical lesions. Because most people have to drink water several times a day, even those without regular dental care and regular use of fluoride toothpaste benefit from water fluoridation. Therefore, water fluoridation should be recommended in all populations with relatively high caries prevalence and without access to organized preventive programs or daily use of fluoride toothpaste. The only limitations to its use are a reliable and controllable water supply, ie, centralized reticulation, which may not be feasible in some developing countries or in rural areas of industrialized countries.

To reduce the risk of dental fluorosis, the supply of fluoride from other sources must be known to adjust water fluoride levels appropriately. The recommended water fluoride concentration in temperate climates such as the United States is 0.7 to 1.2 mg/L; in warm to hot subtropical and tropical regions, only 0.5 to 0.7 mg/L is recommended to prevent the development of esthetically unacceptable fluorosis.

Results from early studies of fluoridated water showed caries reductions of about 50% in the permanent dentition and 40% in the primary dentition, compared to control areas. Significant reductions in root caries were also observed. At that time, the caries prevalence was high in the United States as well as in Europe, where the studies were run, and few topical fluoride agents such as toothpaste and mouthrinses were available. Today, the supplementary effect of fluoridated drinking water would be only 5% to 25% in most European countries and the United States because of improved oral hygiene and daily use of fluoride toothpaste and other topical fluoride agents, which have resulted in very significant reductions in both caries prevalence and incidence. However, about a 50% caries reduction can still be expected by water fluoridation in regions with relatively high caries prevalence, limited dental resources, and limited daily use of fluoride toothpaste.

Some studies of fluoride tablets have shown caries reductions ranging from 30% to 60%. However, the use of fluoride tablets should not be regarded as a public dental health measure, except for some at-risk age groups, such as 5.5 to 7.0 and 11.5 to 13.0 years, the ages at which the molars are erupting. To reduce the risk of fluorosis, fluoride tablets should not be used before the age of 5 years and absolutely not before the age of 3 years. Another problem with administration of fluoride tablets in public dental health programs for children is the limited compliance.

Because the aim of using fluoride tablets is to achieve a supplementary posteruptive cariostatic effect similar to other topical fluoride agents such as toothpaste, only slow-release lozenges, which prolong the fluoride clearance time in the oral fluids, should be recommended. An optimal effect should be achieved if the lozenges are used as a "dessert," ie, directly after meals, particularly in adults with reduced salivary secretion rates.

For very caries-susceptible patients (C2 or C3), fluoride chewing gum should be the preferred "systemic" agents, to be used for 15 to 20 minutes directly after every meal. It is recommended primarily for caries-susceptible adults with reduced salivary secretion rates and for caries-susceptible children and young adults, especially during the eruption of the molars.

Indications for fluoridated salt are limited, despite promising cariostatic results from controlled early studies in countries where the majority of the population did not use fluoride toothpaste daily (Colombia and Hungary). With the current understanding of the cariostatic effects of fluoride, fluoridated salt should be restricted to populations in disadvantaged regions in the tropics and subtropics, where caries prevalence is relatively high and water fluoridation is not feasible.

There is only limited documentation available on fluoridated milk, and indications for its use are restricted. If used at all, fluoridated milk should be restricted to daily use in elementary schools in children with erupting or newly erupted permanent teeth in regions where fluoridated toothpaste is not used daily and fluoride mouthrinse programs cannot be organized.

Topical fluorides for self-care

Although fluoride toothpaste is by far the most widespread fluoride agent, it is used by only about 10% of the world's population. The following six major factors were recently proposed by leading caries research scientists as probable causes for the decline in caries in 20- to 25-year-old individuals in most developed countries over the last 30 years: *(1)* use of fluoride toothpaste; *(2)* improved plaque control; *(3)* school-based preventive programs; *(4)* reduced frequency of sugar intake; *(5)* use of fissure sealants; and *(6)* reduced sugar intake (Bratthall, 1996). Fluoride toothpaste should be recommended as a public caries control program for everyone, regardless of age.

The most common fluoride compounds in toothpastes are NaF and SMFP, followed by SnF_2 and AmF.

The cariostatic effect of fluoride toothpastes is dose related. However, European Union regulations limit the concentration of fluoride in toothpastes to a maximum of 0.15%. Longitudinal double-blind studies of similarly formulated silica-based toothpastes have shown that, at the same fluoride concentration, NaF is about 6% more effective than SMFP. There are no available data from double-blind studies comparing NaF with SnF_2 and AmF. The average caries reduction achieved in various 2- to 3-year clinical studies is about 20% to 30%. However, the cumulative effect over a life span is estimated to be 50% or more.

The cariostatic effects of fluoride toothpastes are also related to accessibility and fluoride clearance in the oral fluids. Accessibility is improved by:

1. Frequent mechanical removal of dental plaque, particularly on the approximal surfaces of the posterior teeth.
2. Deliberate application of fluoride toothpaste to the posterior interdental spaces before approximal cleaning.
3. Thorough swishing with the remaining toothpaste slurry after cleaning, followed only by one brief rinse with water.

The following measures may prolong fluoride clearance time from the oral fluids:

1. Use of as high a fluoride concentration as possible.
2. Increased daily frequency of application of fluoride toothpaste.
3. Use of the toothpaste technique previously recommended.
4. Application of fluoride toothpaste to fill the posterior interdental spaces after cleaning at bedtime. This method is especially recommended to adult high-risk caries patients (C3).

In order to minimize ingestion and the risk of fluorosis, for children younger than 6 years, a pea-sized amount (5 mm) of toothpaste containing 0.10% to 0.15% fluoride is recommended. This is particularly important in areas with 1 mg of fluoride or less in the drinking water.

Toothpastes containing fluoride as well as chemical plaque control agents should be recommended in particular to caries-susceptible patients with a high plaque formation rate (PFRI score 4 or 5), periodontitis, or gingivitis. To date, the following formulations are the most successful:

1. NaF + triclosan + copolymer + SLS (silica based)
2. SMFP + triclosan + zinc citrate + SLS (silica based)
3. NaF + CHX + zinc lactate (silica based)

Toothpastes containing SnF_2 or AmF also have documented antiplaque effects. However, to date the cariostatic effect of these formulations has not been compared with NaF and SMFP in a double-blind clinical trial, as earlier discussed. Even more

efficient formulations are currently under development: The combination of mechanical and chemical plaque control plus fluoride in the same self-care procedure is too appealing and cost effective not to be optimally investigated.

Oral hygiene aids that not only remove plaque mechanically but also release fluoride to the most caries-susceptible tooth surfaces in the dentition, ie, the approximal surfaces of the posterior teeth, would be most appropriate. In recent years, several brands of fluoridated toothpicks, dental tape, and dental floss have been introduced commercially.

Although to date no longitudinal clinical trials have been conducted to evaluate the effect of these new products on approximal caries, some in situ studies show that, at least, use of fluoridated wooden toothpicks will elevate the salivary fluoride concentration to levels comparable with or even higher than those achieved by fluoride mouthrinses or lozenges; interproximally, the fluoride concentration should be considerably higher. However, both fluoride concentration and release differ significantly among the various brands.

To optimize the release of fluoride from wooden toothpicks, moistening in saliva for a few seconds just before application is recommended.

Because toothpicks offer optimal accessibility to the key-risk posterior approximal surfaces, wooden toothpicks impregnated with both NaF and CHX would be an appropriate aid for prevention and control of caries, gingivitis, and periodontitis.

Weekly school-based mouthrinsing with 10-mL neutral 0.2% NaF solutions for 1 minute should still be considered a very cost-effective program for caries control in areas with high caries prevalence and inadequate water fluoride concentration, in the absence of good oral hygiene habits and daily use of fluoride toothpaste. As mentioned earlier, about 90% of the world's population does not use fluoride toothpaste. On the other hand, in areas with very low caries prevalence, a uniformly high standard of oral hygiene, and daily use of fluoride toothpaste, the supplementary caries-inhibiting effect of weekly fluoride mouthrinses is negligible and not cost effective.

Rinsing with 10-mL fluoride solutions (0.025% F) for 1 minute after every toothcleaning procedure is an efficient supplement for caries control in caries-susceptible patients. Fluoride mouthrinses containing chemical plaque control agents (eg, triclosan and SLS, CHX, and AmF and SnF_2) should be more cariostatic than pure neutral NaF solutions. However, in high–caries-risk (C3) patients and particularly those with impaired salivary function (stimulated salivary flow of < 0.7 mL/min and xerostomia), the combined use of fluoride and CHX chewing gums as a dessert for 15 to 20 minutes after every meal is a much more efficient cariostatic alternative than fluoride mouthrinsing to supplement daily use of fluoride toothpaste.

As with other fluoride and chemical plaque control agents, the effect of fluoride gels is related to factors such as the concentration, time of application, and accessibility. Although not recommended as a public health measure for self-care, for selected adult caries-risk patients (C2 or C3), fluoride gels must be regarded as efficient cariostatic supplements to fluoride toothpaste for daily use.

Most commercially available fluoride gels for daily use by self-care contain about 0.5% fluoride in the form of neutral NaF, APF, SnF_2, or AmF plus NaF. The last two also have documented antiplaque effects. A gel containing both NaF plus CHX is available on prescription. To improve the effect of the gels, the recommended application time is 4 to 5 minutes.

For patients with xerostomia, artificial saliva containing NaF is available to reduce the physical and subjective symptoms of dry mouth and to reduce the risk of rampant caries in these extremely high-risk patients. In these patients, meticulous mechanical and chemical plaque control and combinations of the most efficient fluoride agents are essential. Fluoridated artificial saliva is formulated either as a gel or as a spray. Patient acceptance is generally higher for the spray, which is usually applied 20 to 30 times a day.

Topical fluorides for professional application

The following systems are used for professional fluoride application: aqueous solutions for painting; gels; prophylaxis pastes; and slow-release agents such as varnishes, slow-release devices, and glass-ionomer cements. Most systems are used to supplement self-care, based on the predicted risk and not as public health programs.

The fluoride concentration in agents for professional use, 1% to 2% fluoride, is usually much higher than in agents for self-care. The effect of the fluoride agent is strongly correlated to the fluoride concentration, exposure time, and accessibility.

The most common aqueous fluoride solutions for painting are neutral 2% NaF (1% F), 8% SnF_2 (2% F), and APF (1.23% F). Although this method is rarely used today, in patients susceptible to root caries, the combined fluoride and antimicrobial effect of 8% SnF_2 is still recommended as cost effective.

Fluoride gels for professional use are similar to those for self-care, namely, neutral NaF, APF, SnF_2, AmF plus NaF, and, on prescription, NaF plus CHX. To optimize the accessibility, plaque should be removed by PMTC. The gel should be syringed into the posterior interproximal areas and then applied three times for 4 to 5 minutes in a customized tray. The gels are recommended for selected caries-risk patients (C2 or C3), strictly at needs-related intervals. Gels containing SnF_2, AmF plus NaF, and particularly NaF plus CHX have combined fluoride and antiplaque effects.

Prophylaxis pastes are used mainly for PMTC but also for finishing and polishing. For polishing and PMTC on root surfaces and composite restorations, an abrasiveness of RDA 40 to 90 is appropriate; an RDA of 100 to 200 should be used for PMTC and an RDA of more than 200 for finishing. Although all prophylaxis pastes should contain fluoride, the fluoride effect of the pastes should not be overestimated, and, in caries-susceptible patients, fluoride varnishes are recommended to supplement PMTC.

Semi–slow-release and slow-release fluoride agents such as varnishes and glass-ionomer cements are rapidly gaining popularity because of the high cost effectiveness of the slow-release system.

Three different systems of fluoride varnish are commercially available: Duraphat (5% NaF, 2.3% F), Fluor Protector (silane fluoride, 0.1% F), and Bifluorid 12 (6% NaF and 6% CaF_2, about 6% F). Based on clinical studies, the caries reduction with the use of these agents ranges from about 20% to 70%. To date, the three different fluoride varnishes have not been compared in a well-controlled longitudinal clinical study. Experimental studies indicate that Bifluorid 12 should be more efficient than the others, at least in relieving hypersensitivity of dentin. The caries-reducing effect of fluoride varnishes is the end result of mechanical removal of plaque by PMTC, protection of the cleaned caries-susceptible tooth surfaces from direct reaccumulation of plaque, and slow release of fluoride as long as the varnish is retained.

The recommended procedure in caries-risk patients is initially to repeat varnish treatment three times within 7 to 10 days in order to heal gingivitis and thereby reduce plaque formation rate (PFRI) and to arrest enamel caries by sealing the outer micropore surface as soon as possible. Thereafter, the varnish should be reapplied, at needs-related intervals, two to four times per year.

Two specific types of intraoral slow-release devices have been developed: the copolymer membrane device and the fluoride glass device. These systems are currently under evaluation in the United States and the United Kingdom. The first human clinical study using a glass fluoride slow-release device (a 2-year, double-blind study) resulted in as much as 60% to 75% caries reduction.

The assortment of restorative materials, sealants, liners, and cements that contain fluoride and act as slow-release fluoride agents is continuously increasing. The significantly greatest fluoride release is from pure glass-ionomer cements, followed by resin-modified glass-ionomer cements, glass-ionomer cement–modified resin composites (compomers) and fluoridated resin composites and amalgams.

In addition to the slow release of fluoride, particularly from glass-ionomer cement materials, a further advantage is the possibility of replenishing the diminished fluoride reservoir from such sources as daily use of fluoride toothpastes, fluoride lozenges, and fluoride chewing gums. The most efficient method is probably the application of a varnish with a high fluoride concentration, eg, Bifluorid 12. Both experimental in situ and clinical studies have shown glass-ionomer cement restorative material to be very efficient in preventing recurrent caries in caries-susceptible patients. Low-viscosity glass-ionomer cement is also recommended as a slow-release fluoride sealant for erupting molars, superseding acid-etch–retained resin sealants.

CHAPTER 6

USE OF
FISSURE SEALANTS

The pattern of carious and restored tooth surfaces varies significantly in both the permanent and the primary dentitions. In a toothbrushing population, caries susceptibility in the permanent dentition may be ranked in the following order:

1. Fissures of the molars
2. Mesial and distal surfaces of the first molars
3. Mesial surfaces of the second molars and distal surfaces of the second premolars
4. Distal and mesial surfaces of the maxillary first premolars and mesial surfaces of the maxillary second premolars
5. Distal surfaces of the canines and mesial surfaces of the mandibular first premolars
6. Approximal surfaces of the maxillary incisors

The pattern of carious and restored surfaces is related to the buccolingual width of the crown and to sites where dental plaque stagnates, that is, to toothbrush inaccessibility. In nontoothbrushing populations, or in individuals with poor and irregular oral hygiene habits (precluding regular use of fluoride toothpaste) and a high intake of sticky sugary food, cervical lesions may also develop on the buccal surfaces of the maxillary teeth and on the mandibular molars and premolars.

Some decades ago, almost 100% of all the occlusal surfaces of the molars in young adults were restored or carious in most industrialized countries. The main reason for that was the combination of a very high general caries prevalence, in particular on these key-risk surfaces, and the old Black's rule of extension for prevention. As a result, many intact fissures were opened by prophylactic odontomy and restored with amalgam.

The World Health Organization's (WHO) continuously updated global databank on caries prevalence among 12-year-old children shows a very significant decline in most industrialized countries during the last three decades. However, data show a relative increase of caries prevalence on the occlusal surfaces of the permanent molars. Although these eight occlusal surfaces represent only about 6% of the 128 permanent tooth surfaces at risk (third molars excluded), it is estimated that they exhibit about 60% of the total number of decayed or filled surfaces (DFSs) (Bratthall et al, 1996).

Even in developing countries with relatively low caries prevalence, the occlusal surfaces of the permanent molars are carious more frequently than are the approximal surfaces. Fissure caries is partly attributed to the extremely plaque-retentive morphology of the fissure systems. To prevent the accumulation of cariogenic plaque in the depths of the fissures, and thereby prevent the development of caries, so-called fissure sealants were introduced to level out the original occlusal morphology.

Morphology of the Occlusal Surfaces of the Molars

Viewed in a stereomicroscope or scanning electron microscope, the occlusal surface of a permanent molar appears as a convoluted landscape, with high mountains separated by valleys, some of which are deep rifts while others resemble open river beds (Fig 320).

The fissure pattern of permanent molars varies among racial groups. Some teeth have deeply convoluted fissures prone to caries (Taylor, 1978); in other teeth, the fissures are shallow and almost imperceptible. The distribution of fissure patterns will vary among population groups, and even within racial groups, rendering teeth more or less susceptible to dental caries. For example, the permanent first molars of the Inuit vary widely in their fissure patterns and therefore also in their susceptibility to caries (Mayhall, 1977).

One congenital condition resulting in characteristically small teeth and few molar fissures is Down syndrome (Brown and Schodel, 1976). The pattern of cusp form and the fissure pattern are genetically determined. In an individual patient, tooth morphology may be an aid in deciding whether or not a tooth should receive a fissure sealant. Attempts to classify the shape of fissures so that they might be assessed as more or less caries prone were made many years ago (Bossert, 1937).

Furthermore, fissure pattern and its relation to structure within the depth of enamel is highly variable (Mortimer, 1964). Based on the cross-sectional shape of the fissures of first molars, it has been found that most teeth (almost 90%) have so-called normal fissures in cross section. They have a relatively wide opening, followed by a narrow cleft (approximately 1.0 mm deep and 0.1 mm wide) that reaches almost to the dentinoenamel junction (Fig 321). The caries lesion usually starts as an enamel lesion on both sides of the entrance of the fissure, which is visible and accessible to a probe. However, some atypical fissures (fewer than 10%) with a narrow opening and a bulbous widening at the base should be regarded as risk fissures (Fig 322) because a lesion can start at the base as well as at the entrance to the fissure. Fortunately, from a diagnostic point of view, there is a strong correlation between steep cuspal inclination and such sticky risk fissures.

Even extreme risk fissures with irregularities such as horizontal tunnels can be kept free of caries, as shown in Fig 323. These first molar fissures were successfully maintained free of caries, without fissure sealants, by diligent oral hygiene and daily use of fluoride toothpaste. All of her teeth were still caries free at the age of 42 years (Figs 324 and 325). These teeth should remain caries free because they have reached secondary as well as tertiary maturation and because of the high standard of oral hygiene. Figures 326 and 327 exhibit the caries-free teeth of a 40-year-old man with normal molar fissures.

Sticky risk fissures and shallow nonrisk fissures in maxillary and mandibular first molars are exemplified in Figs 328 to 331. Figure 332 shows a caries-free mandibular dentition with excellent shape, size, and position of the teeth.

The occlusal surface morphology of the permanent first molars has been illustrated in drawings by Carlsen (1987). Figures 333 and 334 show the morphology of the occlusal surface of maxillary and mandibular first molars. Both exhibit mesial, central, and distal fossae (1, 2, 3) but different numbers of interlobal grooves: seven in the maxillary molars and five in the mandibular molars. The mandibular permanent second molars also exhibit mesial, central, and distal fossae. However, in the maxillary second molars, the mesial fossa is very shallow and sometimes nonexistent—just a flat surface.

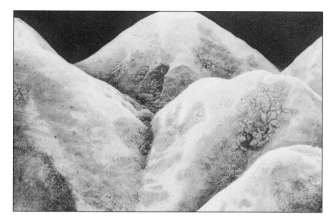

Fig 320 Occlusal morphology of a permanent molar.

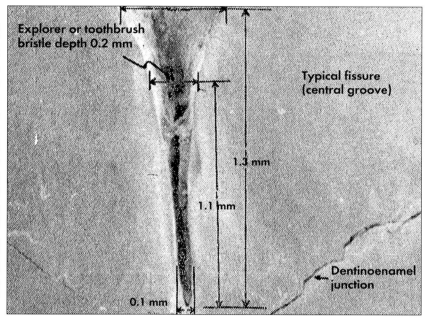

Explorer or toothbrush
bristle depth 0.2 mm

Typical fissure
(central groove)

1.3 mm

1.1 mm

Dentinoenamel
junction

0.1 mm

Fig 321 Cross section of a normal molar fissure.

Fig 322 Atypical fissure with a narrow opening and a bulbous widening at the base. This type of fissure should be considered a risk fissure.

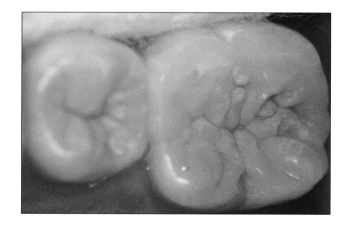

Fig 323 Fissures with horizontal tunnels in a 10-year-old girl. These irregularities put the first molar at extreme risk of developing caries.

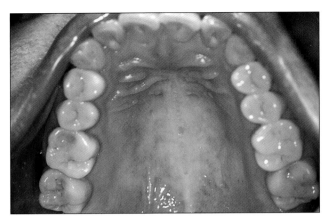
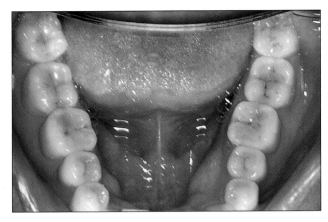

Figs 324 and 325 Maxillary and mandibular teeth of the individual in Fig 323, at the age of 42 years. The dentition is still caries and gingivitis free.

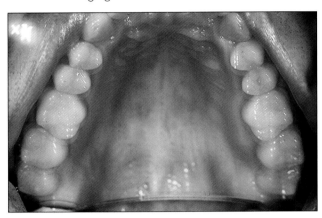
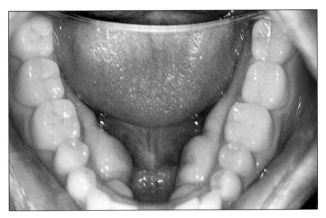

Figs 326 and 327 A 40-year-old man also exhibiting 32 caries- and gingivitis-free teeth.

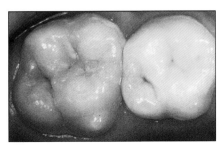
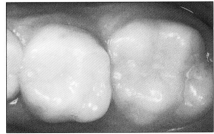
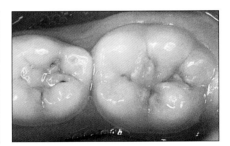

Fig 328 Sticky, at-risk fisures in a maxillary first molar. (From Mejáre and Malmgren, 1994. Reprinted with permission.)

Fig 329 Maxillary first molar with shallow, nonrisk fissures. (From Mejáre and Malmgren, 1994. Reprinted with permission.)

Fig 330 Mandibular first molar exhibiting sticky, at-risk fissures. (From Mejáre and Malmgren, 1994. Reprinted with permission.)

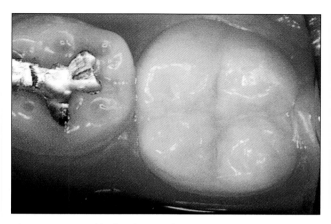

Fig 331 Mandibular first molar with shallow, nonrisk fissures.

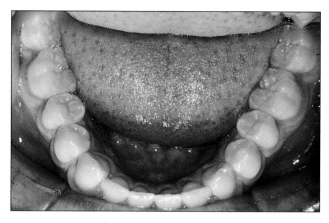

Fig 332 Caries-free mandibular dentition with excellent shape, size, and position of the teeth.

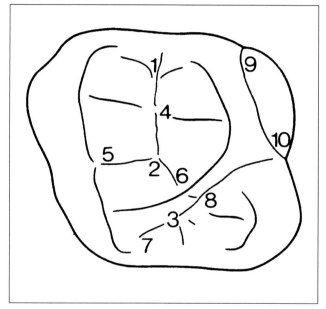

Fig 333 Morphology of the permanent maxillary first molar. The numbers correspond to fossae (1 to 3) and interlobal grooves (4 to 10): (1) Mesial fossa; (2) central fossa; (3) distal fossa; (4) mesial interlobal groove; (5) facial interlobal groove; (6) central interlobal groove; (7) distal interlobal groove; (8) lingual interlobal groove; (9) Carabelli groove; (10) foramen caecum. (Modified from Carlsen, 1987, by Ekstrand et al, 1993. Reprinted with permission.)

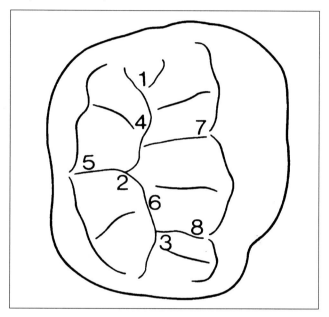

Fig 334 Morphology of the permanent mandibular first molar: (1) Mesial fossa; (2) central fossa; (3) distal fossa; (4) mesial interlobal groove; (5) lingual interlobal groove; (6) distal interlobal groove; (7) mesiofacial interlobal groove; (8) distofacial interlobal groove. (Modified from Carlsen, 1987, by Ekstrand et al, 1993. Reprinted with permission.)

DEVELOPMENT OF OCCLUSAL CARIES

The initiation and development of occlusal caries are strongly correlated to the morphology, the eruption stage, and the functional wear of the occlusal surface. Caries can be successfully prevented or at least significantly modified by plaque control, topical use of fluorides, and application of fissure sealants, in particular during eruption.

It is a common clinical observation that caries on occlusal surfaces does not involve the entire fissure system with the same intensity, but rather is a localized occurrence. Each tooth type in the dentition has its own specific occlusal surface anatomy, and caries is usually detected in relation to the same specific anatomic configuration in identical tooth types. In maxillary molars, for example, the central and distal fossae are sites that typically accumulate plaque and hence are also the sites at which caries most often occurs. In general, occlusal caries is initiated at sites where bacterial accumulations are well protected against functional wear.

Role of eruption stage

In studies by Carvalho et al (1989, 1991, 1992), discussed in chapters 1 and 2, the accumulation of plaque and development of occlusal caries were evaluated in relation to the morphology and eruption stage of the occlusal surface of the permanent first molars in 56 children aged 6 to 8 years. Table 18 shows classification of the children based on the eruption stage of their right first molars at the baseline examination and after 1 year.

Based on the data from these studies, it may be estimated that about five times more plaque will reaccumulate on the occlusal surfaces of erupting molars compared to fully erupted molars with occlusal function 48 hours after professional mechanical toothcleaning (PMTC). Most plaque reaccumulates in the distal and central fossae of

Table 18 Classification of children according to stage of eruption of permanent right first molars at baseline examination and after 1 year in studies by Carvalho et al*

Stage of eruption	Baseline	After 1 y
Only one tooth partially erupted	18	0
Both teeth partially erupted but more than half of facial surface covered with gingival tissue	6	0
Both teeth partially erupted and less than half of facial surface covered with gingival tissue	13	3
Full occlusion	20	53
Total	**57**	**56**

*From Carvalho et al, 1991. Reprinted with permission.

the erupting molars (see Fig 54 in chapter 1). As a consequence, most occlusal caries lesions are initiated in the distal and central fossae during eruption (see Fig 117 in chapter 2).

Figure 335 shows the accumulated plaque on the occlusal surface of a maxillary right first molar during eruption. Figure 336 shows the accumulated plaque on the same tooth after it had erupted and been subjected to functional wear (see also Figs 109 and 110 in chapter 2). The plaque control and topical use of fluorides were intensified during eruption, as discussed in chapter 2.

Ekstrand et al (1993) also studied the pattern of dental plaque and caries on permanent first molar occlusal surfaces in relation to sagittal occlusion in 72 children aged 7 to 10 years before loss of primary second molars. The pattern of de novo 48-hour plaque reaccumulation and caries was recorded in the 10 maxillary and 8 mandibular sites of the occlusal surfaces as shown in Figs 333 and 334 (according to Carlsen, 1987). Similar patterns of reaccumulated plaque (Fig 337) and dental caries (Fig 338) were found on unrestored occlusal surfaces. In similarity with the studies by

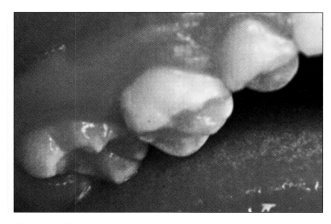

Fig 335 Plaque accumulation on the occlusal surface of a partially erupted maxillary first molar without friction from chewing function. (Courtesy of A. Thylstrup.)

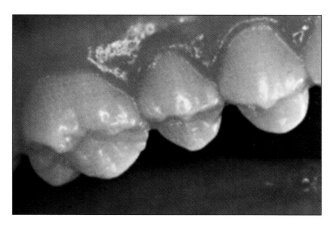

Fig 336 Plaque accumulation on the occlusal surface of the same molar, now completely erupted. There has been gradual physiologic detachment of the gingiva from the surfaces of the tooth, increasing the exposure of the clinical crown. These changes favor mechanical removal or suppression of cariogenic plaque. (Courtesy of A. Thylstrup.)

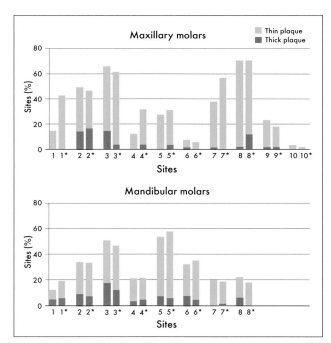

Fig 337 Percentage distribution of plaque scores in relation to macromorphology sites. Numbers correspond to those in Figs 333 and 334. *Plaque observation at 48 hours. (Modified from Ekstrand et al, 1993 with permission.)

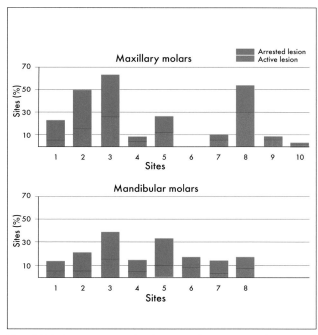

Fig 338 Percentage distribution of active and arrested caries lesions in relation to macromorphology sites. Numbers correspond to those in Figs 333 and 334. (Modifed from Ekstrand et al, 1993 with permission.)

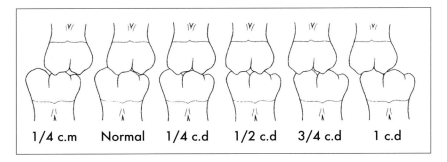

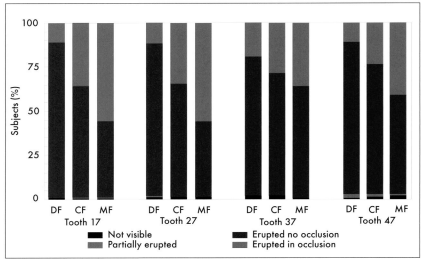

Fig 339 Classification system of sagittal occlusion types. (c) Cusp; (m) Mesial; (d) Distal. (From Ekstrand et al, 1993. Reprinted with permission.)

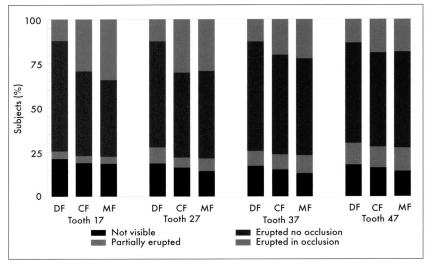

Fig 340 Stage of eruption of second molars in 13-year-old boys (Fédération Dentaire Internationale [FDI] tooth-numbering system). (DF) Distal fossae; (CF) Central fossae; (MF) Mesial fossae. (From Axelsson et al, 1997c. Reprinted with permission.)

Fig 341 Stage of eruption of second molars in 16-year-old boys (FDI tooth-numbering system). (DF) Distal fossae; (CF) Central fossae; (MF) Mesial fossae. (From Axelsson et al, 1997c. Reprinted with permission.)

Carvalho et al (1989, 1991, 1992), most plaque and thus caries were located in the distal (3) and central (2) fossae, the maxillary distolingual interlobal groove (8), and the mandibular lingual interlobal groove (5). In addition, the study showed that the amount of plaque and localization of active caries were related to functional wear, which was related to sagittal molar occlusion type (Fig 339).

The tremendous difference in plaque accumulation on the occlusal surfaces of erupting mo-

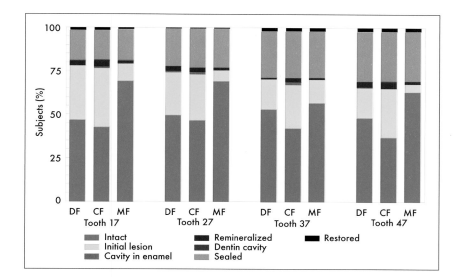

Fig 342 Status at baseline in second molars in 13-year-old boys in the mesial (MF), central (CF), and distal fossae (DF) (FDI tooth-numbering system). (From Axelsson et al, 1997c. Reprinted with permission.)

lars compared with the accumulation on erupted molars with full functional wear explains why almost all occlusal caries in molars is initiated during the extremely long eruption period (12 to 18 months) and why occlusal caries is not common in premolars, which have an eruption time of only 1 to 2 months. This was confirmed in a 5-year longitudinal study by Månsson (1977), who examined first molars every 3 months from the start of eruption. Development of caries occurred, on average, within 11 months of the start of eruption, ie, during eruption (most had decayed within 3 to 9 months). On the other hand, there was virtually no further initiation of occlusal caries beginning 15 months after the start of eruption.

Monitoring and prevention of the development of occlusal caries should therefore be intensified during the eruption period (high-risk period). If the teeth have erupted into natural chewing function without developing occlusal caries, the risk is over; examinations can be more cursory and less frequent. Nor is there any indication for application of fissure sealants.

In addition, the tooth enamel is much more susceptible to dental caries during eruption and until it has been exposed to the oral environment for some years and achieved the so-called secondary maturation (Kotsanos and Darling, 1991). Be-

cause the eruption period is regarded as the key-risk period for development of occlusal caries, it must be noted that, on average, the molars erupt about 6 months earlier in girls than in boys. However, among both genders, there may be individual variations of up to 2 years (Teivens et al, 1996).

A 3-year longitudinal fissure sealant study evaluated, among other things, the eruption stages of the mesial, central, and distal fossae of the permanent second molars in boys from the ages of 13 to 16 years. Figure 340 shows the eruption stage of these three fossae in the maxillary and mandibular second molars in 13-year-old boys. Less than 20% were not visible. The majority of the fossae were erupted but still not in occlusal function, in particular the distal fossae of the maxillary molars. At the age of 16 years, almost 100% of the fossae were erupted, but only about 10% of the maxillary distal fossae were in occlusal function (Fig 341). Figure 342 shows the pattern of intact fossae, initial noncavitated enamel lesions, cavitated enamel lesions, arrested enamel lesions, dentin caries, fissure sealants, and restorations in the fossae of the same second molars at the age of 13 years.

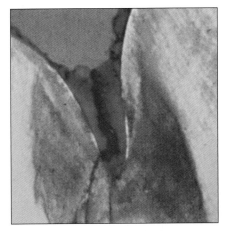

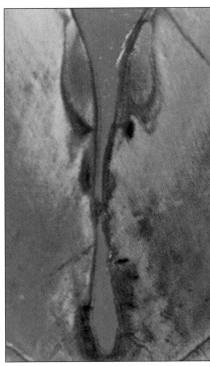

Fig 343 Polarized light micrograph showing, in cross section, an unusually wide, shallow fissure, full of stagnant, cariogenic plaque. An associated noncavitated enamel lesion is located around the entire fissure.

Fig 344 Risk fissure with stagnant, cariogenic plaque located at both the entrance and the base of the fissure. Localized, noncavitated enamel lesions have developed on both sides of the entrance and around the base of the fissure.

Role of tooth-specific anatomy

The caries lesion usually starts in the enamel on either side of the entrance to the fissures and is visible as a noncavitated white-spot enamel lesion. Gentle probing with a sharp explorer will damage the surface zone of such a lesion and initiate cavitation to the lesion body. A rule of thumb is to use sharp eyes and a blunt probe (or no probe at all) and to arrest the lesion by plaque control and fluoride.

Most clinical and scientific concern with respect to occlusal caries has been over the possible events in deep and inaccessible fissures. However, caries always starts in the surface enamel as a result of the metabolic activity of bacterial accumulations on the surface. It is reasonable to assume that evolution of plaque with cariogenic potential requires space that, in this context, is available only above the entrance to the narrow fissures, ie, in the grooves. This assumption is suported by ultrastructural studies indicating that, in contrast to the vital bacteria found at the entrance, nonvital bacteria or different stages of calculus formation are ususally harbored by the depths of the fissures (Ekstrand, 1988; Ekstrand and Björndal, 1997; Theilade et al, 1976).

Figure 343 shows an unusually wide, shallow fissure, full of stagnant, cariogenic plaque and an associated noncavitated enamel lesion around the entire fissure. Figure 344 shows a so-called risk fissure with stagnant, cariogenic plaque in the entrance as well as at the base of the fissure. In this case, localized, noncavitated enamel lesions have developed on both sides of the entrance and around the bulbous base of the fissure.

Using erupting human third molars, Ekstrand and Björndal (1997) investigated:

1. The relationship between the morphology of the interlobal groove and the histologic features of caries.
2. The ultrastructural features of the interlobal groove and content features of caries.
3. Whether the morphology of interlobal grooves influences the viability of the microorganisms.

Figs 345 and 346 Microradiographs of fissurelike (Fig 345) and groovelike (Fig 346) interlobal grooves with noncavitated enamel caries (score 1). The horizontal line indicates the point of entrance (E) to the interlobal groove. The vertical interrupted line indicates the distance from the entrance to the base (B) of the interlobal groove, making it possible to divide the fissure into three (arrows) and the groove into two (arrow) equal parts. The depth of lesion penetration in relation to the different parts is measured along the direction of the prisms, the principal directions of which are indicated on the microradiographs. (From Ekstrand et al, 1997. Reprinted with permission.)

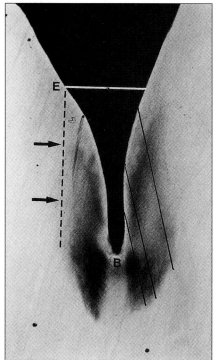
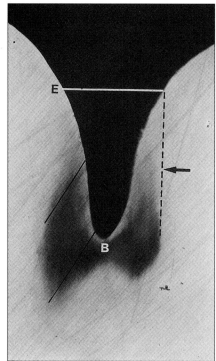

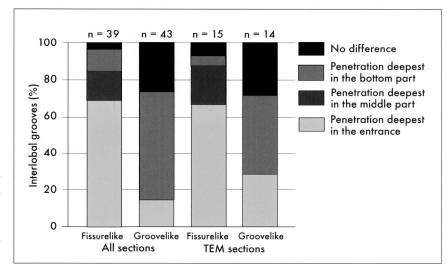

Fig 347 Percentage distribution of lesion penetration depth in relation to localization in fissurelike and groovelike interlobal grooves. (TEM) Transmission electron microscopy. (From Ekstrand et al, 1997. Reprinted with permission.)

The teeth were prefixed and postfixed, and buccolingual sections were prepared. The mesial or distal interlobal groove was classified as fissurelike (Fig 345) or groovelike (Fig 346). All sections were embedded and ground, and microradiographs were made. The interlobal groove contents were analyzed on two sections from each tooth, one with a fissurelike and one with a groovelike morphology; more than 80% of the teeth had evidence of caries. The severity of caries was associated with the length of posteruption time. Fissures were not more prone to caries than grooves. In about 70% of the fissures, caries had penetrated deepest at the entrance rather than in deeper parts (Fig 347).

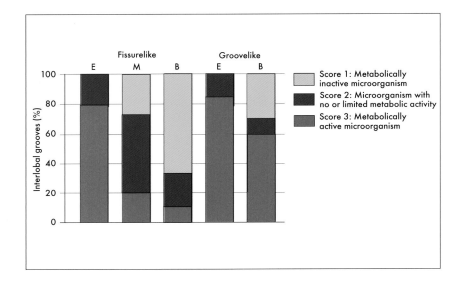

Fig 348 Percentage distribution of scores characterizing the viability of the microorganisms in relation to localization in fissurelike and groovelike interlobal grooves. (E) Entrance; (M) Middle; (B) Bottom. (Modified from Ekstrand et al, 1997 with permission.)

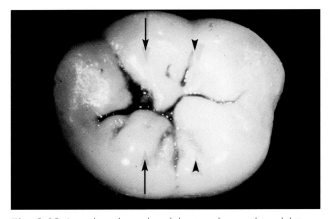

Fig 349 Localized occlusal lesion (arrowheads) in a mandibular molar with a discolored, cavitated lesion (arrows) in the distal fossa. (From Thylstrup et al, 1989. Reprinted with permission.)

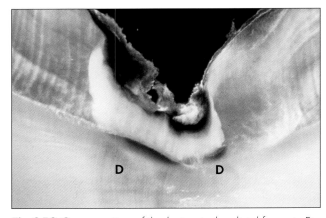

Fig 350 Cross section of the lesion in the distal fossa in Fig 349. There is superficial enamel breakdown with cavitation into about 50% of the enamel but no cavitation into the dentin. Thus the cavity could have been treated with resin-modified glass-ionomer cement without drilling into the dentin, despite obvious demineralization of dentin (D) corresponding to the direction of the enamel rods of the lesion. (From Thylstrup et al, 1989. Reprinted with permission.)

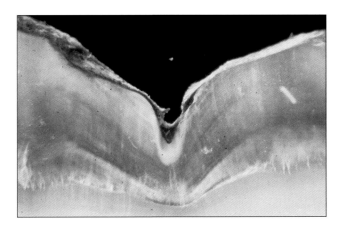

Fig 351 Cross section of the anterior fissure lesion indicated in Fig 349. There is no progressive demineralization. (From Thylstrup et al, 1989. Reprinted with permission.)

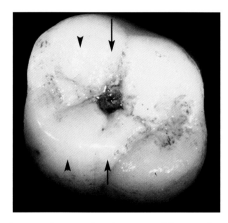

Fig 352 Localized cavity (arrows) in the central fossa of a maxillary first molar. (arrowheads) Fissures. (From Thylstrup et al, 1989. Reprinted with permission.)

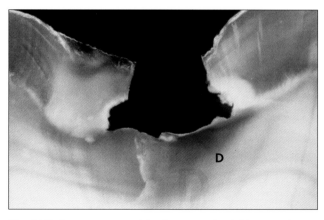

Fig 353 Cross section of the lesion in Fig 352, revealing a truncated cavity. The superficial zone of destruction and the zone of dentinal demineralization (D) are confined to the involved enamel. (From Thylstrup et al, 1989. Reprinted with permission.)

Viable microorganisms were found at the entrance as well as at the bottom part of the grooves. In fissures, viable microorganisms were primarily observed at the entrance, while in deeper parts the microorganisms were less viable or dead (Fig 348). A moderate association was observed between the viability of the microorganisms and differences in the depths of caries penetration in the fissures. Thus, the internal morphology of the interlobal grooves influenced the conditions for bacterial growth, and this determined the location of caries progression within the groove-fossa system.

Progressive destruction of the occlusal surface begins as a local process in the groove-fossa system, as a result of accumulation of bacterial plaque. In this area, which is already sheltered from physical wear, the formation of a microcavity further improves the potential for bacterial attachment and colonization. This accelerates demineralization and destruction, enhancing local conditions for bacterial growth.

Figures 349 to 351 show different stages of localized progressive occlusal lesions in a mandibular molar with a discolored, cavitated lesion in the distal fossa. The cross section of the lesion in the fossa shows superficial enamel breakdown with cavitation into about 50% of the enamel but no cavitation into the dentin; ie, there is no bacterial invasion of the dentinal tubules, and the lesion could be arrested (Fig 350). However, the enamel lesion (demineralized area) is approaching the dentinoenamel junction, and there is demineralized dentin in the contact area, corresponding to the direction of the rods. The fissure in Fig 349 is shown in cross section in Fig 351. There is no progressive demineralization.

Figure 352 shows a localized cavity in the central fossa of a maxillary first molar. The cross section of the lesion shows that the cavity is truncated and that the superficial zone of destruction and the zone of dentinal demineralization are confined to the involved enamel (Fig 353). The cross section of the fissures shows that these are filled with calcified materials, indicating total absence of cariogenic plaque (Fig 354). It is therefore assumed that the dark brown cavity is arrested or stagnant.

In people living in communities without dental health care, the natural progression of occlusal caries is rapid because of the particular anatomic configuration of the occlusal surface where caries is initiated. Occlusal caries usually begins in a fossa, ie, a depression where two or more interlobal grooves meet. Several surfaces are involved in the initial dissolution, and the process

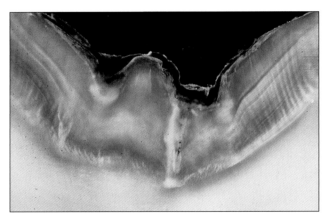

Fig 354 Cross section of the fissures indicated in Fig 352. They are filled with calcified material, indicating the total absence of cariogenic plaque. Thus, the superficial dark brown noncavitated lesions of the fissures should be regarded as inactive and arrested. (From Thylstrup et al, 1989. Reprinted with permission.)

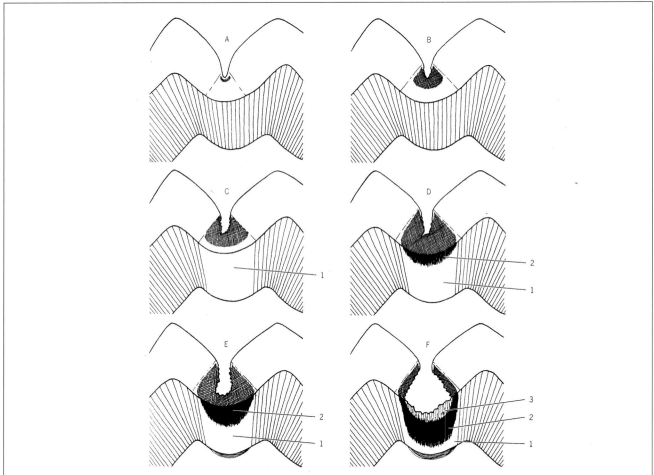

Fig 355 Progressive stages of lesion formation in an occlusal fossa: (A) Noncavitated enamel lesion; (B) Superficial cavitated enamel lesion; (C) Superficial cavitated enamel lesion with some reaction of dentin; (D) Superficial cavitated enamel lesion with noncavitated dentin lesion; (E) Superficial cavitated enamel lesion with noncavitated dentin lesion and some pulpal reaction; (F) Cavitated dentin lesion and more chemical reaction of the pulp. (1) Translucent zone into the pulp; (2) Zone of demineralized dentin; (3) Bacterial invasion and destruction of dentin. (Modified from Ekstrand et al, 1995 with permission.)

is therefore three dimensional. Because enamel demineralization always follows the rods, the enamel lesion initiated in a fossa gradually assumes the three-dimensional shape of a cone, with its base toward the dentinoenamel junction (although a section through such a lesion has the two-dimensional appearance of two separate, independent lesions). The response by the dentin corresponds to the direction of the involved enamel rods. Although textbooks have traditionally emphasized the undermining character of occlusal or so-called hidden caries, the pattern of lesion growth in these areas is not particularly surprising in the context of the structural arrangement of rods in the occlusal groove-fossa system.

As enamel destruction proceeds, a true cavity forms, the outline reflecting the arrangement of rods in the areas. The cavity has the shape of a truncated cone. The particular anatomic configuration of the occlusal surface at the site of caries initiation explains why the opening of occlusal cavities is always smaller than the base. The closed nature of the process obviously favors undisturbed growth of bacteria and hence accelerated destruction of the tissue. Occlusal enamel breakdown is the result of further demineralization from an initially established focus, rather than general demineralization involving the entire fissure system.

Figure 355 illustrates the progressive stages of lesion formation in an occlusal fossa, from the earlier noncavitated enamel lesion to cavitation into the dentin with a zone of bacterial invasion and dentinal destruction, where excavation and restoration are indicated. However, at the second-to-last stage (E), no such invasive intervention is indicated, despite a considerable zone of demineralized dentin and a sclerotic and translucent zone into the pulp. The method of choice would be placement of a fissure sealant or a minimally invasive sealant restoration with a resin-based glass-ionomer material or "compomer."

In conclusion, two factors have been considered of importance for plaque accumulation and caries initiation on occlusal surfaces: *(1)* stage of eruption or functional usage of teeth, and *(2)* tooth-specific anatomy.

DIAGNOSIS OF OCCLUSAL CARIES

If intensified plaque control (self-care and need-related intervals of PMTC), topical use of fluorides, and needs-related use of fluoride-releasing glass-ionomer sealants keep the occlusal surfaces of the molars caries free during eruption (12 to 18 months), until they have reached full occlusion and normal functional wear, future occlusal caries diagnostic problems are solved. That is because the risk for development of cavitated caries lesions is over. However, in new patients with fully erupted molars—for example, first molars in 8- to 11-year-old children, second molars in 14- to 16-year-old children, and third molars in young adults and adults—the occlusal surfaces may already be carious into the dentin with or without cavitation and thus offer delicate differential diagnostic problems.

It might be expected that occlusal caries lesions would be fairly easy to diagnose because, unlike approximal and subgingival root surfaces, these surfaces are readily accessible for visual inspection. However, clinically (visual or visual-tactile by probing) or radiographically, diagnosis of occlusal lesions is a delicate problem because of the complicated three-dimensional shape of the occlusal surfaces, which incorporate fossae and grooves with a great range of individual variations.

Diagnostic methods

In typical fissures, and particularly in atypical sticky fissures (see Figs 322 and 344), most of the early stages of the lesion are hidden from the naked eye, although in a clean, dry typical fissure, it might be possible to observe active noncavitated white-spot lesions on the walls of the entrance. Soon after eruption, most of these lesions are arrested (Figs 356 and 357) and take up a brown stain from items in the diet. This diagnostic problem was recognized many years ago by G. V. Black (1908), who wrote, "Very many pits and fissures show evidence of some slight soft-

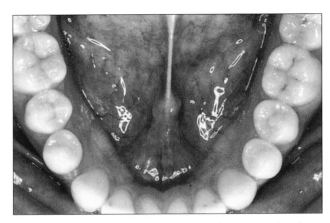

Fig 356 Arrested and stained enamel caries in the deep fissures of the molars in an 18-year-old individual. (From Mejáre and Malmgren, 1994. Reprinted with permission.)

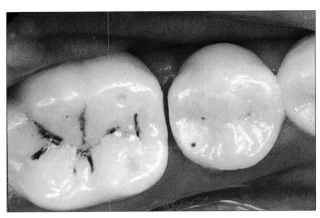

Fig 357 Arrested and stained fissure caries in a mandibular first molar in a 14-year-old individual. (From Mejáre and Malmgren, 1994. Reprinted with permission.)

ening in early youth, which is stopped by the coming of immunity or some change of local conditions. These become dark in color and so remain without further change. These should not be interfered with, as they are just as safe without any filling whatever."

To this day, dentists continue to have difficulty in differentiating active from arrested lesions and usually base their decision on clinical judgment, which should include an assessment of a patient's past caries experience, oral hygiene and dietary habits, salivary function, and likelihood of compliance with a preventive regimen.

None of the aforementioned diagnostic methods has gained universal acceptance, particularly for detection of early occlusal caries. Because occlusal caries today constitutes a major portion of all new lesions in children, its diagnosis assumes considerable importance. Several studies have reported that clinically undetected occlusal lesions extending well into the dentin are a serious problem in many communities. In one comparative study of molar lesions that had been clinically undetected (by mirror and probe), 10% more lesions were found with the aid of bitewing radiographs in 1974, whereas 32% more were found in 1982 (Sawle and Andlaw, 1988).

Radiographs, although better than visual inspection alone, are inaccurate in estimating the extent of the caries lesion or in detecting enamel occlusal lesions (D_1). (For classification of enamel and dentin caries lesions see chapter 5 in volume 2.) Radiographically, occlusal lesions appear as large, subtle, diffuse, dark radiolucent areas in dentin, centrally located under the fissure. Figure 358 shows a limited lesion in dentin (D_3) in the mandibular left second molar, and Fig 359 shows a very advanced lesion in dentin, probably involving the pulp (D_4), in the mandibular right first molar. The sensitivity of both visual and clinical-tactile (probing) diagnosis for noncavitated occlusal enamel and dentin lesions is low, but the sensitivity of combined meticulous clinical and visual examination is somewhat better. However, all the clinical diagnostic methods have high specificity.

Noncavitated occlusal caries lesions are by far the most prevalent lesions in children and young adults; traditionally, these have been restored three times more frequently than have similar lesions on buccal and lingual surfaces. Ideally, noncavitated lesions should be arrested instead of restored. Of great importance is differential diagnosis between active and inactive lesions and between cavitated and noncavitated lesions.

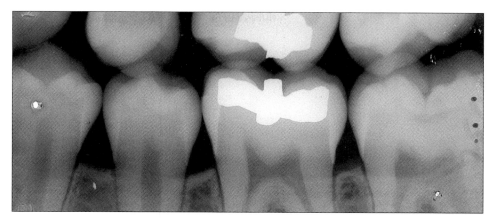

Fig 358 Limited lesion in dentin (D_3) in the mandibular left second molar.

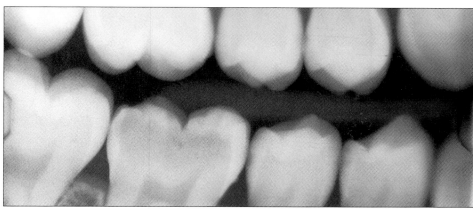

Fig 359 Advanced lesion in the dentin, probably involving the pulp (D_4), in the mandibular right first molar.

Probing

During the past 10 years, the role of explorers in caries detection has become a controverisal issue. Historically, an explorer was essential; if the tip caught in a pit or fissure or a cavity, a restoration was indicated. There is no place for such a procedure in modern caries management: Today a noncavitated lesion is managed by remineralization or by minimally invasive techniques, such as sealants or microrestoratons.

There is also consistent evidence that explorers do not improve the accuracy of caries diagnosis. Applied with slight force, an explorer could damage a tooth surface, converting a white-spot lesion into a cavity. Noncavitated lesions in narrow pits are particularly vulnerable (Figs 360 and 361). Explorers should not be used to probe any pit or fissure or any other tooth surface: Their use in stained or noncavitated pits and fissures is unethical.

When required, a blunt periodontal probe may be used to remove plaque and debris from the tooth surface prior to examination and to check the surface texture of a lesion atraumatically.

Lussi (1991) demonstrated the limitations of the explorer as an aid to accurate diagnosis of occlusal lesions and treatment decisions in an in vitro study. Thirty-four dentists were instructed to provide diagnoses for 61 teeth and recommend treatment. The teeth were then histologically prepared and assessed (the gold standard). The agreement between histologic and clinical diagnoses was determined.

The results showed no difference in diagnostic accuracy between explorer and visual inspection only. Sensitivity (62%) and specificity (84%) showed that the dentists were more likely not to treat carious teeth than they were to restore sound teeth. Approximately 42% of teeth were diagnosed correctly, but the percentage of clini-

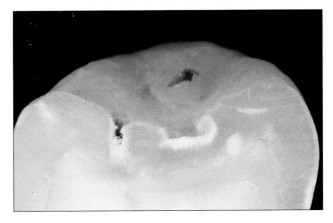

Fig 360 Noncavitated white-spot enamel fissure caries in a deep sticky fissure. (Courtesy of I. Espelid.)

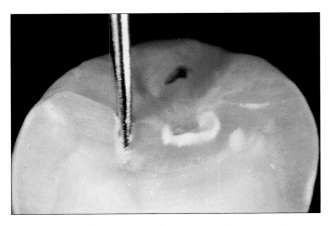

Fig 361 Possible creation of a cavity with pressure from an explorer. (Courtesy of I. Espelid.)

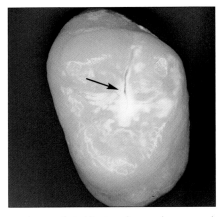

Fig 362 Noncavitated superficial lesion (arrow) surrounding the entrance of the fissure. It is unlikely that the lesion would have been detectable on a conventional bitewing radiograph. (From Lussi, 1991. Reprinted with permission.)

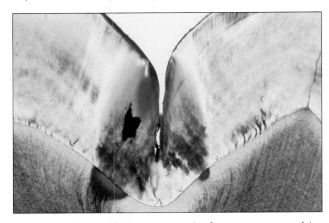

Fig 363 Polarized light micrograph of a cross section of the fissure in Fig 362 shows a noncavitated enamel lesion as well as a dentin lesion. (From Lussi, 1991. Reprinted with permission.)

cally correct treatment decisions was 73%. It was concluded that, compared to visual inspection alone, the use of an explorer does not improve the validity of the diagnosis of fissure caries (Lussi, 1991).

Figures 362 to 370 show the clinical appearance of the occlusal lesions and the histologic sections of the lesions. Because most of the extracted teeth were third molars that were only partially erupted or without chewing function, they exhibited active noncavitated enamel lesions in the fossae surrounding the fissures.

Radiographs

Although conventional bitewing radiographs will significantly improve the potential for detecting noncavitated occlusal lesions in dentin (D_3) (see Figs 358 and 359), there is some risk of a false-positive diagnosis. However, the clinical visual diagnosis may be improved by supplementary radiographs (Figs 371 to 374).

In a Norwegian study (Espelid et al, 1994), 640 dentists were asked to diagnose the occlusal surface of an intact molar based on a color photograph of the occlusal surface (Fig 375) and a radiograph (Fig 376). Of the participants, 15%

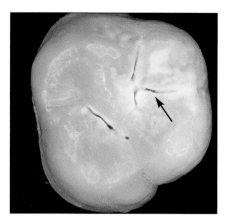

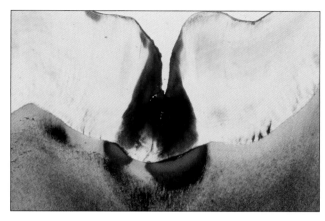

Fig 364 Noncavitated occlusal lesion *(arrow)* in the distal fossa of an extracted third molar. This lesion should have been detectable on a conventional bitewing radiograph. (From Lussi, 1991. Reprinted with permission.)

Fig 365 Polarized light micrograph of a cross section of the lesion in Fig 364. Note the superficial demineralization (red) around the entrance of the fissure, down to the dentino-enamel junction, and extending into the outer third of dentin. Even at this stage of lesion development, invasive therapy is contraindicated. Professional mechanical toothcleaning and fissure sealant with slow release of fluoride should be appropriate. At most, a pointed diamond tip might be used to open the entrance of the fissure, to confirm the diagnosis. In the absence of cavitation at the base of the fissure, a minimally invasive adhesive restoration would be placed. (From Lussi, 1991. Reprinted with permission.)

Fig 366 *(left)* Close-up of the dentinal involvement of the lesion in Fig 364. Note the dentinal tubules in the inner part of the lesion. (From Lussi, 1991. Reprinted with permission.)

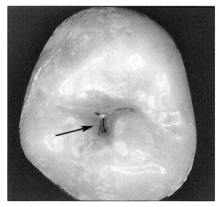

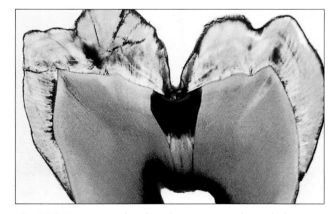

Fig 367 Noncavitated lesion *(arrow)* in the occlusal surface.

Fig 368 Corresponding histologic section through the area marked by the arrow in Fig 367, revealing the extent of the caries (original magnification ×3.2). (Courtesy of A. Lussi.)

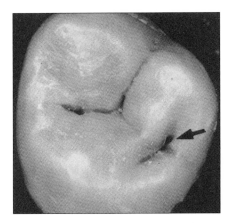

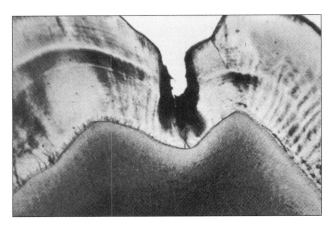

Fig 369 Obvious cavity *(arrow)* in the distal fossa. (From Lussi, 1991. Reprinted with permission.)

Fig 370 Polarized light micrograph of a cross section of the lesion in Fig 369. The cavitation is limited to the outer surface of the enamel around the wide, shallow fissure. The demineralized zone of enamel has not progressed to the dentinoenamel junction. (From Lussi, 1991. Reprinted with permission.)

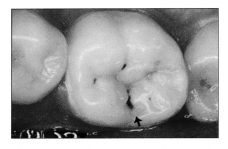

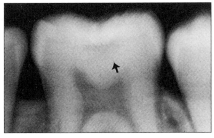

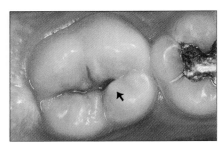

Fig 371 Occlusal cavitated caries lesion in a fissure with a narrow opening *(arrow)*. (From Mejáre and Malmgren, 1994. Reprinted with permission.)

Fig 372 Radiograph showing that the lesion in Fig 371 has progressed into the dentin *(arrow)*. (From Mejáre and Malmgren, 1994. Reprinted with permission.)

Fig 373 Cavitated occlusal caries lesion *(arrow)* in a mandibular molar. (From Mejáre and Malmgren, 1994. Reprinted with permission.)

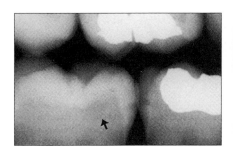

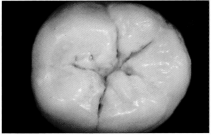

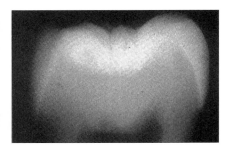

Fig 374 Radiograph confirming that the lesion in Fig 373 has penetrated into the dentin. (From Mejáre and Malmgren, 1994. Reprinted with permission.)

Fig 375 Occlusal surface of an intact molar, used in a study of diagnostic accuracy. (From Espelid et al, 1994. Reprinted with permission.)

Fig 376 Radiograph of the molar in Fig 375. (From Espelid et al, 1994. Reprinted with permission.)

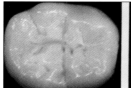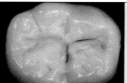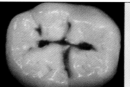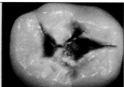

Fig 377 Illustration of grades 1 to 5 *(left to right)* on the scale for diagnosis of occlusal caries presented in Box 14. (From Espelid et al, 1994. Reprinted with permission.)

diagnosed a dentin lesion, 53% considered the occlusal surface to be intact, and 32% were uncertain. They were also asked to suggest therapy appropriate for a 20-year-old patient, with average oral hygiene and dietary habits, who presented with such a lesion. For this individual, 22% of respondents proposed no treatment at all, 23% proposed fluoride treatment, 19% proposed fissure sealants, and 36% proposed restoration.

In a similar study in vivo (Elderton and Nutall, 1983), 18 subjects underwent clinical and (bitewing) radiographic diagnosis by 15 dentists. The suggestions for treatment were wide ranging: from 20 to 153 occlusal restorations. Only two occlusal surfaces received an identical diagnosis by all 15 dentists. This study highlighted the need for considerable improvement in the competence of clinicians, not only with respect to diagnostic skills, but also in the management of occlusal caries. To that end, a five-point scale has been proposed to standardize the criteria for diagnosis based on meticulous clinical visual examination and radiographs (Box 14 and Fig 377) (Espelid et al, 1994; Tveit et al, 1994).

The specificity of a clinical examination exceeds 90%; ie, sound surfaces are most often correctly recognized. This is very important in light of today's relatively small prevalence of occlusal caries. Surfaces that can be maintained unrestored for years with adequate preventive measures must not be treated prematurely by restorative dentistry.

However, the sensitivity at the D_3 level (dentin caries), ie, the ability to recognize diseased surfaces with clinical examination methods, varies between 10% and 80%. The significant differ-

---

Box 14 Five-point scale for diagnosis of occlusal caries

- Grade 1: Noncavitated white-spot or slightly discolored caries lesion in enamel, not detected on the radiograph
- Grade 2: Some superficial cavitation in the entrance of the fissures, some noncavitated mineral loss in the surface of the enamel surrounding the fissures, and/or a caries lesion in enamel, detected on the radiograph
- Grade 3: Moderate mineral loss with limited cavitation in the entrance of the fissure and/or a lesion into the outer third of the dentin, detected on the radiograph
- Grade 4: Considerable mineral loss with cavitation and/or a lesion into the middle third of the dentin, detected on the radiograph
- Grade 5: Advanced cavitation and/or a lesion into the inner third of the dentin, detected on the radiograph

---

ences in the performance of the various classic clinical methods are related to different conditions of occlusal surfaces. Dentin caries in teeth with a macroscopically intact occlusal surface is difficult to detect in everyday practice. Indeed, the lowest sensitivities (10% to 20%) have been found with teeth that exhibit macroscopically intact occlusal surfaces but reveal underlying dentin caries histologically (Figs 378 to 380; see also Figs 362 to 365).

It is not possible with bitewing radiography to detect occlusal enamel lesions reliably. The sensitivity of bitewing radiography for the detection of dentin caries beneath a macroscopically intact surface (hidden caries) is 45%. Visual inspection combined with bitewing radiographs may increase the sensitivity for detection of noncavitated caries lesions in dentin to about 50% and for cavitated caries lesions in dentin to about 90% (Lussi, 1996).

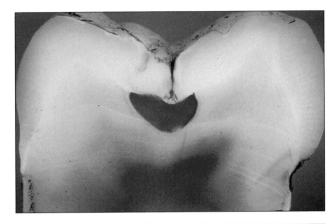

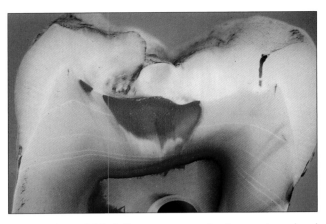

Figs 378 to 380 Different stages of advanced occlusal dentin caries lesions without cavitation into dentin. (Courtesy of A. Lussi.)

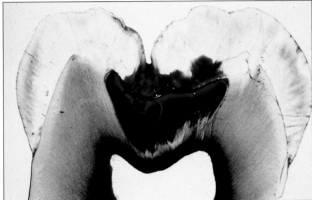

These data highlight the need for improved diagnostic methods for detection of noncavitated caries lesions in dentin. However, from a therapeutic point of view, only cavitated dentin lesions have to be conservatively restored.

New diagnostic methods

Electrical detection systems. Electrical detection methods are believed by many to have the greatest potential for significantly improving diagnostic performance in the years to come. For occlusal surfaces, the commercially available fixed-frequency devices have repeatedly shown high sensitivity and specificity for enamel lesions. For lesions in dentin, sensitivity is high, although specificity is only moderate.

The electrical resistance of a tooth is dependent on its conditions. Sound enamel and dentin lose their insulation properties with ongoing disintegration and replacement by a medium with better conductivity. This is the basis of the electric caries detectors. Electrical conductivity measurements are particularly suitable for the assessment of occlusal lesions with macroscopically intact surfaces. The electrical conductivity of a tooth changes with demineralization, even when the surface remains macroscopically intact. Several studies have demonstrated in vitro and in vivo the favorable performance of fixed-frequency conductance measuring devices in the detection of occlusal dentin caries.

In a recent in vitro study, Ekstrand et al (1997) compared the reproducibility and accuracy of a new visual scoring system, a new electrical conduction tool (Electronic Caries Meter; Lode, Groningen, the Netherlands; Figs 381 and 382), and conventional radiographs for assessment of the depth of demineralization on the occlusal surface. After hemisectioning of the teeth, histologic evaluation was used as the gold standard.

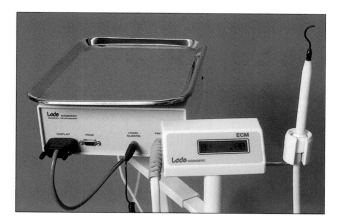

Fig 381 Electronic Caries Meter. (Courtesy of A. Lussi.)

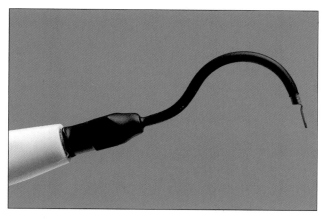

Fig 382 Tip of the Electronic Caries Meter. Note the facility for drying the occlusal surface. (Courtesy of A. Lussi.)

The new visual system appears promising, but takes time to learn and requires the teeth to be clean. The reproducibility and accuracy of the electric conduction method were acceptable, while early occlusal lesions were not detectable on radiographs. The results of this study indicate that both the visual method and the electronic conduction method, used in conjunction with other relevant clinical observations, can improve the accuracy of diagnosis of occlusal caries.

A variable-frequency method also appears to be promising for future development. The disadvantage of the fixed-frequency type of electrical detection system is the somewhat difficult measuring procedure. The sensitivities reported in various in vivo studies have been between 93% and 96%, which is significantly higher than those obtained with any traditional method. The respective specificities (71% to 77%) are typically smaller than those found with visual inspection (Lussi et al, 1995; Verdonschot et al, 1992). These relatively small values for specificity indicate that 23% to 29% of sound teeth may be erroneously diagnosed as diseased and consequently at risk of being treated by invasive methods.

Diagnostic light systems. Two widely used diagnostic light systems are currently available. The one based on laser light is best used for the detection of occlusal and smooth-surface lesions. Among the fluorescence- and light-scattering-based methods, there is presently only one system that is suitable for use in everyday practice. It is based on the detection of fluorescence radiation.

Usually fluorescence is excited by (ultra)violet, blue, or green light, producing fluorescence radiation with longer wavelengths that may be analyzed in the visible part of the spectrum. Spectral investigations on teeth with caries, however, reveal that a better contrast between sound and diseased regions can be achieved when fluorescence is excited in the red and detected in the near infrared region. Under such conditions, fluorescence is much more intense within caries than within sound dental tissues. This allows optical probing into the tooth.

On the basis of the above findings, a relatively simple instrument for caries detection (Diagnodent; KaVo, Biberach, Germany), containing a laser diode as the excitation light source and a photo diode combined with a long-pass filter as the detector, has been developed. The excitation light is transmitted by an optical fiber to the tooth, and a bundle of nine fibers arranged concentrically around it serves for detection. The digital display shows quantitatively the detected fluorescence intensity (in units related to a calibration standard) according to real time and a maximum

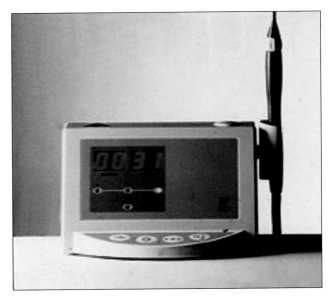

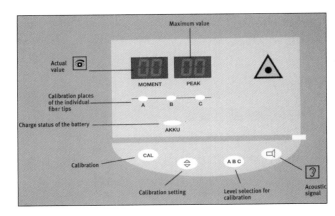

Figs 383 and 384 Diagnodent including real time and maximum (peak) digital display. The device consists of a probe, a fiber-optic lead, and a unit containing electronics and a laser diode.

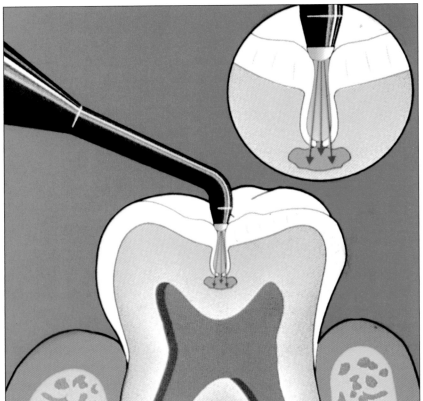

Fig 385 Diagnodent offers the advantage of fluorescence measurement in fissure areas, where the laser is reflected through minute access routes. (Courtesy of A. Lussi.)

Figs 386 and 387 Histologic sections with the tip of the Diagnodent, illustrating the importance of tilting of the tip to detect the carious process. (Courtesy of A. Lussi.)

value (Figs 383 and 384). A tapered fiber-optic tip has been designed for the detection of fissure caries (Fig 385).

The assessment of a tooth with the laser fluorescence system is performed as follows: After calibration with a ceramic standard, the fluorescence of a sound area on the smooth surface of the tooth is measured to provide a baseline value. This value is then subtracted electronically from the fluorescence of the site to be assessed. To determine the extent of the caries, the instrument has to be worked around the site being assessed (Figs 386 and 387). This ensures that the tip picks up fluorescence from the slopes of the fissures walls, where caries is often initiated. A rising tone starting with a value of 10 helps the examiner to find the maximum fluorescence value for the site under assessment.

Combined with the high caries detection performance of Diagnodent, the excellent reproducibility of this device indicates that the laser-based method may be suitable for the longitudinal monitoring of caries and, as a consequence, for assessing carious activity and the outcome of preventive interventions. Its potential application extends beyond locating dentin lesions requiring operative intervention to facilitating the preventive-based management of dental caries.

Diagnostic recommendations

A patient always should be examined clinically, first and foremost with a mirror and a three-in-one syringe without a probe. Drying makes decalcification visible. If there is doubt as to the status of a particular site, further assessment should involve the use of the newer forms of diagnostic devices. This approach allows the dentist to combine the advantages of the greater specificity and speed of clinical examination with the greater sensitivity of the new devices. In no case should the detection of early caries be an indication for operative intervention.

Neither should noncavitated caries lesions in dentin be excavated and restored. The dilemma is that none of these new diagnostic methods or radiographs can differentiate between hidden noncavitated and cavitated dentin lesions, in particular in sticky, atypical fissures (see Figs 322 and 356). In such sites, the entrance of the fissures has to be opened down to the bottom by the use of a pointed rotating diamond tip to verify whether or not the carious dentin is cavitated. If the lesion is noncavitated, it can be successfully arrested by restoration of the opened fissure with a fluoride-releasing glass-ionomer material. Such therapy is termed *fissure blocking*.

MATERIALS AND METHODS

Development

Historically, several methods have been used for prevention and preventive treatment of occlusal caries. In fact there had been a long-standing interest in sealing as a method of preventing occlusal caries for more than 100 years.

One approach was proposed by Hyatt (1923) and Miller (1950). They attempted to fill the occlusal fissure with a sealant material that, by blocking up the fissure, would prevent bacteria and their substrate from coming into contact with that part of the tooth. Clearly, if successfully retained on the tooth, this material would have a good chance of preventing caries of the underlying enamel. The difficulty was to ensure the retention of the sealing material.

Hyatt (1923) recommended that the occlusal fissures of the erupting tooth be sealed with zinc phosphate cement as soon as possible and that, when the tooth was sufficiently erupted, a minimal Class I cavity be prepared and the tooth restored with amalgam before it became carious. In other words, he advocated extension for prevention, according to Black's principles. There was considerable resistance to these proposals from the dental profession, which objected to cutting cavities in caries-free teeth. A certain amount of consumer resistance to the idea of having a tooth filled before it had a cavity in it might also be expected. From the patient's point of view, it was better to wait until caries developed, because the operative procedures were the same.

Hyatt's argument was that it was almost inevitable that the permanent first molar would develop occlusal caries. The tooth should be restored before progression of this caries made restoration difficult, and there was no time like the present. Hyatt's concept, sometimes called *prophylactic odontotomy*, never gained wide acceptance, probably because the procedure involved drilling the child's teeth.

Miller (1950) tested the preventive action of black copper cement when used as a fissure sealant. The copper cement was compared with silver nitrate, and neither material was found to be effective in preventing caries when contrasted with a group of control teeth that received no treatment. However, the copper cement was not retained on the occlusal surface.

While investigating different methods of improving the marginal seal of acrylic resin restorative materials, Buonocore (1955) decided to test the effect of etching the tooth surface with an acid solution before application of the restorative material. This alteration in technique had a dramatic effect on the adhesion of the resin to the tooth, and acid-etching techniques were soon introduced to the field of fissure sealing.

The etched or conditioned enamel improves the marginal adaption and bond strength of plastic resin to enamel. Acid conditioning removes old and fully reacted enamel, increases the surface area, and enhances surface porosity. Presently, it is accepted that phosphoric acid is the conditioner of choice, although several other acids—such as pyruvic, citric, and lactic—have also been used to etch enamel prior to the application of resin (Brauer and Termini, 1972; Retief et al, 1976).

Various concentrations of phosphoric acid, from 30% to 85%, have been used to increase surface topography and retention of sealants. Chow and Brown (1972) showed that concentrations less than 30% are unacceptable because the reaction by-product left on the enamel surface reduces bond strength. Studies by Gwinnett and Buonocore (1965) and Silverstone (1975) showed that resulting tissue porosity was greater with 30% acid than it was with higher concentrations of phosphoric acids. However, from the standpoint of clinical retention, these lower and higher acid concentrations are comparable. It appears that concentrations of 37% to 50% phosphoric acid solutions produce the changes in surface topography that are most conducive to bonding (Dennison and Craig, 1978).

According to the original recommendations by Buonocore (1955), acid conditioning is usually done for 1 minute, although further studies have shown that improvement may result if etching is extended to 2 minutes for primary teeth. The majority of clinical studies have used 1 minute as a minimum time necessary for appropriate etching and subsequent retention. However, a study by Eidelman and coworkers (1974) showed that the retention of sealants is not dramatically affected when the etching time is reduced to 20 seconds.

Requirements for an ideal occlusal sealant include sustained bonding to enamel surface, simple application by the dentist and/or hygienist, biocompatibility with oral tissues, a free-flowing low-viscosity quality capable of entering narrow fissures, and low solubility in the oral environment. One of the first materials clinically tested at Eastman Dental Center as a sealant was methyl 2-cyanoacrylate (Eastman 910 adhesive) in combination with a filler (Cueto and Buonocore, 1967). Although cyanoacrylate initially bonded well to enamel, by the end of 1 year, only one third of the sealants were fully intact.

The second group of materials considered for clinical use as sealants came from the polyurethane family of compounds, some containing fluoride. Polyurethane was originally marketed as a sealant and then as a prolonged fluoride-releasing resin. However, several studies demonstrated their poor retention as sealants, and their use was discontinued (Newbrun et al, 1974; Rock, 1974).

A major breakthrough in sealants was the development by Bowen (1963, 1970) of a new cross-linking, thermosetting dimethacrylate monomer (bis-GMA), a substance currently present in most resin composite materials and sealants. When bis-GMA is diluted with methyl methacrylate or other co-monomers, the flow characteric is improved and it may be used as a sealant. This product mixture exhibits less polymerization shrinkage and a lower coefficient of thermal expansion than does methyl methacrylate alone and is, therefore, more likely to bond with enamel. In addition, bis-GMA can be polymerized with ultraviolet or visible lights, depending on the chemical catalyst added to the mixture. Most commercial sealants today are bis-GMA dimethacrylate– or urethane dimethacrylate–based products.

Commercially available sealants differ in whether they are free of inert fillers or semifilled and whether they are clear, tinted, or opaque. A principal difference is the manner in which polymerization is initiated. The first marketed sealants, called *first-generation sealants*, were activated with an ultraviolet light source. *Second-generation sealants* are autopolymerizing and set on mixing with a chemical catalyst–accelerator system. The *third-generation sealants* are phooinitiated with visible light. Another recent innovation is the sale of fluoride-containing resin-based sealants and different glass-ionomer materials, which should be regarded as slow-release fluoride agents. Today, the third generation of fissure sealants, with or without fluoride, and chemical or light-cured glass-ionomer sealants are the materials of choice for fissure sealing.

Resin-based fissure sealants

Table 19 shows a comparison of some resin-based fissure sealant materials on the market. Delton is the most commonly used light-cured fissure sealant material. Among the fluoride-containing light-cured sealant materials, Helioseal-F seems to be the most frequently used (Fig 388).

Application procedures

It quickly became apparent that application technique is very important to the success of a fissure sealant, but it is difficult to determine from clinical data which aspects of the application technique are most important. Most of the research work carried out on the strength of the bond between resin and tooth and the effect of changing various aspects of the application technique has been done in the laboratory, where it is easier to test the effect of changing one factor in isolation.

Table 19 Characteristics of commercially available resin-based fissure sealants

Product	Manufacturer	Fluoride containing	Color	Polymerization	Application	Handling	Economy
Helioseal F	Vivadent, Liechtenstein	Yes	White	Light-curing	Disposable syringe	Easy; fast	Good
Fissurit F	Voco, Cuxhaven, Germany	Yes	White and transparent	Light-curing	Disposable brush	Slower	Not as good
Fluoroshield	Dentsply/Caulk, Milford, DE	Yes	White and transparent	Self-curing	Disposable brush	Slower	Not as good
Ultraseal XT	Ultradent, South Jordan, UT	Yes	White and transparent	Self-curing	Disposable syringe	Easy; fast	Good
Delton	Johnson & Johnson, New Brunswick, NJ	No	Pink and transparent	Light-curing	Disposable cannula	Easy; fast	Good

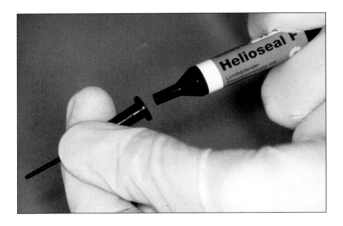

Fig 388 Light-cured resin-based fissure sealant with fluoride supplementation (Helioseal F).

The application procedure is more or less the same for all the modern materials, which depend on acid etching for their retention. The manufacturer will usually recommend that the tooth surface first be mechanically cleaned and washed, isolated, and etched with the material provided (usually phosphoric acid) for a specific time. The acid is then rinsed off the tooth, while the tooth is kept isolated from contact with saliva, and the tooth is dried with compressed air. The sealing material is placed on the occlusal surface, which must be kept isolated from moisture until the material has polymerized. The most critical parts of the application procedure are rinsing of the etched enamel, drying of the tooth surface, and maintaining the isolation of the teeth until the sealant material has polymer-

ized. Variation in the concentration of the etching material appears to have little effect: Provided that the concentration of orthophosphoric acid (which is the most widely used etching material) lies between 30% and 50% by weight, small variations in the concentration do not appear to effect the quality of the etched surface (Silverstone, 1974).

Variation in the time for which the tooth enamel is exposed to etching solution is more important, provided that the enamel is exposed to the etching material for 30 seconds; however, there will be sufficient demineralization to allow adequate retention of the sealant. Even in areas where there is an optimal level of fluoride in the water supply, it appears there is little to be gained by increasing the etching time (Barkmeier et al,

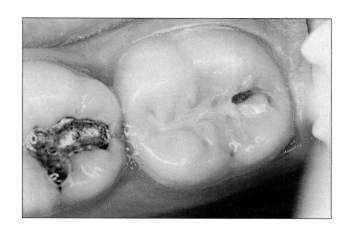

Fig 389 Cavitated caries lesion in dentin in the distal fossae of the mandibular right first molar resulting from the microleakage and loss of the resin-based fissure sealant. (From Mejáre and Malmgren, 1994. Reprinted with permission.)

1985; Beech and Jalaly, 1980; McCabe and Storer, 1980). It may be more difficult to gain adequate retention when the enamel of primary teeth is etched, but clinical studies (Simonsen, 1979) suggest that it may not be necessary to increase the etching time when primary molars are sealed because perfectly adequate retention values are obtained with the etching times recommended for permanent teeth.

Relatively small variations in the time for which the etched enamel is rinsed have a more marked effect on the strength of the bond between resin and enamel (Williams and von Fraunhofer, 1977), and the operator applying sealant resins should take particular care during the rinsing and drying of the tooth surface before applying the resin. The enamel should be rinsed for a full 20 to 25 seconds before being dried.

The acid-etching technique is very sensitive to salivary contamination. For this reason, the retention (success) of fissure sealing depends almost entirely on the efficiency of the clinical moisture control. Teeth to be sealed must remain completely dry. If the etched enamel is contaminated with saliva, the surface should be carefully rinsed with water and dried, and only after new moisture-control measures are taken should the etching procedure be repeated.

The problem regarding moisture control is that almost 100% of occlusal caries is initiated during the extremely long period of eruption of the first

and second molars (12 to 18 months), as discussed earlier. The risk of caries development is over if the occlusal surfaces are maintained caries free until they have reached full occlusal function. Therefore, fissure sealants should be used as early as possible in caries-risk patients and sticky risk fissures (see Figs 322, 323, 328, 331, and 344). However, moisture control of the occlusal surfaces of erupting molars and in particular the distal fossae is very complicated. Isolation of such surfaces with rubber dam is almost impossible. The use of the etching–resin sealant technique in the distal fossae of such teeth is very risky and may initiate caries if microleakage occurs between the etched enamel and the resin (Fig 389).

Retention

The effect of the eruption status of permanent first and second molars on sealant retention was investigated by Dennison et al (1990). Teeth in different stages of eruption were treated. In the earlier stage, an operculum covered the distal marginal ridge; in the most advanced stage, there was complete vertical eruption and the gingival tissue was located below the distal marginal ridge. The number of sealant re-treatments during the 36-month observation period was associated with the eruption status at the time the teeth were originally treated. Molars with an opercu-

Fig 390 Enamel surface etched with 35% phosphoric acid (original magnification ×3,000). (From Mejáre and Malmgren, 1994. Reprinted with permission.)

Fig 391 Resin that has penetrated into the roughness of the etched enamel surface, forming so-called tags (original magnification ×3,000). (From Mejáre and Malmgren, 1994. Reprinted with permission.)

lum covering the distal marginal ridge had twice the probability for re-treatment as did those with gingival tissue located below the distal marginal ridge (Table 20).

The phosphoric acid treatment causes microscopic etching of enamel that allows resinous sealant materials to penetrate into the enamel. The bond formed is purely mechanical; its strength is approximately 60 to 100 N/m², depending on the histology of the surface enamel.

Figure 390 shows the surface structure of tooth enamel etched with 35% phosphoric acid for 1 minute. The resin-based sealant penetrates the microroughness of the etched enamel and forms so-called tags (Fig 391), which significantly increase the mechanical retention of the sealant. Even if most of the sealant material seems to be worn out, at least such tags may remain as a sealant of the enamel surface. There are no indications that noncavitated active incipient caries lesions will reduce the mechanical retention if the surface is mechanically cleaned and etched (Handelman et al, 1987). Rather, the microporous surface will increase the retention.

Figure 392 shows an assortment of useful materials and aids for moisture control and application of sealant materials. Figures 393 and 394 show useful holders for the lips and cotton rolls

Table 20 Sealant re-treatment during a 36-month observation period related to the stage of molar eruption at the time of initial treatment*

Eruption status at initial treatment	No. of surfaces	Re-treatments No.	%
Tissue operculum over distal marginal ridge	28	18	53.6
Distal gingival tissue level with distal marginal ridge	31	8	25.8
Distal gingival tissue below distal marginal ridge	16	0	0.0

*From Dennison et al, 1990. Reprinted with permission.

for moisture control during the sealant procedure. An alternative method for moisture control in erupted teeth could be application of rubber dam.

Figure 395 shows an almost fully erupted maxillary first molar with deep risk fissures, in particular the distal fossae, in a selected 6-year-old caries-risk child. This is a typical indication for fissure sealant therapy. After comprehensive mechanical cleaning of the occlusal surface with

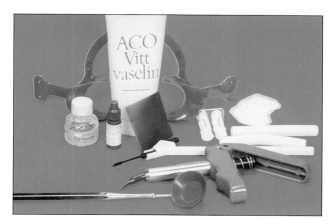

Fig 392 Useful materials and aids for moisture control and application of sealants. (From Mejáre and Malmgren, 1994. Reprinted with permission.)

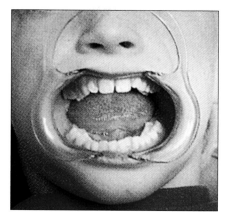

Fig 393 Holder for lips. (From Mejáre and Malmgren, 1994. Reprinted with permission.)

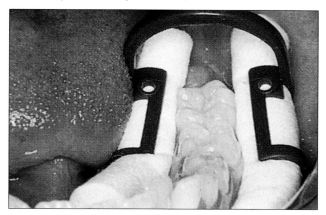

Fig 394 Holder for cotton rolls to facilitate moisture control. (From Mejáre and Malmgren, 1994. Reprinted with permission.)

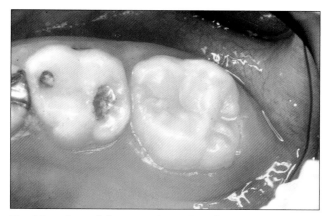

Fig 395 Almost fully erupted maxillary first molar with deep risk fissures, in particular the distal fossae, in a selected 6-year-old caries-risk child. This is a typical indication for fissure sealant therapy. (From Mejáre and Malmgren, 1994. Reprinted with permission.)

a rotating pointed brush and pumice (Fig 396), the fissure system is etched for 30 seconds with a 35% phosphoric acid gel. Thereafter, the surface is washed thoroughly for 30 seconds with water spray and dried with compressed dry air (Fig 397). After careful application of moisture control measures, light-cured resin-based fissure sealant material is placed in the fissure system with a special applicator and light cured for 30 seconds (Fig 398). Finally, the complete retention of the sealant is verified with a probe. The occlusion is checked by the use of a green

occlusal wax. If necessary, a pear-shaped finishing bur is used to adjust the occlusion with the wax in situ.

Misuse

Figures 399 to 401 show how fissure sealant can be misused if the dentist does not know why, when, and how to use it properly for caries prevention. It may be assumed that the brown fissures of these two abraded mandibular occlusal surfaces with almost no cuspal inclination had

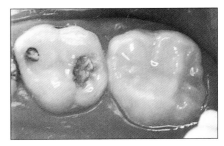

Fig 396 Professional mechanical tooth-cleaning of the occlusal surface with a pointed rotating brush and pumice. (From Mejáre and Malmgren, 1994. Reprinted with permission.)

Fig 397 Etched occlusal surface. (From Mejáre and Malmgren, 1994. Reprinted with permission.)

Fig 398 Fissure system sealed with a light-cured resin-based fissure sealant after careful moisture control. (From Mejáre and Malmgren, 1994. Reprinted with permission.)

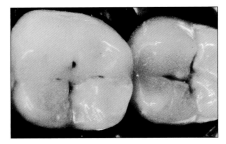

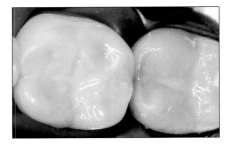

Figs 399 to 401 Misuse of fissure sealant in two mandibular molars that exhibit stained and arrested enamel caries in a young adult. Such occlusal surfaces, with abraded flat cusps and shallow fossae and fissures, should remain caries inactive for a lifetime.

noncavitated caries lesions during the eruption of the first molar, about 15 years ago, and the second molar, 6 to 9 years ago. However, after full eruption, the lesions were successfully arrested and stained. Such fissures should remain caries inactive for a lifetime. Only for esthetic reasons might such fissures be sealed.

Glass-ionomer and resin-modified glass-ionomer fissure sealants

To date, resin-based sealant materials have most frequently been used in molars and premolars with full occlusal function and, consequently, long after eruption. The risk for initiation of occlusal caries is over by that time. However, during the eruption of the molars (12 to 18 months), when the occlusal caries lesions are initiated, the use of resin-based fissure sealants is limited because of the high degree of moisture control that

has to be achieved. Therefore, low-viscosity glass-ionomer and resin-modified glass-ionomer materials are attractive alternatives to resin-based fissure sealants because these materials can be used as slow-release fluoride sealants during the eruption of the molars.

Conventional glass-ionomer materials

The first glass-ionomer cement, developed by Wilson and Kent (1972), was a product of an acid-base reaction between basic fluoroaluminosilicate glass powder and polycarboxylic acid in the presence of water. The set cement comprised an organic-inorganic complex with high molecular weight. Therefore, *glass-ionomer cement* can be defined as a water-based material that hardens following an acid-base reaction between fluoroaluminosilicate glass powder and an aqueous solution of polyacid.

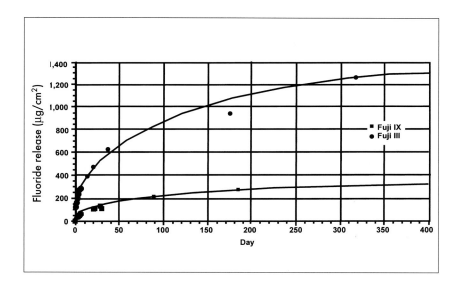

Fig 402 Mean cumulative fluoride release from a glass-ionomer sealant (Fuji III) and a regular glass-ionomer cement (Fuji IX).

The glass was developed on the basis of dental silicate cements and was prepared by melting alumina (Al_2O_3), silica (SiO_2), metal oxides, metal fluorides, and metal phosphates at temperatures higher than 1,100°C. The melted glass is crushed, milled, and powdered to fine particles. The particle size and distribution of the glass powder is of great importance for controlling the setting characteristics of the cement.

Fluoroaluminosilicate glass possesses a unique feature—it releases fluorine without adding other fluoride compounds to the cement. The main source of this release from the set cement is thought to be the cement matrix; however, some of the fluorine is believed to orginate in the glass core. Because fluorine does not exist in the skeletal structure of the glass, fluoride in the glass core probably diffuses throughout the cement matrix and is then released slowly. The physical properties of glass-ionomer cement do not deteriorate even after fluoride release.

Many studies have suggested that the ability of glass-ionomer cement to recharge fluoride is a feature of the fluorine movement within the cement matrix (for review, see Creanor et al, 1995; Damen et al, 1996; and Gao, 2000). In other words, when the level of fluoride ions increases in the proximity of the glass-ionomer cement restoration, fluoride ions go into and accumulate in the cement.

When the level of fluoride ions in the environment decreases, the accumulated fluoride ions are released again. This feature tries to maintain constant levels of fluorine in the oral environment. Figure 402 shows the levels of fluoride ion released from Fuji IX and Fuji III (GC Corp, Tokyo, Japan), which are usable as fissure sealants, as measured with fluoride electrodes. These results show the ability of a glass-ionomer cement to release small amounts of fluoride over an extended period.

The polyacid that reacts with the fluoroaluminosilicate glass in the glass-ionomer system is usually a polycarboxylic acid. Recently, polyvinyl phosphonic acid also has been introduced to this system. The polyacid either is part of the liquid as an aqueous solution or is incorporated into the cement powder as a freeze-dried powder. In the latter case, the liquid is simply water in which the freeze-dried polyacid dissolves on mixing. Thanks to the presence of these particular acids, the glass-ionomer cement has the ability to adhere to tooth structure or metals without the additional step of a special substrate treatment. The setting reaction of conventional glass-ionomer cements starts when the fluoroaluminosilicate glass powder and the aqueous solution of polyacrylic acid are combined, producing an acid-base reaction with the powdered fluoroaluminosilicate glass as the base:

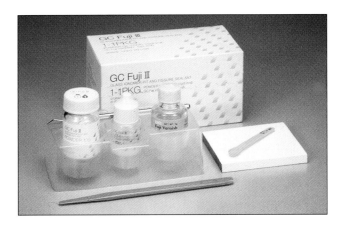

Fig 403 Glass-ionomer fissure sealant (Fuji III).

fluoroaluminosilicate glass (base) + polyacid (acid) → polyacid matrix (salts)

The low-viscosity glass-ionomer cements have been developed as liners, fisure protection materials, sealing materials for hypersensitive cervical areas, and endodontic materials. Such materials are designed with low powder-liquid ratios and are highly flowable. The fluoride released from glass-ionomer materials may act locally as a caries-preventive agent on the enamel of the fissures as well as in the entire oral cavity (see chapter 5). The low powder-liquid ratio of the materials is expected to result in gradual dissolution from the occlusal surface when they are applied as fissure protection materials (see Fig 402).

Low-viscosity glass-ionomer cements are in particular suitable as fissure protection materials during the eruption period of the molars (the high-risk period). Figure 403 shows Fuji III, which is the most commonly used conventional glass-ionomer cement specifically composed for this purpose.

Resin-modified glass-ionomer materials

Resin modification of glass-ionomer cement was designed to produce favorable physical properties similar to those of resin composites and resin cements, while retaining the basic features of the conventional glass-ionomer cement. This goal was achieved by incorporating water-soluble resin monomers into an aqueous solution of polyacrylic acid. In this way, the system undergoes polymerization of the resin monomer while the acid-base reaction continues simultaneously. The resulting resin-modified glass-ionomer cements exhibit many advantages of both resin cements and glass-ionomer cements. The *resin-modified glass-ionomer cement* is defined as a material that undergoes both a polymerization reaction and an acid-base reaction.

The basic composition of the cement liquid is polycarboxylic acid, water, and 2-hydroxyethyl methacrylate (HEMA). It may also contain a small amount of cross-linking material. Some products, such as Vitremer (3M, St Paul, MN), are said to contain a polycarboxylic acid modified with pendant methacrylate groups. The composition and structure of the fluoroaluminosilicate glass for resin-modified glass-ionomer cements are basically similar to those of conventional glass-ionomer cements.

The essential acid-base reaction between the fluoroaluminosilicate glass and the polycarboxylic acid is initiated when the powder and liquid are mixed. At the same time, the polymerization of HEMA and cross-linking material

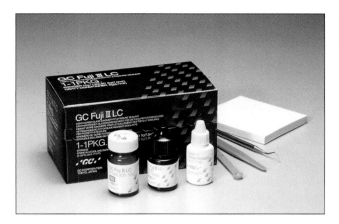

Fig 404 Light-cured resin-modified glass-ionomer fissure sealant. (Fuji III LC).

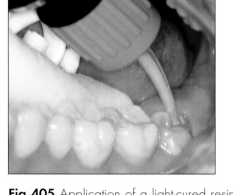

Fig 405 Application of a light-cured resin-modified glass-ionomer fissure sealant (Fuji III LC) via a capsule in the fissures of a partially erupted third molar.

is started by an oxidation-reaction or a photopolymerizing catalyst. This forms a hardened mixture, in which HEMA polymer and polycarboxylic acid are supposed to be linked by hydrogen bonding.

One of the main disadvantages of conventional glass-ionomer cement is that when it comes in contact with water during the early stage of setting, the setting reaction is inhibited, damaging the surface of the cement. Water sensitivity could be reduced by the incorporation of photopolymerization, which promotes faster setting, into the setting reaction. In other words, water no longer inhibits the setting reaction by the time photopolymerization is completed.

Fuji III LC is a light-cured, low-viscosity, resin-modified glass-ionomer cement, specifically prepared as a slow-release fluoride sealant material; it can be used from the beginning of the eruption of the molars. This material shows a high fluidity and ability to penetrate the fissures. It is available both hand mixed (Fig 404) and in capsules. The capsule design optimizes accessibility and is particularly comfortable for dental hygienists who are working without an assistant (Fig 405).

Although resin-modified glass-ionomer materials are retained in the fissure for a required period of time, they do not have to be retained as

long as resin sealant material. Presently, glass-ionomer and resin-modified glass-ionomer cements are the only materials that increase the acid resistance of the tooth.

Adhesion to the tooth surface

The reaction between glass-ionomer cements and the dental tissues is chemical. The strength of the bond formed is approximately 20 N/m^2, which is thus weaker than that obtained with the acid-etching technique and resins, which is mechanical retention. However, the retention of conventional glass-ionomer sealants can be improved by about 50% if the enamel surface is conditioned with polyacrylic acid. It is presumed that the ionic reactivity of a resin-modified glass-ionomer cement to the surface is lower than that of a conventional glass-ionomer cement. However, the reactivity can be markedly increased by treatment of the tooth surface with an acid conditioner.

Treating the enamel surface with phosphoric acid does not decrease the bond of either type of cement, and the bond is stabilized. An aqueous solution of citric acid–ferric chloride or polyacrylic acid–aluminum chloride also is effective in increasing either cement's adhesiveness. It is

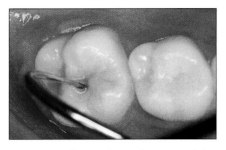

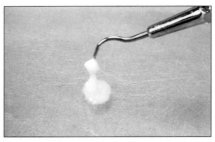

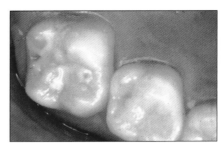

Fig 406 Chemical conditioning of the fissures of a maxillary first molar with polyacrylic acid and mechanical removal of plaque in the fissures with a probe. (From Mejáre and Malmgren, 1994. Reprinted with permission.)

Fig 407 Application of a glass-ionomer fissure sealant by the use of a probe. (From Mejáre and Malmgren, 1994. Reprinted with permission.)

Fig 408 Fissures sealed with a glass-ionomer sealant. (From Mejáre and Malmgren, 1994. Reprinted with permission.)

probably best to use conditioning materials that exhibit acidity at the same level as that of an aqueous solution of polyacrylic acid and aluminum chloride when both enamel and dentin are treated simultaneously. The bond strength of a resin-modified glass-ionomer cement increases because its tensile strength improves with the treatment (for reviews on the composition and characteristics of glass-ionomer and resin-modified glass-ionomer materials, see Braem, 1999; de Gee, 1999; Saito et al, 1999).

Application procedures

Before application of conventional or resin-modified glass-ionomer sealant, careful PMTC of the occlusal surface is performed with a rotating pointed brush and pumice (see Fig 396). The occlusal surface is then thoroughly washed with water spray and dried.

After moisture control is applied, 25% polyacrylic acid is applied with a pointed disposable brush. During conditioning of the surface, a probe is used for removal of debris in the fissures (Fig 406). Then the occlusal surface is washed for 30 seconds with water spray and dried with clean compressed air. Careful moisture control is re-established. After conditioning with polyacrylic acid, the surface is not etched and rough, but the

surface energy is reduced, which increases the wettability and the penetration of the sealant in the fissures.

Fuji III is mixed and applied in the fissure system with a probe (Figs 407 and 408). To protect the glass-ionomer sealant materials from moisture and dehydration during the primary sensitive setting period, occlusal wax is pressed on the occlusal surface for about 5 minutes (Fig 409). This measure also improves the adaptation of the sealant in the fissure and reduces excess. About 80% of the setting strength is attained during the first 15 minutes. However, the strength of the material may continue to increase up to 24 hours after application. Therefore it is recommended that the surface be protected with resin after removal of the wax.

During the last few years, the light-cured resin-modified glass-ionomer fissure sealant (Fuji III LC) has more or less replaced the conventional glass-ionomer sealant. Among the reasons are its reduced sensitivity to moisture and dehydration, increased retention, and much faster and easier procedure. In addition, the retention may be improved further by etching of the enamel surface if optimal moisture control can be achieved.

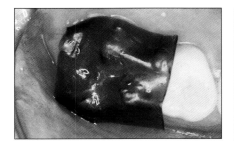

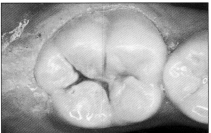

Fig 409 Placement of occlusal wax to protect the sealant from moisture and dryness during the first 5 minutes of setting. (From Mejáre and Malmgren, 1994. Reprinted with permission.)

Fig 410 Sticky fissures with suspected cavitated caries lesions in the distal and central fossae of a mandibular first molar. (From Mejáre and Malmgren, 1994. Reprinted with permission.)

Fig 411 Pointed diamond bur used to open the fissures (odontotomy) down to the bottom in order to achieve a correct caries diagnosis.

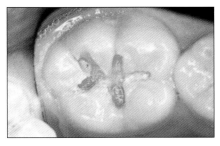

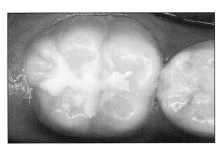

Fig 412 Opened distal and central fossae and related fissures. No cavitated caries lesion was discovered in the bottom of the fossae or fissures. (From Mejáre and Malmgren, 1994. Reprinted with permission.)

Fig 413 Opened fossae and fissures blocked with a light-cured resin-modified glass-ionomer cement (type II), resulting in a so-called extended fissure sealant or fissure blocking. (From Mejáre and Malmgren, 1994. Reprinted with permission.)

Extended fissure sealing (fissure blocking)

Deep and narrow sticky fissures are regarded as caries-risk surfaces, as discussed earlier. Such surfaces are difficult to seal properly because the accessibility for elimination of plaque and debris from the deepest parts of the fissures is limited. In addition, correct diagnosis of caries is very complicated. The consequences of errors in diagnosis of enamel caries before the use of fissure sealants are limited. In most cases, the cavitated caries lesions in dentin should not have developed during the eruption of the molars. However, in new patients with fully erupted first molars (7 to

11 years) and second molars (13 to 16 years) noncavitated or cavitated dentin lesions may be suspected or diagnosed, for example, on bitewing radiographs. The extended fissure sealing (so-called fissure blocking) technique (Le Bell and Forsten, 1980) is very useful for solving these problems. Thereby optimal accessibility for mechancial cleaning of the bottom of the fissure as well as correct diagnosis of caries is achieved.

Figure 410 shows a mandibular left first molar already exhibiting sticky caries lesions in the distal and central fossae that are suspected to be cavitated. The fissure is opened with a rotating pointed diamond tip (Fig 411) and water spray. Figure 412 shows the fissure opened to the bottom in the

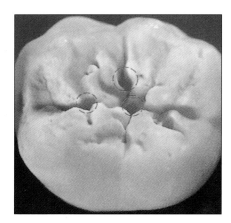

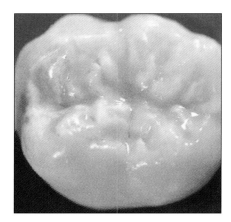

Fig 414 Sticky risk fissures of a mandibular molar, opened for visual caries diagnosis. No cavitated caries lesion into the dentin was discovered.

Fig 415 The opened fissure system in Fig 414 was etched and blocked with a flowable compomer.

distal and central fossae. Because no cavitation into the dentin is discovered, the fissure is washed with water, dried, and conditioned with 25% polyacrylic acid for 10 seconds. The fissure is washed again with water for 30 seconds and dried. The fissure is sealed with a type II light-cured resin-modified glass-ionomer cement (Fig 413).

Figure 414 shows a very irregular fissure system in a mandibular molar that has been opened with a pointed diamond tip for optimal accessibility and etched. The fissure is blocked with flowable compomer (Fig 415).

Minimally invasive preventive restorations

One approach to dealing with the problem of early fissure caries is to use a procedure that employs a minimally invasive resin composite or resin-modified glass-ionomer cement restoration (Houpt et al, 1994; Raadal, 1978; Simonsen, 1980; Walls et al, 1988). The technique involves making a very small, local cavity preparation in the immediate area of the fissure system at which the presence of caries is suspected. No attempt is made to extend the cavity beyound the immediate area affected by caries. The defect in

the occlusal surface is restored with sealant, a resin composite, or a type II resin-modified glass-ionomer restorative material, depending on the size of the defect. Following this, the etched occlusal surface of the tooth may be sealed over the top of any minirestoration with resin composite or resin-modified glass-ionomer material.

The advantage of this approach is that the absolute minimum of tooth substance is removed. Walls et al (1988) reported that the occlusal amalgam restorations in their study occupied, on average, 25% of the occlusal surface of the tooth, while the minimal resin composite restorations occupied 5%. In addition, the procedure avoids the unfortunate consequences of an error in diagnosis. If a healthy tooth is investigated, little harm is done, for it quickly becomes evident that no caries is present and the resulting cavity is very small. If the caries is more extensive than was originally supposed, this will become apparent during the procedure, and appropriate action can be taken.

Figures 416 to 422 show fissure caries in the occlusal surfaces of a mandibular and a maxillary first molar treated with this minimally invasive preparation and the etching–flowable resin composite technique.

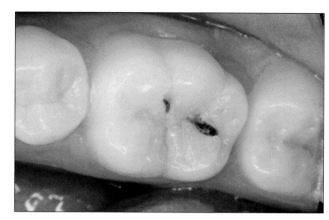

Fig 416 Occlusal caries in a mandibular molar of a young patient. (From Toffenetti, 2001. Reprinted with permission.)

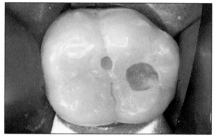

Fig 417 Conservative minipreparation, dictated only by the location and extent of the caries lesions. (From Toffenetti, 2001. Reprinted with permission.)

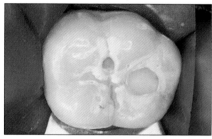

Fig 418 Etched cavity preparation and fissures. (From Toffenetti, 2001. Reprinted with permission.)

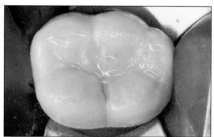

Fig 419 Cavities after restoration with a resin-based composite. The remaining fissures have been sealed with a resin-based fissure sealant. (From Toffenetti, 2001. Reprinted with permission.)

Fig 420 Localized caries lesion in the distal fossae of a maxillary first molar. On a bitewing radiograph, the lesion shows radiolucency beyond the dentinoenamel junction, reaching the outer third of the underlying dentin. (From Lasfargues et al, 2000. Reprinted with permission.)

Fig 421 Minimally invasive preparation. The preparation was initially limited to the caries lesion, and then the entrances of related fissures were opened further. All of the fissure system was etched. (From Lasfargues et al, 2000. Reprinted with permission.)

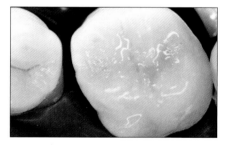

Fig 422 Minicavity after restoration with a microhybrid resin composite. The fissures have been sealed with a resin-based sealant. (From Lasfargues et al, 2000. Reprinted with permission.)

An attractive alternative to flowable resin composite or compomers is type II light-cured resin-modified glass-ionomer cement. Figures 423 to 425 show the restoration of the central fossae of a mandibular second molar after minipreparation with a pointed diamond bur. A type II light-cured resin-modified glass-ionomer cement is used. At the same visit, the central fossa and related fissures of the first molar are blocked with the same material.

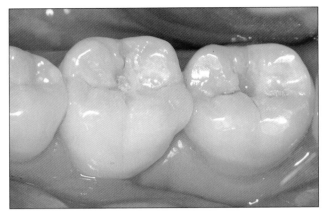

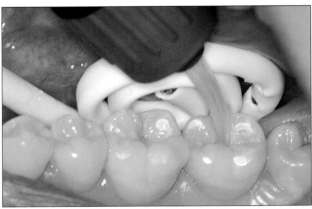

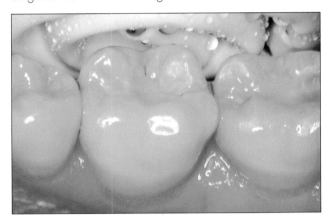

Fig 423 Sticky central fossae in first and second mandibular molars, opened with a minimal preparation because of cavitated enamel caries lesions and suspected dentin lesions. No cavitation was discovered in the dentin.

Figs 424 and 425 *(bottom left and right)* Minimally invasive preparations in the central fossae after restoration with a light-cured resin-modified glass-ionomer cement

CARIES-PREVENTIVE EFFECTS
Resin-based sealants

Sealants are effective as long as they remain firmly adherent to the tooth. Subsequently, the evaluation of their effectiveness involves the determination of reductions in occlusal caries. Early clinical studies with sealants date back to the 1970s, when the split-mouth technique was used; noncarious contralateral teeth were selected for study. Sealant was applied to the occlusal pits and fissures of one tooth, while the other tooth was monitored for carious activity as an untreated control.

The results have shown excellent caries reduction during the first years after treatment: The reduction in caries varied from 36% to 100%, depending solely on the retention of the sealants (Meurman, 1977). For example, greater than 50%

caries reduction was observed 7 years after application of the sealant in a study where the retention was 66% (Mertz-Fairhurst et al, 1984). Fifteen years after a single application of sealants to the permanent first molars, 28% were still completely retained. Only 31% of the originally sealed surfaces were carious or restored, whereas 83% of surfaces were carious or restored in the nonsealed controls (Simonsen, 1991). Results of these studies are based on a single application of the sealants, while in routine dental health care today repeated applications are recommended in cases of failure. Thus, a 100% caries prevention in pits and fissures is to be expected with properly applied and maintained sealants.

However, caries prevalence and incidence were very high during the 1970s in the so-called developed countries, where almost all of the fissure sealant studies were carried out. For example, in Sweden, the caries incidence in children was about 10 times higher in the 1970s than it was in the 1990s (Axelsson, 1998). Therefore, the caries-preventive effect of fissure sealants compared to unsealed teeth in a split-mouth designed study should be very limited in such a population in the twenty-first century.

On average, 12-year-old children in the county of Värmland, Sweden, exhibit only about 0.5 DFSs, and about 80% of these children are caries free. In populations such as this, from a cost-effectiveness point of view, fissure sealants should only be used in erupting molars with sticky risk fissures in selected caries-risk individuals. Based on the significant caries reduction in sealed teeth compared to unsealed teeth in the same mouth in the early 1970s, sealants were judged to be caries preventive as long as they remained adhered to the teeth. Thereafter, the design of sealant studies changed: Nontreated control teeth were no longer acceptable, and the longevity of sealant coverage, ie, clinical retention, became the measure of sealant success. Subsequent studies evaluated the retention of one sealant system over time, or compared the retention of two or more sealant systems.

In a review of the effectiveness of pit and fissure sealants, Ripa (1993b) classified the first generation of successful sealants as those activated with an ultraviolet light source. Second-generation sealants were autopolymerizing sealants that used a chemical-catalyst–accelerator system. The third generation of sealants were photoinitiated with visible light. Although first-generation sealants had results comparable with those of the second generation for the first 3 years after application, their effectiveness subsequently declined. The long-term results of second-generation sealants up to 15 years are highly favorable. The clinical results for up to 5 years of third-generation sealants are similar to those of the second-generation sealants (Ripa, 1993b).

Three longitudinal studies have evaluated the long-term retention of chemically initiated sealants for 10 years or longer. In each program, chemically polymerized sealants were used to treat permanent first molars. Simonsen (1987) re-

ported that 57% of the treated fissures were completely covered 10 years after a single application of sealant and 28% were covered after 15 years.

Romcke et al (1990) reported the results of a sealant program in Canada in which treatments were performed in a mobile dental van. After 10 years, they reported 41% complete retention and 8% partial retention on teeth requiring no sealant re-treatment during the observation period. However, they considered the overall success rate to be 85%, including completely and partially covered teeth that they felt did not require re-treatment and teeth that had maintenance sealant replacement during the observation period.

Wendt and Koch (1988) reported 80% complete retention after 8 years and 94% combined partial and complete retention after 10 years but did not list complete retention separately. They attributed their high success rate to the fact that the sealants were applied under optimal dental office conditions rather than less ideal field conditions.

Recently, Wendt et al (2001a) presented follow-up data on the retention and effectiveness of chemically polymerized sealants on the occlusal fissures and buccal pits of permanent first and second molars after 20 and 15 years, respectively. The population consisted of 72 children, each of whom had their first molars sealed between 1977 and 1980. At the annual examinations, all caries-free, newly erupted second molars were sealed. When sealant was applied to the second molars, the first molars were checked and sealant was reapplied to those that had deficient sealants.

At the follow-up, when the subjects were 26 to 27 years of age, 27 members of the original group had moved from the community. Thus, the present result is based on 45 subjects with 153 sealed first molars and 161 sealed second molars available for inspection. At the follow-up examination of the first molars 20 years after sealant had been applied, 65% showed complete retention, 22% showed partial retention without caries, and 13% had caries or restorations in the occlusal fissures or buccal pits. At the 15-year follow-up of the sec-

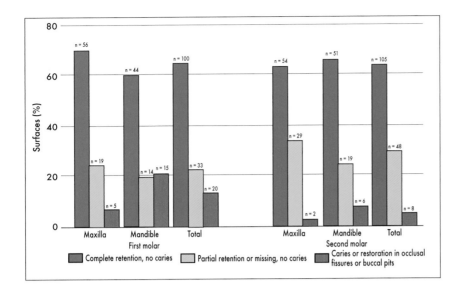

Fig 426 Evaluation of fissure sealants and occlusal caries in permanent first molars after 20 years and in permanent second molars after 15 years in 26- to 27-year-old subjects. (Modified from Wendt et al, 2001a with permission.)

ond molars, the corresponding figures were 65%, 30%, and 5%, respectively (Fig 426). Of the restored or carious molars, significantly more were found in the mandible than in the maxilla ($P <$ .001).

This longitudinal study showed that pit and fissure sealants, applied during childhood, have a long-lasting caries-preventive effect. No matched control group existed in the study by Wendt et al (2001a), but recent Swedish data on caries development in subjects from 11 to 20 years of age indicate a considerably higher caries attack rate on the occlusal surfaces of the molars than were found in this test group (Mejáre et al, 1998).

The experiences from this study were implemented in a large-scale fissure-sealing program. In 1986 to 1987, all dental personnel in Jönköping, Sweden, took part in theoretical and clinical courses in fissure sealing. In connection with this education, the recommendation was given to seal all caries-free pits and fissures on the surfaces of newly erupted permanent first molars. The aim of the study was to evaluate the long-term results of this systematic fissure-sealing program by evaluating data on the occlusal surfaces of the permanent first molars in the 15-year-old patients at seven public dental service clinics in Jönköping, Sweden.

A total of 815 patients (54 to 184 children per clinic) were treated in 1997 at the seven clinics; of these, 745 had been treated annually at the clinics since the eruption of their permanent first molars and were thus included in the present investigation. Of the 2,980 occlusal surfaces of the permanent first molars, 312 (11%) were restored immediately after eruption, and 212 (7%) were not sealed during the study period (Fig 427). Thus, 2,456 (82%) of the occlusal surfaces of the permanent first molars were sealed with an autopolymerizing sealant (Delton) when the children were at a mean age of 7 years 3 months.

In connection with the theoretical and clinical courses, the operators were advised to apply the sealants with chairside assistance. About 45 different operators applied the sealants: Dentists placed 92%, and dental hygienists or dental assistants placed 8%, of the sealants.

The results are presented in Fig 427. When the patients were 15 years old, 134 (6%) of the original 2,456 sealed permanent first molars had undergone a Class II restoration. Thus, 2,322 sealed occlusal surfaces were available for inspection. Of these surfaces, 1,806 (78%) were judged as caries free at the examination by the child's ordinary dentist. Restoration or caries in the occlusal pits and fissures was registered in 516

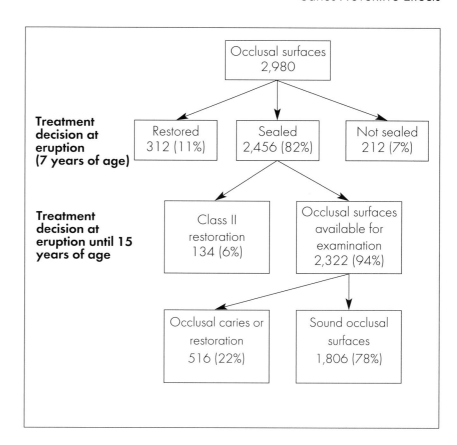

Fig 427 Course of treatment of the occlusal surfaces of the permanent first molars from eruption at approximately 7 years of age until 15 years of age. (From Wendt et al, 2001b. Reprinted with permission.)

(22%) teeth. Of these, 76 (3%) had only caries or restorations in the buccal pits of the mandibular first molar or in the palatal fissure of the Carabelli cusp of the maxillary first molar but not on the occlusal fissures. There was no statistically significant difference between the maxillary and mandibular molars in the distribution of sound surfaces. New sealant material had been reapplied to 15% (271) of the 1,806 sound occlusal surfaces. (Wendt et al, 2001b.)

Bravo et al (1997) compared Delton visible-light–cured fissure sealant with Duraphat (Rorer, Cologne, Germany) fluoride varnish in the prevention of occlusal caries in permanent first molars. A 48-month clinical trial was carried out in three groups of 6- to 8-year-old Spanish schoolchildren: a sealant group (104 children), in which Delton was used; a varnish group (112 children), in which Duraphat was used; and a control group (128 children). Sealant or varnish was applied to

all sound permanent first molars, according to the group. Replacement (sealant) and reapplication (varnish) were carried out every 6 months.

A survival analysis was used to describe the molar failures over time in the three groups. A Cox proportional hazards regression model was built to test the influence of group on molar failure. The median survival times were 28.6 months for the control molars and more than 48.0 months for both sealed and varnished molars. The Cox model indicated a hazard ratio of 0.177 for the sealant versus control comparison, 0.463 for varnish versus control, and 0.382 for sealant versus varnish (Bravo et al, 1997).

In another recent Swedish study, Holst et al (1998) evaluated whether placement of light-cured resin sealant is a preventive method appropriate for delegation to auxiliary personnel. After a 2-day course, providing theoretical as well as practical knowledge, 77 dental assistants working at 12 dental clinics sealed 3,218 permanent first and second molars in children. Indications for sealant were deep fissures and/or a high caries risk in the individual. The material used was the light-cured resin Delton Opaque. The sealants were examined once a year by the children's ordinary dentists, using internationally accepted criteria.

The result showed a total retention rate of 91% after 1 year and 69% after 5 years. There were differences in success rate among the clinics; the majority demonstrated a 5-year total retention rate between 70% and 94%. It was evident that the retention of sealants was influenced by the age of the children when the first molars were sealed but not by the position of teeth in the mouth. Thus, the compliance of the children and the eruption stage were important. Most failures occurred during the first year (35%), and failures were essentially the result of fractures (64%). At the end of the 5-year period, 8% of all teeth present had become carious.

The dental assistants sealed the fissures without any assistance, and none of them had any previous experience of fissure sealants. The authors concluded that sealing of fissures is a procedure well suited for delegation to dental assistants after proper education but that the patients should be followed up, because the success rate showed a great variation (Holst et al, 1998).

The reported incidence of sealants placed according to the etching-resin technique needing replacement or repair in contemporary studies averages between 5% and 10% per year (reviewed by Feigal, 1998). The greatest risk of sealant failure occurs soon after tooth eruption, when the distal marginal ridge has just cleared the soft tissue, thereby leaving the occlusal surface at risk for moisture contamination during the sealing procedure (Dennison et al, 1990). In addition, sealant failure on the buccal and lingual fissures of molars continues to be a recognized challenge, with failure rates higher than those on occlusal surfaces (Barrie et al, 1990; Cooney and Hardwick, 1994; Futatsuki et al,

1995; Messer et al, 1997). Higher failure rates in these difficult clinical situations are most likely the result of inadvertent moisture contamination.

Recent work on improved bonding of sealants to saliva-contaminated enamel may help to improve clincan confidence in sealant success, even when circumstances of application are less than ideal. Hitt and Feigal (1992) first reported the benefit of adding a dentin bonding agent between the etched enamel and the sealant as a way of optimizing bond strength in the face of moisture and salivary contamination. Other studies have confirmed the benefit of placing bonding agents under sealants on contaminated enamel to increase bond strength (Choi et al, 1997; Fritz et al, 1998), reduce microleakage (Borem and Feigal, 1994), enhance flow of resin into fissures (Symons et al, 1996), and improve short-term clinical success (Feigal et al, 1993).

In a 5-year clinical study, Feigal et al (2000) recently scored sealants using a split-mouth design: Half received sealant alone, and half received bonding agent plus sealant. Application procedures followed accepted protocols as published by the manufacturers: a low-speed, dry-brush cleaning of the surface, cotton roll isolation, a 30-second phosphoric acid gel etch, a 15-second rinse, air drying, application of sealant (Fluoroshield), and a 40-second light polymerization. The placement of bonding agent prior to sealant application was performed as follows: Once the surface had been etched, rinsed, and dried, a layer of bonding agent was applied to the surface with a hand-held brush. This layer of agent was then air thinned across the surface. The sealant was immediately applied over the bonding agent. Both materials were photocured together in one curing cycle of 40 seconds. Sealants were applied on the occlusal and buccal or lingual surfaces at the same visit if the buccal or lingual fissures required sealant treatment.

The treatment effects and potential risk factors for sealant failure were tested by means of a Cox regression model. A single-bottle dentin bonding agent was successful in reducing risk of sealant failure, with a hazard ratio (HR) of 0.53 (P = .014) for occlusal and 0.35 (P = .006) for buccal and lingual sealants. Variables that affected success differed between occlusal and buccal or lingual sealants, suggesting that failures on these two surfaces may be dependent on differing factors. Early eruption stage was a significant risk factor for both surfaces (occlusal HR = 2.91, P < .001; buccal and lingual HR = 1.52, P = .015). Behavior (HR = 1.96, P < .001), salivary problems (HR = 1.73, P = .002), and visually apparent variations in enamel (HR = 1.51, P = .018) were significant risk factors for occlusal sealants only. In addition to completing detailed analyses of risk factors for sealant survival, the study showed that single-bottle bonding agents protect sealant survival, yielding half the usual risk of failure for occlusal sealants and one third the risk of failure for buccal or lingual sealants (Feigal et al, 2000).

One concern of some dental practitioners is the possible danger of placing sealants over incipient caries lesions. Studies have shown, however, that the retention of sealant placed over incipient lesions is better than retention of sealant placed on those surfaces classified as caries free (Handelman et al, 1987). Lesions over which sealant material remains intact have not shown signs of progression, suggesting that the carious process has been arrested. Bacterial counts of material taken from these lesions after intervals as long as 2 years have shown huge reductions in the numbers of cultivable bacteria. It appears that as long as the sealant is intact over the pit and fissure, the viability of the bacteria will decrease and ultimately the caries lesion may become sterile (Arana, 1974; Going et al, 1978). Based on radiographic evaluation, it has also been shown that caries lesions in the fissures do not progress as long as the sealant is retained (Handelman et al, 1981, 1985, 1986).

Glass-ionomer sealants

The main potential advantages of glass-ionomer cement compared with the conventional resin-based fissure sealants are its ability to adhere to untreated enamel surface and the continuous release of fluoride from the cement. Fissure sealing with a glass-ionomer cement was introduced in 1974 by McLean and Wilson, who, in a clinical study, reported that 84% of sealants were completely retained after 1 year and 78% were retained after 2 years. In their study, fissures more than 100 µm in width were selected for sealing to allow for a thick bulk of cement. Later clinical trials on the retention of glass-ionomer cements have given contradictory results. While McKenna and Grundy (1987) showed a retention rate of 82.5% after 1 year, others reported a total or almost total loss of sealants within 6 months of placement (Boksman et al, 1987).

In a 5-year longitudinal Swedish study, Mejáre and Mjör (1990) evaluated the retention rate and caries-preventive effect of a hand-mixed glass-ionomer sealant (Fuji III) and two resin-based fissure sealants (Delton and Concise WS [3M]). Four dentists sealed 208 occlusal fissures of permanent molars and premolars with Fuji III, Delton, or Concise WS. Clinical assessments and a replica-scoring technique were used to register the performance of the sealants at baseline, after 6 to 12 months, and then yearly up to 5 years.

As judged clinically, 61% of the glass-ionomer sealants were lost within 6 to 12 months and 84% after 30 to 36 months. Although total loss was recorded clinically for the majority of the glass-ionomer sealants, some retained sealant was observed in 93% of their tooth replicas. The clinical evaluation of the resin-based sealants showed an average complete retention rate of 90% after 4.5 to 5 years. The corresponding figure with the replica technique was 58%.

Caries was recorded in 5% of the resin-based sealed surfaces (six Delton-sealed and two Concise-sealed teeth). In six of them, caries was registered after 6 to 12 months while the other two lesions were diagnosed 2.5 and 4 years after sealing. Caries was not observed on the glass-ionomer cement–sealed surfaces. The differences in retention rates between the replica-scoring technique and the clinical scoring of the resin-based sealants probably reflect the difficulty in clinically judging the retention of a degraded or worn sealant. The replica technique was more sensitive in this respect. The replica technique was also more sensitive in detecting areas with small amounts of remaining sealant. Although 84% of the glass-ionomer sealants were judged clinically to be totally lost, the replicas revealed areas of retained sealants in almost all of the teeth (Mejáre and Mjör, 1990).

This finding, in combination with the well-documented fluoride-release from glass-ionomer materials, may explain why a caries-preventive effect has been observed even after the glass-ionomer sealant has been clinically registered as lost. Later studies have confirmed the experiences of this study (Arrow and Riordon, 1995; Forss et al, 1994; Forss and Halme, 1998; Komatsu et al, 1994).

Arrow and Riordon (1995) compared the retention and caries-preventive effects of fissure sealants based on a glass-ionomer material (Ketac-Fil; Espe, Seefeld, Germany) with those of a traditional resin-based fissure sealant (Delton) placed under field clinical conditions in a split-mouth longitudinal study.

In Perth, western Australia, 465 schoolchildren with a mean age of 7 years (0.72 SD) received sealants on the occlusal surfaces of sound homologous permanent first molar pairs. Test (glass-ionomer cement) and control (resin) sealants were systematically allocated to left and right sides, based on the child's month of birth. Sealants were placed by dental therapists. After 3.64 (± 0.11) years, 415 children were examined by different clinicians, and the clinical status of the teeth and the extent of sealant retention were recorded. Sealants were deemed retained when at least two thirds of the fissure pattern was still sealed. In 252 tooth pairs, neither sealant was retained to this extent. In 71 pairs, the glass-ionomer cement was not retained and the resin sealant was retained. In 40 pairs, the reverse occurred.

Of the 824 teeth examined, only 37 became carious. Six glass-ionomer cement–sealed and 31 resin-sealed teeth developed caries. The effectiveness of the glass-ionomer cement was 80.6% (95% confidence interval: 59.6%, 90.7%). The net gain in caries reduction was 6.1% (95% confidence interval: 3.3%, 8.9%). The relative risk was 0.19 (95% confidence interval: 0.09, 0.40). A relative risk value of less than 1.00 indicates a protective effect of the test agent; this means that glass-ionomer cement–sealed teeth had a 19% chance of becoming carious compared with resin-sealed teeth.

Only 4.5% of the teeth included in this study developed caries during the study period. The study was conducted among children who resided in a fluoridated area, where fluoride toothpaste predominates, and who, in addition, received fissure sealants. The net gain in caries reduction was thus low, and the appropriateness of a general policy of fissure sealing all permanent first molars in this area would have to be questioned. Nevertheless, fissure sealants are likely to be considered necessary for some at-risk children, and then the apparent ability of glass-ionomer cement sealants to prevent caries more effectively than a resin material is of interest.

The authors speculated that retention of small amounts of glass-ionomer cement sealant may provide adequate protection to the entire pit and fissure system of a tooth either by sealing the deepest part of the fissures or by maintaining a locally elevated fluoride concentration. Whether the prevention of caries is due to obturation of the fissures or to both modes of action, it would appear that long-term retention of glass-ionomer cement fissure sealants is not a prerequisite for caries prevention and such treatment should perhaps be regarded more as a very prolonged fluoride application rather than as a sealing of fissures. This study suggests that complete retention of glass-ionomer cement sealant is not necessary for caries prevention in newly erupted permanent first molars (Arrow and Riordon, 1995).

Under field clinical conditions, both resin and glass-ionomer cement sealant materials may be poorly retained over longer periods, but glass-ionomer cement sealants may offer advantages because their caries-preventive effect continues after the bulk of the sealant has been lost. In addition, the risk for initiation of caries should be over if the occlusal surfaces are maintained caries free until full functional wear is established, as discussed earlier. Williams and Winter (1981) compared an early glass-ionomer cement material with a first-generation unfilled resin fissure sealant and found that, after 4 years, retention was similarly poor for both materials (about one third of each group of teeth retained the sealant) but significantly less caries had occurred in teeth sealed with the glass-ionomer cement material.

Forss et al (1994) compared the retention and caries-preventive effect of a glass-ionomer sealant (Fuji III) and Delton light-cured resin sealant. Three health center dentists applied the sealants to 166 children; glass-ionomer sealants were placed on one side and resin-based sealants were placed on the contralateral side of the mouth. After 2 years, one pair of molar teeth in each of 151 children was compared. The average age of the children was 11 years, 3 months (range: 5 years, 8 months to 14 years, 11 months) at the time of sealant application.

At the end of the study, 26% of glass-ionomer and 82% of resin-based sealants were totally present ($P < .001$). During the 2 years, in both groups, 4.6% of the sealed surfaces became carious. The results showed that the retention of glass-ionomer sealants is inferior to that of the resin-based sealants. In this study, however, no difference in caries incidence on the sealed surfaces was observed (Forss et al, 1994). This may be due to the different mechanism of caries prevention of the two sealant materials, as discussed earlier.

The same persons were invited to a dental checkup 6.1 to 7.8 (mean of 7.1) years after the application of sealants. Of the original group, 111 persons (66.8%) participated in this study. The retention of sealants and the caries status of occlusal surfaces and adjacent proximal surfaces were recorded. On the sealed occlusal surfaces, 10% of glass-ionomer and 45% of light-cured resin sealants were totally present and 9% of glass-ionomer cement and 20% of resin sealants were partially present. It was found that 23.0% of the occlusal surfaces sealed with glass-ionomer cement and 16.5% of those sealed with light-cured resin were carious or restored (Forss and Halme, 1998).

Considerably better results for glass-ionomer cement sealants (Fuji III) were shown in a Japanese study in 5- to 7-year-old children (mean of 6.6 years). Dental examinations were conducted twice a year over 3 years. Where there was partial or total sealant loss, sealant was reapplied. When teeth with reapplied sealant (33.0% to 65.7% of teeth at each examination) were included, the retention of sealants after 3 years was 70.3%. The caries reduction on occlusal surfaces of molar teeth, compared to no-treatment controls, was 66.5%. The caries-preventive efficacy of glass-ionomer cement sealants was clinically significant (Komatsu et al, 1994).

In a recent 3-year field trial in Zimbabwe, Frencken et al (1998) treated schoolchildren (mean age of 13.9 years) with atraumatic restorative treatment or fissure sealant. Glass-ionomer cement (restorative type II, 1) was used as the restorative and sealant material. After primitive mechanical cleaning of the tooth with a periodontal probe and wet cotton pellets, the sealant was placed using the press-finger technique. Sealants were placed only on surfaces diagnosed as early enamel lesions and on some small dentin lesions. After 3 years, 50.1% (95% confidence interval: 55.1%, 45.1%) of the fully and partially retained sealants survived; the range was 68.5% to 25.9% among the operators. Regardless of the low rate of retention, the sealed surfaces had a four times lower chance of developing caries than did unsealed surfaces with early enamel lesions over the 3-year period.

Resin-modified glass-ionomer sealants

As discussed earlier, almost 100% of occlusal caries is initiated in the molars during the extremely long eruption time, 12 to 14 months for first molars and 14 to 18 months for second molars, because the accumulation of plaque is very rapid and undisturbed in the absence of occlusion and functional wear. This is particularly the case in the distal and central fossae and related fissures, which is the reason why these sites exhibit higher caries prevalence than the mesial fossae. On average, the permanent teeth erupt 6 months earlier in girls than in boys.

The tooth enamel is much more caries susceptible during eruption and the following year, until the so-called secondary maturation is achieved. That is why the caries-preventive effect of fluorides on such teeth is double that found on teeth that have been exposed to the oral environment for years. However, if the occlusal surfaces are maintained caries free until they have reached full occlusion and functional wear, the risk for initiation of caries is more or less over. That is why the occlusal surfaces of the premolars, which have only a 1- to 2-month eruption time, rarely are carious.

The aforementioned facts were the reasons why 260 subjects with the highest caries prevalence and caries risk were selected for a 3-year longitudinal fissure sealant study from among more than 700 boys aged 13 years at the elementary schools in the city of Karlstad, Sweden (Axelsson et al, 1997c). The aim was to compare the caries-preventive effects of a new light-cured resin-modified glass-ionomer sealant and a conventional light-cured resin sealant in the mesial, central, and distal fossae of erupting or newly erupted second molars.

At baseline, the eruption stage of the mesial, central, and distal fossae was recorded. Figures 340 and 341, earlier in this chapter, show the eruption stage of the three fossae in the four second molars (teeth 17, 27, 37, and 47) at baseline and at the 3-year reexamination, respectively. On average, only about 20% of the fossae had reached occlusion at baseline (lowest in the distal fossae and highest in the maxillary mesial and central fossae). The highest percentage of fossae were erupted without occlusion. Green occlusal wax was used to determine whether occlusion was reached.

After baseline examination, the subjects were randomly assigned to the test group or positive-control group. In the test group, the occlusal surfaces to be sealed were mechanically cleaned with a pointed rotating brush and pumice. The surfaces were then washed, dried, and chemically cleaned for 20 seconds with a conditioner (Cavity Conditioner; GC Corp), which is an aqueous solution of 20% polyacrylic acid and 3% aluminum chloride. Thereafter, the surfaces were washed with water spray and dried, and moisture control was established. A new hand-mixed light-cured resin-modified glass-ionomer sealant (Fuji III LC) was placed in the fossae and related fissures with a probe and light cured for 20 seconds.

The occlusal surfaces of the second molars to be sealed in the positive-control group were mechanically cleaned, washed, and dried in similarity to the test group. The surfaces were etched for 30 seconds with 35% phosphoric acid, washed for 30 seconds with water spray, and dried with clean compressed air. When moisture control was established, an opaque, light-cured conventional resin-based sealant (Delton Opaque) was placed in the fossae and related fissures.

In the minority of teeth that had full occlusion, green occlusal wax was used as a guide to check possible sealant excess. Excess was eliminated with a pear-shaped finishing bur.

All the sealant procedures were carried out by a dental hygienist. Teeth that were already restored or sealed at baseline were excluded from the study. In the test group (Fuji III LC), 383 teeth were sealed: 100 maxillary right second molars (tooth 17), 96 maxillary left second molars (tooth 27), 95 mandibular left second molars (tooth 37), and 92 mandibular right second molars (tooth 47). In the positive-control group (Delton Opaque), 421 teeth were sealed: 105 in tooth 17, 108 in tooth 27, 109 in tooth 37, and 99 in tooth 47.

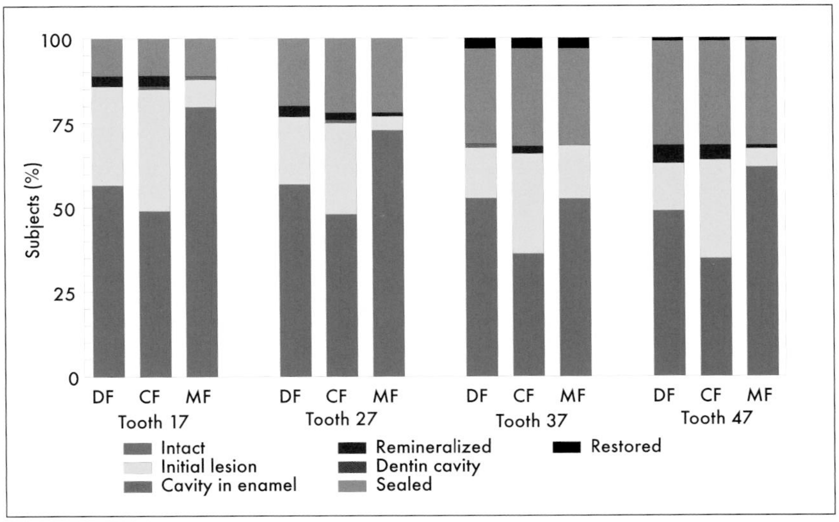

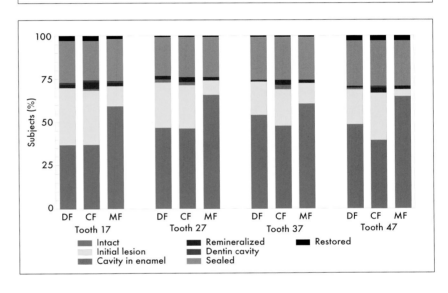

Fig 428 Baseline status in a 13-year-old test group with light-cured resin-modified glass-ionomer sealant (Fuji III LC) placed in the distal fossae (DF), central fossae (CF), and mesial fossae (MF) of the four second molars (teeth 17, 27, 37, and 47). (From Axelsson et al, 1997c. Reprinted with permission.)

Fig 429 Baseline status in a 13-year-old positive-control group with light-cured conventional resin-based sealant (Delton Opaque) placed in the distal fossae (DF), central fossae (CF), and mesial fossae (MF) of the four second molars (teeth 17, 27, 37, and 47). (From Axelsson et al, 1997c. Reprinted with permission.)

Reexaminations were carried out after 1 and 2 years to assess retention and caries control. The final reexamination after 3 years was performed as a blinded assessment by one dentist using visual inspection, probing, and the aid of bitewing radiographs. Figures 428 and 429 show the status at baseline in the distal, central, and mesial fossae of the four second molars in the test group and the positive-control group, respectively. About 25% of the teeth were already sealed, confirming that early use of fissure sealant is a well-established preventive measure in selected caries-risk patients. On the other hand, between 25% and 35% of the central and distal fossae already exhibited noncavitated initial enamel lesions, which should be arrested or at least prevented from progression until the tooth has reached full occlusion and function.

The status after 3 years is shown in Figs 430 (Fuji III LC) and 431 (Delton). Although almost 75% of the light-cured resin-modified glass-ionomer sealants were partially or totally lost after 3 years, only two mandibular second molars had been restored (for unknown reasons) and no caries lesions in dentin were discovered. On the other hand, almost 100% of the enamel (incipi-

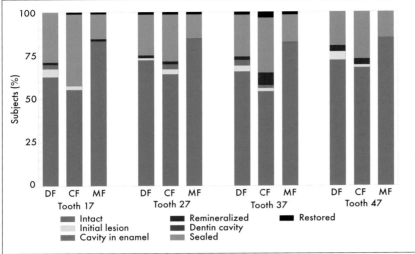

Fig 430 Status after 3 years in the test group (Fuji III LC). Note the high percentage of intact fossae. (DF) Distal fossae; (CF) Central fossae; (MF) Mesial fossae. (From Axelsson et al, 1997c. Reprinted with permission.)

Fig 431 Status after 3 years in the positive-control group (Delton). Note the high percentage of retained sealants. (DF) Distal fossae; (CF) Central fossae; (MF) Mesial fossae. (From Axelsson et al, 1997c. Reprinted with permission.)

ent) caries lesions at baseline were remineralized and diagnosed as intact after 3 years. Because the teeth were fully erupted, the risk for new development of caries should be over. Thus, it may be concluded that close to 100% of the initially sealed occlusal surfaces should remain caries free, even if the subjects were selected caries-risk individuals.

In the positive-control group (Delton Opaque), about 80% to 90% of the sealed surfaces exhibited retained sealants. Only one maxillary second molar had been restored during the 3-year period and no caries lesions in dentin de-

veloped. Because the study was blind for the examiner, opaque sealant was used. Therefore, it was impossible to know what happened in the sealed fossae that were initially carious. However, earlier studies have shown that lesions normally do not progress as long as the sealant is fully retained.

In conclusion, this study revealed that the use of either light-cured resin-modified glass-ionomer sealant or light-cured conventional resin-based sealant resulted in nearly 100% caries prevention for 3 years in initially erupting or newly erupted permanent second molars in selected caries-risk subjects (Axelsson et al, 1997c). However, about 25% of the occlusal surfaces were sealed, about 25% exhibited noncavitated enamel (initial) lesions, and a few were restored at baseline. This finding indicated that fissure sealants should be used as early as possible during the eruption of the molars in selected caries-risk individuals. However, moisture control is difficult to achieve during the early eruption stage, particularly in the distal fossae. Light-cured resin-modified glass-ionomer sealants are less sensitive to moisture than are conventional resin-based fissure sealants, which require etching of the enamel surface for mechanical retention. In addition, resin-modified glass-ionomer sealants are regarded as slow-release fluoride agents that can be recharged with fluoride from other topical fluoride agents. Therefore, light-cured resin-modified glass-ionomer sealants are recommended as the first choice during early eruption of molars in selected caries-risk patients.

Fluoride release and uptake of glass-ionomer materials

The source of fluoride ions from glass-ionomer cements are the calcium fluoride (CaF_2), strontium fluoride (SrF_2), lanthanum fluoride (LaF_2), sodium hexafluoroaluminate (Na_3AlF_6), and aluminum trifluoride (AlF_3), inclusions of the low-temperature glasses used to formulate the powder components. The primary role of these fluorides is to lower the glass fusion temperature during the manufacturing process, although they also improve handling properties and increase the strength and translucency of the cement. During setting, the fluoride ions produced form strong soluble aluminofluoride complexes such as AlF^{2+}, which prevent premature gelation of polyions by aluminum ions. In fully set cements, fluoride is located in the partially degraded glasses that form the glass core and in the polysalt matrix, primarily in the form of aluminum complexes (Wilson et al, 1993).

When a fully set glass-ionomer cement is exposed to neutral aqueous solutions, it absorbs water and releases ions such as sodium, silica, calcium, and fluoride. Of the total amount of fluoride in the set cement, only a small fraction is available for release. This amount is not related to the total fluoride content of the cement but depends on the amount of sodium that maintains the cement's electron neutrality after fluoride release. Only the fluoride present in the matrix is available for elution at neutral conditions.

Although many studies have reported a sustained release of fluoride from set glass-ionomer cements in distilled water over a long period of time (Forsten, 1990; Schwartz et al, 1984; Wilson et al, 1985), the mechanism of fluoride release is still not completely understood. It is generally accepted that two reactions are involved: a short-term reaction of high fluoride release, corresponding to initial elution resulting from the postsetting maturation process, and a long-term reaction of low release, attributed to equilibrium diffusion processes.

The fact that the fluoride released from conventional products is mostly sodium fluoride (NaF), along with the observation that this is not a critical salt for the formation of the polysalt matrix, explains the lack of anticipated weakening of the materials following fluoride release (Wilson and McLean, 1988). A significant increase in fluoride release is observed at low pH values, suggesting that an erosive mechanism is activated at these conditions from the preferential dissolution of the glass particles in the matrix (Cranfield et al, 1982; Forss, 1993).

Another factor governing fluoride release is its concentration in the immersion solution. In the presence of an inverse fluoride concentration gradient, glass-ionomer cements may absorb fluoride from the environment and release it again under specific conditions (Forsten, 1991, 1998). Thus, the concept of fluoride recharging of old glass-ionomer restorations was introduced.

Creanor et al (1994, 1995) have shown that various glass-ionomer materials are able to take up ionic fluoride and release it subsequently.

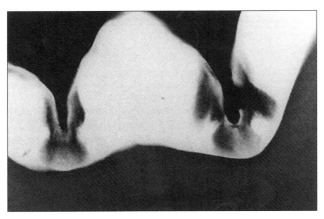

Fig 432 Cross-section microradiograph showing noncavitated enamel lesions around two groovelike fissures. (From Pearce et al, 1999. Reprinted with permission.)

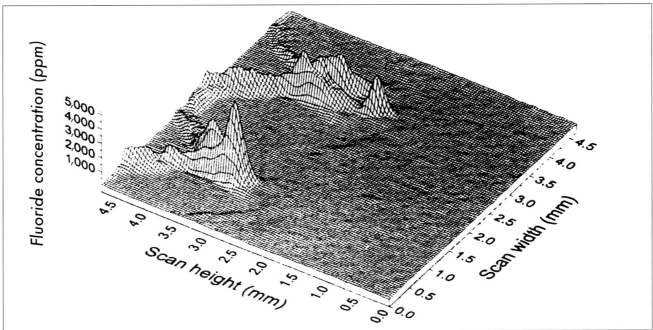

Fig 433 Surface plot of fluoride concentration from the occlusal surface of the tooth shown in Fig 432. Note the very high fluoride concentration in the two carious, groove-shaped fissures. (From Pearce et al, 1999. Reprinted with permission.)

The uptake and release of fluoride were correlated to the concentration and frequency of the fluoride solution used. Seppä et al (1993, 1995) showed that topical application of a fluoride gel on glass-ionomer cements significantly increased the fluoride release and reduced the acid production of plaque and *Streptococcus mutans*. These data further indicate the possibilities of recharging glass-ionomer materials with fluoride by using fluoride varnishes with high fluoride concentration and a slow-release mecha- nism, such as Duraphat (5% NaF) and Bifluorid 12 (6% NaF + 6% CaF_2) (Voco).

The important question of whether the fluoride released from glass-ionomer cements is sufficient to increase the fluoride concentration in saliva was addressed in several studies during the early 1990s. These studies provided evidence of elevated fluoride concentraion in saliva for a period of a few weeks after application of glass-ionomer cement, implying that these materials may be effective short-term intraoral fluoride-releasing devices (Forss et al, 1995; Hallgren et al, 1990; Hatibovic-Kofman and Koch, 1991; Hattab et al, 1991; Koch and Hatibovic-Kofman, 1990).

Recent studies by Pearce et al (1999) have shown that the fluoride concentration in fissures with intact enamel ranges from 1,800 to 4,200 ppm in the outer 60 µm of the surface at the bottom of the fissures and into 120 µm at the entrance of the fissures, while it ranges from 2,700 to 10,000 ppm in fissures with advanced enamel lesions (Figs 432 and 433). Because the most significant caries-preventive effects of fluorides are to act as catalysts for remineralization and inhibitors of demineralization, it is obvious that active placement of a fluoride-releasing and highly flowable light-cured resin-modified sealant to the bottom of the fissures should be very efficient for prevention of fissure caries as well as remineralization of enamel caries and noncavitated dentin lesions in the fissures.

However, the fluoride release and uptake is highest in conventional glass-ionomer cements and gradually reduced in resin-modified glass-ionomer materials and compomers in relation to the content of resins and composite. Resins and composites cannot be recharged with fluoride in the mouth.

The anticariogenic effect of glass-ionomer materials is well documented experimentally and in vivo (Benelli et al, 1993; Mukai et al, 1993; Nagamine et al, 1997; Qvist et al, 1997; Skartveit et al, 1990; Tam et al, 1997; Ten Cate and van Duinen, 1995; Weerheijm et al, 1993, 1999; for reviews, see Gao et al, 2000; Mount, 1999; and Randall and Wilson, 1999). A new approach to further improve the anticariogenic effects of glass-ionomer materials is the addition of chlorhexidine gluconate (CHX) to a fissure sealant (Hoszek et al, 1998). The addition of 10% CHX and 11% tartaric acid mixed in distilled water to glass-ionomer powder delayed the setting, improved penetration into the fissures, and significantly increased the antimicrobial effect in the fissures compared to a conventional glass-ionomer sealant.

Alternative Methods for Prevention of Occlusal Caries

Intensified plaque control and topical use of fluoride during eruption of the molars

Studies of occlusal caries in erupting first permanent molars have shown that a needs-related, noninvasive preventive program arrested or inactivated about 80% to 90% of active enamel lesions (Carvalho et al, 1992). After 3 years, only 2% of the occlusal surfaces had to be sealed and none had to be restored (see Fig 117 in chapter 2). No progression of lesions into dentin was observed on bitewing radiographs. The parents were taught to clean the occlusal surfaces of their children's erupting molars using a specific toothbrushing technique and fluoride toothpaste (see Fig 108 in chapter 2). In selected children with higher caries risk, home care was supplemented at needs-related intervals by PMTC and application of 2% NaF solution. In a matched control group, despite intensive use of topical fluoride and fissure sealing of 70% of the occlusal surfaces, 2% of surfaces had to be restored (Carvalho et al, 1992).

This low-cost, noninvasive program was subsequently implemented on a large scale in Nexö, Denmark. Over a 10-year period, the percentage of caries-free 12-year-old children increased from about 30% to almost 90% without the use of fissure sealants (Thylstrup et al, 1997). The large-scale implementation of this strategy in Nexö resulted in continuously reduced values for decayed, missing, or filled surfaces (DMFSs) in 8-year-old children in spite of little use of fissure sealants (Fig 434). The area is approaching an almost 100% caries-free population (Thylstrup et al, 1997).

Figure 118 in chapter 2 shows the mean DMFSs by age in Nexö compared with the national figures in Denmark in 1987 and 1994. The 1994 Nexö data represent children who have followed this program from the age of 6 years. In this con-

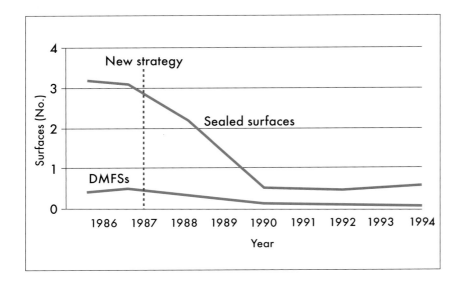

Fig 434 Cross-sectional decayed, missing, or filled surfaces (DMFSs) and sealed surfaces in 8-year-old children from 1986 to 1994. (Modified from Thylstrup et al, 1997 with permission.)

text, epidemiologic data from the WHO Global Data Bank (1997) reveals that the lowest caries prevalence in 12-year-old children is reported in Denmark, Sweden, Finland, Switzerland, Australia, and some countries in Central Africa. In populations with high caries prevalence and incidence, however, there is still an indication for very early application of fluoride-releasing fissure sealants in the permanent first and second molars as a highly efficient caries control measure.

The same concept of intensified plaque control, targeting especially the fissures of erupting molars, was applied in a longitudinal study of 6- to 8-year-old children (first molars) and 11- to 13-year-old children (second molars) in Solntsevsky, Moscow, Russia (Kuzmina, 1997). Because there was no school-based preventive program, randomized negative-control groups were selected. Figures 119 and 120 in chapter 2 show the mean caries incidence, in new DMFSs, from 1994 to 1996 in the study groups, control groups, and corresponding age groups in Denmark. The results for the test group compared well with data from Denmark, confirming that a low-cost, needs-related plaque control program is efficient even in populations with a very high incidence of caries.

These studies show that if the fissures of the molars are maintained free of caries during the risk period, ie, throughout the entire eruption period to the stage of occlusal function, the risk for development of fissure caries is past. In the occlusal surfaces of fully erupted, caries-free molars, general use of fissure sealants represents costly overtreatment. Among children with relatively low caries prevalence, the procedure should be reserved for very sticky, plaque-retentive fissures in the molars of selected children at high caries risk.

The main drawback to the use of sealants compared to the low-cost and low-technology plaque control fluoride program is that this technique calls for specific materials and highly educated personnel to apply the fissure sealant individually to each tooth in question. Further, frequent follow-up is needed to achieve the greatest benefit. In spite of more than 25 years of routine use, the specific diagnostic criteria for when to seal and when not to seal have not been made clear because they ultimately depend on the clinical judgment of the individual dentist or dental hygienist. In addition, the use of sealants for caries prevention is not without problems from a cost-effectiveness point of view. Thus, nonoperative, low-cost techniques, used in combination with individualized application of sealants for occlusal caries prevention, should be further developed (Axelsson et al, 1993a; Axelsson, 1994, 1998; Carvalho et al, 1992; Selwitz et al, 1995).

Use of fluoride and chlorhexidine varnishes

In the very early stages of eruption, fluoride varnish (Duraphat, Fluor Protector [Vivadent], and Bifluorid) and chlorhexidine varnish (Cervitec; Vivadent) may be used until accessibility and moisture control allow the use of fissure sealant in the entire fissure system, including the distal fossae. However, it must again be stressed that etching and use of resin-based fissure sealants are very risky in erupting teeth if full moisture control cannot be achieved. Therefore, light-cured resin-based glass-ionomer sealants should be the first choice of material, placed as early as possible, and regarded as a slow-release fluoride agent during this high-risk period.

In a 48-month clinical study, Bravo et al (1997) showed that the survival of intact occlusal surfaces of first molars in 6- to 8-year-old Spanish children was 48 months in teeth sealed with a light-cured sealant (Delton) as well as those treated with Duraphat fluoride varnish; the survival was 28 months in the teeth of a control group. Every 6 months, partial or totally lost sealants and fluoride varnish were reapplied.

The effects of one and two applications of 40% CHX varnish on the numbers of mutans streptococci in human dental fissure plaque from molar and premolar teeth were evaluated by Ie and Schacken (1993). Twenty-nine subjects (aged 20 to 30 years) participated in the study and were randomly assigned to one of three groups. In each subject in group 1 (control group), two fissures were treated with a placebo varnish containing no CHX. Fissures in group 2 received a single application of 40% CHX varnish. Fissures in group 3 received an additional chlorhexidine varnish application 1 week after the first application.

Fissure plaque samples were taken prior to the first application of chlorhexidine varnish and 1, 2, and 4 months thereafter. Compared with levels of mutans streptococci found in the fissures in the control group, mutans streptococci was significantly suppressed in plaque from group 2 for up to 2 months and in plaque from group 3 for up to 4 months after application. Mutans streptococci were suppressed more strongly in premolars than in molars and more strongly and for a longer period of time in fissures of premolar teeth treated twice than in fissures of premolars treated once (Ie and Schacken, 1993).

Bratthall et al (1995) evaluated the possibility of reducing development of fissure caries using an antimicrobial varnish (Cervitec). Children aged 7 to 8 years and 12 to 13 years, 251 in each age group, were selected in Bangkok, Thailand. Subjects had at least two sound contralateral permanent molars. A split-mouth method was used with one test and one control tooth within the same arch. At baseline and after 2 years, all children were investigated for DMFSs and DMFT. In addition, the size of any cavities was estimated. From 200 children, plaque samples of test and control occlusal surfaces were collected at baseline and after 1 year and processed to estimate the number of mutans streptococci. Cervitec varnish, containing 1% CHX and 1% thymol, was applied at baseline, after 3 to 4 months, and after 8 to 9 months.

Cervitec varnish reduced development of fissure caries significantly. The levels of salivary mutans streptococci at baseline were significantly correlated with caries status at baseline and with total caries increment over the 2-year period. Caries development in a fissure was significantly correlated to the level of plaque mutans streptococci at that same site. Three months after the last application of varnish, a certain reduction of mutans streptococci in plaque could be observed in the test teeth. When the sizes of the lesions were compared, more large cavities were found in the untreated teeth. The authors concluded that varnishes should be considered as further options for prevention of fissure caries, possibly in more individualized programs or in combination with already established methods (Bratthall et al, 1995).

These studies show that both fluoride varnishes and CHX varnishes may be useful alternatives at least to etching and use of resin-based fissure sealants during the eruption of molars. Mixing of a compatible fluoride varnish (Fluor Protector) and CHX varnish (Cervitec) in a 1:1 ratio would achieve a slow release of fluoride as well as an efficient antimicrobial agent.

NEEDS-RELATED PROGRAMS FOR PREVENTION OF OCCLUSAL CARIES

Clinical studies comparing amalgam restorations with sealant therapy have shown a positive efficacy of sealant application in a caries-susceptible population (Straffon and Dennison, 1988). Sealants can be applied in less time than amalgam and resin composite restorations and can be placed by dental auxilliaries. The cost of a sealant application has been estimated to be half or less than half of the cost for a one-surface amalgam or resin composite restoration. However, the practical application of sealants must be meticulous to minimize the need for reapplication. Furthermore, amalgam or resin composite restorations and sealants should not be regarded as alternative methods because sealants are meant to prevent caries, while restorations are used to restore carious teeth. The value of maintained health is difficult to assess solely on a monetary basis.

Mitchell and Murray (1989) concluded that, despite the large number of clinical trials demonstrating the effectiveness of fissure sealants in preventing occlusal caries, it has been suggested that the fundamental principles of cost effectiveness have not been applied to the use of fissure sealants. There is little in the literature that allows up-to-date judgments to be made. A review of the literature suggests that, while the limitations in design of some of the earlier fissure sealant studies make it difficult to draw firm conclusions, the results of later studies, particularly those that evaluated fissure sealants as an integral part of routine dental care, are encouraging. A number of

factors, such as caries prevalence of the population, individual caries risk, selection of teeth (molars, eruption stage, and occlusal morphology), dental care standard, personnel resources (eg, dentists, dental hygienists, and dental assistants), materials, and methods, influence the economic viability of fissure sealants. These factors must be considered and controlled if sealants are to be used cost effectively.

For example, in the early 1970s, in Scandinavia, as in most industrialized countries, there were practically no unrestored occlusal surfaces of permanent first molars in 11- to 12-year-old children. This was due to the combined effect of a very high caries incidence and the universal application of G. V. Black's principle of extension for prevention. Many caries-free fissures were preventively sealed with amalgam restorations. During such times, a more generous use of fissure sealants should have been cost effective. However, in a 5-year Swedish longitudinal study, Månsson (1977) followed caries progression in the fissures of the permanent first molars from eruption. Despite the prevailing very high caries incidence in Sweden, only 50% of the occlusal surfaces became carious; ie, 100% general successful use of fissure sealants would have resulted in maximum caries prevention in 50 of 100 occlusal surfaces. This was also confirmed in the case-control split-mouth study designed by Simonsen (1991) during the same period. After 15 years, he found a 62% caries-preventive effect of fissure sealants.

If restorative therapy were limited to caries with cavitation into dentin, at the current very low caries incidence in Scandinavia, for example, restorations would be indicated in only about 5% of the occlusal surfaces. Cavitation confined to the enamel would be managed by minimal intervention: so-called preventive sealant restorations (fissure blocking) or fissure sealants. Widespread use of fissure sealants cannot be justified as cost effective in the Swedish population today because of the existence of very efficient school-based, needs-related caries-preventive programs. Thus 100% success of gen-

Type of fissure	Caries risk	Treatment	%
	No or low	PMTC + fluoride varnish	70
	High or very high	PMTC + fissure sealant (glass-ionomer)	10
	No or low	PMTC + fissure sealant (glass-ionomer)	10
	High or very high	PMTC + fissure sealant (glass-ionomer)	10

Fig 435 Preventive program for the occlusal surfaces of the molars, selected at the beginning of molar eruption. The aims are to maintain caries-free occlusal surfaces and, at the most, to use a fissure sealant or so-called fissure blocking.

eral use of fissure sealants would have prevented only 5 of 100 sealed occlusal surfaces from restoration of cavitated dentin lesions.

The caries prevalence in Switzerland is also low. Thus, Marthaler (1995) found that only 10% of sound fissures in Zurich schoolchildren became carious within 4 years; therefore, 90% should not have been sealed. He recommended only the preventive restorative sealing of discolored fissures, 27% to 47% of which became carious within the same period.

From a cost-effectiveness point of view, Fig 435 shows an overall simple model for prevention of occlusal caries by needs-related use of fissure sealants (light-cured resin-modified glass-ionomer sealant) in erupting or newly erupted molars with sticky, deep risk fissures and/or in selected caries-risk individuals. In populations with high caries prevalence, a higher percentage of erupting as well as fully erupted caries-free molars should be sealed. Caries-free erupted molars with full occlusion and functional wear in individuals and populations with low caries prevalence should not be sealed. In such individuals and populations, placement of preventive restorative sealants (so-called fissure blocking), is recommended. This procedure is restricted only to sticky risk fissures with suspected caries lesions and to achieve accessibility for visual in-

spection and correct diagnosis. If the lesion is cavitated into the dentin, a minipreparation is performed. This may be supplemented with noninvasive elimination of infected carious dentin with the Cariosolv technique (Medi-Team, Gothenburg, Sweden).

For fissure blocking, light-cured resin-modified glass-ionomer sealants or etching and light-cured resin-based fissure sealants are recommended. Type II light-cured resin-modified glass-ionomer cement or etching, bonding, and restoration with compomer materials are recommended for restoration of noninvasive cavities.

In populations with relatively low caries prevalence, well-organized school-based dental care systems, and sufficient personal resources (eg, Scandinavia), a more detailed program based on early and comprehensive risk prediction is recommended for prevention of fissure caries. Figure 436 exemplifies criteria for early prediction of caries risk, ranging from no risk (C0) to high risk (C3), in 5- to 6-year-old children before eruption of the permanent first molars and in 11- to 12-year-old children before eruption of the permanent second molars. Girls, on average, start 6 months before boys. Recommended variables for risk prediction are caries prevalence (primary teeth in 5- to 6-year-old chil-

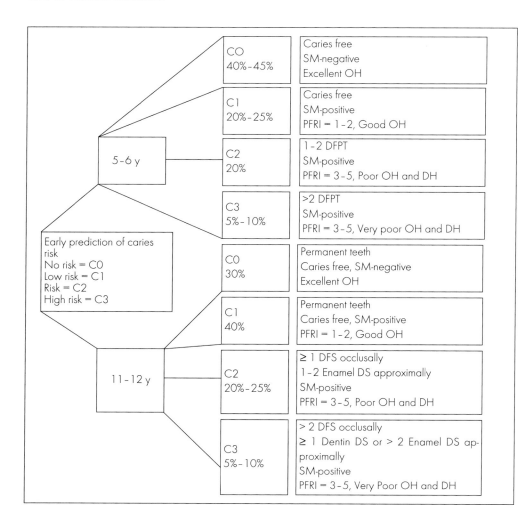

Fig 436 Schedule for early prediction of fissure caries in 5- to 6-year-old children (first molars) and 11- to 12-year-old children (second molars). (SM) Streptococcus mutans; (OH) oral hygiene; (PFRI) Plaque Formation Rate Index; (DFPT) decayed and filled primary teeth; (DH) dietary habits; (DFS) decayed and filled surface; (DS) decayed surface.

dren and on occlusal and approximal surfaces in permanent teeth in 11- to 12-year-old children), salivary screening for *S mutans*, Plaque Formation Rate Index (PFRI), standard of oral hygiene, and dietary habits. Based on this individual risk prediction, a needs-related preventive program is recommended (Fig 437).

Key-risk age group: Age 5 to 7 years

All the parents of 5- to 7-year-old children are educated to observe, and report to the responsible dentist or dental hygienist, when the different permanent first molars emerge. The parents are shown how to clean the occlusal surfaces with a specific toothbrushing technique (see Fig 108 in chapter 2) and fluoride dentifrice twice daily. In selected caries-risk patients (C2 and C3), semiprofessional mechanical toothcleaning could be performed in the occlusal fissure systems by the parents, using an automatic electric toothbrush with a rotating pointed brush (Rota-dent; Pro-Dentec, Batesville, AR). However, the parents are responsible for cleaning all tooth surfaces, not only the occlusal surfaces, twice daily. In no-risk and low-risk children (C0 and C1), supplementary PMTC should be followed by application of a fluoride varnish as early as possible during the eruption of the first molars—once in C0 children and twice in C1 children.

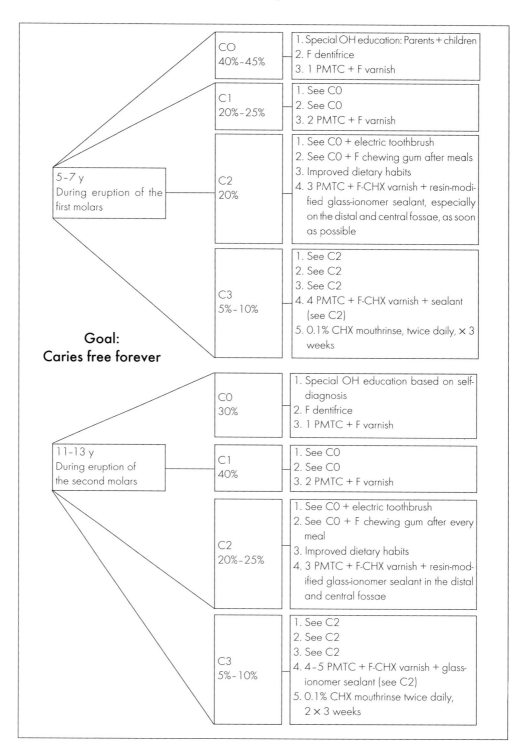

Fig 437 Schedule for cost-effective prevention of fissure caries, during the eruption of the molars in 5- to 7- and 11- to 13-year-old children based on early risk prediction. (OH) Oral hygiene; (PMTC) Professional mechanical toothcleaning; (F) Fluoride; (CHX) Chlorhexidine.

In selected caries-risk children (C3), the home care program is supplemented with the use of fluoride chewing gum for 20 minutes after every meal, dietary control by reduced sugar clearance time, and rinsing with 0.1% CHX twice daily for 3 weeks during the eruption of the molars. Frequent PMTC, followed by application of fluoride-CHX varnish (mixed 1:1), is carried out during eruption (three times in C2 patients and four times in C3 patients). As early as possible after the emergence of the molars, a light-cured resin-modified glass-ionomer cement is placed in the distal and central fossae and sticky risk fissures, not only as a fissure sealant but also as a slow-release fluoride agent during the eruption.

Key-risk age group: Age 11 to 13 years

Materials, methods, and intervals in relation to predicted caries risk (C0 to C3) are similar in 11- to 13- and 5- to 7-year-old children, with a few exceptions (see Fig 437). The daily toothcleaning is performed by the older children themselves and based on self-diagnosis in accordance with the principles described in chapter 2. As soon as the posterior permanent teeth reach approximal contacts, flat fluoridated dental tape and fluoride dentifrice are used to supplement the toothbrush for approximal toothcleaning in molars and premolars.

The intervals in groups C0 and C1 and groups C2 and C3 will be prolonged to once per year and two or three times per year, respectively, when the first molars are erupted and in full occlusal function (6 to 7 years of age) until the second molar emerge and erupt (11.5 to 12.5 years of age). When the second molars are erupted with full occlusal function and the approximal surfaces are maintained caries free, the intervals are prolonged again, strictly in accordance with the individual's predicted risk (one to four times per year).

The long-term outcome of such a program should be caries-free young adults, who, it is hoped, will maintain that status for the rest of their lives.

CONCLUSIONS

Data on the decline of caries prevalence among children and young adults in most industrialized countries over the last two decades show a relative increase in the proportion of caries on the occlusal surfaces of the permanent molars. Even in developing countries with relatively low caries prevalence, the occlusal surfaces of the permanent molars are carious more frequently than are the approximal surfaces. Fissure caries is partly attributable to the extremely plaque-retentive morphology of the fissure systems. To prevent the accumulation of cariogenic plaque in the depths of the fissures, and thereby prevent the development of caries, so-called fissure sealants were introduced to level out the original occlusal morphology.

Morphology of the occlusal surfaces of the molars

Viewed in a stereomicroscope, the occlusal surface of a permanent molar appears as a convoluted landscape, with high mountains (cusps) separated by valleys, some of which are deep rifts (fissures), while others resemble open river beds (grooves and fossae).

There are three more or less distinct and deep fossae in the molars—a mesial, a central, and a distal fossa—and five to seven related fissures and grooves. In cross section, most fissures have a wide opening, followed by a narrow cleft, approximally 1.0 mm deep (width: 0.1 mm) and reaching almost to the dentinoenamel junction. However, there are some (less than 10%) atypical fissures with a narrow opening and a bulbous widening at the base. Such fissures are regarded as sticky risk fissures. Fortunately, from a diagnostic point of view, there is a strong correlation between steep cuspal inclination and such sticky risk fissures.

Development of occlusal caries

The initiation and development of occlusal caries is strongly correlated to the morphology, eruption stage, and functional wear of the occlusal surface. About five times more plaque reaccumulates in erupting molars, particularly in the distal and central fossae and related fissures, than in erupted molars with full occlusion and functional wear. That is the reason why occlusal caries is initiated in molars during the extremely long period of eruption (12 to 18 months) and why the occlusal surfaces of the premolars, which have only a 1- to 2-month eruption time, rarely are carious.

As a consequence, if the occlusal surfaces of the molars are maintained caries free until they have reached occlusion and functional wear by efficient preventive measures (plaque control, fluorides, and fissure sealants) the risk for initiation and development of caries should be over. Occlusal caries lesions start as noncavitated enamel lesions at both sides of the entrance of the normal-shaped fissures. However, in the sticky atypical fissures, caries lesions may also be initiated in the bottom of the fissure. In similarity with the pathogenesis of smooth-surface lesions, cavities into the dentin always are preceeded by noncavitated dentin lesions.

Diagnosis of occlusal caries

In normal fissures and particularly in atypical sticky fissures, most of the early stages of the caries lesions are hidden from the naked eye, although in a clean, dried fissure, it might be possible to observe active noncavitated white-spot enamel lesions on the walls of the entrance of the fissures. Soon after eruption, most of these lesions are arrested and take up a brown stain from items in the diet. Such arrested fissures should remain caries inactive because functional occlusal wear will not permit any reaccumulation of cariogenic plaque. Therefore, diagnosis of occlusal caries should be simple if the occlusal surfaces were maintained free of caries until the surfaces reached full occlusal and functional wear by the aforementioned efficient preventive measures. However, in new patients with fully erupted molars, the occlusal surfaces may already be carious into the dentin with or without cavitation and thus offer delicate differential diagnostic problems.

The most common methods for diagnosis of occlusal caries are visual and tactile (ie, probing) inspection, and radiographs. However, because of the complicated three-dimensional shape of the occlusal surfaces, incorporating fossae, grooves, and fissures with a great range of individual variations, all these diagnostic methods have the potential to result in errors. New diagnostic tools, based on laser fluorescence or electrical conduction, used in combination with clinical visual inspection, tactile probing of cleaned occlusal surfaces, and bitewing radiographs, can improve the accuracy of diagnosis of occlusal enamel and dentin caries.

In sticky atypical fissures, none of these methods can accurately differentiate between noncavitated and cavitated caries lesions in dentin until advanced, open cavities have developed, which can be verified visually and by probing. It is of great importance to differentiate between noncavitated and cavitated dentin lesions, because the former can be arrested and treated by odontotomy and preventive restoration with fluoride-releasing resin-modified glass-ionomer sealants (so-called fissure blocking). Thus, for optimized accessibility and correct diagnosis of dentin lesions suspected to be cavitated, the fissures are opened to the bottom with a pointed diamond bur.

Materials and methods

There are mainly two groups of materials and methods for the fissure-sealing technique. The first group is based on etching of the enamel surfaces with a phosphoric acid (about 35%) and the use of a cross-linking, thermosetting dimethacrylate monomer (bis-GMA), which is diluted with methyl methacrylate or other co-monomers to increase the flow characterstics to reach the depth of narrow fissures. Etching is necessary to increase the roughness of the enamel surface because the retention of the resin-based sealant is mechanical. The resin forms so-called tags into the microroughness of the etched enamel surface to achieve mechanical retention.

There are two different principles for setting of the resin-based sealant materials: self-curing (the second generation) and light-cured (the third generation). During the last decade, the light-cured fissure sealants have been most frequently used. Clear, tinted, and white opaque resin-based sealants are available.

The technique for use of resin-based fissure sealants is as follows:

1. A rotating pointed brush and pumice are used to carry out PMTC of the occlusal surface.
2. The surface is washed with water spray and dried with compressed air.
3. Moisture control measures are applied, and the fissure system is etched for more than 30 seconds with 35% phosphoric acid.
4. The surface is rinsed with water spray for 30 seconds and dried with clean, compressed air, and careful moisture control is reapplied.
5. Light-cured resin sealant is applied with a special applicator and light cured for about 30 seconds.
6. The occlusion is checked with green occlusal wax. If necessary, adjustment is performed with a pear-shaped finishing bur.

This technique is very sensitive to moisture. Therefore, it may be risky to use it during the eruption of the molars, until the distal fossae and distal margin of the occlusal surface are free of the gingiva. Unfortunately, almost all occlusal caries lesions are initiated during this period.

During the last few years, resin-based fissure sealants supplemented with fluorides have also become available. However, most fluoride is lost during the first few days, and the material, unlike glass-ionomer materials, cannot be recharged with fluoride.

Glass-ionomer cement is a water-based material that hardens following an acid-base reaction between fluoroaluminosilicate glass powder and an aqueous solution of polyacid. The reaction between glass-ionomer cements and the dental fissures is chemical. This chemical retention is somewhat weaker than the mechanical retention between resin and an etched enamel surface. However, conditioning of the enamel surfaces with polyacrylic acid can improve the retention of conventional glass-ionomer sealants by about 50%.

A specific caries-preventive effect of glass-ionomer cements, not provided by resin-based sealants, is the release of fluoride; the cement can be recharged with fluorides from topical fluoride agents, such as gels and varnishes. Most of the released fluoride is NaF.

During the last few years, glass-ionomer sealants have been replaced with resin-modified glass-ionomer sealants, which achieve favorable physical properties similar to those of resin composites and resin-based sealants while they retain the basic features of the conventional glass-ionomer material, such as fluoride release and chemical adhesion. This type of material was created by incorporating water-soluble resin monomers into an aqueous solution of polyacrylic acid.

Autopolymerized as well as light-cured resin-modified glass-ionomer sealants are available. However, the recently introduced light-cured resin-modified glass-ionomer sealant (Fuji III LC) is most frequently used. It is popular because of its fast setting reaction and because it is less sensitive to moisture than the etching-resin technique. This is of great importance bcause the light-cured resin-modified glass-ionomer sealants can be used as early as possible during the eruption of the molars as a combination of fissure sealant and a slow-release fluoride agent.

Light-cured resin-modified glass-ionomer sealants are used according to the following method:

1. Careful PMTC is performed as in the use of resin-based sealants.
2. The occlusal surface is chemically conditioned with 25% acrylic acid for 30 seconds.
3. Debris in the fissures is removed with a probe.
4. The occlusal surface is washed for 30 seconds and dried.
5. Moisture control measures are carried out, and sealant is applied with a probe or a capsule.
6. The sealant is light cured for 30 seconds.
7. Occlusion is adjusted in a way similar to that described for resin-based sealants.

In deep, narrow, sticky fissures, there may be suspected hidden caries in molars that are fully erupted. To achieve accessibility for correct diagnosis, such fissures are opened to the bottom with a pointed diamond bur. If no cavitated lesion into the dentin is present, the opened fissure is sealed with either the etching-resin technique or light-cured resin-modified glass-ionomer sealant. Alternative materials could be flowable compomer or type II light-cured resin-modified glass-ionomer cement. This type of therapy is called *fissure blocking* or *extended fissure sealing*.

If a cavitated lesion in the dentin is discovered when the fissure is opened, a minimally invasive preparation is performed to minimize loss of tooth material. Such minicavities may be restored with type II light-cured resin-modified glass-ionomer cement, compomer, or resin composite materials.

Caries-preventive effects

Fissure sealants placed with the etching-resin technique are effective as long as they remain firmly adherent to the tooth. Early split-mouth studies of this technique resulted in about a 50% caries reduction in populations with a high incidence of caries. However, in populations with a low incidence of caries, less caries prevention

may be expected. Thus, general use of fissure sealants in populations with a low prevalence of caries is not cost effective. In such populations, the use of fissure sealants should be restricted to caries-risk patients and molars with atypical sticky fissures as an integrated measure in needs-related caries-preventive programs.

In studies comparing the effect of glass-ionomer or light-cured resin-modified glass-ionomer sealants with resin-based sealants, the former have resulted in at least the same caries-preventive effect as the latter, in spite of their shorter period of retention. This may be attributed to the release of fluoride from the glass-ionomer sealants.

From a caries-preventive point of view, the optimal effect should be achieved by the use of sealants as early as possible during the eruption because almost 100% of occlusal caries is initiated during the extremely long eruption period (12 to 18 months) of the molars. If the occlusal surfaces of the molars are maintained caries free until they have reached full occlusal and functional wear, the risk for initiation of caries is over. Therefore, light-cured resin-modified glass-ionomer sealants should be used as early as possible during the eruption of the molars as the first choice of sealants. Because of their combined sealant and slow-release fluoride effect and lower sensitivity to moisture, light-cured resin-modified glass-ionomer materials should be regarded as a very efficient fissure protector rather than a long-lasting fissure sealant.

Alternative methods for prevention of occlusal caries

Studies have shown that with intensified plaque control via a special daily toothbrushing technique combined with fluoride toothpaste, needs-related PMTC, and topical application of 2% NaF solution during the 12- to 18-month eruption period, it is possible to achieve close to 100% caries prevention and arrest of enamel caries (Carvalho et al, 1992; see chapter 1). Other studies have shown that repeated use of fluoride varnish is as efficient as application of resin-based sealants. In addition, use of CHX varnish has proven to provide a significant reduction in occlusal caries and in the amount of the cariogenic mutans streptococci in fissure plaque. Such varnishes can be used in the earliest stage of eruption until moisture control can be achieved and thus fissure sealants can be placed.

Needs-related programs for prevention of occlusal caries

From a cost-effectiveness point of view, use of fissure sealants should be based on detalied risk prediction as an integrated measure in needs-related caries-preventive programs. However, there are some general principles for the use of sealants:

1. Light-cured resin-modified glass-ionomer sealants should be used as early as possible during the eruption of the molars as a fissure protector, that is, at 5.5 to 6.5 years (first molars) and 11.5 to 12.5 years (second molars), and, on average, 6 months earlier in girls than in boys.
2. Fissure sealants should be used more frequently in populations with high caries prevalence than in those with low caries prevalence.
3. Fissure sealants should be used more frequently in caries-risk individuals and molars with sticky atypical fissures.

(For reviews on fissure sealants, see Association of State and Territorial Dental Directors et al, 1995; Feigal, 1998; Gao et al, 2000; Gordon, 1989; Handelman and Shey, 1996; Meurman and Thylstrup, 1994; and Ripa, 1993a.)

CHAPTER 7

INTEGRATED PREVENTION AND CONTROL OF DENTAL CARIES IN CHILDREN AND YOUNG ADULTS

Dental caries is an infectious, transmissible disease. As early as 1954, Orland et al demonstrated that, although germ-free animals do not develop caries, even with frequent sugar intake, all animals in the group rapidly develop caries lesions when human cariogenic bacteria (mutans streptococci) are introduced in the mouth of one animal. Specific bacteria (acidogenic and aciduric) that colonize the tooth surfaces are recognized as etiologic factors in dental caries (see chapter 1).

However, dental caries is also a multifactorial disease, ie, a disease that may be modified by several factors. For example, frequent intake of fermentable carbohydrates, such as sugar, is regarded as an external (environmental) modifying risk factor or prognostic risk factor. In the presence of these and other external risk factors, the outcome may be modified by internal host factors, such as the quality of the teeth and the amount and quality of saliva (Fig 438):

1. Microflora: acidogenic bacteria that colonize the tooth surface
2. Host: eg, quantity and quality of saliva, quality and shape of the tooth
3. Diet: intake of fermentable carbohydrates, especially sucrose, but also starch
4. Time: total exposure time to organic acids produced by the bacteria of the dental plaque

The development of a clinical lesion involves a complicated interplay among a number of factors in the oral environment and the dental hard tissues. The carious process is initiated by bacterial fermentation of carbohydrates, leading to the formation of a variety of organic acids and a fall in pH. Initially, H^+ will be taken up by buffers in plaque and saliva; when the pH continues to fall (H^+ increases), however, the fluid medium will be depleted of OH^- and PO_3^{3-}, which react with H^+ to form H_2O and HPO_2^{2-}. On total depletion of these compounds, the pH can fall below the critical value of 5.5, at which point the aqueous phase becomes undersaturated with respect to hydroxyapatite. Therefore, whenever surface enamel is covered by a microbial deposit, the ongoing metabolic processes within this biomass cause fluctuations in pH, and occasional steep falls in pH, which may result in dissolution of the mineralized surface (see Fig 278 in chapter 5).

In their classic study of experimental caries in humans, von der Fehr et al (1970) showed that, in the absence of oral hygiene (ie, with free accumulation of plaque and rinsing nine times a day with a sucrose solution), clinical signs of enamel caries develop within 3 weeks (Fig 439). When the same research team repeated the study, but introduced chemical plaque control (rinsing twice a day with 0.2% chlorhexidine so-

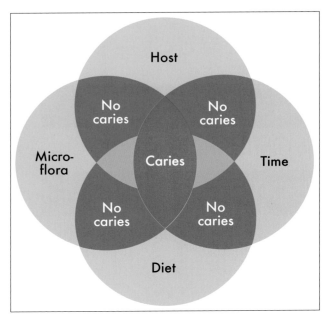

Fig 438 Illustration of the interaction among etiologic risk factors (microflora), external modifying risk factors (diet), internal modifying risk factors (host), and time in the development of dental caries. (Modified from Keyes, 1960.)

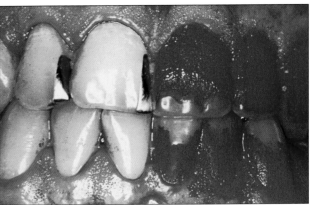

Fig 439 Typical appearance of an experimental subject after refraining from oral hygiene procedures for 3 weeks. On the left side, plaque has been disclosed. On the right side, plaque has been removed, revealing the development of white-spot lesions on the buccocervical surfaces, where plaque accumulation was greatest. (From von der Fehr et al, 1970. Reprinted with permission.)

lution), the subjects did not develop caries, even though they rinsed with sucrose solution nine times a day for 3 weeks (Löe et al, 1972). In other words, when the etiologic factor was suppressed or eliminated, the precondition for caries did not exist, and no lesions developed, despite the subjects' very frequent exposure to sucrose.

Like the inflammation induced in the gingival soft tissues adjacent to the gingival plaque, caries lesions of enamel develop on individual tooth surfaces beneath the undisturbed bacterial plaque (see Fig 439). This explains why combinations of different nonspecific plaque control programs have been so effective against caries,

gingivitis, and periodontitis (for review, see Axelsson, 1994 and 1998, and chapters 2 and 3 in this volume).

More recently, there has been intense interest in the role plaque (amount, formation rate, and ecology) and specific cariogenic microflora play in the etiology of dental caries. However, dental caries is also a multifactorial disease, as already discussed. Thus the caries lesion represents the net result of an extraordinarily complex interplay among "harmless" and "harmful" bacteria, antagonistic and synergistic bacterial species, their metabolic products, and their interaction with the many salivary and other host factors. Cost-effec-

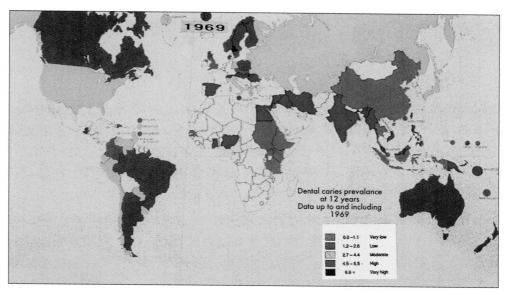

Fig 440 Worldwide caries prevalence among 12-year-old children in 1969. In Sweden, the black area (county of Värmland) has the highest caries prevalence. (From World Health Organization, 1994. Reprinted with permission.)

tive prevention and control of dental caries must therefore supplement different plaque control measures by integration of diet control, salivary stimulation, use of fluorides, and application of fissure sealants based on risk prediction at group, individual, and tooth-surface levels, ie, needs-related and integrated prevention and control of dental caries.

EFFECT OF A NEEDS-RELATED AND POPULATION-BASED CARIES-PREVENTIVE PROGRAM IN CHILDREN AND YOUNG ADULTS: RESULTS AFTER 20 YEARS

According to the World Health Organization's (WHO) first global caries databank (World Health Organization, 1994), caries prevalence in 12-year-old children from Sweden and some other industrialized countries was among the highest in the world in 1969 (Fig 440). However, during the following decades, caries prevalence in 12-year-old children has been reduced significantly in some industrialized countries—particularly in Scandinavia (World Health Organization, 1997) (Figs 441 and 442). The county of Värmland presented the highest caries prevalence in Sweden, and the county of Uppsala had the lowest. According to data collected by the Swedish Board of Health and Welfare (1979), the national average for caries-free 3-, 5-, and 7- to 16-year-old children was 29%, 13%, and 2%, respectively, compared to only 18%, 7%, and 1% in the county of Värmland and 55%, 25%, and 3.5% in the county of Uppsala. Thirty years ago, preventive programs in the schools were either rudimentary or nonexistent. In the county of Värmland, supervised toothbrushing with 0.2% sodium fluoride (NaF), once a month, was carried out in the schools. Very few children brushed their teeth daily.

Under the national dental insurance scheme, needs-related dental care, including preventive dentistry, is provided for children up to 20 years of age, free of charge; about 95% of this care is provided by the Public Dental Health Service. Sweden's adult population is subsidized by the national dental insurance scheme; yet 60% of adults are treated by private dentists. Sweden has no artificially fluoridated drinking water. Less than 5% of the Swedish population has access to naturally fluoridated drinking water containing

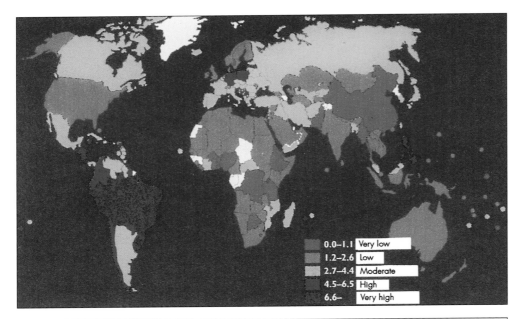

Fig 441 Worldwide caries prevalence in 1993 among 12-year-old children. (From World Health Organization, 1997. Reprinted with permission.)

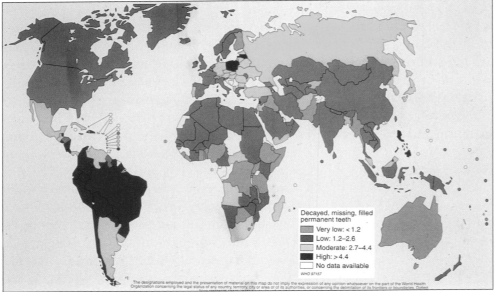

Fig 442 Worldwide caries prevalence in 1997 among 12-year-old children. (From World Health Organization, 1997. Reprinted with permission.)

more than 0.7 mg of fluoride per liter. In the county of Värmland, less than 2% of the population uses such drinking water. The fluoride concentration in drinking water is less than 0.1 mg/L in the city of Karlstad, where approximately 40% of the county's population lives. However, about 80% of Uppsala county's population uses naturally fluoridated drinking water containing approximately 1.0 mg/L of fluoride.

Program development

The so-called Karlstad studies were initiated in 1971 to evaluate and re-evaluate the individual, as well as the combined, effects of different plaque control programs, such as oral hygiene training, professional mechanical toothcleaning (PMTC), and chemical plaque control on dental plaque, caries, gingivitis, and periodontitis in children and adults (Axelsson, 1978; Axelsson and

Lindhe, 1974, 1977, 1978, 1981a, 1981c, 1987; Axelsson et al, 1976, 1987b, 1991). These studies consistently showed that high standards of mechanical plaque control through frequent PMTC and oral hygiene almost completely prevented the initiation and progression of caries. The effects of different topical fluoride programs were also evaluated or combined with plaque control programs (Axelsson and Lindhe, 1975; Axelsson et al, 1976, 1987a; Zickert et al, 1982a). In addition, the effects of mechanical and chemical plaque control on the cariogenic microflora were specifically evaluated (Axelsson et al, 1987b; Emilson et al, 1982; Kristoffersson et al, 1984), as were the relative effects of specific cariogenic microflora and diet on dental caries (Axelsson et al, 1987b; Buischi et al, 1989; Kristoffersson et al, 1986). New methods of predicting caries risk were developed, with special reference to cost effectiveness (Axelsson, 1991).

In most of the earlier studies, a dental assistant was given 1 month's intensive training in the preventive measures to be used in the studies. When the results of the first studies were published, the author was asked by the Swedish Board of Health and Welfare to design a training program for preventive dentistry assistants. In 1975, the first preventive dentistry assistants in Sweden were trained in Värmland. These students had previously undergone the standard 2-year dental assistant training program. The additional demanding, 1-month preventive dentistry assistant program resulted in the following expertise: optimal self-care on an individual and a group basis, execution of PMTC procedures, topical applications of fluoride gels and varnishes, use of fissure sealants, administration of salivary and oral microbiology tests, dietary evaluation, and counseling. Swedish dental hygienists are even better trained in all these procedures, as well as scaling and root planing, finishing restorations, removing overhangs, recontouring in cases of furcation involvement, diagnosis, epidemiology, and administering local anesthetics (Axelsson et al, 1993b) (for further details, see chapter 8).

In the county of Värmland, for every dentist there are, on average, one preventive dentistry assistant and one dental hygienist practicing preventive dentistry, which is the highest ratio in Sweden and probably in the world. In 1975, prophylactic dental clinics were gradually introduced in the elementary schools of Värmland, enabling preventive dentistry assistants or dental hygienists to practice individualized and needs-related preventive dentistry. The county of Värmland is the only province in Sweden with such preventive dentistry clinics in the schools.

For successful prevention and control of dental caries in children and young adults, some basic principles must be adopted, based on the caries incidence of the targeted population. The higher the risk of developing caries for most of the population, the greater the effect of one single preventive measure. For example, 30 to 40 years ago, caries prevalence in Sweden and some other industrialized countries was extremely high (see Fig 440), and almost every child developed several new caries lesions every year, mainly because of very poor oral hygiene. Regular toothbrushing was not an established habit, and there was no effective fluoride toothpaste. Under the prevailing conditions, well-organized Swedish school-based programs, in which a fluoride mouthrinse (0.2% NaF solutions) was administered once every 1 or 2 weeks, resulted in caries reductions of 30% to 50% (Forsman, 1965; Torell and Ericsson, 1965). These results were comparable to the effects of fluoridated drinking water in other industrialized countries (the United States and the United Kingdom) with high caries prevalence at the time. In countries or districts with high caries prevalence and poor oral hygiene, the introduction of a single caries-preventive measure for all children, such as fluoridated drinking water, school-based fluoride mouthrinsing, or daily toothbrushing with fluoride toothpaste, would still result in a very significant reduction in caries.

In the late 1970s, school-based programs involving administration of fluoride mouthrinses once every 1 or 2 weeks were still recommended by the Swedish Board of Health and Welfare.

However, among the population of Värmland, which demonstrated high standards of oral hygiene, regular use of fluoride toothpaste, and low incidence of caries, the supplementary cariostatic effect of the school programs was questionable. Therefore, from 1977 to 1980, a 3-year double-blind study was conducted to evaluate the supplementary effect of fluoride rinsing on caries incidence in 12- to 15-year-old subjects who used fluoride toothpaste twice a day. There was no difference in results between weekly rinsing with 0.05% NaF solution and rinsing with distilled water (Axelsson et al, 1987a). The school-based mouthrinsing programs were consequently withdrawn, without detriment to caries prevalence or incidence. Thus, in populations with low or moderate incidence of caries, well-established self-care habits, and well-organized oral health care, a single preventive measure administered to all subjects in the population, regardless of predicted risk, will not be cost effective. In such populations, individual risk prediction and needs-related combinations of preventive measures are necessary. To ensure a high sensitivity of risk prediction, several etiologic and modifying risk factors have to be combined.

This is exemplified by the following: The Vipeholm study confirmed about 50 years ago that prolonged sugar clearance time is an external modifying risk factor for caries development in mentally handicapped people with heavy accumulation of plaque in the absence of oral hygiene or fluoride (Gustafsson et al, 1954). However, over the past two decades studies have repeatedly failed to find any correlation between the intake of sugar products and caries prevalence in Sweden (Sundin et al, 1983; Kristoffersson et al, 1986). That is because caries prevalence and caries incidence have declined dramatically as a result of the integration of caries-preventive measures by self-care, supported by needs-related professional treatment (Axelsson et al, 1993a). In a totally integrated caries-preventive program, however, external modifying risk factors, such as high frequency of sugar intake, should also be addressed.

Box 15 Goals of the integrated caries-preventive program in Värmland, Sweden

- To have no approximal restorations
- To have no occlusal amalgam restorations
- To have no approximal loss of periodontal attachment
- To motivate and encourage individuals to assume responsibility for their own oral health

The risk for caries development varies significantly for different age groups, individuals, teeth, and surfaces. Caries-preventive measures must be integrated and based on predicted risk from age groups down to the individual tooth surfaces. In other words, a medium-sized suit does not fit all the men in the world; it would be a reasonable fit for, at most, 40% but too small for 30% and too large for the remaining 30%.

Goals

Based on this philosophy and experience from continuously ongoing research evaluating and re-evaluating separate and integrated caries-preventive measures, as well as methods for prediction of caries risk, an integrated caries-preventive program for children up to the age of 19 years was introduced in the county of Värmland, Sweden, in 1979. The goals for the subjects following the program from birth to the age of 19 years are shown in Box 15. It was hoped that these goals would be attained for 19-year-old participants by 1999. The effect of the program has been evaluated once every year on almost 100% of all 3- to 19-year-old participants in a computer-aided epidemiologic program since 1979 (Axelsson et al, 1993a).

Materials and methods

Key-risk age groups

Recent studies have shown that caries lesions are initiated more frequently at specific ages. This applies particularly to children but also to adults. In children, the key-risk periods for initiation of caries seem to be during eruption of the perma-

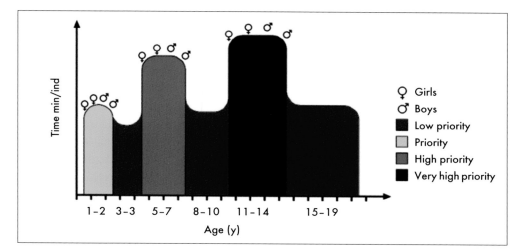

Fig 443 Timing of cost-effective caries-preventive programs, related to age and gender. (min/ind) Minutes per individual. (From Axelsson, 1998. Reprinted with permission.)

nent molars and the period during which the enamel is undergoing secondary maturation. In adults, most root caries develops in the elderly, partly because of the higher prevalence of exposed root surfaces and reduced salivary flow as a side effect of using large amounts of medication.

Key-risk age group 1: Ages 1 to 2 years. Studies by Köhler and colleagues (Köhler and Bratthall, 1978; Köhler et al, 1982) showed that mothers with high salivary levels of mutans streptococci (MS) frequently transmit the bacteria to their babies as soon as the first primary teeth erupt, leading to greater development of caries. Other studies have shown that 1-year-old babies with plaque and gingivitis develop several dental caries lesions during the following years, while babies with clean teeth and healthy gingivae, maintained by regular daily cleaning by their parents, remain caries free (Wendt et al, 1994). It also has been shown that the practice of giving infants sugar-containing drinks in nursing bottles at night increases the development of caries (Wendt and Birkhed, 1995).

In another investigation, Grindefjord et al (1995) studied the relative risk (odds ratio) that 1-year-old infants would develop caries by the age of 3.5 years: Those with poor oral hygiene, poor dietary habits, salivary MS, little or no exposure to fluoride, and parents with a low educational level or an immigrant background were at 32 times

greater risk than were children without the corresponding etiologic and external risk factors. The importance of establishing good habits as early as possible, and of postponing or preventing poor habits, should not be underestimated. In addition, the enamel of erupting and newly erupted primary and permanent teeth is at its most caries-susceptible stage until completion of secondary maturation (Kotsanos and Darling, 1991). In 1- to 3-year-old infants, the specific immune system, particularly immunoglobulins in saliva, is immature. Poor oral hygiene will therefore favor the establishment of cariogenic microflora such as MS.

On average, the permanent teeth in particular erupt 6 to 12 months earlier in girls than they do in boys (Teivens et al, 1996). On this basis, the first-priority key-risk age groups are expectant mothers and 1- to 2-year-old children, starting with girls (Fig 443). To prevent postnatal transmission of cariogenic bacteria and poor dietary habits from mother to child, expectant mothers who are at risk should be offered a special preventive program comprising intensified plaque control (mechanical and chemical) and reduction of sugar intake to reduce the number of cariogenic microflora.

Key-risk age group 2: Ages 5 to 7 years (eruption of first molars). The pattern and amount of de novo plaque reaccumulation on the occlusal surfaces of the permanent first molars,

48 hours after PMTC, was studied in relation to eruption stage by Carvalho et al (1989). Plaque reaccumulation is heavy on the occlusal surfaces of erupting maxillary and mandibular molars, particularly in the distal and central fossae and related fissures. This is in sharp contrast to the fully erupted molars, which are subjected to normal chewing friction (see Fig 54 in chapter 1). Abrasion from normal mastication significantly limits plaque formation, and this explains why almost all occlusal caries in molars begins in the distal and central fossae during the extremely long eruption period of 14 to 18 months. In contrast, fissure caries is very rare in premolars, which have a brief eruption period of only 1 to 2 months. In addition, the enamel of erupting and newly erupted teeth is considerably more susceptible to caries until secondary maturation is completed, more than 2 years after eruption. Therefore, the caries-reducing effect of fluoride is about 50% greater in erupting and newly erupted teeth than it is in teeth that have undergone secondary maturation.

The next high-risk age is, therefore, from 5 to 7 years, during eruption of the first molars (the key-risk teeth), starting with girls (see Fig 443). Intensified mechanical plaque control twice a day with fluoride toothpaste should be performed by the children's parents, particularly on the erupting first molars (see chapter 2). Home care should be supplemented at needs-related intervals by professional mechanical tooth-cleaning and application of fluoride varnish. In the most caries-susceptible children, resin-modified glass-ionomer cement should be used in the fissures as a slow-release fluoride agent.

Key-risk age group 3: Ages 11 to 14 years (eruption of second molars). Normally, the second molars start to erupt at the age of 11 to 11.5 years in girls and at around the age of 12 years in boys. The total eruption time is 16 to 18 months. During this period, the approximal surfaces of the newly erupted posterior teeth are undergoing secondary maturation of the enamel and are also at their most susceptible to caries. Therefore, 11- to 14-year-old children have not only, by far,

the highest number of intact tooth surfaces but also the greatest number of surfaces at risk.

Integrated plaque control measures and use of fluoride agents should therefore be intensified on the approximal surfaces of all the posterior teeth and the occlusal surfaces of the second molars, starting with girls aged 11 to 11.5 years (see Fig 443), to protect intact tooth surfaces and to remineralize noncavitated caries lesions. If this program is maintained throughout the secondary maturation period, and needs-related self-care habits are established, there is a high probability that the remaining intact tooth surfaces will remain intact for the individual's entire life.

Key-risk individuals

In children, caries prevalence and caries incidence related to the age group and the combination of Plaque Formation Rate Index (PFRI) plus salivary MS levels will give the highest sensitivity value for prediction of caries risk. The percentage of selected key-risk individuals should also be related to age. In other words, the highest percentage of key-risk individuals should be selected from among the 11- to 14-year-old children and the lowest percentage from among the 3- to 4-, 8- to 10-, and 15- to 19-year-old children (see Fig 443).

Numerous cross-sectional as well as longitudinal studies have shown significant correlations between salivary MS levels and caries prevalence and caries incidence (for review, see Bratthall, 1991; Bratthall and Ericsson, 1994). At the surface level, even more significant correlations between colonization by MS and caries incidence have been found (Axelsson et al, 1987b; Kristoffersson et al, 1984).

Most of the early salivary studies of MS were carried out in child populations with relatively high caries prevalence (Sweden in the 1970s), and at that time more than 1 million colony-forming units (CFUs) of MS per 1 mL of saliva was shown to be a good predictor of caries risk (Klock and Krasse, 1977; Zickert et al, 1982b). However, since then, caries prevalence in Sweden and many other industrialized countries has decreased significantly. The correlation between one single etiologic factor, such as salivary MS levels, and caries prevalence and caries incidence tends to be weaker in such populations because dental caries is a multifactorial disease.

More recent cross-sectional studies in Swedish schoolchildren have repeatedly found that the cutoff for correlation between salivary MS counts and caries prevalence is whether the individual tests negative or positive for MS rather than 1 million or more CFUs of MS per 1 mL of saliva (Kristoffersson et al, 1986). However, the dilemma is that only 10% to 30% of individuals are MS negative in most populations (higher among young children and lower among the elderly). The question is how to select 5% to 25% high-risk and very high-risk individuals from among the 70% to 90% of MS-positive subjects. The answer is combining salivary MS tests and the PFRI for the prediction of caries risk.

Enamel caries lesions develop only on the specific tooth surfaces where thick plaque with a high percentage of acidogenic and aciduric bacteria remains too long; its "acid slag products" demineralize the underlying tooth surface. The quantity of plaque that forms on clean tooth surfaces during a given time represents the net result of interactions among etiologic factors, many internal and external risk indicators and risk factors, and protective factors. This observation was the rationale for the development of the PFRI (Axelsson, 1987, 1991). The index, based on the amount of plaque freely accumulated (de novo) 24 hours after PMTC, is described in more detail in chapter 1.

An earlier 30-month longitudinal study showed a very strong correlation between the development of approximal caries lesions and the level of MS colonization (Axelsson et al, 1987b). Other studies have shown that salivary MS counts are correlated to the number of tooth surfaces that are colonized by MS (Lindquist et al, 1989b; for review, see Bowden and Edwardsson, 1994; Bowden, 1997; and Bratthall and Ericsson, 1994). Therefore, it seems reasonable that MS-positive individuals with high and very high PFRI scores (4 and 5, respectively) should be more susceptible to caries than MS-negative individuals or MS-positive individuals with very low or low PFRI scores (1 and 2, respectively). That is because the total number of the most cariogenic bacteria (MS) should be significantly higher on tooth surfaces in subjects with a PFRI score of 4 or 5 than in subjects with a PFRI score of 1 or 2, if the percentage of MS in their plaque biofilm is the same.

In 1984 a combined cross-sectional and longitudinal study was initiated in 667 children aged 14 years. The study had the following objectives:

1. To determine the distribution of the PFRI in a large number of schoolchildren and the distribution characteristics of plaque formation on the individual tooth surfaces in the dentition
2. To determine whether a combination of the MS level and the PFRI score is more related to caries prevalence than are the variables individually
3. To determine whether caries development can be predicted by a combination of salivary MS levels and the PFRI

An analysis of caries prevalence (mean number of decayed or filled surfaces [DFSs]) related to different PFRI scores indicated a threshold for caries risk between PFRI scores 2 and 3. This finding was subsequently confirmed in the longitudinal part of the study, over 5 years (Axelsson, 1987, 1991). For MS this critical level was between 0 and 100,000 CFUs/mL saliva.

In the longitudinal part of the study, MS-positive individuals with PFRI scores of 3 to 5 developed five times more approximal caries lesions in dentin per individual per 5 years than did MS-negative individuals. Similar data were also reported in a 3-year longitudinal study in Polish 13- to 16-year-old children (Axelsson et al, 2003).

Based on the experiences from these studies, the following guidelines for selection of no caries risk (C0), low caries risk (C1), caries risk (C2), and high caries risk (C3) were developed for the combination of salivary MS test and PFRI. A salivary MS test screens out MS-negative subjects (about 20%) as not being at risk. Of the remaining 80% or so (MS-positive subjects), those with a PFRI score of 3 or greater are selected as risk patients (ap-

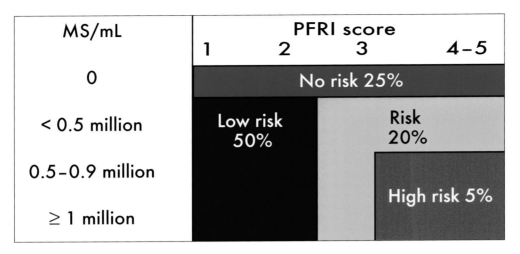

MS/mL	PFRI score
	1　　2　　3　　4–5
0	No risk 25%
< 0.5 million	Low risk 50%　　Risk 20%
0.5–0.9 million	
≥ 1 million	High risk 5%

Fig 444 Four-point scale for prediction of caries risk based on MS levels and PFRI. (Modified from Axelsson, 1991 with permission.)

proximately 25%). From among these subjects, an extremely high-risk group may be further selected: those with a PFRI score of 4 or 5 and an MS score of 2 (0.5 to 0.9 million MS/mL) or 3 (≥ 1 million MS/mL) (around 5%). These guidelines are illustrated in Fig 444.

More detailed criteria for selection of C0, C1, C2, and C3 individuals in different age groups based on caries prevalence, caries incidence, etiologic factors, external and internal modifying factors, and preventive factors are presented in chapter 8.

Key-risk surfaces

Depending on the age and caries prevalence of the population, there may be pronounced variations in the pattern of both lost teeth and carious or restored surfaces. The molars are the key-risk teeth. In a toothbrushing population, the key-risk surfaces are the fissures of the molars and the approximal surfaces from the mesial aspect of the second molars to the distal aspect of the first premolars (see Fig 52 in chapter 1). Integration of mechanical plaque control by self-care and the use of fluoride toothpaste, supplemented at needs-related intervals by PMTC and application of fluoride varnish and chlorhexidine varnish, should therefore target these key-risk teeth and surfaces, according to the principles discussed in earlier chapters.

The distal surface of the mandibular right first molar is the most frequently carious (see Fig 53 in chapter 1). This is probably because most people are right handed, and it is well known that in right-handed people the mandibular right linguoapproximal surfaces of molars show the greatest tendency to plaque accumulation and gingivitis. That the distal surfaces of the second premolars constitute a relatively high percentage of carious surfaces may be explained as follows: The wide mesial surfaces of the first molars are frequently carious and exposed to cariogenic microflora when the second premolars erupt. In caries-susceptible (C2 or C3) individuals, it is difficult to achieve successful arrest of such enamel lesions during the short period of eruption (1 to 2 months) of the second premolars, and lesions are sometimes unrestored. Until completion of secondary maturation of the enamel, the environment is extremely unfavorable for the newly erupted distal surfaces of the second premolars.

The approximal surfaces of molars and second premolars also exhibit the highest scores of MS colonization (see Fig 51 in chapter 1). The lowest approximal MS colonization and prevalence of approximal DFSs are found in the mandibular incisors. Thus the risk for development of approximal caries lesions seems to be correlated to the buccolingual width of the tooth crown, particularly in toothbrushing populations. That is because of the limited accessibility of the toothbrush to the wide approximal surfaces of the molar and premolars, which are blocked by buccal and lingual papillae.

Preventive programs

Ages 0 to 3 years. Dental hygienists or preventive dentistry assistants provide prenatal counseling on an individual and a group basis. To prevent postnatal transmission of cariogenic microbes and poor dietary habits from mother to child, selected key-risk individuals are offered a special preventive program at the public dental health centers. At the child welfare centers, dental hygienists or preventive dentistry assistants counsel parents on good oral hygiene and dietary habits for their children as well as the importance of early introduction of the use of fluoride toothpaste.

Special oral hygiene aids for these age groups are described in chapter 2. A pea-sized amount of fluoride toothpaste is recommended with the toothbrush to reduce side effects of swallowed toothpaste. However, no systemic fluorides (tablets, etc) are used (see chapter 5). The reasons for that are twofold: *(1)* No preeruptive caries-preventive effects may be expected, and *(2)* almost all visible fluorosis on the incisal third of the incisors is induced at the age of 20 to 24 months. The overall goals in this age group are to establish good habits as early as possible and to postpone or prevent the introduction of bad habits as long as possible, because established habits are trusted, regardless of whether they are good or bad.

Ages 3 to 5.5 years. In kindergarten, preventive dentistry assistants or dental hygienists carry out a preventive program with the teachers' assistance. The preventive program includes supervised toothbrushing with a fluoride toothpaste and games based on oral health education. In addition, at the age of 3 years, all children are given gentle PMTC by preventive dentistry assistants or dental hygienists at the public dental health center. This early introduction is not only for pre-

vention of disease but also to familiarize the child with dental care, as a means of avoiding dental fear in the future. The selected key-risk individuals (about 10%) receive PMTC and fluoride varnish treatment two to four times per year. Special efforts are focused on education of the parents to be responsible for the daily cleaning of their children's teeth.

Ages 5.5 to 7.5 years. In this age group, the caries-preventive methods are focused on maintaining caries-free fissures in the permanent first molars until the teeth are fully erupted and exposed to normal functional chewing wear. Thereafter, the risk for development of occlusal caries should be past. Therefore the parents are educated to observe when the first molars begin to erupt and how to intensify cleaning the fissures twice a day with a special toothbrushing technique and fluoride toothpaste. As soon as the mesial surfaces of the first molars are in contact with the distal surfaces of the primary second molars, the parents are responsible for daily cleaning of these surfaces with fluoridated dental tape in a special holder.

Based on the predicted risk, the daily toothbrushing performed by the parents is supplemented with needs-related intervals of PMTC, use of fluoride and chlorhexidine varnishes, and application of fissure sealants with light-cured resin-modified glass-ionomer cements.

Ages 7.5 to 11.5 years. In this relatively low-risk age group, the children are educated by preventive dental assistants or dental hygienists in the preventive clinics of the elementary schools and at the public dental health clinics. The children are educated so they can gradually take over the responsibility from their parents for the daily cleaning of their own teeth. About 10% of this group, selected key-risk individuals, may receive supplemented needs-related professional caries-preventive treatment.

Ages 11.5 to 14 years. This age group is the key-risk age group and should receive the most generous and intensive caries-preventive program from the cost-effectiveness point of view. They have the highest number of still caries-free but highly caries-susceptible permanent tooth surfaces until the second molars are fully erupted and the so-called secondary maturation of the enamel is achieved.

Dental hygienists or preventive dentistry assistants give lessons on preventive dentistry, as well as self-care training in the elementary schools. They cooperate with schoolteachers, school nurses, dietary consultants, psychologists, school physicians, and dentists as a health team to optimize oral health as well as general health from a holistic point of view. For example, dental hygienists and prophylactic dental assistants have been the pioneers to successfully prevent the debut of smoking from the age of 12 years.

Based on the experiences from the Brazilian study (Axelsson et al, 1994a) described in chapter 2, needs-related oral hygiene habits are established by the age of 12 years. Starting with girls from the age of 11.5 years (see Fig 443) special education is focused on the cleaning of the fissures of erupting second molars. In particular, 12- to 13-year-old children are given careful education in the application of fluoride dentifrice on dental tape to clean the approximal surfaces of the molars and premolars prior to brushing.

Use of the PFRI is recommended for establishing needs-related oral hygiene habits. The PFRI score indicates how frequently the teeth should be cleaned and which tooth surfaces should be cleaned first; special attention should be paid to surfaces with heaviest reaccumulation of plaque. For individuals with a PFRI score of 4 or 5 and high caries risk, it is recommended that they clean their teeth just *before* each meal and take a fluoride chewing gum or lozenges flavored with xylitol or sorbitol right *after* each meal. For optimal topical effect, this special lozenge should be allowed to dissolve in the mouth.

Based on the predicted caries risk, supplementary professional caries-preventive measures, such as PMTC, topical application of fluoride varnish and chlorhexidine varnish, and

application of fissure sealants with resin-modified glass-ionomer cements as a slow-release fluoride agent in erupting second molars, are performed by dental hygienists or prophylactic dental assistants at needs-related intervals at the preventive clinic in the schools or at the public dental health clinics. With complete implementation of such an intensified and needs-related program up to the age of 14 years, it should be possible to maintain all tooth surfaces free of caries for the rest of the individual's life. Tooth enamel exposed to the oral environment, including fluoride, is significantly more caries resistant than newly erupted enamel, as discussed earlier.

Ages 14 to 19 years. In this age group, needs-related self-care habits should already have been established and the so-called secondary maturation is achieved. Thus these ages are regarded as a low-risk group. However, special attention must be focused on erupting third molars, which face problems similar to those of erupting first and second molars. In addition, an unhealthy lifestyle, such as poor dietary habits, may occur in individuals who leave their homes early to continue their studies. In these patients and other selected caries-risk individuals (about 10%), supportive professional caries-preventive measures are necessary to maintain the earlier achieved oral health condition.

Caries diagnosis and evaluation

In the Värmland studies, caries diagnosis was carried out by dentists in the public dental health service in a standardized fashion (according to the Swedish Board of Health and Welfare's recommendations for collecting epidemiologic data). Buccal and lingual surfaces showing caries lesions with cavitation by probing were regarded as decayed surfaces (DSs). Caries lesions on the approximal surfaces of the molars and premolars and the occlusal surfaces of the molars were diagnosed by bitewing radiographs as well as by probing. All lesions in dentin diagnosed on radio-

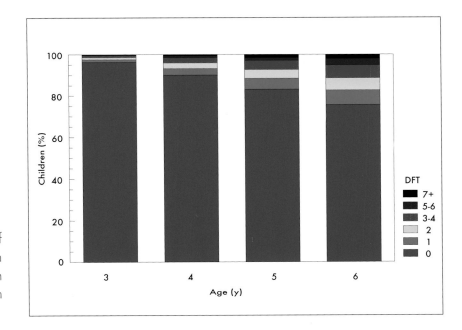

Fig 445 Frequency distribution of decayed and filled primary teeth (DFT) in 3- to 6-year-old children in the county of Värmland, Sweden, in 2001.

graphs were regarded as DSs even without clinically evident cavitations. For comparison, the caries diagnosis criteria by WHO for national epidemiologic surveys are based only on clinically visible cavities and by probing. In the Värmland studies, all filled surfaces were regarded as filled surfaces. Missing surfaces were not included, because no teeth were missing as a result of caries in the permanent dentition of these 6- to 19-year-old individuals; all missing teeth had been extracted for orthodontic reasons.

Since 1979, the effects of the needs-related preventive programs have been evaluated annually in almost 100% of 3- to 19-year-old patients at surface, tooth, individual, clinic, and county levels in a computerized epidemiological system.

Results

Caries prevalence in primary teeth

The preventive program was gradually introduced at the health centers for expectant mothers and the child welfare centers in the 1970s. The percentage of caries-free 3-year-old children

increased from 35% in 1973 to 97% in 1993 and is still maintained at this near-100% level. In 1999 the mean values of decayed and filled primary teeth in 3-, 4-, and 5-year-old children were only 0.07, 0.3, and 0.4, respectively. Fig 445 shows the frequency distribution of decayed and filled primary teeth in 3- to 6-year-old children in 2001. The percentages of caries-free 3-, 4-, 5- and 6-year-old children were 97%, 90%, 83%, and 76%, respectively.

Caries prevalence in permanent teeth

Figure 446 presents the caries prevalence in the county of Värmland, expressed as mean DFSs for all surfaces and approximal surfaces per individual in 1979 and 1999, in all age groups from 7 to 19 years. The average caries reduction ranged from 85% to 95%. The mean total number of DFSs per individual aged 12, 16, and 19 years, respectively, declined from 6.0, 13.0 and 22.3 DFSs in 1979 to 0.3, 1.2 and 2.1 in 1999. In the same age groups, DFSs on the approximal surfaces, representing 20% to 25% of the total in 1979, declined from 1.1, 3.0, and 5.0 DFSs in 1979 to 0.1, 0.6, and 1.1 in 1999.

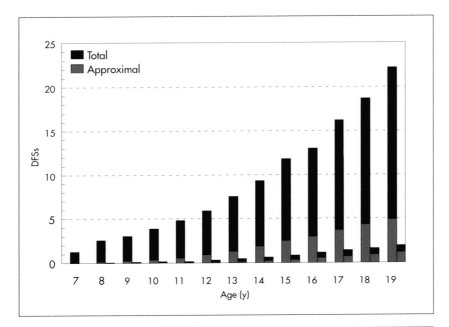

Fig 446 Mean number of DFSs for all surfaces and approximal surfaces in 1979 (*left bar*) and 1999 (*right bar*) in 7- to 19-year-old individuals in the county of Värmland, Sweden.

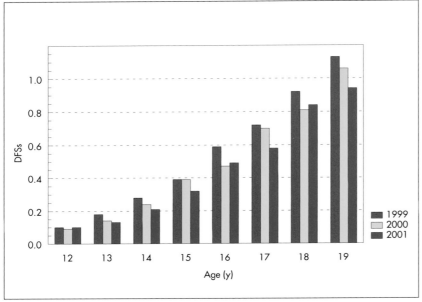

Fig 447 Mean number of approximal DFSs in 1999, 2000, and 2001 in 7- to 19-year-old individuals in the county of Värmland, Sweden.

The mean number of DFSs per individual on the approximal surfaces in 1999, 2000, and 2001 confirms the continuing improvement in dental health status in all age groups (Fig 447). In 2001, the mean number of approximal DFSs in 19-year-old residents of the county of Värmland was only 0.9, while the number for all Sweden was 1.4.

The frequency distribution of DFSs in the different age groups in 1999 reveals that 98%, 96%, 94%, 92%, 91%, 86%, 81%, 77%, 72%, 71%, 66%, 66%, and 60% of individuals aged 7 through 19 years, respectively, were caries free (Fig 448). Figure 449 shows the frequency distribution of DFSs on the approximal surfaces in 2001. Among 12- and 19-year-old individuals, 95% and 65%, respectively, were caries free.

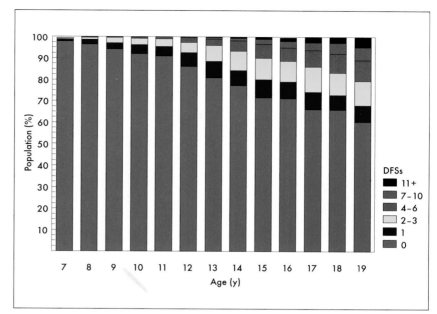

Fig 448 Frequency distribution of DFSs in 7- to 19-year-old individuals in the county of Värmland, Sweden, in 1999.

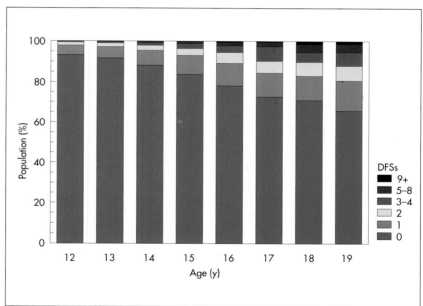

Fig 449 Frequency distribution of approximal DFSs in 7- to 19-year-old individuals in the county of Värmland, Sweden, in 2001.

Caries incidence in permanent teeth

Compared to 1979, the mean number of new DSs per individual per year was also reduced between 85% and 95% in 1999, including all surfaces, as well as the approximal surfaces, in the different age groups. In 7-, 12-, and 19-year-old individuals, respectively, the caries incidence for all surfaces dropped from 0.9, 1.2, and 2.0 new DSs per indi-

vidual in 1979 to 0.02, 0.06, and 0.2 in 1999 (Fig 450). In the same groups, caries incidence for the approximal surfaces declined from 0.0, 0.2, and 0.9 new DSs per individual in 1979 to 0.0, 0.03, and 0.2 in 1999. The frequency distribution of new DSs shows that 88% to 98% of children in the different age groups did not develop any new DSs in 1999 (Fig 451). Among 7-year-old children, 98% developed no new DSs, 1% developed one new DS, and

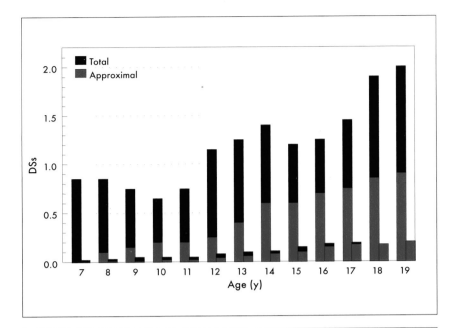

Fig 450 Mean number of total new DSs and approximal new DSs in 1979 (left bar) and 1999 (right bar) in 7- to 19-year-old individuals in the county of Värmland, Sweden.

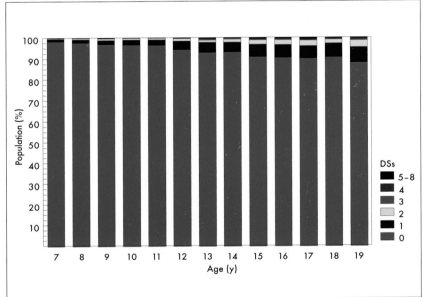

Fig 451 Frequency distribution of new DSs in 7- to 19-year-old individuals in the county of Värmland, Sweden, in 1999.

0.4% developed two new DSs. The corresponding figures for 12-year-old children were 94%, 4%, and 1%, respectively. Among 19-year-old individuals, 88% developed no new DSs, 8% had one new DS, 3% had two new DSs, 1% had three or four new DSs, and 0.1% had five or six new DSs.

Treatment time and cost effectiveness

According to data collected annually from all counties in Sweden by the Swedish Board of Health and Welfare, the mean treatment time by a dentist per child in 1979 was 1.75 hours in the county of Värmland; and the national average was 1.70 hours. However, as an effect of the needs-related preventive program by preventive dental assistants and dental hygienists, the mean treatment time by a dentist was dramatically reduced to only 20 minutes per child per year in 1999, which is the outstanding lowest value among the Swedish counties. Most of the dentists' time was spent on examinations because the need for restorations was minimized.

The total costs per child per year (2001), including the needs-related preventive program by dental hygienists or preventive dental assistants, treatment by dentists, and orthodontic treatment by specialized orthodontists, was about 90 euros (US $100); the average for all of Sweden was 100 euros (US $110). Because the average caries prevalence in children and young adults living in the county of Värmland is the lowest in Sweden, it must be concluded that this needs-related caries-preventive program is very cost effective.

The cost-benefit ratio of such a program is also very high because intact teeth and healthy gingiva are beautiful, functional, and appealing and may positively influence general health. As a contrast, unreasonable "drilling, filling, killing (the pulp), and billing" should have gone out of fashion many decades ago.

Validity of the data

According to the WHO, 12-year-old children are selected as the indicator age group for epidemiologic surveys on caries prevalence in children worldwide. The decayed, missing, or filled teeth (DMFT) index is used for such studies. However, in some countries with limited dental care resources, one DMFT may represent a first molar extracted because of untreated caries lesions on most surfaces or a first molar with all its surfaces decayed, which means five decayed, missing, or filled surfaces (DMFSs). On the other hand, one DMFT in Sweden and particularly in the county of Värmland mostly means only one overtreated fissure in a first molar (one filled surface). In addition, no permanent teeth are lost because of dental caries in individuals up to the age of 20 years. Therefore, the DFS index is used to provide

a more detailed evaluation of the effect of the caries-preventive programs.

In addition, all noncavitated approximal dentin caries lesions diagnosed on bitewing radiographs are included in the Swedish data. In the early 1980s it was shown that none of 58 radiographically diagnosed lesions penetrating approximately halfway through dentin exhibited cavitation into dentin (Fig 452). Only 50% had cavitation in the enamel. This was confirmed by carefully opening the lesions for visual diagnosis, the gold standard of diagnosis (Bille and Thylstrup, 1982). Other studies have shown similar results (Mejàre and Malmgren, 1986; Pitts and Rimmer, 1992).

The teeth of children in Värmland are extremely clean, which facilitates clinical examination, and optimal light conditions as well as fiber-optic light are available during the examination. Since 1979, all dentists in the county of Värmland have been working in cooperation, using standardized methods for caries diagnosis according to the study criteria. Data from close to 100% of each age group from 3 to 19 years are collected annually since 1979 in a computer-aided epidemiologic system.

According to the WHO's criteria, only open cavities, diagnosed visually and by probing, are registered at the tooth level in caries epidemiology. In addition, most national surveys are carried out as primitive field studies. These facts should be considered when the data from the Värmland studies are compared with the WHO's global data.

National and international comparison of data in 12-year-old children

Since 1969, the WHO has compiled a world map of caries prevalence among 12-year-old children, expressed in DMFT. At the age of 12 years, the five-level scale varied from 0.0 to more than 6.5 DMFT (Table 21).

The WHO's goals for caries prevalence in 12-year-old children in the years 2000, 2010, and 2025 are 3.0, 2.0, and 1.0 DMFT, respectively. On the WHO global maps from 1969, 1973, and

Fig 452 Comparison of radiographic and clinical scoring of approximal caries lesions. Radiographic score (Möller and Poulsen, 1973; Gröndahl et al, 1977): (0) No radiographic changes in enamel; (1) Radiographic changes in enamel; (2) Radiographic lesion that has reached the dentino-enamel junction; (3) Radiolucent lesion penetrating approximately halfway through dentin; (4) Radiolucent lesion close to the pulp. Clinical score: (1, 2) Progressive changes in enamel; (3) Changes in dentin without cavitation in the enamel; (4, 5) Changes in dentin and progressive cavitation in the enamel, ie, still no bacterial invasion of the dentinal tubules and no indication for invasive tooth preparation; (6) Cavitation involving dentin—possible indication for tooth preparation and restoration. (Modified from Bille and Thylstrup, 1982 with permission.)

Radiographic score	Clinical score 1	2	3	4	5	6	Total
0	2	4					6
1	6	21	16	5	2		50
2	1	11	16	7			35
3	1	5	22	23	7		58
4				1	6	2	9

Table 21 Scale of caries prevalence on WHO maps

Color	Level	Range of DMFT
Green	Very low caries prevalence	0.0–1.1
Blue	Low caries prevalence	1.2–2.6
Yellow	Moderate prevalence	2.7–4.4
Red	High prevalence	4.5–6.5
Brown	Very high caries prevalence	> 6.5

1997, caries prevalence in 12-year-old children has already decreased from a very high level to low levels in several industrialized countries during the last three decades (see Figs 440 to 442).

In 1969, there were sharp contrasts: The number of DMFT was very high, high, or moderate in the industrialized countries and generally very low, low, or occasionally moderate in the developing countries. During the following 20 to 30 years, there was a downward trend in caries prevalence in most of the industrialized countries, including a particularly dramatic improve-

ment (from very high to low prevalence) in the Scandinavian countries, Australia, and New Zealand. From 1993 to 1997, the following decreases in levels of DMFT were achieved: Canada, France, Spain, Italy, Greece, and Iceland from moderate to low; Brazil, Peru, and Paraguay from very high to high; Germany and the Balkan countries from high to moderate; and Australia and Finland from low to very low. On the other hand, during the same period, prevalence increased from very low to low in most of China and from low to moderate in South Africa.

Regionally, some areas improved even more dramatically; for example, in the county of Värmland, in southwestern Sweden, prevalence declined from among the highest in the world (40.0 DFSs in 1964) to very low (less than 1.0 DFS in 1994). Caries prevalence has decreased from very high or high levels to moderate or low levels in most industrialized countries, while it has increased from very low or low levels to low or moderate levels in many developing countries. Figure 453 shows the mean level of DMFT in 12-year-old children in industrialized and develop-

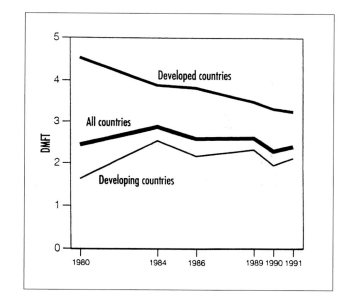

Fig 453 Mean level of DMFT among 12-year-old children in developing and industrialized countries, as well as globally, from 1980 to 1991. (From World Health Organization, 1994. Reprinted with permission.)

ing countries compared to the global means from 1980 to 1991. The trend in developing countries is similar to the global trend, because the developing countries represent most of the world's population, with about 40% living in China, India, and Indochina. In the developing countries, the general trend is for caries prevalence to increase, except where preventive programs have been established.

According to the most recent data from 2001 by the WHO, the mean global DMFT for 12-year-old children is 1.7 DMFT. The lowest and highest values in developing and industrialized countries and areas are shown in Table 22. Based on these data, it may be concluded that the mean number of 0.3 DFSs in 12-year-old children in the county of Värmland in 1999 (see Fig 446) is among the lowest values in the world, particularly among industrialized countries.

Recently, the Significant Caries (SiC) Index, together with a proposal for a new global oral health goal for 12-year-old children, was introduced by D. Bratthall of the WHO Collaborating Centre, Malmö, Sweden (Bratthall, 2000). Attention is drawn to the skewed distribution of dental caries within a given population, indicating that there are still large groups of individuals who have considerably more caries than the WHO and Fédération Dentaire Internationale (FDI) target level of 3.0

Table 22 Mean DMFT among 12-year-old children in developing and industrialized countries

Level	Country	DMFT	Year
Developing countries			
Very low	Ghana	0.1	1991
	Tanzania	0.4	1977
	Liberia	0.3	1994
High	Paraguay	5.9	1983
	Lebanon	5.7	1994
	El Salvador	5.1	1989
Very high	Guatemala	8.1	1987
	Peru	7.0	1990
Industrialized countries			
Very low	Australia	0.8	1998
	Denmark	0.9	2001
	Sweden	1.0	2000
	Finland	1.1	1997
Moderate	Bulgaria	4.2	1998
	Canada	3.0-3.7	1989-1991
	Czech Republic	3.4	1998
	Hungary	3.8	1996
	Latvia	4.2	1998
	Poland	4.0	1998
	Slovakia	4.3	1998
	Ukraine	4.4	1992
	Russian Federation	3.7	1985-1995
	Serbia	2.9-7.8	1994
High	Bosnia and Herzegovina	6.1	2001
Very high	Romania	7.3	1998

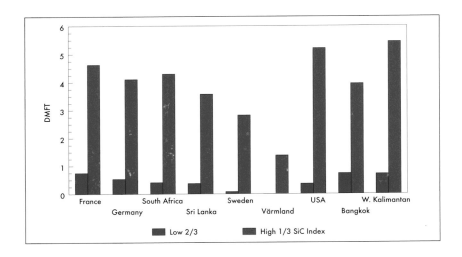

Fig 454 The SiC Index for 12-year-old children in 1999 in the county of Värmland, Sweden, compared to France, Germany, South Africa, Sri Lanka, Sweden, the United States, Bangkok, and West Kalimantan.

DMFT by the year 2000. The index is calculated as follows: Individuals are sorted according to their DMFT values. The one third of the population with the highest caries score is selected, and the mean DMFT for this subgroup is calculated. This value constitutes the SiC Index. The SiC Index can be easily calculated and used as a measure for future oral health goals. The SiC Index should be less than 3.0 DMFT in the 12-year-old children in a given population, and it is hoped that this global oral health goal is reached at the latest by the year 2015.

Figure 454 shows the SiC Index based on recent epidemiologic data from some countries compared with data from 1999 in 12-year-old children in the county of Värmland, where more than 85% were caries free (see Fig 448). Among the 15% with 1.0 DFS or more, the mean value of DFSs was only 1.4.

FACTORS AFFECTING RATES OF CARIES PREVALENCE

What lies behind the spectacular fall in caries prevalence in some countries? How can the prevalence rate be prevented from rising again? How can the deteriorating situation in other countries be halted? The answer to all three questions is one and the same: *prevention*. The factors underlying the unprecedented public health success story in the industrialized world are the promotion of oral hygiene, the widespread use of fluoride toothpaste, the fluoridation of water (or, in some countries, the introduction of fluoridated salt), nutritional counseling (no sweets between meals, etc), and the establishment of well-organized, school-based preventive programs (particularly in Scandinavia). In other words, integrated caries prevention.

Apart from the fluoridation of water, salt, and milk, which requires more advanced technology and supervised central administration, the aforementioned methods use simple techniques, cost little, and are perfectly suited to implementation at the primary health care level. As a result of the progress made in the last 25 years, the developing countries now have the knowledge and means of prevention to enable them to avoid the costly problems that the industrialized countries have had to face and indeed are still facing. Dental care, as currently practiced, falls into one of five categories (Box 16). Table 23 shows the distribution of the different forms of dental care for children and young adults in the county of Värmland, Sweden, from 1900 to 1990, underlying the dramatic reduction of DFSs in 12-year-old children from 1964 (40.0 DFSs) to 1994 (less than 1.0 DFS).

Table 23 Percentages of forms of dental care provided to children and young adults (aged 0 to 19 years) in the county of Värmland, Sweden

Period	"Primary" primary prevention	Primary prevention	Secondary prevention	Tertiary prevention	Relief of pain
1900–1929	0	0	0	25	75
1930–1949	0	0	2	48	50
1950–1959	0	0	10	70	20
1960–1969	0	5	20	70	5
1970–1979	5	15	40	40	0*
1980–1989	15	45	30	10	0

*Number is approximate.

Box 16 Categories of dental care

- "Primary" primary prevention: preventive measures for all expectant mothers, to prevent postnatal transmission from mother to child of cariogenic microbes and poor dietary habits
- Primary prevention: maintenance of the intact dentition, ie, prevention of dental caries, gingivitis, and periodontitis in a completely healthy mouth
- Secondary prevention: prevention of recurrence of disease (dental caries, gingivitis, and periodontitis) after successful symptomatic treatment
- Tertiary prevention: symptomatic treatment of dental caries, gingivitis, and periodontitis (ie, restorations, scaling, and periodontal surgery)
- Relief of pain: extractions and endodontic treatment

As an effect of "primary" primary prevention and primary prevention, available to expectant mothers and 1- to 3-year-old children through prenatal, maternal, and child welfare centers, the percentage of caries-free 3-year-old children in the county of Värmland, Sweden, has increased from only 35% in 1973 to 97% in 1993.

The reasons that Sweden, particularly in the county of Värmland, has succeeded in such a significant reduction in caries prevalence and caries incidence among children and young adults will be discussed in more detail in the following sections.

Sugar consumption

Caries prevalence was much higher 50 years ago than it was 30 years ago. At that time daily toothbrushing was almost nonexistent among the Swedish population, especially among children and the mentally retarded. No topical fluoride agents were used, and less than 5% of the Swedish population had access to drinking water with a fluoride concentration of more than 0.7 mg/L. At that time, the famous Vipeholm study in southern Sweden (Gustafsson et al, 1954) was conducted on mentally retarded individuals with no established oral hygiene habits; ie, they had heavy amounts of plaque on the teeth. The study showed a very significant correlation between estimated sugar clearance time and caries incidence.

National statistics from the Swedish sugar and sweets companies show that the results from the Vipeholm study have not been successfully applied. The daily national average consumption of sugar, about 115 g per day per individual, remained unchanged from 1960 to 1990. Unfortunately, the proportion of indirect consumption has increased from about 20% in 1960 to 65% in 1990; ie, consumption of sweets, soft drinks, marmalade, and cake have increased significantly (Fig 455). In 1990, the average annual consumption of sugar-containing sweets was 10

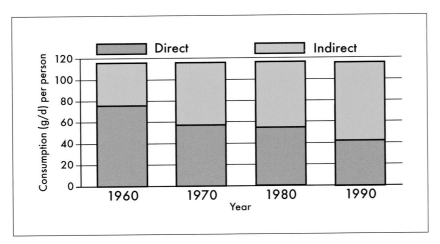

Fig 455 Mean sugar consumption (g/d) per person in Sweden from 1960 to 1990.

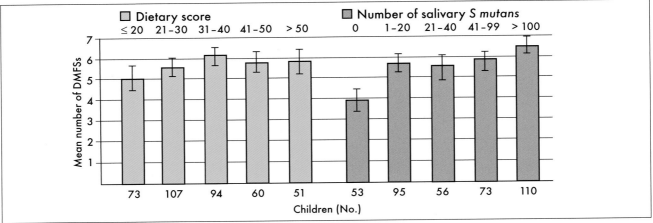

Fig 456 A study of 13- to 14-year-old schoolchildren in Karlstad with good oral hygiene and relatively low caries prevalence shows no correlation between caries prevalence and different dietary scores and different levels of salivary *Streptococcus mutans*. The cutoff is *S mutans* negative or positive. (From Kristoffersson et al, 1986. Reprinted with permission.)

kg per individual (at a retail cost of about US $200). The most recent data show that Denmark and Sweden have the highest consumption of sweets in the world: 17 and 15 kg per individual per year, respectively.

In spite of that fact, Denmark and Sweden have the lowest caries prevalence in 12-year-old children among industrialized countries: 0.9 and 1.0 DMFT, respectively. Therefore, the reduction in caries cannot be attributed to reduced consumption of sweets. This was also confirmed in other studies that found no correlation between an estimated sugar clearance score and caries prevalence in 13- to 14-year-old schoolchildren (Kristoffersson et al, 1986) (Fig 456). Similar results were also obtained in other Swedish studies (Birkhed et al, 1989; Stecksen-Blicks et al, 1985; Sundin et al, 1983). The estimated sugar clearance time in the Swedish adult population is not as important a modifying risk factor on caries prevalence as a smoking habit is on periodontal disease prevalence (Axelsson et al, 1998). In populations with high caries prevalence and poor oral hygiene or absence of oral hygiene habits, on the other hand, high sugar clearance values remain a powerful external modifying risk factor.

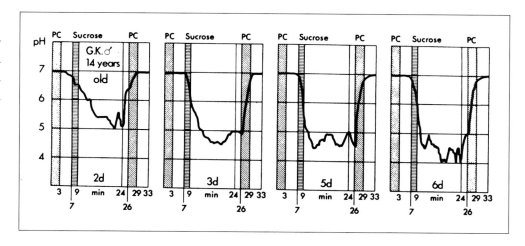

Fig 457 Telemetrically recorded pH of 2-, 3-, 5- and 6-day-old interdental plaque in a 14-year-old boy during and 15 minutes after a sucrose rinse (15 mL, 10%). (PC) 3 minutes of paraffin chewing. (From Imfeld, 1978. Reprinted with permission.)

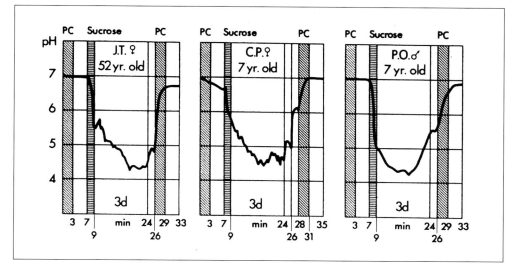

Fig 458 Comparison of telemetrically recorded pH values of 3-day-old interdental plaque in a 52-year-old woman and two 7-year-old children during and 15 minutes after a 2-minute sucrose rinse (15 mL, 10%). (PC) 3 minutes of paraffin chewing. (From Imfeld, 1978. Reprinted with permission.)

Plaque amount and plaque formation rate

Colonization of the tooth surfaces by bacteria-forming plaque is the key etiologic factor of dental caries. The telemetric method has been used to evaluate the influence of plaque age on pH, following a 2-minute rinse with 10% sucrose solution. Figure 457 shows the fall in pH in 2-, 3-, 5-, and 6-day-old interdental plaque biofilms in a 14-year-old boy.

Irrespective of the subject's age, and in experiments in the same test subject over a 2-year period, it seems that a critical fall in pH (to less than

5) occurs only in 3-day-old plaque biofilms. In a toothbrushing population, such mature plaque biofilms would be found, if at all, only on the approximal surfaces of the molars and premolars. This explains why, in such a population, these surfaces are the most susceptible to caries.

Figure 458 shows the fall in pH in a 3-day-old plaque biofilm in a 52-year-old woman, a 7-year-old girl, and a 7-year-old boy after they rinsed with sucrose (Imfeld, 1978, 1983). On the other hand, maximum 12-hour-old plaque will only result in a marginal drop in pH (see Fig 6 in chapter 1). By using the wire telemetric method, Igarashi et al (1989) showed that after a 1-minute

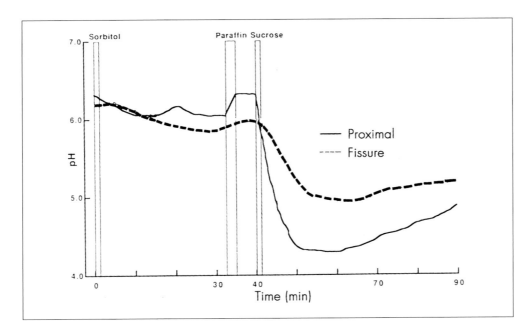

Fig 459 Results of the wire telemetric method show the low pH of approximal plaque compared to that of fissure plaque. (From Igarashi et al, 1989. Reprinted with permission.)

rinse with 10% sucrose solution, the pH was much lower in 4-day-old approximal plaque than in the corresponding fissure plaque (Fig 459).

Firestone et al (1987) used the same telemetric test in vivo, measuring the drop in pH after subjects rinsed with a 10% sucrose solution. Four different sites on molars with approximal plaque were compared to plaque-free approximal surfaces. The authors concluded: "Removing plaque from interdental surfaces significantly reduced the exposure of the surfaces to plaque acids following sucrose rinse. This further supports mechanical removal of plaque from interdental surfaces as a means of reducing dental caries."

The basic principle of the nonspecific plaque hypothesis is that a thick plaque biofilm on the tooth surface, if left undisturbed for long periods, allows the total amount of acid produced within this plaque to initiate the development of a caries lesion. Accordingly, very fast plaque formers (PFRI scores 4 and 5) would be expected to develop more caries lesions than slow or very slow plaque formers (PFRI scores 1 and 2) if the standard of oral hygiene and the composition of the microflora were the same in the two groups.

In addition, caries lesions tend to develop on the particular tooth surfaces on which most plaque reaccumulates between toothcleaning procedures (mesiolingual and distolingual surfaces of the mandibular molars and mesiobuccal and distobuccal surfaces of the maxillary molars), and, in a toothbrushing population, where the toothbrush has limited access (the approximal surfaces of the molars and premolars). This is confirmed in studies on the pattern of caries prevalence in different populations (Axelsson et al, 1990, 1993a; Forsling et al, 1999).

Not only the frequency but also the main target of needs-related oral hygiene procedures should be based on the score and the pattern of the PFRI. Because the quantity of plaque that forms on clean tooth surfaces during a given time represents the net result of interactions among etiologic factors, many internal and external risk factors, and protective factors, future research should be directed toward methods of identifying the major factors that cause rapid plaque formation in the individual patient. If possible, these factors should be reduced or eliminated. One of the most important factors for a high PFRI score is inflamed gingiva and as

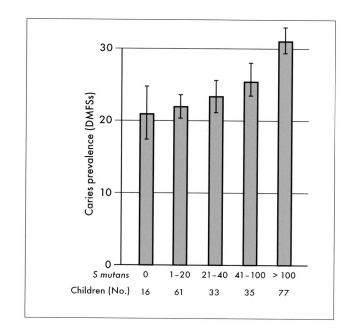

Fig 460 A study of 12-year-old Brazilian children shows a correlation between very high salivary *S mutans* levels and very high caries prevalence. (From Buischi et al, 1989. Reprinted with permission.)

a consequence increased gingival exudate flow, which is conducive to the growth of bacteria, since it contains glucose and many important amino acids. Thus, because of the high standard of gingival plaque control in children in the county of Värmland, the majority exhibit low or very low PFRI compared with children in areas with a poorer quality of oral hygiene.

Mutans streptococci

Several cross-sectional studies in human populations with relatively high caries prevalence have shown a correlation between very high salivary MS levels and very high caries prevalence (Buischi et al, 1989; Klock and Krasse, 1977; Salonen et al, 1990; Zickert et al, 1982b). This is exemplified in Fig 460, which is based on data from a study of 12-year-old Brazilian children (Buischi et al, 1989). However, in populations with relatively low caries prevalence and high standards of oral hygiene, the threshold value of more than 1 million CFUs of MS per 1 mL of saliva no longer seems to apply, as exemplified in Fig 456, which is based on data from a study of 13- to 14-year-old children in Karlstad, Sweden (Kristoffersson et al, 1986). In this population, the critical difference was between MS-negative and MS-positive subjects. Thus, the lower the prevalence and incidence of caries in a population, the more difficult it is to demonstrate a significant correlation for one single etiologic or modifying factor.

Because MS requires a hard, solid, nondesquamating surface for colonization (Berkowitz et al, 1975; Stiles et al, 1976), infants do not harbor MS until some time after tooth eruption: The major source of the infection is thought to be maternal. Several studies have suggested that the extent of colonization by MS and, to some degree, subsequent carious activity experienced by a child may be correlated with the mother's salivary level of MS: Mothers with high levels of MS tend to have children with high levels and vice versa (Caufield et al, 1988; Köhler and Bratthall 1978; Köhler et al, 1984; van Houte et al, 1981).

While correlations between caries or MS levels in mothers and those in their children may be explained in part by common genetic or environmental factors, some have suggested that a child's degree of colonization may be directed by the mother's levels of MS at the time of transmission.

457

In a landmark study, Köhler and coworkers (1983, 1984) selected mothers with initially high levels of MS in saliva and determined the effects of various preventive and treatment regimens aimed at reducing MS below a predetermined threshold level. Children of these mothers were monitored for initial acquisition of MS and, subsequently, for carious activity over a 3-year period. A statistically significant difference was observed between control and experimental groups in terms of when a child acquired MS, the levels of MS harbored by the mother and child, and the child's caries outcome. It was shown that the earlier the colonization by MS, the higher the caries prevalence at 4 years of age. These findings have confirmed the importance of the preventive programs that have been provided so successfully to pregnant women and babies for more than 25 years in the county of Värmland, Sweden.

Longitudinal human clinical studies have shown correlations between high salivary MS counts and high caries incidence. In preschool-aged children (primary dentition), correlations between salivary MS counts and caries incidence have been shown by Alaluusua et al (1990), Köhler et al (1988), Roeters et al (1995), Thibodeau and O'Sullivan (1996), and Twetman et al (1996). In the permanent dentition, Klock and Krasse (1978, 1979) showed that 9- to 12-year-old children with very high salivary MS levels developed significantly more new carious surfaces than did children with low levels of MS during a 2-year period. However, when a test group of children with more than 1 million CFUs of MS per 1 mL of saliva received high-quality plaque control by frequent PMTC, they developed fewer new carious surfaces than did the control groups with either high or low salivary MS levels (0.9 new carious surfaces versus 4.3 and 2.2 new carious surfaces, respectively). These studies in humans confirm the earlier animal studies by Fitzgerald and Keyes (1960), which indicated that dental caries is an infectious disease, transmissible by MS. Clinical studies have also confirmed that MS can be isolated from dental plaque covering active caries lesions in enamel (Axelsson et al, 1987b; Kristoffersson et al, 1984).

In Mölndal, Sweden, Zickert et al (1982b) also found a significant correlation between the prevalence of MS in saliva and the incidence of new caries lesions. During a 3-year period, children with high levels of salivary MS (more than 10^6 CFUs/mL) developed about three times as many new caries lesions as did control groups with lower levels of MS. Subjects in test groups following a treatment program including chlorhexidine developed significantly fewer cavities.

In particular, controlled intraindividual longitudinal studies monitoring the microflora at the tooth-surface level have clarified the cariogenic potential of MS (Axelsson et al, 1987b; Kristoffersson et al, 1984; Macpherson et al, 1990). An advantage of such studies is that several other external and internal modifying factors, such as diet, fluoride, and saliva, are equal for test and control sites. These studies have clearly shown that, in the same mouth, a tooth surface colonized by MS is at greater risk for caries than is a similar surface without MS.

In 13- to 14-year-old children, the highest levels of colonization by MS are found on the approximal surfaces of the molars and second premolars. In fact, the previously mentioned study of more than 600 children aged 14 years (Axelsson, 1991) showed that the same surfaces also had the highest PFRI scores. These observations explain why, in toothbrushing populations, these approximal surfaces have the highest prevalence of DFSs. For optimal caries prevention in such populations, plaque control and topical application of fluorides should target these key-risk surfaces.

Plaque control

Plaque control can be mechanical or chemical and performed by self-care or by professionals (dentists, dental hygienists, and preventive dentistry assistants). In the most successful preventive programs based on plaque control, self-care has been combined with PMTC at needs-related intervals. Chemical plaque control by self-care and/or by professionals is normally used only in selected high-risk individuals.

As discussed earlier, germ-free animals do not develop caries in spite of frequent intake of sucrose (Orland et al, 1954). If cariogenic bacteria from human dental plaque is inoculated in the mouth of one subject of such a group of germ-free animals, however, rampant caries develops in all the groups (Fitzgerald and Keyes, 1960).

When the experimental caries study in humans was repeated with sucrose mouthrinsing nine times per day for 3 weeks, the test subjects did not develop any enamel caries lesions: Chemical plaque control was maintained with 0.2% chlorhexidine digluconate mouthrinse twice a day (Löe et al, 1972). In contrast, almost every subject abstaining from oral hygiene for 3 weeks developed enamel caries lesions (von der Fehr et al, 1970). From these classic studies in animals and humans, it may be concluded that plaque control should be central to caries prevention because the dental plaque biofilm is the key etiologic factor in dental caries. In other words, it is a question of the quality of cleaning of every tooth surface with needs-related intervals: "Clean teeth never decay."

The caries-preventive effect of daily toothbrushing has been questioned. In the hundreds of studies based on supervised toothbrushing, the effect has been compared with positive controls who already had an established habit of daily toothbrushing, sometimes more frequent and efficient than the test groups; ie, they were not negative-control groups (for review, see Bellini et al, 1981; Hotz, 1998). To evaluate the separate caries-preventive effects of daily toothbrushing versus fluoride dentifrice, a 3-year longitudinal study should be carried out in a nontoothbrushing population of 12-year-old children, with a maximum number of tooth surfaces at risk, high caries incidence, and without fluoridated drinking water. Two test groups and one negative-control group would be randomly selected. Test group 1 would

have their teeth brushed daily in the school with placebo toothpaste by a dental assistant. The quality of the toothcleaning would be checked with a plaque-disclosing agent. Test group 2 would have their teeth brushed daily with fluoride toothpaste by the same dental assistant. By comparing test group 1 with the negative-control group, the separate effect of daily toothbrushing would be evaluated on all buccal and lingual surfaces and the approximal surfaces of the incisors—those easily accessible to the toothbrush. Between test groups 1 and 2, the separate effect of fluoride toothpaste would be evaluated on the same surfaces. For ethical reasons, no preventive measures would be withdrawn in test group 1 and the negative-control group, but of course, these groups would receive appropriate needs-related preventive measures at the end of the study. Unfortunately, it is probably too late to run such a study, because toothbrushing with fluoride toothpaste is such a widespread, well-established caries-preventive measure, that it could not ethically be withheld from caries-susceptible subjects for such a long experimental period.

However, experimentally, Dijkman et al (1990) carried out a crossover study in which teeth with experimentally induced 100-μm-deep enamel lesions were placed in situ for 3 months. In the control teeth, which were not brushed, 50% of the lesions progressed. In the test teeth, which were brushed daily with a placebo toothpaste, none of the lesions progressed, and no new lesions developed. In a test in which the teeth were brushed daily with a fluoride dentifrice, 40% of the experimentally induced lesions decreased in depth, ie, an additional caries reduction. The authors concluded that the remineralizing efficacy of fluoride toothpaste is the result of the cleaning effect of the brushing by the toothpaste (presumably on the pellicle) as well as the fluoride effect on mineral nucleation and growth. As a measure directed against the caries etiologic factor (the dental plaque biofilm), cleaning prevents the development of new caries lesions and the progression of enamel lesions, while fluoride acts as a catalyst for remineralization.

Even in early toothbrushing studies with positive-control groups, a significant caries reduction has been achieved, at least on buccal and lingual surfaces, where the toothbrush has accessibility (Ainamo, 1971; Finn et al, 1955; Fogels et al, 1982; Fosdick, 1950; Granath et al, 1976, 1978; Strålfors et al, 1967). In a longitudinal clinical study (Fogels et al, 1982), one test group of 8- to 10-year-old children received a 1-minute supervised toothbrushing per day at school using a nonfluoride toothpaste (for school as well as for home use). The quality of the toothcleaning was checked every time. In a matched control group, 99% of the children brushed their teeth with a fluoride toothpaste one to three times per day, according to their parents. The control group developed significantly more new DSs than the test group, 2.3 DSs compared to 0.4 DSs per individual, respectively ($P < .001$). The authors concluded that their study suggests that children who participate in a 1-minute daily brushing regimen with a nonfluoride toothpaste may show improved dental health. Their increment of dental caries (DSs) over a 1-year period was found to be lower than the increment obtained from a comparable group of children whose oral hygiene regimen was not intentionally altered or influenced and who used their regular fluoride toothpaste at home.

The fact that 98% of Swedish schoolchildren brush their teeth one to three times per day as an effect of the school-based preventive program may therefore explain the very significant caries reduction in Swedish schoolchildren during the last 20 years, especially on all buccal and lingual surfaces and the approximal surfaces of the incisors (Kuusela et al, 1997). During the 1960s, very few Swedish schoolchildren brushed their teeth daily. In addition, the percentage of fluoride toothpastes on the market has increased from about 80% in 1971 to almost 100% in 1990. However, only 10% of the children use dental tape daily (Riise et al, 1991) and fewer than 1% use a combination of dental tape and fluoride toothpaste.

Thus there is still an unexploited caries-preventive effect by self-care on the approximal surfaces of the posterior teeth; in a split-mouth, 2-year study, daily supervised flossing resulted in a more than 50% caries reduction in a toothbrushing population using fluoride toothpaste (Wright et al, 1979). If self-diagnosis is included as a motivating factor and behavior modification principles are used to establish good habits, the caries-preventive effects of mechanical plaque control by self-care can be significantly improved (for details, see chapter 1).

This was proven in a 3-year longitudinal study (Axelsson et al, 1994a). Twelve-year-old children in São Paulo, Brazil, were randomly placed into two test groups and one positive-control group. All the children had a well-established habit of using a toothbrush and fluoride toothpaste daily. In addition, all children were exposed to fluoridated drinking water. Children in test group 1 were trained and instructed to identify sites with inflamed gingivae, which could heal, and sites with enamel caries, which could remineralize. By using the PFRI as a guideline, they understood how frequently they needed to clean their teeth and which surfaces required special attention during toothcleaning procedures. According to the "linking method" (Weinstein and Getz, 1978), they were motivated to initially apply fluoride toothpaste on the approximal surfaces of the molars and premolars. These surfaces were then meticulously cleaned with dental tape and the "rubbing technique." Afterward they routinely cleaned the buccal and lingual surfaces with a toothbrush and fluoride toothpaste. After the three initial visits, the children were recalled monthly during the first 4 months and then were recalled every third month for reevaluation of the results based on self-diagnosis. The children in test group 2 were trained, both on models and in their mouths, to clean tooth surfaces meticulously by using dental tape, fluoride toothpaste dentifrice, and toothbrush. They were recalled for reinstruction at the same intervals as test group 1. During the 3-year period, the children in test group 1 developed 60% and 75% fewer new approximal dentin caries lesions per individual on the molars and premolars than did children in the test group 2 and the control group, respectively.

The reductions in plaque and gingivitis were also significantly greater in test group 1 than in test group 2 and the control group (Albandar et al, 1994a). Because all three groups had the same exposure to fluoride, from toothpaste and drinking water, the greater reduction in approximal caries in test group 1 than in test group 2 and the control group must be the effect of improved interdental plaque control.

Following are the conclusions derived from the study (Axelsson et al, 1994a):

1. In a toothbrushing population, using fluoride toothpaste and fluoridated drinking water, a highly significant reduction in the incidence of approximal caries will be achieved with an oral hygiene training program based on self-diagnosis and the linking method.
2. In such a population, frequent repetition of meticulous oral hygiene training is almost useless.

The cost effectiveness (cost savings) of large-scale implementation of the system used in this low-cost, low-technology self-care study is almost inestimable. Since the late 1970s, these principles have gradually been introduced in the self-care education programs of schoolchildren in the county of Värmland. This is one important reason for the dramatic caries reduction and improved standard of oral hygiene during the last two decades, not only in comparison with other countries, but also with other Swedish counties.

The caries-preventive effects of PMTC are well documented. In the early 1970s, studies showed that the reduction in caries generated by frequent PMTC ranged from 70% to 90% (Axelsson and Lindhe, 1974, 1978, 1981a, 1981c; Axelsson et al, 1976; Gisselsson et al, 1983; Klimek et al, 1985; for review, see Axelsson, 1981, 1994; Bellini et al, 1981; and Hotz 1998). In addition, when fluoride was replaced with placebo in the dentifrice and prophylaxis paste, similar caries-preventive effects were shown (Axelsson et al, 1976). However, it must be pointed out that use of a rotating rubber cup combined with a prophylaxis paste

in a population brushing with fluoride toothpaste one to three times per day is not really *professional* mechanical toothcleaning. In a toothbrushing population, PMTC must focus on:

1. The approximal surfaces of the molars and premolars.
2. The lingual and buccal surfaces of the mandibular molars and the buccal surfaces of the maxillary molars (for details, see chapter 3).

These facts may explain the range of the caries-preventive effects of the so-called PMTC in different studies. Since 1975, needs-related intervals of PMTC have been implemented in integrated preventive programs for children and young adults in the county of Värmland (for review, see Axelsson, 1978, 1981, 1994, 1998 and Hotz, 1998).

Chlorhexidine digluconate is regarded as the gold standard among the chemical plaque control agents for self-care as well as for professional use. In addition, chlorhexidine has a specific antimicrobial effect on MS. For self-care, chlorhexidine is most frequently used as a 0.2% or 0.1% mouthrinse twice a day, intermittently for 2 to 3 weeks, in selected caries-risk individuals with high levels of salivary MS. Chlorhexidine is also available for self-care in gels and for professional use in gels and varnishes.

A meta-analysis of randomized controlled clinical studies on the caries-inhibiting effect of plaque control by chlorhexidine revealed an overall caries reduction of 46% (van Rijkom et al, 1996). For fluoride toothpastes and other fluoride agents, similar analyses show a caries reduction of only 20% to 25%. Thus chlorhexidine is twice as effective because it targets the cause of dental caries (the cariogenic plaque) directly, while fluoride is an external factor that modifies enamel demineralization and remineralization.

However, the effect of chlorhexidine as well as other chemical agents such as fluoride is related not only to the concentration and substantivity (clearance time) but also to accessibility. Therefore the effects of mouthrinses and gels applied by self-care are limited on the approximal surfaces of the molars and premolars, which are blocked by buccal and lingual papillae. During the last decades, promising results have been published on the effects of chlorhexidine- and chlorhexidine-thymol–releasing varnishes, ie, professional chemical plaque control. The depressant effect on MS colonizing the approximal surfaces persists for up to 6 months, and significant effects on remineralization of enamel caries lesions are reported (Huizinga et al, 1992; Petersson et al, 1991b; for details, see chapter 4). Therefore it seems reasonable from a caries-preventive perspective that PMTC should be supplemented with such a varnish on the proximal surfaces of the molars and premolars and the fissures of the molars, in accordance with the individual assessment of caries risk.

Mechanical plaque control for the key-risk individuals, teeth, and surfaces by PMTC, as well as by self-care, has been practiced more intensively in Värmland county than any other county in Sweden. These facts may explain why during the last 20 to 25 years the caries prevalence there has been significantly reduced compared to the rest of Sweden. This can be exemplified in the WHO's key reference group for global caries prevalence, 12-year-old children. In 1964, the mean caries prevalence in the county of Värmland was about 40.0 DMFSs, generally located on the approximal and occlusal surfaces of the molars and premolars (see Fig 52 in chapter 1). Also, some buccal and lingual surfaces of the molars and approximal surfaces of the incisors were carious, and one mandibular first molar was extracted.

During the following 10 years, the use of the toothbrush and fluoride toothpaste were introduced. As an effect of these two preventive measures, the DFSs disappeared, ie, "were brushed away" from the approximal surfaces of the thin incisors and the readily accessible buccal and lingual surfaces of the molars and premolars. The numbers dropped to about 25.0 DFSs. The separate effects of the toothbrush versus fluoride toothpaste are difficult to estimate.

In 1975, a needs-related plaque control program (both professional and home care) was introduced, concentrating on the key-risk tooth surfaces of schoolchildren. As a result, the DFSs disappeared from the approximal surfaces of the molars and the premolars. The number of DFSs dropped to 3.0 (this value represents mainly overtreatment of first molar fissures) in 1984. The preventive program for the occlusal surfaces of the molars was initiated in 1984. Caries prevalence in 12-year-old children was less than 1.0 DFSs in 1994 and only 0.3 DFSs in 1999, and about 85% of children were caries free (see Figs 447 and 448).

Fluoride

Today, there is general agreement that almost 100% of the caries-preventive effect of systemic fluoride from drinking water, salt, and milk is posteruptive, ie, topical (for review, see Burt and Fejerskov, 1996; and Fejerskov and Clarkson, 1996). It is also generally accepted that the most significant caries-preventive effect of fluoride is its active role in the remineralizing process of early enamel caries lesions and retarding the demineralization of enamel (for review, see ten Cate and Featherstone, 1996).

The enamel of erupting and newly erupted teeth is much more susceptible to caries than is enamel exposed to the oral environment for years (Kotsanos and Darling, 1991). In addition, about five times more dental plaque reaccumulates on erupting than on fully erupted teeth (Carvalho et al, 1989). Therefore, plaque control and the use of fluoride should be intensified during eruption. These facts explain the significant caries reduction of about 50% obtained from fluoridation of drinking water during the 1950s and 1960s, when daily toothbrushing was not routine among children and toothpaste did not contain fluoride (for review, see Burt and Fejerskov, 1996). The same effects may still be expected in populations with poor oral hygiene habits and, therefore, almost no use of fluoride toothpaste and, as a consequence, a high prevalence of caries.

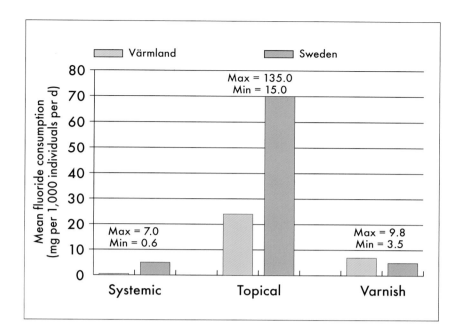

Fig 461 Use of fluoride (mg per 1,000 individuals per day) in the county of Värmland compared to Sweden as a whole. (From data by Friberg et al, 1989.)

The Swedish Board of Health and Welfare did not recommend fluoride toothpaste for children younger than 4 years of age until about 15 years ago, when a toothpaste containing 0.02% fluoride was first introduced for 1- to 3-year-old children. Fewer than 10% of the Swedish 2- to 3-year-old children used fluoride tablets on a regular basis. Thus, in 1979, in the county of Uppsala, which has naturally fluoridated water, 78% of 3-year-old children and 43% of 5-year-old children were caries free, compared to only 68% and 35% of 3- and 5-year-old children, respectively, in all of Sweden.

A study on the consumption of fluorides, expressed in milligrams of fluoride per 1,000 individuals per day, showed that the county of Värmland used only 0.6 mg of systemic fluorides (ie, fluoride tablets), the lowest value in Sweden. The mean value for all of Sweden was 5.0 mg, and the county of Halland used 7.0 mg, the highest value in Sweden. The consumption of topical fluoride agents by self-care in such vehicles as toothpaste, gels, and mouthrinses was 24.0 mg in the county of Värmland, 70.0 mg in all Sweden, 135.0 mg in the county of Halland (maximum in Sweden),

and 15.0 mg in the county of Uppsala (the minimum value in Sweden). The consumption of fluoride varnish was 6.0 mg in the county of Värmland. The national average was 5.0 mg; the average was 9.8 mg in the county of Halland (the maximum value for Sweden) and 3.5 mg in the county of Uppsala (the minimum for Sweden) (Friberg et al, 1989) (Fig 461).

Thus, the role of fluorides may not explain why the county of Värmland has reduced caries prevalence among children and young adults from the highest values to the lowest values among the Swedish counties and in the world during the last two decades. The main reason is rather that the county has focused on cause-related caries-preventive measures, because caries-preventive effects achieved in studies based on high-quality plaque control by self-care and PMTC range from 50% to 90%, while the effects of fluorides in toothbrushing populations range from 20% to 25%, as discussed earlier. Of course, the use of fluorides based on individual need represents an important measure in integrated caries-preventive programs, and fluoride toothpaste should be used daily by everyone.

During the 1960s, caries prevalence and caries incidence were extremely high in Sweden; the highest values were found in the county of Värmland. At that time, very few children brushed their teeth daily. This is why fluoride mouthrinsing once every 1 to 2 weeks resulted in about a 40% caries reduction during the early 1960s (Forsman, 1965; Torell and Ericsson, 1965). Fluoride toothpaste was gradually introduced: Usage increased from 0% in 1965, to about 80% in 1970, and to almost 100% in the 1990s. In a 3-year longitudinal double-blind study in 12- to 15-year-old children who brushed daily with fluoride toothpaste, a mouthrinse containing 0.05% NaF weekly proved ineffective when compared to distilled water used as placebo (Axelsson et al, 1987a). In the same study, PMTC and fluoride varnish were applied four times per year in another randomized test group, which resulted in a significant caries reduction.

Subsequently, all mouthrinsing programs in the elementary schools in Värmland were discontinued in 1980, contravening the Swedish Board of Health and Welfare's recommendations. Instead, in selected risk and high-risk individuals, the daily self-care program was supplemented with a needs-related program of PMTC and application of fluoride varnish to key-risk surfaces. Despite termination of the fluoride mouthrinsing program in the schools, the needs-related program continued to have significant effects on dental caries. In 1990, 12-year-old children in the county of Värmland had 1.1 DFSs compared with 2.0 DFSs in the county of Uppsala, which has natural fluoride in the drinking water, and the national average of 2.2 DFSs.

It is surprising that 65% of Swedish schoolchildren still follow a weekly fluoride mouthrinsing program. On the other hand, school fluoride mouthrinsing programs may still be very cost effective in countries with no fluoridated drinking water, a high prevalence and incidence of caries, and poor or absent oral hygiene habits and therefore no fluoride tooth-

---

Box 17 Recommendations for the use of fluorides in a self-care program

- Ages 2 to 19 years: brushing with fluoride toothpaste one to three times per day, based on individual needs (PFRI)
- Ages 12 to 19 years: use of dental tape with fluoride toothpaste one to two times per day on the approximal surfaces of the molars and premolars, based on predicted risk
- Selected high-risk individuals: cleansing of all tooth surfaces before meals; consumption of fluoride lozenges that are allowed to dissolve in the mouth or use of fluoride chewing gum immediately after meals for optimal topical and saliva-stimulating effects.
- Selected high risk individuals: intermittent use of a combined antimicrobial fluoride mouthrinse (eg, 0.1% chlorhexidine plus 0.05% NaF or stannous fluoride [SnF$_2$] plus amine fluoride [AmF] solution) twice a day for 3 weeks

---

paste. The most recent recommendations for the use of fluorides in the Värmland self-care program are shown in Box 17.

The program recommendations for professional use are: PMTC plus fluoride varnish, eg, Fluor Protector (Vivadent, Liechtenstein), Duraphat (Rorer, Cologne, Germany), or Bifluorid 12 (Voco, Cuxhaven, Germany), on the key-risk surfaces (the occlusal surfaces of the molars, the approximal surfaces of the molars and premolars, and other smooth surfaces showing signs of active enamel caries lesions). In high-risk individuals, this procedure is repeated every 6 months, two applications at 2- to 3-day intervals, or three to six single treatments per year. Studies have shown that fluoride varnishes provide additional benefit even when daily fluoride toothpaste or fluoridated water is used (Axelsson et al, 1987a; Seppä et al, 1981). (For details on fluorides, see chapter 5.)

Fissure sealants

In the early 1970s, a 27-month longitudinal study was conducted in Sweden to evaluate caries progression from the time of eruption in the permanent first molars. The teeth were examined every 3 months from the start of eruption. The results showed that only 50% of the molars were carious. Most of the lesions had developed already within 3 to 9 months and no further initiation of caries lesions occurred after 15 months. That means the caries lesions were initiated during the 12- to 14-month eruption of the first molars. In premolars, which have an extremely short eruption time of only 1 to 2 months, caries lesions will not develop.

At that time, almost 100% of the occlusal surfaces in 11- to 12-year-old Swedish children were restored, representing an enormous overtreatment that followed Black's recommendation of "extension for prevention." Now, when the caries incidence in Sweden and the county of Värmland is more than 10 times lower than it was in the early 1970s, only about 5% of the molars would be restored if the indication for restoration was a caries lesion with cavitation into the dentin. Therefore, from a cost-effectiveness perspective, the general use of fissure sealants cannot be justified.

The use of fissure sealant was almost nonexistent in Sweden during the 1970s. However, during the 1980s, this procedure has increasingly been used. In some regions of Sweden, the general use of fissure sealant is common, a tremendous overtreatment. When fissure sealants are used in fully erupted teeth, 70% to 95% "success" is reported in longitudinal studies; that is, 70% to 95% of the sealants remain and/or 5% to 30% of the surfaces were carious. In partially erupted teeth, the reported success rate has been only about 50%. How many of these surfaces, if left unsealed, would have developed a caries lesion with cavitation?

As mentioned earlier, it has been shown that the amount, as well as the formation rate, of plaque is about five times higher—especially in the distal and central fossae of the occlusal surfaces—in partially erupted maxillary and mandibular first molars than in the fully erupted teeth. In studies by Carvalho et al (1989, 1991, 1992), 90% of the central and 50% to 60% of the distal fossae already exhibited active enamel caries lesions during eruption. The parents of these children were given instruction in toothcleaning, especially the occlusal surfaces, immediately before meals and bedtime. In selected risk individuals, PMTC and painting with 2% NaF solution was performed at needs-related intervals during the eruption of the teeth.

One year later, 95% of the molars were fully erupted. As a result of this noninvasive preventive program, about 80% to 90% of the active enamel lesions were arrested or inactivated. During the following 2 years, when the molars were fully erupted, in functional occlusion, and subject to natural attrition, PMTC was carried out at less frequent intervals in the risk individuals. After 3 years, only 2% of the occlusal surfaces had to be sealed because of active enamel lesions and none had to be restored. No caries lesions progressed into the dentin according to the bitewing radiographs. In a matched control group, 2% of the occlusal surfaces had to be restored despite massive use of topical fluoride and fissure sealing of 70% of the surfaces (Carvalho et al, 1989, 1991, 1992).

In the early 1980s, a special needs-related program for prevention of occlusal caries in the molars was introduced in the county of Värmland. Use of fissure sealants was restricted to erupting molars with "sticky" risk fissures and selected high–caries-risk individuals from a cost-effectiveness point of view. Gradually, resin-based fissure sealants were replaced with glass-ionomer and light-cured resin-modified glass-ionomer sealant materials, which also serve as a slow-release fluoride agent during the eruption of the molars.

Based on the author's own research and the studies by Carvalho et al (1989, 1991, 1992), a more comprehensive preventive program for the occlusal surfaces of the molars was introduced in the county of Värmland some years ago. Thus it is estimated that only 10% to 20% of molars have to be sealed in order to achieve nearly 100% caries prevention (for details, see chapter 6).

Educational level

A computerized and analytical oral epidemiology system showed that in randomized samples of 50- and 65-year-old subjects with only elementary school education, 8% and 26%, respectively, were edentulous, compared to 0% and 2% for individuals in the same age groups with higher educational levels. The mean number of remaining teeth in 50-year-old individuals with low, middle, and high levels of education were 19.4, 23.8, and 25.0 teeth, respectively (excluding third molars). Loss of periodontal attachment was almost twice as high in 50-year-old subjects with a low educational level than it was in those with a high educational level. In addition, the latter exhibited 15% to 20% more intact tooth surfaces. These facts indicate that dental health status is strongly related to the educational level in the same age group and population (Axelsson et al, 1990). In 1988, 70% of the 65-year-old, 50% of the 50-year-old, and 22% of the 35-year-old adult population had only an elementary school educational level.

A more recent study in more than 1,000 seventh grade children in Goa, India, showed that poor oral hygiene was the outstanding number one risk factor for developing enamel as well as dentin caries, followed by a low educational level of the mothers, which was more important than lack of fluoride toothpaste use (Mascarenhas, 1998). At present, in Värmland's integrated preventive program, the well-educated 35-year-old adults in 1988 are the parents of 1- to 19-year-old children, and some are grandparents of 1- to 7-year-old children. In addition, most of the young parents in the county of Värmland today have personal experience of the needs-related preventive program for up to 15 to 18 years, which should positively influence their competence to be responsible for their young children's daily oral hygiene. Nearly 100% of these young parents have a higher educational level than elementary school. These facts may explain some of the successful promotion of oral health among the children—especially during the ages of 1 to 10 years, when parents accompany their children to their dental appointments, as well as to counseling sessions at the maternal and child health centers.

About 80% of the population in the county of Uppsala live in the city of Uppsala and its suburbs. Because the city, dominated by the University of Uppsala, has been a center of learning for more than 500 years, the population in the county has the highest educational level in Sweden. This difference was even more significant 30 years ago. That means the grandparents as well as the parents of the 1- to 19-year-old children in Uppsala in the early 1970s were well educated. In combination with naturally fluoridated drinking water in Uppsala, this explains the significantly higher percentage of caries-free 3-, 5-, and 6-year-old children in Uppsala in 1971, compared to all of Sweden and especially the rural, sparsely populated county of Värmland (about 250 km from north to south and 150 km from east to west).

In spite of these facts, the county of Värmland today exhibits the lowest caries prevalence in 1- to 19-year-old children in Sweden and among all industrialized countries. That is because the integrated dental care programs have focused so much on cause-related caries-preventive measures: in other words, plaque control by self-care supplemented with needs-related PMTC. That is a rational approach, because periodontal diseases are prevented and controlled as well.

CONCLUSIONS

When the needs-related and integrated caries-preventive program for children and young adults was introduced and the computer-aided baseline examination was carried out, the following goals were set up for those who had followed the program from birth to the age of 19 years:

1. To have no approximal restorations
2. To have no occlusal amalgam restorations
3. To have no approximal loss of periodontal attachment
4. To motivate individuals to assume responsibility for their own health

As a result of this program, caries prevalence in 19-year-old individuals was reduced from more than 24.0 DFSs in 1979 to only 2.0 DFSs in 1999. Of these, 1.0 DFS was occlusal. For 9 years, dentists have not been allowed to use amalgam in Sweden in 1- to 19-year-old children. Therefore, very few occlusal amalgam restorations currently exist because of caries in 19-year-old individuals; in a few years there will be none. As an average there was only 1.0 approximal DFS in 1999 and less than 0.5 was filled, because this number included noncavitated dentin caries lesions diagnosed on bitewing radiographs, which should be arrested by "prevention instead of extension" or at least "prevention before extension."

Needs-related self-care habits based on self-diagnosis were established as early as possible. In addition, supplemental PMTC and chemical plaque control were used in key-risk individuals at needs-related intervals. As a consequence, approximal loss of periodontal attachment was prevented. Thus, it can be concluded that close to 100% of the goals stated 20 years ago had been achieved. In addition, it seems realistic that 20-year-old individuals with no lost teeth, only 2.0 DFSs, no loss of periodontal attachment, and well-established, excellent self-care habits, should maintain at least 25 healthy, natural teeth for the rest of their lives. This seems a realistic prediction, as has already been proven in a 30-year longitudinal study in adults. In this study, 51- to 65- and 66- to 80-year-old subjects who participated in a needs-related preventive program performed by a dental hygienist from 1972 to 2002 on average lost only 0.4 and 0.7 teeth, respectively, did not lose periodontal attachment, and developed only two new carious surfaces per subject over 30 years (Axelsson et al, 2004; for details, see chapter 3).

Caries prevalence in the county of Värmland is still being reduced. From 1999 to 2001, approximal DFSs in 19-year-old individuals was reduced from 1.0 to 0.9. Because of implementation of continuously ongoing clinical research on improving and testing new preventive methods and materials and methods to predict risk for oral diseases, as well as new knowledge from other researchers and continued education of the well-trained and highly motivated dental professionals involved in the program, it seems reasonable that caries prevalence in the children and young adults in this region can still be improved at an even lower cost. Finally, the evaluation of new methods, as well as the reevaluation of established methods, by clinical research in the local population before large-scale implementation has great benefits in contrast to the hazards of direct implementation from animal and in vitro experiments. Although there are "evergreens," preventive measures should be tailored to reflect trends in the pattern of dental disease in a population.

CHAPTER 8

ORAL HEALTH PROMOTION AND NEEDS-RELATED PREVENTIVE PROGRAMS

TRENDS IN ORAL HEALTH

In many industrialized and several developing countries, 30% to 50% of the population older than 65 years is edentulous (40% in the United States; almost 60% in the United Kingdom). Many 50- to 65-year-old adults have lost most of their molars—the most efficient teeth for chewing. Among 35- to 50-year-old adults in many industrialized countries, 35 to 50 tooth surfaces are restored with esthetically unacceptable materials, such as amalgam or gold. However, among adults in many industrialized countries, the percentage of edentulous people has decreased and the number of remaining teeth has increased in recent decades. Several factors have shaped oral health trends globally, affecting the rates of caries and periodontal disease.

Dental caries

Caries prevalence has decreased significantly from a high or very high level among children and young adults in most industrialized countries during the last two decades, as discussed in chapter 7. On the other hand, caries prevalence in these

age groups is increasing from a low or very low level in many developing countries, where the majority of the world population is living.

What lies behind the spectacular drop in caries prevalence in some countries? How can the prevalence be prevented from rising again? How can the deteriorating situation in other countries be halted? The answer to all three questions is one and the same: *prevention, more prevention, and still more prevention.*

In industrialized countries, the promotion of oral hygiene, the widespread use of fluoride toothpastes, the introduction of fluoride into drinking water, or salt in some countries, and the availability of advice on nutrition (no sweets between meals, etc) are the factors behind an unprecedented public health success story, as discussed in chapter 7.

Apart from the fluoridation of water, salt, and milk, which requires more advanced technology and supervised central administration, the aforementioned methods use simple techniques, cost little, and are perfectly suited to implementation at the primary level of oral health care. As a result of the progress made in the last 25 years, developing countries now have the knowledge and means of prevention to enable them to avoid the costly problems that industrialized countries have had to face and indeed are still facing.

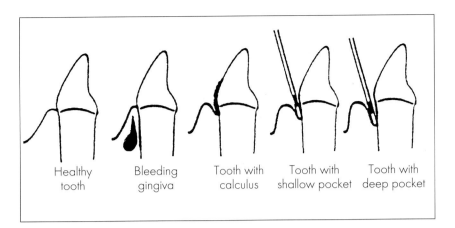

Healthy tooth   Bleeding gingiva   Tooth with calculus   Tooth with shallow pocket   Tooth with deep pocket

Fig 462 Clinical signs of periodontal disease recorded by the CPITN.

Periodontal diseases

There are few international surveys on the prevalence of periodontal diseases expressed in loss of periodontal support (probing attachment loss). The Community Periodontal Index of Treatment Needs (CPITN) was recommended in the early 1980s by a joint working group from the World Health Organization (WHO) and the Fédération Dentaire Internationale (FDI) (Ainamo et al, 1982). The CPITN records four clinical signs of periodontal diseases (Fig 462):

1. Bleeding gingiva
2. Calculus
3. Shallow periodontal pockets (4 to 5 mm)
4. Deep periodontal pockets (> 5 mm)

For measurement of periodontal status, the mouth is divided into six parts, or sextants. A specially designed probe is used to test the condition of the gingiva around the tooth selected as the index tooth for each sextant. If several clinical signs are present simultaneously, the most severe is selected.

The WHO has compiled data on more than 100 surveys carried out in individuals aged 35 to 44 years (Miyazaki et al, 1991a, 1991b, 1992). These data should be treated with caution, because very few of the surveys provide a national estimate. In addition, other data show that treatment needs indicated by the CPITN are greatly overestimated

at the sextant and even the tooth level compared to the surface level, that is, the "true" treatment need (Axelsson, 1998; Axelsson et al, 2000). Nevertheless, the data are of great interest because they consistently show a similar pattern of frequency and severity of involvement, challenging some generally accepted ideas about the distribution and the etiologic process of periodontal disease.

The data show that the percentage of people who have deep pockets (> 5 mm) and the mean number of sextants per person displaying deep pockets are low to very low; ie, the severe forms of periodontal disease, requiring complicated treatment, are far from common (Miyazaki et al, 1991a, 1991b, 1992). Moreover, there seems to be no difference in frequency between industrialized countries and developing countries for the severe forms of periodontal disease. On the other hand, the initial forms (bleeding and calculus) are much more prevalent in developing countries because fewer dental care resources are available and oral hygiene habits are poorer.

In light of these data, it may be stated that generalized periodontal destruction is rare in 40-year-old adults (Miyazaki et al, 1991b). Where some signs of such destruction are present, only a limited part of the dentition is affected. Except in a minority of patients, the initial forms (bleeding and calculus) do not seem to progress inevitably to the advanced stages of the disease (Norderyd et al, 1999).

Importance of Changing Attitudes Toward Oral Health

If the goal of oral health care is to maintain a natural dentition throughout life, the loss of all teeth is the ultimate failure, closely followed by the high percentage of people with only 20 or fewer remaining teeth. The relevant question is why such a high failure rate has been accepted by the public and the dental profession. Intact teeth and healthy gingiva are simply beautiful, attractive, functional parts of the body and should be much more highly regarded by the population. On the other hand, carious teeth, swollen, red, bleeding gingiva, and foul-smelling breath are most unattractive.

Under similar conditions, patients would never accept destruction of other parts of the body: An ugly false nose or other part of the body, even one normally covered by clothes, would never be accepted. Why should anyone accept false teeth? Patients would not accept having even 1% of the nose replaced by amalgam or gold. Imagine having to amputate a finger once every 5 years and replace it with a gold finger, in spite of regular checkups once or twice a year, because of an infectious disease—in an age when this disease, with a well-known etiology, could be successfully prevented.

It is the duty of dental professionals to educate and motivate the public, health personnel, and politicians to regard intact teeth and healthy gingiva as highly as, for example, a healthy nose, eyes, or ears and a justifiable external mode of dress. It is all a matter of changing attitudes and priorities. Famous clothing designers change fashion annually, and people accept the extra costs without hesitation.

A healthy and well-maintained mouth facilitates communication and human relationships. In addition, the boost in health, well-being, and self-confidence not only is very important for quality of life but also contributes at a very basic biologic level to protection from systemic infection and other general health problems. For ex-ample, analytical studies have disclosed a clear relationship between periodontal diseases and cardiovascular diseases (for review, see Beck et al, 1996 and Kolltveit and Eriksen, 2001), as well as preterm low–birth weight deliveries (Offenbacher et al, 1996, 1998, 1999), diabetes mellitus (Grossi and Genco, 1998), and other general health problems. Therefore, when oral health is compromised, overall health and quality of life are also compromised.

Patient responsibility

Motivation is defined as readiness to act or the driving force behind a person's actions. Greater responsibility has been described as the motivating factor of longest duration. Optimized responsibility may sometimes result in lifelong motivation, in contrast to the limited durations of encouragement provided by, for example, commendation or a salary increase.

Adults should believe, "No dentist or dental hygienist should accept more responsibility for my oral status than I do myself, because it is my mouth." However, in many industrialized countries with well-organized social health and welfare systems, the population is more or less passive; patients regard the dentist and dental hygienist as responsible for their oral health, the physician as responsible for their general health, and the politicians as responsible for their social welfare.

With the current level of knowledge about the etiology, prevention, and control of dental caries and periodontal diseases, it has been shown that patients who are well motivated and well educated in self-diagnosis and self-care can prevent and control these diseases by themselves. Much more important to general health, quality of life, and costs for health and welfare are the following examples: It is estimated that, among external (environmental) carcinogenic factors, an unhealthy diet accounts for about 30% of cancers, smoking for about 20%, and viruses for about 10%. (The simple recommendation for diet is reduction of animal fat and increased intake of fiber-rich vegetables and fruits, which are the cheapest and most accessible food products in tropical and subtropical climates, where most of the world population lives.) For cardiovascular diseases, unhealthy diet and smoking, along with physical inactivity and chronic infectious diseases such as periodontal diseases, are also highly ranked as external (environmental) risk factors. Physical inactivity may also result in skeletal disorders, particularly back pain. The important question is: "Who is responsible for what you eat or whether you smoke, exercise, and clean your teeth?" Health maintained and controlled by self-diagnosis and self-care is not only cost effective but also an important factor in quality of life to maintain independence and health.

Practitioner responsibility

The principles of *leges artis* require clinicians to practice dentistry according to modern science and established, well-tried methods, ie, the state of the art. From experimental and well-controlled longitudinal clinical studies in humans, the following conclusions may be drawn about etiologic and modifying factors of dental caries and periodontal diseases and efficient methods of prevention and control (for review, see Axelsson 1994 and 1998):

1. Dental caries and periodontal diseases can successfully be prevented and controlled by self-care supplemented by needs-related professional preventive measures.
2. Noncavitated caries lesions affecting enamel, root, and even dentin can be arrested successfully.
3. Regeneration of periodontal attachment is a reality.

The profession is obliged to be continuously updated and implement the state of the art (the most recent level of science- and evidence-based methods). Therefore, the profession must concentrate on prevention, control, and arrest of dental caries

and periodontal diseases. For dental caries, "prevention instead of extension," or at least "prevention before extension," should be given priority. By the same token, aggressive treatment of dental caries with extractions and "drilling, filling, killing (the pulp), and billing" and periodontal diseases with aggressive scaling, extensive flap surgery, extractions, and replacement of teeth with very expensive implant therapy must be regarded as outdated and more or less unjustified.

Oral health and general health are strongly correlated with the level of education. All over the world, the level of education is improving. Eventually, increasingly well-educated patients will learn the implications of high-quality dentistry according to the principles of *leges artis* and will request more preventive dentistry. Dentists who are not willing to comply with their patients' requests will find that their practices decline.

SYSTEMS OF ORAL CARE

Needs analysis

Public health leaders (ministers of health, chief dental officers, etc) in all countries confront the same challenge: how to design, set up, regularly evaluate, and modify services that cover needs and meet demand. To add to the complexity of the task, all decisions are subject to budgetary constraints, which often pay very little attention to genuine public health considerations.

In the field of oral health, there is a set of methods enabling each country to analyze the needs, the demand, and the response of the oral health care system. These methods have been developed over the last 20 years, in the course of international collaborative studies on oral health care systems, and promoted by the WHO. The study is performed on a random population sample of each key age group for oral diseases: 12 years, 35 to 44 years, and 65 years and older. The oral health status of each individual is recorded by clinical examination, and the findings are used to calculate the current health status and treatment needs. Moreover, a questionnaire completed by the subjects provides information on attitudes, behavior, level of satisfaction, and other factors relating to demand for and consumption of care.

The study is supplemented by a survey of a sample of the personnel concerned, including dentists, dental auxiliaries, dental hygienists, dental assistants, heads of school health services, and so on. In addition to these data on the consumers and providers of care, the study provides structural, macroeconomic, and sociologic data.

On completion of an international collaborative study, a full description of the health system of the sector is available, together with an estimate of the levels of disease and the treatment needs. It is thus possible to analyze the weaknesses and strengths of the system and the way in which these are perceived by the population. All these data have to be taken into account whenever it is intended, for example, to introduce a new system of fees, to promote hygiene and prevention programs, or to redesign personnel-training programs. If repeated periodically, the exercise makes it possible to follow trends in disease and to estimate the growth of demand and treatment coverage and thus to make the necessary adjustments and plan personnel resources for the coming decades.

Oral diseases differ from one country to another, but only in terms of prevalence or severity, not in type. The principles of prevention are likewise universal: improved plaque control and controlled use of fluorides through self-care, supplemented by professional care, and a low-sugar diet. Types of treatment comply, theoretically, with the principle that identical problems require an identical response. Despite these similarities, oral health care systems differ markedly from one country to another. It is therefore useful to compare the various systems and to see what aspects, structures, or approaches seem to provide the best response to a given situation.

To date, international collaborative studies have overturned some generally accepted ideas:

1. Oral health status does not seem to depend on the type of personnel or on access to care.
2. Care provided in the school setting is very effective in the treatment of childhood oral diseases but does not appear to improve the oral health of adults in the long term, except when reinforced by the education of both patients and oral health personnel in preventive techniques.
3. The most decisive factors for oral health are linked to the ability of oral health personnel and governments to promote preventive activities.

In other words, facilitating access to care and increasing the number of dentists are not necessarily the best responses in the oral health field. When oral health programs are planned, priority should always be given to prevention, whatever the level of resources available, and each situation should be analyzed in its own right and in comparison with others.

Improving access to oral care

Traditional systems for oral care are based on various combinations of public-salaried services and private practice. The public services are usually responsible for prevention and for care of schoolchildren and disadvantaged groups. Private practitioners provide a wide range of treatment to the general public. All these systems are oriented in such a way that the dentist provides most of the care.

As discussed earlier, the level of dental caries in developing countries has rarely been as high as in industrialized countries, except for Latin America (see Figs 440 to 442 in chapter 7). In some developing countries, successful preventive programs have been implemented. However, in many places there is still a threat of increasing caries prevalence, related to changing diet and lifestyles in combination with poor oral hygiene and lack of use of fluoride toothpaste. In many communities, the oral care systems do not meet even the basic needs of the public. Most public services have only very low coverage; communities in low-income rural and urban areas cannot afford private oral care. Further, developing countries cannot afford to establish, staff, and run education facilities for dentists or hope to provide adequate employment opportunities for dentists trained abroad.

In all countries, economic restraints, changes in demand for oral health care, political pressures to extend services to underprivileged groups, and concerns about quality, costs, and effectiveness of care demand that alternative ways of organizing oral health care be examined and implemented. Cost and lack of access for underprivileged and low-income groups constrain all oral health care systems.

What actions can be taken to combat this neglect, break down the barriers of costs, and improve access to oral health care? Alternative oral care systems must be developed so that a maximum number of people can have access to and can afford oral health care. Several recent advances give great scope for the transformation of the delivery and quality of oral care:

1. New educational technologies via interactive training (eg, via the Internet) that make learning of both knowledge and skills simpler and faster for all types of personnel
2. Simplified and logical design of oral clinics that improves the workplace and substantially reduces the capital cost of equipment and the need for maintenance
3. Better materials that are easier and simpler to use

Based on these technological advances, three types of care can be defined:

1. Simple, low technology; very cost effective
2. Moderate level of technology; rather expensive
3. High technology; often extremely expensive

The first level of oral care includes education of the population in self-diagnosis and self-care on an individual and a group basis; professional me-

chanical toothcleaning (PMTC); use of fluoride varnish and fissure sealants; scaling, root planing, and debridement; and low-technology treatment of single-surface caries lesions with the so-called atraumatic restorative technique, which has the potential to revolutionize the type of care that can be given in developing countries with a low ratio of dentists.

For small carious cavities, hand instruments can be used to scrape out the diseased tissue. The damaged area can be repaired with glass-ionomer cement, which also has a preventive effect, functioning as a slow-releasing fluoride agent that can be recharged with fluoride from other agents, such as fluoride toothpaste. The aim of the first-level (low-technology) oral care is to prevent the need for more traditional and costly invasive (moderate-technology) oral care.

The second level of care (moderate technology) includes multiple-surface restorations, extractions, simple periodontal surgery, and removable prostheses—that is, the traditional invasive oral care practiced by most dentists worldwide.

The third and most complex, high-technology level of oral care includes precision prosthetics, implants, laminates and ceramic inlays and onlays, orthodontics, regenerative periodontal treatment, and complex oral surgery and medicine—in other words, costly and complex procedures that require highly qualified specialists or "oral physicians." Therefore, in any society, the availability of high-technology oral care will be limited.

A rational, health-promoting, affordable mix of oral care should be planned and implemented in all countries. Emphasis on prevention and control of oral diseases will minimize the need for intervention at the moderate- and high-technology levels. As a consequence of improving oral health in most industrialized countries, the need for moderately complex care is decreasing. With further emphasis on prevention, the need and demand for first-level interventions will increase slightly, while the need for high-technology care will probably increase for sev-

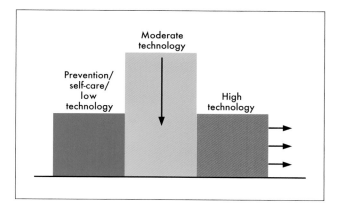

Fig 463 Changes in distribution of tasks in oral health care from past to present in highly industrialized countries. (Modified from WHO, 1994. Reprinted with permission.)

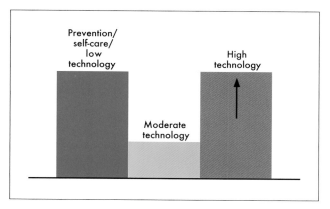

Fig 464 Distribution of tasks in oral health care in highly industrialized countries. (Modified from WHO, 1994. Reprinted with permission.)

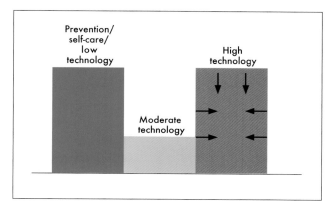

Fig 465 Future distribution of tasks in oral health care in highly industrialized countries. (Modified from WHO, 1994. Reprinted with permission.)

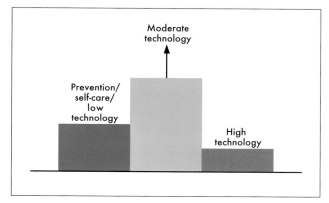

Fig 466 Typical distribution of tasks in oral health care in developing countries. (Modified from WHO, 1994. Reprinted with permission.)

eral decades because of the desire to preserve natural teeth and the increasing numbers of elderly people who have some natural teeth and edentulous people who want implant treatment.

In developing countries, the first (low-technology, noninvasive) level of oral care will continue to be the major need. In those developing countries where the prevalence of caries is increasing, a rising demand for moderate-technology care will continue over the next few decades. High-technology oral care must, on the other hand, still be limited.

The proportion of the population taking advantage of oral health care services varies from country to country, but in only a few might it be considered optimum. For example, in Sweden, almost 100% of 1- to 19-year-old individuals receive well-organized and regular oral care, predominantly preventive dentistry, free of charge (Axelsson et al, 1993a). In the county of Värmland, Sweden, 95% of the adult population visits oral health care clinics regularly for oral care, including preventive care (Axelsson et al, 2000). In most countries, total coverage is an unrealistic goal, but steps should be taken to ensure that oral care is available to all those who need it (Figs 463 to 467).

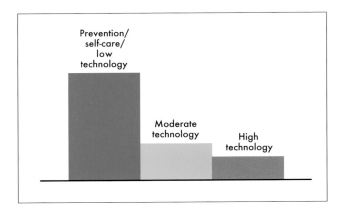

Fig 467 Eventual future distribution of tasks in oral health care in all countries. (Modified from WHO, 1994. Reprinted with permission.)

INFLUENCE OF DENTAL INSURANCE SYSTEMS ON ORAL HEALTH

There is no doubt that different financing and insurance systems significantly influence oral health status. Some of the different approaches to financing oral care are quality control guidelines, fixed fee agreements, capitations schemes, health maintenance organizations, rewarding increased preventive care, and public health funding.

Quality control guidelines

Using information about the duration of treatment methods and acceptable care products, quality control guidelines are being prepared, indicating the average number of years each type of care should last. If care procedures do not last the specified time, the clinician is then obliged to provide retreatment free of charge. Such guidelines are intended to reduce unnecessary treatment that results in progressive destruction of tooth substance and periodontal support and higher costs for oral care.

Fixed fees

In some countries, for most procedures, dentists may charge only fixed fees agreed upon by the health authorities and the professionals. These fees may be exceeded only for special treatment and after a review of the diagnosis and proposed procedure. In countries using this system, the costs of oral care are not rising, and costs are even decreasing in some countries.

However, dental insurance systems based on fixed fees for different specified treatment procedures, such as one anterior resin composite restoration versus complicated multisurface restorations, crowns, and endodontic treatment, are not a guarantee of improved oral health status among the population. For example, in Germany and Japan, the dental insurance systems are heavily concentrated on restorative dentistry, ie, so-called tertiary prevention. For items such as crowns and restorations, the dentists are relatively well reimbursed, but there is virtually no reimbursement for preventive dentistry according to the schedules issued by the national dental insurance schemes. As a consequence, dental practices concentrate on "drilling, filling, and billing" to survive.

This may explain why, in contrast to other industrialized countries, such as the Scandinavian countries, Canada, and Australia, Germany and Japan maintained persistently between moderate and high caries prevalence, respectively, in 12-year-old children from 1969 to 1997 (see Figs 440 to 442 in chapter 7). This is surprising, because both Germany and Japan have well-educated dentists, very high scientific standards in medicine, physics, chemistry, and economics, and well-educated populations with a high standard of living. On the other hand, it indicates the very powerful impact of the prevailing dental insurance system on the oral health status of the population.

Capitation schemes

Capitation schemes pay the dentist a fixed sum for each person enrolled as a patient in the practice. For this fixed annual fee, a dentist contracts to maintain the oral health of all the enrolled patients. However, patients must undertake to attend checkups on a regular basis, or they lose their rights and have to pay for treatment needed to restore their oral health. It seems likely that costs will be reduced and the oral health status improved by this type of program.

Health maintenance organizations

Health maintenance organizations contract with a group of oral care professionals to provide care to a group of communities or individuals, at agreed fees. Health maintenance organizations are usually organized and managed by companies that specialize in health insurance. This has proved an effective way to limit the costs of providing comprehensive oral care.

Rewarding preventive care

In some countries, projects to encourage preventive care give dental care managers a financial reward if levels of disease are reduced in the patients in their geographic area.

Public health funding

During the last decades, in most Scandinavian countries, oral health care programs for children and young adults, including school-based pre-

ventive programs, have been organized free of charge by the department of health. This is one reason for the vast reduction of caries prevalence in Scandinavian children from 1969 to 1997. For example, in Sweden, the Public Dental Health Service has been granted a fixed annual allowance by the Department of Health to carry out needs-related dental care for all individuals up to 20 years of age. This has encouraged and motivated the Public Dental Health Service to focus on preventive dentistry and to delegate preventive treatment to dental hygienists and preventive dentistry assistants, to minimize costly restorative dentistry by dentists.

Particularly in the county of Värmland, Sweden, needs-related programs practiced by dental hygienists and preventive dentistry assistants have been successfully implemented under this capitation system (see chapter 7).

Proposed oral health insurance system

In 1973, a national dental insurance was introduced for all adults in Sweden, irrespective of whether they chose the Public Dental Health Service (40%) or private practice (60%). About 80% to 90% of the Swedish adult population visits dental clinics regularly for maintenance programs. Until 1999, restorative dentistry, including crowns and prostheses, had been based on an itemized fee schedule, but preventive dentistry and periodontal treatment were based on an hourly fee, with different rates for specialized periodontists, general dental practitioners, dental hygienists, and preventive dental assistants. In 1999 the insurance system was reviewed. A capitation system based on the individual's predicted risk combined with analytical epidemiology for quality control was introduced parallel to a modified version of the earlier system.

Oral health professionals must practice according to modern scientific principles and established, well-tried methods, as earlier discussed. The cause of both dental caries and periodontal diseases are known, as are efficient preventive measures, which have been proven for up to 30 years, irrespective of age (see chapter 3; Axelsson et al, 2004). Therefore high-quality oral health care must focus on prevention and control of dental caries, periodontal diseases, and other oral diseases, ie, primary and secondary prevention. Existing treatment needs (tertiary prevention) must also be addressed.

In this context, it seems that all national dental health insurance schemes and financing systems should promote prevention and control of oral diseases and include provisions for analytical epidemiology for quality control and analyses of cost effectiveness by the ministry of health, in accordance with national and global oral health goals. In other words, dental professionals should expect to be well paid for successful prevention and control of oral diseases. Ancient Chinese doctors were said to be very well paid as long as they were able to keep their patients healthy. If they failed, they received no reimbursement at all. An increasingly well-educated and well-motivated population will undoubtedly prefer an oral health care program that gives priority to prevention and control of oral diseases to extractions, large flaps, extensive restorations, and costly implant therapy, with the possible consequence impaired general health and reduced quality of life.

The following proposal for an oral health insurance system is based on the state of the art. The insurance would involve a capitation system based on the predicted risk, including needs-related primary, secondary, and tertiary prevention. The individual annual fee would be combined with a negative fee based on the effect of the needs-related maintenance program on disease progression. For example, if the patient developed a new caries lesion, the patient's annual fee would be increased by US $40 and the dentist's reimbursement would be reduced by US $40. Likewise the annual fee would be increased by a certain amount for the patient, and reduced by the same amount for the dentist, if the patient exhibited further loss of periodontal support.

For quality control of such an oral health insurance system, a computer-aided analytical oral epidemiologic system should be introduced (discussed further in chapter 9). The dentist could present beautiful graphs in the waiting room to show the success of the practice in improving the oral health status and eliminating treatment need among the patients. It would significantly increase the dentist's reputation and good will. Patients would feel privileged to belong to the practice and pleased with the effects of their own efforts in self-care; they would be willing to pay a reasonable fee for the high quality of oral health care they were being offered. Old-fashioned colleagues who are focused on drilling, filling, and billing fear that they will cut off the tree branch they are sitting on if they practice preventive dentistry. On the contrary, such dentists will lose patients as the public becomes better educated about the benefits of prevention. Undoubtedly, no single preventive measure or combination of preventive measures would have such a significant impact on the improvement of oral health status as the aforementioned proposal:

1. Patients will be motivated to learn and improve self-diagnosis and self-care to improve their oral health status and prevent increased annual fees for oral health care.
2. Dentists will have a reasonable opportunity to practice state-of-the-art dentistry, and, as a consequence, their annual reimbursement will not be reduced.

GLOBAL AVAILABILITY OF ORAL HEALTH PERSONNEL

The most important precondition for successful implementation of efficient needs-related preventive programs is an adequate number of well-educated and well-informed oral health personnel, not only dentists but also dental hygienists and preventive dentistry assistants, who are specially trained and committed to practicing preventive dentistry.

Ratio of oral health personnel to population

There has been a dramatic decrease in the previously high levels of dental caries among children in most industrialized countries during the last three decades (WHO, 1997). In many developing countries, however, caries prevalence has increased from previous levels. In addition, the prevalence of moderate periodontal treatment needs (CPITN scores 1 to 3) is much higher in developing countries than in industrialized countries, despite a similar prevalence of advanced periodontal treatment needs (CPITN score 4). However, the dilemma is that the developing countries represent the absolute majority of the world's population.

Figure 468 shows the percentage of the world's population by WHO regions in 1986. Asia alone (SEARO and WPRO) contains more than 50% of the world's population. China and India alone have 2.5 billion inhabitants (40%). According to the WHO's Data Bank, Europe has more than 50% of the world's oral health personnel resources (eg, dentists, dental hygienists, dental assistants, technicians), although it has only 17.7% of the world's population (Figs 468 and 469) (WHO, 1994). The extremes of the ratio of dentists to inhabitants are found in Scandinavia (1:1,000) and in China and India (1:120,000).

Increasing the availability of oral health personnel

Improving dental schools worldwide

Overcoming this imbalance in the ratio of oral health personnel, particularly of dentists, to the population in different countries is a major challenge. Many developing countries are experiencing either an actual increase in caries prevalence or an inability to reduce the existing level. Although the prevalence of disease is, or threatens to be, only moderate, its impact is much more serious than would be the case in industrialized countries because of the almost total lack of effective care.

These countries have responded to the problems by attempting to develop new dental schools with curricula adapted from those that were common, but are now changing, in industrialized countries. More serious still is the fact that, irrespective of which curricula these countries may have chosen, there is a crippling lack of qualified teaching staff. In practical terms, this shortfall results in failure to deliver, often by a wide margin, whatever curricula are offered in faculty handbooks.

This predicament is not limited to developing countries. There is a wide divergence of dental school performance across the range of industrialized countries and within any country in which there are a number of dental schools. Even where schools may be judged to be similar in quality, strength and creativity may vary markedly from subject to subject. The profession is thus facing a challenge to minimize undesirable variation and to provide up-to-date and high-quality education for all.

An answer lies in broad exploitation of new communication media, which are so versatile in keeping information at hand and in offering opportunities for interactive use by undergraduates and graduates, as well as teaching staff. An obvious outcome will be the development of a new type of student as well as new types of teachers. Major elements of this new experience will be:

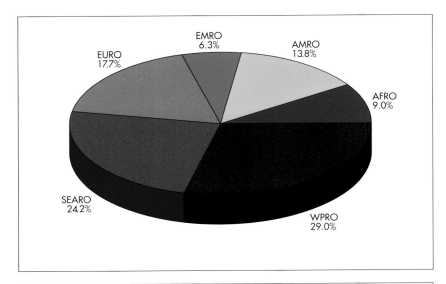

Fig 468 World population by World Health Organization regions. (EURO) Regional Office for Europe; (EMRO) Regional Office for the Eastern Mediterranean; (AMRO) Regional Office for the Americas; (AFRO) Regional Office for Africa; (WPRO) Regional Office for the Western Pacific; (SEARO) Regional Office for Southeast Asia. (Modified from WHO, 1994. Reprinted with permission.)

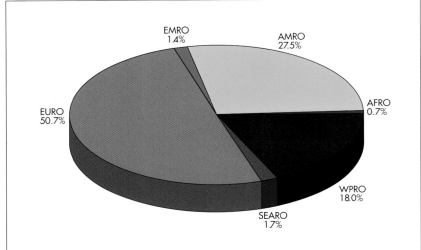

Fig 469 Oral health personnel by World Health Organization regions. (EURO) Regional Office for Europe; (EMRO) Regional Office for the Eastern Mediterranean, (AMRO) Regional Office for the Americas; (AFRO) Regional Office for Africa; (WPRO) Regional Office for the Western Pacific; (SEARO) Regional Office for Southeast Asia. (Modified from WHO, 1994. Reprinted with permission.)

1. Much greater ease in pooling intellectual resources.
2. Speedy exchange of information, especially in relation to effective new methodologies and materials.
3. Development of common assessment methods.
4. Access to a wide variety of databanks.
5. Greater opportunity to adapt careers over an extended period of time based on the speed of change in society and the job market.

Preparation of a broad set of modules, serving the oral health stream within a health sciences structure, will demonstrate what might be achieved and thus initiate a snowballing effect toward the ultimate objective of a complete computer-assisted curriculum. Already there are several centers of excellence around the globe producing the type of interactive materials needed for computer-assisted training. The assortment of such programs is steadily increasing and available as CD-ROM and on the Internet. There is a need to direct that excellence toward the overall objective. This is what the WHO hopes to do, in collaboration with the International Federation of Dental Education Associations, which is enthusiastic about this approach (WHO, 1994).

Succinctly, empowerment of the learner is the principal aim of this concept, in both a variation on the theme of universities without walls and an attempt to make excellence available whatever the constraints of quantity and quality in teaching staff. It will benefit dental schools where the resources needed to deliver an adequate and appropriate curriculum are very scarce. Every school will need a certain core of staff to administer, coordinate, guide, evaluate, and conduct practical exercises. The specific teaching and learning, however, would come largely from the student, the graduate, and the teachers interacting with the computer programs at their disposal.

The excellence of those programs would depend on having a "brain trust" of the very best experts responsible for devising the programs. In this way the best available courses would be offered to all students, everywhere, rather than the wide range of messages, often contradictory or outdated, that students currently receive. Whatever the definitive choice of approaches, an initiative of this type will streamline the training of health personnel, with sufficient flexibility to ensure that training of health personnel will keep pace with the leading edge of science more clearly than is the case today.

Expanding the role of dental hygienists

In addition to the overwhelming need for dental treatment by dentists, particularly in developing countries in Asia, Africa, and Latin America, which are home to the majority of the world's population, there is a clear indication for training of dental hygienists, who are committed exclusively to prevention and control of dental caries and periodontal diseases and periodontal treatment. In addition, their work is very cost effective (Axelsson et al, 1991, 2004; Axelsson, 1998).

The role and supply of dental hygienists are of increasing interest worldwide, mainly because of a growing acknowledgment of the importance of oral health as a part of general health, renewed emphasis on setting and attaining health policy goals, and recognition of dental hygienists as a major resource for attaining those goals. Dental hygienists today constitute one of the largest and fastest growing groups in oral health service. They practice, in collaboration with other health professionals, primarily as clinicians and health educators. Their work involves the use of preventive and therapeutic methods to promote good health and to prevent and control oral diseases.

According to FDI surveys (FDI, 1990), dental hygienists are now trained in more than 25 countries. The pioneers were the United States (1906), followed by Norway (1924), Great Britain (1943), Canada (1947), Japan (1949), Nigeria (1958), Sweden (1968), and the Netherlands (1968). The United States has the highest ratio of dental hygienists per head of the population (1:2,500), followed by Japan, Sweden, and Canada (about 1:4,000) (Fig 470). The United States also has the highest ratio of dental hygienists to dentists (0.75:1), followed by Japan (0.50:1), South Korea (0.50:1), Canada (0.50:1), and Sweden (0.30:1) (Fig 471).

Figure 472 shows the ratio of dental hygienists to dentists in Sweden from 1975 to 2000 (Axelsson et al, 1993b). This was estimated in 1993 based on the ongoing training program and was in fact achieved in 2000. However, in some areas of Sweden the ratio is much higher. For example, in the county of Värmland, there is one dental hygienist and one preventive dentistry assistant for each dentist. In the author's private clinic, dental hygienists have spent six to eight times more hours providing care than dentists during the last three decades.

In most countries, the length of the dental hygienist training program is 2 years, but it may range from 1 to 4 years. The entrance requirement is qualification as a dental assistant or matriculation from high school. The occupation is held predominantly by women (more than 90%).

According to a survey in 13 countries (Australia, Canada, Denmark, Italy, Japan, Korea, the Netherlands, Nigeria, Norway, Sweden, Switzerland, Great Britain, and the United States), which

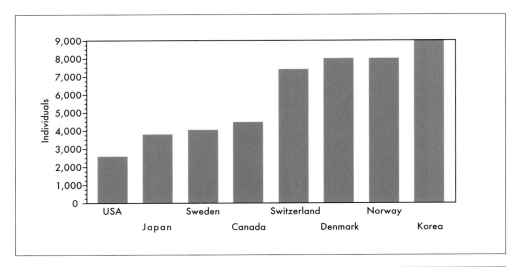

Fig 470 Population per dental hygienist.

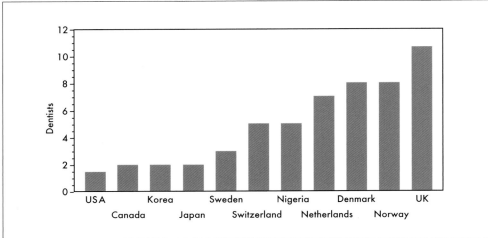

Fig 471 Dentists per dental hygienist.

covers most of the world's dental hygienists, the legal scope of their clinical practice is remarkably similar (FDI, 1990). It is characterized by a common set of procedures and activities, including treatment planning for the dental hygiene stages of care; history taking on general health, socioeconomic, and oral health aspects; explanation of optimized self-care for individuals and groups; scaling; root planing; debridement, and PMTC; topical application of fluoride gels and varnishes; application of fissure sealants; finishing of restorations and removal of overhangs, dietary evaluations and counseling; and administration of salivary and oral microbiology tests. In some countries, including Sweden, the training program also emphasizes behavioral science. In Sweden, dental hygienists are also trained in administration of local anesthesia (infiltration and regional block).

In Sweden, the training program for dental assistants also includes practical preventive dentistry. They are responsible for most of the preventive measures so successfully carried out in children and young adults (see chapter 7). Because the training program of these preventive dentistry assistants takes less than half the time required for the dental hygiene program, this personnel category could well be appropriate in other countries.

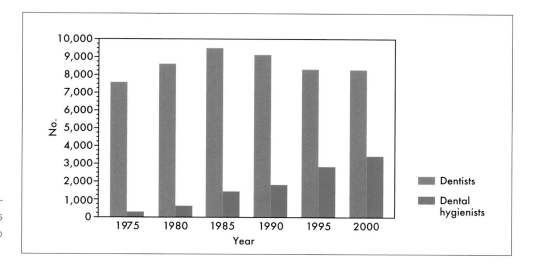

Fig 472 Number of dentists and dental hygienists in Sweden from 1975 to 2000.

Future of oral health personnel

For too long, the oral cavity has been separated from the rest of the body and "donated" to dentists, a profession more or less independent from the general medical professions. At the beginning of this century, dentistry centered around extraction of teeth and making complete and partial dentures, with the assistance of technicians. Then came an era of drilling, filling, killing (the pulp), and billing in most industrialized countries and aggressive exploration of the tooth crown because of dental caries. In recent decades, in many industrialized countries, the roots of the teeth have also been exploited by subspecialists in periodontology and endodontics, and teeth have been moved around for functional and esthetic reasons by specialists in orthodontics.

Recently, successful reconstructions with implant technology has been achieved through the teamwork of oral surgeons, periodontists, prosthodontists, and dental technicians. Specialists in periodontology now offer regeneration of periodontal support.

Integrated education

Modern knowledge and experience obliges dental professionals to focus on prevention and control of dental caries, periodontal diseases, and other oral diseases concurrently with elimination of existing treatment needs.

Increasing numbers of individuals worldwide will have intact tooth crowns and no loss of periodontal support. Therefore, the tooth crowns and the roots have to be treated together, and the oral cavity must be considered in the context of the body as a whole. This requires a more holistic approach, centered on the owner of the oral cavity. Preventive dentistry and oral health promotion have to be integrated with general health promotion, in collaboration with general health personnel. Dental education has to be comprehensively reoriented to serve the changing needs of oral health, linked more closely to the requirements of the whole health sector.

As an alternative to existing schools, which are specific to medicine, dentistry, nursing, and pharmacy, the aim should be to create an integrated system for educating all health personnel within a health sciences school structure. Elimination of separate training of auxiliary and professional personnel categories is also desirable. A "ladder" system of education is envisaged, in which, for the oral health team, the oral physician, instead of today's dentist, is seen as the highest level category, at parity with the other medical specialties. Other oral health personnel categories, such as dental assistants, preventive dentistry assistants, and dental hygienists would leave this ladder at various levels, corresponding to defined lists of duties, as would be the case for personnel focusing on any other health area. A number of dental schools, predominantly but not exclusively in Europe, have already begun to move toward this type of structure.

Integrated health teams

At the community and city levels, integrated local health teams should be established to improve oral health as well as general health among the population. Such a team should consist of well-trained professionals, highly experienced in prevention and health promotion rather than treatment:

1. General physician: the leader of the team; the most highly qualified and experienced in leadership and communication, general medicine, general health promotion, and epidemiology
2. Psychologist: highly qualified and experienced in behavioral science, with special reference to establishing good, healthy lifestyle habits as early as possible, and preventing or eliminating bad habits, such as smoking
3. Nutritionist: highly qualified and experienced in the concept of "input and output" related to eating at the cellular level as well as the individual level; emphasis should be placed on the consequences of healthy versus unhealthy dietary habits
4. Sociologist: well educated and experienced in the influence of socioeconomic conditions on health status
5. Physiotherapist: well educated and experienced in how to prevent the development

(primary prevention) and the recurrence (secondary prevention) of musculoskeletal orthopedic problems (such as back pain), as well as diagnosis and early rehabilitation of such problems
6. Physical education teacher: well educated and experienced in the consequences of physical training and activity versus physical inactivity for general health and how to activate all members of the population
7. Engineer: well educated and experienced in the effects of the external and internal environment on health and how to diagnose and eliminate unhealthy environments
8. Oral physician: leader of the oral health personnel team; highly qualified and experienced in preventive dentistry and oral health promotion as an integrated part of general health promotion
9. General health nurse: responsible for education of the population in self-diagnosis and self-care, with special reference to prevention and control of diseases
10. Dental hygienist: well educated and experienced in optimizing oral self-diagnosis and self-care for prevention and control of dental caries, periodontal diseases, and other oral diseases, supplemented by needs-related professional preventive measures

The local health team should be supported, at the county level, by a central resource team with a great variety of specialists and subspecialists. Most should have a scientific background at the doctoral level:

1. Physicians, all specialties
2. Psychologists
3. Sociologists
4. Nutritionists
5. Oral physicians
6. Epidemiologists
7. Statisticians
8. Economists

The remainder should have extensive research experience:

1. Physiotherapists
2. Sports trainers
3. Engineers
4. General nurses
5. Dental hygienists
6. Public relations officers (art director and copy-writers)

The effects of such teams on the health status of the population and the costs should be repeatedly evaluated in terms of the following:

1. Absence from work because of disease
2. Internationally (WHO) accepted, disease-specific variables
3. Cost effectiveness and cost-benefit ratio

> **Box 18** World Health Organization's international oral health goals for 2000
>
> - The World Health Organization's databases for oral health will be further developed. Coordinated national database systems will be established. A World Health Organization personal computer card for oral health recordings will be defined and produced.
> - Fifty percent of 5- to 6-year-old children will be caries free.
> - Children will have no more than three decayed, missing, or filled teeth at 12 years.
> - Eighty-five percent of the population will have all their teeth left at 18 years.
> - The levels of edentulousness at ages 35 to 44 years will be 50% less than the levels reported in 1969.
> - The levels of edentulousness at ages 65 years and older will be 25% less than the levels reported in 1969.

GLOBAL GOALS FOR ORAL HEALTH

Governments and communities should recognize the need to develop and maintain preventive programs for oral diseases. Communities will be responsible for these activities. All communities should be able to afford and manage basic, health-promoting oral care so that adult teeth will be retained throughout life.

Realistic goals for the effects of all preventive dental care programs should be set up at global as well as national, county, district, and clinic levels. Such goals should also be updated in accordance with improved oral health status at certain intervals. The goals set up in the county of Värmland, Sweden, in 1979, for those who had followed the needs-related program from birth to the age of 20 years, were almost 100% achieved in 1999 (see chapter 7).

Oral health for life should be approaching reality for all. To promote improved oral health worldwide, the WHO's Oral Health Unit drew up international oral health goals for severity of oral disease involvement at various key ages-12 years, 35 to 44 years, and 65 years and older-by 2000 (Box 18).

In 1988, the first International Conference on Preventive Dentistry and Epidemiology was held in Karlstad, Sweden, in collaboration with the WHO, the FDI, and the Swedish Board of Health and Welfare. Among other topics discussed at the workshops were realistic goals by the years 2010 and 2025, to follow the WHO's goals for the year 2000; what was known about already existing preventive measures; and the future potential of large-scale implementation of recent clinical research (Axelsson et al, 1988). The goals proposed for the years 2010 and 2025 (Box 19) involve:

1. Creation of gradually more well-developed, computerized analytical epidemiologic systems for quality control of oral and general health programs and cost effectiveness.
2. Focusing on education of all the population in self-diagnosis and self-care, which is the most cost-effective form of oral health care.

Box 19 International oral health goals for the years 2010 and 2025

Goals for the year 2010

- A complete electronic global, nation-based WHO database for oral health and a coordinated general health database will be established.
- Ninety percent of 5-year-old children will be caries free.
- Children will have no more than two decayed, missing, or filled teeth at 12 years of age.
- Seventy-five percent of 20-year-old adults will be caries inactive.
- Seventy-five percent of 20-year-old adults will not develop destructive periodontal disease.
- More than 75% of all children and young adults will have sufficient knowledge of the etiology and prevention of oral diseases to motivate self-diagnosis and self-care.

Goals for the year 2025

- A global electronic database for automatic oral and general health evaluation, including possibilities for health economy analysis, will be established.
- Ninety percent of 5-year-old children will be caries free.
- Children will have no more than one decayed, missing, or filled tooth at 12 years of age.
- Ninety percent of 20-year-old adults will be caries inactive.
- Ninety percent of the whole population will not develop destructive periodontal diseases.
- More than 75% of the total population will have sufficient knowledge of the etiology and prevention of oral diseases to motivate self-diagnosis and self-care.

3. Prevention and control of the development and progression of caries and periodontitis in the vast majority of the adult population (secondary prevention), parallel with efficient primary prevention in children, so that about 90% of young children will be caries free.

More detailed and tougher goals can be set up at national, county, city, and clinic levels, depending on the actual oral health status and dental care resources.

ESTABLISHMENT OF NEEDS-RELATED PREVENTIVE PROGRAMS BASED ON RISK PREDICTION AND RISK PROFILES

In countries with high prevalence and incidence of dental caries and/or periodontal diseases and limited resources of dentists, a so-called whole-population strategy for preventive programs, performed by other oral health personnel such as dental hygienists and preventive dentistry assistants, should still be cost effective. For example, if all the countries with nonfluoridated water, high or very high caries prevalence and incidence, little or no daily toothcleaning, and no use of fluoride toothpaste were to introduce not only water fluoridation, but also daily supervised toothbrushing with a fluoride toothpaste in the schools, it would prove very cost effective for caries prevention and control in children.

For successful prevention and control of dental caries in both children and adults, some basic principles must be adopted: for example, the higher the risk of developing caries (new carious surfaces) in most of the population, the greater the effect of one single preventive measure. This may be illustrated by the Swedish experience. In Sweden, 35 to 40 years ago, caries prevalence was extremely high. Almost every child developed several new lesions every year, mainly because of very poor oral hygiene. Regular toothbrushing was not an established habit, and no effective fluoride toothpaste was available. Under the prevailing conditions, well-organized, school-based fluoride mouthrinse programs in which 0.2% sodium fluoride solution was provided once every 1 or 2 weeks resulted in caries reductions of 30% to 50% (Forsman, 1965; Torell and Ericsson, 1965).

Twenty years later, a 3-year double-blind study revealed no benefit from weekly rinsing with sodium fluoride solutions compared to rinsing with distilled water (Axelsson et al, 1987a). There had been a dramatic fall in both caries prevalence and caries incidence from 1964 to 1984, and particularly after 1974, following the introduction of needs-related preventive programs. As an analogy, a raincoat is very cost effective for a week in London in November but not for a visit to the Sahara Desert.

Likewise, in populations with poor standards of oral hygiene and limited oral health care resources, most children have gingivitis, and most adults have both gingivitis and untreated chronic periodontitis. Under these conditions, until the existing treatment needs are met, a whole-population strategy for general oral health promotion should be applied. School-based education in self-diagnosis and daily oral hygiene should be very cost effective, because periodontitis always is preceded by gingivitis. In addition, such a program is very rational, because clean teeth never decay.

Similarly, at the population level, it is easy to find a positive correlation between one single risk factor and caries incidence in populations with high caries prevalence and incidence, where almost 100% of individuals develop new caries lesions every year. However, in populations with low or moderate caries incidence, well-established self-care habits, and well-organized oral health care, administration of one single preventive measure to all subjects in the population, irrespective of predicted risk, will not be cost effective; individual risk prediction and needs-related combinations of preventive measures are necessary. To ensure high sensitivity of risk prediction, several etiologic and modifying risk factors must be combined. For cost effectiveness, the so-called high-risk strategy would be the method of choice.

There are several similarities between periodontal diseases and dental caries. Both are multifactorial infectious diseases. Besides etiologic, preventive, and control factors, many other factors may modify the prevalence, onset, and progression of the diseases. These modifiers may be divided into external (environmental) and internal (endogenous) factors.

Probability statements

For cost effectiveness, the methods used to select and predict groups and individuals at "true" risk for disease development should be as sensitive as possible. The optimal *sensitivity* for a diagnostic risk test is 100%; that is, of 100 individuals selected as "risk individuals," all are true risk individuals. Similarly, for methods used to select true "nonrisk individuals," *specificity* should be as high as possible; that is, of selected nonrisk individuals, 100% are truly nonrisk.

Usually, the higher sensitivity, the lower the specificity. Thus, the clinician is usually forced to choose a test based on the consequences of making an error. A test method with high sensitivity and low specificity is likely to err in the direction of false positives. For a disease such as periodontitis, the implications are not usually serious when people are incorrectly identified as having active disease: They do not suffer extreme anxiety or undergo radical treatment. They may undergo intensified preventive treatment and incur increased treatment costs.

The consequence of a false-negative result, when a person with disease is incorrectly identified as healthy, is that the disease may progress further before it is diagnosed at a subsequent examination. This may be serious in the case of aggressive disease and of major consequence when there are prolonged intervals between examinations. For periodontal disease, as well as dental caries, the consequence of a false-positive diagnosis are less serious than those of a false-negative diagnosis. Alternatively, multiple tests, one highly sensitive and one highly specific, may be combined.

The probability of development of dental caries or periodontitis, given the result of a test method, is known as the *predictive value*. The positive predictive value is the probability that a patient with a positive test result actually has active caries or periodontitis. Similarly, the negative predictive value is the probability that a patient with a negative test result actually has inactive dental caries or periodontal disease.

Likelihood ratios can be used to evaluate the performance of a diagnostic test that is dichotomous or has interval properties. In addition, likelihood ratios can be used to calculate the probability of disease after a positive or negative test result.

Sensitivity, specificity, and predictive values are probability statements, representing the proportion of people with disease who have a positive test. Likelihood ratios are based on odds, which is the ratio of two probability values that contain the same information but express it differently. The relationship between the two is expressed in the following formulas:

$$Odds = \frac{Probability\ of\ event}{1 - Probability\ of\ event}$$

$$Probability = \frac{Odds}{1 + Odds}$$

The likelihood ratio for any value of a diagnostic test method is the probability of getting that test result when disease is present, divided by the probability of the result when disease is absent. Thus, likelihood ratios express how many times more or less likely a test result is to be found in diseased test subjects than it is in nondiseased test subjects.

This type of calculation can be done with different values of an interval scale diagnostic test. The resulting distribution of likelihood ratios can then provide the clinician with information on the likelihood of disease for a range of values.

Risk categories

For the clinician, accurate prediction of patients or sites at high risk of developing caries or periodontitis is of fundamental importance. Because of the particular nature of dental caries as well as periodontal diseases, prediction of disease progression is important. Diagnostic tests differentiate whether or not a person has a specific disease

at the time. *Risk indicators* (RIs) are factors that have proved to be significantly associated with the occurrence of a specific disease but only in cross-sectional studies.

Risk factors (RFs), on the other hand, are those that significantly increase the likelihood that people without disease, if exposed to these factors, will succumb to the disease within a specified time interval. Although diagnostic tests and RIs can be evaluated by cross-sectional research design, longitudinal studies are necessary to confirm RFs.

The term *risk factor* is rather loosely used and can refer to an attribute or exposure associated with increased probability of disease (not necessarily causal), any type of determinant (cause), or a determinant that can be modified. This loose terminology may cause confusion when multivariate aspects of diseases are considered. Use of the term should be restricted as follows: risk factors are characteristics of the person or environment that, when present, directly result in an increased likelihood that a person will get a disease and, when absent, directly result in a decreased likelihood.

Exposure to a risk factor means that a person has been exposed to or has manifested the factor prior to the onset of the disease. There may be continuous exposure, an isolated episode, or multiple exposures over a period of time. Clinicians must recognize that risk factors, like etiologic factors (the periopathogens and the cariogenic microflora), are based on the current state of knowledge about a direct relationship: In light of further knowledge, a current risk factor for a specific disease may in the future be excluded. Because dental caries and periodontal diseases are the result of multiple etiologic factors (different species of the cariogenic microflora and periopathogens) and modifying factors, removal of a risk factor, such as the cessation of smoking in periodontal diseases and reduced sugar intake in dental caries, does not necessarily cure the disease. It should reduce the likelihood of disease development, but, once a person has the disease, removal of the risk factor may or may not result in cure.

Prognostic risk factors (PRFs) are factors that increase the risk that an already existing disease will progress. For example, in patients with existing periodontal disease, smoking will increase the risk for progression of the disease. Thus in adults it is more practical to assess PRFs instead of RFs.

Risk markers and *risk predictors* (eg, advanced attachment loss and deep diseased pockets in periodontal disease and carious or restored surfaces in dental caries) are usually biologic markers that indicate either disease or disease progression but currently are thought not to be causal or represent historical evidence of the disease, such as the number of missing teeth, carious or restored surfaces, or past evidence of periodontal disease. If a risk predictor is more strongly associated with the disease than a risk factor, and the risk predictor and risk factor are also associated with each other, then the risk predictor will appear in the multivariate model instead of the risk factor. For example, baseline periodontal status and number of decayed, missing, or filled surfaces (DMFSs) are usually strongly associated with the occurrence of new disease, because they measure past disease and thus are risk predictors.

Microorganisms are etiologic risk factors and are also associated with new disease. Microorganisms are also associated with baseline periodontal status and caries prevalence (DMFSs) because they are partially responsible for this status. Thus, baseline periodontal status or DMFSs may replace microorganisms in the multivariate model. Having a risk predictor in the model results in a prediction model rather than a risk model.

Finally, dental caries and periodontal diseases have multiple levels of measurement: person, tooth, and site. A person can have disease, although many teeth and even more tooth sites may be disease free. Thus, the same person can have sites with onset of disease as well as established sites of disease, exhibiting progression. If the risk factor for disease onset is different than the prognostic factors for disease outcome, established sites should be evaluated under prog-nosis and not under risk prediction. Currently, however, both types of sites tend to be considered together when risk factors are delineated. In the future it may be useful to consider the two as separate entities.

In addition to etiologic and preventive factors, many other factors may modify the prevalence, onset, and progression (incidence) of dental caries and periodontal diseases. Such factors are divided into external (environmental) and internal (endogenous) factors.

External modifiers

External modifying risk indicators, risk factors, and prognostic risk factors for dental caries are frequent intake of sticky-sugar containing products (prolonged sugar clearance time), poor oral hygiene, irregular dental care habits, regular intake of medicines with saliva-depressive side effects, and low socioeconomic (particularly low educational) level.

Examples of external modifying RIs, RFs, and PRFs for periodontal diseases are smoking, use of smokeless tobacco, irregular dental care, low socioeconomic (particularly low educational) level, infectious and other acquired diseases, side effects of medication, and poor dietary habits. The most important are smoking, poor oral hygiene, and irregular attendance habits.

Therefore, poor oral hygiene habits, irregular dental care habits, and low educational level are jointly external RIs, RFs, and PRFs for dental caries and periodontal diseases.

Internal modifiers

The most important internal modifying RIs, RFs, and PRFs for dental caries are reduced salivary flow, buffering capacity, and antimicrobial effects; impaired host response; atypical plaque-retentive shape of the tooth crowns (particularly the fissures of the molars); and low quality of the tooth enamel.

489

Internal RIs, RFs, and PRFs related to periodontal disease include genetic factors, impaired host factors, chronic diseases, and reduced salivary flow and quality. Most studies of genetic factors in periodontal disease have concerned the aggressive forms of disease. Family studies suggest that susceptibility to the aggressive forms of disease, particularly in prepubertal and adolescent children, is at least in part influenced by host genotype. Inherited phagocytic cell deficiencies appear to confer risk for aggressive periodontitis in prepubertal children. The prevalence and distribution of aggressive periodontitis in affected families are most consistent with an autosomal-recessive mode of inheritance. Comparisons between adult monozygous twins reared together and reared apart indicate that early family environment has no appreciable influence on probing depth and attachment loss in adults.

A number of apparently genetically determined syndromes or diseases appear to carry an associated increased risk for periodontal destruction. Defects of phagocytic cell function, especially polymorphonuclear leukocytes (PMNLs), but also mononuclear phagocytes, is common in patients with aggressive periodontitis.

Genetic polymorphism of the proinflammatory cytokine interleukin 1 (1L-1) has been shown to be an RF and a PRF for periodontal diseases (Kornman et al, 1997). It has also been shown that the combination of genetic polymorphism of IL-1 with a smoking habit has a synergistic negative effect (Axelsson et al, 2001; McGuire and Nunn, 1999). It is estimated that about 80% of the most severe forms of periodontal diseases could be explained by these two joint factors.

Of the chronic diseases, poorly controlled diabetes (type 1 as well as type 2) is the most important RF and PRF for periodontal disease.

Individual caries-risk assessment

Caries risk may be evaluated by combining etiologic factors; caries prevalence (experience); caries incidence (increment); external and internal modifying risk indicators, risk factors, and prognostic risk factors; and preventive factors, at the individual level, as no risk (C0), low risk (C1), risk (C2), and high risk (C3). These conditions may vary in different age groups. Therefore, the criteria for C0, C1, C2, and C3 should be defined for at least the following general groups: preschool children (primary teeth), schoolchildren (permanent teeth), adults, and the elderly.

Caries prevalence and caries incidence as well as socioeconomic conditions may vary considerably among different populations and countries. The criteria listed for caries prevalence and incidence in children (C0 to C3) apply to Scandinavian conditions (low caries prevalence). Boxes 20 through 23 exemplify criteria for C0 to C3 in preschool children, children, adults, and the elderly, respectively. The criteria include etiologic factors; caries prevalence; caries incidence; external modifying RIs, RFs, and PRFs; internal modifying RIs, RFs, and PRFs; and preventive factors. The more factors that can be identified in the individual subject, the greater the validity of the predicted risk evaluation.

Individual periodontitis risk assessment

The relative risk for developing periodontal disease can be evaluated by combining clinical examination; preventive factors; etiologic factors; the absence or presence of environmental (external) and host-related (internal) risk indicators, risk factors, and prognostic risk factors; risk markers; and risk predictors. Periodontal risk increases in accordance with the severity and number of these factors to which the individual is exposed. Criteria for grading individual risk into one of four classes, from no risk to high risk (P0 to P3), have been proposed for children, young adults, adults, and the elderly (Table 24). The colors, from green to red, symbolize escalated risk. The criteria are based on history taking, established clinical diagnostic criteria, and supplementary bacterial sampling and laboratory tests, where indicated.

Box 20 Prediction of caries risk in preschool children*

No caries risk (C0 [green])
- Etiologic factors:
  - *Streptococcus mutans* negative
  - Low levels of salivary lactobacilli (< 10,000 CFU/mL)
  - Very low or low plaque formation rate (PFRI 1 or 2)
- Caries prevalence: Caries free
- Caries incidence: No incidence
- External modifying risk indicators and risk factors: None
- Internal modifying risk indicators and risk factors: None
- Preventive factors:
  - Excellent standard of oral hygiene: Well-motivated and well-educated parents
  - Regular use of fluoride toothpaste
  - Excellent dietary habits
  - Regular preventive dental care habits

Low caries risk (C1 [blue])
- Etiologic factors:
  - Streptococcus mutans positive (< 100,000 CFU/mL)
  - Low levels of salivary lactobacilli (< 10,000 CFU/mL)
  - Very low or low plaque formation rate (PFRI 1 or 2)
- Caries prevalence: Caries free (no caries lesions in dentin)
- Caries incidence: No new caries lesions in dentin
- External modifying risk indicators and risk factors: None
- Internal modifying risk indicators and risk factors: None
- Preventive factors:
  - Fairly good standard of oral hygiene: Regular cleaning by well-motivated parents
  - Regular use of fluoride toothpaste
  - Fairly good dietary habits
  - Regular preventive dental care habits

Caries risk (C2 [yellow])
- Etiologic factors:
  - *Streptococcus mutans* positive (> 100,000 CFU/mL)
  - High levels of salivary lactobacilli (100,000 CFU/mL)
  - Moderate or high plaque formation rate (PFRI 3 or 4)
- Caries prevalence: High (approximal caries lesions in dentin or restored surfaces on the primary molars)
- Caries incidence: High (more than one new caries lesion in dentin per year)
- External modifying risk indicators, risk factors, and prognostic risk factors:

  - High frequency of intake of sugar-containing products (prolonged sugar clearance time)
  - Low socioeconomic background
- Internal modifying risk indicators, risk factors, and prognostic risk factors:
  - Low salivary buffering effect
  - Reduced immune response
- Preventive factors:
  - Poor standard of oral hygiene
  - Irregular use of fluoride toothpaste
  - Poor dietary habits
  - Irregular preventive dental care habits

High caries risk (C3 [red])
- Etiologic factors:
  - *Streptococcus mutans* positive (> 1 million CFU/mL)
  - Very high levels of salivary lactobacilli (> 100,000 CFU/mL)
  - High or very high plaque formation rate (PFRI 4 or 5)
- Caries prevalence: Very high (caries lesions in dentin or restorations on most approximal surfaces and fissures and some active caries lesions in enamel on the buccal surfaces)
- Caries incidence: Very high (more than two new caries lesions in dentin per year)
- External modifying risk indicators, risk factors, and prognostic risk factors:
  - Very high frequency of intake of sugar-containing products (extremely prolonged sugar clearance time)
  - Low or very low socioeconomic background
- Internal modifying risk indicators, risk factors, and prognostic risk factors:
  - Very low salivary buffering effect
  - Reduced immune response
- Preventive factors:
  - Very low standard of oral hygiene, without assistance by the parents
  - Irregular or no use of fluoride toothpaste
  - Very poor dietary habits
  - No preventive dental care and irregular dental care

*(CFU) colony-forming units; (PFRI) Plaque Formation Rate Index

Box 21 Prediction of caries risk in children from 6 to 19 years of age

No caries risk (C0 [green])

- Etiologic factors:
  - *Streptococcus mutans* negative
  - Low levels of salivary lactobacilli (< 10,000 CFU/mL)
  - Very low or low plaque formation rate (PFRI 1 or 2)
- Caries prevalence: Caries free
- Caries incidence: No incidence
- External modifying risk indicators and risk factors: None
- Internal modifying risk indicators and risk factors: None
- Preventive factors:
  - Excellent standard of oral hygiene
  - Regular use of fluoride toothpaste
  - Excellent dietary habits
  - Regular preventive dental care habits

Low caries risk (C1 [blue])

- Etiologic factors:
  - *Streptococcus mutans* positive (< 100,000 CFU/mL)
  - Low levels of salivary lactobacilli (< 10,000 CFU/mL)
  - Very low or low plaque formation rate (PFRI 1 or 2)
- Caries prevalence: No caries lesions in dentin or restored surfaces
- Caries incidence: No new caries lesions in dentin
- External modifying risk indicators and risk factors: None
- Internal modifying risk indicators and risk factors: None
- Preventive factors:
  - Good standard of oral hygiene
  - Regular use of fluoride toothpaste
  - Good dietary habits
  - Regular preventive dental care habits

Caries risk (C2 [yellow])

- Etiologic factors:
  - *Streptococcus mutans* positive (> 100,000 CFU/mL)
  - High levels of salivary lactobacilli (100,000 CFU/mL)
  - Moderate or high plaque formation rate (PFRI 3 or 4)
- Caries prevalence: High
  - At 6 to 11 years of age: Fissure caries in permanent first molars and approximal caries lesions in dentin or restorations in primary molars
  - At 12 to 19 years of age: Fissure caries in most molars, caries lesions in enamel, and a few caries lesions in dentin on the approximal surfaces of some molars and premolars
- Caries incidence: High (more than one new caries lesion in dentin or some new enamel caries lesions per year)
- External modifying risk indicators, risk factors, and prognostic risk factors:
  - High frequency of intake of sugar-containing products (prolonged sugar clearance time)

  - Low socioeconomic background
- Internal modifying risk indicators, risk factors, and prognostic risk factors:
  - Reduced stimulated salivary secretion rate (< 0.7 mL/min)
  - Low salivary buffering effect
  - Reduced immune response
- Preventive factors:
  - Poor standard of oral hygiene
  - Irregular use of fluoride toothpaste
  - Poor dietary habits
  - Irregular preventive dental care habits

High caries risk (C3 [red])

- Etiologic factors:
  - *Streptococcus mutans* positive (> 1 million CFU/mL)
  - Very high levels of salivary lactobacilli (> 100,000 CFU/mL)
  - High or very high plaque formation rate (PFRI 4 or 5)
- Caries prevalence: Very high
  - At 6 to 11 years of age: Occlusal and mesial surfaces of the permanent first molars are carious (active enamel or dentin caries) or restored; most primary molars are restored or lost; and some active caries lesions in enamel may be present on the surfaces of the permanent incisors
  - At 12 to 19 years of age: Occlusal surfaces of the permanent molars are restored; active caries lesions in enamel, dentin, or restorations are present on most of the approximal surfaces of the molars and premolars and some incisors; and some active caries lesions in enamel are present on the buccal surfaces of the posterior teeth and the lingual surfaces of the mandibular molars
- Caries incidence: Very high (more than two new caries lesions in dentin and several new caries lesions in enamel per year)
- External modifying risk indicators, risk factors, and prognostic risk factors:
  - Very high frequency of intake of sugar-containing products (extremely prolonged sugar clearance time)
  - Low or very low socioeconomic background
- Internal modifying risk indicators, risk factors, and prognostic risk factors:
  - Reduced stimulated salivary secretion rate (< 0.7 mL/min)
  - Very low salivary buffering effect
  - Severely compromised immune response
- Preventive factors:
  - Very low standard of oral hygiene
  - Irregular or no use of fluoride toothpaste
  - Very poor dietary habits
  - No preventive dental care and irregular dental care

Box 22 Prediction of caries risk in adults (*continued on next page*)

No caries risk (C0 [green])

- Etiologic factors:
  - *Streptococcus mutans* negative
  - Low levels of salivary lactobacilli (< 10,000 CFU/mL)
  - Very low or low plaque formation rate (PRFI 1 or 2)
- Caries prevalence: Caries free or only occlusal carious or restored surfaces in the molars
- Caries incidence: No incidence
- External modifying risk indicators, risk factors, and prognostic risk factors: None
- Internal modifying risk indicators, risk factors, and prognostic risk factors: None
- Preventive factors:
  - Excellent standard of oral hygiene
  - Regular use of fluoride toothpaste
  - Excellent dietary habits
  - Regular preventive dental care habits

Low caries risk (C1 [blue])

- Etiologic factors:
  - *Streptococcus mutans* positive (< 100,000 CFU/mL)
  - Low levels of salivary lactobacilli (< 10,000 CFU/mL)
  - Very low or low plaque formation rate (PFRI 1 or 2)
- Caries prevalence:
  - At 20 to 35 years of age: A few occlusal carious or restored surfaces in the molars
  - At 36 to 50 years of age: Only occlusal carious or restored surfaces
  - At 51 to 65 years of age: Occlusal carious or restored surfaces and fewer than four approximal carious surfaces or restored surfaces
- Caries incidence: No more than one new caries lesion every 5 years
- External modifying risk indicators, risk factors, and prognostic risk factors: Few or none
- Internal modifying risk indicators, risk factors, and prognostic risk factors: Few or none
- Preventive factors:
  - Good standard of oral hygiene
  - Regular use of fluoride toothpaste
  - Good dietary habits
  - Regular preventive dental care habits

Caries risk (C2 [yellow])

- Etiologic factors:
  - *Streptococcus mutans* positive (> 100,000 CFU/mL)
  - High levels of salivary lactobacilli (100,000 CFU/mL)

- Moderate or high plaque formation rate (PFRI 3 or 4)
- Caries prevalence: High
  - At 20 to 35 years of age: Caries lesions or restorations on most occlusal and some posterior approximal surfaces
  - At 36 to 50 years of age: More than one tooth lost directly or indirectly due to caries; caries lesions or restorations on most occlusal and posterior approximal surfaces
  - At 51 to 65 years of age: More than two teeth lost directly or indirectly due to caries; caries lesions or restorations on most occlusal and approximal surfaces (also in the maxillary incisors) as well as some buccal surfaces (in industrialized countries, most caries lesions are recurrent caries)
- Caries incidence: High
  - At 20 to 50 years of age: More than one new caries lesion in dentin per year
  - At 51 to 65 years of age: More than two new caries lesions per year (more than 75% recurrent caries)
- External modifying risk indicators, risk factors, and prognostic risk factors:
  - High frequency of intake of sugar-containing products (prolonged sugar clearance time)
  - Low socioeconomic background (particularly low educational level)
- Internal modifying risk indicators, risk factors, and prognostic risk factors:
  - Reduced stimulated salivary secretion rate (< 0.7 mL/min)
  - Low salivary buffering effect
  - Reduced immune response
- Preventive factors:
  - Poor standard of oral hygiene
  - Irregular use of fluoride toothpaste
  - Poor dietary habits
  - Irregular preventive dental care habits

High caries risk (C3 [red])

- Etiologic factors:
  - Very high salivary *Streptococcus mutans* level (> 1 million CFU/mL)
  - Very high levels of salivary lactobacilli (> 100,000 CFU/mL)
  - High or very high plaque formation rate (PFRI 4 or 5)
- Caries prevalence: Very high
  - At 20 to 35 years of age: More than one tooth lost due to caries, directly or indirectly (endodontic reasons or root fracture because of a post); carious or restored occlusal surfaces; most approximal surfaces, including the maxillary incisors, and some buccal surfaces carious or restored

Box 22 Prediction of caries risk in adults (*continued*)

High caries risk (C3 [red]) (continuation)

- At 36 to 50 years of age: More than two teeth lost directly or indirectly due to caries; caries lesions or restorations on most posterior tooth surfaces and the approximal surfaces of the incisors; multiple recurrent caries lesions and some root caries lesions
- At 51 to 65 years of age: More than three teeth lost directly or indirectly due to caries; caries lesions or restorations on almost all posterior tooth surfaces and the approximal surfaces of the incisors; multiple recurrent caries lesions as well as root caries lesions

- Caries incidence: Very high
  - At 20 to 35 years of age: More than two new caries lesions per year
  - At 36 to 50 years of age: More than three new caries lesions per year (more than 85% recurrent caries)
  - 51 to 65 years of age: More than four new caries lesions per year (more than 90% recurrent and root caries)

- External modifying risk indicators, risk factors, and prognostic risk factors:
  - Very high frequency of intake of sugar-containing products (excessive sugar clearance time)
  - Low or very low socioeconomic level (particularly very low educational level)
  - Regular use of medicines with saliva-depressive effects
- Internal modifying risk indicators, risk factors, and prognostic risk factors:
  - Very low stimulated salivary secretion rate or xerostomia (0.0 to 0.4 mL/min)
  - Very low salivary buffering effect
  - Chronic diseases resulting in xerostomia (Sjögren syndrome, etc)
  - Severely compromised immune response
- Preventive factors:
  - Very low standard of oral hygiene
  - Irregular or no use of fluoride toothpaste
  - Very poor dietary habits

Box 23 Prediction of caries risk in the elderly (*continued on next page*)

No caries risk (C0 [green])

- Etiologic factors:
  - *Streptococcus mutans* negative or low salivary levels of *Streptococcus mutans* (< 10,000 CFU/mL)
  - Low levels of salivary lactobacilli (< 10,000 CFU/mL)
  - Very low or low plaque formation rate (PRFI 1 or 2)
- Caries prevalence: Very low
  - At 66 to 80 years of age: No teeth lost due to caries; caries lesions or restorations only on the occlusal surfaces of molars
  - At 81+ years of age: No teeth lost to caries; caries lesions or restorations on the occlusal surfaces and a few posterior approximal surfaces
- Caries incidence: None
- External modifying risk indicators, risk factors, and prognostic risk factors: None
- Internal modifying risk indicators, risk factors, and prognostic risk factors: None
- Preventive factors:
  - Very good standard of oral hygiene.
  - Regular use of fluoride toothpaste and other fluoride agents, such as fluoride chewing gum
  - Excellent dietary habits
  - Regular preventive dental care habits

Low caries risk (C1 [blue])

- Etiologic factors:
  - Low salivary *Streptococcus mutans* levels (< 100,000

CFU/mL)
  - Low levels of salivary lactobacilli (< 10,000 CFU/mL)
  - Very low or low plaque formation rate (PFRI 1 or 2)
- Caries prevalence:
  - At 66 to 80 years of age: Fewer than two teeth lost due to caries; caries lesions or restorations on the occlusal surfaces and some posterior approximal surfaces
  - At 81+ years of age: Fewer than four teeth lost due to caries; caries lesions or restorations on the occlusal surfaces and several posterior approximal surfaces
- Caries incidence:
  - At 66 to 80 years of age: No more than one new caries lesion every 5 years (recurrent or root caries)
  - At 81+ years of age: No more than one new caries lesions every 3 years (recurrent or root caries)
- External modifying risk indicators, risk factors, and prognostic risk factors: Few or none
- Internal modifying risk indicators, risk factors, and prognostic risk factors: Few or none
- Preventive factors:
  - Good standard of regular oral hygiene
  - Regular use of fluoride toothpaste and other fluoride agents, such as fluoride chewing gum
  - Good dietary habits
  - Regular preventive dental care habits

Box 23 Prediction of caries risk in the elderly (*continued*)

Caries risk (C2 [yellow])

- Etiologic factors:
  - High salivary *Streptococcus mutans* levels (> 100,000 CFU/mL)
  - High levels of salivary lactobacilli (100,000 CFU/mL)
  - High or very high plaque formation rate (PFRI 4 or 5)
- Caries prevalence:
  - At 66 to 80 years of age: Four to six teeth lost directly or indirectly due to caries; caries lesions or restorations on more than 70% of the posterior tooth surfaces and more than 50% of the surfaces of the maxillary incisors (mainly approximally); some root surface caries lesions
  - At 81+ years of age: Six to eight teeth lost directly or indirectly due to caries; caries lesions or restorations on more than 80% of the posterior tooth surfaces and more than 50% of the surfaces of the maxillary incisors; several root caries lesions
- Caries incidence:
  - At 66 to 80 years of age: More than one new caries lesion per year (more than 80% recurrent and root caries)
  - At 81+ years of age: More than two new caries lesions per year (about 20% recurrent caries and more than 70% root caries)
- External modifying risk indicators, risk factors, and prognostic risk factors:
  - High frequency of intake of sugar-containing products (prolonged sugar clearance time) and sticky, cooked starch products
  - Low socioeconomic background (particularly low educational level)
  - Regular use of medicines with saliva-depressive effects
- Internal modifying risk indicators, risk factors, and prognostic risk factors:
  - Reduced stimulated salivary secretion rate (< 0.7 mL/min)
  - Low salivary buffering effect
  - Reduced immune response
- Preventive factors:
  - Poor standard of oral hygiene
  - Irregular use of fluoride toothpaste
  - Poor dietary habits
  - Irregular preventive dental care habits

High caries risk (C3 [red])

- Etiologic factors:
  - Very high salivary *Streptococcus mutans* level (> 1 million CFU/mL)

  - Very high levels of salivary lactobacilli (> 100,000 CFU/mL)
  - High or very high plaque formation rate (PFRI 4 or 5)
- Caries prevalence: Very high
  - At 66 to 80 years of age: Six to 10 teeth lost directly or indirectly due to caries; caries lesions or restorations on more than 90% of the posterior tooth surfaces, more than 60% of the surfaces of the maxillary incisors, and more than 30% of the surfaces of the mandibular incisors; many recurrent and root caries lesions
  - At 81+ years of age: More than 10 teeth lost directly or indirectly due to caries; caries lesions or restorations on more than 90% of the posterior tooth surfaces and most of the surfaces of the incisors; caries lesions on most exposed root surfaces
- Caries incidence: Very high
  - At 66 to 80 years of age: More than three new caries lesions per year (more than 90% recurrent and root caries)
  - At 81+ years of age: More than three new caries lesions per year (about 20% recurrent caries and more than 75% root caries)
- External modifying risk indicators, risk factors, and prognostic risk factors:
  - Very high frequency of intake of sugar-containing products (excessive sugar clearance time) and sticky, cooked starch products
  - Low or very low socioeconomic level (particularly very low educational level)
  - Regular use of medicines with saliva-depressive effects
- Internal modifying risk indicators, risk factors, and prognostic risk factors:
  - Very low stimulated salivary secretion rate (0.0 to 0.4 mL/min)
  - Very low unstimulated salivary secretion rate and xerostomia (0.0 to 0.1 mL/min)
  - Very low salivary buffering effect
  - Chronic diseases or irradiation resulting in xerostomia (Sjögren syndrome, etc)
  - Senile dementia
  - Physical disability
  - Severely compromised host response
- Preventive factors:
  - Very low standard of oral hygiene
  - Irregular or no use of fluoride toothpaste
  - Very poor dietary habits
  - No preventive dental care and irregular dental care

Table 24 Criteria for evaluating individual periodontal risk in children, young adults, adults, and the elderly *(continued on next page)*

No periodontal risk (P0; green)		Low periodontal risk (P1; blue)	
I	Children	III	Adults
I:1	Healthy gingivae	III:1	Fewer than five diseased approximal pockets > 3 mm (CPITN 3–4)
I:2	Excellent oral hygiene habits	III:2	Good oral hygiene habits
I:3	No approximal probing loss of attachment	III:3	Mean approximal probing loss < 1 mm
I:4	No internal or external risk indicators or risk factors	III:4	No internal or external risk indicators or risk factors

No periodontal risk (P0; green)		Low periodontal risk (P1; blue)	
II	Young adults	IV	Elderly
II:1	Healthy gingivae	IV:1	Chronic periodontitis: fewer than five diseased approximal pockets > 5 mm (CPITN 4)
II:2	Excellent oral hygiene habits	IV:2	Good oral hygiene habits
II:3	No approximal probing loss of attachment	IV:3	Mean approximal probing loss of attachment < 2 mm
II:4	No internal or external risk indicators or risk factors	IV:4	No tooth loss caused by periodontal disease
		IV:5	No internal or external risk indicators or prognostic risk factors

No periodontal risk (P0; green)		Periodontal risk (P2; yellow)	
III	Adults	I	Children
III:1	No diseased periodontal pockets	I:1	Gingival bleeding index < 20% (CPITN 1)
III:2	Excellent oral hygiene habits	I:2	Poor oral hygiene habits
III:3	No approximal probing loss of attachment	I:3	Internal risk indicators, risk factors, and prognostic risk factors
III:4	No internal or external risk indicators or risk factors	I:4	External risk indicators, risk factors, and prognostic risk factors (low educational level of parents, etc)

No periodontal risk (P0; green)		Periodontal risk (P2; yellow)	
IV	Elderly	II	Young adults
IV:1	No diseased periodontal pockets	II:1	One to five diseased approximal pockets > 3 mm (CPITN 3–4)
IV:2	Excellent or good oral hygiene habits	II:2	Poor oral hygiene habits
IV:3	Mean approximal probing loss of attachment < 1 mm	II:3	Mean approximal probing loss of attachment < 1 mm
IV:4	No internal or external risk indicators or risk factors	II:4	Internal risk indicators, risk factors, and prognostic risk factors
		II:5	External risk indicators, risk factors, and prognostic risk factors (low socioeconomic level, smoking etc)

Low periodontal risk (P1; blue)		Periodontal risk (P2; yellow)	
I	Children	III	Adults
I:1	Gingival bleeding index < 10% (CPITN 1)	III:1	Chronic periodontitis: more than five diseased approximal pockets > 5 mm (CPITN 4)
I:2	Good oral hygiene habits		
I:3	No approximal probing loss of attachment		
I:4	No internal or external risk indicators or risk factors		

Low periodontal risk (P1; blue)	
II	Young adults
II:1	Gingival bleeding index < 10% (CPITN 1)
II:2	Good oral hygiene habits
II:3	No approximal probing loss of attachment
II:4	No internal or external risk indicators or risk factors

During the last 20 to 25 years in the county of Värmland, almost 100% of the children and young adults (1 to 20 years of age) have been assigned to one of the four caries-risk categories for the needs-related caries-preventive program, as discussed in chapter 7.

During the last 5 years, nearly 100% of the adult patients at the public dental health clinics, representing about 50% of the adult population in the county of Värmland, Sweden, and the majority of the adult patients in private clinics have been assigned to one of four classes of general risk, periodontal risk, caries risk, and iatrogenic

Table 24 Criteria for evaluating individual periodontal risk in children, young adults, adults, and the elderly (continued)

III:2	Poor oral hygiene habits
III:3	Mean approximal probing loss of attachment > 2 mm
III:4	Internal risk indicators, risk factors, and prognostic risk factors (type 1 or 2 diabetes, etc)
III:5	External risk indicators, risk factors, and prognostic risk factors (smoking, low educational level, irregular dental care, etc)

Periodontal risk (P2; yellow)

IV	Elderly
IV:1	Chronic periodontitis: more than 15 of the approximal sites (CPITN 4)
IV:2	Poor oral hygiene habits
IV:3	Mean approximal probing loss of attachment > 4 mm
IV:4	More than six teeth lost because of periodontal disease
IV:5	Internal risk indicators, risk factors, and prognostic risk factors (type 2 diabetes, etc)
IV:6	External risk indicators, risk factors, and prognostic risk factors (smoking, low educational level, etc)

High periodontal risk (P3; red)

I	Children
I:1	Localized or generalized aggressive periodontal diseases (0.1% to 0.3% of the Scandinavian population)
I:2	Very poor oral hygiene
I:3	Most diseased sites infected with bacteria associated with aggressive periodontitis (Actinobacillus actinomycetemcomitans, etc)
I:4	Internal risk indicators, risk factors, and prognostic risk factors, such as genetic interleukin 1 polymorphism, dysfunction of the polymorphonuclear leukocytes, reduced immunoglobulin G2 response, type 1 diabetes, Down syndrome, and leukemia
I:5	External risk indicators, risk factors, and prognostic risk factors (low education level of the parents, acquired immunodeficiency syndrome, irregular dental care, etc)

High periodontal risk (P3; red)

II	Young adults
II:1	Localized or generalized aggressive periodontal diseases
II:2	High periodontal incidence (annual probing loss of attachment) and several sites related to aggressive periodontitis
II:3	Very poor oral hygiene
II:4	Most diseased sites infected with bacteria associated with aggressive periodontitis (Actinobacillus actinomycetemcomitans, Porphyromonas gingivalis, etc)
II:5	Internal risk indicators, risk factors, and prognostic risk factors, such as genetic interleukin 1 polymorphism, dysfunction of the polymorphonuclear leukocytes, reduced immunoglobulin G2 response, type 1 diabetes, leukemia, etc
II:6	External risk indicators, risk factors, and prognostic risk factors (smoking, low educational level, acquired immunodeficiency syndrome, irregular dental care, etc)

High periodontal risk (P3; red)

III	Adults
III:1	Aggressive periodontitis: high periodontal incidence (annual probing loss of attachment) and several sites with aggressive periodontitis
III:2	More than four teeth lost because of periodontal disease
III:3	Very poor oral hygiene and most diseased sites infected by Porphyromonas gingivalis, Bacteroides forsythus, and other periopathogens
III:4	Internal risk indicators, risk factors, and prognostic risk factors (type 1 or type 2 diabetes, genetic interleukin 1 polymorphism, etc)
III:5	External risk indicators, risk factors, and prognostic risk factors (smoking, low educational level, acquired immunodeficiency syndrome, irregular dental care, etc)

High periodontal risk (P3; red)

IV	Elderly
IV:1	Aggressive periodontitis (periodontitis gravis and complicata involving most teeth)
IV:2	More than 10 teeth lost because of periodontal disease
IV:3	Very poor oral hygiene; most diseased sites infected by Porphyromonas gingivalis, Bacteroides forsythus, and other periopathogens
IV:4	Internal risk indicators, risk factors, and prognostic risk factors (type 1 or type 2 diabetes, genetic interleukin 1 polymorphism, etc)
IV:5	External risk indicators, risk factors, and prognostic risk factors (smoking, low educational level, irregular dental care, etc)

risk: 0 = no risk; 1 = low risk; 2 = risk; 3 = high risk. This assessment was developed to improve the needs-related preventive program for adults, which was initiated in 1985.

General risk includes general health problems, physiologic compliance problems, physical disabilities, and so on. Iatrogenic risk includes risk for cusp fractures due to complicated restorative dentistry and root fractures due to posts and complicated fixed prosthodontics. Figure 473 shows the mean scores for general risk, periodontal risk, caries risk, and iatrogenic risk for the adult population of the county of Värmland, Sweden. Iatrogenic risk exhibited the high-

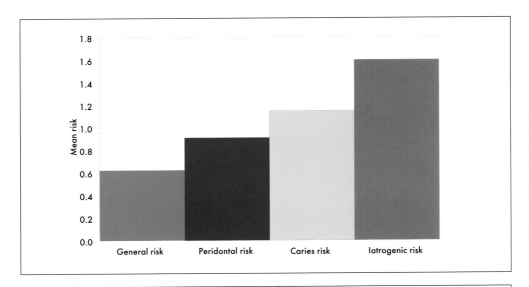

Fig 473 Mean risk scores for general risk, periodontal risk, caries risk, and iatrogenic risk in the adult population, county of Värmland, Sweden. Risk was assessed with the following scale: 0 = no risk; 1 = low risk; 2= risk; and 3 = high risk.

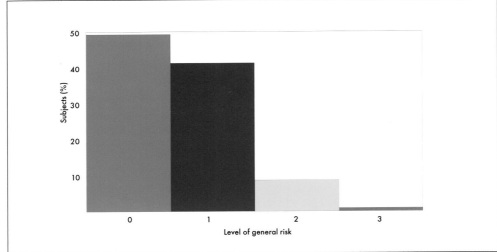

Fig 474 Frequency distribution of general risk scores (0 = no risk; 3 = high risk) in adult patients, aged 20 to 80 years, in the county of Värmland, Sweden.

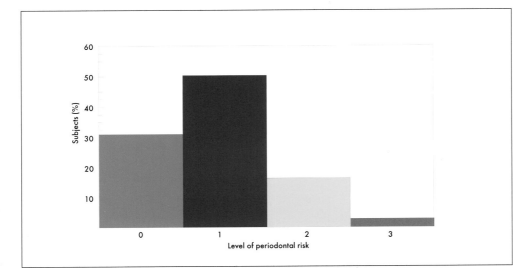

Fig 475 Frequency distribution of periodontal risk scores in adult patients, aged 20 to 80 years, in the county of Värmland, Sweden. Results indicated that 32% had no periodontal risk (0); 50% had low periodontal risk (1); 16% had periodontal risk (2); and 2% had very high periodontal risk (3).

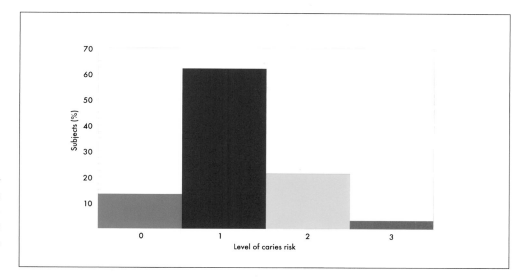

Fig 476 Frequency distribution of caries-risk scores (0 = no risk; 3 = high risk) in adult patients, aged 20 to 80 years, in the county of Värmland, Sweden.

est mean value. That is because Sweden, along with Japan, has the highest percentage of elderly in the world, and the elderly in the county of Värmland have the highest number of remaining teeth in Sweden. However, in spite of the high percentage of elderly individuals, the general risk exhibited the lowest mean value. Figure 474 shows the frequency distribution of general risk scores 0 to 3. The frequency distribution of the four classes of periodontal risk (P0 to P3) in 20- to 80-year-old adults is shown in Fig 475.

The frequency distributions of no periodontal risk, low periodontal risk, periodontal risk, and high periodontal risk, respectively, were 61%, 36%, 2%, and 1% in 20-year-old adults; 24%, 59%, 15%, and 2% in 40-year-old adults; 11%, 58%, 27%, and 4% in 60-year-old adults; and 12%, 55%, 28%, and 4% in 80-year-old adults. That means the frequency distribution of high-risk patients is less than 5% even in 80-year-old adults. In developing countries with limited dental care resources as well as in industrialized countries, the percentage of the more severe (aggressive) forms of periodontitis is estimated to be between 5% to 15%.

Figure 476 shows the frequency distribution of the four classes of caries risk in the adult population. Only 3% to 4% of the population is classified as high risk. However, the percentage is high-

er in 60- to 80-year-old than in 20- to 30-year-old adults because older patients tend to have reduced stimulated salivary flow, exposed root surfaces (increased risk for root caries), higher frequency of physical and mental disabilities, and other general health problems.

The frequency distribution of the four classes of iatrogenic risk is shown in Fig 477. Because about 50% of the Swedish adult population is more than 50 years old, adults as a group exhibit a high prevalence of restored tooth surfaces and teeth with posts. When the adult population became older, Sweden had among the highest caries prevalences in the world. That is the main reason why more than 40% of those older than 50 years are classified with iatrogenic risk and almost 15% with high iatrogenic risk. On the other hand, the majority of 20- to 30-year-old adults have no iatrogenic risk.

The fee per year for needs-related maintenance, or so-called supportive care, is based on the sum of general risk, periodontal, caries, and iatrogenic risk classes.

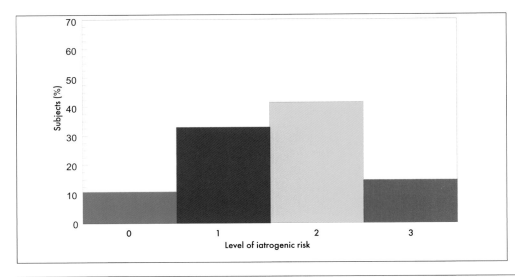

Fig 477 Frequency distribution of iatrogenic risk scores (0 = no risk; 3 = high risk) in adult patients, aged 20 to 80 years, in the county of Värmland, Sweden.

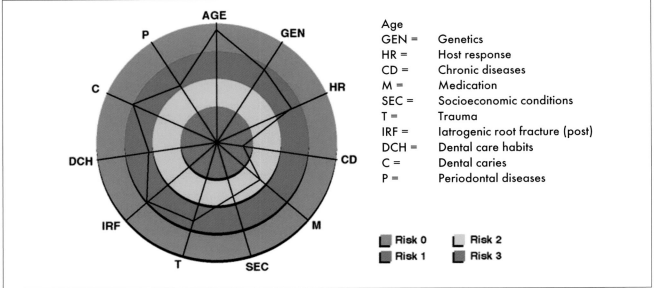

Age
GEN = Genetics
HR = Host response
CD = Chronic diseases
M = Medication
SEC = Socioeconomic conditions
T = Trauma
IRF = Iatrogenic root fracture (post)
DCH = Dental care habits
C = Dental caries
P = Periodontal diseases

Risk 0   Risk 2
Risk 1   Risk 3

Fig 478 Risk profile for tooth loss.

Risk profiles for tooth loss, dental caries, and periodontal diseases can be presented graphically, by combining symptoms (risk markers) of disease (prevalence, incidence, treatment needs, etc); etiologic factors, external modifying risk indicators, risk factors, and prognostic risk factors; internal modifying risk indicators, risk factors, and prognostic risk factors; and preventive factors. This can be done manually or by computer. Degrees of risk, 0, 1, 2, or 3, are displayed in green, blue,

yellow, or red, respectively. The graphs are very useful tools for communication with the patient when oral health status, etiology, modifying factors, prevention, possibilities, responsibilities, and reevaluations are discussed.

Risk profiles for tooth loss

A profile of risk for future tooth loss can be compiled by combining several RIs, RFs, and PRFs (Fig 478). Among these are age, estimated risk for

(Fig 478). Among these are age, estimated risk for periodontal diseases (P0 to P3) and dental caries (C0 to C3), poor socioeconomic conditions, chronic diseases, iatrogenic root fractures, trauma, genetics and impaired host response, medication, and irregular dental care habits.

In many industrialized countries, elderly people have heavily restored dentitions because of high caries incidence 30 to 50 years previously, when dental treatment was invasive. In such populations, the most frequent reason for tooth loss would be iatrogenic root fractures caused by posts in endodontically treated teeth. On the other hand, in developing countries with very limited oral health care resources, the main reasons for tooth loss would be untreated periodontal diseases and dental caries among elderly people and untreated dental caries and trauma among young people.

Detailed risk profiles for dental caries

If a patient is at high risk predominantly for either caries or periodontal disease, a more detailed risk profile is available for the specific disease. Box 24 shows a list of abbreviations for the most important variables related to caries risk.

Figure 479 illustrates how a high-risk patient (C3) was transformed to a low-risk patient (C1) by improved self-care supplemented by professional preventive measures. The greater the difference between the solid line (baseline) and the dotted line (maintenance phase), the greater the improvement. The absence of any change suggests that this particular factor cannot be influenced (eg, genetic factors, age, and some chronic diseases).

The patient in question was a 40-year-old woman with the following clinical diagnosis and anamnestic data at the first visit.

Clinical variables related to caries

- The caries prevalence was very high. All occlusal surfaces, most approximal surfaces, and some buccal surfaces were restored. There were several recurrent caries lesions.

Box 24 Abbreviations for caries-risk variables

Caries risk
C0 = No risk
C1 = Low risk
C2 = Risk
C3 = High risk

Symptoms of dental caries
CP = Caries prevalence
CI = Caries incidence

Etiologic factors (EF)
PFRI = Plaque Formation Rate Index
PI = Plaque Index
MS = Salivary mutans streptococci count
LBC = Salivary lactobacilli count

External modifying risk indicators, risk factors, and prognostic risk factors (EMRIRF)
SEC = Socioeconomic conditions
ID = Infectious diseases
CD = Chronic diseases
M = Medication
DCH = Dental care habits
OSCT = Oral sugar clearance time
DHI = Dietary Habit Index

Internal modifying risk indicators, risk factors, and prognostic risk factors (IMRIRF)
CD = Chronic diseases
SSR = Stimulated salivary secretion rate
SBE = Salivary buffering effect

Preventive factors (PF)
GEN = Genetic factors
ED = Educational level
RPDC = Regular preventive dental care habits
CO = Compliance
SOH = Standard of oral hygiene
F = Use of fluorides
DC = Dietary control
SS = Salivary stimulation

- The caries incidence was very high; the patient was developing more than three new caries lesions (more than 85% recurrent caries) per year.

Etiologic factors. Values for etiologic factors were extremely high:

- The PFRI was very high (score 5).
- The amount of plaque was excessive (Plaque Index = 93%).

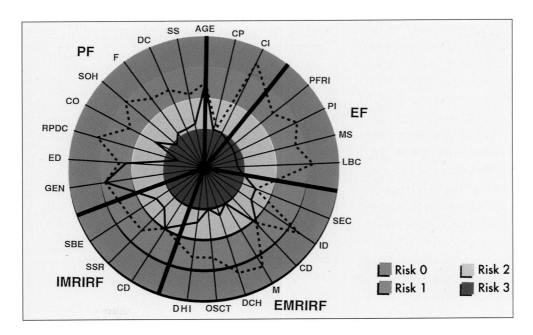

Fig 479 Risk profile for dental caries. (*solid line*) Initial presentation, C3; (*dotted line*) 1-year recall, C1. (For explanation of abbreviations, see Box 24.)

- The level of salivary mutans streptococci was very high (> 1 million colony-forming units [CFU]/mL).
- The salivary lactobacillus level was very high (> 500,000 CFU/mL).

External modifiers

- Her socioeconomic condition was low.
- The presence of ongoing infectious disease required medication with saliva-depressive effects.
- She had mild rheumatoid arthritis, which occasionally required medication with saliva-depressive side effects.
- To date, her dental attendance habits were very irregular.
- Her intake of sticky, sugar-containing products was very frequent, which resulted in extremely prolonged sugar clearance time.
- Her dietary habits were poor; she had negligible intake of fiber-rich fresh vegetables and fruits, accounting for the low Dietary Hygiene Index.

Internal modifiers

- The immune response was chronically impaired.
- The stimulated salivary secretion rate was reduced (0.5 mL/min).
- The salivary buffering effect was low.

Preventive factors

- She had no known genetic defects of tooth shape, saliva, etc.
- Her educational level was moderate.
- Preventive dental care habits were absent.
- Her level of cooperation was poor.
- The standard of oral hygiene was very low.
- She did not use fluoride toothpaste or other fluoride agents.
- Dietary control was very poor.
- Added salivary stimulation from chewing of fiber-rich food was lacking.

During case presentation, the risk profile was used as a tool for communication with the patient. Concurrently, the patient was educated in self-diagnosis to confirm the diagnosis of her own oral health status and treatment needs. Thereafter, an agreement was reached with respect to a treatment strategy, in which responsibility for the patient's oral health was shared between the patient and the oral health personnel at the clinic.

This was followed by an initial intensive preventive period, including education in self-care based on self-diagnosis; elimination of plaque-retentive factors; semipermanent restoration of recurrent caries with resin-modified glass-ionomer material; so-called complete-mouth disinfection, comprising PMTC, tongue cleaning, and chlor-hexidine (CHX) therapy (varnish, gels, toothpaste, or mouthrinse); and applications of fluoride varnish. The first reevaluation was carried out after 3 months. Thereafter, the patient began a maintenance program tailored to her individual requirements.

The first detailed reexamination was carried out after 1 year. Most important at this reexamination was that the patient was activated in self-evaluation. Again, the risk profile was used as a tool for communication with the patient, to supplement self-evaluation in the mouth and on radiographs. Figure 479 shows how successfully the patient and the dental personnel had carried out their responsibilities.

The etiologic factors were dramatically reduced by improved mechanical plaque control and intermittent use of CHX by self-care, supplemented by needs-related intervals of PMTC and CHX varnish:

- The PFRI was reduced from score 5 to score 2.
- The Plaque Index was reduced from 93% to 8%.
- The mutans streptococci count was reduced from > 1 million to < 10,000 CFU/mL.
- The lactobacilli count was reduced from > 500,000 to < 10,000 CFU/mL.

Marked reductions in sugar clearance time and Dietary Hygiene Index were achieved by:

- Elimination of sticky, sugar-containing products from the diet.
- Reduction of the total number of meals and snacks from nine to four per day.

- An increase in the intake of fiber-rich vegetables and fruits, to stimulate salivation by chewing.
- An increase in the intake of vegetable fat and proteins and a reduction in the intake of animal fat and proteins.

The salivary secretion rate was increased from 0.6 to 1.0 mL/min and the buffering effect of saliva was improved from low to normal by:

- Use of fluoride chewing gum for 20 minutes after every meal.
- An improvement in dietary habits, particularly an increase in her intake of fiber-rich products that require chewing, eg, fresh vegetables and fruits. The chewing stimulates salivation.
- Use of cheese and fresh fruit as dessert.
- Elimination of medicines with saliva-depressive effects.

Fluoride supplementation, a modifying caries-preventive factor intended to retard demineralization, enhance remineralization, and modify falls in plaque pH, was achieved by:

- Regular use of fluoride toothpaste.
- Use of fluoride chewing gum for 20 minutes after every meal.
- Application of fluoride varnish after PMTC, at needs-related intervals.
- Placement of glass-ionomer cement restorations, which function as slow-release agents for fluoride and can be recharged with fluoride from varnish, toothpaste, chewing gum, etc.

As an effect of the above improvements by self-care and dental visits at needs-related intervals, for professional preventive measures and self-evaluation, the caries incidence was 0 after 1 year; no new lesions had developed. If there are no new lesions after a further 2 years of excellent self-care habits in combination with the needs-related maintenance program, the patient will be classified as low risk (C1).

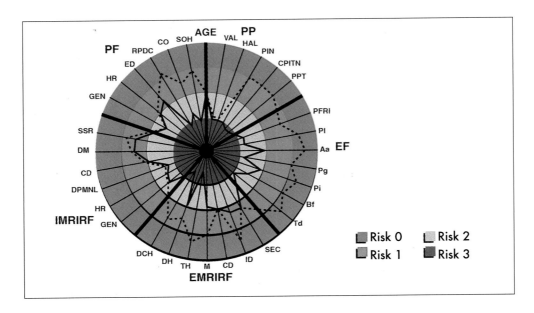

Fig 480 Risk profile for periodontal diseases. (*solid line*) Baseline, P3; (*dotted line*) 2 years later, P1. (For explanation of abbreviations, see Box 25.)

Box 25 Abbreviations related to periodontal risk

Periodontal risk
P0 = No periodontal risk
P1 = Low periodontal risk
P2 = Periodontal risk
P3 = High periodontal risk

Clinical periodontal diagnosis
PP = Periodontitis prevalence (experience)
VAL = Vertical attachment loss
HAL = Horizontal attachment loss (furcation involvement)
PIN = Periodontitis incidence (activity)
PPT = Periodontal pocket temperature
CPITN = Community Periodontal Index of Treatment Needs

Etiologic factors (EF)
PFRI = Plaque Formation Rate Index
PI = Plaque Index (scores and pattern)
Aa = *Actinobacillus actinomycetemcomitans*
Pg = *Porphyromonas gingivalis*
Pi = *Prevotella intermedia*
Bf = *Bacteroides forsythus*
Td = *Treponema denticola*

External modifying risk indicators, risk factors, and prognostic risk factors (EMRIRF)
SEC = Socioeconomic conditions
ID = Infectious diseases
CD = Chronic diseases
M = Medication
TH = Tobacco habits
DH = Dietary habits
DCH = Dental care habits

Internal modifying risk indicators, risk factors, and prognostic risk factors (IMRIRF)
GEN = Genetic factors
HR = Host response
DPMNL = Defective PMNL function
CD = Chronic diseases
DM = Diabetes mellitus
SSR = Stimulated salivary secretion rate

Preventive factors (PF)
HR = Host response
ED = Educational level
RPDC = Regular preventive dental care habits
CO = Compliance
SOH = Standard of oral hygiene

Detailed risk profiles for periodontal diseases

A detailed risk profile for patients at high risk predominantly for periodontal diseases is shown in Fig 480. The abbreviations in Fig 480 are explained in Box 25. The risk profile illustrates graphically how high periodontal risk has been reduced to low risk (from P3 to P1) by improved needs-related plaque control measures, via self-care and supplementary professional treatment. The greater the difference between the solid line (baseline) and the dotted line (maintenance phase), the greater the improvement. The absence of any change suggests that this particular factor cannot be influenced (for example, genetic factors, host response, and some chronic diseases).

The patient in question was a 50-year-old man with the following clinical diagnosis and anamnestic data at his initial assessment.

Clinical variables related to periodontal disease

- Eight teeth had been lost because of periodontal disease (four maxillary molars, two maxillary premolars, and two mandibular molars).
- The mean vertical attachment loss on the approximal surfaces was 3 mm more than the average for his age group. In addition, several posterior teeth had two- and three-wall infrabony pockets. All of the remaining molars had grade I to II furcation involvement (horizontal attachment loss).
- Retrospective radiographs and diagnoses of vertical attachment loss from the referring dentist showed irregular but advanced loss of periodontal support during the last few years (periodontal incidence).
- More than 60% of the approximal sites were diseased, with greater than 5-mm probing depths (CPITN score 4). Purulent exudate was frequent. Analysis of the gingival crevicular fluid showed high levels of prostaglandin E_2, IL-1b, aspartate aminotransferase, and other endogenous metalloproteinases, particularly from PMNLs, which together indicated active lesions with advanced breakdown of periodontal tissues.
- The periodontal pocket temperature was elevated in all pockets deeper than 3 mm, which also indicated active lesions.

Etiologic factors

- He was a very fast plaque former (PFRI 5).
- The standard of oral hygiene was very poor (Plaque Index = 76%).
- DNA probe analyses from the deepest pockets showed the following values, on a scale of 0 to 5: *Actinobacillus actinomycetemcomitans* = score 3 (> 10^5); *Porphyromonas gingivalis* = score 5 (> 10^6); *Prevotella intermedia* = score 3 (> 10^5); *Bacteroides forsythus* = score 4 (10^6); and *Treponema denticola* = score 5 (> 10^6)

External modifiers

- His socioeconomic condition, including education, was about average.
- He had a history of urinary infection (infectious diseases).
- He had diagnosed hypertension and had experienced some minor heart infarctions (chronic diseases).
- He was taking medication for his cardiovascular disease.
- He had smoked more than 20 cigarettes a day since the age of 15 years (more than 35 pack years).
- His dietary habits were poor, including frequent snacks between meals, sweets, and sweet drinks. His body mass index was high.
- His dental care habits were very irregular.

Internal modifiers

- Use of the new Periodontal Susceptibility Test revealed that he was positive for the polymorphic IL-1 gene cluster; ie, he was genetically impaired. It has been shown that this genetic defect is strongly correlated to increased susceptibility to periodontal diseases (Kornman et al, 1997), particularly in combination with smoking (Axelsson et al, 2001; McGuire and Nunn, 1999).
- His host response was also reduced because of defective PMNL function, an effect of regular smoking. The importance of aggressive, phagocytosing PMNLs as the first line of non-

specific defense in periodontal pockets should not be underestimated.

- As stated previously, he had a diagnosis of cardiovascular disease, which could be attributable to the presence of several diseased pockets, from which gram-negative microorganisms and their lipopolysaccharides continuously entered the connective tissue and the vascular system. Other contributing factors could be 35 years of smoking, poor dietary habits, hereditary factors, and physical inactivity.
- He occasionally experienced symptoms of type 2 diabetes.
- Because of regular medication with saliva-depressive effects, his stimulated salivary secretion rate was low (< 0.7 mL/min)

Preventive factors

- He was genetically impaired.
- Instead of having an effective host response, his first line of defense was impaired because of smoking.
- His educational level was slightly above average.
- He sought preventive dental care only irregularly.
- His compliance on oral hygiene, smoking, and dietary habits was very poor, resulting in a very low standard of oral hygiene. He used a toothbrush and toothpaste only irregularly.

During the case presentation, a graphic illustration (see Fig 480) was used as a tool for communication with the patient. Concurrently, the patient was educated in self-diagnosis to confirm the diagnosis of his own oral health status and treatment needs. Thereafter, an agreement was reached by the patient and the oral health personnel with respect to a treatment strategy in which responsibility for the patient's oral health was shared between the patient and the oral health personnel (dentist and dental hygienist) at the clinic. This was followed by an initial intensive preventive period, including education in needs-related plaque control measures, based on self-diagnosis.

The dentist and dental hygienist, working in cooperation, eliminated all supragingival and subgingival plaque-retentive factors. Conservative, nonaggressive methods were used for scaling and root planing to achieve smooth root cementum without exposing dentinal tubules, which would have led to bacterial invasion. The subgingival biofilm and nonattaching microflora were comprehensively removed by nonaggressive debridement and powered irrigation with bactericidal chemical plaque control agents (iodine solution). During this initial intensive period, the entire oral cavity was treated according to the so-called one-stage complete-mouth disinfection strategy: three times in 1 week, the dental hygienist cleaned the tongue and all tooth surfaces (supragingival as well as subgingival), both mechanically (PMTC) and chemically with chemical plaque control agents (CHX and iodine).

Thereafter, the patient practiced needs-related plaque control measures twice a day, based on self-diagnosis and self-evaluation. Plaque disclosure before and after cleaning was performed every day during the first week and weekly thereafter. Needs-related plaque control measures included use of selected mechanical toothcleaning aids and a tongue scraper as well as a toothpaste that contained triclosan. For the first 4 weeks, the patient also used a CHX mouthrinse twice a day. Because CHX is cationic, the patient was instructed not to use toothpastes containing anions (eg, sodium lauryl sulfate and monofluorophosphate) within an hour before or after rinsing with CHX.

The first reevaluation was carried out after 3 months. Thereafter, the patient began a maintenance program tailored to his individual requirements. Maintenance included needs-related intervals of clinical evaluation, PMTC, nonaggressive debridement of diseased pockets, and control of the oral hygiene standard.

The 1-year recall assessment involved comprehensive clinical examination, digitized computer-aided radiographs, DNA probe analyses, pocket temperature measurement, and gingival crevicular fluid analysis. It was confirmed that only three remaining deep pockets (> 5 mm) exhibited signs of activity: prostaglandin E_2, IL-1, and aspartate aminotransferase levels in gingival crevicular fluid were still high, and the pocket temperature was elevated. The levels of *A actinomycetemcomitans, P gingivalis*, and *B forsythus* remained high. Use of millimeter-graded probes in combination with the digitized radiographs disclosed the presence of two-wall infrabony pockets at all three active sites.

At this stage, the patient was highly motivated (prepared to act) and his standard of oral hygiene was excellent. After a "case presentation," including reevaluation of the risk profile, the patient decided to stop smoking if the remaining three active lesions could be healed and arrested by regenerative therapy.

One week before regenerative therapy, any remaining subgingival biofilms were mechanically removed by nonaggressive debridement, followed by comprehensive powered irrigation with iodine solution. Because the sites contained high levels not only of the anaerobes *P gingivalis* and *B forsythus* but also the exogenous pathogen *A actinomycetemcomitans,* a fiber that delivered controlled, slow release of tetracycline was placed in the pockets for 1 week. In addition to the needs-related mechanical plaque control measures and the use of fluoride toothpaste that contained triclosan, the patient began a CHX rinsing program 1 week before surgery.

Tailor-made miniflap surgery was used both to gain accessibility to the three different periodontal lesions and for regenerative therapy at the one surgical session. After nonaggressive mechanical cleaning of the root surfaces with curettes (used with negative angle) and PER-IO-TOR reciprocating instruments (Dentatus, Hägersten, Sweden), followed by chemical cleaning and surface conditioning with ethylenediaminetetraacetic acid gel (PrepGel; Biora, Lund, Sweden), a new matrix-guided regenerative material (Emdogain gel; Biora) was placed on the root surfaces. The miniflaps were resutured.

For the first postoperative month, only chemical plaque control by rinsing twice a day with CHX solution and the use of an extrasoft toothbrush was allowed around the treated sites, to prevent disruption of healing by mechanical trauma, particularly from interdental cleaning aids. After 4 weeks, the patient resumed needs-related mechanical plaque control measures and the needs-related maintenance program, based on evaluations, PMTC, and nonaggressive debridement at sites where subgingival biofilms had reformed, despite the concerted efforts at gingival plaque control by both patient and hygienist.

The second detailed reexamination was carried out after 2 years. At these reexaminations it is most important that the patient be activated in self-evaluation. Digitized radiographs (Fig 481), an intraoral camera (Fig 482), and a lighted mouth mirror are very useful tools for this purpose.

The risk profile (see Fig 480) was again used as a tool for communication with the patient, to supplement self-evaluation in the mouth and on radiographs. The changes in the risk profile (dotted line) shows how successfully the patient and the dental personnel fulfilled their responsibilities:

- Etiologic factors were dramatically reduced by improved mechanical plaque control and intermittent use of CHX by self-care, supplemented at needs-related intervals by PMTC and debridement.
- The PFRI was reduced from 5 to 2 (indicating that the gingivae had healed following the establishment of meticulous gingival plaque control).
- The PI was reduced from 76% to 8%.
- The exogenous periopathogens *A actinomycetemcomitans* and *P gingivalis,* as well as the opportunistic periopathogens *B forsythus, P intermedia*, and *T denticola*, were almost totally eliminated.
- The urinary infection and the periodontal pockets healed. The patient stopped smoking and improved his dietary habits. The need for medication for infection as well as for cardiovascular disease was reduced.
- For 2 years, the patient had participated in a needs-related maintenance program, which included regular dental care habits and regular professional preventive care.

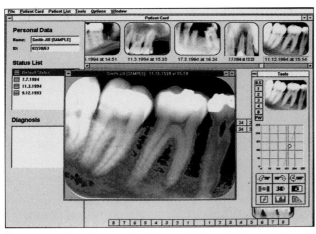

Fig 481 Digital intraoral radiography. With the use of computer-aided digital subtraction radiography (Digora; Soredex Orion, Finland), bone loss or bone regeneration can be measured and visualized in color within 2 to 4 months.

Fig 482 Intraoral camera (Vista Cam2; Dürr Dental, Bietigheim-Bissingen, Germany). These cameras are very useful for case presentation and education of the patient in self-diagnosis.

- The reduced need for medication for cardiovascular disease eased the saliva-depressive effects of the drugs. Together with changes in dietary habits (increased intake of fiber-rich vegetables, etc), this led to improved salivary function: The stimulated salivary secretion rate increased from < 0.7 to 1.2 mL/min.
- Because the patient stopped smoking, the PMNL function and thereby the host response seemed to improve.
- The patient's educational level, that is, knowledge of dental diseases, self-diagnosis, and self-care, increased considerably over 2 years.
- High motivation, based on self-diagnosis, knowledge, and training resulted in establishment of excellent needs-related plaque control habits and compliance.

The outcome of these efforts and improvement by self-care and needs-related professional preventive treatment was that there was no further loss of periodontal support during the 2-year period. Instead, a mean 6-mm gain of vertical attachment was achieved at three sites, as a result of successful regenerative therapy. All periodontal sites were healthy (CPITN 0), and no increased periodontal pocket temperature was observed.

Periodontal risk was therefore reassessed: Although the risk lessened, the patient tested positive (Periodontal Susceptibility Test) for the polymorphic IL-1 gene cluster; he therefore will continue in a maintenance program, with recall at needs-related intervals. The aim will be to gradually prolong the intervals between recalls.

This example illustrates how useful the risk profile is as a tool for:

1. Case presentation and communication with the patient
2. Establishment of needs-related self-care habits
3. Detailed evaluation of self-care and professional preventive treatment, even in individuals assessed as high risk according to current knowledge

Schedules for needs-related preventive programs based on risk prediction

In countries with ample oral health personnel, oral health resources, and relatively high standards of living and oral health, by far the most cost-effective strategy for improvement of oral

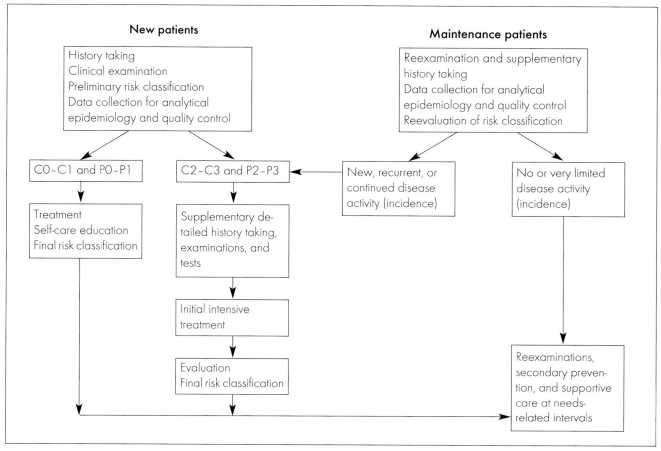

Fig 483 Schedule for needs-related preventive and maintenance (supportive) care.

health status is needs-related self-care based on self-diagnosis, supplemented with PMTC and topical application of fluoride agents at needs-related intervals. Figure 483 illustrates briefly a schedule for needs-related preventive and maintenance (supportive) care for new and maintenance patients.

According to the state of the art, all new patients should be introduced to a needs-related preventive program in the following order.

Screening and history taking

The goals at the very first appointment are:

1. To obtain a brief overview of oral health status by screening diagnosis, supplemented with necessary radiographic examinations

2. To gain an impression of the owner of the oral cavity by taking an oral and a general history

For the oral history, an evaluation of dental care, oral hygiene, fluoride use, dietary and smoking habits, and attitudes toward oral health are of great importance. The most important variables for general history evaluations are level of education, occupation, lifestyle, general diseases, use of medicines, attitudes toward general health, and body mass index.

It is important to remember that the diagnosis reveals only the present oral status, but the history discloses the reasons for that status. On the basis of the results of screening and history taking, appropriate detailed supplementary examinations and tests are selected.

Examinations and tests

Supplementary examinations and tests are carried out with the following goals:

1. To obtain detailed information on oral health status
2. To obtain detailed information on etiologic and modifying factors related to the patient's oral health status

The most important clinical variables related to periodontal diseases are vertical probing attachment loss (mm); horizontal attachment loss (furcation involvement, grades 0 to III); Gingival Index; probing depth; alveolar bone level, shape, and structure, based on complete-mouth radiographs, conventional or computer-aided digitized; bleeding on probing; purulent exudate; periodontal pocket temperature; amount and content of gingival crevicular fluid; and subgingival plaque-retentive factors (eg, calculus, rough root surfaces, overhangs of restorations). If retrospective data are available, the incidence of periodontal attachment loss is estimated. From an etiologic point of view, PFRI, Plaque Index, and the occurrence of subgingival plaque biofilms and specific periopathogens such as *A actinomycetemcomitans*, *P gingivalis*, *B forsythus*, *T denticola*, and *P intermedia* (diagnosed via DNA-DNA checkerboard technique), are of great importance. The most important external and internal modifying factors for periodontal diseases are oral and dental care habits, smoking habits, educational level, systemic diseases (particularly type 1 and type 2 diabetes), and genetic susceptibility. For diagnosis of dental caries, the most important clinical variables are caries prevalence, expressed in DMFSs, the location and size of noncavitated and cavitated enamel and dentin caries lesions, the color and texture (active or inactive) of noncavitated and cavitated root caries lesions, and the location and size of caries lesions diagnosed on radiographs (conventional bitewing radiographs or computer-aided digitized radiographs). If retrospective data are available, the rate of caries progression (incidence) is evaluated. Stimulated and unstimulated salivary flow (mL/min) and the buffer capacity of the saliva are evaluated. In the near future, antimicrobial and genetic tests of saliva should be available. The most important etiologic factors related to dental caries are PFRI and specific cariogenic bacteria, in particular the salivary counts of mutans streptococci and lactobacilli. Important modifying factors related to dental caries are dietary habits (particularly estimated sugar clearance time), oral hygiene, fluoride and dental care habits, educational level, systemic diseases, use of medicines with negative effects on salivary flow and quality, and the shape and morphology of the teeth.

Risk classification

Based on all data from the diagnosis and history taking, the patient is classified according to risk for dental caries and periodontal diseases: no risk (C0 and P0, respectively), low risk (C1 and P1), risk (C2 and P2), or high risk (C3 and P3). The individual risk profile is established as a tool for case presentation and communication with the patient as discussed earlier.

Initial intensive treatment

The number of visits and the materials and methods used are strictly related to the patient's classification and predicted risk. The initial intensive treatment has the following main goals:

1. To establish needs-related self-care habits based on self-diagnosis and education
2. To heal diseased periodontal tissues as soon as possible, resulting in a dramatic reduction in plaque formation rate (PFRI)
3. To achieve arrest of caries lesions without cavitation
4. To eliminate plaque-retentive factors, such as restoration overhangs, unpolished restorations, calculus, rough root surfaces, and cavitated caries lesions

For nonrisk (C0 and P0) and low-risk (C1 and P1) patients without any treatment need, the initial preventive treatment may be limited to one or two visits. For risk (C2 and P2) and high-risk (C3 and P3) patients, the initial intensive treatment may range from three to six visits. These visits will be concentrated in as short a period as possible (7 to 10 days) at about 2-day intervals. However, the first two should take place at exactly a 24-hour interval for evaluation of PFRI and establishment of needs-related oral hygiene habits based on PFRI and education in self-diagnosis (for details, see chapters 1 and 2).

To facilitate mechanical plaque control by self-care as well as PMTC, plaque-retentive factors will be eliminated as early as possible. The most severe plaque-retentive factors are remaining roots that are untreatable because of caries and teeth that are untreatable because of advanced periodontal disease. Such teeth should be extracted as soon as possible. Overhangs and undercuts of restorations and crowns must be eliminated, and rough surfaces of restorations finished and polished. Cavitated caries lesions should be excavated and restored at least semipermanently with glass-ionomer cement or resin-modified glass-ionomer cement. This should be especially beneficial in caries-risk patients, because glass-ionomer cement may act as a slow-release fluoride agent, which can be recharged with fluoride by daily use of fluoride toothpaste and professional use of fluoride varnish and gels.

Subgingival as well as supragingival calculus is eliminated and rough root surfaces are planed with a nonaggressive technique without exposure of the root's dentinal tubules. With these procedures subgingival biofilms should be removed as well. Recontouring can reduce the plaque retention in root grooves and furcation-involved teeth.

The so-called one-stage complete-mouth disinfection method (see chapter 4) should be implemented in caries-risk (C2 and C3) as well as periodontal-risk (P2 and P3) patients during this initial intensive treatment period. For C2 and C3 patients this means mechanical cleaning of *all*

tooth surfaces and the dorsum of the tongue and rinsing afterward with 0.1% or 0.2% CHX solution twice a day. At every visit, PMTC and tongue scraping are performed; these are followed with CHX mouthrinsing and interproximal placement of CHX gel or the use of CHX varnish on the key-risk tooth surfaces. In P2 and P3 patients, the complete-mouth disinfection also includes subgingival mechanical removal of the biofilm in combination with irrigation with iodine solution (bactericidal). Depending on the outcome of the analyses of the subgingival microflora, topical use of doxycycline gels or systemic use of antibiotics (amoxicillin and/or metronidazol) may be indicated in especially susceptible patients with impaired general health (eg, severe cardiovascular problems, poorly controlled type 1 or 2 diabetes).

As an effect of this intensive combination of mechanical and chemical plaque control by self-care as well as professionals (dentist or dental hygienist) the periodontal tissues—particularly the free gingivae—will heal. In addition, the PFRI will be dramatically reduced and thus facilitate daily oral hygiene procedures. By supplementation of this treatment with fluoride chewing gum for 20 minutes after every meal and the use of fluoride varnish on the key-risk surfaces at every visit, noncavitated enamel and dentin caries in C2 and C3 patients may be arrested, and active root caries lesions may be transformed to inactive lesions.

Reevaluation

Reevaluation has the following goals:

1. To evaluate the results of the initial intensive treatment
2. To assess patient compliance
3. To evaluate supplementary need for regenerative therapy due to lost periodontal support and restorative treatment
4. To determine the materials and methods needed for the maintenance period and the optimal recall intervals

Patients assessed as C2–C3 and P2–P3 should be recalled within 3 to 6 months after the initial intensive treatment for evaluation of the effect on the symptoms of active caries lesions, gingivitis, and periodontitis that the patient exhibited at the baseline examination. Noncavitated caries lesions may have been arrested, and active root caries lesions should be inactive in C2 and C3 patients. Inflamed gingival sites should be healed, and probing depths significantly reduced in P2 and P3 patients. Repeated oral microbiology tests may be indicated in C2 and C3 patients with very high counts of salivary mutans streptococci and lactobacilli and P2 and P3 patients with high levels of periopathogens in deep, diseased pockets at baseline.

At this reevaluation, the dentist decides whether supplementary restorative dentistry or regenerative periodontal therapy is indicated. Through interviews and questionnaires, the patient's knowledge, attitude, and compliance are evaluated. The quality of the patient's self-performed oral hygiene is evaluated by plaque disclosure. It is of great importance to motivate the patient to participate in the evaluation by self-diagnosis at every visit. Based on the outcome of the initial intensive treatment and risk classification, the materials and methods needed for the maintenance period (supportive care) and the optimal recall intervals are determined.

Maintenance period (supportive care)

Intervals, materials, and methods of the maintenance program are strictly related to the patient's classification and predicted risk. The maintenance program has these optimal goals:

1. To ensure that healthy individuals with no experience of oral diseases remain healthy (primary prevention)
2. To prevent recurrence of oral disease (secondary prevention) after initial successful symptomatic treatment (tertiary prevention); ie, to ensure no new caries lesions and no further loss of periodontal attachment

3. To encourage continuous improvement in self-care habits to prolong the intervals and reduce the need for professional preventive measures.

All these goals have been proven to be successfully achieved by a dental hygienist in adult patients for 30 years (see Axelsson et al, 2004, discussed in chapter 3) and by dental hygienists as well as preventive dentistry assistants in patients from birth to 20 years of age (see chapter 7).

However, if the effect of the maintenance (supportive) care is failing and the patient develops new caries lesions or exhibits new or further loss of periodontal attachment, he or she will undergo new detailed history taking, clinical diagnosis, and testing to evaluate the reasons for new continued disease activity (incidence). Based on the outcome of these detailed analyses, another needs-related intensive treatment will be carried out according to the flowchart in Fig 483.

Recall examination

Recall intervals, diagnosis, and supplementary history taking are strictly related to the patient's classification and predicted risk. The reexamination has several goals:

1. To evaluate the effect of the maintenance program
2. To prolong the intervals of the maintenance program and the recall examinations, provided that the patient's self-care has improved and there is no oral disease activity
3. To evaluate whether new treatment needs have occurred
4. To repeat an intensive treatment period and introduce a more comprehensive maintenance program if the previous program has proved unsuccessful

The recall examination should always be carried out by a dentist. In P3 patients, and perhaps in P2 patients, the recall examination should be carried out by a dentist specializing in periodontal diseases.

Implementation

These principles for needs-related preventive programs are based on an ideal ratio of dental professionals, well-developed oral health care systems, high economic and educational standards among the population, and recent knowledge about diagnosis and preventive materials and methods. Local conditions may limit full implementation of this state-of-the-art program.

For example, in Sweden, almost all children and young adults and 80% to 90% of adults are so-called recall patients; they receive some kind of regular maintenance program, but only a minority receive a strictly needs-related preventive program. In many other countries, however, maintenance programs are still nonexistent, at least for adults.

Table 25 shows schedules of needs-related preventive programs for different age groups strictly based on risk classification, an ideal ratio of dental professionals, and well-developed oral health care systems in a well-educated population with high economic standards. To simplify the schedules of the preventive program, abbreviations were created for many dental terms (Box 26).

Table 25 Schedules for needs-related preventive programs (continued on following pages; see page 520 for explanation of abbreviations)

Screening		History taking			Examinations/tests		Patient category Predicted risk	Initial intensive treatment			Reevaluation		Maintenance period				Reexamination		
EX	DP	HT	DP		EX	DP		PMM	DP	Intervals	EX	DP	EX	PMM	DP	Intervals	EX	DP	Intervals
Pregnant women (PW)																			
EX	D	GHT OHT	D	DH	EX, CCITN, CPITN	D DH	PW: C0	SCE, SDE, OHMT, OHMD, SFD, PMTC	D DH	1xD			SDE	SCE, PMTC	DH	1Y	SDE, EX, CI	D	2–3 Y
EX	D	GHT OHT	D	DH	SDE, EX, CCITN, CPITN	D DH	PW: C1	SCE, OHMT, OHMD, SFD, PMTC, PFV	D DH	2xD			SDE, EX	SCE, PMTC, PFC, PCV	DH	6M	SDE, EX, CI, CCITN	D	1–2 Y
EX	D	GHT OHT	D		SDE, DEX, CPR, CCITN, SSR, SBC, CPITN, OMT, PFRI, DHE	D	PW: C2	SCE, OHMT, OHMD, SFD, DHR, PMTC, PFV	D DH	2xD + 1x2D	DEX, 3M	D	SDE, EX	SCE, PMTC, PFV, PCV	DH	3M	SDE, DEX, CI, CCITN, CPITN	D	1Y
EX	DSC	GHT OHT	D		SDE, DEX, CPR, CCITN, CI, CPITN, SSR, SBC, OMT, PFRI, DHE	DSC	PW: C3	SCE, OHMT, OHMP, SFD, DHR, OHC, PMTC, SC, PCV, PFV, FCG	DSC DH	2xD 2x2D	DEX, 3M	DSC	SDE, DEX	SCE, PMTC, PFV, PCV, FCG	DH	3M	SDE, DEX, CI, CCITN, CPITN, PI, OMT	D, DSC	1Y
EX	D	GHT OHT	D		SDE, DEX, CPR, CCITN, PP, CPITN, SSR, SBC, OMT, PFRI, DHE, PST	D	PW: C2,P2	SCE, OHMT, OHMP, SFD, DHR, OHC, PMTC, SC, PCI, PCV, PFV	D DH	2xD + 2x2D	DEX, 3M	D	SDE, DEX	SCE, DEB, PMTC, PFV, PCV	DH	3M	SDE, DEX, CI, CCITN, CPITN, PI, OMT	D	1Y
Children (CH)																			
EX	DH	GHT OHT	D	DH	PC-SDE, EX, CPR	DH	CH 0–4 Y: C0	PC-SCE, SFD, OHMT, PMTC	PDA DH	1xD			EX	PC-SCE, PMTC	PDA DH	1Y	PC-SDE, EX	D	2Y
EX	DH	GHT OHT	D	DH	PC-SDE, EX, CPR	DH	CH 0–4 Y: C1	PC-SCE, SFD, OHMT, PMTC, PFV	PDA DH	2xD			EX	PC-SCE, PMTC, PFV, PCV	PDA DH	6M	PC-SDE, EX, CI	D	1Y
EX	D	GHT OHT	D		PC-SDE, EX, CPR, OMT, PFRI, DHE	D	CH 0–4 Y: C2	PC-SCE, DHR, OHMT, PMTC, SFD, PFV, PCV	PDA DH	2xD			PC-SDE, EX	PC-SCE, PMTC, PFV, PCV	PDA DH	3M	PC-SDE, EX, CI	D	1Y

Screening EX	Screening DP	History taking HT	History taking DP	Examinations/tests EX	Examinations/tests DP	Patient category / Predicted risk	Initial intensive treatment PMM	Initial DP	Initial Intervals	Reevaluation EX	Reevaluation DP	Maintenance EX	Maintenance PMM	Maintenance DP	Maintenance Intervals	Reexamination EX	Reexamination DP	Reexamination Intervals
EX	D	GHT OHT	D	PC-SDE, EX, CPR, OMT, PFRI, DHE	D	CH 0–4 Y: C3	PC-SCE, SFD, DHR, OHMT, PMTC, PCV, PFV	PDA DH	2xD + 1x2D	DEX, 3M, PC-SDE	D	PC-SDE DEX	PC-SCE, PMTC, PFV, PCV	DH	3M	PC-SDE, DEX, CI, CCITN	D	1Y
EX	DH	GHT OHT	DH	PC-SDE, EX, CPR	DH	CH 5–7 Y: C0	PC-SCE, OHMT, SFD, PMTC, PFV	PDA DH	1xD			PC-SDE, EX	PC-SCE, PMTC, PFV	PDA DH	6M	PC-SDE, Ex	D	1Y
EX	DH	GHT OHT	DH	PC-SDE, EX, CPR, CCITN	D DH	CH 5–7 Y: C1	PC-SCE, OHMT, SFD, PMTC, PFV	PDA DH	2xD			PC-SDE, EX	PC-SCE, PMTC, PFV	PDA DH	6M	PC-SDE, EX, CI	D	1Y
EX	D	GHT OHT	D	PC-SDE, DEX, PFRI, DHE, OMT, CPR, CCITN, PLI	D	CH 5–7 Y: C2	PC-SCE, OHME, SFD, DHR, PMTC, PFV, PCV, SFT, PFS-PFGI	DH	2xD + 1x2D	EX, 2M	D	PC-SDE, DEX	PC-SCE, PMTC, PFV, PCV, SFT, PFS-PFGI	DH	3M	PC-SDE, DEX, CI, CCITN, OMT	D	1Y
EX	D	GHT OHT	D	PC-SDE, DEX, PFRI, DHE, OMT, CPR, CCITN, PLI	DSC	CH 5–7 Y: C3	PC-SCE, OHME, SFD, DHR, PMTC, PCV, PFV, SFT, PFS-PFGI	D DH	2xD + 1x2D	DEX, 2M	D	PC-SDE, DEX	PC-SCE, PMTC, PFV, PCV, SFT, PFS-PFGI	DH	2–3 M	PC-SDE, DEX, CI, CCITN, OMT, DHE, PLI	D	1Y
EX	DH	GHT OHT	DH	SDE, EX, CPR, CCITN	DH	CH 8–11 Y: C0	SCE, OHMT, SFD, PMTC, PFV	PDA DH	1xD			SDE, EX	SCE, PMTC, PFV	PDA DH	1Y	SDE, EX, CI, CCITN	D	2Y
EX	DH	GHT OHT	DH	SDE, EX, CPR, CCITN	DH	CH 8–11 Y: C1	SCE, OHMT, SFD, PMTC, PFV	PDA DH	2xD			SDE, EX	SCE, PMTC, PFV	PDA DH	6M	SDE, EX, CI, CCITN	D	2Y
EX	D	GHT OHT	D	SDE, DEX, PFRI, DHE, OMT, CPR, CCITN, PLI	D	CH 8–11 Y: C2	SCE, OHMT, SFD, DHR, PMTC, PCV, PFV, SFT	DH	2xD + 1x2D	EX, 3M	D	SDE, EX	SCE, PMTC, PCV, PFV, SFT	DH	3M	SDE, EX, CI, CCITN, OMT, DHE, PLI	D	1–2 Y

Table 25 Schedules for needs-related preventive programs (continued; see page 520 for explanation of abbreviations)

Screening EX	History taking HT	DP	Examinations/tests EX	DP	Patient category Predicted risk	Initial intensive treatment PMM	DP	Intervals	Reevaluation EX	DP	Maintenance EX	Maintenance PMM	DP	Intervals	Reexamination EX	DP	Intervals
D	GHT OHT	D	SDE, DEX, PFRI, DHE, OMT, CPR, CCITN, PLI	D	CH 8–11 Y: C3	SCE, OHMT, SFD, DHR, PMTC, PCV, PFV, SFT	DH	2xD + 1x2D	DEX, 3M	D	SDE, DEX	SCE, PMTC, PCV, PFV, SFT	DH	3M	SDE, DEX, CI, CCITN, OMT, DHE, PLI	D	1Y
D	GHT OHT	PDA DH	SDE, EX, CPR, CI	PDA DH	CH 12–13 Y: C0	SCE, OHMT, SFD, PMTC, PFV	PDA DH	1xD			SDE, EX	SCE, PMTC, PFV	PDA DH	6M	SDE, EX, CI, CCITN	D	1–2 Y
D	GHT OHT	DH	SDE, EX, CPR, CI, CCITN, PFRI	DH	CH 12–13 Y: C1	SCE, OHMT, OHMD, SFD, PMTC, PFV	DH D	2xD			SDE, EX	SCE, PMTC, PFV	PDA DH	4–6 M	SDE, EX, CI, CCITN	D	1–2 Y
D	GHT OHT	D	SDE, DEX, PFRI, DHE, OMT, CPR, CI, CCITN, PLI	D	CH 12–13 Y: C2	SCE, OHMT, OHMD, SFD, DHR, FCG, PMTC, PCV, PFV, PFS-PFGI	DH D	2xD + 2x2D	SDE, EX, 3M	D DH	SDE, EX	SCE, PMTC, SFT, PCV, PFV, PFS-PFGI	DH	3M	SDE, DEX, CI, CCITN, OMT, DHE, PLI	D	1Y
D	GHT OHT	D	SDE, DEX, PFRI, DHE, OMT, CPR, CI, CCITN, CPITN, PLI, SBC, SSR	D	CH 12–13 Y: C3	SCE, OHMT, OHMD, SFD, DHR, SFCM, PMTC, PCV, PFV, FCG, PFS-PFGI	DH D	2xD + 2x2D	SDE, DEX, 2 M	D DH	SDE, DEX	SCE, SFT, SFCM, PMTC, PCV, PFV, PFS-PFGI	DH	2–3 M	SDE, DEX, CI, CCITN, OMT, DHE, PLI	D	1Y
D	GHT OHT	D	SDE, DEX, PLI, PFRI, OMT, PP, PI, CPITN, CPR, CI, CCITN, PST	D	CH 12–13 Y: P2, P3	SCE, OHMT, OHMD, SFD, OHC, OHME, DEB, PMTC, PFV, PCV	DH D	2xD + 1x2D	SDE, DEX, 3M	D	SDE, DEX	SCE, OHC, PMTC, DEB, PLI, TAB?, SAB?, PCV, PFV	DH	3M	SDE, DEX, CI, PLI, CCITN, PI, CPITN, OMT	D	1Y
D	GHT OHT	PDA DH	SDE, EX, CPR, CI, CCITN	PDA DH	CH 14–19 Y: C0	SCE, OHMT, SFD, PMTC, PFV	PDA D DH	1xD			SDE, EX	SCE, PFV, PMTC	PDA DH	6M–1Y	SDE, EX, CI	D	2Y
D	GHT OHT	DH	SDE, EX, CPR, CI, CCITN	DH	CH 14–19 Y: C1	SCE, OHMT, OHMD, SFD,	PDA D DH	2xD			SDE, EX	SCE, PMTC, PFV	PDA DH	6M	SDE, EX, CI, CCITN	D	1–2 Y

Screening EX	Screening DP	History taking HT	History taking DP	Examinations/tests EX	Exam DP	Patient category Predicted risk	Initial intensive treatment PMM	Initial DP	Initial Intervals	Reevaluation EX	Reeval DP	Maintenance EX	Maintenance PMM	Maint DP	Maint Intervals	Reexamination EX	Reexam DP	Reexam Intervals
EX	D	GHT OHT	D	SDE, DEX, PLI, PFRI, DHE, OMT, CPR, CI, CCITN, SBC	D	CH 14–19 Y: C2	SCE, OHMT, OHMD, SFD, DHR, FCG, PMTC, PCV, PFV	DH, D	2xD + 1x2D	SDE, DEX, 3M	D, DH	SDE, EX	SCE, FCG, PMTC, PCV, PFV	DH	3–4 M	SDE, DEX, CI, CCITN, PLI, DHE, OMT	D	1Y
EX	D	GHT OHT	D	SDE, DEX, PLI, PFRI, DHE, OMT, SSR, SBC, CPR, CI, CCITN	D	CH 14–19 Y: C3	SCE, OHMT, OHMD, SFD, DHR, SFCM, FCG, PMTC, PCV, PFV	DH, D,	2xD + 1x2D	SDE, DEX, 3M	D, DH	SDE, DEX	SCE, FCG, SFCM, PMTC, PCV, PFV	DH	2–3 M	SDE, DEX, CI, CCITN, PLI, DHE, OMT	D	1Y
EX	D	GHT OHT	D	SDE, DEX, PLI, PFRI, OMT, PP, PI, CPITN	D	CH 14–19 Y: P2	SCE, OHMT, OHMD, SFD, OHC, SC, RP, PCI, PMTC	DH, D	2xD + 1x2D	SDE, DEX, 3M	D, DH	SDE, DEX	SCE, DEB, PCI, PMTC, PFV	DH	3M	SDE, DEX, CI, PI, CPITN, PLI, OMT, PPT	D	1Y
EX	DSP	GHT OHT	DSP	SDE, DEX, PLI, PFRI, OMT, PP, PI, CPITN, PPT, PST		CH 14–19 Y: P3	SCE, OHMT, OHMD, SFD, OHC, SC, RP, PCI, PMTC, TAB?, SAB?	DH, D	2xD + 1x2D	SDE, DEX, 3M	DSP	SDE, DEX	SCE, DEB, PCI, OHC?, PMTC, PFV, TAB?, SAB?	DH	3M	SDE, DEX, CI, PI, CPITN, PLI, OMT, PPT	DSP	1Y

Adults (AD)

Screening EX	Screening DP	History taking HT	History taking DP	Examinations/tests EX	Exam DP	Patient category Predicted risk	Initial intensive treatment PMM	Initial DP	Initial Intervals	Reevaluation EX	Reeval DP	Maintenance EX	Maintenance PMM	Maint DP	Maint Intervals	Reexamination EX	Reexam DP	Reexam Intervals
EX	D	GHT OHT	D	SDE, EX, CPR, CCITN, CPITN	D	AD: CO	SCE, OHMT, SFD, PMTC, PFV	DH	1xD			SDE, EX	SCE, PMTC	PDA, DH	1Y	SDE, EX, CI, CCITN	D	2–3 Y
EX	D	GHT OHT	D	SDE, EX, CPR, CI, CCITN, CPITN, PFRI	D	AD: C1	SCE, OHMP, SFD, DHR, DEB, PMTC, PFV	DH	2xD	SDE, EX, 6M	DH	SDE, EX	SCE, PMTC, PFV	PDA, DH	6M – 1Y	SDE, EX, CI, CCITN, CPITN, PLI	D	2Y
EX	D	GHT OHT	D	SDE, DEX, CPR, CI, CCITN, PFRI, PLI, DHE, OMT, SSR, CPITN	D	AD: C2	SCE, OHMT, SFD, DEB, PMTC, PCV, PFV, DHR, SFCM, FCG, SS?	DH	2xD + 2x2D	SDE, EX, 3M	D	SDE, DEX	SCE, FCG, DHR, PMTC, PCV, PFV, DEB, SS	DH	4–6 M	SDE, DEX, CI, CCITN, PLI, CPITN, OMT, SSR, DHE	D	1Y

Table 25 Schedules for needs-related preventive programs (continued; see page 520 for explanation of abbreviations)

Screening		History taking		Examinations/tests		Patient category	Initial intensive treatment			Reevaluation		Maintenance period				Reexamination		
EX	DP	HT	DP	EX	DP	Predicted risk	PMM	DP	Intervals	EX	DP	EX	PMM	DP	Intervals	EX	DP	Intervals
EX	D	GHT OHT	DSC	SDE, DEX, CPR, CI, CCITN, PFRI, PLI, DHE, OMT, SSR, SBC, CPITN	DSC	AD: C3	SCE, OHMT, OHMP, SFD, DEB, PMTC, PCV, PFV, DHR, SFCM, FCG, SS?	D DH	2xD + 2x2D	SDE, DEX, 3M	DSC	SDE, DEX	SCE, FCG, SFCM, DHR, DEB, PMTC, PCV, PFV, SS?	DH	3M	SDE, DEX, CI, CCITN, PLI, CPITN, OMT, DHE, SSR, SBC	DSC	1Y
EX	D	GHT OHT	D	SDE, EX, CPR, CCITN, CPITN	D	AD: P0	SCE, OHMT, OHMP, SFD, DEB, PMTC	DH	1xD			SDE, EX	SCE, DEB, PMTC	PDA DH	1Y	SDE, EX, CPITN	D DH	2 – 3 Y
EX	D	GHT OHT	D	SDE, EX, PLI, PFRI, CPR, PP, CPITN, CCITN	D	AD: P1	SCE, OHMT, OHMP, SFD, SC, RP, DEB, PMTC	DH	2xD	SDE, EX, 6M	DH	SDE, EX	SCE, DEB, PLI, PMTC, OHC	DH	6M	SDE, EX, PI, CPITN, CCITN, PLI	D DH	2Y
EX	D	GHT OHT	D	SDE, DEX, PP, PI, CPITN, PLI, PFRI, OMT, CPR, CCITN, PPT	D	AD: P2	SCE, OHMT, OHMP, SFD, SC, RP, DEB, PCI, PMTC, PFV, OHC	DH	2xD + 2x2D	SDE, DEX, 3M	D	SDE, EX	SCE, OHC, PCI, DEB, PMTC	DH	4–6 M	SDE, DEX, PI, CPITN, PLI, CCITN	D	1Y
EX	DSP	GHT OHT	D DSP	SDE, DEX, PP, PI, CPITN, PLI, PFRI, OMT, CPR, CCITN, PPT, PST	D DSP	AD: P3	SCE, OHMT, OHMP, OHMI, OHME, SFD, SC, RP, DEB, PCI, PMTC, PFV, OHC, TAB?, SAB?	DH DSP	2xD + 2 – 3x 2D	SDE, DEX 3M	D DSP	SDE, DEX	SCE, OHC, DEB, PCI, PMTC, TAB?, SAB?	DH DSP	3–4 M	SDE, DEX, PI, CPITN, PLI, CCITN, OMT, PPT	DSP	1Y

Screening		History taking		Examinations/tests		Patient category	Initial intensive treatment			Reevaluation		Maintenance period				Reexamination		
EX	DP	HT	DP	EX	DP	Predicted risk	PMM	DP	Inter-vals	EX	DP	EX	PMM	DP	Inter-vals	EX	DP	Inter-vals
	D	GHT OHT	D	SDE, DEX, PP, CPR, PI, CI, CPITN, CCITN, PFRI, PLI, OMT, DHE, SSR, SBC, PPT	D	AD: C2, P2	SCE, OHMT, OHMP, OHMI, OHME, SFD, SC, RP, DEB, PCI, PMTC, PCV, PFV, SFCM, FCG, DHR, SS	D	2xD + 2x2D	SDE, DEX, 3M	DSP, D, DH	SDE, DEX	SCE, FCG, SFCM, DEB, PCI, PMTC, PCV, PFV, SS, DHR	DH, D	4M	SDE, DEX, PI, CI, CPITN, CCITN, PLI, PFRI, DHE, SSR, SBC, PPT, OMT	D	1Y
	DSC DSP	GHT OHT	DSC DSP	SDE, DEX, PP, CPR, CI, PI, CPITN, CCITN, PFRI, PLI, OMT, DHE, SSR, SBC, PPT, PST	D DSP DSC	AD: C3, P3	SCE, OHMT, OHMP, OHMI, OHME, SFD, SC, RP, DEB, PCI, PMTC, PCV, PFV, SFCM, FCG, SS, DHR, TAB?, SAB?	DSC DSP	2xD + 3x2D	SDE, DEX, 2-3 M	DSC, DSP, D	SDE, DEX	SCE, FCG, SFCM, DEB, PCI, PMTC, PCV, PFV, SS, DHR, SAB?, TAB?	DH, D, DSC, DSP	2-3 M	SDE, DEX, PI, CI, CPITN, CCITN, PLI, PFRI, DHE, SSR, SBC, PPT, OMT	DSC, DSP	6M-1Y

Box 26 Abbreviations for schedules of needs-related preventive programs (see Table 25)

Dental professionals (DP)

D = Dentist
DH = Dental hygienist
DSC = Dentist specialized in cariology
DSP = Dentist specialized in periodontology
PDA = Preventive dentistry assistant

Patient category

AD = Adults
CH = Children
PW = Pregnant women

Predicted risk

C0 = No caries risk
C1 = Low caries risk
C2 = Caries risk
C3 = High caries risk
P0 = No periodontal risk
P1 = Low periodontal risk
P2 = Periodontal risk
P3 = High periodontal risk

History taking (HT)

GHT = General history taking
OHT = Oral history taking

Examinations and tests (EX)

CCITN = Community Caries Index of Treatment Needs
CI = Caries incidence (decayed surfaces/year)
CPITN = Community Periodontal Index of Treatment Needs
CPR = Caries prevalence (decayed, missing, or filled surfaces)
DEX = Detailed examination
DHE = Dietary habits evaluation
EX = Examination
OMT = Oral microflora test
PC-SDE = Self-diagnosis education for parents of the child
PFRI = Plaque Formation Rate Index
PI = Periodontitis incidence (attachment loss/year)
PLI = Plaque index
PP = Periodontitis prevalence (attachment loss)
PPT = Periodontal pocket temperature
PST = Genetic periodontal susceptibility test
SBC = Salivary buffer capacity test
SDE = Self-diagnosis education
SSR = Salivary secretion rate

Intervals

1xD = One single visit
2xD = One visit per day on 2 consecutive days
2–3x2D = Two to three visits at 2-day intervals
1x2D = One visit 2 days after the first two visits
2x2D = Two visits at 2-day intervals after the first two visits
3x2D = Three visits at 2-day intervals after the first two visits
2M = Every 2 months
3M = Every 3 months
4M = Every 4 months
6M = Every 6 months
1Y = Every year
2Y = Every 2 years
3Y = Every 3 years

Preventive materials and methods (PMM)

DEB = Debridement
DHR = Dietary habit recommendation
FCG = Fluoride chewing gum
OHC = Oral hygiene by chemical plaque control
OHE = Oral health education
OHMD = Oral hygiene by mechanical plaque control with fluoridated dental tape
OHME = Oral hygiene by mechanical plaque control with electric toothbrush
OHMI = Oral hygiene by mechanical plaque control with interdental brush
OHMP = Oral hygiene by mechanical plaque control with fluoridated wooden toothpick
OHMT = Oral hygiene by mechanical plaque control with toothbrush
PCI = Antimicrobial irrigation of diseased pockets
PC-SCE = Self-care education for parents of the child
PCV = Chlorhexidine varnish
PFCG = Fluoride antimicrobial gels
PFGI = Glass-ionomer cement
PFS = Fissure sealant
PFV = Fluoride varnish
PMTC = Professional mechanical toothcleaning
RP = Root planing
SAB = Systemic use of antibiotics
SC = Scaling
SCE = Self-care education
SFCM = Fluoride antimicrobial mouthwash
SFD = Fluoride dentifrice
SFT = Fluoride tablets (topical by lozenges)
SS = Salivary stimulation
TAB = Topical use of antibiotics
? = Must be decided on a case-by-case basis

CONCLUSIONS

Trends in oral health

In many industrialized countries, the percentage of edentulousness among people older than 65 years has been high (30% to 50%). However, this percentage is continuously decreasing; furthermore, the number of remaining teeth is significantly increasing in the elderly. These changes are effects of the prevention programs initiated several decades ago for schoolchildren as well as continuously better organized maintenance care for the adult populations in many industrialized countries. As a consequence of preventive programs for schoolchildren, caries prevalence in children has decreased from high or very high levels to moderate and low levels in many industrialized countries during the last three decades. However, in the adult population, the caries prevalence (decayed, missing, or filled teeth or surfaces) is still relatively high because most caries lesions developed when these adults were children. The prevalence of chronic periodontitis has decreased during the last decades because of improved oral hygiene and more efficient maintenance care, but the percentage of severe or aggressive periodontitis has been unaltered (about 5% to 15%).

In many developing countries, on the other hand, the percentage of edentulousness among the elderly is continuously increasing, as is caries prevalence in children, from very low levels to low and moderate levels. The prevalence of chronic periodontitis among adults is also rising. These changes are consequences of changed lifestyles (particularly dietary habits), poor oral hygiene, and very limited dental care resources. However, the prevalence of severe or aggressive periodontitis is about the same as in industrialized countries.

Importance of changing attitudes toward oral health

For successful promotion of oral health and improvement of the oral health levels, dental professionals have to change attitudes toward oral health among patients, oral health personnel, and politicians.

Intact teeth, healthy gingiva, and functional natural teeth should be as much appreciated as other healthy and intact parts of the body. Patients must be well educated in self-diagnosis and self-care, which will enable them to be responsible for their own oral health. This approach also has cost-effectiveness and quality-of-life benefits.

Oral health personnel are obliged to follow the so-called *leges artis* principles, which dictate that dentists and their staff have to be continuously updated about the most recent science- and evidence-based materials and methods, not only to treat but even more importantly to prevent oral diseases. Likewise, departments of health should prioritize prevention of oral diseases.

Systems of oral care

Three different levels of dental care can be identified: The first level of oral care includes education of the population in self-diagnosis and self-care on an individual and a group basis, PMTC, professional use of fluorides and fissure sealant, nonsurgical treatment of periodontal diseases and low-technology treatment of single-surface cavitated caries lesions (atraumatic restorative technique). Because most of this first-level care is focused on prevention and could be successfully performed by a dental hygienist, it is very cost effective from an oral health point of view.

The second level of dental care is based on moderate technology and is rather expensive. It includes multisurface restorations, endodontics, extractions, simple periodontal surgery, and removable prostheses—that is, traditional invasive oral care, practiced by most dentists worldwide.

The third, high-technology level of oral care includes precision prosthetics; implants; laminates and ceramic inlays and onlays; orthodontics; regenerative periodontal treatment; and complex oral surgery and medicine—in other words, costly and complex procedures that require highly qualified specialists. Therefore, in any society, the availability of high-technology oral care will be limited.

A rational, health-promoting, affordable mix of oral care should be planned and implemented in all countries. Emphasis on prevention and control of oral diseases will minimize the need for intervention at the moderate- and high-technology levels. As a consequence of improving oral health in most industrialized countries, the need for moderately complex care is decreasing. With further emphasis on prevention, the need and demand for first-level interventions will increase slightly, while the need for high-technology care will probably increase for several decades, because of the desire to preserve natural teeth and the increasing numbers of elderly people who have some natural teeth and edentulous people who want implant treatment.

In developing countries, the first (low-technology, noninvasive) level of oral care will continue to be the major need. In those developing countries where the prevalence of caries is increasing, a rising demand for moderate-technology care will continue over the next few decades. High-technology oral care must, on the other hand, still be limited.

Influence of dental insurance systems on oral health

A great variety of dental insurance and fee systems exist. Among these are quality control guidelines, fixed fees, capitation schemes, health maintenance organizations, rewarding increased preventive care, and public health funding.

Because the dental insurance and fee system has a great impact on the content of dental care, it should be specially designed to focus on preventive dentistry. Therefore, a yearly fee based on risk prediction has been proposed, combined with fee penalties if the outcome of the individualized preventive program is failing (if the patient still develops caries or periodontitis).

Global availability of oral health personnel

Europe, representing only 17% to 18% of the world's population, has more than 50% of the world's oral health personnel resources (dentists, dental hygienists, dental assistants, technicians, etc). In China and India, representing about 40% of the world's population, the ratio of dentists to inhabitants is only 1:120,000 compared to 1:1,000 to 1:2,000 in most European countries. Thus, it seems unrealistic to attempt needs-related distribution of dentists, because the resources of teachers and dental schools are very limited in most developing countries. New technology and global interactive training programs may be helpful to overcome this problem.

Training programs for dental hygienists and preventive dentistry assistants are much shorter and less expensive than are training programs for dentists. In addition, dental hygienists and preventive dentistry assistants are completely focused on preventive dentistry, including education of the population in self-care as well as low-cost, low-technology professional preventive treatment and nonsurgical periodontal treatment. Therefore, training programs for these two categories of oral health personnel should be prioritized not only in developing countries but also in most industrialized countries to improve the oral health status of the world's population as soon as possible.

From future general health point of view, local integrated teams of physicians, psychologists, sociologists, nutritionists, oral physicians (dentists), physiotherapists, physical education teachers, general health nurses, and dental hygienists should be established.

Global goals for oral health

Realistic goals for the effects of all preventive dental care programs should be set up at global, national, county, district, and clinic levels. Such goals should also be updated in accordance with improved oral health status and dental care resources at certain intervals.

Oral health for life should be approaching reality for all. To promote improved oral health worldwide, the WHO's Oral Health Unit drew up international oral goals for severity of oral disease involvement at various key ages—12 years, 35 to 44 years, and 65 years and older—by the year 2000. At the global level, the WHO should be responsible for new updated goals.

Establishment of needs-related preventive programs based on risk prediction and risk profiles

In countries with high prevalence and incidence of dental caries and/or periodontal diseases and limited resources of dentists, a so-called whole population strategy for preventive programs performed by other oral health personnel such as dental hygienists and preventive dentistry assistants, should still be cost effective.

On the other hand, in countries with relatively good oral health status and well-organized dental care systems and resources, needs-related preventive programs based on predicted risk have to be implemented from a cost-effectiveness point of view. Criteria for classification of caries risk as well as periodontal risk in different age groups should be set up in relation to the oral health status and dental care resources. Such criteria should be reevaluated at certain intervals.

Evaluation of the individual patient's risk profile is of great importance. The risk profile is a very useful tool for case presentation and communication with the patient; establishment of needs-related self-care habits; and detailed evaluations of self-care and professional preventive treatment.

All new patients should be introduced to a needs-related preventive program, in the following order:

1. Screening, history taking, and preliminary risk classification
2. Supplementary examinations, history taking, and tests in selected risk and high-risk individuals
3. Final risk classification
4. Initial intensive treatment (number of visits, materials, and methods to be determined by the individual risk classification)
5. In risk and high-risk patients, evaluation of the effect of the initial intensive treatment after 3 to 6 months, to determine the intervals, materials, and methods of the maintenance program and the intervals of reexaminations (see Fig 483 and Table 25).

CHAPTER 9

ANALYTICAL COMPUTERIZED ORAL EPIDEMIOLOGY FOR QUALITY CONTROL OF PREVENTIVE PROGRAMS AND TREATMENT

A great deal of time, effort, and money are spent on oral health care each year, so it is reasonable that the government agency responsible for health care has an audit system that regularly evaluates the total effect of the national oral care and dental insurance systems on the oral health status and treatment needs of the population. At the same time, this review encourages and motivates dentists to continuously evaluate the efficiency of their own preventive program at surface, tooth, and individual levels, as well as for their total patient population. Furthermore, a national dental insurance scheme should promote preventive dentistry and analytical epidemiology for quality control.

According to the Swedish text *Medical Terminology* (Lindskog and Zetterberg, 1975), the definition of *epidemiology* is "the medical science of the spread, etiology, and prevention of the epidemic (infectious) diseases." Because certain types of transmissible microorganisms that colonize the tooth surfaces are implicated in the etiology of both caries and periodontal diseases, these diseases are regarded as epidemic diseases.

As discussed in chapter 8, the World Health Organization (WHO), in collaboration with the Fédération Dentaire Internationale (FDI), international dental associations, and ministries of health, established goals for the level of oral health to be attained by the year 2000 for selected indicator age groups in children and adults. One goal recommended that computer-based epidemiologic systems be established to monitor whether these goals are being attained. Epidemiologic surveys in randomized samples of the population are recommended; these should be repeated at 5-year intervals.

Very powerful personal computers are now available, and portable computers are suitable for field surveys. Today, large volumes of epidemiologic data may be collected, and direct statistical evaluation and graphic presentation of the results are readily accomplished with computer processing.

ORAL HEALTH–RELATED EPIDEMIOLOGIC DATA

Based on the experiences of a longitudinal preventive clinical study in adults in Karlstad, county of Värmland, Sweden (Axelsson and Lindhe, 1978, 1981a; Axelsson et al, 1991, 2004; for details, see chapter 3), a needs-related preventive program for the adult population of the county of Värmland was designed in 1985. This program is continuously updated in accordance with the principles described in chapter 8.

In 1988, a new computer-aided analytical epidemiologic system was designed to evaluate the effects of this preventive program at the population, individual, tooth, and surface levels, as well as the role of etiologic and modifying factors (Axelsson et al, 1988, 1990, 2000). In contrast to other medical disciplines, dentistry has well-established and measurable variables for the evaluation of oral health. These variables should be stratified according to their importance. The main reasons for loss of teeth are dental caries and periodontal diseases. Variables associated with these conditions should be given priority in oral health epidemiology. The masticatory efficiency of the dentition and the condition of the oral mucosa should also be included.

Tooth loss

The final outcome of untreated caries and periodontal disease is total edentulousness. According to the WHO's goals, edentulousness in 35- to 44-year-old and 65-year-old adults should have been reduced from 1969 levels by 50% and 25%, respectively, by the year 2000.

In field surveys, retrospective determination of the reason that teeth are missing is frequently difficult and uncertain. Information with respect to which teeth are most frequently missing and the reason for extraction is important for planning appropriate preventive measures. Important questions that arise are:

1. Why are the maxillary molars the most frequently missing teeth?
2. Why are maxillary premolars missing more frequently than mandibular premolars?
3. Why are the mandibular canines the most resistant of all the teeth?

Loss of occlusal contacts

Mastication is the primary function of the teeth. Masticatory efficiency, that is, chewing capacity,

may be expressed in terms of the Eichner index, which is based on the number of occlusal contacts in the molar and premolar areas (Eichner, 1955; Österberg and Landt, 1976).

Dental caries and periodontal diseases

There is a strong correlation between oral health status and the occurrence of dental caries and periodontal diseases. The prevalence of these diseases is therefore the most important dental health variable. Many parallels may be drawn between these two diseases: The etiologies are known in both diseases; pathogenic microorganisms that colonize the tooth surfaces are implicated.

Both diseases are site related, that is, unevenly distributed among the teeth and tooth surfaces. For example, the difference in prevalence of both caries and marginal periodontitis between the distal surface of the maxillary first molars and distal surface of the mandibular canines is usually more significant than the difference in total prevalence between individuals: There are specific, highly susceptible, key-risk teeth and surfaces. If the standard of oral health is to be improved by preventive measures, such facts must be acknowledged and the mechanisms explained.

The prevalence of both caries and periodontal disease should be presented at the individual level as well as at tooth-surface levels. The prevalence of both diseases represents the end result of all incidences and does not progress linearly. In other words, prevalence represents the results of unpredictable site-specific exacerbations and periods of disease quiescence.

Apical periodontitis

In most countries, the prevalence of apical periodontitis has not been determined because complete-mouth intraoral radiographs or orthopantomograms are required for diagnosis.

Endodontic treatment

Data on the prevalence of endodontic treatment should also be collected. In an adult population, the number of root fractures is strongly correlated with the number of endodontically treated teeth with posts. Most coronal fractures also occur in endodontically treated teeth.

Mucosal diseases

From the oral health aspect, diagnosis of and collection of data on diseases of the oral mucosa are very important. In many countries, the prevalence of serious diseases, such as precancerous and cancerous lesions and human immunodeficiency virus–associated lesions, is increasing.

Treatment needs

To allow planning and organization of the resources necessary to meet the need for oral treatment, an estimate must be made of treatment needs, not only for marginal periodontal diseases but also for caries, apical periodontitis, malocclusion, oral mucosal diseases, and bone diseases. Some new indices for treatment needs have therefore been designed: the Community Caries Index of Treatment Needs (CCITN) and Apical Periodontitis Index of Treatment Needs (APITN) (Axelsson, 1988a, 1988b). These indices are analogous to the well-established Community Periodontal Index of Treatment Needs (CPITN) (Ainamo et al, 1982).

Etiologic and modifying factors

In addition to epidemiologic data on prevalence and treatment needs, epidemiologic studies should also include causal and modifying factors, in terms of the previously mentioned definition of epidemiology. This is also confirmed by the definition of *epidemiology* in *Encyclopedia Britannica* (1963): "The medical science concerned with the description of factors and conditions that are significantly associated with the occurrence of an infectious process, disease or abnormal physiological state in a human community with elucidation of the manner in which these factors and conditions operate in the causative complex."

MATERIALS AND METHODS

Materials

In 1988, the baseline examination was carried out in randomized samples of 35-, 50-, 65-, and 75-year-old residents (N = 1,086) in the county of Värmland, Sweden. They were stratified into living areas (50% rural area and 50% urban area), gender, and dental care system (Public Dental Health Service, private practice, and dental nonattendance). The largest portion of the study group was 50-year-old subjects (n = > 400), the mean age of the adult population; these individuals were followed longitudinally for 10 years to the age of 60 years. New cross-sectional studies were scheduled for every 5 years.

Variables

A specially designed computer program was used for collecting the data. Table 26 shows, in ranking order, the variables included in the new analytical oral epidemiology system.

Questionnaire

Prior to the clinical examination, the participants answered a questionnaire about their dental care habits, oral hygiene habits, dietary and smoking habits, systemic diseases and use of medicines, socioeconomic background, lifestyle, knowledge of causes and prevention of oral diseases, and other factors (Fig 484). In addition, complete-mouth radiographs were taken.

Table 26 Variables included in the analytical computer-based oral epidemiology system

Code	Variable		Code	Variable
1	Oral epidemiology		4	Etiologic factors
1:1	Percent edentulous		4:1	Nonspecific oral microflora: Plaque Index (PI) and Plaque Formation Rate Index (PFRI) (Axelsson, 1987, 1991)
1:2	Number of teeth			
1:3	Function of teeth: Modified Eichner index		4:2	Specific microflora
2	Prevalence		5	Modifying factors
2:1	Dental caries, DMFT, DFT, DMFSs, DFSs		5:1	External modifying risk indicators, risk factors, and prognostic risk factors: poor oral hygiene and dietary habits, smoking and snuffing habits (and other unhealthy lifestyle habits), socioeconomic background (particularly low educational level), use of medicines, presence of infectious and other acquired diseases, etc
2:2	Marginal periodontitis			
2:2:1	Vertical loss of attachment (mm)			
2:2:2	Horizontal loss of attachment (furcation involvement grade 0, I, II, III)			
2:3	Apical periodontitis index (Axelsson, 1988b)			
2:4	Oral mucosal lesions and bone diseases		5:2	Internal modifying risk indicators, risk factors, and prognostic risk factors: chronic diseases (diabetes mellitus, cardiovascular diseases, Sjögren syndrome, etc), impaired host response (particularly reduced polymorphonuclear leukocyte [PMNL] function), genetic susceptibility to periodontal diseases, reduced salivary secretion rate, etc
2:5	Malocclusion			
3	Treatment needs			
3:1	Dental caries (CCITN) (Axelsson, 1988a)			
3:2	Marginal periodontitis (CPITN)			
3:3	Apical periodontitis (APITN) (Axelsson, 1988b)			
3:4	Oral mucosal lesions and bone diseases			
3:5	Malocclusion			

Clinical examination

The clinical examination was carried out by well-trained and coordinated examiners, and the data were entered directly into a portable computer by a dental assistant. This system has been used in three large-scale cross-sectional epidemiologic studies in randomized adult samples in the county of Värmland, Sweden (1988, 1993, and 1998), and in a longitudinal study of 50-year-old adults from 1988 to 1998 (Axelsson and Paulander, 1994; Axelsson et al, 1988, 1990, 1998, 2000, 2001; Paulander et al, 2003, 2004a, 2004b, 2004c).

Edentulousness and missing teeth

At the clinical examination, it was registered whether one or both arches were edentulous. The number and pattern of remaining teeth were examined as well as replacement of missing teeth with complete dentures, partial dentures, or fixed prosthodontics. In this system, the third molars were excluded. Thus, the maximum number of natural teeth in each individual was 28.

Tooth function

The chewing function of the teeth was diagnosed in accordance with the modified Eichner index, which is based on the occlusal contacts in the molar and premolar areas (Eichner, 1955; Österberg and Landt, 1976). Figure 485 shows the criteria for the 10 different classes (A1 to A3, B1 to B4, and C1 to C3) of the modified Eichner index.

Dental caries: Prevalence and treatment needs

A caries lesion should be regarded as a symptom of the disease dental caries. Table 27 shows the clinical diagnosis related to the type, localization, size, depth, and shape of the caries lesion. Caries lesions are diagnosed by the following methods, used separately or in combination:

QUESTIONNAIRE RELATED TO YOUR ORAL AND GENERAL HEALTH

Name: . . . .
Address: . . . .
Home telephone: . . . .
Work telephone: . . . .

Please answer the following questions as accurately as you can. If you have any difficulties in understanding the questions, please ask our personnel.

1a. What is your education? Please mark your highest level. Choose one.
  ❏ 1. Elementary school, normally 6 to 8 years
  ❏ 2. Vocational training, at least 1 year after elementary school or the equivalent (vocational school, domestic science school)
  ❏ 3. Comprehensive school, secondary modern school (9 years)
  ❏ 4. Upper secondary school/high school
  ❏ 5. Education at least 1 year after high school (college or university without graduation)
  ❏ 6. Graduation from university/college
  ❏ 7. Other education. Specify . . . .
  ❏ 8. If you have immigrated to Sweden, what year did you come? . . . .
1b. Profession: . . . .

2a. What occupation do you have at the moment? Choose one.
  ❏ 1. Permanent appointment
     Full time ( ) Part time( )
  ❏ 2. Self-employed person
  ❏ 3. Unemployed
  ❏ 4. Student
  ❏ 5. Disabled (more than 3 months)
  ❏ 6. Relief work
  ❏ 7. Parental leave
  ❏ 8. Homemaker
  ❏ 9. Other: . . . .
  ❏ 10. Early retirement
     Profession before retirement? . . . .
  ❏ 11. Retiree
2b. Does your work involve a night shift?
  Yes ❏ No ❏
  Swing shift? Yes ❏ No ❏

2c. If you are unemployed, how long have you been unemployed?
  ❏ 1. Less than 1 month
  ❏ 2. 1 to 4 months
  ❏ 3. 5 to 8 months
  ❏ 4. 9 to 12 months
  ❏ 5. More than 1 year
2d. Current civil status:
  ❏ 1. Married or living together
  ❏ 2. Widowed
  ❏ 3. Single parent How many children? . . . .
  ❏ 4. Single

3. Do you breathe mainly through your mouth when you sleep? Choose one.
  ❏ 1. Never
  ❏ 2. Rarely
  ❏ 3. Often
  ❏ 4. Always
  ❏ 5. Don't know

4. Do you suffer from any of the following diseases?
  ❏ Heart problems   ❏ Asthma
  ❏ High blood pressure ❏ Allergy
  ❏ Diabetes   ❏ Rheumatism
  ❏ Gastric ulcer
  ❏ Repeated/long infectious diseases.
     Specify. . . . .
  ❏ Other diseases. Specify. . . . .

5. Do you have a dry mouth? Choose one.
  ❏ 1. Never
  ❏ 2. Rarely
  ❏ 3. Often
  ❏ 4. Always

6. Do you regularly take any medicines?
  ❏ 1. No
  ❏ 2. Yes
  ❏ 3. Medicine 1 . . . .
  ❏ 4. Medicine 2 . . . .
  ❏ 5. Medicine 3 . . . .

7. How often do you eat every day (including between-meal snacks)? . . . .

Fig 484 Questionnaire given to the participants in the epidemiologic study of the effectiveness of the needs-related preventive program for adults.

8. How often do you eat sweets (for example, chewing gum with sugar, candy, chocolate)? Choose one.
   ❑ 1. Never/rarely
   ❑ 2. Every week
   ❑ 3. Two to three times/week
   ❑ 4. Four to seven times/week
   ❑ 5. More than seven times/week

9. How often do you eat products containing sugar (for example, cake, marmalade, honey, juice, fruit syrup, soft drinks, coffee/tea with sugar)? Choose one.
   ❑ 1. Never/rarely
   ❑ 2. Every week
   ❑ 3. Two to three times/week
   ❑ 4. Four to seven times/week
   ❑ 5. More than seven times/week

10. How often do you eat potato chips, cheese puffs, or something similar? Choose one.
    ❑ 1. Never/rarely
    ❑ 2. Every week
    ❑ 3. Two to three times/week
    ❑ 4. Four to seven times/week
    ❑ 5. More than seven times/week

11. How often do you eat fresh fruit? Choose one.
    ❑ 1. Never/rarely
    ❑ 2. Every week
    ❑ 3. Two to three times/week
    ❑ 4. Four to seven times/week
    ❑ 5. More than seven times/week

12. How often do you eat fresh vegetables? Choose one.
    ❑ 1. Never/rarely
    ❑ 2. Every week
    ❑ 3. Two to three times/week
    ❑ 4. Four to seven times/week
    ❑ 5. More than seven times/week

13. What type of fat do you use on your sandwiches? Choose one.
    ❑ 1. Butter
    ❑ 2. Margarine
    ❑ 3. Reduced-fat margarine
    ❑ 4. None at all

14. What type of fat do you use for cooking? Choose one.
    ❑ 1. Butter
    ❑ 2. Margarine
    ❑ 3. Oil
    ❑ 4. Liquid reduced-fat margarine

15. What type of bread do you mainly eat? Choose one.
    ❑ 1. Soft white bread
    ❑ 2. Dark wheat bread
    ❑ 3. Hard white bread
    ❑ 4. Hard dark bread

16. How often do you eat the following products?
    Pasta:
    ❑ 1. Daily
    ❑ 2. A few times/week
    ❑ 3. Rarely or never
    Root crops (for example, carrots, turnips):
    ❑ 1. Daily
    ❑ 2. A few times/week
    ❑ 3. Rarely or never
    Beans:
    ❑ 1. Daily
    ❑ 2. A few times/week
    ❑ 3. Rarely or never
    Potatoes:
    ❑ 1. Daily
    ❑ 2. A few times/week
    ❑ 3. Rarely or never
    Rice:
    ❑ 1. Daily
    ❑ 2. A few times/week
    ❑ 3. Rarely or never

17. Have you ever smoked?   ❑ Yes   ❑ No
    Have you previously smoked but stopped?
    ❑ Yes   ❑ No
    Started (year): . . . . . . Stopped (year): . . . .
    Number of cigarettes/day: . . . .
    Are you still smoking now?   ❑ Yes   ❑ No
    What year did you start to smoke regularly? . . . . .
    Number of cigarettes/day: . . . .
    Number of packs/year: . . . .
    Type of cigarettes:
    ❑ Mild/light   ❑ Regular cigarettes

Fig 484 (continued)

Has your smoking consumption been unchanged during the whole smoking period?

❑ Yes   ❑ No   If no:

❑ Smoked more earlier   ❑ Smoked less earlier

Have you at any time stopped smoking?

❑ Yes   ❑ No   If yes, when (year)? . . . .

How long until you started smoking again (months)?

. . . .

Have you at any time reduced your amount of smoking?   ❑ Yes   ❑ No

If yes, when (year)? . . . .

For how long did you smoke less (months)? . . . .

Number of cigarettes/day:. . . . .

Have you at any time increased your smoking?

❑ Yes   ❑ No

If yes, when (year)? . . . .

For how long did you smoke more (months)? . . .

Number of cigarettes/day: . . . .

Have you at any time used snuff (smokeless tobacco) instead of smoking cigarettes?   ❑ Yes   ❑ No

If yes, when did you use snuff (year)?

18. Do you use snuff regularly? Choose one.

❑ Yes   ❑ No

19. Do you exercise regularly? Choose one.

❑ Yes   ❑ No

If yes, what type of exercise?

How many times/week?

20a. How tall are you? . . . . (cm)

20b. How much do you weigh? . . . . (kg)

21. What is your Body Mass Index?. . . . .

22a. Do you go to a dentist regularly?

❑ Private dentist

❑ Public Dental Health Service dentist

❑ No

22b. How often do you go to your dentist (note: not dental hygienist)?

❑ Three to four times/year

❑ Twice/year

❑ Once/year

❑ Once every second year

❑ Once every third or fourth year

23. How are you called to your dentist?

❑ I make an appointment myself

❑ The dentist calls me

❑ Both

24. How far from home is the nearest dentist? . . . . .(km)

25. When did you last go to your dentist (year)?

26. Have you changed dentists during the last 5 years?

❑ Yes   ❑ No   If yes, when (year)?

❑ Changed from private dentist to the Public Dental Health Service

❑ Changed from the Public Dental Health Service to private dentist

27. How much does your dental care cost per year? . . . . . (SEK)

28. Do you find this to be:

❑ Inexpensive

❑ Expensive

❑ Too expensive

❑ Reasonable price

29. How do you feel about going to the dentist? Choose all that apply.

❑ Nothing special   ❑ Relaxed

❑ Uncomfortable   ❑ Stressed

❑ Necessary   ❑ Scared

❑ Positive   ❑ Terrified

❑ Other

30. Why do you think some people do not go for dental care? Choose all that apply.

❑ It is expensive.

❑ It is difficult to get an appointment.

❑ They are afraid.

❑ Other reasons. Specify. . . . .

31. How would you define the condition of your teeth, on a scale from 1 to 5?

1 = Very poor   5 = Very good

❑   ❑   ❑   ❑   ❑

1   2   3   4   5

Fig 484 (continued)

32. How well do you look after your teeth, on a scale from 1 to 5?

1 = Very poorly   5 = Very well

❑	❑	❑	❑	❑
1	2	3	4	5

33. How important do you think your teeth are, on a scale from 1 to 5?

1 = Completely unimportant   5 = Very important

❑	❑	❑	❑	❑
1	2	3	4	5

34. How do you rate the dental care you have received so far, on a scale from 1 to 5?

1 = Very poor   5 = Very good

❑	❑	❑	❑	❑
1	2	3	4	5

35. What do you like most about the dental care you have received so far? . . . .
. . . .
. . . .

36. What are you most dissatisfied with about the dental care you have received so far? . . . .
. . . .
. . . .

37. What do you feel is missing most from the dental care you have received so far? . . . .
. . . .
. . . .

38. What do you like most of all when you go to your dentist? . . . .
. . . .
. . . .

39. What do you dislike most of all when you go to your dentist? . . . .
. . . .
. . . .

Questions for those who go to a dental hygienist

40. What do you like most of all when you go to a dental hygienist? . . . .
. . . .
. . . .

41. What do you dislike most of all when you go to a dental hygienist? . . . .
. . . .

42. If you go regularly to a dental hygienist:
Since when (year)? . . . .
How often per year? . . . .
❑ Three to four times   ❑ Twice   ❑ Once

43a. How often do you use any of the following dental hygiene aids?

	Never	One to four times/ week	Daily, only in the morning	Daily, only in the evening	Daily, mornings and evenings	After meals
Toothbrush	❑	❑	❑	❑	❑	❑
Toothpick	❑	❑	❑	❑	❑	❑
Dental tape	❑	❑	❑	❑	❑	❑
Interdental brush	❑	❑	❑	❑	❑	❑
Interspace brush	❑	❑	❑	❑	❑	❑

43b. If you use several of the aids, which one do you start with? . . . .

43c. Do you use toothpaste together with any of the aids apart from the toothbrush?
❑ Yes   ❑ No   If yes, which one? . . . .

44. Are you left- or right-handed?
❑ Right   ❑ Left
Do you use fluoride toothpaste?
❑ Yes   ❑ No

45. There can be several different difficulties with the oral cavity or teeth. Do you have any of the following problems? Please think carefully about the questions and answer.

	No problems	Some problems	Quite troubled	Big problem
Color of the teeth	❑	❑	❑	❑
Shape of the teeth	❑	❑	❑	❑
Inclined teeth	❑	❑	❑	❑
Horizontal or vertical overlap	❑	❑	❑	❑

Fig 484 (continued)

	No prob-lems	Some prob-lems	Quite trou-bled	Big prob-lem
Teeth too narrow	❏	❏	❏	❏
Sore mouth	❏	❏	❏	❏
Wounds or blisters	❏	❏	❏	❏
Changes in taste	❏	❏	❏	❏
Pain around the man-dibular/maxillary joint	❏	❏	❏	❏
Temporomandibular joint problem (sound)	❏	❏	❏	❏
Difficulties in opening the mouth	❏	❏	❏	❏
Bruxism	❏	❏	❏	❏
Bleeding from the gingiva	❏	❏	❏	❏
Bad breath	❏	❏	❏	❏
Trouble with material from dental fillings	❏	❏	❏	❏
Shooting pain	❏	❏	❏	❏

46. Do you have any problems or difficulties other than those mentioned above? ❏ Yes   ❏ No
    Specify. . . . .

47. The following statements and opinions are rather common. Please indicate your opinion.

	I agree com-pletely	I some-what agree	I some-what dis-agree	I dis-agree com-pletely
"To have beautiful and perfect teeth is very important for how other people treat you."	❏	❏	❏	❏
"Minor faults are not important, as long as the teeth work."	❏	❏	❏	❏
"To have teeth missing is something to be ashamed of."	❏	❏	❏	❏
"It doesn't matter what your teeth look like, as long as you can chew the food you like."	❏	❏	❏	❏

48. How often during the past 3 to 4 weeks have you been bothered by any of the following problems?

	Daily	A few times per week	A few times	Never
Stomach trouble	❏	❏	❏	❏
Headache	❏	❏	❏	❏
Palpitation of the heart	❏	❏	❏	❏
Sleeping problems	❏	❏	❏	❏
Breathing difficulties	❏	❏	❏	❏
Shortness of breath	❏	❏	❏	❏
Chest pains or pressure	❏	❏	❏	❏
Dizziness	❏	❏	❏	❏
Depression	❏	❏	❏	❏
Restlessness	❏	❏	❏	❏
Nervousness or worry	❏	❏	❏	❏
Tiredness and weakness	❏	❏	❏	❏
Difficulty relaxing	❏	❏	❏	❏

49. How much are the above-mentioned problems ob-structing your everyday life?

	Very obstruct-ing	Rather obstruct-ing	Obstruct-ing in some ways	Not obstruct-ing at all
Stomach problems	❏	❏	❏	❏
Headache	❏	❏	❏	❏
Palpitation of the heart	❏	❏	❏	❏
Sleeping problems	❏	❏	❏	❏
Breathing difficulties	❏	❏	❏	❏
Shortness of breath	❏	❏	❏	❏
Chest pains or pressure	❏	❏	❏	❏
Dizziness	❏	❏	❏	❏
Depression	❏	❏	❏	❏
Restlessness	❏	❏	❏	❏
Nervousness or worry	❏	❏	❏	❏
Tiredness and weakness	❏	❏	❏	❏
Difficulty relaxing	❏	❏	❏	❏

Fig 484 (continued)

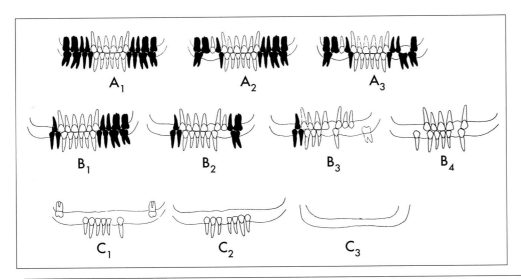

Fig 485 Modified Eichner index for assessing the chewing function of the teeth. Criteria for the 10 different classes (A_1 to A_3, B_1 to B_4, and C_1 to C_3) are illustrated. Black teeth indicate posterior teeth with occlusal contacts. (From Österberg and Landt, 1976. Reprinted with permission.)

Table 27 Diagnosis of clinical caries lesions related to type, localization, size, depth, and shape

Type		Localization	Size/depth	Shape
Primary caries	Crown	Occlusal surfaces Free smooth surfaces (buccal and lingual) Approximal surfaces Supragingival or subgingival	Enamel caries (incipient) Dentin caries (manifest)	"Smooth" surfaces "Rough" surfaces Cavitation Without cavitation Cavitation into the enamel Cavitation into the dentin Cavitation into the pulp
Secondary caries (recurrent)	Crown	Occlusal surfaces Free smooth surfaces (buccal and lingual) Approximal surfaces Supralingual or subgingival	Enamel caries (incipient) Dentin caries (manifest)	"Smooth" surface "Rough" surface Cavitation Without cavitation Cavitation into the enamel Cavitation into the dentin Cavitation into the pulp
Primary caries	Root	Buccal, lingual, mesial, or distal Supragingival or subgingival	Cementum caries (surface or incipient) Dentin caries (manifest)	Soft surface (active lesion) Arrested surface Without cavitation Cavitation
Secondary caries (recurrent)	Root	Buccal, lingual, mesial, or distal Supragingival or subgingival	Cementum caries (surface or incipient) Dentin caries (manifest)	Soft surface (active lesion) Arrested surface Without cavitation Cavitation

1. Clinically, with visual inspection and probing
2. Radiographically, with a standardized bitewing view (specifically, the approximal surfaces of the posterior teeth)
3. Radiographically, with digitized radiographs
4. Visually, with fiber-optic transillumination of the approximal surfaces of the anterior teeth

For many years, caries prevalence has been expressed in terms of decayed, missing, or filled teeth (DMFT) and decayed, missing, or filled tooth surfaces (DMFSs). In retrospect, it is difficult to ascertain whether the teeth were lost because of caries. In most industrialized countries, the only reason permanent teeth are missing in children and young adults is orthodontic treatment. Caries prevalence should therefore mainly be based on the numbers of decayed or filled teeth (DFT) and decayed or filled surfaces (DFSs).

During the last few years, caries prevalence has been separated into coronal caries and root caries, particularly among the elderly. At least in longitudinal epidemiologic studies, the depth of the caries lesion is differentiated as enamel caries and dentinal caries. Cavitated versus noncavitated lesions should also be considered as well as primary and recurrent (secondary) caries (see Table 27).

For estimating caries treatment needs, a new index, analogous to the CPITN, was designed (Table 28). The rationale is that estimation of treatment needs should encompass more than the restorative need. The emphasis should be on prevention. Caries lesions in enamel should be remineralized. Studies by Bille and Thylstrup (1982), Pitts and Rimmer (1992), and Lussi (1991) showed that very few dentinal caries lesions on smooth surfaces or in fissures exhibited cavitation into the dentin. Therefore, most such lesions can be arrested. Nyvad and Fejerskov (1986) showed that, in response to improved oral hygiene, active root caries can also be successfully converted to inactive lesions. These findings further emphasize the importance of *prevention instead of extension*, in contrast to the traditional concept of extension for prevention.

The clinician must address the following questions in making a decision about invasive intervention:

1. How quickly has the caries lesion progressed?
2. Why did the patient's self-care and my professional preventive measures fail to prevent development of the lesion?
3. How can we improve our combined preventive efforts to reverse, or at least arrest, the lesions?
4. How long should we wait to evaluate the results of these preventive efforts?

Table 28 Community Caries Index of Treatment Needs*

Score	Diagnosis	Treatment needs
0	Intact enamel	Prevention
1	Primary active enamel caries	Prevention
2:1	Primary dentin caries; no cavitation into dentin	Prevention
2:2	Recurrent (secondary) caries; no cavitation into dentin	Prevention
3:1	Primary dentin caries; with cavitation into dentin	Prevention + Restoration
3:2	Recurrent (secondary caries); with cavitation into dentin	Prevention + Restoration
4:1	Primary (active) root caries; no cavitation into dentin	Prevention
4:2	Recurrent (active) root caries; no cavitation into dentin	Prevention
5:1	Primary root caries; with cavitation into dentin	Prevention + Restoration
5:2	Recurrent root caries; with cavitation into dentin	Prevention + Restoration

*From Axelsson et al (1988, 1990). Reprinted with permission.

For caries prevalence as well as treatment needs, a special matrix was designed (Fig 486). The kind of restorative materials already present and the current caries treatment needs can be analyzed in detail at a surface level in the system, allowing the pattern of these variables in the dentition to be shown.

Marginal periodontitis: Prevalence and treatment needs

The prevalence of marginal periodontal diseases refers to vertical loss of periodontal attachment in all teeth; horizontal loss of attachment with furcation involvement may also occur in multirooted teeth.

Because marginal periodontal diseases are site related, measurement of loss of alveolar bone on radiographs is not an acceptable substitute for

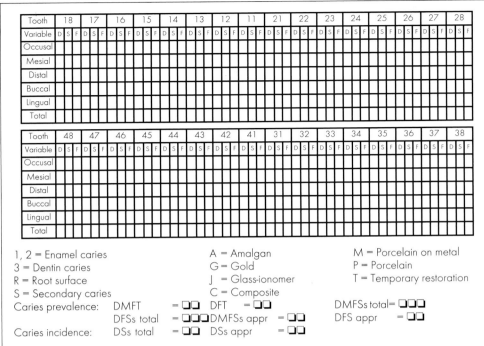

Fig 486 Special matrix for direct recording of data related to dental caries in the computer at individual, tooth, and surface levels (FDI tooth-numbering system). (D) Primary decayed surface; (S) Secondary decayed surface; (F) Filled surface; (DMFT) Decayed, missing, or filled teeth; (DFSs) Decayed or filled surfaces; (DSs) Decayed surfaces; (DFT) Decayed or filled teeth; (DMFSs) Decayed, missing, or filled surfaces; (appr) Approximal. (From Axelsson et al, 1988, 1990. Reprinted with permission.)

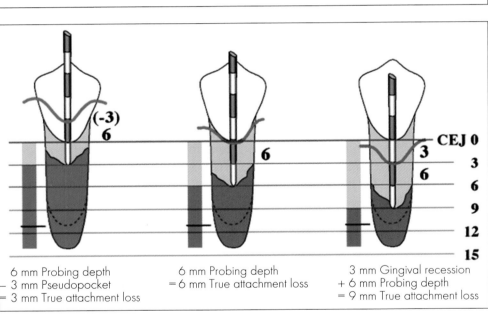

Fig 487 Relationship of probing depth to attachment loss. (Modified from Rateitschak et al, 1989 with permission.)

assessment of clinical loss of attachment on every tooth surface. Buccal and lingual destruction of alveolar bone cannot be diagnosed on radiographs. In fact, the correlation between approximal bone loss diagnosed from radiographs and clinical loss of attachment is very weak (Goodson et al, 1984). Therefore, most epidemiologic surveys and longitudinal clinical studies on the prevalence and incidence of marginal periodontal disease are based on clinical loss of attachment.

Probing depth, unlike probing attachment level, does not disclose long-term failure or success of a maintenance program after periodontal therapy. This is exemplified in Fig 487; despite

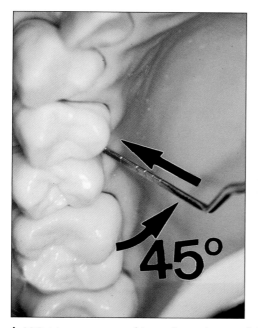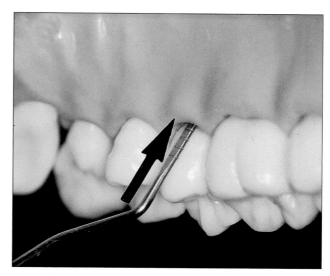

Figs 488 and 489 Measurement of loss of attachment. If the gingival margin is located coronal to the CEJ, the probing depth is exceeding probing attachment loss (PAL).

three different levels of periodontal attachment loss (3, 6, and 9 mm), the probing depth is 6 mm.

Vertical loss of attachment is the distance between the cementoenamel junction (CEJ) and the base of the pocket. It can be measured manually with a millimeter-graded probe. During measurement, use of an intraoral mirror with fiber-optic illumination is recommended as an aid.

When the CEJ is located subgingivally, loss of attachment is measured as shown in Figs 488 and 489. The probe is held with a light pencil grasp so that it can be moved and directed with minimal force. The end of the probe is then placed against the enamel surface coronal to the margin of the gingiva, so that the angle formed by the working end of the probe and the long axis of the tooth crown is approximately 45 degrees. With slightly decreased probe-crown angle, the distance between the free gingival margin and the CEJ is measured. The distance between the CEJ and the base of the pocket is measured as the loss of attachment level. If the free gingival margin is apical to the CEJ or crown margin, attachment loss is measured directly from the visible CEJ to the bottom of the

pocket. The probe is used as parallel as possible to the long axis of the root. It is important that the point of the probe continuously follow the root surface to prevent penetration of the pocket epithelium and connective tissue, resulting in underestimation of attachment loss.

The mesial surface is assessed mesiobuccally and mesiolingually, and the highest value is registered as representing mesial loss of attachment. The distal surface is measured only distobuccally. The buccal and lingual surfaces are measured on the most prominent part of the root surfaces. In multirooted teeth, maxillary and mandibular molars, the highest buccal value is registered. Electronically computerized periodontal probes have also been introduced (eg, the Florida Probe [Florida Probe, Gainesville, FL]).

Radiographs are also useful for diagnosis of loss of periodontal support and for quality control of a maintenance program: A standardized technique with long-cone and attached film holders should be used for periapical and vertical bitewing radiographs. However, studies by Goodson et al (1984) showed that advanced vertical probing attachment

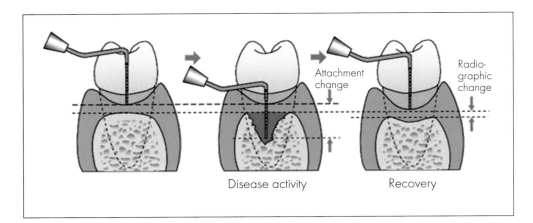

Fig 490 Relationship of probing attachment levels to radiographic change. (Modified from Goodson et al, 1984. Reprinted with permission.)

> **Box 27** Classification of furcation involvement
> - Grade I: Horizontal loss of supporting tissues not exceeding one third of the width of the tooth
> - Grade II: Horizontal loss of supporting tissues exceeding one third of the width of the tooth but not encompassing the total width of the furcation area
> - Grade III: Horizontal through-and-through destruction of the supporting tissues in the furcation

loss (PAL) may occur interproximally in the molar area before loss of alveolar bone is discernible radiographically (Fig 490). This is due to the marked difference in the degree of mineralization between spongiform and compact bone. Changes of less than 50% in mineral loss are difficult to observe with conventional radiographic techniques. This problem may be overcome to some extent with the more sensitive computer-aided subtraction radiography.

Advanced periodontal disease around multirooted teeth will destroy the supporting structures of the furcation area. Treatment is often complicated. Therefore, the precise identification of the presence and extent of periodontal tissue breakdown within the furcation area of each multirooted tooth is of importance for proper diagnosis and treatment planning.

Three grades of furcation involvement may be classified (Box 27). For accurate diagnosis, a slim, curved instrument is necessary. A slim double-ended curette, such as the Goldman-Fox No. 3 (Hu-Friedy, Chicago, IL), is appropriate. Special diagnostic probes for assessing furcation involvement are also available, eg, the flexible disposable tip of the TPS probe (Vivadent, Lichtenstein) and the Furcation Probe (LM Dental, Finland). Figure 491 shows cross sections of maxillary and mandibular teeth and variations in furcation topography on different teeth.

Mesial furcation involvement of the maxillary molars has to be diagnosed in the mesiolingual-apical direction (Fig 492). On the distal aspect, furcation involvement must be assessed in a distobuccal-apical direction (Fig 493). Eccentric radiographs should also be used to detect furcation involvement. The levels of the entrance of the furcation area should be compared with the most coronal margin of the alveolar bone. Thereafter, the diagnosis is verified by probing.

The CPITN was developed jointly by the FDI and the WHO in 1977 (Ainamo et al, 1982). The CPITN is now a recognized index to indicate levels of periodontal conditions in populations for which specific interventions might be considered. The major features of the CPITN method include:

1. Use of the specially designed periodontal probe.
2. Division of the dentition into sextants.
3. Assignment of scores for all teeth or index teeth in each sextant.

Fig 491 Cross sections of maxillary and mandibular teeth and variations in furcation topography on different teeth. (Modified from Rateitschak et al, 1989 with permission.)

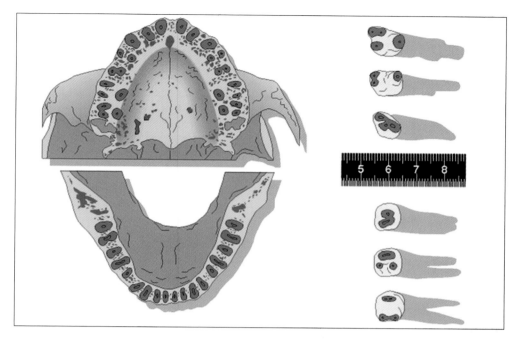

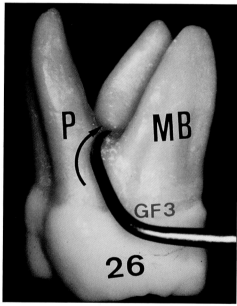

Fig 492 Mesial furcation involvement of the maxillary first molar. This must be diagnosed in a mesiolingual-apical direction. (MB) Mesiobuccal root; (P) Palatal root; (GF3) Goldman-Fox No. 3 curette.

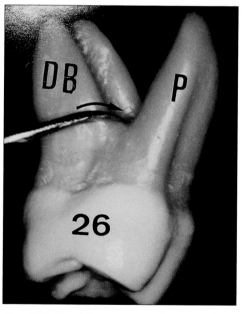

Fig 493 Distal furcation involvement of the maxillary first molar. This must be diagnosed in a distobuccal-apical direction. (DB) Distobuccal root; (P) Palatal root.

Only the highest score for each sextant is recorded. The code numbers recorded indicate the types of periodontal treatment needs (Box 28).

In individual subjects as well as in populations, "true" periodontal treatment needs are considerably overestimated when CPITN is used at the sextant level rather than the surface level. In addition,

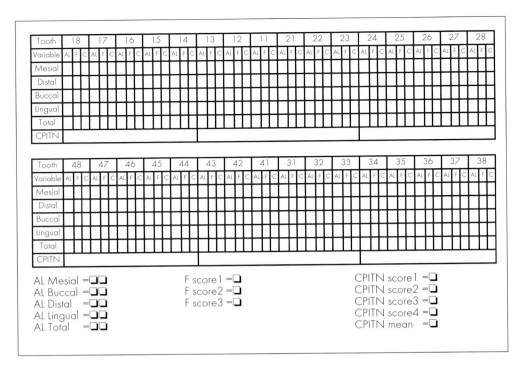

Fig 494 Matrix for computer-aided analysis (FDI tooth-numbering system). (AL) Probing attachment level; (F) Furcation involvement; (C) CPITN. (From Axelsson et al, 1988, 1990. Reprinted with permission.)

Box 28 CPITN scores

- Code 0 (healthy gingiva): There is no need for periodontal treatment.
- Code 1 (gingival bleeding on probing): Oral hygiene education is required.
- Codes 2 (calculus) and 3 (diseased 4- to 5-mm pockets): Scaling and root planing are required in addition to oral hygiene education. Scaling and root planing are also supplemented with elimination of plaque-retentive margins of restorations and crowns.
- Code 4 (> 5-mm-deep pockets): Complex treatment is required in addition to scaling and root planing and oral hygiene education.

Table 29 Apical periodontal prevalence*

Score	Size	Margin
1:1	1–2 mm	Diffuse
1:2	1–2 mm	Well-defined
2:1	3–5 mm	Diffuse
2:2	3–5 mm	Well-defined
3:1	> 5 mm	Diffuse
3:2	> 5 mm	Well-defined

*From Axelsson (1988b). Reprinted with permission.

Apical periodontitis: Prevalence and treatment needs

missing teeth and sites should be excluded when the true treatment need is estimated. At the surface level, the CPITN is also very useful for estimation of true treatment needs in the individual patient.

A special matrix was also designed for the diagnosis of marginal periodontal prevalence and treatment needs. Vertical as well as horizontal loss of attachment was diagnosed at the surface and the individual levels. The CPITN was diagnosed at the surface, sextant, and individual levels (Fig 494).

A new score for the prevalence of apical periodontitis was designed, based on the size and the structure of the margin of the apical periodontal lesion (Table 29). Based on the diagnosis of apical periodontitis, the existence of previous endodontic treatment, and the anticipated complexity of existing treatment needs for endodontic therapy, a new APITN was also designed (Table 30).

Table 30 Apical Periodontal Index of Treatment Needs*

Score	Diagnosis[†]	Treatment needs
1:1	AP in incisors and premolars, no RF	RF
1:2	AP in molars, no RF	RF
2:1	AP + RF in incisors and premolars	RF + retrograde RF?
2:2	AP + RF in molars	RF + retrograde RF?
3:1	AP + RF + POST in incisors and premolars	Retrograde RF
3:2	AP + RF + POST in molars	Retrograde RF or extraction

*From Axelsson (1988b). Reprinted with permission.
[†](AP) Apical periodontitis; (RF) Root filling; (POST) Post-and-core restoration.

Mucosal diseases: Prevalence and treatment needs

A simplified system was used for the scoring of prevalence of red lesions, white lesions, and ulcerative lesions. Their presence was noted on the right and left sides of the palate, the tongue, the floor of the mouth, the vestibule, and the lips. The WHO's more detailed diagnosis can be used in our system as well (WHO, 1987).

The treatment need for lesions of the oral mucosa was based on more detailed analysis of duration, location, size, color, and structure of the lesions, supplemented with information from detailed history taking (see Fig 484).

Malocclusion: Prevalence and treatment needs

In this study, only the prevalence of horizontal and vertical overlap, measured in millimeters, and the size of mouth opening were evaluated, but many other variables can be used. In this study, the treatment needs of patients with malocclusion can so far be based only on measurements of overbite and overjet.

Etiologic factors

Only the PI, based on the percentage of stained plaque surfaces, was determined. The analytical program can, however, also include the PFRI (Axelsson, 1987, 1991) and specific oral microflora associated with the etiology of caries and periodontal diseases as well as oral mucosa lesions.

Modifying factors

Many external (environmental) and internal (endogenous) modifying risk indicators (RIs), risk factors (RFs), and prognostic risk factors (PRFs), related in particular to dental caries and periodontal diseases, were evaluated in this study (see Fig 484). In addition, the role of reduced stimulated salivary secretion rate in dental caries and the role of genetic polymorphism of the proinflammatory cytokine interleukin 1 (IL-1α and IL-1β) in susceptibility to periodontal diseases were evaluated.

RESULTS OBTAINED BY THE ANALYTICAL EPIDEMIOLOGY SYSTEM

The main objective of the analytical epidemiology system was to evaluate the effect of the county's preventive program for adults on tooth loss, dental caries, and periodontal diseases. Therefore, the results from 1988 (baseline) to 1998, based on cross-sectional studies in randomized samples of 35-, 50-, 65-, and 75-year-old subjects and the 10-year longitudinal study on a randomized sample of 50-year-old subjects up to the age of 60 years, are focused on variables related to tooth loss, dental caries, and periodontal disease. However, some other data from the baseline examination, such as the function of the teeth (modified Eichner index), the presence of apical periodontitis, and the presence of oral mucosa lesions, are also presented.

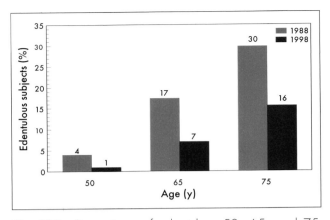

Fig 495a Percentage of edentulous 50-, 65-, and 75-year-old subjects in the county of Värmland, Sweden, in 1988 and 1998. (From Axelsson et al, 2000. Reprinted with permission.)

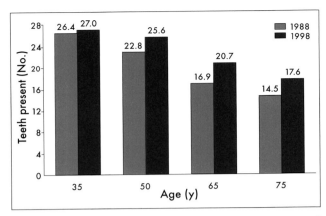

Fig 495b Mean number of teeth (excluding third molars) by age group in the county of Värmland, Sweden, in 1988 and 1998. (From Axelsson et al, 2000. Reprinted with permission).

During the study, enormous amounts of data were collected, and full analysis has not been completed. Most analyses so far have been related to periodontal diseases.

Edentulousness and missing teeth

Percentage of edentulous subjects in 1988 and 1998

The percentage of edentulous 50-, 65-, and 75-year-old adults decreased from 4%, 17%, and 30%, respectively, in 1988 to 1%, 7%, and 16%, respectively, in 1998 (Fig 495a). One of the WHO's global oral health goals for the year 2000 was that the percentage of edentulousness in 35- to 44- and 65-year-old adults be reduced by 50% and 25%, respectively, from the levels in 1969. In the county of Värmland, edentulousness is nonexistent in 35- to 44-year-old adults and is found in only 1% of 50-year-old adults. Edentulousness declined in 65-year-old adults by about 60%, from 17% to 7%, in the 10 years from 1988 to 1998. Estimations based on these data indicate a further decline during the following 10 years, from 1998 to 2008, to about 0%, 2%, and 8% in 50-, 65-, and 75-year-old adults, respectively.

Mean number of teeth in 1988 and 1998

The mean number of teeth (third molars excluded) increased from 26.4, 22.8, 16.9, and 14.5 in 1988 to 27.0, 25.6, 20.7, and 17.6 in 35-, 50-, 65-, and 75-year-old adults, respectively, in 1998 (Fig 495b). Thus, from 1988 to 1998, the mean number of remaining teeth in 35-, 50-, 65-, and 75-year-old adults increased by 0.6, 2.8, 3.8, and 3.1, respectively. For reference, the mean number of teeth in randomized national samples of 50-year-old adults increased by only 1.7 teeth per individual from 1974 to 1985 (Håkansson, 1978, 1991). Excluding third molars, the fully dentate adult has 28 teeth. Estimates based on the data from the two cross-sectional studies from 1988 and 1998 and the data from the 10-year longitudinal study indicate that during the next 10 years, the mean number of remaining teeth in 65- and 75-year-old adults will increase to 25.0 and 20.0, respectively.

Lost teeth from 50 to 60 years of age

On average, the randomized sample of 50-year-old subjects (n = 313) in 1988 lost fewer than 0.7 teeth to the age of 60 years (Fig 495c). Sixty-

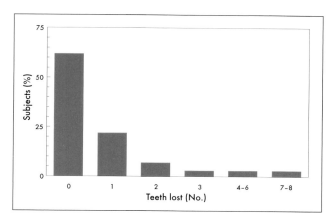

Fig 495c Frequency distribution of tooth loss from the age of 50 years (1988) to the age of 60 years (1998) in a randomized sample of 313 subjects. (From Axelsson et al, 2000. Reprinted with permission.)

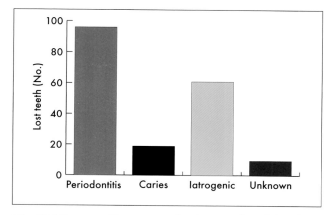

Fig 496 Estimated frequency distribution of tooth loss from the age of 50 years (1988) to the age of 60 years (1998) in a randomized sample of 313 subjects, according to cause: periodontitis, caries, iatrogenic reasons (mainly root fractures in teeth with a post), and unknown reasons. (Modified from Axelsson et al, 2000 with permission.)

two percent of the subjects did not lose one single tooth, and 22% lost only one tooth. Most of the teeth were lost because of periodontal disease (fewer than 100). The next most common cause of tooth loss was iatrogenic reasons, mainly because of root fractures in teeth with posts, which should be regarded as an indirect effect of advanced caries at a younger age (Fig 496).

For reference, in a national, randomized sample of individuals of the same age, the mean number of lost teeth from 1974 to 1985 was 2.5 to 3.0 (Håkansson, 1991). In a randomized sample of 46- to 55-year-old adults from the county of Stockholm, Sweden, on average 4.0 teeth were lost from 1970 to 1990 (Jansson et al, 2002).

Frequency distribution of teeth at baseline

Figures 497 to 500 show the frequency distribution of remaining teeth in the randomized sample of 35-, 50-, 65-, and 75-year-old subjects at the baseline examination in 1988. Forty-five percent of the 35-year-old subjects exhibited the maximum number of remaining teeth.

Pattern of remaining teeth at baseline

Figures 501 to 504 show the pattern of remaining teeth in the randomized sample of 35-, 50-, 65-, and 75-year-old subjects at the baseline examination in 1988. Apart for a small percentage of lost premolars, 100% of the teeth remained in the 35-year-old subjects. The only reason for lost premolars was orthodontic treatment. However, in the 50-year-old subjects, it was obvious that the molars were the key-risk teeth. In 65- and 75-year-old subjects, after the molars, the maxillary premolars exhibited the lowest percentage of remaining teeth, while the mandibular canines were the most resistant.

Function of the teeth according to the modified Eichner index

The chewing function of the teeth was examined according to the modified Eichner index (see Fig 485) in the randomized samples of 35-, 50-, 65-, and 75-year-old subjects at the baseline examination in 1988 (Fig 505). Almost 75% of 35-year-old subjects exhibited no missing occlusal contacts in the molar and premolar regions (A_1). The edentulous individuals (C_3) were found mainly among the 65- and 75-year-old (20% to 30%) and a few 50-year-old subjects (fewer than 5%).

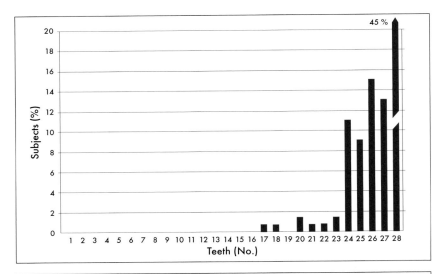

Fig 497 Frequency distribution of remaining teeth in a randomized sample of 35-year-old subjects in 1988 (FDI tooth-numbering system). (From Axelsson et al, 1988, 1990. Reprinted with permission.)

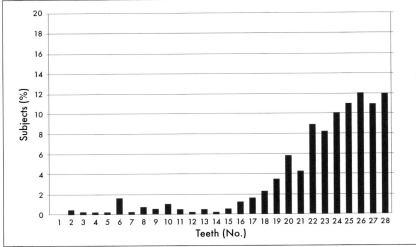

Fig 498 Frequency distribution of remaining teeth in a randomized sample of 50-year-old subjects in 1988 (FDI tooth-numbering system). (From Axelsson et al, 1988, 1990. Reprinted with permission.)

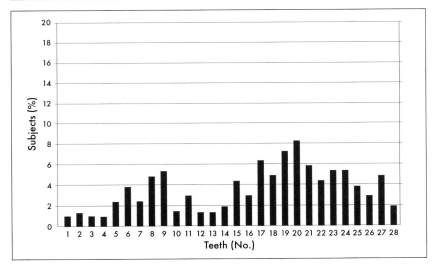

Fig 499 Frequency distribution of remaining teeth in a randomized sample of 65-year-old subjects in 1988 (FDI tooth-numbering system). (From Axelsson et al, 1988, 1990. Reprinted with permission.)

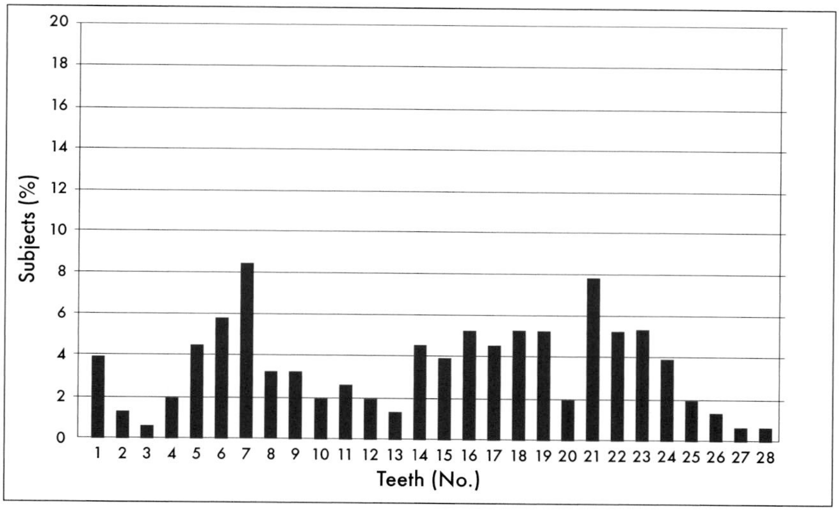

Fig 500 Frequency distribution of remaining teeth in a randomized sample of 75-year-old subjects in 1988 (FDI tooth-numbering system). (From Axelsson et al, 1988, 1990. Reprinted with permission.)

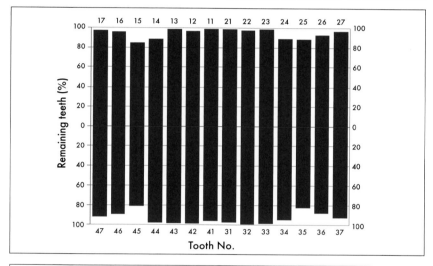

Fig 501 Pattern of remaining teeth (third molars excluded) in 35-year-old subjects in 1988 (FDI tooth-numbering system). (From Axelsson et al, 1988, 1990. Reprinted with permission.)

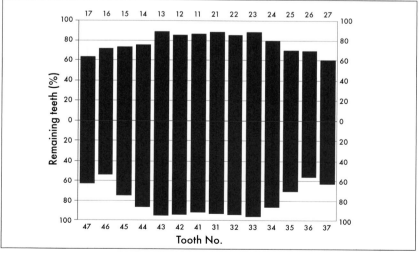

Fig 502 Pattern of remaining teeth (third molars excluded) in 50-year-old subjects in 1988 (FDI tooth-numbering system). (From Axelsson et al, 1988, 1990. Reprinted with permission.)

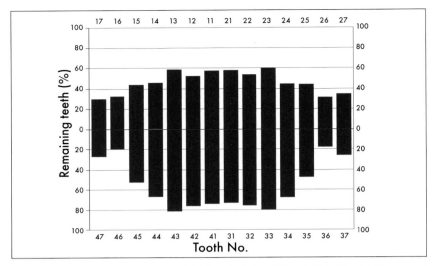

Fig 503 Pattern of remaining teeth (third molars excluded) in 65-year-old subjects in 1988 (FDI tooth-numbering system). (From Axelsson et al, 1988, 1990. Reprinted with permission.)

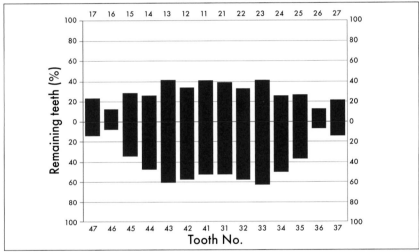

Fig 504 Pattern of remaining teeth (third molars excluded) in 75-year-old subjects in 1988 (FDI tooth-numbering system). (From Axelsson et al, 1988, 1990. Reprinted with permission.)

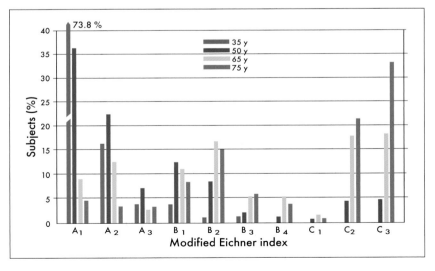

Fig 505 Chewing function in randomized samples of 35-, 50-, 65-, and 75-year-old subjects in 1988 according to the 10 classes of the modified Eichner index, illustrated in Fig 485. (From Axelsson et al, 1988, 1990. Reprinted with permission.)

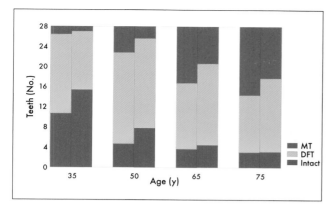

Fig 506a Frequency distribution of the mean number of intact teeth, decayed and filled teeth (DFT), and missing teeth (MT) in randomized samples of 35-, 50-, 65-, and 75-year-old subjects in 1988 compared with 1998. (Modified from Axelsson et al, 2000 with permission.)

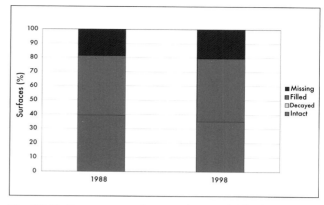

Fig 506b Percentage of intact, decayed, filled, and missing tooth surfaces in a randomized sample of subjects at the age of 50 years (1988) and at the age of 60 years (1998). (Modified from Axelsson et al, 2000 with permission.)

These data will be compared to the new data from 1998 in a new randomized sample of 35, 50-, 65, and 75-year-old subjects. The dramatic reduction of edentulousness and increased number of teeth from 1988 to 1998 indicate that particularly scores C_1 to C_3 and B_2 to B_4 should have been significantly reduced from 1988 to 1998 in 65- and 75-year-old subjects.

Dental caries: Prevalence, incidence, and treatment needs

Prevalence

From 1988 to 1998, particularly in the 35- and 50-year-old subjects, the mean number of intact teeth increased significantly (Fig 506a). The number of missing teeth decreased most obviously in the 65- and 75-year-old subjects.

Incidence

Only marginal changes in the frequency distribution of intact surfaces and DMFSs in the randomized sample of 50-year-old subjects occurred during the 10-year period from 1988 to 1998 (Fig 506b). In 1988 as well as 1998, carious surfaces were almost nonexistent. That is be-

cause more than 95% of this age group received regular maintenance care. Furthermore, the absolute majority of DMFSs were carious and restored before the subjects were 16 years old, when Sweden exhibited among the highest caries prevalence in the world among children (see Fig 440 in chapter 7).

From 1988 to 1998, the mean number of intact surfaces was reduced by 4.3 surfaces, while the mean number of filled (restored) and missing surfaces was increased by 2.5 and 1.8 surfaces, respectively (Fig 507). However, that does not mean that caries incidence was 4.3 carious surfaces (ie, the reduced number of intact surfaces) per individual per 10 years, because many tooth surfaces may have been restored or lost as a result of cuspal or root fractures or periodontal disease.

For comparison, it was recently shown that randomized samples of 55-, 65-, and 75-year-old individuals in the city of Gothenburg, Sweden, developed 6.0, 3.8, and 1.8 new coronal DFS and 5.3, 8.1, and 14.3 new root surface DFS, respectively, per subject per 10 years (Fure, 2003).

Treatment needs

Figure 508 shows the mean number of dentin, recurrent (secondary), and root caries lesions in the randomized sample of 35-, 50-, 65-, and 75-year-

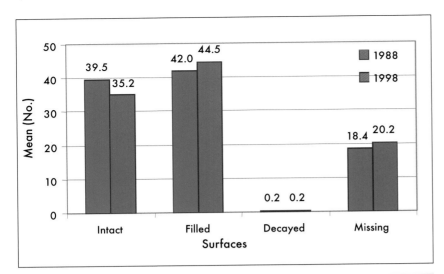

Fig 507 Mean numbers of intact, decayed, filled, and missing tooth surfaces in a randomized sample of subjects at the age of 50 years (1988) and at the age of 60 years (1998). (Modified from Axelsson et al, 2000 with permission.)

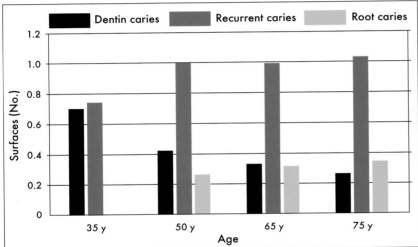

Fig 508 Mean numbers of dentin, recurrent (secondary), and root caries lesions in randomized samples of 35-, 50-, 65-, and 75-year-old subjects in 1988. (From Axelsson et al, 1988, 1990. Reprinted with permission.)

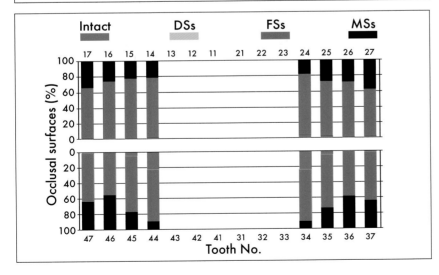

Fig 509 Caries prevalence in 50-year-old subjects: Frequency distribution of intact, decayed (DSs), filled (FSs), and missing surfaces (MSs) occlusally (FDI tooth-numbering system). (From Axelsson et al, 1988, 1990. Reprinted with permission.)

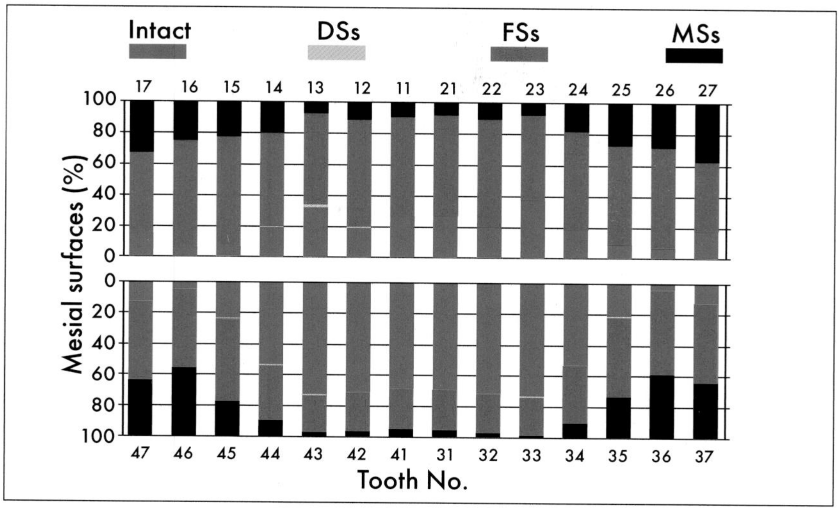

Fig 510 Caries prevalence in 50-year-old subjects: Frequency distribution of intact, decayed (DSs), filled (FSs), and missing surfaces (MSs) mesially (FDI tooth-numbering system). (From Axelsson et al, 1988, 1990. Reprinted with permission.)

Fig 511 Caries prevalence in 50-year-old subjects: Frequency distribution of intact, decayed (DSs), filled (FSs), and missing surfaces (MSs) distally (FDI tooth-numbering system). (From Axelsson et al, 1988, 1990. Reprinted with permission.)

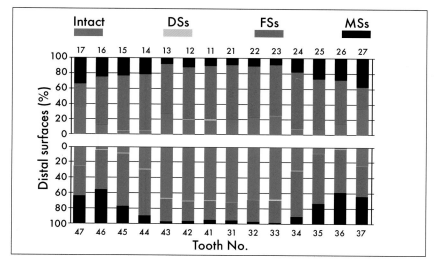

old subjects at the baseline examination in 1988. Because the majority of these lesions were non-cavitated, the existing restorative treatment needs in 1988 were already very limited.

Pattern of caries prevalence at baseline

The patterns of intact, decayed, filled, and missing surfaces in the randomized sample of 50-year-old subjects at the baseline examination are shown in Figs 509 to 513. The highest percentage of intact surfaces were found buccally and lingually, particularly in the mandibular incisors, while the key-risk surfaces were the occlusal surfaces of the molars and the approximal surfaces of molars and premolars.

Marginal periodontitis: Prevalence, incidence, and treatment needs

Prevalence in 1988 and 1998

The mean PAL, measured mesiobuccally, mesiolingually, buccally, distobuccally, and lingually on all teeth, decreased from 0.9, 1.8, 2.4, and 2.9 mm in 1988 to 0.8, 1.5, 1.8, and 2.2 mm in 1998 in randomized samples of 35-, 50-, 65-, and 75-year-old subjects, respectively (Fig 514a). That represents about a 25% improvement of the periodontal status in the 65- and 75-year-old subjects over 10 years. In addition, the number of remaining teeth increased 20% to 25% during the same period. For the following 10-year period, estimates

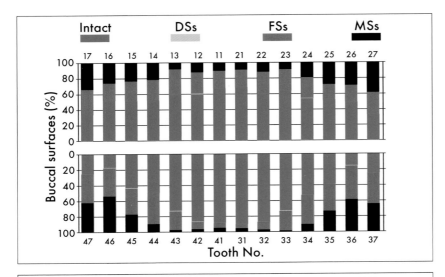

Fig 512 Caries prevalence in 50-year-old subjects: Frequency distribution of intact, decayed (DSs), filled (FSs), and missing surfaces (MSs) buccally (FDI tooth-numbering system). (From Axelsson et al, 1988, 1990. Reprinted with permission.)

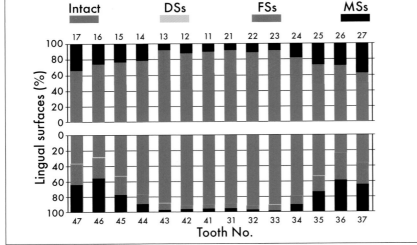

Fig 513 Caries prevalence in 50-year-old subjects: Frequency distribution of intact, decayed (DSs), filled (FSs), and missing surfaces (MSs) lingually (FDI tooth-numbering system). (From Axelsson et al, 1988, 1990. Reprinted with permission.)

based on the data from the longitudinal study indicate continued reductions in average PAL to less than 1.0 mm, 1.6 to 1.7 mm, and less than 2.0 mm for 50-, 65-, and 75-year-old subjects, respectively.

Figure 514b shows the frequency distribution of mean mesial PAL per individual in randomized samples of 35-, 50-, 65-, and 75-year-old subjects in 1988 and 1998. To exclude iatrogenic PAL, particularly on the buccal surfaces but also on lingual surfaces, the frequency distribution of PAL on the mesial surfaces is presented at individual as well as site levels. Because the prevalence of PAL represents the sum of PAL longitudinally, only 3% of the 35-year-old subjects

exhibited a mean PAL of more than 2.0 mm in 1998, compared to about 60% of the 75-year-old subjects. However, the proportion of individuals with more than 2.0 mm of mean PAL was reduced 25% to 50% in the different age groups from 1988 to 1998. In 1998, only about 10% of the 65-year-old subjects had more than 3.0 mm of mean PAL.

For comparison, among randomized samples of more than 7,000 adults in the United States, about 40% exhibited more than 3.0 mm of attachment loss and about 15% exhibited more than 5 mm of attachment loss (Brown et al, 1996).

In all age groups in the Swedish study, there was a considerable reduction of mesial sites

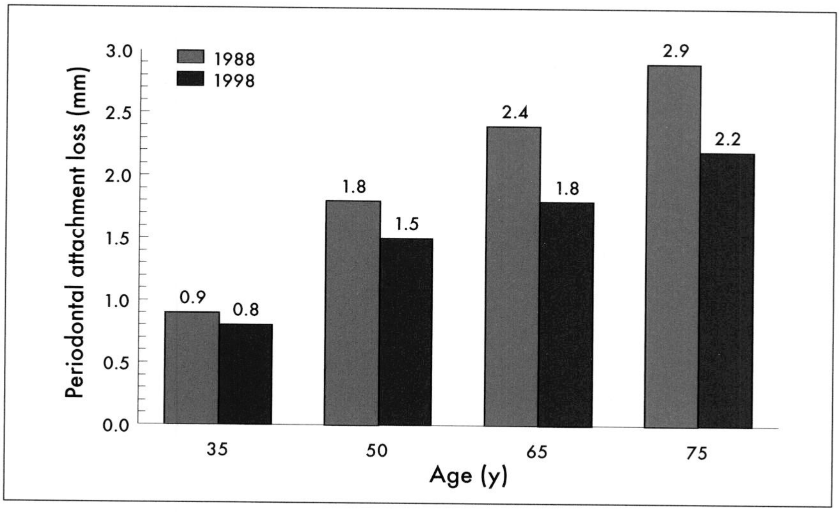

Fig 514a Mean loss of probing attachment by age group in the county of Värmland, Sweden, in 1988 and 1998. (From Axelsson et al, 2000. Reprinted with permission).

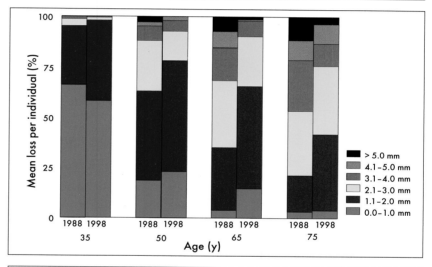

Fig 514b Frequency distribution of different mean levels of PAL at mesial sites per individual, by age group, in 1988 and 1998. (From Axelsson et al, 2000. Reprinted with permission.)

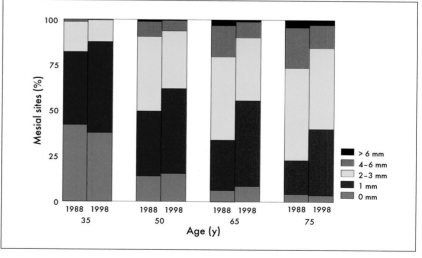

Fig 515 Frequency distribution of PAL at mesial sites, by age group, in 1988 and 1998. (From Axelsson et al, 2000. Reprinted with permission.)

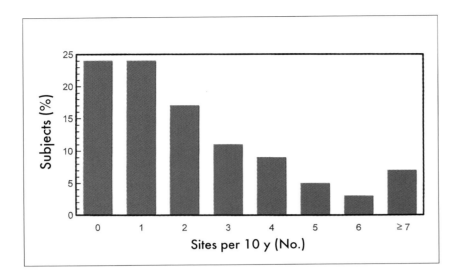

Fig 516 Frequency distribution of individuals who lost 2.0 mm or more of periodontal attachment mesially from 1988 to 1998. (From Axelsson et al, 2000. Reprinted with permission.)

with 2.0 mm or more of PAL from 1988 to 1998 (Fig 515). In 65-year-old subjects, only about 10% of the mesial sites had 4.0 mm or more of PAL in 1998 compared to more than 20% in 1988.

Incidence from 1988 to 1998

The mean PAL in a randomized sample of 50-year-old subjects from 1988 to 1998 was very limited (about 0.2 mm), because more than 95% of the individuals had received regular maintenance care during the same period.

For comparison, data from the studies, "Natural History of Periodontal Disease," showed that the mean annual PAL was 0.1 mm in Norwegian adult academics receiving regular dental care and 0.25 mm in Sri Lankan tea-workers receiving no dental care (Löe et al, 1978, 1986). That means PAL of 1.0 and 2.5 mm, respectively, per 10 years. Two 20-year longitudinal Swedish studies in randomized samples of adults in the counties of Jönköping (Hugoson and Laurell, 2000) and Stockholm (Jansson et al, 2002) showed an average annual alveolar bone loss of 0.1 mm, which means 1.0 mm per 10 years.

Figure 516 shows the frequency distribution of mesial sites with 2.0 mm or more of attachment loss per individual per 10 years. Of the individuals in the study, 24% did not exhibit one single site with 2.0 mm or more of attachment loss, and an-

other 24% exhibited only one such site. About 15% of the individuals had five or more sites with 2.0 mm or more of attachment loss.

Pattern of attachment loss

In analytical epidemiology, it is important to identify not only individuals at risk but also teeth and surfaces at risk. The pattern of tooth loss and attachment loss in different age groups should therefore be analyzed cross-sectionally as well as longitudinally.

Most buccal PAL in subjects up to the age of 50 years is attributable to toothbrush abrasion, whereas almost all approximal PAL is unquestionably caused by the gingival microflora. From a preventive aspect, it is therefore of interest to compare the pattern of approximal PAL with the pattern of tooth loss in randomly selected age groups (see Figs 501 to 504). Figures 517 and 518 show the pattern of PAL on the mesial and distal surfaces in 35- and 75-year-old subjects, respectively. In 35-year-old subjects, there is already twice as much PAL on the distal surfaces of the maxillary first molars as there is on the mesial surfaces. This is of consequence for the risk of furcation involvement on the "inaccessible" distal surfaces and could have been prevented if these surfaces had been the target of

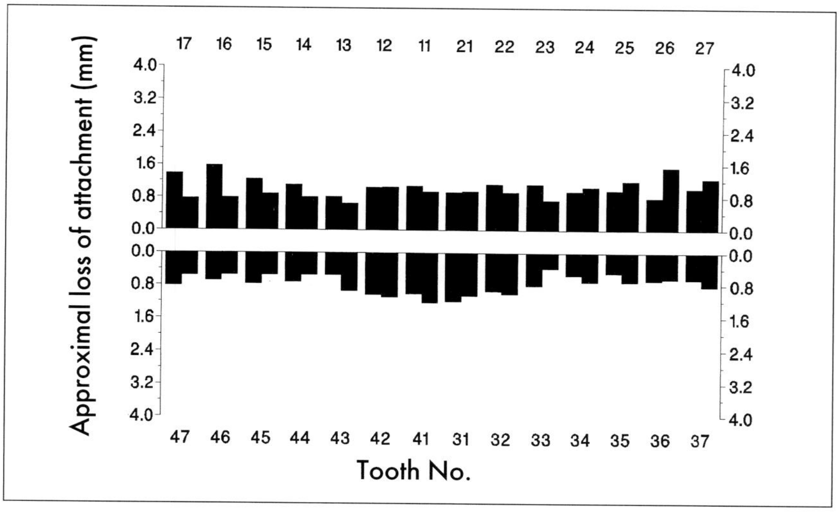

Fig 517 Prevalence and pattern of approximal loss of attachment in 35-year-old subjects (FDI tooth-numbering system). (From Axelsson et al, 1988, 1990. Reprinted with permission.)

Fig 518 Prevalence and pattern of approximal loss of attachment in 75-year-old subjects (FDI tooth-numbering system). (From Axelsson et al, 1988, 1990. Reprinted with permission.)

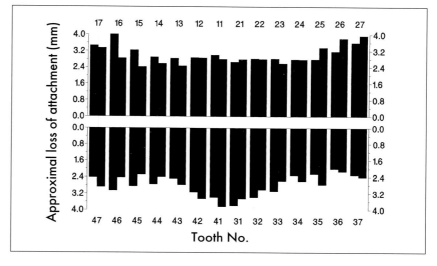

gingival plaque control from early adulthood. In 75-year-old subjects, PAL was still greater on the distal surfaces of the remaining maxillary first molars, despite 90% loss of these teeth (see Fig 504). The increased PAL on the mandibular central and lateral incisors in 35-year-old subjects may be partly explained by local factors, such as plaque retention associated with heavy supragingival calculus formation, abnormal frenum, smoking, thin alveolar bone, and crowded and rotated teeth.

The patterns of PAL on the buccal and lingual surfaces in 35-year-old subjects are shown in Figs 519 and 520. The greater PAL on the buccal sur-

faces is attributable to toothbrush abrasion and is in accordance with the results of early studies on the pattern of abrasive defects (Sangnes, 1975).

Treatment of teeth with furcation involvement is much more complicated than is treatment of diseased pockets in single-rooted teeth. Therefore, diagnosis and analytical epidemiology of the pattern of furcation involvement are important. Figures 521 and 522 show the frequency distribution of uninvolved and furcation-involved surfaces in randomly selected 50-and 75-year-old subjects. In 50-year-old subjects, furcation involvement grades II and III are most frequent on the distal surfaces of the maxillary first molars, while grade I is

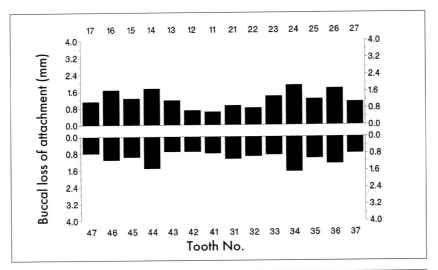

Fig 519 Prevalence and pattern of buccal loss of attachment in 35-year-old subjects (FDI tooth-numbering system). (From Axelsson et al, 1988, 1990. Reprinted with permission.)

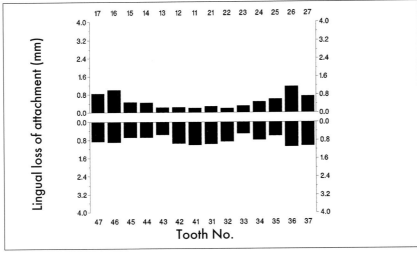

Fig 520 Prevalence and pattern of lingual loss of attachment in 35-year-old subjects (FDI tooth-numbering system). (From Axelsson et al, 1988, 1990. Reprinted with permission.)

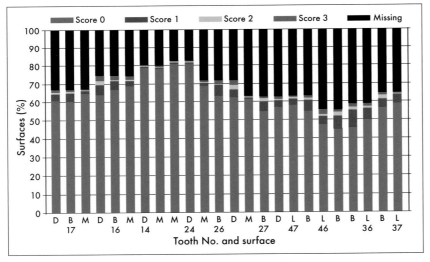

Fig 521 Pattern and frequency distribution of furcation involvement in 50-year-old subjects (FDI tooth-numbering system). Score 0 represents uninvolved surfaces; scores 1 to 3 represent varying grades of furcation involvement. (D) Distal; (B) Buccal; (M) Mesial; (L) Lingual. (From Axelsson et al, 1988, 1990. Reprinted with permission.)

Fig 522 Pattern and frequency distribution of furcation involvement in 75-year-old subjects (FDI tooth-numbering system). Score 0 represents uninvolved surfaces; scores 1 to 3 represent varying grades of furcation involvement. (D) Distal; (B) Buccal; (M) Mesial; (L) Lingual. (From Axelsson et al, 1988, 1990. Reprinted with permission.)

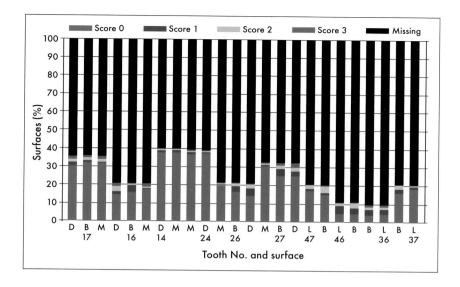

most common on the buccal surfaces of the mandibular first molars (Axelsson et al, unpublished data, 1990).

In a previously mentioned study, Löe et al (1978) compared the prevalence of PAL in a randomly selected group of Norwegian students and academics with that in Sri Lankan tea workers. The maxillary molars of the Sri Lankan group were the key-risk teeth. Most of the maxillary molars exhibited grade III (through-and-through) furcation involvement.

Treatment needs

The WHO recommends the use of the CPITN at individual and sextant levels for estimating periodontal treatment needs in population surveys (Ainamo et al, 1982; Cutress et al, 1987). To date, 200 to 300 surveys have been undertaken in more than 100 different countries. At both the individual and sextant levels, the most frequent findings from almost 100 CPITN surveys of 35- to 40-year-old subjects, in more than 50 countries, were calculus and shallow pocketing (scores 2 and 3). With few exceptions, both the percentage of subjects and the mean number of sextants per subject with score 4 were low (Miyazaki et al, 1991b).

In this context, it is important to recognize that using the highest CPITN score at the individual, sextant, and even tooth levels, according to the hierarchical principle, grossly overestimates periodontal treatment need. Therefore, in the present study, CPITN scores were evaluated and compared at the sextant level as well as the surface level. Analogous with DMFT and DMFS indices in caries epidemiology, sextants and sites were stratified as CPITN scores 1, 2, 3, 4, or missing. Edentulous sextants and lost teeth will have had no periodontal treatment needs for several years. By using the CPITN at the site level and excluding missing sites, true periodontal treatment needs can be evaluated in individuals as well as in groups and populations. Figure 523 shows the frequency distribution of CPITN scores 0 to 4 and missing teeth at the sextant level in 1988 and 1998 in randomized samples of 35-, 50-, 65-, and 75-year-old subjects. For comparison, Fig 524 shows the frequency distribution of more than 110,000 sites, diagnosed according to CPITN scores 0 to 4 and missing sites, representing true periodontal treatment needs.

The existing periodontal treatment need, based on CPITN, was reduced from 1988 to 1998 at both the sextant and the site levels. However, at the sextant level, treatment needs are overestimated, even

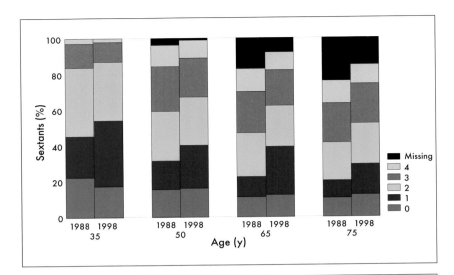

Fig 523 Frequency distribution, at the sextant level, of CPITN scores 0 to 4 and missing teeth, by age group, in 1988 and 1998. (From Axelsson et al, 2000. Reprinted with permission.)

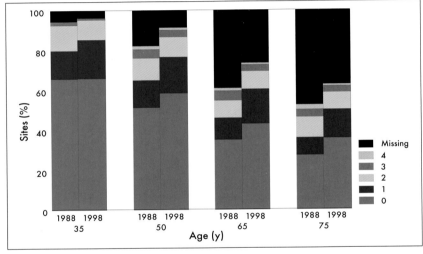

Fig 524 Frequency distribution, at the site level, of CPITN scores 0 to 4 and missing teeth, by age group, in 1988 and 1998. (From Axelsson et al, 2000. Reprinted with permission.)

when edentulous sextants are excluded. An unexpected finding was that only 2% to 3% of all sites in 75-year-old subjects had a CPITN score of 4, and about 60% of the remaining sites were healthy (see Fig 524). In 1998, about 95%, 70%, 65%, and 60% of the 35-, 50-, 65-, and 75-year-old subjects, respectively, had no single site deeper than 5.0 mm (CPITN score 4), reaffirming the high quality of the county's needs-related maintenance program for both younger adults and the elderly (Fig 525).

The randomized sample of 50-year-old subjects in 1988 was followed longitudinally and re-examined at the age of 60 years. The percentage of individuals without any site with a CPITN score

of 4 had increased during the 10-year period, while the number of individuals with 7 to 12 sites with a CPITN score of 4 was reduced by about 50% (Fig 526).

Pattern of treatment needs

To plan an efficient preventive strategy, it is essential to have data on the pattern of disease in the dentition at the surface level. Figures 527 to 529 show the pattern of CPITN scores and missing teeth on different tooth surfaces in the randomized sample of 50-year-old subjects in 1988. Score 4 is almost negligible, and score 3 is limited

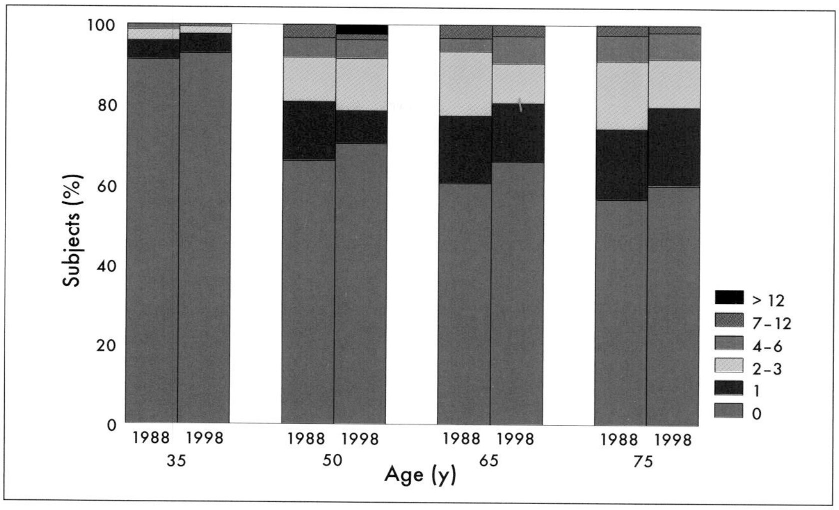

Fig 525 Frequency distribution of individuals with 0, 1, 2 to 3, 4 to 6, 7 to 12, and more than 12 sites with CPITN score 4, by age group, in 1988 and 1998. (From Axelsson et al, 2000. Reprinted with permission.)

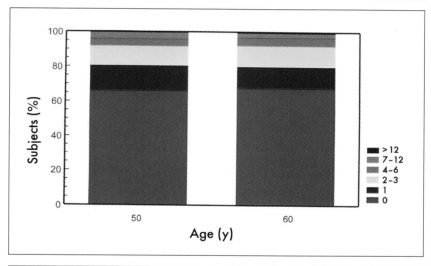

Fig 526 Frequency distribution of individuals with 0, 1, 2 to 3, 4 to 6, 7 to 12, and more than 12 sites with CPITN score 4 at 50 years of age (1988) and 60 years of age (1998). (From Axelsson et al, 2000. Reprinted with permission.)

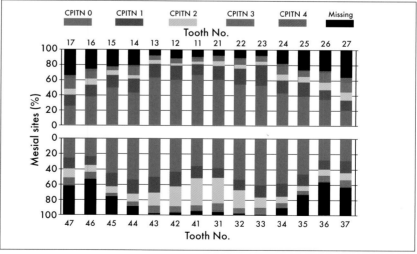

Fig 527 Frequency distribution of missing teeth and mesial surfaces with CPITN scores 0 to 4 in 50-year-old subjects (FDI tooth-numbering system). (From Axelsson, 1998. Reprinted with permission.)

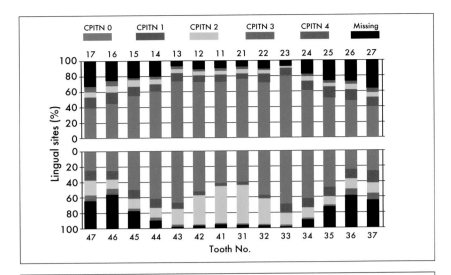

Fig 528 Frequency distribution of missing teeth and lingual surfaces with CPITN scores 0 to 4 in 50-year-old subjects (FDI tooth-numbering system). (From Axelsson, 1998. Reprinted with permission.)

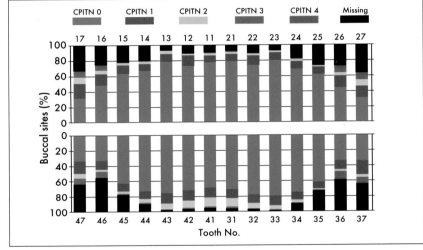

Fig 529 Frequency distribution of missing teeth and buccal surfaces with CPITN scores 0 to 4 in 50-year-old subjects (FDI tooth-numbering system). (From Axelsson, 1998. Reprinted with permission.)

mainly to the mesial and distal surfaces of the maxillary molars, the key-risk surfaces. The mesial and lingual surfaces of the mandibular incisors have the highest percentages of score 2, indicating supragingival calculus that requires removal.

Apical periodontitis: Prevalence and treatment needs

Prevalence

Figure 530 shows the frequency distribution of the mean number of apical periodontal lesions in randomized samples of 50-, 65-, and 75-year-old subjects at the baseline examination in 1988. The lesions are stratified according to the six different scores suggested by Axelsson (1988b), which are based on the size and structure of the margin of the lesions (see Table 29). In all age groups, 1- to 2-mm lesions with a diffuse margin were most frequent, followed by 3- to 5-mm lesions with a diffuse margin. However, the total number of apical lesions was low. The mean values per individual was 0.5, 0.4, and 0.3, respectively, in 50-, 65-, and 75-year-old subjects. The mean value per tooth, however, should have been reduced significantly from 1988 to 1998 as an effect of continuously reduced caries prevalence in all age groups.

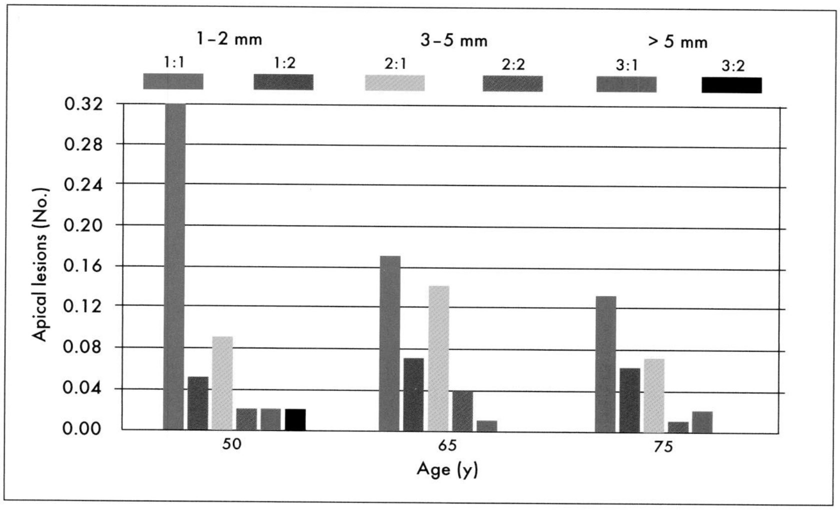

Fig 530 Mean number of apical periodontitis lesions per subject in randomized samples of 35-, 50-, 65-, and 75-year-old subjects in 1988. The lesions are stratified according to the index proposed by Axelsson (1988b) (see Table 29). (From Axelsson et al, 1988, 1990. Reprinted with permission.)

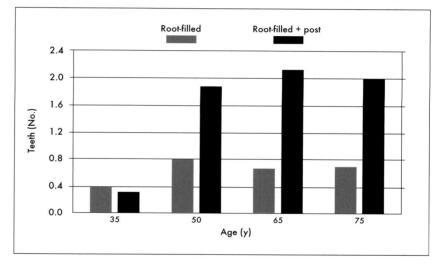

Fig 531 Mean number of endodontically treated teeth, with and without posts, per individual in randomized samples of 35-, 50-, 65-, and 75-year-old subjects in 1988. (From Axelsson et al, 1988, 1990. Reprinted with permission.)

Prevalence of endodontic treatment

The mean number of endodontically treated teeth, with and without posts, per individual in the randomized sample of 35-, 50-, 65-, and 75-year-old subjects at the baseline examination in 1988 is shown in Fig 531. In 35-year-old subjects, only 0.7 tooth per individual was endodontically treated (0.3 with a post). In 50-, 65-, and 75-year-old subjects, about 2.5 teeth per individual were endodontically treated (about 2.0 with post), as a consequence of the extremely high caries prevalence in Sweden when these subjects were children and young adults.

The highest mean value of endodontic treatment per tooth was found in 75-year-old subjects, followed by the 65-year-old subjects, because they have fewer remaining teeth than the 50-year-old subjects. Thus, in the new cross-sectional study in 1998, in randomized samples of 35-, 50-, 65-, and 75-year-old subjects, the mean number of endodontic treatment per tooth was reduced significantly.

Treatment needs

The APITN, developed by Axelsson (1988b), is based on the estimated complexity of treatment

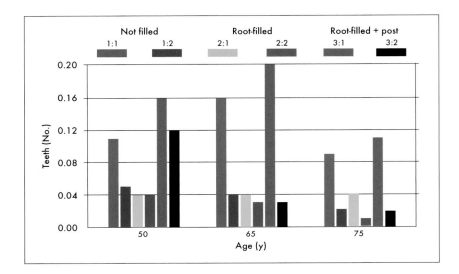

Fig 532 Mean number of teeth per individual with apical periodontal treatment needs in randomized samples of 50-, 65-, and 75-year-old subjects in 1988. The treatment needs are stratified according to the APITN proposed by Axelsson (1988b; see Table 30). (From Axelsson et al, 1988, 1990. Reprinted with permission.)

of the apical lesion (see Table 30). This is related to single or multirooted teeth, untreated roots, and endodontically treated roots with or without a post. Figure 532 shows the mean values of the six different scores per individual in the randomized samples of 50-, 65-, and 75-year-old subjects at the baseline examination in 1988. In all age groups, endodontically treated incisors and premolars with a post and an apical lesion (score 3.1) exhibited the highest mean value. Such lesions are mostly treated with surgery and retrograde root fillings.

Data on the prevalence of and treatment needs for apical periodontitis in randomized samples of 35-, 50-, 65-, and 75-year-old subjects in 1998 later will be compared to data from 1988 at the individual level as well as the pattern in the dentition.

Oral mucosal lesions: Prevalence

The mean values of oral mucosal lesions in men were compared with those of women in the randomized sample of 35-, 50-, 65-, and 75-year-old subjects in 1988 (Fig 533). There was 0.4 lesion per individual among 35- and 50-year-old men but fewer than 0.2 lesion among women. Figure 534 shows the mean number of red, white, and ulcerous lesions per individual in the four age groups.

Particularly in 35-year-old subjects, but also in 50- and 65-year-old subjects, white lesions dominate, while red lesions dominate in 75-year-old subjects.

The mean values of oral mucosal lesions in the maxillary and mandibular arches, vestibule, tongue, floor of the mandible, and lips per individual in the four age groups are shown in Fig 535. The highest mean values of lesions are found in the maxillary arches of all age groups, followed by inside the maxillary lips in 35- and 50-year-old subjects. White lesions located in the maxillary vestibular mucosa and inside the maxillary lips were mainly found in 35- and 50-year-old men as a local effect of snuffing (smokeless tobacco). The higher frequency of red lesions in 75-year-old subjects mainly was the result of denture stomatitis in the maxilla.

Modifying factors

Risk indicators for edentulousness

Based on the data from the cross-sectional study in randomized samples of 35-, 50-, 65-, and 75-year-old subjects in 1988, some important RIs for edentulousness were identified. In particular, low educational level, smoking, irregular dental care habits, and living in rural areas were shown to be important RIs for edentulousness.

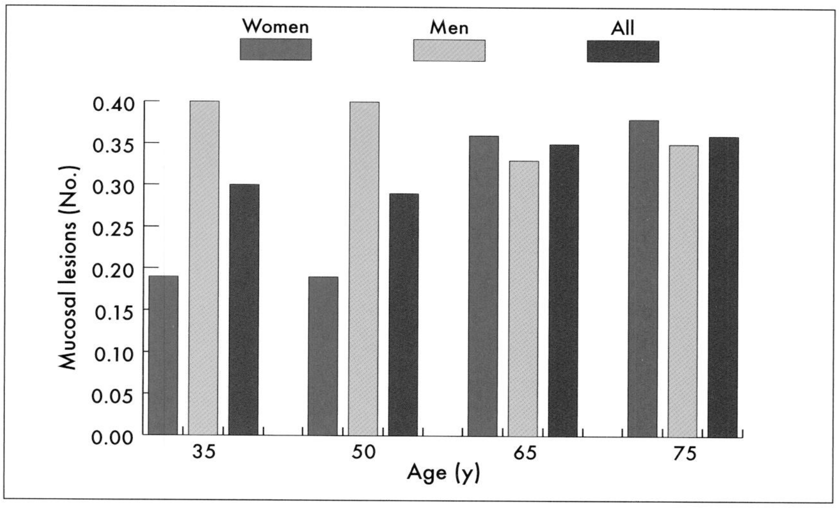

Fig 533 Comparison of mean numbers of oral mucosal lesions in men and women in randomized samples of 35-, 50-, 65-, and 75-year-old subjects. (From Axelsson et al, 1988, 1990. Reprinted with permission.)

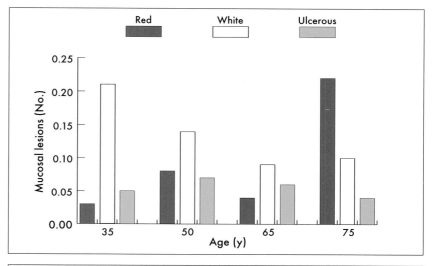

Fig 534 Mean number of red, white, and ulcerous oral mucosal lesions per subject in randomized samples of 35-, 50-, 65-, and 75-year-old subjects in 1988. (From Axelsson et al, 1988, 1990. Reprinted with permission.)

Fig 535 Mean number of oral mucosal lesions in the maxilla, mandible, vestibule, tongue, floor of the mouth, and lips in randomized samples of 35-, 50-, 65-, and 75-year-old subjects in 1988. (From Axelsson et al, 1988, 1990. Reprinted with permission.)

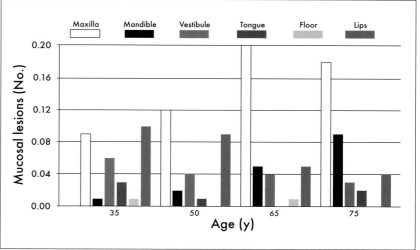

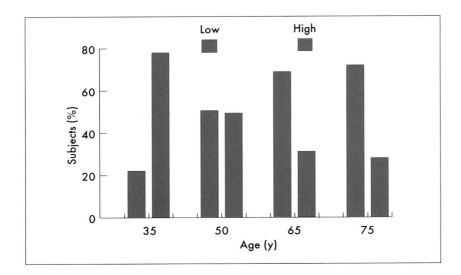

Fig 536 Percentage of subjects with low educational level (elementary school only) and high educational level (higher than elementary school) in randomized samples of 35-, 50-, 65-, and 75-year-old subjects in 1988. (From Axelsson et al, 1988, 1990. Reprinted with permission.)

Figure 536 shows the percentage of individuals in the four age groups who had low educational level (only elementary school) and high educational level (higher level than elementary school). In 1988, 22% of the 35-year-old subjects had only elementary school level compared to 72% of the 75-year-old subjects. Because of continuously on-going education even of the adult population in Sweden, it is estimated that in 2003 at least 75% of 50-year-old adults, 60% of 65-year-old adults, and about 50% of 75-year-old adults have an educational level higher than elementary school.

The role of low educational level as an RI for edentulousness is shown in Fig 537. As a matter of fact, in 1988 no edentulousness existed in the county of Värmland, Sweden, in 35-year-old subjects, irrespective of educational level. In 50-, 65-, and 75-year-old subjects, the percentage of edentulousness was 8.5%, 25.4%, and 42.0%, respectively, among individuals with low educational level compared to only 0.5%, 3.6% and 13.6%, respectively, among individuals with higher educational level.

In 1988, the percentages of smokers, former smokers, and snuffers were 30%, 11%, and 11%, respectively, in 35-year-old subjects; 30%, 13%, and 9%, respectively, in 50-year-old subjects; 21%, 9%, and 3%, respectively, in 65-year-old

subjects; and 6%, 9%, and 4%, respectively, in 75-year-old subjects (Fig 538). The overall frequency of smokers was 27.4%.

Approximately 32% to 37% of the 35-, 50-, and 65-year-old men, but only 16% of the 75-year-old men, were smokers. In women, the highest frequency of smokers was found among the 35-year-old subjects (37%) and the lowest among the 75-year-old subjects (8%). Significantly ($P <$ .001) more men than women in the 65-year-old category were smokers (34% versus 15%).

In 1998, the percentages of smokers in new randomized samples of 35-, 50-, 65-, and 75-year-old subjects were reduced to 19%, 29%, 15%, and 8%, respectively, as an effect of smoking-preventive programs presented in the elementary schools for more than two decades and among adults during the last decade.

The mean number of pack years (a composite value of the number of packs of cigarettes smoked per day, multiplied by the number of years smoking; ie, half a pack of cigarettes per day for 4 years would equal 2 pack years) among the smokers in the four age groups was 10, 15, 29, and 22, respectively, and the mean number of cigarettes per day was 11, 10, 12, and 8, respectively, in 1998. In 1988, 4%, 19%, and 36% of 50-, 65-, and 75-year-old nonsmokers, respectively, were edentulous. The corresponding

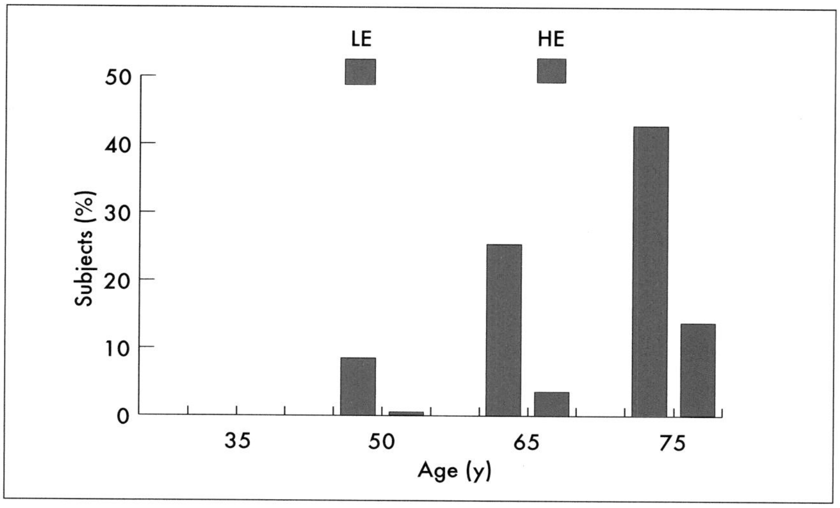

Fig 537 Baseline values for edentulism among subjects with low educational (LE) and high educational (HE) levels, defined as elementary school (6 to 9 years) and more than elementary school, respectively. (From Axelsson et al, 1988, 1990. Reprinted with permission.)

Fig 538 Tobacco habits in 35-, 50-, 65-, and 75-year-old subjects in 1988. (From Axelsson et al, 1998. Reprinted with permission.)

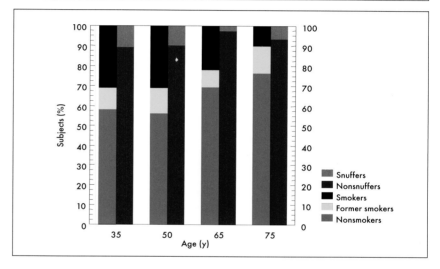

rates of edentulousness among 50-, 65-, and 75-year-old smokers were 7%, 30%, and 41%, respectively.

In the study group, 50% of the 35-, 50-, 65-, and 75-year-old subjects were randomized from rural areas and 50% from urban areas (the city of Karlstad). In 1988, among 50- and 65-year-old subjects, respectively, there were almost 2.5 and 3.0 times more edentulous people living in rural areas than in urban areas (Fig 539). There are several reasons for that difference: The rural population has a lower income, lower educational level, and lesser access to dental care than do those living in urban areas.

In the randomized samples, 51% were women and 49% were men. Figure 540 compares the percentage of edentulousness in women in 1988 to edentulousness in men and in the sample as a whole. Among the 50- and 65-year-old subjects, there is a remarkable difference between women and men.

Risk indicators, risk factors, and prognostic risk factors for tooth loss

Risk indicators. The role of low educational level as an RI for tooth loss in randomized samples of 35-, 50-, 65-, and 75-year-old subjects in 1988 is

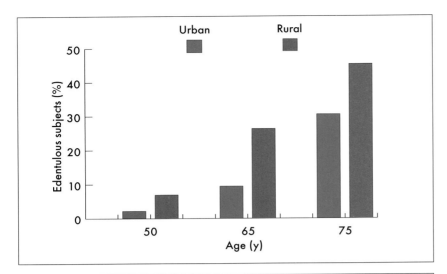

Fig 539 Comparison of edentulousness in randomized samples of 50-, 65-, and 75-year-old subjects from rural and urban areas in 1988. (From Axelsson et al, 1988, 1990. Reprinted with permission.)

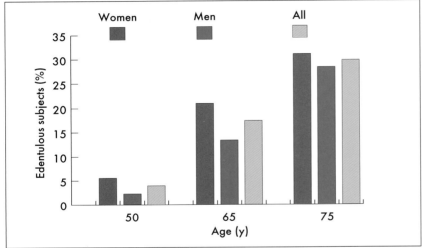

Fig 540 Comparison of edentulousness in women and men and all individuals in randomized samples of 50-, 65-, and 75-year-old subjects in 1988. (From Axelsson et al, 1988, 1990. Reprinted with permission.)

shown in Fig 541. Among 50-, 65-, and 75-year-old subjects, individuals with only an elementary school education had, on average, lost more than three more teeth than had those with a higher educational level.

The role of smoking as an RI for tooth loss in the same age groups is shown in Fig 542. In 65- and 75-year-old subjects, smokers had on average lost 3.5 and 5.8 more teeth, respectively, than nonsmokers. Figures 543 to 546 compare the patterns of lost teeth in 35-, 50-, 65-, and 75-year-old smokers and nonsmokers. Smoking seems to have a local effect on tooth loss. Particularly among 65- and 75-year-old subjects, smokers had lost a much

higher percentage of maxillary incisors than had nonsmokers.

The role of living in rural areas as an RI for tooth loss in the same randomized sample is shown in Fig 547. On average, 65-year-old subjects in rural areas had lost almost three more teeth than had those living in urban areas. Figure 548 compares the mean number of teeth in women and men in 1988. Among 50-year-old subjects, women had one more remaining tooth compared with men, but this trend is reversed among 65- and 75-year-old subjects, among whom men had one or two more remaining teeth, respectively, compared with women.

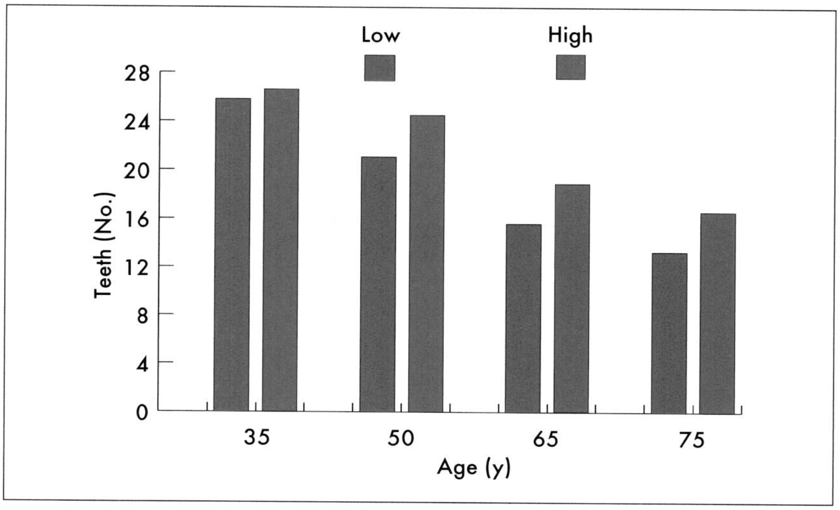

Fig 541 Baseline values for mean numbers of teeth among subjects with low education (LE) and high education (HE), defined as elementary school (6 to 9 years) and more than elementary school, respectively. (From Axelsson et al, 1988, 1990. Reprinted with permission.)

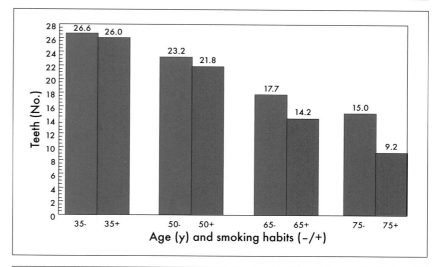

Fig 542 Mean number of teeth in smokers (+) and nonsmokers (-). (From Axelsson et al, 1998. Reprinted with permission.)

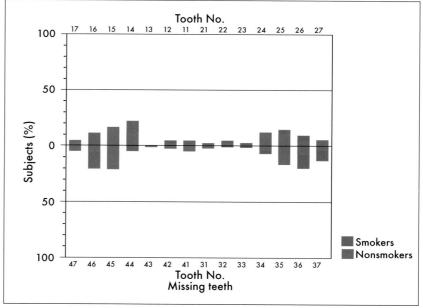

Fig 543 Pattern of tooth loss (%) in a randomized sample of 35-year-old smokers and nonsmokers (FDI tooth-numbering system). (From Axelsson et al, 1998. Reprinted with permission.)

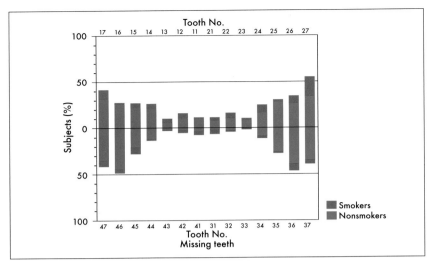

Fig 544 Pattern of tooth loss (%) in a randomized sample of 50-year-old smokers and nonsmokers (FDI tooth-numbering system). (From Axelsson et al, 1998. Reprinted with permission.)

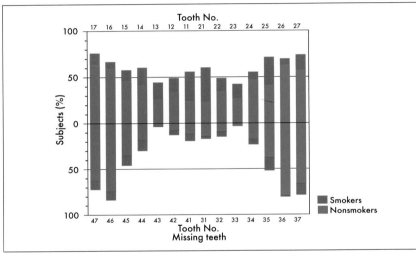

Fig 545 Pattern of tooth loss (%) in a randomized sample of 65-year-old smokers and nonsmokers (FDI tooth-numbering system). (From Axelsson et al, 1998. Reprinted with permission.)

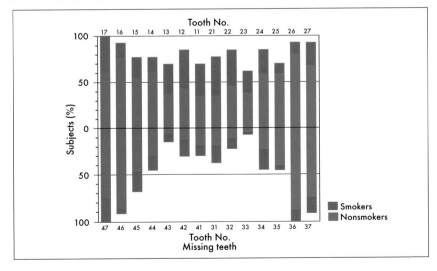

Fig 546 Pattern of tooth loss (%) in a randomized sample of 75-year-old smokers and nonsmokers (FDI tooth-numbering system). (Axelsson et al, 1998. Reprinted with permission.)

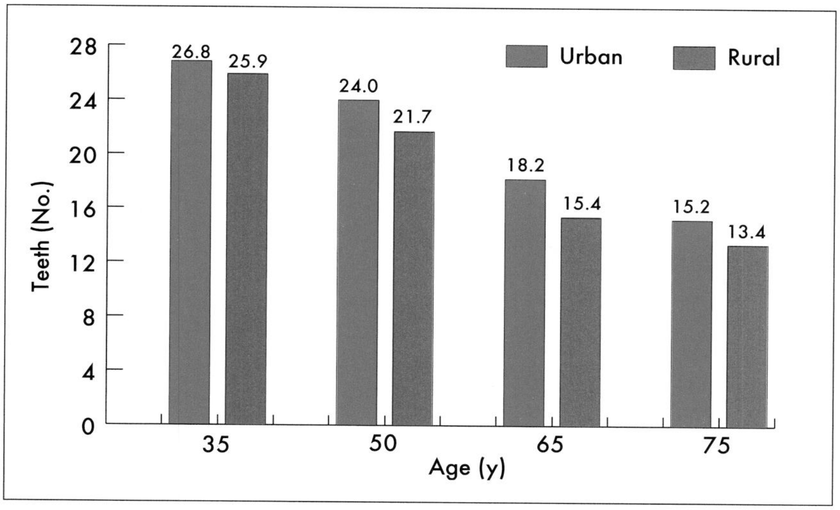

Fig 547 Comparison of mean number of teeth per individual in randomized samples of 35-, 50-, 65-, and 75-year-old subjects living in urban areas and rural areas. (From Axelsson et al, 1988, 1990. Reprinted with permission.)

Fig 548 Comparison of mean number of teeth per individual in women and men in randomized samples of 35-, 50-, 65-, and 75-year-old subjects. (From Axelsson et al, 1988, 1990. Reprinted with permission.)

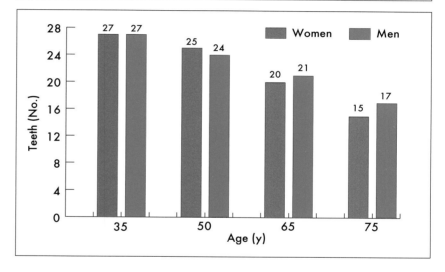

Stimulated salivary flow of less than 0.7 mL/min is regarded as an internal RF for dental caries. Normal mean values for women and men are 1.5 and 2.0 mL/min, respectively. In the Värmland study, about 15%, 20%, and 25% of the 50-, 65-, and 75-year-old subjects, respectively, exhibited only 0.0 to 0.7 mL/min of stimulated salivary secretion (Fig 549). Figure 550 shows the role of reduced stimulated salivary secretion rate as an RI for tooth loss in randomized samples of 50-, 65-, and 75-year-old subjects. In 50- and 65-year-old subjects, those with a secretion rate of 0.0 to 0.7 mL/min had lost about two and three more teeth than had those with secretion rates of more than 1.5 mL/min, probably because of dental caries.

In another cross-sectional analytical epidemiologic study in 1993, a randomized sample of more than 600 residents in the county of Värmland, Sweden, aged 50 to 55 years, were examined (Axelsson and Paulander, 1994). Adults living in urban areas had, on average, 24.0 remaining teeth, while those in rural areas had only 22.0 (Fig 551). Third molars were excluded. Women had 1.5 more remaining teeth than did men. Patients receiving regular dental care in private practice or the Public Dental Health Service had 23.8 and 22.5 remaining teeth, respectively; those receiving care sporadically had only 17.0 remaining teeth. People with a higher educational level had 4.0 more teeth than did

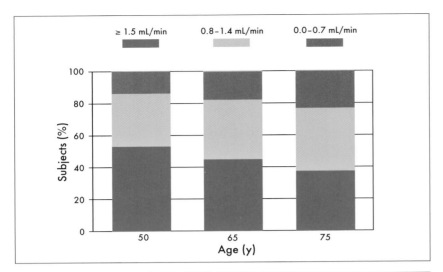

Fig 549 Frequency distribution of stimulated whole salivary secretion rates, by age. (From Axelsson et al, 1990. Reprinted with permission.)

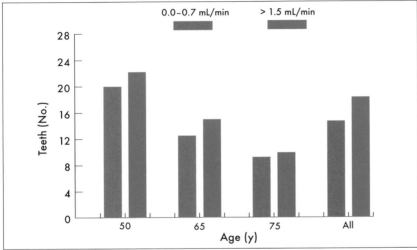

Fig 550 Mean numbers of teeth (excluding third molars) in individuals with low (0.0 to 0.7 mL/min) and high (> 1.5 mL/min) stimulated salivary secretion rates, by age. (From Axelsson et al, 1990. Reprinted with permission.)

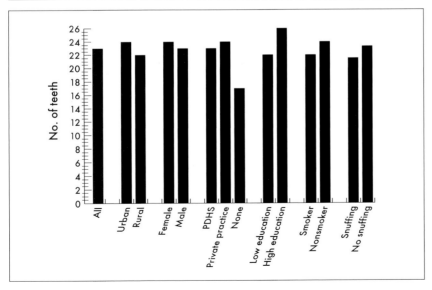

Fig 551 Mean number of remaining teeth in 50- to 55-year-old individuals, related to living area, gender, dental care habits (regular care in Public Dental Health Service [PDHS] or in a private practice or no regular dental care), educational level, smoking habits, and tobacco snuffing habits. (From Axelsson and Paulander, 1994. Reprinted with permission.)

Fig 552 Mean number of teeth lost per individual (from 50 to 60 years of age) per 10 years, related to smoking habits, PI, sex, educational level, living area, and dental care habits. (From Axelsson et al, 2000. Reprinted with permission.)

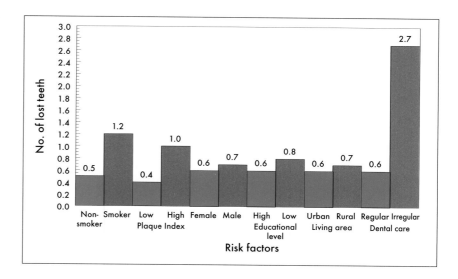

those with a low educational level. Nonsmokers had 2.0 more teeth than did smokers, and nonsnuffers had, on average, 1.5 more remaining teeth than did snuff users. From this study, it may be estimated that a 50- to 55-year-old male smoker with a low educational level who is receiving irregular dental care and living in a rural area, could have, on average, about 50% fewer teeth than a female nonsmoker with a higher educational level who is receiving regular dental care and living in an urban area.

Risk factors and prognostic risk factors. In the previously discussed 10-year longitudinal study of a randomized sample of 50-year-old subjects who were followed from 1988 to 1998 (Axelsson et al, 2000), when the subjects were 60 years old, the mean number of lost teeth was very low (0.7) because more than 95% of the adult population in the county of Värmland receives regular needs-related maintenance care. The mean number of lost teeth per individual per 10 years was assessed in relation to smoking habits, PI, gender, educational level, rural or urban living environment, and regularity of dental care (Fig 552). Because of the aforementioned well-maintained population, only smoking and high PI were significant RFs for tooth loss. However, the mean value of lost teeth was 0.8 in individuals with low

educational levels and 0.6 in those with higher educational levels. The small percentage (less than 5%) of subjects with irregular dental care lost 2.7 teeth, while those with regular dental care lost less than 0.6 teeth.

In the same study, among other things, the role of genetic polymorphism of the proinflammatory cytokine IL-1 (IL-1α and IL-1β) was evaluated as an RF or a PRF for tooth loss and loss of periodontal support in smokers and nonsmokers. A new periodontal risk test (PRT) set and a specific DNA primer for genetic IL-1 polymorphism were used. The results showed that 44.5% of the subjects tested positive (PRT-positive) for genetic polymorphism of IL-1, 32% were heterozygotic PRT-positive for IL-1α and IL-1β, and 37 were homozygotic negative (PRT-negative). More than 95% of the subjects had received regular maintenance care at needs-related intervals during the 10-year period. Therefore, the mean number of teeth lost because of periodontal diseases was less than 0.4 per subject per 10 years. During the 10-year period, PRT-negative nonsmokers, PRT-positive nonsmokers, PRT-negative smokers, and PRT-positive smokers lost 0.16, 0.30, 0.43, and 0.95 tooth per subject, respectively (Fig 553). Figure 554 shows the frequency distribution of lost teeth in the four subgroups. Genetic polymorphism of IL-1 in combination with

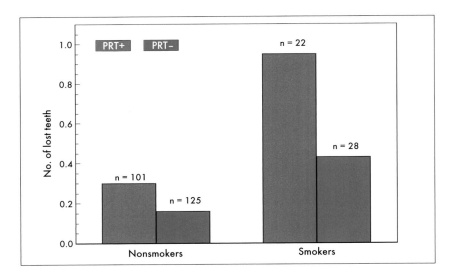

Fig 553 Mean number of lost teeth per subject per 10 years in smokers and nonsmokers testing positive (PRT+) or negative (PRT–) for genetic polymorphism of IL-1. (From Axelsson et al, 2001. Reprinted with permission.)

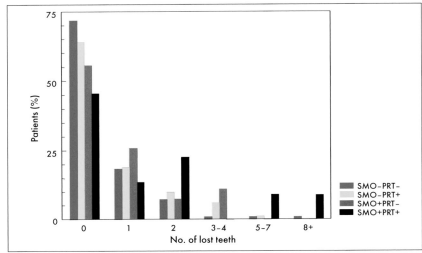

Fig 554 Frequency distribution of lost teeth in nonsmokers testing negative (SMO–PRT–), nonsmokers testing positive (SMO–PRT+), smokers testing negative (SMO+PRT–), and smokers testing positive (SMO+PRT+) for genetic polymorphism of IL-1. (From Axelsson et al, 2001. Reprinted with permission.)

smoking synergistically increased the risk of tooth loss because of periodontal diseases. This finding is in agreement with that of a recent study by McGuire and Nunn (1999).

Risk indicators, risk factors, and prognostic risk factors for dental caries

Substantial data about known or possible RIs, RFs, and PRFs for dental caries have been collected from these cross-sectional and longitudinal analytical epidemiologic studies in randomized samples of 35-, 50-, 65-, and 75-year-old subjects (see Fig 484). For example, the roles of

dental care, oral hygiene, dietary and smoking habits, educational level, occupation, body mass index, lifestyle, and stimulated salivary secretion rate, have been or will be evaluated. A few of these factors are presented.

Risk indicators. The role of educational level as an RI for dental caries is shown through a comparison of the percentages of intact, decayed, filled, and missing surfaces in randomized samples with a low educational level (elementary school) and a higher educational level (more than elementary school level) (Fig 555). Subjects with a low educational level exhibited a

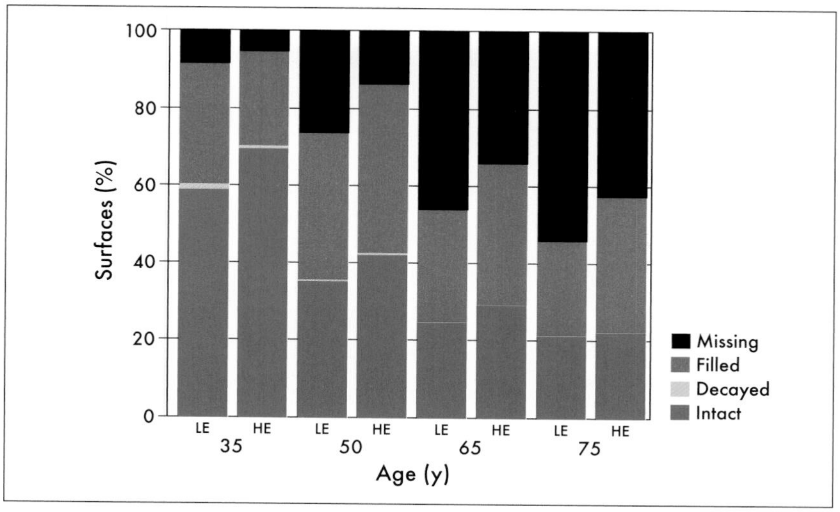

Fig 555 Frequency distribution of intact, decayed, filled, and missing surfaces in randomized samples of 35-, 50-, 65-, and 75-year-old subjects with low educational (LE) level (elementary school) and high educational (HE) level (higher level than elementary school). (Modified from Paulander et al, 2003. Reprinted with permission.)

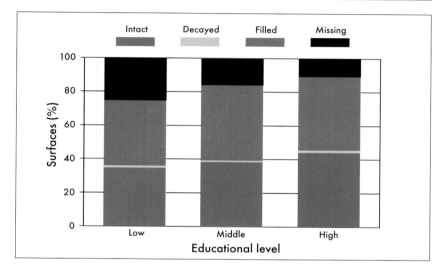

Fig 556 Frequency distribution of intact and decayed, missing, and filled surfaces in 50-year-old subjects in relation to their level of education: low (elementary school), middle (high school), or high (tertiary education). (From Axelsson et al, 1988, 1990. Reprinted with permission.)

higher percentage of missing surfaces in all age groups and fewer intact surfaces, particularly among 35- and 50-year-old subjects.

Figure 556 shows the percentage of sound surfaces and DMFSs in 50-year-old subjects in 1988 in relation to educational level. Subjects with higher levels of education had a greater percentage of intact surfaces and a lower percentage of missing surfaces than did those with less education. However, the percentage of carious surfaces was almost negligible, indicating that the available resources for provision of dental care were adequate, at least with respect to treatment of caries.

The results of this large-scale, analytical, cross-sectional study show that low educational level is a very significant RI for tooth loss, dental caries, and periodontal diseases, not necessarily because highly educated people are more intelligent or wealthier. In Sweden, there are very limited differences in net income, after tax, between occupational categories such as poorly educated laborers and well-educated teachers. The difference in dental health status is attributable to the fact that highly educated people know how to learn from written information, to seek information about health promotion, and to apply theoretical information, eg, to self-care.

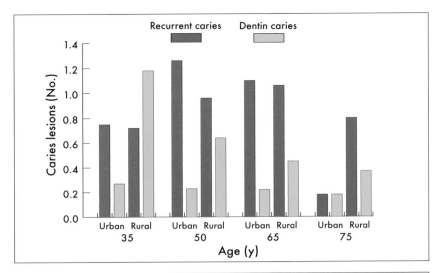

Fig 557 Comparison of mean number of primary dentin and recurrent caries lesions per individual in randomized samples of 35-, 50-, 65-, and 75-year-old subjects living in urban and rural areas. (From Axelsson et al, 1988, 1990. Reprinted with permission.)

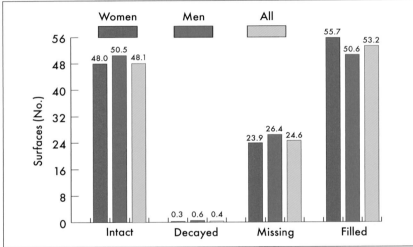

Fig 558 Comparison of mean number of intact, decayed, missing, and filled surfaces per individual in women, men, and both in a randomized sample of 50-year-old subjects. (From Axelsson et al, 1988, 1990. Reprinted with permission.)

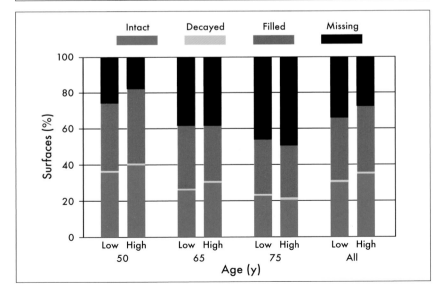

Fig 559 Frequency distribution of intact, decayed, filled, and missing surfaces in individuals with low (0.0 to 0.7 mL/min) and high (> 1.5 mL/min) stimulated salivary secretion rates, by age. (From Axelsson et al, 1990. Reprinted with permission.)

Fig 560 Frequency distribution of all manifest caries lesions, by type of dental care, among a randomized sample of 50-year-old subjects in the county of Värmland, Sweden. (From Axelsson et al, 1988, 1990. Reprinted with permission.)

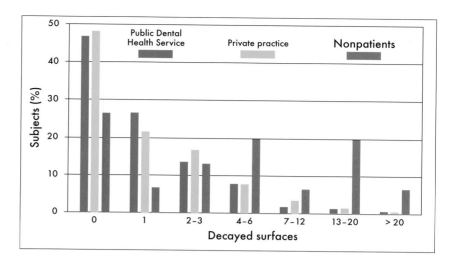

The mean values of primary dentin and recurrent caries lesions in randomized samples of 35-, 50-, 65-, and 75-year-old subjects living in urban or rural areas in 1988 are shown in Fig 557. In 35- and 50-year-old subjects, the mean number of primary dentin caries lesions is approximately four and three times higher, respectively, in subjects living in rural areas than in those living in urban areas. This indicates, at least, that treatment needs are higher in rural areas than in urban areas.

The mean numbers of intact, decayed, filled, and missing surfaces in a randomized sample of 50-year-old women and men are compared in Fig 558. On average, men exhibited 2.5 more intact surfaces and missing surfaces and 5.1 fewer restored surfaces than did women.

The percentages of intact surfaces and DMFSs in subjects with low stimulated salivary secretion rate (0.0 to 0.7 mL/min) were compared to those in subjects with high stimulated salivary secretion rate (more than 1.5 mL/min) in a randomized sample of 35-, 50-, 65-, and 75-year-old dentate subjects and 75-year-old edentulous subjects (Fig 559). The 50- and 65-year-old subjects with a low secretion rate exhibited lower percentages of intact surfaces and higher percentages of missing surfaces than did those with high secretion rate. However, in 75-year-old subjects, dentate as well as edentulous, there

was no difference. That is probably because such a high percentage of the surfaces were missing as the result of reasons other than dental caries.

In another cross-sectional study of a randomized sample of more than 600 subjects aged 50 to 55 years, among many clinical and anamnestic variables, caries prevalence was related to stimulated whole salivary secretion rate and regular use of medicines with known systemic effects on salivary secretion rate (Axelsson and Paulander, 1994). Of the subjects, 29% were taking medication regularly, and 22% used medicines that impair salivary secretion rate. The data showed conclusively that the salivary secretion rate is an important factor in caries severity and should be considered when caries risk is assessed (see Fig 298 in chapter 5). Very low stimulated salivary secretion rate, or hyposialosis (less than 0.7 mL/min, and particularly less than 0.4 mL/min), results in a high risk of caries. Clinically, it is therefore important to determine whether salivary secretion rate is normal or impaired.

The frequency distribution of all the caries treatment needs (decayed surfaces) in 50-year-old subjects in 1988 was stratified according to their regular attendance at the Public Dental Health Service or a private practice or their status as nonpatients (Fig 560). Almost 50% of all

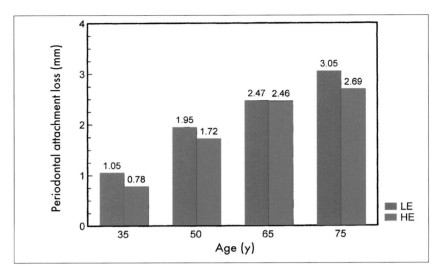

Fig 561 Mean periodontal attachment loss at mesial sites among subjects with low educational (LE) and high educational (HE) levels, defined as elementary school (6 to 9 years) and more than elementary school, respectively. (Modified from Paulander et al, 2003. Reprinted with permission.)

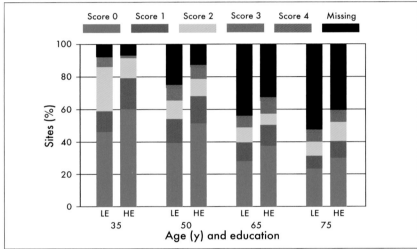

Fig 562 Frequency distribution of CPITN scores among subjects with low educational (LE) and high educational (HE) levels, defined as elementary school (6 to 9 years) and more than elementary school, respectively. (Modified from Paulander et al, 2003. Reprinted with permission.)

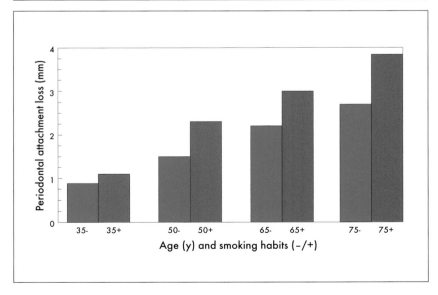

Fig 563 Mean periodontal attachment loss in smokers (+) and nonsmokers (−). (From Axelsson et al, 1998. Reprinted with permission.)

Fig 564 Pattern of mean mesial attachment loss in 50-year-old smokers and nonsmokers (FDI tooth-numbering system). (From Axelsson et al, 1998. Reprinted with permission.)

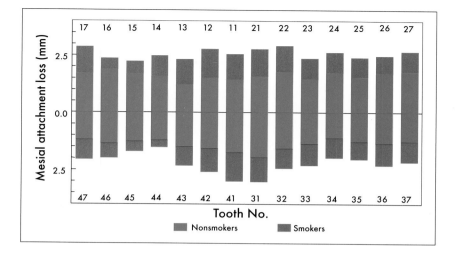

the patients in the Public Dental Health Service as well as private practice had no dentin caries, recurrent caries, or root caries. Individuals with 4 to 6, 7 to 12, 13 to 20, and more than 20 carious surfaces were found mainly among the nonpatients.

Risk indicators, risk factors, and prognostic risk factors for periodontal diseases

Thus far, the role of RIs, RFs, and PRFs related to periodontal diseases has been evaluated more extensively than that of RIs, RFs, and PRFs related to dental caries. However, substantial amounts of data remain to be evaluated.

Risk indicators. The role of low educational level as an RI for periodontal diseases was examined in a randomized sample of 35-, 50-, 65-, and 75-year-old subjects (Figs 561 and 562). Figure 561 shows the mean loss of probing periodontal attachment on the mesial surfaces in the four age groups. To exclude iatrogenic recession on the buccal surfaces, only the mesial values are presented. With the exception of the 65-year-old subjects, poorly educated subjects had significantly more PAL than did highly educated subjects.

The frequency distribution of sites with CPITN scores 1 to 4 and missing sites in the four age groups is presented in Fig 562. In all age groups, subjects with less education had fewer healthy sites and more missing sites than did those with a higher level of education. Particularly among 35-year-old subjects, poorly educated subjects had more calculus that required removal (CPITN score 2) and diseased 4- to 5-mm pockets (CPITN score 3). The less pronounced differences in existing treatment needs between poorly and highly educated subjects in the other age groups are probably attributable to the extraction of the most diseased teeth (particularly molars) in the poorly educated subjects. This cross-sectional study revealed that low educational level is a powerful RI for tooth loss and periodontal diseases as well as dental caries (Axelsson et al, unpublished data, 1990; Paulander et al, 2003).

Probing attachment loss and CPITN were compared in adult smokers and nonsmokers (Axelsson et al, unpublished data, 1990; Axelsson et al, 1998). The subjects were randomly selected 35-, 50-, 65-, and 75-year-old subjects (N = 1,086) from the county of Värmland, Sweden. Probing attachment loss and the CPITN score were measured at all mesial, buccal, distal, and lingual sites (N = almost 100,000). The mean PALs in smokers were significantly greater (P = .001) than those in nonsmokers in every age group (Fig 563). Among poorly educated 50-year-old women, nonsmokers had 50% less PAL than did smokers. The pattern of attachment loss in smokers and nonsmokers on mesial and lingual surfaces is presented in Figs 564

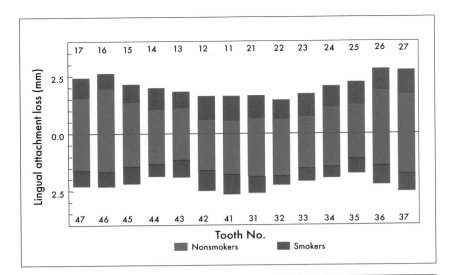

Fig 565 Pattern of mean lingual attachment loss in 50-year-old smokers and nonsmokers (FDI tooth-numbering system). (From Axelsson et al, 1998. Reprinted with permission.)

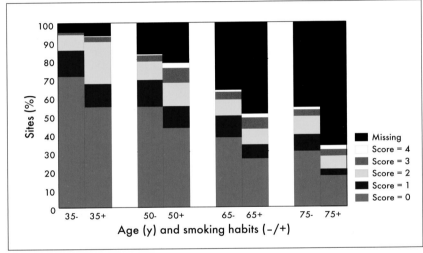

Fig 566 Frequency distribution of CPITN scores in smokers (+) and nonsmokers (–). (From Axelsson et al, 1998. Reprinted with permission.)

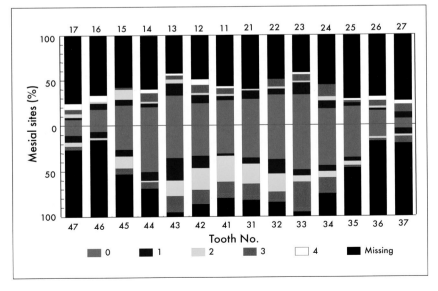

Fig 567 Frequency distribution of CPITN scores at mesial sites in 65-year-old smokers (FDI tooth-numbering system). (From Axelsson et al, 1998. Reprinted with permission.)

Fig 568 Frequency distribution of CPITN scores at mesial sites in 65-year-old nonsmokers (FDI tooth-numbering system). (From Axelsson et al, 1998. Reprinted with permission.)

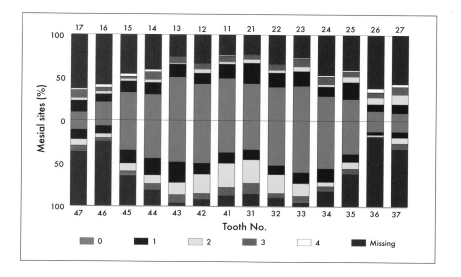

and 565, respectively. Particularly on the maxillary lingual surfaces and the approximal surfaces of the maxillary and mandibular incisors, smoking seemed to have a local effect.

The frequency distribution of CPITN scores in smokers and nonsmokers is presented in Fig 566. The patterns of CPITN scores and missing sites on mesial surfaces in 65-year-old smokers and nonsmokers are shown in Figs 567 and 568, respectively. The highest percentages of CPITN score 4 and missing sites were found on the maxillary mesial sites of the molars in smokers. On the other hand, smokers had a lower percentage of CPITN score 1 (gingival bleeding on probing) than did nonsmokers.

It was also notable that, in this randomized, large-scale, analytical epidemiologic study, no differences were found in the oral hygiene, dietary habits, or dental attendance habits of smokers and nonsmokers. Thus, this study showed that smoking is an extremely powerful environmental (external) RI for loss of teeth, loss of periodontal probing attachment, and periodontal treatment needs, correlated with the total, cumulative exposure to smoking (Axelsson et al, unpublished data, 1990; Axelsson et al, 1998).

In the previously discussed cross-sectional study in a randomized sample of more than 600 residents aged 50 to 55 years in the county of Värmland, Sweden, the role of living area, gender, dental care habits, educational level, and smoking and snuffing habits as RIs for periodontal attachment loss was evaluated in similarity to their role as RIs for tooth loss (see Fig 551).

Figure 569 shows the mean probing loss of attachment on the mesial surfaces. (Attachment loss on the mesial surfaces is caused by bacteria, while most buccal attachment loss is a side effect of toothbrushing.) Irregular dental care seemed to be the strongest RI for PAL, followed by smoking and low educational level. However, fewer than 5% of subjects in this age group had irregular dental care.

In the same sample, the characteristics of the subjects with limited and advanced periodontal attachment loss were evaluated from the available data, because it is important to learn not only why the most diseased subjects are diseased but also why the healthiest subjects remain healthy. Therefore, the characteristics of those with the 25% lowest mesial attachment loss (lowest quartile [LQ]) were compared to those of the subjects with the 25% most advanced mesial PAL (highest quartile [HQ]). Mean mesial PAL was 1.9 mm. In the healthiest quartile, the mean mesial PAL ranged from only 0.1 to 1.1 mm while in the most diseased quartile it ranged from 2.2 to 7.2 mm (Fig 570). The mean mesial PAL for LQ and HQ was 0.9 and 3.25 mm, respectively.

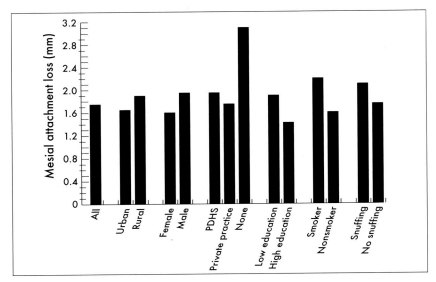

Fig 569 Mean periodontal attachment loss mesially in 50- to 55-year-old individuals related to living area, gender, dental care habits (regular care in Public Dental Health Service [PDHS] or in a private practice or no regular dental care), educational level, smoking habits, and tobacco snuffing habits. (From Axelsson and Paulander, 1994. Reprinted with permission.)

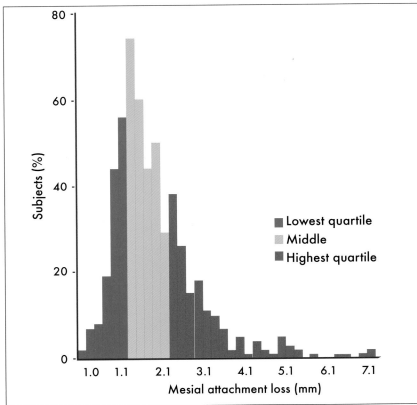

Fig 570 Frequency distribution of the mean mesial PAL per subject in different quartiles of a randomized sample of 50- to 55-year-old subjects. Lowest quartile (green), Highest quartile (red). (Modified from Paulander et al, 2004b with permission.)

Figures 571 and 572 detail the presence or absence of various RIs in subjects in the lowest and highest quartiles of mesial PAL. In the HQ, the percentage of smokers is almost 60%, while in the LQ, only 15% are smokers. Among the HQ, the high percentage of men and individuals with low educational level, living in rural areas, was also observed (similar to the data shown in Fig 569). No individuals with irregular dental care were found in the LQ. Only about 15% of patients in the LQ used medicines regularly, compared to about 35% in the HQ.

Figs 571 and 572 Presence or absence of various RIs in the lowest (LQ) and highest (HQ) quartiles of mesial PAL. (Modified from Paulander et al, 2004b with permission.)

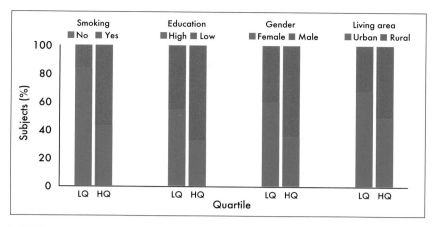

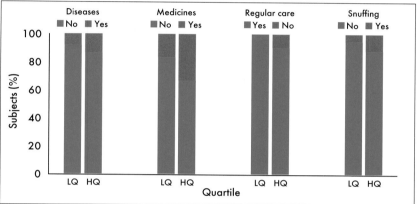

Table 31 shows the mean mesial PAL related to smoking, educational level, gender, living area, cardiovascular disease and/or diabetes, regular medication, regular dental care visits, and snuffing in the LQ and HQ as well as the lowest and highest deciles (10%). The percentage of differences in mean mesial PAL is also shown. The difference between smokers and non-smokers is higher in the LQ than in the HQ, which indicates that the most severe high-risk group is exposed to several other important known RIs, such as genetic susceptibility, lifestyle, stress, and irregular dental care.

Figure 573 compares the mean number of teeth, mean number of cigarettes smoked per day, mean plaque percentage, and mean body mass index in the LQ and the HQ. The HQ, on average, smoked four times more cigarettes per day, had lost more than five more teeth, had almost a 30% higher plaque percentage, and had

a higher body mass index than did the LQ. The odds ratio for the HQ, related to various RIs, is shown in Fig 574.

Statistical analyses of the enormous amount of data remain to be completed. However, significant differences in mesial PAL were found to be related to smoking, educational level, gender, living area, general health status, dental fear, dental attendance, smokeless tobacco, medication, and regular exercise (Paulander et al, 2004b).

Risk factors and prognostic risk factors. In the previously discussed 10-year longitudinal study in a randomized sample of 50-year-old subjects, who were reexamined at the age of 60 years, several RFs and PRFs for tooth loss, loss of periodontal support, and other conditions were evaluated. Because more than 95% of the subjects received regular maintenance care, the fre-

Table 31 Mean mesial periodontal attachment loss (PAL) in subjects in the lowest and highest quartiles and deciles of loss, according to various risk indicators

Risk indicator	Lowest quartile of attachment loss		Highest quartile of attachment loss		Lowest decile of attachment loss		Highest decile of attachment loss	
	PAL (mm)	Difference (%)	PAL (mm)	Difference (%)	PAL (mm)	Difference (%)	PAL (mm)	Difference (%)
Non-smoking	0.89	+9.0	3.15	+4.4	0.64	+15.6	4.09	+2.4
Smoking	0.97		3.29		0.74		4.19	
Low education	0.97	−13.4	3.34	−7.8	0.73	−17.8	4.25	−4.2
High education	0.84		3.08		0.60		4.07	
Female	0.90	+1.1	3.11	+6.8	0.63	+7.9	4.17	+1.0
Male	0.91		3.32		0.68		4.21	
Urban	0.85	+21.2	3.19	+3.8	0.63	+23.8	4.09	+5.9
Rural	1.03		3.31		0.78		4.33	
No CVD/DM*	0.90	0.0	3.28	−5.8	0.65	+10.8	4.15	+15.4
CVD/DM*	0.90		3.09		0.72		4.79	
No regular medication	0.92	−12.0	3.28	−7.0	0.68	−17.6	4.22	−6.2
Regular medication	0.81		3.05		0.56		3.96	
Irregular dental attendance†	N/A	N/A	4.82	−34.6	N/A	N/A	5.75	−30.3
Regular dental attendance	0.90		3.15		0.65		4.01	
No snuffing	0.91	−3.3	3.23	+7.4	0.65	+1.5	4.16	+7.5
Snuffing	0.88		3.47		0.66		4.47	

*(CVD/DM) Cardiovascular diseases/diabetes mellitus.
†There were no subjects with irregular dental attendance in the lowest decile and quartile of attachment loss

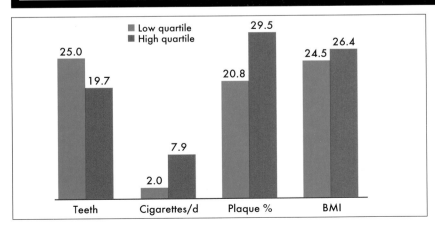

Fig 573 Risk indicators for periodontal attachment loss: Mean number of teeth, cigarettes smoked per day, mean plaque percentage, and mean body mass index (BMI) in subjects in the lowest and highest quartiles of mean mesial attachment loss. (Modified from Paulander et al, 2004b with permission.)

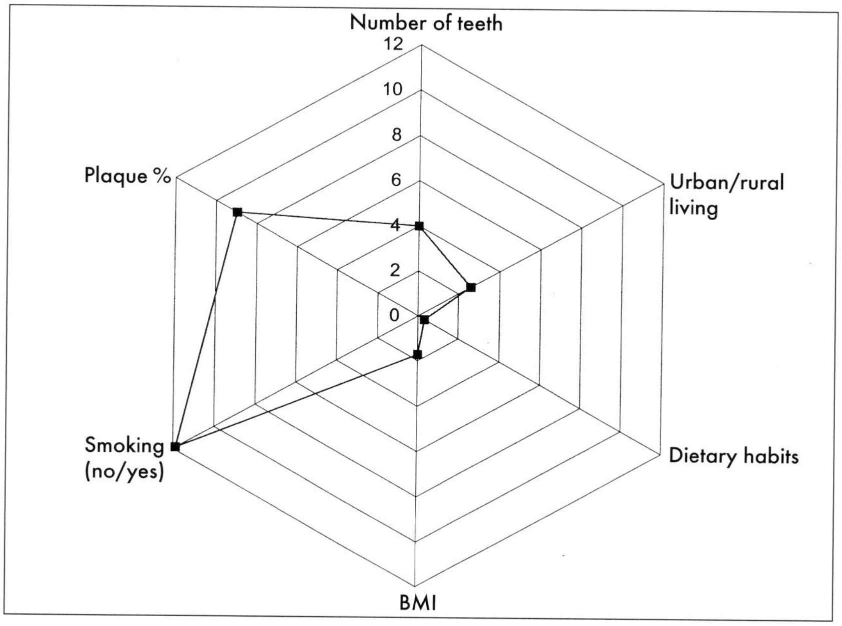

Fig 574 Odds ratios for subjects in the highest quartile of mean mesial periodontal attachment loss related to number of teeth, living area (urban or rural), dietary habits, body mass index (BMI), smoking habits (yes or no), and plaque percentage. (Modified from Paulander et al, 2004b with permission.)

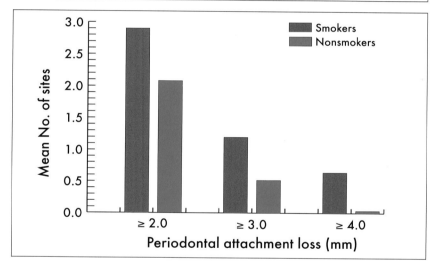

Fig 575 Mean number of mesial sites per individual per 10 years with ≥ 2.0 mm, ≥ 3.0 mm, and ≥ 4.0 mm of periodontal attachment loss in smokers and nonsmokers. (From Axelsson et al, 2000. Reprinted with permission.)

quency distribution of sites with 2.0 mm or more PAL was very limited (see Fig 516). Nevertheless the mean number of sites with 2.0 mm or more, 3.0 mm or more, and 4.0 mm or more of periodontal attachment loss was significantly greater in smokers than in nonsmokers (Axelsson et al, 2000) (Fig 575).

In the same sample, the role of genetic polymorphism of the proinflammatory cytokine IL-1α and IL-1β in smokers and nonsmokers was evaluated as an RF and a PRF for tooth loss (see Figs 553 and 554) and alveolar bone loss. The aforementioned genetic periodontal risk test was used. The

mean alveolar bone loss during the 10-year period in nonsmokers testing negative, nonsmokers testing positive, smokers testing negative, and smokers testing positive was approximately 0.3, 0.3, 0.6, and 1.2 mm, respectively (Fig 576). Figure 577 shows the frequency distribution of different levels of alveolar bone loss in the four subgroups (Axelsson et al, 2001). The combination of genetic polymorphism of IL-1α and IL-1β and smoking was shown to be a synergistic, powerful RF and PRF for alveolar bone loss. Further analyses of other RFs and PRFs from the collected data will be carried out.

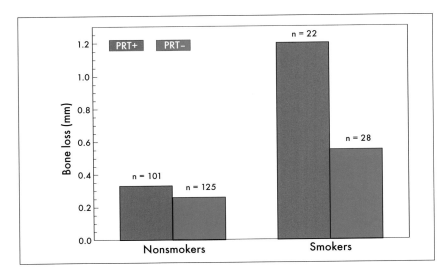

Fig 576 Mean alveolar bone loss per subject per 10 years in smokers and nonsmokers testing positive (PRT+) or negative (PRT–) for genetic polymorphism of interleukin 1. (From Axelsson et al, 2001. Reprinted with permission.)

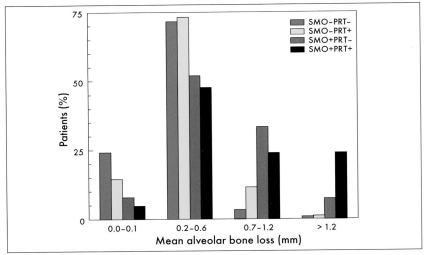

Fig 577 Frequency distribution of mean alveolar bone loss from 1988 to 1998 in nonsmokers testing negative (SMO–PRT–), nonsmokers testing positive (SMO–PRT+), smokers testing negative (SMO+PRT–), and smokers testing positive (SMO+PRT+) for genetic polymorphism of interleukin 1. (From Axelsson et al, 2001. Reprinted with permission.)

CONCLUSIONS

According to the WHO's goals for the year of 2000, ministries of health worldwide should support computer-aided oral epidemiology programs to collect data on the oral health status of the population at a national level. Surveys are recommended in randomized samples of specific age groups at, for instance, 5-year intervals to evaluate the effect of national oral health programs and to facilitate international comparison. In every county or district, similar surveys should be carried out for comparison at a national level.

However, every dentist should also be eager to monitor the oral health status of his or her own patients at the group, individual, tooth, and surface level for quality control. Otherwise, the optimistic practitioner will just remember the successful cases, and the pessimistic practitioner will remember only the failures. If a needs-related preventive program is presented to patients according to the main principles described in chapter 8, simplified graphs can be hung in the waiting room as a feedback mechanism for patients. The data will show how successfully the practice and the patients, together, continuously improve patients' oral health status. Patients will tell their relatives and friends about the successful outcome of the practice. As a consequence, the practice will be overwhelmed with new patients and may have to expand with more dental hygienists, if possible, and perhaps more dentists. The author has had the very best personal experience with a ratio of 6 to 8 dental hygienist hours per dentist hour for more than 30 years in his private practice, specializing in preventive dentistry and periodontology.

Objectives

A detailed analytical epidemiologic program was designed to evaluate the effect of a needs-related preventive program for adults that was introduced in the county of Värmland, Sweden, in 1985 (see chapter 8). In contrast to other medical disciplines, dentistry has well-established and measurable variables for evaluation of oral health. The main reasons for loss of teeth are dental caries and periodontal diseases. The masticatory efficiency of the dentition and the condition of the oral mucosa should also be included in the assessment.

Materials and methods

In ranking order of importance, data on the following variables were recorded:

1. Number and percentage of edentulous subjects
2. Mean number and frequency distribution of teeth
3. Chewing function of the teeth
4. Prevalence and treatment needs of dental caries and periodontal diseases
5. Apical periodontitis: prevalence and treatment needs
6. Oral mucosa lesions: prevalence and treatment needs

Epidemiology should also include etiologic and other modifying factors.

Results and discussion

Baseline examination was carried out in 1988, in randomized samples of 35-, 50-, 65-, and 75-year-old subjects. A number of 50-year-old subjects were followed longitudinally to the age of 60 years. The outcome after 10 years in new randomized samples of 35-, 50-, 65-, and 75-year-old subjects showed, among other things:

1. The percentage of edentulous subjects was reduced by more than 50%.
2. The mean number of remaining teeth increased by about three teeth in the 50-, 65-, and 75-year-old subjects.
3. The mean number of intact teeth in 35- and 50-year-old subjects increased by more than 5.0 and 2.5 teeth, respectively.
4. Caries treatment needs were almost nonexistent.
5. Probing periodontal attachment loss was reduced by about 25% in 65- and 75-year-old subjects.
6. Among the 35-, 50-, 65-, and 75-year-old subjects, 95%, 70%, 65%, and 60% of the subjects, respectively, did not exhibit one single deep periodontal pocket (CPITN score 4).

Those who were 50-year-old subjects in 1988 lost approximately 0.7 teeth (0.35 because of periodontitis, 0.15 because of caries, and 0.20 for iatrogenic reasons) and 0.4 mm of alveolar bone per subject per 10 years to the age of 60 years. The mean number of intact surfaces was reduced by only four surfaces (less than two because of caries) during the 10-year period.

Among modifying RIs, RFs, and PRFs, poor oral hygiene, poor dietary habits, irregular dental care habits, reduced stimulated salivary flow, and low educational level proved to be important for increased caries risk. In particular, smoking in combination with genetic polymorphism of the proinflammatory cytokines IL-1α and IL-1β, poor oral hygiene, irregular dental care habits, and low educational level increased the risk for periodontal disease.

To maintain a healthy dentition, free from dental caries as well as periodontal diseases, daily cleaning of *all* tooth surfaces, use of fluoride toothpaste, healthy dietary habits, no smoking, and regular preventive dental care visits with needs-related materials, methods, and intervals would be successful for at least 90% of a population. In addition, a high level of education with an understanding of how to implement health information regarding self-diagnosis and self-care would be beneficial.

REFERENCES

Abello R, Buitrago C, Prate CM, DeVizio W, Bakar SK (1990). Effect of mouthrinse containing triclosan and a copolymer on plaque formation in the absence of oral hygiene. Am J Dent 3:57–61.

Abelson DC, Barton JE, Maietti GM, Cowherd MG (1981). Evaluation of interproximal cleaning by two types of dental floss. Clin Prev Dent 3:19–21.

Abelson DC, Barton J, Mandel ID (1990). The effect of chewing sorbitol-sweetened gum on salivary flow and cemental plaque pH in subjects with low salivary flow. J Clin Dent 2:3–5.

Abrahams L, Yonese M, Higuchi W, et al (1980). In vivo remineralization using a sustained topical fluoride delivery system. J Dent Res 59:583–587.

Adams D, Addy M (1994). Mouthrinses. Adv Dent Res 8(2):291–301.

Addy M, Wright R (1978). Comparison of the in vivo and in vitro antibacterial properties of povidone iodine and chlorhexidine gluconate mouthrinses. J Clin Periodontol 5:198–205.

Addy M, Moran J, Griffiths A, Wills-Wood NJ (1985). Extrinsic tooth discolouration by metals and chlorhexidine. I. Surface protein denaturation or dietary precipitation? Br Dent J 159:331–334.

Addy M (1986). Chlorhexidine compared with other locally delivered antimicrobials: A short review. J Clin Periodontol 13:957–964.

Addy M, Drummer P, Griffiths G, Hicks R, Kingdom A, Shaw W (1986). Prevalence of plaque, gingivitis and caries in 11–12-year-old children in South Wales. Community Dent Oral Epidemiol 14:115–118.

Addy M, Hassen H, Moran J, Wade W, Newcombe R (1988). Use of antimicrobial containing acrylic strips in the treatment of chronic periodontal disease. J Periodontol 59:557–564.

Addy M, Jenkins S, Newcombe R (1989a). Studies on the effect of toothpaste rinses on plaque regrowth. (I). Influence of surfactants on chlorhexidine efficacy. J Clin Periodontol 16:380–384.

Addy M, Wade WG, Jenkins S, Goodfield S (1989b). Comparison of two commercially available chlorhexidine mouthrinses: I. Staining and antimicrobial effects in vitro. Clin Prev Dent 11(5):10–14.

Addy M, Moran J, Wade W (1994). Chemical plaque control in the prevention of gingivitis and periodontitis. In: Lang N, Karring T (eds). Proceedings of the 1st European Workshop on Periodontology. London: Quintessence:244–257.

Addy M (1997). Antiseptics in periodontal therapy. In: Karring T, Lang N (eds). Clinical Periodontology and Implant Dentistry. Copenhagen: Munksgaard:461–487.

Addy M, Moran JM (1997a). Evaluation of oral hygiene products: Science is true; don't be misled by the facts. Periodontol 2000 15:40–51.

Addy M, Moran JM (1997b). Clinical indications for the use of chemical adjuncts to plaque control: Chlorhexidine formulations. Periodontol 2000 15:52–54.

Addy M, Renton-Harper P (1997). The role of antiseptics in secondary prevention. In: Lang NP, Karring T, Lindhe J (eds). Proceedings of the 2nd European Workshop on Periodontology—Chemicals in Periodontics. Berlin: Quintessence:152–173.

Adriaens PA, de Boever JA, Loesche WJ (1988a). Bacterial invasion in root cementum and radicular dentine of periodontally diseased teeth in humans: A reservoir of periodontopathic bacteria. J Periodontol 59:222–230.

References

Adriaens PA, Edwards CA, de Boever JA, Loesche WJ (1988b). Ultrastructural observations on bacterial invasion in cementum and radicular dentine of periodontally diseased human teeth. J Periodontol 55:493–503.

Adriaens PA, Gjermo P (1997). Anti-plaque and anti-gingivitis efficacy of toothpastes. In: Lang NP, Karring T, Lindhe J (eds). Proceedings of the 2nd European Workshop on Periodontology—Chemicals in Periodontics. Berlin: Quintessence:204–220.

Adriaens PA (2000). Local delivery of antimicrobials: Evidence of efficacy in the treatment of gingivitis and periodontitis. J Parodontol Implantol Orale 19:159–194.

Afflitto J, Fakhry-Smith S, Gaffar A (1989). Salivary and plaque triclosan levels after brushing with a 0.3% triclosan/copolymer/NaF dentifrice. Am J Dent 2:207–210.

Afseth J (1983). Some aspects of the dynamics of Cu and Zn retained in plaque as related to their effect on plaque pH. Scand J Dent Res 91:169–174.

Agerbaek N, Poulsen S, Melsen B, Glavind L (1977). Effect of professional tooth cleaning every third week on gingivitis and dental caries in children. Community Dent Oral Epidemiol 6:40–41.

Ainamo J (1970). Concomitant periodontal disease and dental caries in young adult males. Suom Hammaslaak Toim 66:303–366.

Ainamo J (1971). The effect of habitual toothcleaning on the occurrence of periodontal disease and dental caries. Suom Hammaslaak Toim 67:63–70.

Ainamo J (1979). Occurrence of plaque, gingivitis and caries as related to self reported frequency of toothbrushing in fluoride areas in Finland. Community Dent Oral Epidemiol 7:142–146.

Ainamo J, Barmes D, Beagrie G, Cutress T, Martin J, Sardo-Infirri J (1982). Development of the WHO community periodontal index of treatment needs (CPITN). Int Dent J 32:281–291.

Ainamo J, Paloheimo L, Nordblad A, Murtomaa H (1986). Gingival recession in schoolchildren at 7, 12 and 17 years of age in Spoco, Finland. Community Dent Oral Epidemiol 14:283–286.

Ainamo J, Etemadzadeh H (1987). Prevention of plaque growth with chewing gum containing chlorhexidine acetate. J Clin Periodontol 14:524–527.

Ainamo J, Xie Q, Ainamo A, Kallio P (1997). Assessment of the effect of an oscillating/rotating electric toothbrush on oral health. A 12-month longitudinal study. J Clin Periodontol 24:28–33.

Alaluusua S, Kleemola-Kujala E, Grönros L, Evälathi M (1990). Salivary caries-related tests as predictors of future caries increment in teenagers. A 3-year longitudinal study. Oral Microbiol Immunol 5:77–81.

Alaluusua S, Malmivirta R (1994). Early plaque accumulation—A sign for caries risk in young children. Community Dent Oral Epidemiol 22:273–276.

Albandar JM, Axelsson P, Buischi Y, Mayer M (1994a). Long-term effect of two preventive programs on the incidence of plaque and gingivitis in adolescents. J Periodontol 65:605–610.

Albandar JM, Gjermo P, Preus HR (1994b). Chlorhexidine use after two decades of over-the-counter availability. J Periodontol 65:109–112.

Al-Mubarak S, Ciancio S, Aljada A, et al (2001). Comparative evaluation of an adjunctive oral irrigation in diabetics [abstract 846]. J Dent Res 80:141.

Althenhofen E, Lutz F, Guggenheim B (1989). Mikrobiologischer In-Vitro-Vergleich plaquehemmender Mundspülmittel. Schweiz Monatsschr Zahnmed 99:13–18.

American Dental Association (1986). Guidelines for acceptance of chemotherapeutic products for the control of supragingival dental plaque and gingivitis. J Am Dent Assoc 112:529–532.

American Dental Association (1988a). Council on dental therapeutics accepts Listerine. J Am Dent Assoc 117:515–516.

American Dental Association (1988b). Council on dental therapeutics accepts Peridex. J Am Dent Assoc 117:516–517.

American Dental Association (1988c). Report of workshop aimed at defining guidelines for caries clinical trials: Superiority and equivalency claims for anticaries dentifrices. Council on Dental Therapeutics. J Am Dent Assoc 117:663–665.

Amjad Z, Nancollas GH (1979). Effects of fluoride on the growth of hydroxyapatite and human dental enamel. Caries Res 13:250–258.

Ammari AB, Bloch-Zupan A, Ashley PF (2003). Systematic review of studies comparing the anticaries efficacy of children's toothpaste containing 600 ppm of fluoride or less with high fluoride toothpastes of 1,000 ppm or above. Caries Res 37:85–92.

Anderson RL, Vess RW, Panlilio AL, Favero MS (1990). Prolonged survival of Pseudomonas cepacia in commercially manufactured povidone-iodine. Appl Environ Microbiol 56:3598–3600.

Angelillo IF, Nobile CGA, Pavia M (2002). Evaluation of the effectiveness of a pre-brushing rinse in plaque removal: A meta-analysis. J Clin Periodontol 29:301–309.

Apiou J, Gueguen M, Doleux S, Bonnaure-Mallet M (1994). Evaluation of a new toothbrush concept with regard to bacterial elimination. J Clin Periodontol 21:347–350.

Arai T, Kinoshita S (1977). A comparison of plaque removal by different toothbrushes and toothbrushing methods. Bull Tokyo Med Dent Univ 24:177–188.

Arana EM (1974). Clinical observations of enamel after acid-etch procedure. J Am Dent Assoc 89:1102–1106.

Araujo A, Naspitz G, Chelotti A, Cai S (2002). Effect of Cervitec on mutans streptococci in plaque and on caries formation on occlusal fissures of erupting permanent molars. Caries Res 36:373–376.

Arends J, Jongebloed WL (1979). Crystallite dimensions of enamel. J Biol Buccale 6:161–171.

Arends J, Nelson DGA, Dijkman AG, Jongebloed WL (1984). Effect of various fluorides on enamel structure and chemistry. In: Guggenheim B (ed). Cariology Today. Basel: Karger:245–258.

Arends J, Christoffersen J (1990). Nature and role of loosely bound fluoride in dental caries. J Dent Res 69:601–605.

Arends J, Ruben J (1993). Chlorhexidine CHX release by dentine after varnish treatment [abstract 88]. Caries Res 27:231–232.

Arno A, Waerhaug J, Lövdal A, Schei O (1958). Incidence of gingivitis as related to sex, occupation, tobacco consumption, toothbrushing and age. Oral Surg Oral Med Oral Pathol 11:587–595.

Arnold FA, McClure FJ, White CL (1960). Sodium fluoride tablets for children. Dent Prog 1:8–12.

Arnold FA Jr, Likins RC, Russel AL, Scott DB (1962). 15th year of the Grand Rapids fluoridation study. J Am Dent Assoc 65:780–785.

Arrow P, Riordan P (1995). Retention and caries preventive effects of a GIC and resin-based fissure sealant. Community Dent Oral Epidemiol 23:282–285.

Ashley FP, Sainsbury RH (1981). The effect of a school-based plaque control programme on caries and gingivitis. Br Dent J 150:41–45.

Asikainen S, Sandholm L, Sandman S, Ainamo J (1984). Gingival bleeding after chlorhexidine rinses with or without mechanical oral hygiene. J Clin Periodontol 11:87–94.

Asikainen S, Alaluusua S, Saxén L (1991). Recovery of A. actinomycetemcomitans from teeth, tongue, and saliva. J Periodontol 62:203–206.

Asikainen S, Chen C, Slots J (1995). Actinobacillus actinomycetemcomitans genotypes in relation to serotypes and periodontal status. Oral Microbiol Immunol 10:65–68.

Association of State and Territorial Dental Directors, the New York State Health Department, the Ohio Department of Health and the School of Public Health, University of Albany, State University of New York (1995). Workshop on guidelines for sealant use: Recommendations. J Public Health Dent 55:263–273.

Ast DB (1943). The caries-fluorine hypothesis and a suggested study to test its application. Public Health Rep 58:857–879.

Ast DB, Finn SB, McCaffrey I (1950). The Newburgh-Kingston caries-fluorine study. I. Dental findings after 3 years of water fluoridation. Am J Public Health 40:716–724.

Ast DB, Fitzgerald B (1962). Effectiveness of water fluoridation. J Am Dent Assoc 65:581–588.

Ast DB, Cons NC, Pollard ST, Garfinkel J (1970). Time and cost factors to provide regular periodic dental care for children in a fluoridated and a non-fluoridated area: Final report. J Am Dent Assoc 80:770–776.

Attin T, Koidl U, Buchalla W, Schaller H, Kielbassa A, Hellwig E (1997). Correlation of microhardness and wear in differently eroded bovine dental enamel. Arch Oral Biol 42:243–250.

Attin T, Kocabiyik M, Buchalla W, Hannig C, Becker K (2003). Susceptibility of enamel surfaces to demineralization after application of fluoridated carbamide peroxide gels. Caries Res 37:93–99.

Attwood D, Blinkhorn AS (1989). Reassessment of the effect of fluoridation on cost of dental treatment among Scottish schoolchildren. Community Dent Oral Epidemiol 17:79–82.

Attwood D, Blinkhorn AS (1991). Dental health in schoolchildren 5 years after water fluoridation ceased in south-west Scotland. Int Dent J 41:43–48.

Axelsson P (1969). The Eva-system. Sver Tandläkarförb Tidn 61:1086–1104.

Axelsson P, Lindhe J (1974). The effect of a preventive programme on dental plaque, gingivitis and caries in schoolchildren. Results after one and two years. J Clin Periodontol 1:126–138.

Axelsson P, Lindhe J (1975). The effect of fluoride on gingivitis and dental caries in a preventive programme based on plaque control. Community Dent Oral Epidemiol 3:156–160.

Axelsson P, Lindhe J, Wäseby J (1976). The effect of various plaque control measures on gingivitis and caries in schoolchildren. Community Dent Oral Epidemiol 4:232–239.

Axelsson P, Lindhe J (1977). The effect of a plaque control programme on gingivitis and dental caries in schoolchildren. J Dent Res 56 (special issue): C142–C148.

Axelsson P (1978). The Effect of Plaque Control Procedures on Gingivitis, Periodontitis and Dental Caries [thesis]. Gothenburg: Univ of Gothenburg.

Axelsson P, Lindhe J (1978). Effect of controlled oral hygiene procedures on caries and periodontal disease in adults. J Clin Periodontol 5:133–151.

Axelsson P (1981). Concept and practice of plaque control. Pediatr Dent 3:101–113.

Axelsson P, Lindhe J (1981a). Effect of controlled oral hygiene procedures on caries and periodontal disease in adults. Results after six years. J Clin Periodontol 8:239–248.

Axelsson P, Lindhe J (1981b). The significance of maintenance care in the treatment of periodontal disease. J Clin Periodontol 8:281–294.

Axelsson P, Lindhe J (1981c). Effect of oral hygiene and professional tooth-cleaning on gingivitis and dental caries. Community Dent Oral Epidemiol 6:251–255.

Axelsson P (1987). Placknybildningsindex PFRI—Indikator för karies- och parodontitprevention, munhygien-frekvens och ytrelaterad munhygien. Tandlakartidningen 79:387–391.

Axelsson P, Lindhe J (1987). Efficacy of mouthrinses in inhibiting dental plaque and gingivitis in man. J Clin Periodontol 14:205–212.

Axelsson P, Paulander J, Nordqvist K, Karlsson R (1987a). The effect of fluoride containing dentifrice, rinsing and varnish on interproximal dental caries. A 3-year clinical trial. Community Dent Oral Epidemiol 15:177–180.

Axelsson P, Kristoffersson K, Karlsson R, Bratthall D (1987b). A 30-month longitudinal study of the effects of some oral hygiene measures on Streptococcus mutans and approximal dental caries. J Dent Res 66:761–765.

Axelsson P (1988a). The Community Caries Index of Treatment Needs (CCITN). Presented at the 1st International Conference on Preventive Dentistry and Epidemiology, Karlstad, Sweden.

Axelsson P (1988b). The Apical Periodontal Index of Treatment Needs (APITN). Presented at the 1st International Conference on Preventive Dentistry and Epidemiology, Karlstad, Sweden.

Axelsson P, Paulander J, Tollskog G (1988). A new computer-based oral epidemiology system. Presented at the 1st International Conference on Preventive Dentistry and Epidemiology, Karlstad, Sweden.

Axelsson P, Paulander J, Tollskog G (1990). A new computer-based oral epidemiology system. Presented at the 2nd International Conference on Preventive Dentistry and Epidemiology, Karlstad, Sweden.

Axelsson P (1991). A four-point scale for selection of caries risk patients, based on salivary S. mutans levels and plaque formation rate index. In: Johnson NW (ed). Risk Markers for Oral Diseases Caries, vol 1. Cambridge, NY: Cambridge UP:158–170.

Axelsson P, Lindhe J, Nyström B (1991). On the prevention of caries and periodontal disease. Results of a 15-year-longitudinal study in adults. J Clin Periodontol 13:182–189.

Axelsson P (1993a). New ideas and advancing technology in prevention and nonsurgical treatment of periodontal disease. Int Dent J 43:223–238.

Axelsson P (1993b). Current role of pharmaceuticals in prevention of caries and periodontal disease. Int Dent J 43:473–482.

Axelsson P, Paulander J, Svärdström G, Tollskog G, Nordensten S (1993a). Integrated caries prevention: The effect of a needs-related preventive program on dental caries in children—County of Värmland, Sweden—Results after 12 years. Caries Res 27(suppl 1):83–84.

Axelsson P, Rolandsson M, Bjerner B (1993b). How Swedish dental hygienists apply their training program in the field. Community Dent Oral Epidemiol 21:297–302.

Axelsson P, Östlund L, Hontwedt M, Paulander J (1993c). The effect of mouthrinses twice a day on gingivitis, plaque and plaque formation rate index (PFRI)—A four month double-blind study [abstract]. Scand J Dent Res 101.

Axelsson P, Birkhed D, Paulander J, Östlund L, Hontwedt M (1993d). The effect of amine fluoride/stannous fluroide chlorhexidine and NaF mouthrinse on plaque formation rate (PFR) and salivary levels of mutans streptococci and lactobacilli—A 4 month double-blind study. Presented at the 4th World Congress on Preventitive Dentistry, Helsinki, Finland.

Axelsson P (1994). Mechanical plaque control. In: Lang N, Karring T (eds). Proceedings of the 1st European Workshop on Periodontics, 1993. London: Quintessence:219–243.

Axelsson P, Paulander J (1994). The Oral Health Status in 50–55-Year-Olds in the County of Värmland. Värmland, Sweden: County Council of Värmland.

Axelsson P, Buischi YAP, Barbosa MFZ, Karlsson R, Pradi MCB (1994a). The effect of a new oral hygiene training program on approximal caries in 12–15-year-old Brazilian children: Results after three years. Adv Dent Res 8:278–284.

Axelsson P, Östlund L, Hontwedt M, Paulander J (1994b). The effect of amine fluoride/stannous fluoride, chlorhexidine and NaF mouthrinse on plaque, plaque formation rate (PFRI) and gingivitis—A 4 month randomized double-blind study. Presented at the Scandinavian Division of Dental Research Meeting, Gothenburg, Sweden, 19–21 Aug.

Axelsson P, Paulander J, Hontwedt M, Östlund L, Engström A (1997a). The effect of F-chewing gum on salivary secretion rate, plaque (PI), plaque formation rate (PFRI), salivary mutans streptococci (MS) and oral mucosa in subjects with reduced salivary secretion rate—A 6-month longitudinal study. Presented at the 5th World Congress on Preventive Dentistry, Cape Town, South Africa.

Axelsson P, Kocher T, Vivien N (1997b). Adverse effects of toothpastes on teeth, gingiva and oral mucosa. In: Lang NP, Karring T, Lindhe J (eds). Proceedings of the 2nd European Workshop on Periodontology. Chemicals in Periodontics. Berlin: Quintessence:259–261.

Axelsson P, Paulander J, Engström A, Hontwedt M, Östlund L, Svärdström G (1997c). The effect of a new light-cured resin-modified glass-ionomer fissure sealant material on occlusal caries in erupting and newly erupted second molars. A 3-year longitudinal study. Presented at the 5th World Congress on Preventive Dentistry, Cape Town, South Africa.

Axelsson P (1998). Needs-related plaque control measures based on risk prediction. In: Lang NP, Attström R, Löe H (eds). Proceedings of the European Workshop on Mechanical Plaque Control. Berlin: Quintessence:190–247.

Axelsson P, Paulander J, Lindhe J (1998). Relationship between smoking and dental status in 35-, 50-, 65-, and 75-year-old individuals. J Clin Periodontol 25:297–305.

Axelsson P, El Tabakk S (1999a). Caries prevalence in 12-year-old Egyptian schoolchildren related to prevalence of fluorosis, salivary mutans streptococci levels and dietary habits [abstract]. Göteborg, Sweden: Dept of Periodontology Research Congress, Göteborg University.

Axelsson P, El Tabakk S (1999b). Caries incidence in Egyptian schoolchildren related to prevalence of fluorosis, salivary mutans streptococci levels and dietary habits [abstract]. Göteborg, Sweden: Dept of Periodontology Research Congress, Göteborg University.

Axelsson P, Paulander J, Svärdström G, Kaijser H (2000). Effects of population based preventive programs on oral health conditions. J Parodontol Implantol Orale 19:255–269.

Axelsson P, Paulander J, Nordström L, Jonsson AS, Appel B (2001). The Role of Genetic Interleukin-1 Polymorphism on Tooth Loss and Periodontal Support Loss in 50- to 60-year-old Smokers and Non-smokers [abstract]. Gothenburg, Sweden: Scientific Congress, Dept of Periodontology, Univ of Gothenburg.

Axelsson P, Struzycka I, Wojcieszek D, Wierzbicka M (2003). Prediction of caries risk based on salivary mutans streptococci (MS) levels and plaque formation rate index (PFRI). Caries Res (in press).

Axelsson P, Nyström B, Lindhe J (2004). The long-term effect of a plaque control program on tooth mortality, caries and periodontal disease in adults. Results after 30 years of maintenance. J Clin Periodontol (in press).

Azmak N, Atilla G, Luoto H, Sorsa T (2002). The effect of subgingival controlled-release delivery of chlorhexidine chip on clinical parameters and matrix metalloproteinase-8 levels in gingival crevicular fluid. J Periodontol 73:608–615.

Baab DA, Johnson RH (1989). The effect of a new electric toothbrush on supragingival plaque and gingivitis. J Periodontol 60:336–341.

Backer-Dirks O (1966). Posteruptive changes in dental enamel. J Dent Res 45:503–511.

Bader H, Williams R (1997). Clinical and laboratory evaluation of powered electric toothbrushes: Comparative efficacy of 2 powered brushing instruments in furcations and interproximal areas. J Clin Dent 8:91–94.

Bader JD, Sugars DA, Bonito AJ (2001). A systematic review of selected prevention and management methods. Community Dent Oral Epidemiol 29:399–411.

Badersten A, Egelberg J (1972). Den plaque-avlägsnande effekten av tandkräm vid tandborstning. Tandlakartidningen 64:770–773.

Badersten A, Nilvéus R, Egelberg J (1984a). Effect of nonsurgical periodontal therapy. II. Severely advanced periodontitis. J Clin Periodonol 11:63–76.

Badersten A, Nilvéus R, Egelberg J (1984b). Effect of nonsurgical periodontal therapy. III. Single versus repeated instrumentation. J Clin Periodontol 11:114–124.

Balanyk T, Sandham H (1985). Development of sustained-release antimicrobial dental varnishes effective against Streptococcus mutans in vitro. J Dent Res 64:1356–1360.

Banoczy J, Zimmermann P, Pinter A, Hadas E, Bruszt V (1985). Effect of fluoridated milk on caries: 5 year results. J R Soc Health 105:99–103.

Banoczy J, Szóke J, Kertész P, Tóth Z, Zimmermann P, Gintner Z (1989). Effect of amine fluoride/stannous fluoride-containing toothpaste and mouthrinsings on dental plaque, gingivitis, plaque and enamel F-accumulation. Caries Res 23:284–248.

Banoczy J, Nemes J (1991). Effect of amine fluoride (AmF)/Stannous fluoride (SnF$_2$) toothpaste and mouthwashes on dental plaque accumulation, gingivitis, and root-surface caries. Proc Finn Dent Soc 87:555–559.

Banoczy J, Szóke J, Nasz I (1995). Effect of an antibacterial varnish and amine-fluoride/stannous fluoride (AmF/SnF$_2$) toothpaste on streptococcus mutans counts in saliva and dental plaque of children. J Clin Dent 6:131–134.

Banting D, Bosma W, Bollmer B (1989). Clinical effectiveness of a 0.12% chlorhexidine mouthrinse over 2 years. J Dent Res 68(special issue):1716–1718.

References

Barkmeier WM, Gwinnett AJ, Shaffer SE (1985). Effects of enamel etching time on bond strength and morphology. J Clin Orthod 19:36–38.

Barkvoll P, Rölla G, Lagerlöf F (1988). Effect of sodium lauryl sulphate on the deposition of alkali soluble fluoride on enamel in vitro. Caries Res 22:139–144.

Barkvoll P, Rölla G, Svendsen K (1989). Interaction between chlorhexidine digluconate and sodium lauryl sulfate *in vivo*. J Clin Periodontol 16:593–595.

Barkvoll P (1991). Actions and Interactions of Sodium Lauryl Sulfate and Chlorhexidine in the Oral Cavity [thesis]. Oslo: Univ of Oslo.

Barkvoll P, Rölla G (1995). Triclosan reduces the clinical symptoms of the allergic patch test reaction (APR) elicited with 1% nickel sulphate in sensitised patients. J Clin Periodontol 22:485–487.

Barnes GP, Parker W, Lyon T, Fultz R (1986). Indices used to evaluate signs, symptoms and etiologic factors associated with diseases of the periodontium. J Periodontol 57:643–651.

Barnes CM, Weatherford T, Menaker L (1993). A comparison of the Braun Oral-B plaque remover D5 electric and a manual toothbrush in affecting gingivitis. J Clin Dent 4:48–51.

Barrie AM, Stephen KW, Kay EJ (1990). Fissure sealant retention: A comparison of three sealant types under field conditions. Community Dent Health 7:273–277.

Bashir HMS (1988). In Vivo and In Vitro Evaluation of a Fluoride-Releasing Glass [thesis]. Leeds, UK: Univ of Leeds.

Bass CC (1954). An effective method of personal oral hygiene. J La State Med Soc 106:100–112.

Bastiaan RJ (1984). A comparison of the clinical effectiveness of a single and a double headed toothbrush. J Clin Periodontol 11:331–339.

Bates DG, Navia JM (1979). Chemotherapeutic effect of zinc on streptococcus mutans and rat dental caries. Arch Oral Biol 24:799–805.

Bauroth K, Charles C, Mankodi SM, Simmonsi K, Zhao Q (2002). Essential oil mouthrinse vs flossing: Antiplaque/antigingivitis efficacy [abstract 2867]. J Dent Res 81:358.

Bay I, Kardel K, Skouygaard M (1967). Quantitative evaluation of the plaque-removing ability of different types of toothbrushes. J Periodontol 38:526–533.

Bay I, Rölla G (1981). Morphological studies of plaque formation and growth after NaF and SnF$_2$ rinses. In: Rölla G, Sönju T, Embery G (eds). Tooth Surface Interactions and Preventive Dentistry. London: IRL Press:27–32.

Beal JF, James PMC, Bradrock G, Anderson RJ (1979). The relationship between dental cleanliness, dental caries incidence and gingival health. A longitudinal study. Br Dent J 146:111–114.

Beck J, Garcia R, Heiss G, et al (1996). Periodontal disease and cardiovascular disease. J Periodontol 67(suppl):1123–1137.

Beech JL, Jalaly T (1980). Bonding of polymers to enamel: Influence of deposits formed during etching, etching time and period of water immersion. J Dent Res 59:1156–1162.

Bellini HT, Arneberg P, von der Fehr FR (1981). Oral hygiene and caries. A review. Acta Odontol Scand 39:257–265.

Bender GR, Thibodeau EA, Marquis RE (1985). Reduction of acidurance of streptococcal growth and glycolysis by fluoride and gramicidin. J Dent Res 64:90–95.

Benelli E, Serra M, Rodrigues A, Cury J (1993). In situ anticariogenic potential of glass ionomer cement. Caries Res 27:280–284.

Benson B, Henyon G, Grossman E (1993). Clinical plaque removal efficacy of three toothbrushes. J Clin Dent 4:21–25.

Bergenholtz A, Hugoson A, Lundgren D, Östgren A (1969). The plaque-removing ability of various toothbrushes used with the roll technique. Sven Tandlak Tidskr 62:15–25.

Bergenholtz A (1972). Mechanical cleaning in oral hygiene. Oral Hygiene. Munksgaard, Copenhagen: 27–60.

Bergenholtz A, Bjorne A, Vikström B (1974). The plaque-removing ability of some common interdental aids. An intra-individual study. J Clin Periodontol 1160–1165.

Bergenholtz A, Brithon J (1980). Plaque removal by dental floss or toothpicks. An intra-individual comparative study. J Clin Periodontol 7:516–524.

Bergenholtz A, Bjorne A, Glantz P-O, Vikström B (1980). Plaque removal by various triangular toothpicks. J Clin Periodontol 7:121–128.

Bergenholtz A, Olsson A (1984). Efficacy of plaque removal using interdental brushes and waxed dental floss. Scand J Dent Res 92:198–203.

Bergenholtz A, Gustafsson L, Segerlund N, Hagberg C, Östby P (1984). Role of brushing technique and toothbrush design in plaque removal. Scand J Dent Res 92:344–351.

Bergström J, Lavsted S (1979). An epidemiologic approach to toothbrushing and dental abrasion. Community Dent Oral Epidemiol 7:57–64.

Bergström J, Eliasson S (1988). Cervical abrasion in relation to toothbrushing and periodontal health. Scand J Dent Res 96:405–411.

Berkowitz RJ, Jordan HV, White G (1975). The early establishment of *Streptococcus mutans* in the mouths of infants. Arch Oral Biol 20:725–730.

Bille J, Thylstrup A (1982). Radiographic diagnosis and clinical tissue changes in relation to treatment of approximal carious lesions. Caries Res 16:1–6.

Binney A, Addy M, Newcombe R (1992). The effect of a number of commercial rinses compared to toothpaste on plaque regrowth. J Periodontol 63:839–842.

Birkhed D, Sundin B, Westin SI (1989). Per capita consumption of sugar-containing products and dental caries in Sweden from 1960 to 1985. Community Dent Oral Epidemiol 17:41–43.

Bjertness E, Eriksen HM, Hansen BF (1986). Caries prevalence of 35-year-old Oslo citizens in 1973 and 1984. Community Dent Oral Epidemiol 14:277–282.

Bjertness E (1991). The importance of oral hygiene on variation in dental caries in adults. Acta Odontol Scand 49:97–102.

Björn A-L (1974). Dental Health in Relation to Age and Dental Care, supplement 29, Odontologisk Revy Series. Lund, Sweden: Gleerup.

Björn A-L, Andersson U, Olsson A (1981). Gingival recession in 15-year old pupils. Swed Dent J 5:141–146.

Black GV (1908). Operative Dentistry. Chicago: Medico-Dental.

Black GV, McKay FS (1916). Mottled teeth: An endemic developmental imperfection of the enamel of the teeth, heretofore unknown in the literature in dentistry. Dent Cosmos 58:129–156.

Blahut P (1993). A clinical trial of the INTERPLAK powered toothbrush in a geriatric population. Compend Suppl (16):S606–S610.

Blayney JR, Tucker WH (1948). The Evanston dental caries study. J Dent Res 27:279–286.

Blayney JR, Hill IN (1967). Fluorine and dental caries. J Am Dent Assoc 74(special issue):233–302.

Blinkhorn AS, Attwood D, Gavin G, O'Hickey S (1992). Joint epidemiological survey on dental health of 12-year-old schoolchildren in Dublin and Glasgow. Community Dent Oral Epidemiol 20:307–308.

Boksman L, Gratton DR, McCutcheon E, Plotzke OB (1987). Clinical evaluation of a glass ionomer cement as a fissure sealant. Quintessence Int 18:707–709.

Bolden TE, Zambon JJ, Sowinski J, et al (1992). The clinical effect of a dentifrice containing triclosan and a copolymer in a sodium fluoride/silica base on plaque formation and gingivitis: A six-month clinical study. J Clin Dent 4:123–131.

Bollen CML, Mongardini C, Papaioannou W, Van Steenberghe D, Quirynen M (1998). The effect of a one-stage full-mouth disinfection on different intra-oral niches. Clinical and microbiological observations. J Clin Periodontol 25:56–66.

Bonesvoll P, Olsen I (1974). Influence of teeth, plaque and dentures on the retention of chlorhexidine in the human oral cavity. J Clin Periodontol 1:214–221.

Bonesvoll P, Lökken P, Rölla G, Paus PN (1974a). Retention of chlorhexidine in the human oral cavity after mouth rinses. Arch Oral Biol 19:209–212.

Bonesvoll P, Lökken P, Rölla G (1974b). Influence of concentration, time, temperature and pH on the retention of chlorhexidine in the human oral cavity after mouthrinses. Arch Oral Biol 19:1025–1029.

Bonesvoll P, Gjermo P (1978). A comparison between chlorhexidine and some quaternary ammonium compounds with regard to retention, salivary concentration and plaque-inhibiting effect in the human oral cavity after mouthrinses. Arch Oral Biol 23:289–294.

Borem LM, Feigal RJ (1994). Reducing microleakage of sealants under salivary contamination: Digital-image analysis evaluation. Quintesssence Int 25:283–289.

Borutta A, Kunzel W, Rubsam F (1991). The caries-protective efficacy of 2 fluoride varnishes in a 2-year controlled clinical trial [in German]. Dtsch Zahn Mund Kieferheilkd Zentralbl 79:543–549.

Bosman CW, Powell RN (1977). The reversal of localized experimental gingivitis. A comparison between mechanical toothbrushing procedures and a 0.2% chlorhexidine mouth rinse. J Clin Periodontol 4:161–172.

Bossert WA (1937). The relation between the shape of the occlusal surfaces of molars and the prevalence of decay. J Dent Res 16:63–68.

Bouwsma O, Yost K, Baron H (1992). Comparison of a chlorhexidine rinse and a wooden interdental cleaner in reducing interdental gingivitis. Am J Dent 5:143–146.

Bowden G, Ellwood DC, Hamilton IR (1979). Microbial ecology of the oral cavity. In: Alexander M (ed). Advances in Microbial Ecology, vol 3. New York: Plenum:135–177.

Bowden G (1990). Effects of fluoride on the microbial ecology of dental plaque. J Dent Res 69:653–659.

Bowden G, Edwardsson S (1994). Oral ecology and dental caries. In: Thylstrup A, Fejerskov O (eds). Textbook of Clinical Cariology. Copenhagen: Munksgaard:45–69.

Bowden G (1997). Does assessment of microbial composition of plaque/saliva allow for diagnosis of disease activity of individuals? Community Dent Oral Epidemiol 25:76–81.

Bowen RL (1963). Properties of a silica-reinforced polymer for dental restoration. J Am Dent Assoc 66:57–64.

Bowen RL (1970). Crystalline dimethacrylate monomers. J Dent Res 49:810–815.

Boyd RL, Leggott P, Robertson P (1988). Effects on gingivitis of two different 0.4% SnF$_2$ gels. J Dent Res 67:503–507.

Boyd RL, Murray P, Robertson PB (1989a). Effect on periodontal status of rotary electric toothbrushes vs. manual toothbrushes during periodontal maintenance. I. Clinical results. J Periodontol 60:390–395.

Boyd RL, Murray P, Robertson P (1989b). Effect of rotary electric toothbrush versus manual toothbrush on periodontal status during orthodontic treatment. Am J Orthod Dentofacial Orthop 96:342–347.

Boyd RL, Chun Y (1994). 18-month evaluation of the effects of a 0.4% stannous fluoride gel on gingivitis in orthodontic patients. Am J Orthod Dentofacial Orthop 105:35–41.

Boyd RL (1997). Clinical and laboratory evaluation of powered electric toothbrushes: Review of the literature. J Clin Dent 8(special issue):67–71.

Boyde A (1984). Airpolishing effects on enamel, dentine, cement and bone. Br Dent J 156:287–291.

Brady JR, Gray W, Bhaskar S (1973). Electron microscopic study of the effect of water jet lavage device on dental plaque. J Dent Res 52:1310–1315.

Braem MJA (1999). Physical properties of glass-ionomer cements: Fatigue and elasticity. In: Davidson CL, Mjör IA (eds). Advances in Glass-Ionomer Cements. Chicago: Quintessence.

Bratel J, Berggren U (1991). Long-term oral effects of manual or electric toothbrushes used by mentally handicapped adults. Clin Prev Dent 13:5–7.

Bratthall D (1966). Effekten på plaqueutbredningen hos skolbarn efter engångsupplysning med audiovisuell teknik. Sver Tandläkarförb Tidn 58:176–182.

Bratthall D (1991). The global epidemiology of mutans streptococci. In: Johnson NW (ed). Risk Markers for Oral Diseases. Vol 1: Dental Caries. Markers of High and Low Risk Groups and Individuals. Cambridge: Cambridge UP:287–312.

Bratthall D, Ericsson D (1994). Tests for assessment of caries risk. In: Thylstrup A, Fejerskov O (eds). Textbook of Clinical Cariology. Copenhagen: Munksgaard:333–353.

Bratthall D, Serinirach R, Rapisuwon S, et al (1995). A study into the prevention of fissure caries using an antimicrobial varnish. Int Dent J 45:245–254.

Bratthall D (1996). Dental caries: Intervened—Interrupted—Interpreted. Concluding remarks and cariography. Eur J Oral Sci 104:486–491.

Bratthall D, Hänsel-Petersson G, Sundberg H (1996). Reasons for the caries decline: What do the experts believe? Eur J Oral Sci 104:416–422.

Bratthall D (2000). Introducing the Significant Caries Index together with a proposal for a new global oral health goal for 12-year-olds. Int Dent J 50:378–384.

Brauer GM, Termini DJ (1972). Bonding of bovine enamel to restorative resin: Effect of pretreatment of enamel. J Dent Res 34:151–161.

Braun R, Ciancio S (1992). Subgingival delivery by an oral irrigation device. J Periodontol 63:469–472.

Bravo M, Garcia-Anllo I, Baca P, Llodra JC (1997). A 48-month survival analysis comparing sealant (Delton) with fluoride varnish (Duraphat) in 6- to 8-year-old children. Community Dent Oral Epidemiol 25:247–50.

Brecx M, Netuschil L, Reichert B, Schreil G (1990). Efficacy of Listerine, Meridol and chlorhexidine mouthrinses on plaque, gingivitis and plaque bacteria vitality. J Clin Periodontol 17:292–297.

Brecx M, Brownstone E, MacDonald L, Gelsky S, Cheang M (1992). Efficacy of Listerine, Meridol and chlorhexidine mouthrinses as supplements to regular tooth cleaning measures. J Clin Periodontol 19:202–207.

Brecx M, MacDonals L, Legary K, Cheang M, Forgay M (1993). Long-term effects of Meridol and chlorhexidine mouthrinses on plaque, gingivitis, staining and bacterial vitality. J Dent Res 72:1194–1197.

Briner W, Grossman E, Buckner R, et al (1986a). Effect of chlorhexidine gluconate mouthrinse on plaque bacteria. J Periodontal Res 21:44–52.

Briner W, Grossman E, Buckner R, et al (1986b). Assessment of susceptibility of plaque bacteria to chlorhexidine after 6 months oral use. J Periodontal Res 21:53–59.

Brown JP, Schodel D (1976). A review of controlled surveys of dental disease in handicapped persons. J Dent Child 43:313–320.

Brown L, Brunelle J, Kingman A (1996). Periodontal status in the US 1988–1991. Prevalence, extent and demographic variations. J Dent Res 75:672–683.

Brownstein C, Briggs S, Schweitzer K, Briner W, Kornman K (1990). Irrigation with chlorhexidine to resolve naturally occurring gingivitis. A methodologic study. J Clin Periodontol 17:588–593.

Brunelle JA, Carlos JP (1990). Recent trends in dental caries in US children and the effect of water fluoridation. J Dent Res 69:723–727.

Bruun C, Lambrou MJ, Fejerskov O, Thylstrup A (1982). Fluoride in mixed human saliva after different topical fluoride treatments and possible relation to caries inhibition. Community Dent Oral Epidemiol 10:124–129.

Bruun C, Thylstrup A (1984). Fluoride in whole saliva and dental caries experience in areas with high or low concentrations of fluoride in the drinking water. Caries Res 18:450–456.

Bruun C, Givskov H, Thylstrup A (1984). Whole saliva fluoride after toothbrushing with NaF and MFP dentifrices with different F concentrations. Caries Res 18:282–288.

Bruun C, Thylstrup A (1988). Dentifrice usage among Danish schoolchildren. J Dent Res 67:1114–1117.

Bruun C, Givskov H (1991). Formation of CaF_2 on sound enamel and in caries-like enamel lesions after different forms of fluoride application *in vitro*. Caries Res 25:96–100.

Bruun C, Givskov H (1993). Calcium fluoride formation in enamel from semi- or low-concentrated F agents *in vitro*. Caries Res 27:96–99.

Budtz-Jorgensen E, Mojon P, Rentsch A, Roehrich N, von der Muehli D, Baehni P (1996). Caries prevalence and associated predisposing conditions in recently hospitalized elderly persons. Acta Odontol Scand 54:251–256.

Buischi YAP, Axelsson P, Zülske Barbosa M, Mayer M, Carmen M, de Oliviera L (1989). Salivary S. mutans and caries prevalence in Brazilian schoolchildren. Community Dent Oral Epidemiol 17:20–30.

Buischi YAP, Axelsson P, Oliveira LB, Mayer MPA, Gjermo P (1994). The effect of two preventive programs on oral health knowledge and habits among Brazilian schoolchildren. Community Dent Oral Epidemiol 22:41–46.

Buonocore MG (1955). A simple method of increasing the adhesion of acrylic filling materials to enamel surfaces. J Dent Res 34:849–853.

Burt BA, Eklund SA, Loesche WJ (1986a). Dental benefits of limited exposure to fluoridated water in childhood. J Dent Res 61:1322–1325.

Burt BA, Ismail AI, Eklund SA (1986b). Root caries in an optimally fluoridated and a high fluoride community. J Dent Res 65:1154–1158.

Burt BA (1992). The changing patterns of systemic fluoride intake. J Dent Res 71:1228–1235.

Burt BA, Fejerskov O (1996). Water fluoridation. In: Fejerskov O, Ekstrand J, Burt B. Fluoride in Dentistry. Copenhagen: Munksgaard.

Butler WJ, Segreta V, Collins E (1985). Prevalence of dental mottling in school-aged lifetime residents of 16 Texas communities. Am J Public Health 75:1408–1412.

Büyükyilmaz T, Tangugsorn V, Ogaard B, Arends J, Ruben J, Rölla G (1994). The effect of titanium tetrafluoride application around orthodontic brackets. Am J Orthod Dentofacial Orthop 105:293–296.

Büyükyilmaz T, Ogaard B, Rölla G (1997). The resistance of titanium tetrafluoride-treated human enamel to strong hydrochloric acid. Eur J Oral Sci 105:473–477.

Cancro L, Fishman S (1995). The expected effect on oral health of dental plaque control through mechanical removal. Periodontol 2000 8:60–74.

Carey HM, Daly CG (2001). Subgingival debridement of root surfaces with a micro-brush: Macroscopic and ultrastructural assessment. J Clin Periodontol 28:820–827.

Carlsen O (1987). Dental Morphology. Copenhagen: Munksgaard.

Carlsson J, Egelberg J (1965). Effect of diet on early plaque formation in man. Odontol Revy 16:112–125.

Carlsson J, Hamilton I (1994). Metabolic activity of oral bacteria. In: Thylstrup A, Fejerskov O (eds). Textbook of Clinical Cariology. Copenhagen, Munksgaard:71–89.

Carvalho JC, Ekstrand KR, Thylstrup A (1989). Dental plaque and caries on occlusal surfaces of first permanent molar in relation to stage of eruption. J Dent Res 68:773–779.

Carvalho JC, Ekstrand KR, Thylstrup A (1991). Results of 1 year of non-operative caries treatment of erupting permanent first molars. Community Dent Oral Epidemiol 19:23–28.

Carvalho JC, Thylstrup A, Ekstrand KR (1992). Results after 3 years of non-operative occlusal caries treatment of erupting first permanent molars. Community Dent Oral Epidemiol 20:187–192.

Caton JG, Blieden TM, Lowenguth RA, et al (1993). Comparison between mechanical cleaning and an antimicrobial rinse for the treatment and prevention of interdental gingivitis. J Clin Periodontol 20:172–178.

Caufield P, Ratanapridakul K, Allen D, Cutter G (1988). Plasmid-containing strains of Streptococcus mutans cluster within family and racial cohorts: Implications for natural transmission. Infect Immun 56:3216–3220.

Chabanski M, Gillam D, Bulman J, Newman H (1996). Prevalence of cervical dentine sensitivity in a population of patients referred to a specialist periodontology department. J Clin Periodontol 23:989–992.

Chapple I, Walmsley AD, Saxby MS, Moscrop H (1992). Effect of subgingival irrigation with chlorhexidine during ultrasonic scaling. J Periodontol 63:812–816.

Charles C, Sharma JG, Qaqish J, Galustians H, Zhao Q (2002). Antiplaque/antigingivitis efficacy of an essential oil mouthrinse vs flossing [abstract 2866]. J Dent Res 81:358.

Chesters RK, Huntington E, Burchell CK, Stephen KW (1992). Effect of oral care habits on caries in adolescents. Caries Res 26:299–304.

Chestnutt IG, Jones PR, Jacobson AP, Schafer F, Stephen KW (1995). Prevalence of clinically apparent recurrent caries in Scottish adolescents, and the influence of oral hygiene practices. Caries Res 29:266–271.

Choi JW, Drummond JL, Dooley R, Punwani I, Soh JM (1997). The efficacy of primer on sealant shear bond strength. Pediatr Dent 19:286–288.

Chong F, Loo L, Ng O, et al (1985). Relative effectiveness in plaque removal by three toothbrush designs and a new clinical method of interproximal plaque evaluation. Singapore Dent J 10:27–31.

Chow LC, Brown WE (1972). Phosphoric acid conditioning of teeth for pit and fissure sealants. J Dent Res 51:151–156.

Chow LC, Takagi S, Carey CM, Sieck BA (2000). Remineralization effects of a two-solution fluoride mouthrinse: An in situ study. J Dent Res 79:991–995.

Chow LC, Takagi S, Frukhtbeyn S, et al (2002). Remineralization effect of a low-concentration fluoride rinse in an intraoral model. Caries Res 36:136–141.

Christoffersen MR, Christoffersen J, Arends J (1984). Kinetics of dissolution of calcium hydroxyapatite. VII. The effect of fluoride ions. J Cryst Growth 67:107–114.

Christou V, Timmerman M, van der Velden U, van der Weijden F (1998). Comparison of different approaches of interdental oral hygiene: Interdental brushes versus dental floss. J Periodontol 69:759–764.

Churchill HV (1931). Occurrence of fluoride in some waters of the United States. J Ind Eng Chem 23:996–998.

Ciancio SG (1988). Use of mouthrinses for professional indications. J Clin Periodontol 15:520–523.

Ciancio SG, Mather ML, Zambon JJ, Reynolds HS (1989). Effect of a chemotherapeutic agent delivered by an oral irrigation device on plaque, gingivitis and subgingival microflora. J Periodontol 60:310–315.

Ciancio SG (1995). Chemical agents: Plaque control, calculus reduction and treatment of dentinal hypersensitivity. Periodontol 2000 81:75–86.

Ciardi JE, Bowen WH, Rölla G (1978). The effect of antibacterial compounds on glucosyltransferase activity from Streptococcus mutans. Arch Oral Biol 23:301–305.

Clark DC (1982). A review on fluoride varnishes: An alternative topical fluoride treatment. Community Dent Oral Epidemiol 10:117–123.

Clark DC, Stamm JW, Roberts G, Tessier C (1985). Results of a 32-month fluoride varnish study in Sherbrooke and Lac-Mégantic, Canada. J Am Dent Assoc 111:949–953.

Clarkson BH, Wefel JS, Feagin FF (1986). Fluoride distribution in enamel after in vitro caries-like lesion formation. J Dent Res 65:963–966.

Clarkson J (1992). A European view of fluoride supplementation. Br Dent J 172:357.

Clarkson JJ (2000). International collaborative research on fluoride. J Dent Res 79:893–904.

Claydon N, Addy M (1995). The use of planimetry to record and score the modified Navy Index and other area-based plaque indices. J Clin Periodontol 22:670–673.

Clerehugh V, Lennon M, Worthington H (1988). Aspects of the validity of buccal loss of attachment ≥ 1 mm in studies of early periodontitis. J Clin Periodontol 15:207–210.

Cline NV, Layman DL (1992). The effects of chlorhexidine on the attachment and growth of cultured human periodontal cells. J Periodontol 63:598–602.

Cobb C, Rodgers R, Killoy W (1988). Ultrastructural examination of human periodontal pockets following the use of an oral irrigation device in vivo. J Periodontol 59:155–159.

Collaert B, Attström R, de Bruyn H, Movert R (1992a). The effect of delmopinol rinsing on dental plaque formation and gingivitis healing. J Clin Periodontol 19:274–280.

Collaert B, Edwardsson S, Attström R, Hase JC, Aström M, Movert R (1992b). Rinsing with delmopinol 0.2% and chlorhexidine 0.2%: Short term effect on salivary microbiology, plaque and gingivitis. J Periodontol 63:618–625.

Collaert B, Attström R, Edwardsson S, Hase J, Åström M, Movert R (1994). Short-term effect of topical application of delmopinol on salivary microbiology, plaque and gingivitis. Scand J Dent Res 102:17–25.

Cooney PV, Hardwick F (1994). A fissure sealant pilot project in a third party insurance program in Manitoba. J Can Dent Assoc 60:140–141,144–145.

Corpron R, Clark J, Tsai A, et al. (1986). Intraoral effect of a fluoride-releasing device on acid-softened enamel. J Am Dent Assoc 113:383–388.

Corpron R, More F, Beltran E, et al. (1991). In vivo fluoride uptake of human root lesions using a fluoride-releasing device. Caries Res 25:158–160.

Cortellini P, Pino Prato G, Tonetti M (1994). Periodontal regeneration of human infrabony defects. V. Effect of oral hygiene on long-term stability. J Clin Periodontol 21:606–610.

Costerton JW, Lewandowski Z, DeBeer D, Caldwell D, Korber D, James G (1994). Biofilms, the customized microniche. J Bacteriol 176:2137-2142.

Costerton JW, Lewandowski Z, Caldwell D, Korber D, Lappin-Scott H (1995). Microbial biofilms. Ann Rev Microbiol 49:711-745.

Cowsar D, Tarwater O, Tanquary A (1976). Controlled release of fluoride from hydrogels for dental applications. In: Andrade JD (ed). Hydrogels for Medical and Related Applications, vol 31. Washington: American Chemical Society:180-197.

Cranfield M, Kuhn AT, Winter G (1982). Factors relating to the rate of fluoride ion release from glass-ionomer cement. J Dent 10:333-341.

Creanor SL, Strang R, Gilmour W, et al (1992). The effect of chewing gum use on *in situ* enamel lesion remineralization. J Dent Res 71:1895-1900.

Creanor SL, Carruthers L, Saunders W, Strang R, Foye R (1994). Fluoride uptake and release characteristics of glass ionomer cements. Caries Res 28:322-328.

Creanor SL, Saunders W, Carruthers L, Strang R, Foye R (1995). Effect of extrinsic fluoride concentration on the uptake and release of fluoride from two glass ionomer cements. Caries Res 29:424-426.

Crenshaw MA, Bawden JW (1981). Fluoride binding by organic matrix from early and late developing bovine fetal enamel determined by low rate dialysis. Arch Oral Biol 26:437-476.

Crommelin DJ, Higuchi WI, Fox JL, Spooner PJ, Katdare AV (1983). Dissolution rate behaviour of hydroxyapatite-fluorapatite mixtures. Caries Res 17:289-296.

Cronin M, Dembling W (1996). An investigation of the efficacy and safety of a new electric interdental plaque remover for the reduction of interproximal plaque and gingivitis. J Clin Dent 7:74-77.

Cubells AB, Dalmau LB, Petrone ME, Chaknis P, Volpe AR (1991). The effect of triclosan/copolymer/fluoride dentifrice on plaque formation and gingivitis: A six-month clinical study. J Clin Dent 2:63-69.

Cueto E, Buonocore M (1967). Sealing of pits and fissures with an adhesive resin. Its use in caries prevention. J Am Dent Assoc 75:121-128.

Cullinan MP, Westerman B, Hamlet SM, Palmer JE, Faddy MJ, Seymour GJ (2003). The effect of a triclosan-containing dentifrice on the progression of periodontal disease in an adult population. J Clin Periodontol 30:414-419.

Cumming BR, Löe H (1973). Consistency of plaque distribution in individuals without special home care instruction. J Periodontal Res 8:94-100.

Cummins D, Watson GK (1989). Computer model relating chemistry to biological activity of metal anti-plaque agents. J Dent Res 68 (special issue):1702-1705.

Cummins D (1991). Zinc citrate/triclosan: A new anti-plaque system for the control of plaque and the prevention of gingivitis: Short term clinical and mode of action studies. J Clin Periodontol 18:455-461.

Cunea E, Axelsson P (1997). Plackbildungsrateindex bei 3- bis 19-jährigen. Phillip J 7-8:237-239.

Curnow M, Pine CM, Burnside G, Nicholson J, Chesters R, Huntington E (2000). A clinical trial of the efficacy of supervised school toothbrushing in high caries risk children [abstract 120]. Caries Res 34:349.

Cutress TW (1966). Effect of sodium fluoride solutions of varying pH on solubility of bovine enamel. Arch Oral Biol 11:121-130.

Cutress TW, Ainamo J, Sardo-Infirri J (1987). The Community Periodontal Index of Treatment Needs (CPITN) procedure for population groups and individuals. Int Dent J 37:222-233.

Dahlén G, Lindhe J, Sato K, Hanamura H, Okamoto H (1992). The effect of supragingival plaque control on the subgingival microbiota in subjects with periodontal disease. J Clin Periodontol 19:802-809.

Dajean S, Menanteau J (1989). A western-blotting study of enamel glycoproteins in rat experimental fluorosis. Arch Oral Biol 34:413-418.

Dale J (1969). Toothbrushing frequency and its relationship to dental caries and periodontal disease. Aust Dent J 120-123.

Damen J, Buijs M, Ten Cate J (1996). Uptake and release of fluoride by saliva-coated glass-ionomer cement. Caries Res 30:454-457.

Darveau RP, Tanner A, Page R (1997). The microbial challenge in periodontitis. Periodontol 2000 14:12-32.

Davies A (1973). The mode of action of chlorhexidine. J Periodontal Res Suppl 12:68-75.

Davies A, Rooney J, Constable G, Lamp D (1988). The effect of variations in toothbrush design on dental plaque scores. Clin Prev Dent 10:3-9.

Dawes C, Jenkins GN, Tonge CH (1963). The nomenclature of the integuments of the enamel surface of teeth. Br Dent J 16:65-68.

Dawes C (1983). A mathematical model of salivary clearance of sugar from the oral cavity. Caries Res 17:321-334.

Dawes C (1987). Physiological factors affecting salivary flow rate, oral sugar clearance and the sensation of dry mouth in man. J Dent Res 66:648-653.

Dawes C, Weatherell JA (1990). Kinetics of fluoride in the oral fluids. J Dent Res 69:638-644.

Dawes C, Macpherson LMD (1992). Effects on 9 different chewing gums and lozenges on salivary flow rate and pH. Caries Res 26:176-182.

Dean HT (1934). Classification of mottled enamel diagnosis. J Am Dent Assoc 21:1421–1426.

Dean HT, Elvove E (1935). Studies on the minimal threshold of the dental sign of chronic endemic fluorosis (mottled enamel). Public Health Rep 50:1719–1729.

Dean HT, Elvove E (1936). Some epidemiological aspects of chronic endemic dental fluorosis. Am J Public Health 26:567–575.

Dean HT, Jay P, Arnold FA Jr, McClure FJ, Elvove E (1939). Domestic water and dental caries, including certain epidemiological aspects of oral L acidophilus. Public Health Rep 54:862–888.

Dean HT, Jay P, Arnold FA Jr, Elvove E (1941). Domestic water and dental caries. II. A study of 2,832 white children aged 12–14 years, of 8 suburban Chicago communities, including L acidophilus studies of 1,761 children. Public Health Rep 56:761–792.

Dean HT (1942). The investigation of physiological effects by the epidemiological method. In: Moulton FR (ed). Fluorine and Dental Health. Washington, DC: American Association for the Advancement of Science:23–31.

Dean HT, Arnold FA Jr, Elvove E (1942). Domestic water and dental caries. V. Additional studies of the relation of fluoride domestic waters to dental caries experience in 4,425 white children aged 12–14 years of 13 cities in 4 states. Public Health Rep 57:1155–1179.

Dean HT, Arnold FA Jr, Jay P, Knutson JW (1950). Studies on mass control of dental caries through fluoridation of the public water supply. Public Health Rep 65:1403–1408.

Deasy MJ, Singh SM, Rustogi KN, et al (1991). Effect of a dentifrice containing triclosan and a copolymer on plaque formation and gingivitis. Clin Prev Dent 13:12–19.

De Bruyn H (1987). Fluoride Varnishes and Enamel Caries [thesis]. Groningen: Univ of Groningen.

De Bruyn H, Arends J (1987). Fluoride varnishes—A review. J Biol Buccale 15:71–82.

De Bruyn H, Buskes J, Jongebloed W, Arends J (1988a). Fluoride uptake and inhibition of intra-oral demineralization, following the application of varnishes with different concentrations of fluoride. J Biol Buccale 16:81–87.

De Bruyn H, Van Rijn L, Purdell-Lewis D, Arends J (1988b). Influence of various fluoride varnishes on mineral loss under plaque. Dent Res 22:76–83.

De Gee AJ (1999). Physical properties of glass-ionomer cements: Setting shrinkage and wear. In: Davidson CL, Mjör IA (eds). Advances in Glass-Ionomer Cements. Chicago: Quintessence.

De Jager M, Wiedemann W, Klinger H, Meizer B, Sturm D (1998). Plaque removal efficacy of 2 counter-rotational electric toothbrushes [abstract 257]. J Dent Res 77:664.

De la Rosa M, Guerra J, Johnston D, Radike A (1979). Plaque growth and removal with daily toothbrushing. J Periodontol 50:660–665.

De los Santos R, Lin Y, Corpron R, Beltran E, Strachan D, Landry P (1994). In situ remineralization of root surface lesions using a fluoride chewing gum or fluoride-releasing device. Caries Res 28:441–446.

Den Besten PK, Crenshaw MA (1984). The effects of chronic high fluoride levels on forming enamel in the rat. Arch Oral Biol 29:675–680.

Den Besten PK (1986). Effects of fluoride on protein secretion and removal during enamel development in the rat. J Dent Res 665:1272–1277.

Denepitiya J, Fine D, Singh S, DeVizio W, Volpe A, Persson P (1992). Effect upon plaque formation and gingivitis of a triclosan/copolymer/fluoride dentifrice. A 6-month clinical study. Am J Dent 5:307–311.

Dennison JB, Craig RG (1978). Characterization of enamel surfaces prepared with commercial and experimental etchants. J Am Dent Assoc 97:799–805.

Dennison JB, Straffon LH, More FG (1990). Evaluating tooth eruption on sealant efficacy. J Am Dent Assoc 121:610–614.

DePaola PF, Lax M (1968). The caries-inhibiting effect of acidulated phosphate-fluoride chewable tablets: A 2-year double-blind study. J Am Dent Assoc 78:554–557.

DePaola LG, Overholser CD, Meiller TF, Minah GE, Niehaus C (1989). Chemotherapeutic inhibition of supragingival dental plaque and gingivitis development. J Clin Periodontol 16:311–315.

De Soet JJ, Gruythuysen RJM, Bosch JA, Van Amerongen WE (2002). The effect of 6-monthly application of 40% chlorhexidine varnish on the microflora and dental caries incidence in a population of children in Surinam. Caries Res 36:449–455.

De Soete M, Mongardini C, Pauwels M, et al (2001). One-stage full-mouth disinfection. Long-term microbiological results analyzed by checkerboard DNA-DNA hybridization. J Periodontol 72:374–382.

Dijkman A, Huizinga E, Kuben J, Arends J (1990). Remineralization of human enamel in situ after 3 months: The effects of not brushing versus the effect of an F-dentifrice and a F-free dentifrice. Caries Res 263–266.

Dijkman GE, Arends J (1992). Secondary caries in situ around fluoride-releasing light-curing composites: A quantitative model investigation on 4 materials with a fluoride content between 0 and 26 vol%. Caries Res 26:351–357.

Dodds M, Hsieh S, Johnson D (1991). The effect of increased mastication by daily gum-chewing on salivary gland output and dental plaque acidogenicity. J Dent Res 70:1474–1478.

Dörfer CE, Berbig B, von Bethlenfalvy ER, Staehle HJ, Pioch T (2001). A clinical study to compare the efficacy of 2 electric toothbrushes in plaque removal. J Clin Periodontol 28:987–994.

Downer MC, Blinkhorn AS, Attwood D (1981). Effect of fluoridation on the cost of dental treatment among urban Scottish schoolchildren. Community Dent Oral Epidemiol 9:112–116.

Dreizen S, Brown LR (1976). Xerostomia and dental caries. In: Stiles HM, Loesche WJ, O'Brien TC (eds). Microbial Aspects of Dental Caries [Proceedings of a Workshop on Microbial Aspects of Dental Caries, 21–24 June 1976, St Simons Island, GA], vol 1. Washington, DC: IRL:263–273.

Dreizen S, Brown LR, Daly TE, Drane JB (1977). Prevention of xerostomia-related dental caries in irradiated cancer patients. J Dent Res 56:99–104.

Driscoll WS, Heifetz SB, Korts DC (1978). Effect of chewable fluoride tablets on dental caries in schoolchildren. Results after 6 years of use. J Am Dent Assoc 97:820–824.

Driscoll WS, Heifetz SB, Brunell JA (1981). Caries-preventive effects of fluoride tablets in schoolchildren 4 years after discontinuation of treatments. J Am Dent Assoc 103:878–881.

Driscoll WS, Swango PA, Horowitz AM, Kingman A (1982). Caries-preventive effects of daily and weekly fluoride mouthrinsing in a fluoridated community: Final results after 30 months. J Am Dent Assoc 105:1010–1013.

Driscoll WS, Nowjack-Raymer R, Selwitz RH, Li SH, Heifetz SB (1992). A comparison of the caries-preventive effects of fluoride mouthrinsing, fluoride tablets, and both procedures combined: Final results after 8 years. J Public Health Dent 52:111–116.

Duckworth R, Duckworth SC (1978). The ingestion of fluoride in tea. Br Dent J 145:368–370.

Duckworth RM, Morgan SN, Murray AM (1987). Fluoride in saliva and plaque following use of fluoride-containing mouthwashes. J Dent Res 66:1730–1734.

Duckworth RM, Knoop DTM, Stephen KW (1991). Effect of mouthrinsing after toothbrushing with a fluoride dentifrice in human salivary fluoride levels. Caries Res 25:287–291.

Duckworth RM, Morgan SN (1991). Oral fluoride retention after use of fluoride dentifrices. Caries Res 25:123–129.

Duckworth RM, Morgan SN, Ingram GS, Page DJ (1992). Oral fluoride reservoirs and their relationship to anticaries efficacy. In: Embery G, Rölla G (eds). Clinical and Biological Aspects of Dentifrices. New York: Oxford UP.

Duckworth RM (1993). Fluoride in Plaque and Saliva [thesis]. Amsterdam: Univ of Amsterdam.

Duckworth RM, Jones S (1994). The effect of sodium lauryl sulphate on human salivary fluoride levels [abstract]. J Dent Res 73:271.

Duckworth RM, Jones Y, Nicholson J, Jacobson APM, Chestnut IG (1994). Studies on plaque fluoride after use of F-containing dentifrices. Adv Dent Res 8:202–207.

Duckworth RM, Stewart D (1994). Effect of mouthwashes of variable NaF concentration but constant NaF content on oral fluoride retention. Caries Res 28:43–47.

Dudding NJ, Muhler JC (1962). Technique of application of stannous fluoride in a compatible prophylactic paste and as a topical agent. J Dent Child 29:219–224.

Duggal MS, Chawla HS, Curzon MEJ (1991). A study of the relationship between trace elements in saliva and dental caries in children. Arch Oral Biol 36:881–884.

Eakle W, Ford C, Boyd R (1986). Depth penetration in periodontal pockets with oral irrigation. J Clin Periodontol 13:39–44.

Eastoe J, Fejerskov O (1984). Composition of mature enamel proteins from fluorosed teeth. In: Fearnhead RW, Suga S (eds). Tooth Enamel IV. Amsterdam: Elsevier:326–330.

Echeverria JJ (1998). Managing the use of oral hygiene aids to prevent damage: Effects and sequelae of the incorrect use of mechanical plaque removal devices. In: Lang NP, Attström R, Löe H (eds). Proceedings of the European Workshop on Mechanical Plaque Control: Status of the Art and Science of Dental Plaque Control. Castle of Münchenwiler, Berne, Switzerland, 9–12 May 1998. Chicago: Quintessence:268–278.

Egelberg J (1964). Gingival exudate measurements for evaluation of inflammatory changes of the gingivae. Odontol Revy 15:381–398.

Eichner K (1955). Über eine Gruppeneinteilung des Lückengebisses für die Prothetik. Dtsch Zahnärztl Z 10:1831–1834.

Eidelman E, Shapiro J, Haupt M (1974). The retention of fissure sealants using 20 seconds etching time. 3-year follow-up. J Dent Child 89:121–126.

Ekstrand J, Koch G, Petersson L (1980). Plasma fluoride concentration and urinary fluoride excretion in children following application of the fluoride-containing varnish Duraphat. Caries Res 14:185–189.

Ekstrand K (1988). Strukturel undersogelse af det organiske vaev og den bakterielle plaque i okklusalfladers fure-fossa-system i relation til emaljeforandringer. En lysmikroskopisk og ultrastrukturel undersogelse foretaget på ikke-frembrudte og delvis frembrudte visdomstaender [thesis]. Copenhagen: Royal Dental College.

Ekstrand J, Spak C-J, Vogel G (1990). Pharmacokinetics of fluoride in man and its clinical relevance. J Dent Res 9:550–555.

Ekstrand K, Nielsen L, Carvalho J, Thylstrup A (1993). Dental plaque and caries on permanent first molar occlusal surfaces in relation to sagittal occlusion. Scand J Dent Res 101:9–15.

Ekstrand K, Kuzmina I, Björndal L, Thylstrup A (1995). Relationship between external and histologic features of progressive stages of caries in the occlusal fossa. Caries Res 29:243–250.

Ekstrand K, Björndal L (1997). Structural analyses of plaque and caries in relation to the morphology of the groove-fossa system on erupting mandibular third molars. Caries Res 31:336–348.

Ekstrand K, Ricketts D, Kidd E (1997). Reproducibility and accuracy of three methods for assessment of demineralization depth on the occlusal surface: An in vitro examination. Caries Res 31:224–231.

Elderton RJ, Nuttall NM (1983). Variation among dentists in planning treatment. Br Dent J 154:201–206.

Ellingsen J, Svatun B, Rölla G (1980). The effects of stannous and stannic ions on the formation and acidogenicity of dental plaque in vivo. Acta Odontol Scand 38:219–222.

Ellingsen J, Eriksen H, Rölla G (1982a). Extrinsic dental stain caused by stannous fluoride. Scand J Dent Res 90:9–13.

Ellingsen J, Rölla G, Svatun B (1982b). Effect on plaque formation and acidogenicity of stored aqueous solutions of stannous fluoride. Scand J Dent Res 90:429–433.

Ellingsen JE, Rölla G (1994). Dental plaque inhibition by a combination of triclosan and polydimethylsiloxane (silicone oil). Scand J Dent Res 102:26–29.

Elworthy A, Edgar R, Moran J, et al (1995). A 6-month home-usage trial of 0.1% and 0.2% delmopinol mouthwashes. II. Effects on the plaque microflora. J Clin Periodontol 22:527–532.

Emilson CG, Fornell J (1976). Effect of toothbrushing with chlorhexidine gel on salivary microflora, oral hygiene and caries. Scand J Dent Res 84:308–319.

Emilson CG, Krasse B, Westergren G (1976). Effect of a fluoride-containing chlorhexidine gel on bacteria in human plaque. Scand J Dent Res 84:56–62.

Emilson CG (1977a). Susceptibility of various microorganisms to chlorhexidine. Scand J Dent Res 85:255–265.

Emilson CG (1977b). Outlook for Hibitane in dental caries. J Clin Periodontol 4:136–143.

Emilson CG (1981). Effect of chlorhexidine gel treatment on Streptococcus mutans population in human saliva and dental plaque. Scand J Dent Res 89:239–246.

Emilson CG, Axelsson P, Kallenberg L (1982). Effect of mechanical and chemical plaque control measures on oral microflora in schoolchildren. Community Dent Oral Epidemiol 10:111–116.

Emilson CG, Lindquist B, Wennerholm K (1987). Recolonization of human tooth surfaces by Streptococcus mutans after suppression by chlorhexidine treatment. J Dent Res 66:1503–1508.

Emilson CG, Lindquist B (1988). Importance of infection level of mutans streptococci for recolonization of teeth after chlorhexidine treatment. Oral Microbiol Immunol 3:64–67.

Emilson CG (1994). Potential efficacy of chlorhexidine against mutans streptococci and human dental caries. J Dent Res 73:682–691.

Emling R, Yankell S (1994). 30-day evaluation of the action toothbrush for clinical safety and efficacy. J Clin Dent 4:120–124.

Encyclopaedia Britannica, vol 8 (1963). London: William Benton.

Englander HR, Wallace DA (1962). Effects of naturally fluoridated water on dental caries in adults. Public Health Rep 77:887–893.

Englander HR, Keyes PH, Gestwicki M, Sultz HA (1967). Clinical anti-caries effect of repeated topical sodium fluoride applications by mouthpieces. J Am Dent Assoc 75:638–644.

Englander HR, Carlos JP, Senning RS, Mellberg JR (1969). Residual anti-caries effect of repeated topical sodium fluoride applications by mouthpieces. J Am Dent Assoc 78:783–787.

Englander HR, Sherill LT, Miller BG, Carlos JP, Mellberg JR, Senning RS (1971). Incremental rates of dental caries after repeated topical sodium fluoride applications in children with lifelong consumption of fluoridated water. J Am Dent Assoc 82:354–358.

Epstein JB, McBride BC, Stevenson-Moore P, Spinelli J (1991). The efficacy of chlorhexidine gel in reduction of Streptococcus mutans and Lactobacillus species in patients treated with radiation therapy. Oral Surg Oral Med Oral Pathol 71:172–178.

Ericsson Y, Andersson R (1983). Fluoride ingestion with fluoridated domestic salt under Swedish dietary conditions. Caries Res 17:277–288.

Eriksen HM, Gjermo P (1973). Incidence of stained tooth surfaces in students using chlorhexidine-containing dentifrices. Scand J Dent Res 81:533–537.

Espelid I, Tveit AB, Fjelltveit A (1994). Variations among dentists in radiographic detection of occlusal caries. Caries Res 28:169–175.

Evans R, Stam JW (1991). An epidemiologic estimate of the critical period during which human maxillary central incisors are most susceptible to fluorosis. J Public Health Dent 51:251–259.

Fanning EA, Cellier KM, Sommerville CM (1980). South Australian kindergarten children: Effects of fluoride tablets and fluoridated water on dental caries in primary teeth. Aust Dent J 25:259–263.

FDI (1990). Basic Facts 1990: Dentistry around the world. London: FDI.

Featherstone JDB, Zero DT (1992). Laboratory and human studies to elucidate the mechanism of action of fluoride-containing dentifrices. In: Embery G, Rölla G (eds). Clinical and Biological Aspects of Dentifrices. Oxford: Medical Publications:41–50.

Feigal RJ, Hitt J, Splieth C (1993). Retaining sealant on salivary contaminated enamel. J Am Dent Assoc 124:88–97.

Feigal RJ (1998). Sealant and preventive restorations: Review of effectiveness and clinical changes for improvement. Pediatr Dent 20:85–92.

Feigal RJ, Musherure P, Gillespie B, Levy-Polack M, Quelhas I, Hebling J (2000). Improved sealant retention with bonding agents: A clinical study of two-bottle and single-bottle systems. J Dent Res 79:1850–1856.

Fejerskov O, Johnson NW, Silverstone LM (1974). The ultrastructure of fluorosed human dental enamel. Scand J Dent Res 82:357–372.

Fejerskov O, Silverstone LM, Melsen B, Möller IJ (1975). Histological features of fluorosed human dental enamel. Caries Res 9:190–210.

Fejerskov O, Thylstrup A, Larsen MJ (1981). Rational use of fluorides in caries prevention. Acta Odont Scand 39:241–249.

Fejerskov O, Clarkson BH (1996). Dynamics of caries lesion formation. In: Fejerskov O, Ekstrand J, Burt B (eds). Fluoride in Dentistry. Copenhagen: Munksgaard.

Fejerskov O, Baelum V, Richards A (1996a). Dose-response and dental fluorosis. In Fejerskov O, Ekstrand J, Burt B. Fluoride in Dentistry. Munksgaard.

Fejerskov O, Richards A, Den Besten P (1996b). The effect of fluoride on tooth mineralization. In Fejerskov O, Ekstrand J, Burt B. Fluoride in Dentistry. Munksgaard.

Feller R, Kiger R, Triol C (1993). Anticaries efficacy of a triclosan/copolymer dentifrice. J Dent Res 72:248.

Fine DH, Letizia J, Mandel ID (1985). The effect of rinsing with Listerine antiseptic on the properties of developing dental plaque. J Clin Periodontol 12:660–666.

Fine DH (1994). Evaluation of antimicrobial mouthrinses and their bactericidal effectiveness. J Am Dent Assoc 125(suppl 2):11–19.

Fine DH (1995). Chemical agents to prevent and regulate plaque development. Periodontol 2000 8:87–107.

Finkelstein P, Grossman E (1979). The effectiveness of dental floss in reducing gingival inflammation. J Dent Res 58:1034–1039.

Finkelstein P, Grossman E (1984). The clinical quantitative assessment of the mechanical cleaning efficiency of toothbrushes. Clin Prev Dent 6:7–12.

Finn SB, Klapper CE, Volker JF (1955). Intra-oral effects upon hamster caries. In: Sognnaes RF (ed). Advances in Experimental Caries Research: A Symposium Presented on December 29, 1953, at the Boston Meeting. Washington: American Association for the Advancement of Science:152.

Firestone AR, Imfeld T, Schiffer S, Lutz F (1987). Measurement of interdental plaque pH in humans with an indwelling glass pH electrode following a sucrose rinse. A long-term retrospective study. Caries Res 21:555–558.

Fischman SL (1986). Current status of indices of plaque. J Clin Periodontol 13:371–374.

Fitzgerald RJ, Keyes PH (1960). Demonstration of the etiologic role of streptococci in experimental caries in the hamster. J Am Dent Assoc 61:9–19.

Flemmig TF, Newman MG, Doherty FM, Grossman E, Meckel AH, Bakdash MB (1990). Supragingival irrigation with 0.06% chlorhexidine in naturally occurring gingivitis. I. 6-month clinical observations. J Periodontol 61:112–117.

Flötra L, Gjermo P, Rölla G, Waerhaug J (1971). Side-effects of chlorhexidine mouthwashes. Scand J Dent Res 79:119–125.

Fogels HR, Cancro LP, Bianco J, Fischman SL (1982). The anti-caries effect of supervised tooth brushing with a non-fluoride dentifrice. J Dent Child 424–427.

Fomon SJ, Ekstrand J (1996). Fluoride intake. In: Fejerskov O, Ekstrand J, Burt B (eds). Fluoride in Dentistry. Copenhagen: Munksgaard.

Fontana M, Gonzalez-Cabezas C, Haider A, Stookey GK (2002). Inhibition of secondary caries lesion progression using fluoride varnish. Caries Res 36:129–135.

Forabosco A, Baletti R, Spinato S, Colao P, Casolari C (1996). A comparative study of a surgical method and scaling and root planing using the Odontoson. J Clin Periodontol 23:611–614.

Forgie AH, Paterson M, Pine CM, Pitts, NB, Nugent ZJ (2000). A randomized controlled trial of the caries-preventive efficacy of a chlorhexidine-containing varnish in high-caries-risk adolescents. Caries Res 34:432–439.

Forsling JO, Halling A, Lundin S, et al (1999). Proximal caries prevalence in 19-year-olds living in Sweden. A radiographic study in 4 counties. Swed Dent J 23:59–70.

Forsman B (1965). Effect of mouthrinses with sodium fluoride in schools at Växjö, Sweden. Sver Tandlakforb Tidn 57:705–709.

Forss H, Seppä L (1990). Prevention of enamel demineralization adjacent to glass ionomer filling materials. Scand J Dent Res 98:173–178.

Forss H (1993). Release of fluoride and other elements from light-cured glass ionomers in neutral and acidic conditions. J Dent Res 72:1257–1262.

Forss H, Saarni U, Seppä L (1994). Comparison of glass-ionomer and resin-based fissure sealants: A 2-year clinical trial. Community Dent Oral Epidemiol 22:21–24.

Forss H, Näse L, Seppä L (1995). Fluoride concentration, mutans streptococci and lactobacilli in plaque from old glass ionomer fillings. Caries Res 29:50–53.

Forss H, Halme E (1998). Retention of a glass-ionomer cement and a resin-based fissure sealant and effect on carious outcome after 7 years. Community Dent Oral Epidemiol 26:21–25.

Forsten L (1990). Short- and long-term fluoride release from glass ionomers and other fluoride-containing filling materials in vitro. Scand J Dent Res 98:179–185.

Forsten L (1991). Fluoride release and uptake by glass-ionomers. Scand J Dent Res 99:241–245.

Forsten L (1998). Fluoride release and uptake by glass-ionomers and related materials and its clinical effect. Biomaterials 19:503–508.

Fosdick LS (1950). The reduction of the incidence of dental caries. I. Immediate toothbrushing with a neutral dentifrice. J Am Dent Assoc 40:133.

Francetti L, del Fabbro M, Testori T, Weinstein RL (2000). Chlorhexidine spray versus chlorhexidine mouthwash in the control of dental plaque after periodontal surgery. J Clin Periodontol 27:425–430.

Frandsen AM (1970). Oral Hygiene. Copenhagen: Munksgaard.

Frandsen A, Barbano JP, Suomi JD, Chang JJ, Burke AD (1970). The effectiveness of the Charters-, scrub and roll methods of toothbrushing by professionals in removing plaque. Scand J Dent Res 80:267–271.

Frandsen A (1986). Mechanical oral hygiene practices. In: Löe H, Kleinman DV (eds). Dental Plaque Control Measures and Oral Hygiene Practices. Oxford: IRL Press:93–116.

Frencken JE, Makoni F, Sithole WD, Hackenitz E (1998). Three-year survival of one-surface ART restorations and glass-ionomer sealants in a school oral health programme in Zimbabwe. Caries Res 32:119–126.

Friberg A, Johansson R, Malmberg E (1989). Förbrukningen av fluor som karies profylax under en tioårsperiod. Tandläk Tidn 81:772–782.

Fritz UB, Finger WJ, Stean H (1998). Salivary contamination during bonding procedures with one-bottle adhesive system. Quintessence Int 29:567–572.

Full CA, Parkins FM (1975). Effect of cooking vessel composition on fluoride. J Dent Res 54:192.

Fure S, Emilson C (1990). Effect of chlorhexidine gel treatment supplemented with chlorhexidine varnish and resin on mutans streptococci and Actinomyces on root surfaces. Caries Res 24:242–247.

Fure S, Zickert I (1990). Root surface caries and associated factors. Scand J Dent Res 98:391–400.

Fure S (2003). Ten-year incidence of tooth loss and dental caries in elderly Swedish individuals. Caries Res 37:462–469.

Furia TE, Schenkel AG (1968). 2,4,4-Tricloro-2-Hydroxy-diphenyl Ether. New, broad spectrum bacteriostat. Soap Chemical Spec 44:3–17.

Furuichi Y, Lindhe J, Ramberg P, Volpe AR (1992). Patterns of de novo plaque formation in the human dentition. J Clin Periodontol 19:423–433.

Furuichi Y, Ramberg P, Lindhe J, Nabi N, Gaffar A (1996). Some effects of mouthrinses containing salifuor on de novo plaque formation and developing gingivitis. J Clin Periodontol 23:795–802.

Futatsuki M, Kubota K, Yeh YC, Park K, Moss SJ (1995). Early loss of pit and fissure sealant: A clinical and SEM study. J Clin Pediatr Dent 19:99–104.

Gaffar A, Afflitto J, Nabi N (1997). Chemical agents for the control of plaque and plaque microflora: An overview. Eur J Oral Sci 105:502–507.

Galagan DJ, Vermillion JR, Nevitt GA, Stadt ZM, Dart RE (1957). Climate and fluid intake. Public Health Rep 72:484–490.

Gao W, Smales RJ, Yip HK (2000). Demineralisation and remineralisation of dentine caries, and the role of glass-ionomer cements. Int Dent J 50:51–56.

Garcia-Godoy F, DeVizio W, Volpe AR, Ferlauto RJ, Miller JM (1990). Effect of triclosan / copolymer / fluoride dentifrice on plaque formation and gingivitis. Am J Dent 3:15–26.

Garmyn P, Van Steenberghe D, Quirynen M (1998). Efficacy of plaque control in the maintenance of gingival health: Plaque control in primary and secondary prevention. In: Lang NP (ed). Proceedings of the European Workshop on Mechanical Plaque Control. Chicago: Quintessence:107–120.

Geddes DAM, McNee SG (1982). The effect of 0.2% (48 mM) NaF rinses daily on human plaque acidogenicity in situ (Stephan curve) and fluoride content. Arch Oral Biol 27:765–769.

Gelhard TBFM, Arends J (1984). In vivo remineralization of artificial subsurface lesions in human enamel. I. J Biol Buccale 12:49–57.

Gibbons RJ, Van Houte J (1980). Bacterial adherence and the formation of dental plaques. In: Beachey EH (ed). Bacterial Adherence and Recognition. Series B. London: Chapman and Hall:60–104.

Gibson J, Wade A (1977). Plaque removal by bass and roll brushing techniques. J Periodontol 48:456–459.

Giertsen E, Svatun B, Saxton A (1987). Plaque inhibition by hexetidine and zinc. Scand J Dent Res 95:49–54.

Giertsen E, Scheie AA, Rölla G (1988). Inhibition of plaque formation and plaque acidogenicity by zinc and chlorhexidine combinations. Scand J Dent Res 96:541–550.

Giertsen E, Scheie AA, Rölla G (1989a). Dose related effects of zinc chloride on dental plaque formation and plaque acidogenicity in vivo. Caries Res 23:272–277.

Giertsen E, Scheie AA, Rölla G (1989b). Plaque inhibition by a combination of zinc citrate and sodium lauryl sulfate. Caries Res 23:278–283.

Giertsen E (1990). Clinical and Microbiological Aspects of Zn²⁺ and Detergents Related to Dental Plaque [thesis]. Oslo: Univ of Oslo.

Giertsen E, Bowen W, Pearson S (1991). Combined effects of Zn²⁺-chlorhexidine and Zn²⁺-cetylpyridinium chloride on caries incidence in partially desalivated rats. Scand J Dent Res 99:301–309.

Giertsen E, Scheie A (1995). Effects of mouthrinses with chlorhexidine and zinc ions combined with fluoride on the viability and glycolytic activity of dental plaque. Eur J Oral Sci 103:306–312.

Gilbert RJ (1987). The oral clearance of zinc and triclosan after delivery from a dentifrice. J Pharmacol 39:480–483.

Gilbert RJ, Williams P (1987). The oral retention and anti-plaque efficacy of triclosan in human volunteers. Br J Clin Pharmacol 23:579–583.

Gillespie GM, Roviralta G (1985). Salt fluoridation [Scientific Publication no 501]. Washington, DC: Pan American Health Organization (WHO-AMRO).

Gillette WB, Van House R (1980). Three effects of improper oral hygiene procedures. J Am Dent Assoc 101:476–481.

Gisselsson H, Björn AL, Birkhed D (1983). Immediate and prolonged effect of individual preventive measures in caries and gingivitis susceptible children. Swed Dent J 7:13–21.

Gisselsson H, Birkhed D, Björn AL (1988). Effect of professional flossing with chlorhexidine gel on approximal caries in 12- to 15-year-old schoolchildren. Caries Res 22:187–192.

Gisselsson H, Birkhed D, Björn AL (1994). Effect of a 3-year professional flossing program with chlorhexidine gel on approximal caries and cost of treatment in preschool children. Caries Res 28:394–399.

Gjermo P, Flötra L (1970). The effect of different methods of interdental cleaning. J Periodontal Res 5:230–236.

Gjermo P, Baastad KL, Rölla G (1970). The plaque inhibiting capacity of 11 antibacterial compounds. J Periodontal Res 5:102–109.

Gjermo P (1974). Chlorhexidine in dental practice. J Clin Periodontol 1:143–152.

Gjermo P, Bonesvoll P, Rölla G (1974). Relationship between plaque-inhibiting effect and retention of chlorhexidine in the human oral cavity. Arch Oral Biol 19:1031–1034.

Gjermo P, Bonesvoll P, Hjeljord L, Rölla G (1975). Influence of cariation of pH of chlorhexidine mouth rinses on oral retention and plaque-inhibiting effect. Caries Res 9:74–82.

Gjermo P (1986). Promotion of Self Care in Oral Health. Oslo: Dental Faculty, Scandinavian Working Group for Preventive Dentistry.

Gjermo P (1989). Chlorhexidine and related compounds. J Dent Res 68:750–760.

Gjermo P, Saxton C (1991). Antibacterial dentifrices. Clinical data and relevance with emphasis on zinc/triclosan. J Clin Periodontol 18:468–473.

Glantz P (1969). On wettability and adhesiveness. A study of enamel, dentine, some restorative materials and dental plaque. Odontol Revy 20(suppl):17.

Glantz P, Attström R (1986). Tooth-surface altering agents. State-of-the-science review. In: Löe H, Kleinman D (eds). Dental Plaque Control Measures and Oral Hygiene Practices. Oxford: IRL Press:85–194.

Glavind L (1977). Effect of monthly professional mechanical tooth cleaning on periodontal health in adults. J Clin Periodontol 4:100–106.

Glavind L, Zeuner E (1986). The effectiveness of a rotary electronic toothbrush on oral cleanliness in adults. J Clin Periodontol 13:135–138.

Going RE, Loesche WJ, Grainger DA, Syed SA (1978). The viability of micro-organisms in carious lesions 5 years after covering with a fissure sealant. J Am Dent Assoc 97:455–462.

Goldberg AJ, Tanzer J, Munster E, Amara J, Thal F, Birkhed D (1981). Cross-sectional clinical evaluation of recurrent enamel caries, restorational marginal integrity, and oral hygiene status. J Am Dent Assoc 102:635–641.

Goodson JM, Haffajee AD, Socransky SS (1984). The relationship between attachment level loss and alveolar bone loss. J Clin Periodontol 11:348–359.

Goodson JM (1989). Pharmacokinetic principles controlling efficacy of oral therapy. J Dent Res 68(special issue):1625–1632.

Goodson JM (1994). Antimicrobial strategies for treatment of periodontal diseases. Periodontol 2000 5:142–168.

Gordon J, Lamster I, Seiger M (1985). Efficacy of Listerine antiseptic in inhibiting the development of plaque and gingivitis. J Clin Periodontol 12:697–704.

Gordon J (1989). Fissure sealants. In: Murray JJ. The Prevention of Dental Disease, ed 2. New York: Oxford UP:218–238.

Gordon J, Frascella J, Reardon R (1996). A clinical study of the safety and efficacy of a novel electric interdental cleaning device. J Clin Dent 7:70–73.

Gottlow J, Nyman S, Karring T (1992). Maintenance of new attachment gained through guided tissue regeneration. J Clin Periodontol 19:315–317.

Graf H, Mühlemann HR (1966). Telemetry of plaque pH from interdental area. Helv Odontol Acta 19:94–101.

Granath L, Rootzen H, Liljegren F, Holst K (1976). Variation in caries prevalence related to combination of dietary and oral hygiene habits in 6-year-olds. Caries Res 10:308–317.

Granath L, Rootzlen H, Liljegren E, Holst K, Köhler L (1978). Variation in caries prevalence related to combinations of dietary and oral hygiene habits and chewing fluoride tablets in 4-year-old children. Caries Res 12:83–92.

Graves RC, Disney J, Stamm J (1989). Comparative effectiveness of flossing and brushing in reducing interproximal bleeding. J Periodontol 60:243–247.

Greenstein G (1999). Povidine-idodine's effect and role in the management of periodontal diseases: A review. J Periodontol 70:1397–1405.

Grindefjord M, Dahllöf G, Ekström G, Höjer B, Modéer T (1993). Caries prevalence in 2.5-year-old children. Caries Res 27:505–510.

Grindefjord M, Dahllöf G, Nilsson B, Modeer T (1995). Prediction of dental caries development in 1-year-old children. Caries Res 29:343–348.

Gripp VC, Schlagenhauf U (2002). Prevention of early mutans streptococci transmission in infants by professional tooth cleaning and chlorhexidine varnish treatment of the mother. Caries Res 36:366–372.

Groeneveld A (1976). Over het werkingsmechanisme van fluoride in carieus glazuur. Ned Tandartsenbl 31:299–304.

Gröndahl HG, Hollender L, Malmcrona E, Sundquist B (1977). Dental caries and restorations in teenagers. I. Index and score system for radiographic studies of proximal surfaces. Swed Dent J 1:45–50.

Gross A, Tinanoff N (1977). Effect of SnF_2 mouthrinse on initial bacterial colonization of tooth enamel. J Dent Res 56:1179–1183.

Grossi S, Genco R (1998). Periodontal disease and diabetes mellitus: A two-way relationship. Ann Periodontol 3:51–61.

Grossman E, Reiter GP, Sturzenberger OP, de la Rosa M, Dickinson TD, Ferreti GA (1986). 6-month study of the effects of a chlorhexidine mouthrinse on gingivitis in adults. J Periodontal Res 21:33–43.

Grossman E, Dembling W, Proskin H (1995). A comparative clinical investigation of the safety and efficacy of an oscillating/rotating electric toothbrush and a sonic toothbrush. J Clin Dent 6:108–112.

Grossman E, Cronin M, Dembling W, Proskin H (1996). A comparative study of extrinsic tooth stain removal with 2 electric toothbrushes and a manual brush. Am J Dent 9:25–29.

Grossman E, Proskin H (1997). A comparison of the efficacy and safety of an electric and a manual children's toothbrush. J Am Dent Assoc 128:469–474.

Gunbay S, Bicakci N, Parlak H, Guneri T, Kirilmaz L (1992). The effect of zinc chloride dentifrices on plaque growth and oral zinc levels. Quintessence Int 23:619–624.

Gustafsson BE, Quensel CE, Lanke LS, et al (1954). The Vipeholm dental caries study. The effect of different levels of carbohydrate intake on caries activity in 436 individuals observed for five years. Acta Odontol Scand 11:232.

Gwinnett AJ, Buonocore MG (1965). Adhesives and caries prevention. A preliminary report. Br Dent J 119:77–84.

Gwinnett AJ, Golub LM, Kleinberg I (1975). Effect of a repeated prophylaxis on plaque accumulation and gingival crevicular fluid flow [abstract 605]. J Dent Res 54(special issue A).

Haffajee AD, Thompson M, Torresyap G, Guerrero D, Socransky SS (2001). Efficacy of manual and powered toothbrushes. I. Effect on clinical parameters. J Clin Periodontol 28:937–946.

Håkansson J (1978). Dental Care Habits, Attitudes Towards Dental Health and Dental Status Among 20–60-Year-Old Individuals in Sweden [thesis]. Lund, Sweden: Lund Univ.

Håkansson R (1991). Dental Care Habits and Dental Status in 1974–1985 Among Adults in Sweden. Comparative Cross-Sectional and Longitundinal Investigations [thesis]. Lund, Sweden: Lund Univ.

Hallgren A, Oliveby A, Twetman S (1990). Salivary fluoride concentrations in children with glass-ionomer cemented orthodontic appliances. Caries Res 24:239–241.

Hamilton IR (1977). Effects of fluoride on enzymatic regulation of bacterial carbohydrate metabolism. Caries Res 11:262–278.

Hamilton IR (1987). Effects of changing environment on sugar transport and metabolism by oral bacteria. In: Reizer J, Peterkofsky A (eds). Sugar Transport and Metabolism in Gram-positive Bacteria. Chichester: Ellis Horwood:94–131.

Hamilton IR (1990). Biochemical effects of fluoride on oral bacteria. J Dent Res 69:660–667.

Hamilton IR, Bowden G (1996). Fluoride effects on oral bacteria. In: Fejerskov O, Ekstrand J, Burt B. Fluoride in Dentistry. Copenhagen: Munksgaard.

Hamp SE, Lindhe J, Fornell J, Johansson LÅ, Karlsson R (1978). Effect of a field program based on systematic plaque control on caries and gingivitis in schoolchildren after 3 years. Community Dent Oral Epidemiol 6:17–23.

Hamp SE, Bergendal B, Erasmie T, Lindström G, Mellbring S (1982). Dental prophylaxis for youths in their late teens. II. Knowledge about health and diseases and the relation to dental health behavior. J Clin Periodontol 9:35–45.

Hancock E (1996). Prevention. In: Genco RJ, Newman MG (eds). Annals of Periodontology. Chicago: American Academy of Periodontology:223–249.

Handelman SL, Leverett DH, Solomon ES, Brenner CM (1981). Use of adhesive sealants over occlusal carious lesions: Radiographic evaluation. Community Dent Oral Epidemiol 9:256–259.

Handelman SL, Leverett DH, Iker HP (1985). Longitudinal radiographic evaluation of the progress of caries under sealants. J Pedod 9:119–126.

Handelman SL, Leverett DH, Espeland MA, Curzon JA (1986). Clinical radiographic evaluation of sealed carious and sound tooth surfaces. J Am Dent Assoc 113:751–754.

Handelman SL, Leverett DH, Espeland M, Curzon J (1987). Retention of sealants over carious and sound tooth surfaces. Community Dent Oral Epidemiol 15:1–5.

Handelman SL, Shey Z (1996). Michael Buonocore and the Eastman Dental Center: A historic perspective on sealants. J Dent Res 75:529–534.

Hanioka T, Tanaka M, Kataoka K, Ojima M, Shizukuishi S (1995). Clinical evaluation of the plaque removal efficacy of three toothbrushes. J Clin Dent 6:113–116.

Hanke MT (1940). Studies on the local factors in dental caries. I. Destruction of plaques and retardation of bacterial growth in the oral cavity. J Am Dent Assoc 27:1379–1393.

Hannah JJ, Johnson JD, Kuftinec NM (1989). Long-term clinical evaluation of toothpaste and oral rinse containing sanguinaria extract in controlling plaque, gingival inflammation and sulcular bleeding during orthodontic treatment. Am J Orthod Dentofacial Orthop 96:199–207.

Hänsel Petersson G, Twetman S, Bratthall D (2002). Evaluation of a computer program for caries risk assessment in school children. Caries Res 36:327–340.

Hansen F, Gjermo P (1971). The plaque-removing effect of four toothbrushing methods. Scand J Dent Res 79:502–506.

Hansen BF, Johansen J (1977). Periodontal treatment needs of 35-year-old citizens in Oslo. J Clin Periodontol 4:263–271.

Hansen BF, Bjertness E, Gjermo P (1990). Changes in periodontal disease indicators in 35-year-old Oslo citizens from 1973 to 1984. J Clin Periodontol 17:249–254.

Hardwick JL, Teasdale J, Bloodworth G (1982). Caries increments over 4 years in children aged 12 at the start of water fluoridation. Br Dent J 153:217–222.

Hardy J, Newman H, Strahan J (1982). Direct irrigation and subgingival plaque. J Clin Periodontol 9:57–65.

Harper DS, Mueller LF, Fine JB, Gordon J, Laster LL (1990a). Clinical efficacy of a dentifrice and oral rinse containing sanguinaria extract and zinc chloride during 6 months of use. J Periodontol 61:352–358.

Harper DS, Mueller LF, Fine JB, Gordon J, Laster LL (1990b). Effect of 6 months use of a dentifrice and oral rinse containing sanguinaria extract and zinc chloride upon the microflora of the dental plaque and oral soft tissues. J Periodontol 61:359–363.

Harper DS, Gordon J, Fine J, Hovliaras C (1991). The effect of subgingival irrigation as an antiseptic mouthwash on periodontal pocket microflora. J Dent Res 70(special issue):324.

Harrap GJ, Saxton CA, Best JS (1983). Inhibition of plaque growth by zinc salts. J Periodontal Res 18:634–642.

Harrap GJ, Best JS, Saxton CA (1984). Human oral retention of zinc from mouthwashes containing zinc salts and its relevance to dental plaque control. Arch Oral Biol 29:87–91.

Harte DB, Manly R (1975). Effect of toothbrush variables on wear of dentin produced by four abrasives. J Dent Res 54:993–998.

Hatibovic-Kofman S, Koch G (1991). Fluoride release from glass ionomer cement in vivo and in vitro. Swed Dent J 15:253–258.

Hattab FN, el Mowafy OM, Salem NS, el Badrawy WAG (1991). An in vivo study on the release of fluoride from glass-ionomer cement. Quintessence Int 22:221–224.

Hawkins BF, Kohout F, Lainson P, Heckert A (1986). Duration of toothbrushing for effective plaque control. Quintessence Int 17:361–365.

Hayes M, Roden R (1990). The effects of potassium fluoride and potassium laurate on pH gradients in *streptococcus downeii*. Microbial Ecol Health Dis 3:121–128.

Heasman PA, Stacey F, Heasman L, Sellers P, Macgregor ID, Kelly PJ (1999). A comparative study of the Philips HP 735, Braun/Oral B D7 and the Oral B 35 Advantage toothbrushes. J Clin Periodontol 26:85–90.

Heasman PA, Heasman L, Stacey F, McCracken GI (2001). Local delivery of chlorhexidine gluconate (PerioChip) in periodontal maintenance patients. J Clin Periodontol 28:90–95.

Helldén L, Camosci D, Hock J, Tinanoff N (1981). Clinical study to compare the effect of stannous fluoride and chlorhexidine mouthrinses on plaque formation. J Clin Periodontol 8:12–16.

Hellstadius K, Asman B, Gustafsson A (1993). Improved maintenance of plaque control by electrical toothbrushing in periodontitis patients with low compliance. J Clin Periodontol 20:235–237.

Hellström M, Ramberg P, Krok L, Lindhe J (1996). The effect of supragingival plaque control on the subgingival microflora in human periodontitis. J Clin Periodontol 23:934–940.

Hellwig E, Attin T (1994). Fluoride retention in dentin after topical application of fluoride varnishes [abstract 67]. Caries Res 28:199.

Hennessey TD (1973). Some antibacterial properties of chlorhexidine. J Periodontal Res 8(suppl 12):61–67.

Hitt JC, Feigal RJ (1992). Use of a bonding agent to reduce sealant sensitivity to moisture contamination: An in vitro study. Pediatr Dent 14:41–46.

Hjeljord LG, Rölla G, Bonesvoll P (1978). Chlorhexidine protein interactions. J Periodontal Res 8(suppl 12):11–16.

Ho HP, Niederman R (1997). Effectiveness of the Sonicare sonic toothbrush on reduction of plaque, gingivitis, probing pocket depth and subgingival bacteria in adolescent orthodontic patients. J Clin Dent 8:15–19.

Hodge HC (1950). The concentration of fluorides in the drinking water to give the point of minimum caries with maximum safety. J Am Dent Assoc 40:436–439.

Holmen L, Thylstrup A, Ogaard B, Kragh F (1985a). A polarized light microscopic study of progressive stages of enamel caries in vivo. Caries Res 19:348–354.

Holmen L, Thylstrup A, Ogaard B, Kragh F (1985b). A scanning electron microscopic study of progressive stages of enamel caries in vivo. Caries Res 19:355–367.

Holmen L, Méjare J, Malmgren B, Thylstrup A (1988). The effect of regular professional plaque removal on dental caries in vivo. A polarized light and scanning microscope study. Caries Res 22:250–256.

Holst A, Martensson I, Maurin M (1997). Identification of caries risk children and prevention of caries in pre-school children. Swed Dent J 21:185–191.

Holst A, Braune K, Sullivan Å (1998). A 5-year evaluation of fissure sealants applied by dental assistants. Swed Dent J 22:195–201.

Honkala E, Kannas L, Riise J (1991). Oral health habits of schoolchildren in 11 European countries. Int Dent J 15:253–258.

Hoppenbrouwers PM, Driessens FC, Borggreven JM (1987a). The mineral solubility of human tooth roots. Arch Oral Biol 32:319–322.

Hoppenbrouwers PMM, Driessens FCM, Borggreven JMPM (1987b). The demineralization of human dental roots in the presence of fluoride. J Dent Res 66:1370–1374.

Horning GM, Cobb CM, Killoy WJ (1987). Effect of an air-powder abrasive system on tooth surfaces in periodontal surgery. J Clin Periodontol 14:213–220.

Horowitz HS, Kau MCW (1974). Retained anticaries protection from topically applied acidulated phosphate fluoride: 30- and 36- month post-treatment effects. J Prev Dent 1:22–27.

Horowitz H, Ismail A (1996). Topical fluorides in caries prevention. In: Fejerskov O, Ekstrand J, Burt B. Fluoride in Dentistry. Copenhagen: Munksgaard.

Horwitz J, Machtei EE, Peled M, Laufer D (2002). Amine fluoride/stannous fluoride and chlorhexidine mouthwashes as adjuncts to surgical periodontal therapy: A comparative study. J Periodontol 71:1601–1606.

Hoszek A, Piereville F, Schittek M, Ericson D (1998). Fissure penetration and antibacterial effect in vitro of a glass-ionomer cement containing chlorhexidine gluconate. Swed Dent J 22:133–141.

Hotz P (1998). Role of dental plaque control in the prevention of caries. In: Lang NP (ed). Proceedings of the European Workshop on Mechanical Plaque Control. Chicago: Quintessence:35–49.

Houpt M, Fukus A, Eidelman E (1994). The preventive resin (composite resin/sealant) restoration: Nine-year results. Quintessence Int 25:155–159.

Huber B, Rueger K, Hefti A (1985). The effect of the duration of toothbrushing on plaque reduction. Schweiz Monatsschr Zahnmed 95:985–992.

Hugoson A, Koch G, Thilander H, Hoogendoorn H (1974). Lactoperoxidase in the prevention of plaque accumulations, gingivitis and dental caries (III). Effect of mouthrinses with amyloglucosidase and glucoseoxidase in the model system of experimental gingivitis and caries in man. Odontol Revy 25:69–80.

Hugoson A, Koch G, Bergendal T, et al (1986). Oral health of individuals aged 3–80 years in Jönköping, Sweden, in 1973 and 1983. II. A review of clinical and radiographic findings. Swed Dent J 10:175–194.

Hugoson A, Koch G, Bergendal T, et al (1995). Oral health of individuals aged 3-80 years in Jönköping, Sweden, in 1973, 1983 and 1993. I. A review of findings on dental care habits and knowledge of oral health. Swed Dent J 19:225–241.

Hugoson A, Norderyd O, Slotte C, Thorstensson H (1998). Oral hygiene and gingivitis in a Swedish adult population 1973, 1983 and 1993. J Clin Periodontol 25:807–812.

Hugoson A, Laurell L (2000). A prospective longitudinal study on periodontal bone height changes in a Swedish population. J Clin Periodontol 27:665–674.

Huizinga ED, Ruben J, Arends J (1990). Effect of an antimicrobial-containing varnish on root demineralisation in situ. Caries Res 24:130–132.

Huizinga ED, Arends J (1991). The effect of an antimicrobial releasing varnish on root demineralisation in situ. The influence of the demineralisation period. J Biol Buccale 19:29–33.

Huizinga ED, Ruben JL, Arends J (1991). Chlorhexidine and thymol release from a varnish system. J Biol Buccale 19:343–348.

Huizinga ED, Ruben J, Arends J (1992).The effect of an antimicrobial releasing varnish on enamel demineralisation in sites. Caries Res (in press).

Hull PS (1980). Chemical inhibition of plaque. J Clin Periodontol 7:431–442.

Hunter L, Addy M, Moran J, Kohut B, Hovliaras CA, Newcombe RG (1994). A study of a pre-brushing mouthrinse as an adjunct to oral hygiene. J Periodontol 65:762–765.

Hutton WL, Linscott BW, Williams DB (1951). The Brantford fluorine experiment. Interim report after 5 years of water fluoridation. Can J Public Health 42:81–87.

Hutton WL, Linscott BW, Williams DB (1956). Final report of local studies on water fluoridation in Brantford. Can J Public Health 47:89–92.

Hyatt TP (1923). Prophylactic odontotomy. The cutting into the tooth for the prevention of disease. Dent Cosmos 65:234–241.

Ie Y, Schaeken M (1993). Effect of single and repeated application of chlorhexidine varnish on mutans streptococci in plaque from fissures of premolar and molar teeth. Caries Res 27:303–306.

Igarashi K, Lee I, Schachtele C (1989). Comparison of in vivo human dental plaque pH changes within artificial fissures and at interproximal sites. Caries Res 23:417–422.

Imfeld T (1978). In vivo assessment of plaque acid production. A long-term retrospective study. In Guggenheim, Proc. ERGOB Conf Health and Sugar Substitutes. Basel: Karger:218–223.

Imfeld T (1983). Identification of low caries risk dietary components. In: Myers H (ed). Monographs in Oral Science. Basel: Karger.

Ingraham RQ, Williams JE (1970). An evaluation of the utility of application and cariostatic effectiveness of phosphate-fluoride in solutions and gel states. J Tenn Dent Assoc 50:5–12.

Irigoyen ME, Sanchez-Hinojosa G (2000). Changes in dental caries prevalence in 12-year-old students in the State of Mexico after 9 years of salt fluoridation. Caries Res 34:303–307.

Ismail AI, Shveller J, Langille D, MacInnis WA, McNally M (1993). Should the drinking water of Truro, Nova Scotia, be fluoridated? Water fluoridation in the 90s. Community Dent Oral Epidemiol 21:118–125.

Isokangas P, Tiekso J, Alanen P, Mäkinen KK (1989). Long-term effect of xylitol chewing gum on dental caries. Community Dent Oral Epidemiol 17:444–448.

Isokangas P, Makinen KK, Tiekso J, Alanen P (1993). Long term effect of Xylitol chewing gum in the prevention of dental caries: A follow-up 5 years after termination of a prevention program. Caries Res 27:495–498.

Itic J, Serfaty R (1992). Clinical effectiveness of subgingival irrigation with a pulsated jet irrigator versus syringe. J Periodontol 63:174–181.

Itthagarun A, Wei SH, Wefel JS (2000). The effect of different commercial dentifrices on enamel lesion progression: An in vitro pH-cycling study. Int Dent J 50:21–28.

Jacobsen APM, Stephan KW, Strang R (1992). Fluoride uptake and clearance from the buccal mucosa following mouthrinsing. Caries Res 26:56–58.

Jannesson L, Renvert S, Kjellsdotter P, Gaffar A, Nabi N, Birkhed D (2002). Effect of a triclosan-containing toothpaste supplemented with 10% xylitol on mutans streptococci in saliva and dental plaque. A 6-month clinical study. Caries Res 36:36–39.

Jansson L, Lavstedt S, Zimmerman M (2002). Marginal bone loss and tooth loss in a sample from the County of Stockholm—A longitudinal study over 20 years. Swed Dent J 26:21–29.

Jeffcoat MK, Bray KS, Ciancio SG, et al (1998). Adjunctive use of a subgingival controlled-release chlorhexidine chip reduces probing depth and improves attachment level compared with scaling and root planing alone. J Periodontol 69:989–997.

Jeffcoat MK, Palcanis KG, Weatherford TW, Reese M, Geurs NC, Flashner M (2000). Use of a biodegradable chlorhexidine chip in the treatment of adult periodontitis: Clinical and radiographic findings. J Periodontol 71:256–262.

Jenkins GN, Edgar WM, Ferguson DB (1969). The distribution and metabolic effects of human plaque fluorine. Arch Oral Biol 14:105–119.

Jenkins GN, Krebsbach P (1985). Experimental study of the migration of charcoal particles in the human mouth. Arch Oral Biol 30:697–699.

Jenkins S, Addy M, Wade W (1988). The mechanism of action of chlorhexidine. A study of plaque growth on enamel inserts in vivo. J Clin Periodontol 15:415–424.

Jenkins S, Addy M, Newcombe R (1989). Toothpastes containing 0.3% and 0.5% triclosan. I. Effects on 4-day plaque regrowth. Am J Dent 2:211–214.

Jenkins S, Addy M, Newcombe R (1991a). Triclosan and sodium lauryl sulphate mouthrinses. I. Effects on salivary bacterial counts. J Clin Periodontol 18:140–144.

Jenkins S, Addy M, Newcombe R (1991b). Triclosan and sodium lauryl sulphate mouthwashes. II. Effects on 4-day plaque regrowth. J Clin Periodontol 18:145–148.

Jenkins S, Addy M, Newcombe R (1993a). Evaluation of a mouthrinse containing chlorhexidine and fluoride as adjunct to oral hygiene. J Clin Periodontol 20:20–25.

Jenkins S, Addy M, Newcombe R (1993b). The effects of a chlorhexidine toothpaste on the development of plaque, gingivitis and tooth staining. J Clin Periodontol 20:59–62.

Jensen ME, Kohout F (1988). The effect of a fluoridated dentifrice on root and coronal caries in an older adult population. J Am Dent Assoc 117:829–832.

Jepsen S (1998). The role of manual toothbrushes in effective plaque control: Advantages and limitations of manual toothbrushes. In: Lang NP (ed). Proceedings of the European Workshop on Mechanical Plaque Control. Quintessence:121–137.

Joburi WAL, Clark C, Fisher R (1991). A comparison of the effectiveness of two systems for the prevention of radiation caries. Clin Prev Dent 13:101–102.

Johansen JR, Gjermo P, Eriksen H (1975). Effect of 2-years' use of chlorhexidine-containing dentifrices on plaque, gingivitis, and caries. Scand J Dent Res 83:288–292.

Johansson G, Fridlund B (1996). Young adults' views on dental care—A qualitative analysis. Scand J Caring Sci 10:197–204.

Johnson MF (1993). Comparative efficacy of NaF and SMFP dentifrices in caries prevention: A meta-analytic overview. Caries Res 27:328–336.

Johnson BD, McInnes C (1994). Clinical evaluation of the efficacy and safety of a new sonic toothbrush. J Periodontol 65:692–697.

Jolkovsky DL, Waki MY, Newman MG, et al (1990). Clinical and microbiological effects of subgingival and gingival margin irrigation with chlorhexidine gluconate. J Periodontol 61:663–669.

Jones CL, Ritchie J, Marsh P, Van der Ouderaa F (1988a). The effect of long-term use of a dentifrice containing zinc citrate and a non-ionic agent on the oral flora. J Dent Res 67:46–50.

Jones CL, Stephen K, Ritchie J, Huntington E, Saxton C, Van der Ouderaa F (1988b). Long-term exposure of plaque to zinc citrate. Caries Res 22:84–90.

Jones CL, Saxton C, Ritchie J (1990). Microbiological and clinical effects of a dentifrice containing zinc citrate and triclosan in the human experimental gingivitis model. J Clin Periodontol 17:570–574.

Jones CG (1997). Chlorhexidine: Is it still the gold standard? Periodontol 2000 15:55–62.

Joshipura KJ, Kent R, DePaola P (1994). Gingival recession: Intra-oral distribution and associated factors. J Periodontol 65: 864–871.

Jost-Brinkman PG, Heintze SD, Loundos J (1994). Studie zur Wirksamkeit elektrischer Zahnbürsten bei Multiband-Patienten. Kieferorthopädie 8:235–246.

Joyston-Bechal S (1992). Prevention of dental disease following radiotherapy and chemotherapy. Int Dent J 42:47–53.

Kalaga A, Addy M, Hunter B (1989a). Comparison of chlorhexidine delivery by mouthwash and spray on plaque accumulation. J Periodontol 60:127–130.

Kalaga A, Addy M, Hunter B (1989b). The use of 0.2% chlorhexidine as an adjunct to oral health in physically and mentally handicapped adults. J Periodontol 60:381–385.

Källestål C, Matsson L, Holm A-K (1990). Periodontal conditions in a group of Swedish adolescents (I). A descriptive epidemiologic study. J Clin Periodontol 17:601–608.

Källestål C, Uhlin S (1992). Buccal attachment loss in Swedish adolescents. J Clin Periodontol 19:485–491.

Kalsbeek H, Kwant GW, Groeneveld A, Dirks OB, van Eck AA, Theuns HM (1993). Caries experience of 15-year-old children in the Netherlands after discontinuation of water fluoridation. Caries Res 27:201–205.

Kalter PGE, Flissebaalje TD, Groeneveld A (1980). Fluoride retention in human enamel after a single phosphoric acid and mixed phosphoric acid/SnF_2 application in vitro. Arch Oral Biol 25:15–18.

Kandelman D, Gagnon G (1990). A 24-month clinical study of the incidence and progression of dental caries in relation to consumption of chewing gum containing xylitol in school preventive programs. J Dent Res 69:1771–1775.

Karlsson B-S, Larsson Y (1976). Den kariesförebyggande effekten av mekanisk plackkontroll bland 13–16-åriga skolbarn. Tandläkartidningen 68:1085–1086.

Kashani H, Birkhed D, Petersson LG (1995). Uptake and release of fluoride from birch and lime toothpicks. Eur J Oral Sci 103:112–115.

Kashani H, Birkhed D, Petersson LG (1998a). Fluoride concentration in the approximal area after using toothpicks and other fluoride-containing products. Eur J Oral Sci 106:564.

Kashani H, Birkhed D, Arends J, Ruben J, Petersson L, Odelius H (1998b). Effect of toothpicks with and without fluoride on de- and remineralization of enamel and dentine in situ. Caries Res 32:422–427.

Kashani H, Emilson C, Birkhed D (1998c). Effect of NaF-, SnF_2- and chlorhexidine-impregnated birch toothpicks on mutans streptococci and pH in approximal dental plaque. Acta Odontol Scand 56:197–201.

Kashket S, Rodriguez VM, Bunick FJ (1977). Inhibition of glucose utilization in oral streptococci by low concentrations of fluoride. Caries Res 11:301–307.

Kashket S, Kashket ER (1985). Dissipation of the proton motive force in oral streptococci by fluoride. Infect Immun 48:19–22.

Kato T, Iijima H, Ishihara K, Kaneko T, Hirai K, Naito Y (1990). Antibacterial effects of Listerine on oral bacteria. Bull Tokyo Dent Coll 31:301–307.

Katsanoulas T, Renée I, Attström R (1992). The effect of supragingival plaque control on the composition of the subgingival flora in periodontal pockets. J Clin Periodontol 19:760–765.

Katz S (1982). The use of fluoride and chlorhexidine for the prevention of radiation caries. J Am Dent Assoc 104:164–710.

Keene H, Fleming T (1987). Prevalence of caries-associated microflora after radiotherapy in patients with cancer of the head and neck. Oral Surg Oral Med Oral Pathol 64:421–426.

Kelner M (1963). Comparative analysis of the effects of automatic and conventional toothbrushing in mental retardes. Penn Dent J 30:102–108.

Keyes PH (1960). The infectious and transmissible nature of experimental dental caries. Findings and implications. Archs Oral Biol 1:304–320.

Kho P, Smales FC, Hardie JM (1985). The effect of supragingival plaque control on the subgingival microflora. J Clin Periodontol 12:676–686.

Kiger RD, Nylund K, Feller R (1991). A comparison of proximal plaque removal using floss and interdental brushes. J Clin Periodontol 18:681–684.

Kilian M, Larsen MJ, Fejerskov O, Thylstrup A (1979). Effects of fluoride on the initial colonization of teeth in vivo. Caries Res 13:319–329.

Kinane D, Jenkins W, Paterson A (1992). Comparative efficacy of the standard flossing procedure and a new floss applicator in reducing interproximal bleeding. J Periodontol 63:757–760.

Kinane D (1998). The role of interdental cleaning in effective plaque control (need for interdental cleaning in primary and secondary prevention). In: Lang NP, Attström R, Löe H (eds). Proceedings of the European Workshop on Mechanical Plaque Control. Berlin: Quintessence:156–168.

Kitamura M, Kiyak H, Mulligan K (1986). Predictors of root caries in the elderly. Community Dent Oral Epidemiol 14:34–38.

Kjaerheim V, von der Fehr FR, Poulsen S (1980). Two-year study on the effect of professional toothcleaning on schoolchildren in Oppegård, Norway. Community Dent Oral Epidemiol 8:401–406.

Kjaerheim V, Waaler SM (1994). Experiments with triclosan-containing mouthrinses: Dose response and an attempt to locate the receptor site(s) of triclosan in the mouth. Adv Dent Res 8(2):302–306.

Kjaerheim V (1995). Experiments with triclosan [thesis]. Oslo: Univ of Oslo.

Kjaerheim V, Barkvoll P, Waaler S, Rölla G (1995a). Triclosan inhibits histamine-induced inflammation in human skin. J Clin Periodontol 22:423–426.

Kjaerheim V, Röed A, Brodin P, Rölla G (1995b). Effects of triclosan on the rat phrenic nerve-diaphragm preparation. J Clin Periodontol 22:488–493.

References

Kleber CJF, Putt MS (1988). Evaluation of a floss-holding device compared to hand-held floss for interproximal plaque, gingivitis, and patient acceptance. Clin Prev Dent 10:6–14.

Kleemola-Kujala E, Rasanen L (1982). Relationship of oral hygiene and sugar consumption to risk of caries in children. Community Dent Oral Epidemiol 10:224–233.

Klimek J, Prinz M, Mellwig E, Arends J (1985). Effect of a preventive program based on professional toothcleaning and fluoride application on caries and gingivitis. Community Dent Oral Epidemiol 13:295–298.

Klock B, Krasse B (1977). Microbial and salivary conditions in 9–12-year-olds. Scand J Dent Res 85:56–63.

Klock B, Krasse B (1978). Effect of caries-preventive measures in children with high numbers of S. mutans and lactobacilli. Scand J Dent Res 86:221–230.

Klock B, Krasse B (1979). A comparison between different methods for prediction of caries activity. Scand J Dent Res 87:129–139.

Klock B (1984). Long-term effect of intensive caries prophylaxis. Community Dent Oral Epidemiol 12:69–71.

Knutson JW, Armstrong WD (1943). The effect of topically applied sodium fluoride on dental caries experience. Public Health Rep 58:1701–1715.

Knutson J (1948a). Sodium fluoride solution: Technique for applications to the teeth. J Am Dent Assoc 36:37–39.

Knutson JW (1948b). An evaluation of the effectiveness as a caries control measure of the topical application of solutions of fluorides. J Dent Res 27:340–350.

Kobayashi S, Kawasaki K, Takagi O, et al (1992). Caries experience in subjects 18-22 years of age after 13 years' discontinued water fluoridation in Okinawa. Community Dent Oral Epidemiol 20:81–83.

Koch G (1967). Effect of sodium fluoride in dentifrice and mouthwash on incidence of dental caries in schoolchildren [thesis]. Odontol Revy 18(suppl 12).

Koch G, Petersson LG, Rydén H (1979). Effect of fluoride varnish (Duraphat) treatment every 6 months compared with weekly mouthrinses with 0.2% NaF solution on dental caries. Swed Dent J 3:39–44.

Koch G, Bergmann-Arnadottir I, Bjarnason S, Funnbogason S, Höskuldsson O, Karlsson R (1989). A 3-year controlled clinical trial on caries-preventing effect of fluoride dentifrices with and without anticalculus agents. In: ten Cate JM (ed). Recent Advances in the Study of Dental Calculus. Oxford: IRL Press:259–267.

Koch G, Hatibovic-Kofman S (1990). Glass-ionomer cements as a fluoride release system in vivo. Scand J Dent Res 14:267–273.

Köhler B, Bratthall D (1978). Intrafamilial levels of Streptococcus mutans and some aspects of the bacterial transmission. Scand J Dent Res 86:35–42.

Köhler B, Andréen I, Jonsson B, Hultqvist E (1982). Effect of caries preventive measures on Streptococcus mutans and lactobacilli in selected mothers. Scand J Dent Res 90:102–108.

Köhler B, Bratthall D, Krasse B (1983). Preventive measures in mothers influence the establishment of the bacterium Streptococcus mutans in their infants. Arch Oral Biol 28:225–231.

Köhler B, Andréen I, Jonsson B (1984). The effect of caries preventive measures in mothers on dental caries and the oral presence of the bacteria Streptococcus mutans and lactobacilli in their children. Arch Oral Biol 29:879–883.

Köhler B, Andréen I, Jonsson B (1988). The earlier the colonization of mutans streptococci the higher the caries prevalence. Oral Microbiol Immunol 3:14–17.

Kolltveit KM, Eriksen HM (2001). Is the observed association between periodontitis and atherosclerosis causal? Eur J Oral Sci 109:2–7.

Komatsu H, Shimokobe H, Kawakami S, Yoshimura M (1994). Caries-preventive effect of glass ionomer sealant reapplication: Study presents 3-year results. J Am Dent Assoc 125:543–549.

Kornman KS (1986). Antimicrobial agents. In: Löe H, Kleinman DV (eds). Dental Plaque Control Measures and Oral Hygiene Practices. Oxford: IRL Press:121–142.

Kornman KS, Crane A, Wang H-Y, et al (1997). The interleukin-1 genotype as a severity factor in adult periodontal disease. J Clin Periodontol 24:72–77.

Kotsanos N, Darling AI (1991). Influence of posteruptive age of enamel on its susceptibility to artificial caries. Caries Res 25:241–250.

Koulourides T, Cueto H, Pigman W (1961). Rehardening of softened enamel surfaces on human teeth by solutions of calcium phosphate. Nature 189:226.

Koulourides T, Feaging F, Pigman W (1965). Remineralization of dental enamel by saliva in vitro. Ann NY Acad Sci 131:751–757.

Koulourides T, Phantumvanit P, Munksgaard ED, Housch T (1974). An intra-oral model used for studies of fluoride-incorporation in enamel. J Oral Pathol 3:185–195.

Kozai K, Wand D, Sandham J, Phillips H (1991). Changes in strains of mutans streptococci induced by treatment with chlorhexidine varnish. J Dent Res 70:1252–1257.

Krasse B, Emilson CG (1986). Reduction of *Streptococcus mutans* in humans. In: Hamada S, Michalek SM, Kiyono H, Menaker L, McGhee JR (eds). Molecular Microbiology and Immunobiology of *Streptococcus mutans*. Amsterdam: Elsevier:381–389.

Kristoffersson K, Bratthall D (1982). Transient reduction of *Streptococcus mutans* interdentally by chlorhexidine gel. Scand J Dent Res 90:417–22.

Kristoffersson K, Axelsson P, Bratthall D (1984). The effect of a professional tooth-cleaning program on interdentally localized *Streptococcus mutans*. Caries Res 18:385–390.

Kristoffersson K, Axelsson P, Birkhed D, Bratthall D (1986). Caries prevalence, salivary *Streptococcus mutans* and dietary habits in 13-year-old Swedish schoolchildren. Community Dent Oral Epidemiol 14:20–25.

Kula K, Kula T, Davidson W, et al. (1987). Pharmacological evaluation of an intra-oral fluoride-releasing device in adolescents. J Dent Res 66:1538–1542.

Kunisada T, Yamada K, Oda S, Hara O (1997). Investigation on the efficacy of povidone-iodine against antiseptic-resistant species. Dermatology 95 (suppl 2):14–18.

Künzel W (1980). Effect of an interruption in water fluoridation on the caries prevalence of the primary and secondary dentition. Caries Res 14:304–310.

Kuusela S, Honkala E, Kannas L, Tynjala J, Wold B (1997). Oral hygiene habits of 11-year-olds in 22 European countries and Canada in 1993-1994. J Dent Res 76:1602–1609.

Kuzmina I (1997). A caries preventive program among children in a district of Moscow [thesis]. Copenhagen: Univ of Copenhagen.

Lagerweij MD, ten Cate JM (2002). Remineralisation of enamel lesions with daily applications of a high-concentration fluoride gel and a fluoridated toothpaste: An in situ study. Caries Res 36:270–274.

Lamb W, Corpron R, More F, Beltran E, Strachan D, Kowalski C (1993). In situ remineralization of subsurface enamel lesion after the use of a fluoride chewing gum. Caries Res 27:111–116.

Lamster IB, Alfano M, Seiger M, Gordon J (1983). The effect of Listerine antiseptic on reduction of existing plaque and gingivitis. Clin Prev Dent 5:12–16.

Lang NP, Cumming BR, Löe H (1973). Toothbrushing frequency as it relates to plaque development and gingival health. J Periodontol 7:396–405.

Lang NP, Cumming B, Löe H (1977). Oral hygiene and gingival health in Danish dental students and faculty. Community Dent Oral Epidemiol 5:237–242.

Lang NP, Hotz P, Graf H, et al (1982). Effects of supervised chlorhexidine mouthrinses in children. A longitudinal clinical trial. J Periodontal Res 17:101–111.

Lang NP, Brecx M (1986). Chlorhexidine digluconate—An agent for chemical plaque control and prevention of gingival inflammation. J Periodontal Res 21(suppl):16:74–89.

Lang NP, Karring T (1994). Proceedings of the 1st European Workshop on Periodontics, 1993. London: Quintessence.

Lang NP, Ronis DL, Farghaly M (1995). Preventive behaviours as correlates of periodontal health status. J Public Health Dent 55:10–17.

Lang NP, Mombelli A, Attström R (1997). Dental plaque and calculus. In: Lindhe J, Karring T, Lang N, (eds). Clinical Periodontology and Implant Dentistry. Copenhagen: Munksgaard:102–137.

Lang NP (1998). Commonly used indices to assess oral hygiene and gingival and periodontal health and diseases. In: Lang NP, Attström R, Löe H (eds). Proceedings of the European Workshop on Mechanical Plaque Control. Berlin: Quintessence:50–71.

Larsen MJ, Kirkegaard E, Poulse S (1989). Patterns of dental fluorosis in a European country in relation to fluoride concentrations in drinking water. J Dent Res 66:10–12.

Larsen MJ (1990). Chemical events during tooth dissolution. J Dent Res 69:575–580.

Larsen MJ, Bruun C (1994). Caries chemistry and fluoride—Mechanisms of action. In: Thylstrup A, Fejerskov O (eds). Textbook of Clinical Cariology, ed 2. Copenhagen: Munksgaard:231–257.

Lasfargues JJ, Kaleka R, Louis JJ (2000). A new system of minimally invasive preparations: The Si/Sta concept. In: Roulet JF, Degrange M. Adhesion: The Silent Revolution in Dentistry. Quintessence:107–152.

Laster LL, Lobene RR (1990). New perspectives on sanguinaria clinicals: Individual toothpaste and oral rinse testing. J Can Dent Assoc 56:19–30.

Leach SA, Speechley JA, White MJ, Abbott JJ (1986). Remineralisation *in vivo* by stimulating salivary flow with Lycasin: A pilot study. In: Leach SA (ed). Factors Relating to Demineralisation and Remineralisation of the Teeth. Oxford: IRL Press:69–79.

Leach SA, Lee GTR, Edgar WM (1989). Remineralization of artificial caries-like lesions in human enamel *in situ* by chewing sorbitol gum. J Dent Res 68:1064–1068.

Le Bell Y, Forsten L (1980). Sealing of preventively enlarged fissures. Acta Odontol Scand 38:101–104.

Lee H, Swartz M (1971). Sealing of developmental pits and fissures. I. In vitro study. J Dent Res 50:133–140.

Lee H, Ocumpaugh D, Swartz M (1972). Sealing of developmental pits and fissures. II. Fluoride release from flexible fissure sealers. J Dent Res 51:183–190.

Lefkowitz H, William B, Robinson G (1962). Effectiveness of automatic hand brushes in removing dental plaque and debris. J Am Dent Assoc 65:351-361.

Lerner E, Barak M, Landau I, Palmer M, Kolatch B, Soskolne A (1996). Chlorhexidine release profile from a Perio Chip [abstract 3308]. J Dent Res 75(special issue):431.

Lemke CW, Doherty JM, Arra MC (1970). Controlled fluoridation: The dental effects of discontinuation in Antigo, Wisconsin. J Am Dent Assoc 80:782-786.

Leverett D, Adair S, Shields C, et al (1987). Relationship between salivary and plaque fluoride levels and dental caries experience in fluoridated and non-fluoridated communities [abstract 57]. Caries Res 21:179.

Leverett D, Adair SM, Proskin HM (1988). Dental fluorosis among children in fluoridated and non-fluoridated communities [abstract]. J Dent Res 67:230.

Li YH, Bowden GHW (1994). The effect of surface fluoride on the accumulation of biofilms of oral bacteria. J Dent Res 73:1615-1626.

Lie T (1978). Ultrastructural study of early dental plaque formation. J Periodontal Res 13:391-409.

Lie T, Enersen M (1986). Effects of chlorhexidine gel in a group of maintenance-care patients with poor oral hygiene. J Periodontol 57:364-369.

Lin J, Raghavan S, Fuerstenau DW (1981). The adsorption of fluoride ions by hydroxyapatite from aqueous solutions. Colloids Surf 3:357-370.

Lindhe J, Koch G (1967). The effect of supervised oral hygiene on the gingivae of children. J Periodontal Res 2:215.

Lindhe J, Hamp SE, Löe H (1975). Plaque-induced periodontal disease in beagle dogs. J Periodontal Res 10:243-255.

Lindhe J, Nyman S, Karring T (1982). Scaling and root planing in shallow pockets. J Clin Periodontol 9:415-418

Lindhe J, Nyman S (1984). Long-term maintenance of patients treated for advanced periodontal disease. J Clin Periodontol 11:504-514.

Lindhe J, Rosling B, Socransky SS, Volpe AR (1993). The effect of a triclosan-containing dentifrice on established plaque and gingivitis. J Clin Periodontol 20(5):327-334.

Lindquist B, Edward S, Torell P, Krasse B (1989a). Effect of different caries preventive measures in children highly infected with mutans streptococci. Scand J Dent Res 97:330-337.

Lindquist B, Emilson C, Wennerholm K (1989b). Relationship between mutans streptococci in saliva and their colonization of tooth surfaces. Oral Microbiol Immunol 4:71-76.

Lindskog BI, Zetterberg BL (1975). Medicinsk Terminologi. Stockholm, Sweden: Nordiska Bokhandeln.

Listgarten MA, Mayo H, Tremblay R (1975). Development of dental plaque on epoxy resin crowns in man. A light and electron microscopic study. J Periodontol 46:10-26.

Listgarten MA (1976). Structure of the microbial flora associated with periodontal health and disease in man. A light and electron microscopic study. J Periodontol 47:1-18.

Listgarten MA, Lindhe J, Helldén L (1978). Effects of tetracycline and/or scaling on human periodontal disease. Clinical, microbiological and histological observations. J Clin Periodontol 5:246-271.

Listgarten MA, Schifter CC, Sullivan P, George C, Rosenberg ES. (1986). Failure of a microbial assay to reliably predict disease recurrence in a treated periodontitis population receiving regularly scheduled prophylaxes. J Clin Periodontol 13:768-773.

Listgarten MA (1994). The structure of dental plaque. Periodontol 2000 5:52-65.

Lobene R, Soparkar PM, Newman MB (1982). Use of dental floss. Effect on plaque and gingivitis. Clin Prev Dent 4:5-8.

Lobene RR, Battista GW, Petrone DM, Volpe AR, Petrone ME (1991). Clinical efficacy of an anticalculus fluoride dentifrice containing triclosan and a copolymer: A six-month study. Am J Dent 4:83-85.

Löe H, Silness J (1963). Periodontal disease in pregnancy. I. Prevalence and severity. Acta Odontol Scand 21:533-551.

Löe H, Theilade E, Jensen SB (1965). Experimental gingivitis in man. J Periodontol 36:177-187.

Löe H, Schiött CR (1970a). The effects of suppression of the oral microflora upon the development of dental plaque and gingivitis. In: McHugh WD (ed). Dental plaque. Edinburgh: E & S Livingston:247-255.

Löe H, Schiött CR (1970b). The effect of mouthrinses and topical application of chlorhexidine on the development of dental plaque and gingivitis in man. J Periodontal Res 5:79-83.

Löe H, von der Fehr FR, Schiött CR (1972). Inhibition of experimental caries by plaque prevention. The effect of chlorhexidine mouthrinses. Scand J Dent 80:1-9.

Löe H, Rindom Schiött C, Glavind L, Karring T (1976). 2 years oral use of chlorhexidine in man. I. General design and clinical effects. J Periodontal Res 11:135-144.

Löe H, Ånerud Å, Boysen H, Smith M (1978). The natural history of periodontal disease in man. The rate of periodontal destruction before 40 years of age. J Periodontol 49:607-620.

Löe H, Anerud Å, Boysen H (1986). Natural history of periodontal disease in man: Rapid, moderate and no loss of attachment in Sri Lankan laborers 14–46 years of age. J Clin Periodontol 13:431–440.

Löe H, Kleinmann D (1986). Dental Plaque Control Measures and Oral Hygiene Practices. Oxford: IRL Press.

Löe H, Anerud Å, Boysen H (1992). The natural history of periodontal disease in man: Prevalence, severity and extent of gingival recession. J Periodontol 63:489–495.

Lövdal A, Arno A, Schei O, Waerhaug J (1961). Combined effect of subgingival scaling and controlled oral hygiene on the incidence of gingivitis. Acta Odontol Scand 19:537–555.

Lu KH, Ruhlman CD, Chung KL, Sturzenberger OP, Lehnhoff RW (1987). A 3-year clinical comparison of a sodium monofluorophosphate dentifrice with sodium fluoride dentifrices on dental caries in children. J Dent Child 54:241–244.

Lumikari M, Soukka T, Nurmio S, Tenovou J (1991). Inhibition of the growth of *Streptococcus mutans*, *Streptococcus sobrinus* and *Lactobacillus casei* by oral peroxidase systems in human saliva. Arch Oral Biol 36:155–160.

Lundgren M, Emilson CG, Österberg T, Steen G, Birkhed D, Steen B (1997). Dental caries and related factors in 88- and 92-year-olds. Cross-sectional and longitudinal comparisons. Acta Odontol Scand 55:282–291.

Lundström F, Krasse B (1987a). *Streptococcus mutans* and *lactobacilli* frequency in orthodontic patients; the effect of chlorhexidine treatments. Eur J Orthodont 9:109–116.

Lundström F, Krasse B (1987b). Caries incidence in orthodontic patients with high levels of *Streptococcus mutans*. Eur J Orthodont 9:117–121.

Luoma H (1972). The effects of chlorhexidine and fluoride combinations on the potassium, sodium and phosphorus content and acid production of cariogenic streptococci. Arch Oral Biol 17:1431–1437.

Luoma H, Murtomaa H, Nuuja T, et al (1978). A simultaneous reduction of caries and gingivitis in a group of schoolchildren receiving chlorhexidine-fluoride applications. Results after 2 years. Caries Res 12:290–298.

Lussi A (1991). Validity of diagnostic and treatment decisions of fissure caries. Caries Res 25:296–303.

Lussi A, Schaffner M, Hotz P (1991). Dental erosion in a population of Swiss adults. Community Dent Oral Epidemiol 19:286–290.

Lussi A, Firestone A, Schoenberg V, Hotz P, Stich H (1995). In vivo diagnosis of fissure caries using a new electrical resistance monitor. Caries Res 29:81–87.

Lussi A (1996). Impact of including or excluding cavitated lesions when evaluating methods for the diagnosis of occlusal caries. Caries Res 30:389–393.

MacAlpine R, Magnusson, Kiger R, Crigger M, Garrett S, Egelberg J (1985). Antimicrobial irrigation of deep pockets to supplement oral hygiene instruction and root debridement. I. Bi-weekly irrigation. J Clin Periodontol 12:568–577.

Macgregor I, Balding J, Regis D (1998). Flossing behaviour in English adolescents. J Clin Periodontol 25:291–296.

MacNeil SR, Johnson VB, Killoy WJ, Yonke M, Ridenhour L (1999). The time and ease of placement of the chlorhexidine chip local delivery system. Compend Contin Educ Dent 19:1158–1167.

Macpherson LMD, MacFarlane TW, Stephen KW (1990). An intra-oral appliance study of the plaque microflora associated with early enamel demineralization. J Dent Res 69:1712–1716.

Magnusson I, Lindhe J, Yoneyama T, Liljenberg B (1984). Recolonization of a subgingival microbiota following scaling in deep pockets. J Clin Periodontol 11:193–207.

Maia LC, de Souza IP, Cury JA (2003). Effect of a combination of fluoride dentifrice and varnish on enamel surface rehardening and fluoride uptake in vitro. Eur J Oral Sci 111:68–72.

Malmberg E (1976). Tuveforsoket. Tre års profylaktisk behandling av 16–19-åringer. Tandläkartidningen 68:1087–1089.

Maltz M, Zickert I, Krasse B (1981). Effect of intensive treatment with chlorhexidine on number of Streptococcus mutans in saliva. Scand J Dent Res 89:445–449.

Maltz M, Emilson CG (1982). Susceptibility of oral bacteria to various fluoride salts. J Dent Res 61:786–790.

Mandel ID, Kleinberg I (1986). Methods for plaque biochemistry alteration. State-of-the-science review. In: Löe H, Kleinman D (eds). Dental Plaque Control Measures and Oral Hygiene Practices. Oxford: IRL Press:147–179.

Mandel ID (1988). Chemotherapeutic agents for controlling plaque and gingivitis. J Clin Periodontol 15:488–498.

Manji F, Baelum V, Fejerskov O (1986). Dental fluorosis in an area of Kenya with 2 ppm fluoride in the drinking water. J Dent Res 65:659–662.

Mankodi S, Walker C, Conforti N, DeVizio W, McCool JJ, Volpe AR (1992). Clinical effect of a triclosan-containing dentifrice on plaque and gingivitis: A six-month study. Clin Prev Dent 14:4–10.

Mann J, Karniel C, Triol C (1993). Clinical caries study of a triclosan/copolymer dentifrice. J Dent Res 72:248.

References

Manning RH, Edgar WM (1992). Salivary stimulation by chewing-gum and its role in the remineralisation of carieslike lesions in human enamel *in situ*. J Clin Dent 3:71–74.

Manning RH, Edgar WM (1993). PH changes in plaque after eating snacks and meals, and their modification by chewing sugared or sugar-free gum. Br Dent J 174:241–244.

Månsson B (1977). Caries progression in the first permanent molars. A longitudinal study. Swed Dent J 1:185–191.

Margolis HC, Moreno EC, Murphy BJ (1986). Effect of low levels of fluoride in solution on enamel demineralization *in vitro*. J Dent Res 65:23–29.

Marini I, Checchi L, Vecchiet F, Spiazzi L (2000). Intraoral fluoride releasing device: A new clinical therapy for dentine sensitivity. J Periodontol 71:90–95.

Markovic N, Abelson DC, Mandel ID (1988). Sorbitol gum in xerostomics: The effects on dental plaque pH and salivary flow rates. Gerodontology 7:71–75.

Marks RG, Cont AJ, Morehead JE, et al (1994). Results from a 3-year caries clinical trial comparing NaF and SMFP fluoride formulations. Int Dent J 44:275–285.

Marquis R, Burne R, Parsons DT, Casiano-Coion AE (1993). Arginine deiminase and alkali generation in plaque. In: Bowen WA, Tabac L (eds). Cariology for the 90's. Rochester: Univ of Rochester Press:309–317.

Marsh PD, Bradshaw DJ (1990). The effect of fluoride on the stability of oral bacterial communities in vitro. J Dent Res 69:668–671.

Marsh PD (1992). Microbiological aspects of the chemical control of plaque and gingivitis. J Dent Res 71:1431–1438.

Marthaler TM, Schenardi C (1962). Inhibition of caries in children after 5-years use of fluoridated table salt. Helv Odontol Acta 6:1–6.

Marthaler TM, Meija R, Todt C, Vines JJ (1978). Caries-preventive salt in fluoridation. Caries Res 12(suppl 1):15–20.

Marthaler TM (1995). Zahnmedizinische Gruppenprophylaxe in der Schweiz: Beobachtungen und Schlüsse für die Vorbeugung in Deutschland. DAZ Forum 14:211–215.

Martin WE, Kiger RD, Levy S, Feller R (1987). A clinical evaluation of mechanical and conventional toothbrushing by institionalized elderly patients. J Dent Res 66:235.

Mascarenhas AK (1998). Oral hygiene as a risk indicator of enamel and dentin caries. Community Dent Oral Epidemiol 26:331–339.

Mathiesen AT, Ogaard B, Rölla G (1996). Oral hygiene as a variable in dental caries experience in 14-year-olds exposed to fluoride. Caries Res 30:29–33.

Matthijs S, Adriaens PA (2002). Chlorhexidine varnishes: A review. J Clin Periodontol 29:1–8.

Mayfield L, Attström R, Söderholm G (1998). Cost-effectiveness of mechanical plaque control. In: Lang NP (ed). Proceedings of the European Workshop on Mechanical Plaque Control. Chicago: Quintessence: 177–189.

Mayhall JT (1977). The oral health of a Canadian Inuit community: An anthropological approach. J Dent Res 56(special issue):C55–C61.

Maynard JH, Jenkins S, Moran J, Addy M, Newcombe R, Wade W (1993). A 6-month home usage trial of a 1% chlorhexidine toothpaste. II. Effects on the oral microflora. J Clin Periodontol 20:207–211.

McCabe J, Storer R (1980). Adaption of resin restorative materials to etched enamel and the interfacial work of fracture. Br Dent J 148:155–158.

McClure DB (1966). A comparison of toothbrushing techniques for the preschool child. J Dent Child 33:205–210.

McGregor IDM, Rugg-Gunn (1979). A survey of toothbrushing sequence in children and young adults. J Periodontal Res 14:225–230.

McGuire JR, Nunn ME (1999). Prognosis versus actual outcome. IV. The effectiveness of clinical parameters and IL-1 genotype in accurately predicting prognoses and tooth survival. J Periodontol 70:49–50.

McInnes C, Johnson B, Emling R, Yankell S (1994). Clinical and computer-assisted evaluations of stain removal ability of the sonicare electronic toothbrush. J Clin Dent 5:13–18.

McKay FS (1928). The relation of mottled enamel to caries. J Am Dent Assoc 15:1429–1437.

McKay FS (1948). Mass control of dental caries through the use of domestic water supplies containing fluorine. Am J Public Health 38:828–832.

McKenna EF, Grundy GE (1987). Glass ionomer cement fissure sealants applied by operative dental auxiliaries-retention rate after one year. Aust Dent J 32:200–203.

McLean JW, Wilson AD (1974). Fissure sealing and filling with an adhesive glass-ionomer cement. Br Dent J 136:269–276.

McLey L, Boyd R, Sarker S (1997). Clinical and laboratory evaluation of powered electric toothbrushes: Laboratory determination of relative abrasion of three powered toothbrushes. J Clin Dent 8:76–80.

McNabb H, Mombelli A, Lang N (1992). Supragingival cleaning 3 times a week. The microbiological effect in moderately deep pockets. J Clin Periodontol 19:348–356.

Meiller TF, Kutcher MJ, Overholser CD, Niehaus C, DePaola LG, Siegel MA (1991). Effects of an antimicrobial mouthrinse on recurrent aphthous ulcerations. Oral Surg Oral Med Oral Pathol 72:425–429.

Mejáre I, Malmgren B (1986). Clinical and radiographic appearance of proximal carious lesions at the time of operative treatment in young permanent teeth. Scand J Dent Res 94:19–26.

Mejáre I, Mjör IA (1990). Glass ionomer and resin-based fissure sealants: A clinical study. Scand J Dent Res 98:345–350.

Mejáre I, Malmgren B (1994). Molarens ocklusalyta—Tidig diagnostik och behandling. Stockholm, Sweden: Gothia.

Mejáre I, Källestål C, Stenlund H, Johansson H (1998). Caries development from 11 to 22 years of age: A prospective radiographic study. Prevalence and distribution. Caries Res 32:10–16.

Mellberg JR, Chomicki WG (1983). Fluoride uptake by artificial caries lesions from fluoride dentifrices in vivo. J Dent Res 62:540–542.

Mellberg JR, Ripa LW (1983). Fluorides in Preventive Dentistry. Chicago: Quintessence:44.

Mellberg JR, Blake-Haskins J, Petrou ID, Grote NE (1991). Remineralization in situ from a triclosan/copolymer/fluoride dentifrice. J Dent Res 70: 1441–1443.

Melsen B, Rölla G (1983). Reduced clinical effect of monofluorophosphate in the presence of sodium lauryl sulphate. Caries Res 17:549–553.

Menaker L, Weatherford T, Pitts G, Ross NM, Lamm R (1979). The effects of Listerine antiseptic on dental plaque. Ala J Med Sci 16:1.

Merchant A, Pitiphat W, Douglass CW, Crohin C, Joshipura K (2002). Oral hygiene practices and periodontitis in health care professionals. J Periodontol 73:531–535.

Mertz-Fairhurst EJ, Fairhurst CW, Williams JE, Della-Giustina VE, Brooks JD (1984). A comparative clinical study of two pit and fissure sealants: 7-year results. J Am Dent Assoc 109:252–255.

Messer LB, Calache H, Morgan MV (1997). The retention of pit and fissure sealants placed in primary schoolchildren by Dental Health Services, Victoria. Aust Dent J 42:233–239.

Meurman JH (1977). Fissure sealing in occlusal caries prevention. Clinical and experimental studies. Proc Finn Dent Soc 73(8):7–45.

Meurman JH (1988). Ultrastructure, growth and adherence of Streptococcus mutans after treatment with chlorhexidine and fluoride. Caries Res 22:283–287.

Meurman JH, Thylstrup A (1994). Fissure sealants and dental caries. In: Thylstrup A, Fejerskov O(eds). A Clinical Textbook of Clinical Cariology. Copenhagen:Munksgaard.

Meyer-Lueckel H, Satzinger T, Kielbassa AM (2002). Caries prevalence among 6- to 16-year-old students in Jamaica 12 years after the introduction of salt fluoridation. Caries Res 36:170–173.

Mikkelsen L, Börglum-Jensen S, Rindom-Schiött C, Löe H (1981). Classification and prevalences of plaque streptococci after 2 years oral use of chlorhexidine. J Periodontal Res 16:646–658.

Mikkelsen L, Börglum-Jensen S, Löe H (1982). Susceptibility to chlorhexidine of plaque streptococci after 2 years oral chlorhexidine hygiene. J Periodontal Res 17:366–373.

Miller WD (1889). Die Mikroorganismen der Mundhohle. Leipzig: Thieme.

Miller J (1950). A clinical investigation in preventive dentistry. Dent Pract 1:66–75.

Miller WD (1973). The Micro-Organisms of the Human Mouth [unaltered reprint with an introductory essay by K. G. König (Nijmegen)]. Basel: Karger.

Milosevic A, Lennon M, Fear S (1997). Risk factors associated with tooth wear in teenagers: A case control study. Community Dent Health 14:143–147.

Minah GE, DePaola LG, Overholser CD, Meiller TF, Niehaus C, Lamm RA (1989). Effects of 6 months use of an antiseptic mouthrinse on supragingival dental plaque microflora. J Clin Periodontol 16:347–352.

Mirth D, Shern R, Emilson C, et al (1982). Clinical evaluation of an intra-oral device for the controlled release of fluoride. J Am Dent Assoc 105:791–797.

Mirth D, Adderly D, Amsbaugh S, et al (1983). Inhibition of experimental dental caries using an intra-oral fluoride-releasing device. J Am Dent Assoc 107:55–58.

Mitchell L, Murray JJ (1989). Fissure sealants: A critique of their cost-effectiveness. Community Dent Oral Epidemiol 17:19–23.

Miyazaki H, Pilot T, Leclercq MH, Barmes D (1991a). Profiles of periodontal conditions in adolescents measured by CPITN. Int Dent J 41:67–73.

Miyazaki H, Pilot T, Leclercq MH, Barmes D (1991b) Profiles of periodontal conditions in adults, measured by CPITN. Int Dent J 41:74–80.

Miyazaki H, Pilot T, Leclerq M-H (1992). Periodontal Profiles. An Overview of CPITN Data in the WHO Global Oral Data Bank for the Age Group 15–19 Years, 35–44 Years and 65–75 Years. Geneva: WHO.

Möller IJ, Poulsen S (1973). A standardized system for diagnosing, recording and analyzing dental caries data. Scand J Dent Res 81:1–11.

References

Mombelli A, Nicopoulou-Karayianni K, Lang NP (1990). Local differences in the newly formed crevicular microbiota. Schweiz Monatsschr Zahnmed 100:154–158.

Mombelli A (1998). The role of dental plaque in the initiation and progression of periodontal diseases. In: Lang NP, Attström R, Löe H (eds). Proceedings of the European Workshop on the Mechanical Plaque Control. Berlin: Quintessence:85–97.

Mongardini C, van Steenberghe D, Dekeyser C, Quirynen M (1999). One stage full- versus partial-mouth disinfection in the treatment of chronic adult or generalized early-onset periodontitis. I. Long-term clinical observations. J Periodontol 70:632–645.

Moran J, Addy M, Newcombe R (1988). A clinical trial to assess the efficacy of sanguinarine-zinc mouthrinse (Viadent) compared with chlorhexidine mouthrinse (Corsodyl). J Clin Periodontol 15:612–616.

Moran J, Addy M, Roberts S (1992a). A comparison of natural product, triclosan and chlorhexidine mouthrinses on 4-day plaque regrowth. J Clin Periodontol 19:578–582.

Moran J, Addy M, Wade WG, Maynard JH, Roberts SE, Aström M (1992b). A comparison of delmopinol and chlorhexidine on plaque regrowth over a 4-day period and salivary bacterial counts. J Clin Periodontol 19:749–753.

Moran J, Addy M (1995). A comparative study of stain removal with 2 electric toothbrushes and a manual brush. J Clin Dent 6:188–193.

Mörch T, Waerhaug J (1956). Quantitative evaluation of the effect of toothbrushing and toothpicking. J Periodontol 27:183.

Mörch T, Bjorvatn K (1981). Laboratory study of fluoride impregnated toothpicks. Scand J Dent Res 89:499–505.

Mornstad H (1975). Acute sodium fluoride toxicity in rats in relation to age and sex. Acta Pharmacol Toxicol 37:425–428.

Mortimer KU (1964). Some histological features of fissure caries in enamel. Proceedings of the European Organisation for Caries Research (ORCA). 2:85–95.

Mount GJ (1999). Glass-ionomers: A review of their current status. Oper Dent 24:115–124.

Mousqués T, Listgarten MA, Phillips RW (1980). Effect of scaling and root-planing on the composition of the human subgingival microbial flora. J Periodontal Res 15:144–151.

Muhler JC, Radike AW, Nebergall W, et al (1955). Effect of a stannous fluoride-containing dentifrice on caries reduction in children. II. Caries experience after one year. J Am Dent Assoc 50:163–166.

Mukai M, Ikeda M, Yanagihara G, et al (1993). Fluoride uptake in human dentine from glass-ionomer cement in vivo. Arch Oral Biol 38:1093–1098.

Müller LJ, Darby M, Allen D, Tolle S (1987). Rotary electric toothbrushing. Clinical effects on the presence of gingivitis and supragingival dental plaque. Dent Hyg (Chic) 61:546–550.

Müller HP, Müller RF, Lange DE (1990). Morphological compositions of subgingival microbiota in Actinobacillus actinomycetemcomitans-associated periodontitis. J Clin Periodontol 17:549–556.

Murray JJ (1971a). Adult dental health in fluoride and non-fluoride areas. Part 1. Mean DMF values by age. Br Dent J 131:391–395.

Murray JJ (1971b). Adult dental health in fluoride and non-fluoride areas. Part 3. Tooth mortality by age. Br Dent J 131:487–492.

Murray JJ, Rugg-Gunn AJ, Jenkins GN (1992). Fluorides in caries prevention. 3rd ed. London: Wright.

Murray LE, Roberts A (1994). The prevalence of self-reported hypersensitive teeth. Hypersensitive dentine. Arch Oral Biol 39:129.

Murtomaa H, Turtola L, Rytömaa I (1984). Differentiating positively and negatively health oriented Finnish university students by discriminant analyses. Community Dent Oral Epidemiol 12:243–248.

Nabi N, Mukherjee C, Gaffar A, Scmidt R (1989). In vitro and in vivo studies of triclosan/PM/MA copolymer/NaF combination as an antiplaque agent. Am J Dent 2:197–206.

Nagamine M, Itota T, Torii Y, Irie M, Staninec M, Inoue K (1997). Effect of resin-modified glass ionomer cements on secondary caries. Am J Dent 10:173–178.

Nakagawa T, Saito A, Hosaka Y, et al (1990). Bactericidal effects on subgingival bacteria of irrigation with a povidone-iodine solution (Neojodin). Bull Tokyo Dent Coll 31:199–203.

National Research Council (1980). Zinc. In: Recommended Dietary Allowances, ed 9. Washington, DC: National Academy of Sciences:144–147.

Nemes J, Banoczy J, Wierzbicka M, Rost M (1991). The effect of mouthwashes containing amino-fluoride and stannous fluoride on plaque formation and gingivitis in adults. Fogorv Sz 84:233–236.

Nemes J, Banoczy J, Wierzbicka M (1992). Clinical study on the effect of amine fluoride/stannous fluoride on exposed root surfaces. J Clin Dent 3:51–53.

Netuschil L, Weiger R, Preisler R, Brecx M (1995). Plaque bacteria counts and vitality during chlorhexidine, Meridol and Listerine mouthrinses. Eur J Oral Sci 103:355–361.

Newbrun E, Plasschaert AJ, König KG (1974). Progress of caries in fissures of rat molars treated with occlusal sealant. J Am Dent Assoc 89:121–126.

Newbrun E (1985). Chemical and mechanical removal of plaque. Compendium 6:110–116.

Newbrun E (1989). Effectiveness of water fluoridation. J Public Health Dent 49(special issue):279–289.

Newman MG, Flemmig T, Nachnanai S, et al (1990). Irrigation with 0.06% chlorhexidine in naturally occurring gingivitis. II. 6 months microbiological observations. J Periodontol 61:427–433.

Nogueira-Fiho GR, Toledo S, Cury JA (2000). Effect of 3 dentifrices containing triclosan and various additives. An experimental gingivitis study. J Clin Periodontol 27:494–498.

Nolte WA (1973). Oral ecology. In: Nolte WA (ed). Oral Microbiology, ed 2. St Louis: Mosby:21.

Nordbo H, Skogedal O (1982). The rate of cervical abrasion in dental students. Acta Odontol Scand 40:45–47.

Norderyd O, Hugoson A, Grusovin G (1999). Risk for severe periodontal disease in a Swedish adult population. A longitudinal study. J Clin Periodontol 26:608–615.

Nosal G, Scheidt M, O'Neil R, Van Dyke T (1991). The penetration of lavage solution into the periodontal pocket during ultrasonic instrumentation. J Periodontol 62:554–557.

Nyman S, Rosling B, Lindhe J (1975). Effect of professional tooth-cleaning on healing after periodontal surgery. J Clin Periodontol 2:80–86.

Nyman S, Lindhe J, Rosling B (1977). Periodontal surgery in plaque-infected dentitions. J Clin Periodontol 4:240–249.

Nyman S, Westfelt E, Sarhed G, Karring T (1988). Role of "diseased" root cementum in healing following treatment of periodontal disease. A clinical study. J Clin Periodontol 15:464–468.

Nyvad B, Fejerskov O (1986). Active root surface caries converted into inactive caries as a response to oral hygiene. Scand J Dent Res 94:281–284.

Nyvad B, Kilian M (1987). Microbiology of the early colonization of human enamel and root surfaces in vivo. Scand J Dent Res 95:369–380.

Nyvad B, Fejerskov O (1997). Assessing the stage of caries lesion activity on the basis of clinical and microbiological examination. Community Dent Oral Epidemiol 25:69–75.

Nyvad B, ten Cate J, Fejerskov O (1997). Arrest of root surface caries in situ. J Dent Res 76:1845–1853.

O'Beirne G, Johnson R, Persson G, Spektor M (1996). Efficacy of a sonic toothbrush on inflammation and probing depth in adult periodontitis. J Periodontol 67:900–908.

Offenbacher S, Katz V, Fertik G, et al (1996). Periodontal infection as a risk factor for preterm low birth weight. J Periodontol 67(suppl 10):1103–1113.

Offenbacher S, Jared HL, O'Reilly PG, et al (1998). Potential pathogenic mechanisms of periodontitis-associated pregnancy complications. Ann Periodontol 3:233–250.

Offenbacher S, Madianos PN, Champagne CME, et al (1999). Periodontitis-atherosclerosis syndrome: An expanded model of pathogenesis. J Periodontal Res 34:346–352.

Ogaard B, Rölla G, Helgeland K (1983). Alkali-soluble and alkali-insoluble fluoride retention in demineralized enamel in vivo. Scand J Dent Res 91:200–204.

Ogaard B, Arends J, Rolla G, Ekstrand J, Oliveby A (1986). Action of fluoride on initiation of early enamel caries in vivo. Caries Res 20:270–277.

Ogaard B, Rölla G, Dijkman T, Ruben J, Arends J (1991). Effect of fluoride mouthrinsing on caries lesions development in shark enamel: An in situ model study. Scand J Dent Res 99:372–377.

Ogaard B, Cruz R, Rölla G (1992). Fluoride dentifrices, a possible cariostatic mechanism. In: Embery G, Rölla G (eds). Clinical and Biological Aspects of Dentifrices. New York: Oxford University Press.

Ogaard B, Seppä L, Rölla G (1994a). Relationship between oral hygiene and approximal caries in 15-year-old Norgwegian. Caries Res 28:297–300.

Ogaard B, Seppä L, Rölla G (1994b). Professional topical fluoride applications—Clinical efficacy and mechanism of action. Adv Dent Res 8:190–201.

Ogaard B (2001). CaF_2 formation: Cariostatic properties and factors of enhancing the effect. Caries Res 35 (suppl 1):40–44.

O'Leary TJ, Drake RB, Naylor JE (1972). The plaque control record. J Periodontol 43:38–39.

Olsson J, Odham G (1978). Effect of inorganic ions and surface active organic compounds on the adherence of oral streptococci. Scand J Dent Res 86:108–117.

Olsson M, Lindhe J (1991). Periodontal characteristics in individuals with varying form of the upper central incisors. J Clin Periodontol 18:78–82.

Oosterwaal P, Mikx F, van't Hof M, Renggli H (1991a). Comparison of the antimicrobial effect of the application of chlorhexidine gel, amino fluoride gel and stannous fluoride gel in debrided periodontal pockets. J Clin Periodontol 18:245–251.

Oosterwaal P, Mikx F, van't Hof M, Renggli H (1991b). Short-term bactericidal activity of chlorhexidine gel, stannous fluoride gel and amine fluoride gel tested in periodontal pockets. J Clin Periodontol 18:97–100.

Ophaug RH, Singer L, Harland BF (1985). Dietary fluoride intake of 6-month and 2-year-old children in four dietary regions of the US. Am J Clin Nutr 42:701–707.

Ophaug RH, Jenkins GN, Singer L, Krebsbach PH (1987). Acid diffusion analysis of different forms of fluoride in human dental plaque. Arch Oral Biol 32:459–462.

Oppermann R (1979). Effect of chlorhexidine on acidogenicity of dental plaque in vivo. Scand J Dent Res 87:302–308.

Oppermann R (1980a). The effect of some cationic antiseptics on the acidogenicity of dental plaque in vivo. Acta Odontol Scand 38:155–161.

Oppermann R (1980b). The Effect of Organic and Inorganic Ions on the Acidogenicity of Dental Plaque In Vivo [thesis]. Oslo: Univ of Oslo.

Oppermann R, Johansen J (1980). The effect of fluoride and non-fluoride salts of copper, silver and tin on the acidogenicity of dental plaque in vivo. Scand J Dent Res 88:476–480.

Oppermann R, Rölla G (1980). Effect of some polyvalent cations on the acidogenicity of dental plaque in vivo. Caries Res 14:422–427.

Oppermann RV, Rölla G, Johansen JR, Assev S (1980). Thiol groups and reduced acidogenicity of dental plaque in the presence of metal ions in vivo. Scand J Dent Res 88:389–396.

Orland FJ, Blayney JR, Wendell-Harrison R (1954). Use of the germ-free animal technique in the study of experimental dental caries. J Dent Res 33:147–174.

Ostela I, Tenovuo J (1990). Antibacterial activity of dental gels containing combinations of amine fluoride, stannous fluoride, and chlorhexidine against cariogenic bacteria. Scand J Dent Res 98:1–7.

Ostela I, Karhuvaara L, Tenovuo J (1991). Comparative antibacterial effects of chlorhexidine and stannous fluoride-amine fluoride containing dental gels against salivary mutans streptococci. Scand J Dent Res 99:378–383.

Österberg T, Landt H (1976). Index for occlusal status. Tandläkartidningen 68:1216–1223.

Pader M (1988). Oral Hygiene Products and Practice. New York: Marcel Dekker.

Palin-Palokas T, Hansen H, Heinonen O (1987). Relative importance of caries risk factors in mentally retarded children. Community Dent Oral Epidemiol 15:19–23.

Paloheimo L, Ainamo L, Niemi M-L, Viikinkoski M (1987). Prevalence of and factors related to gingival recession in Finnish 15- to 20-year-old subjects. Community Dent Health 4:425–430.

Palomo F, Wantland L, Sanchez A, DeVivio W, Carter W, Baines E (1989). The effect of a dentifrice containing triclosan and a copolymer on plaque formation and gingivitis: A 14-week clinical study. Am J Dent 2:231–237.

Palomo F, Wantland L, Sanchez A, Volpe AR, McCool J, DeVizio W (1994). The effect of three commercially available dentifrices on supragingival plaque formation and gingivitis: A six-month clinical study. Int Dent J 44:75–81.

Park KK, Schemehorn BR, Stookey GK (1993). Effect of time and duration of sorbitol gum chewing on plaque acidogenicity. Pediatr Dent 15:197–202.

Paulander J, Axelsson P, Lindhe J (2003). Relationship between level of education and oral health status in 35-, 50-, 65- and 75-year-olds. J Clin Periodontol 30:697–704.

Paulander J, Wennström J, Axelsson P, Lindhe J (2004a). Risk factors for alveolar bone loss in 50-year-old individuals. A 10-year cohort study. J Clin Periodontol (in press).

Paulander J, Axelsson P, Wennström J, Lindhe J (2004b). Some characterisitics of 50/55-year-old individuals with various experience of destructive periodontal disease: A cross-sectional study. J Clin Periodontol (in press).

Paulander J, Wennström J, Axelsson P, Lindhe J (2004c). Periodontal bone loss related to tooth position: A 10-year cohort study. J Clin Periodontol (in press).

Pearce EIF (1981). The artificial mineralization of dental plaque. In: Ferguson DB (ed). Environment of the Teeth. Vol 3: Frontiers of Oral Physiology Series. Basel: Karger:108–124.

Pearce EIF (1982). Effect of plaque mineralization on experimental dental caries. Caries Res 16:460–471.

Pearce EIF, Moore AJ (1985). Remineralization of softened bovine enamel following treatment of overlying plaque with a mineral-enriching solution. J Dent Res 64:416–421.

Pearce EIF, Dibdin GH (1992). Enzymic hydrolysis of monofluorophosphate by mixed oral organisms. Caries Res 26:225–226.

Pearce EIF, Coote G, Larsen M (1995). The distribution of fluoride in carious human enamel. J Dent Res 74:1775–1782.

Pearce EIF, Larsen M, Coote G (1999). Fluoride in enamel lining pits and fissures of the occlusal groove-fossa system in human molar teeth. Caries Res 33:196–205.

Petersson LG (1976). Fluorine gradients in outermost surface enamel after various forms of topical application of fluorides in vivo. Odontol Revy 27:25–50.

Petersson LG, Koch G, Rasmusson CG, Stanke H (1985). Effect on caries of different fluoride prophylactic programs in preschool children. A 2-year clinical study. Swed Dent J 9:97–104.

Petersson LG, Arthursson L, Östberg C, Jönsson G, Gleerup A (1991a). Caries-inhibiting effects of different modes of Duraphat varnish reapplications: A 3-year radiographic study. Caries Res 25:70–73.

Petersson LG, Maki Y, Twetman S, Edwardsson S (1991b). Mutans streptococci in saliva and interdental spaces after topical application of an antibacterial varnish in schoolchildren. Oral Microbiol Immunol 6:284–287.

Petersson LG, Edwardsson S, Arends J (1992). Antimicrobial effect of a dental varnish, in vitro. Swed Dent J 16:183–189.

Petersson LG (1993). Fluoride mouthrinses and fluoride varnishes. Caries Res 27:35–42.

Petersson LG, Kashani H, Birkhed D (1994). In vitro and in vivo studies of an NaF impregnated toothpick. Swed Dent J 18:69–73.

Petersson LG, Twetman S, Pakhomov GN (1997). Fluoride Varnish for Community-based Caries Prevention in Children. Geneva: Oral Health, Division of Noncommunicable Diseases, World Health Organization.

Petersson LG, Magnusson K, Andersson H, Deijerborg G, Twetman S (1998). Effect of semi-annual applications of a chlorhexidine/fluoride varnish mixture on approximal caries incidence in schoolchildren. Eur J Oral Sci 106:623–627.

Petersson LG, Magnusson K, Anderson H, Almquist B, Twetman S (2000). Effect of quarterly treatments with a chlorhexidine and a fluoride varnish on approximal caries in caries-susceptible teenagers: A 3-year clinical study. Caries Res 34:140–143.

Petersson LG, Arvidsson I, Lynch E, Engstrom K, Twetman S (2002). Fluoride concentrations in saliva and dental plaque in young children after intake of fluoridated milk. Caries Res 36:40–43.

Petit MD, van Steenbergen T, Scholte LM, van der Velden U, de Graff J (1993). Epidemiology and transmission of *Porphyromonas gingivalis* and *Actinobacillus actinomycetemcomitans* among children and their family members. J Clin Periodontol 20:641–650.

Pienihäkkinen K, Söderling E, Ostela I, Leskelä I, Tenovuo J (1995). Comparison of the efficacy of 40% chlorhexidine varnish and 1% chlorhexidine-fluoride gel in decreasing the level of salivary mutans streptococci. Caries Res 29:62–67.

Pitts NB, Rimmer PA (1992). An in vivo comparison of radiographic and directly assessed clinical caries status of posterior approximal surfaces in primary and permanent teeth. Caries Res 26:146–152.

Potter D, Manwell M, Dess R, Levine E, Tinanoff N (1984). SnF$_2$ as an adjunct to toothbrushing in an elderly institutionalized population. Spec Care Dent 4:216–218.

Poulsen S, Agerbæk N, Melsen B, Korts DC, Glavind L, Rölla G (1976). The effect of professional toothcleansing on gingivitis and dental caries in children after 1 year. Community Dent Oral Epidemiol 4:195–199.

Preber H, Ylipaa V, Bergström J, Ryden H (1991). Comparative study of plaque removing efficiency using rotary electric and manual toothbrushes. Swed Dent J 15:229–234.

Pretara-Spanedda P, Grossman E, Curro FA, Generallo CH (1989). Toothbrush bristle density: Relationship to plaque removal. Am J Dent 2:345–348.

Primosch R (1985). A report on the efficacy of fluoridated varnishes in dental caries prevention. Clin Prev Dent 7:12–22.

Quirynen M (1986). Anatomical and Inflammatory Factors Influence Bacterial Plaque Growth and Retention in Man [thesis]. Leuven: Catholic Univ.

Quirynen M, Marechal M, Busscher HJ, Weerkamp AH, Darius PL, van Steenberghe D (1990a). The influence of surface free energy and surface roughness on early plaque formation. An in vivo study in man. J Clin Periodontol 17:138–144.

Quirynen M, Marechal M, van Steenberghe D (1990b). Comparative antiplaque activity of sanguinarine and chlorhexidine in man. J Clin Periodontol 17:223–232.

Quirynen M, Dekeyser C, van Steenberghe D (1991). The influence of gingival inflammation, tooth type, and timing on the rate of plaque formation. J Periodontol 62:219–222.

Quirynen M, Bollen C, Vandekerckhove B, Dekeyser C, Papapanou W, Eyssen H (1995). Full- versus partial-mouth disinfection in the treatment of periodontal infections. Short-term clinical and microbiological observations. J Dent Res 74:1459–1467.

Quirynen M, Mongardini C, Pauwels M, Bollen C, van Eldere J, van Steenberghe D (1999). One-stage full-versus partial-mouth disinfection in the treatment of chronic adult or generalized early-onset periodontitis. J Periodontol 70:646–656.

Quirynen M, Mongardini C, De Soete M, et al (2000). The role of chlorhexidine in the one stage full-mouth disinfection treatment of patients with advanced adult periodontitis. Long-term clinical and microbiological observations. J Clin Periodontol 27:578–589.

Qvist V, Laurberg L, Poulsen A, Teglers PT (1997). Longevity and cariostatic effects of everyday conventional glass-ionomer and amalgam restorations in primary teeth: Three-year results. J Dent Res 76:1387–1396.

Raadal M (1978). Follow-up study of sealing and filling with composite resins in the prevention of occlusal caries. Community Dent Oral Epidemiol 6:176–180.

Rajala M, Selkäinaho K, Paunio I (1980). Relationship between reported toothbrushing and dental caries in adults. Community Dent Oral Epidemiol 8:128–131.

617

References

Ramberg P, Furuichi Y, Lindhe J, Gaffar A (1992). A model for studying the effects of mouthrinses on de novo plaque formation. J Clin Periodontol 19(7):509–520.

Ramberg P, Lindhe J, Dahlén G, Volpe AR (1994). The influence of gingival inflammation on de novo plaque formation. J Clin Periodontal 21:51–56.

Ramberg P, Axelsson P, Lindhe J (1995a). Plaque formation at healthy and inflamed gingival sites in young individuals. J Clin Periodontol 22:85–88.

Ramberg P, Furuichi Y, Sherl D, et al (1995b). The effect of triclosan on developing gingivitis. J Clin Periodontol 22:442–448.

Ramberg P, Furuichi Y, Volpe AR, Gaffar A, Lindhe J (1996). The effects of antimicrobial mouthrinses on de novo plaque formation at sites with healthy and inflamed gingivae. J Clin Periodontol 23:7–11.

Ramfjord SP, Caffesse RG, Morrison EC, et al (1987). 4 modalities of periodontal treatment compared over 5 years. J Clin Periodontol 14:445–452.

Rams TE, Slots J (1996). Local delivery of antimicrobial agents in the periodontal pocket. Periodontol 2000 10:139–159.

Randall RC, Wilson N (1999). Glass-ionomer restoratives: A systematic review of a secondary caries treatment effect. J Dent Res 78:628–637.

Rasmusson CG, Rasmusson L (1994). Kliniska försök med ljushärdande glasjonomercement. Tandläkartidningen 86:117–123.

Rateitschak KH, Wolf H, Hassel T (1989). Color Atlas of Dental Medicine, ed 2. New York: Thieme Verlag.

Ravald N, Birkhed D (1991). Factors associated with active and inactive root caries in patients with periodontal disease. Caries Res 25:377–384.

Ravald N, Birkhed D (1992). Prediction of root caries in periodontally treated patients maintained with different fluoride programmes. Caries Res 26:450–458.

Reddy MS, Jeffcoat MK, Geurs NC, et al (2003). Efficacy of controlled-release subgingival chlorhexidine to enhance periodontal regeneration. J Periodontol 74:411–419.

Reintsema H, Schuthof J, Arends J (1985). An in vivo investigation of fluoride uptake in partially demineralised human enamel from several different dentifrices. J Dent Res 64:19–23.

Reitman WR, Whiteley RT, Robertson PB (1980). Proximal surface cleaning by dental floss. Clin Prev Dent 2:7–10.

Renvert S, Wikström M, Dahlén G, Slots J, Egelberg J (1990). Effect of root debridement on the elimination of Actionobacillus actinomycetemcomitans and Bacteroides gingivalis from periodontal pockets. J Clin Periodontol 17:345–350.

Renvert S, Birkhed D (1995). Comparison between 3 triclosan dentifrices on plaque, gingivitis and salivary microflora. J Clin Periodontol 22:63–70.

Renvert S, Glavind L (1998). Individualized instruction compliance in oral hygiene practices: Recommendations and means of delivery. In: Lang NP, Attström R, Löe H (eds). Proceedings of the European Workshop on Mechanical Plaque Control. Berlin: Quintessence:156–168.

Results of the Michigan Workshop (1989). J Public Health Dent 49(special issue):331–337.

Retief DH, Brischoff J, van der Merve E (1976). Pyruvic acid as an etchant agent. J Oral Rehabil 3:245–265.

Reynolds MA, Lavigne CK, Minah GE, Suzuki JB (1992). Clinical effects of simultaneous ultrasonic scaling and subgingival irrigation with chlorhexidine. J Clin Periodontol 19:595–600.

Richards LF, Westmoreland WW, Tashiro M, McKay CH, Morrison JT (1967). Determining optimum fluoride levels for community water supplies in relation to temperature. J Am Dent Assoc 74:389–397.

Richards A, Kragstrup J, Nielsen-Kudsk F (1985). Pharmacokinetics of chronic fluoride ingestion in growing pigs. J Dent Res 64:425–430.

Riise J, Haugejorden O, Wold B, Aarö LE (1991). Distribution of dental health behaviours in Nordic schoolchildren. Community Dent Oral Epidemiol 19:9–13.

Rindom-Schiött C, Briner W, Löe H (1976). 2-year oral use of chlorhexidine in man. II. The effect on the salivary bacterial flora. J Periodontal Res 11:145–152.

Riordan PJ (1991). Dental caries and fluoride exposure in Western Australia. J Dent Res 70:1029–1034.

Ripa LW, Leske GS, Forte F, Varma A (1988). Effect of a 0.05% neutral NaF mouthrinse on coronal and root caries of adults. Gerodontology 6:131–136.

Ripa LW (1989). Review of the anticaries effectiveness of professionally applied and self-applied topical fluoride gels. J Public Health Dent 49(special issue):297–309.

Ripa LW (1990). An evaluation of the use of professional (operator-applied) topical fluorides. J Dent Res 69:786–796.

Ripa LW (1991). A critique of topical fluoride methods (dentifrices, mouthrinses, operator- and self-applied gels) in an era of decreased caries and increased fluorosis prevalence. J Public Health Dent 51:23–41.

Ripa LW (1993a). The current status of pit and fissure sealants. A review. J Can Dent Assoc 5:367–379.

Ripa LW (1993b). Sealants revisted: An update of the effectiveness of pit-and-fissure sealants. Caries Res 27(suppl 1):77–82.

Robinson C, Shore RC, Kirkham J, Stonehouse NJ (1990). Extracellular processing of enamel matrix proteins and the control of crystal growth. J Biol Buccale 18:355–361.

Robinson C, Kirkham J, Weatherell (1996). Fluoride in teeth and bone. In: Fejerskov O, Ekstrand J, Burt B (eds). Fluoride in Dentistry. Copenhagen: Munksgaard.

Robinson PJ, Maddalozzo D, Breslin S (1997). A 6-month clinical comparison of the efficacy of the Sonicare and the Braun Oral-B electric toothbrushes on improving periodontal health in adult periodontitis patients. J Clin Dent 8:4–9.

Rock WP (1974). Fissure sealants. Further results of clinical trials. Br Dent J 136:317–321.

Rodda JC (1968). A comparison of 4 methods of toothbrushing. New Zealand Dent J 64:162–167.

Roeters F, Van der Hoeven J, Burgersdijk R, Schaeken M (1995). Lactobacilli, mutans streptococci and dental caries: A longitudinal study in 2-year-olds up to the age of 5 years. Caries Res 29:272–279.

Rölla G, Löe H, Rindom-Schiött C (1970). The affinity of chlorhexidine for hydroxyapatite and salivary mucins. J Periodontal Res 5:90–95.

Rölla G, Löe H, Rindom-Schiött C (1971). Retention of chlorhexidine in the human oral cavity. Arch Oral Biol 16:1109–1116.

Rölla G, Melsen B (1975a). Desorption of protein and bacteria from hydroxyapatite by fluoride and monofluoro-phosphate. Caries Res 9:66–73.

Rölla G, Melsen B (1975b). On the mechanism of the plaque inhibition by chlorhexidine. J Dent Res 54(spec issue):B57–B62.

Rölla G (1977). Effects of fluoride on initiation of plaque formation. Caries Res 11:243–261.

Rölla G (1988). On the role of calcium fluoride in the cariostatic mechanism of fluoride. Acta Odont Scand 46:341–345.

Rölla G, Saxegaard E (1990). Critical evaluation of the composition and use of topical fluoride with emphasis on the role of calcium fluoride in caries inhibition. J Dent Res 69:780–785.

Rölla G, Ogaard B, Cruz RA (1991). Fluoride containing toothpastes, their clinical effect and mechanism of cariostatic action—A review. Int Dent J 41:171–174.

Rölla G, Gaare D, Ellingsen J (1993). Experiments with a toothpaste containing polydimethylsiloxan/triclosan. Scand J Dent Res 101:130–132.

Rölla G, Ellingsen JE, Gaare D (1994). Clinical effects and possible mechanisms of action of stannous fluoride. Int Dent J 44:99–105.

Rölla G, Ekstrand J (1996). Fluoride in oral fluids and dental plaque. In: Fejerskov O, Ekstrand J, Burt B (eds). Fluoride in Dentistry. Copenhagen: Munksgaard.

Rölla G, Kjaerheim V, Waaler SM (1997). The role of antiseptics in primary prevention. In: Lang NP, Karring T, Lindhe J (eds). Proceedings of the 2nd European Workshop on Periodontology—Chemicals in Periodontics. Berlin: Quintessence:120–130.

Romcke R, Lewis D, Maze B, Vickerson R (1990). Retention and maintenance of fissure sealants over 10 years. J Can Dent Assoc 56:235–237.

Rosling B (1976). Plaque Control. A Determining Factor in the Treatment of Periodontal Disease [thesis]. Sweden: Univ of Gothenburg.

Rosling B, Nyman S, Lindhe J (1976). The effect of systematic plaque control on bone regeneration in infrabony pockets. J Clin Periodontol 3:38–53.

Rosling B, Slots J, Webber RL, Christersson LA, Genco RJ (1983). Microbiological and clinical effects of topical subgingival antimicrobial treatment on human periodontal disease. J Clin Periodontol 10:487–514.

Rosling BG, Slots J, Christersson LA, Gröndahl HG, Genco RJ (1986). Topical antimicrobial therapy and diagnosis of subgingival bacteria in the management of inflammatory periodontal disease. J Clin Periodontol 13:975–981.

Rosling B, Wannfors B, Volpe AR, Furuichi Y, Ramberg P, Lindhe J (1997a). The use of a triclosan/copolymer dentifrice may retard the progression of periodontitis. J Clin Periodontol 24:873–880.

Rosling B, Dahlén G, Volpe A, Furuichi Y, Ramberg P, Lindhe J (1997b). Effect of triclosan on the subgingival microbiota of periodontitis susceptible subjects. J Clin Periodontol 24:881–887.

Rosling B, Hellström MK, Dolata S, Wannfors B, Serino G, Lindhe J (1998). Antibacterial treatment of periodontitis lesions using nonsurgical technique—10 years evaluation of clinical, radiographical and microbiological variables [abstract].

Rosling B, Hellström M-K, Ramberg P, Socransky SS, Lindhe J (2001). The use of PVP-iodine as an adjunct to non-surgical treatment of chronic periodontitis. J Clin Periodontol 28:1023–1031.

Ross NM, Charles CH, Dills SS (1989). Long-term effects of Listerine antiseptic on dental plaque and gingivitis. J Clin Dent 1:92–95.

Ross NM, Mankodi SM, Mostler KL, Charles CH, Bartles LL (1993). Effect of rinsing time on antiplaque-antigingivitis efficacy of listerine. J Clin Periodontol 20:279–281.

Rugg-Gunn A, MacGregor I (1978). A survey of toothbrushing behaviour in children and young adults. J Periodontal Res 13:3832–3389.

619

Rugg-Gunn A, MacGregor I, Edgar W, Ferguson M (1979). Toothbrushing behaviour in relation to plaque and gingivitis in adolescent shoolchildren. J Periodontal Res 14:231–238.

Russel AL, Elvove E (1951). Domestic water and dental caries. VII. A study of the fluoride-dental caries relationship in an adult population. Public Health Rep 66:1389–1401.

Russel AL (1953). The inhibition of approximal caries in adults with lifelong fluoride exposure. J Dent Res 32:138–143.

Rustogi KN, Petrone DM, Singh SM, Volpe AR, Tavss E (1990). Clinical study of a pre-brush rinse and a triclosan/copolymer mouthrinse: Effect on plaque formation. Am J Dent 3:567–569.

Saito S, Tosaki S, Hirota K (1999). Characteristics of glass-ionomer cements. In: Davidson CL, Mjör IA (eds). Advances in Glass-Ionomer Cements. Chicago: Quintessence.

Saglie FR, Carranza F Jr, Newman M, Cheng L, Lewin K (1982a). Identification of tissue-invading bacteria in human periodontal disease. J Periodontal Res 452–455.

Saglie FR, Newman M, Carranza F, Pattison G (1982b). Bacterial invasion of gingiva in advanced periodontitis in humans. J Periodontol 53:217–222.

Saglie FR, Carranza Jr FA, Newman MG (1985). The presence of bacteria within the oral epithelium in periodontal disease. I. A scanning and transmission electron microscopic study. J Periodontol 56:618–624.

Salonen L, Allander L, Bratthall D, Helldén L (1990). Mutans streptococci, oral hygiene and caries in an adult Swedish population. J Dent Res 69:1469–1475.

Sandham HJ, Brown J, Phillips HI, Chan KH (1988). A preliminary report of long-term elimination of detectable mutans streptococci in man. J Dent Res 67:9–14.

Sandham HJ, Brown J, Chan K, Phillips H, Burgess R, Stokl A (1991). Clinical trial in adults of an antimicrobial varnish for reducing mutans streptococci. J Dent Res 70:1401–1408.

Sandham HJ, Nadeau L, Phillips HI (1992). The effect of chlorhexidine varnish treatment on salivary mutans streptococcal levels in child orthodontic patients. J Dent Res 71:32–35.

Sangnes G (1974). Effectiveness of vertical and horizontal toothbrushing techniques in the removal of plaque. J Dent Child 41:39–43.

Sangnes G (1975). Effectiveness and Adverse Effects of Toothbrushing Procedures [thesis]. Oslo, Norway: Univ of Oslo.

Sangnes G (1976). Traumatization of teeth and gingiva related to habitual tooth cleaning procedures. J Clin Periodontol 3:94–103.

Sangnes G, Gjermo P (1976). Prevalence of oral soft and hard tissue lesions related to mechanical tooth-cleansing procedures. Community Dent Oral Epidemiol 4:77–83.

Sanz M, Vallcorba N, Fabregues S, Müller I, Herkströter F (1994). The effect of a dentifrice containing chlorhexidine and zinc on plaque, gingivitis, calculus and tooth staining. J Clin Periodontol 21:431–437.

Sanz M, Herrera D (1998). Role of oral hygiene during the healing phase of periodontal therapy. In: Lang NP, Attström R, Löe H, et al (eds). Proceedings of the European Workshop on Mechanical Plaque Control. Berlin: Quintessence:248–267.

Sarker S, McLey L, Boyd R (1997). Clinical and laboratory evaluation of powered electric toothbrushes: Laboratory determination of relative interproximal cleaning efficiency of four powered toothbrushes. J Clin Dent 8:81–85.

Sawle RF, Andlaw RJ (1988). Has occlusal caries become more difficult to diagnose? A study comparing clinically undetected lesions in molar teeth of 14–16-year-olds in 1974 and 1982. Br Dent J 164:209–211.

Saxe SR, Greene JC, Bohannan HM, Vermillon JR (1967). Oral debris, calculus and periodontal disease in the beagle dog. Periodontics 5:217–225.

Saxegaard E, Rölla G (1988). Kinetics of acquisition and loss of calcium fluoride by enamel in vivo. Caries Res 23:406–411.

Saxer UP, Mühlemann H (1983). Synergistic antiplaque effects of a zinc fluoride/hexetidine containing mouthwash. A review. Schweiz Monatsschr Zahnheilkd 93:689–704.

Saxton CA (1973). Scanning electron microscope study of the formation of dental plaque. Caries Res 7:102–119.

Saxton CA (1975). The Formation of Human Dental Plaque: A Study by Scanning Electron Microscopy [thesis]. London: Univ of London.

Saxton CA (1976). The effects of dentifrices on the appearance of the tooth surface observed with the scanning electron microscope. J Periodontal Res 11:74–85.

Saxton CA (1986). The effects of a dentifrice containing zinc citrate and 2,4,4`-trichloro-2`-hydroxydiphenyl ether. J Periodontol 57:555–561.

Saxton CA, Lane R, Van der Ouderaa F (1987). The effects of a dentifrice containing a zinc salt and a noncationic antimicrobial agent on plaque and gingivitis. J Clin Periodontol 14:144–148.

Saxton CA, Svatun B, Lloyd A (1988). Anti-plaque effects and mode of action of a combination of zinc citrate and a nonionic antimicrobial agent. Scand J Dent Res 96:212–217.

Saxton CA (1989a). The reduced development of gingivitis after the use of dentifrice. J Dent Res 68(special issue):743.

Saxton CA (1989b). Maintenance of gingival health by a dentifrice containing zinc citrate and Triclosan. J Dent Res 68:1724-1726.

Saxton CA, Van der Ouderaa F (1989). The effect of a dentifrice containing zinc citrate and triclosan on developing gingivitis. J Periodontal Res 24:75-80.

Sbordone L, Ramaglia L, Gulletta E, Iacono V (1990). Recolonization of the subgingival microflora after scaling and root planing in human periodontitis. J Periodontol 61:579-584.

Schaeken M, de Haan P (1989). Effects of sustained-release chlorhexidine acetate on the human dental plaque flora. J Dent Res 68:119-123.

Schaeken M, Van der Hoeven J, Hendriks J (1989). Effects of varnishes containing chlorhexidine on the human dental plaque flora. J Dent Res 68:1786-1789.

Schaeken M, Keltjens H, van der Hoeven JS (1991a). Effects of fluoride and chlorhexidine on the microflora of dental root surfaces and progression of root-surfac caries. J Dent Res 2:150-153.

Schaeken M, Schouten CWA, van der Hoeven JS (1991b). Influence of contact time and concentration of chlorhexidine varnish on mutans streptococci in interproximal dental plaque. Caries Res 25:292-295.

Schaeken MJ, Beckers HJ, van der Hoeven JS (1996). Effect of chlorhexidine varnish on Actinomyces naesludii genospecies in plaque from dental fissures. Caries Res 30:40-44.

Scheie AA, Arneberg P, Krogstad O (1984). The effect of orthodontic treatment on the prevalence of Streptococcus mutans in plaque and saliva. Scand J Dent Res 92:211-217.

Scheie A, Assev S, Rölla G (1985). The effect of SnF2 and NaF on glucose uptake and metabolism in S mutans OMZ176. In: Leach SA (ed). Factors Relating to Demineralisation and Remineralisation of the Teeth. Oxford: IRL Press:99-104.

Scheie A, Kjeilen J (1987). Effects of chlorhexidine, NaF and SnF$_2$ on glucan formation by salivary and culture supernatant GTF adsrobed to hydroxyapatite. Scand J Dent Res 95:532-535.

Scheie A, Assev S, Rölla G (1988). Combined effect of xylitol, NaF and ZnCl$_2$ on growth and metabolism of Streptococcus sobrinus OMZ176. APMIS 96:761-767.

Scheie A (1989). Modes of action of currently known chemical antiplaque agents other than chlorhexidine. J Dent Res 68(special issue):1609-1616.

Scheie A (1994a). Chemoprophylaxis of dental caries. In: Thylstrup A, Fejerskov O (eds). Textbook of Clinical Cariology. Copenhagen: Munksgaard:311-326.

Scheie A (1994b). Mechanisms of dental plaque formation. Adv Dent Res 8:246-253.

Scheinin A, Mäkinen KK, Tammisalo E, Rekola M (1975). Turku sugar studies. XVIII. Incidence of dental caries in relation to 1-year consumption of xylitol chewing gum. Acta Odontol Scand 33:269-278.

Schemehorn BR, Henry G (1996). A laboratory investigation of stain removal from enamel surface: Comparative efficacy of 3 electric toothbrushes. Am J Dent 9:21-24.

Schemehorn BR, Zwart A (1996). Hard tissue abrasivity of an automatic interdental plaque remover. J Clin Dent 7:78-80.

Schiött CR, Löe H, Jensen SB, Kilian M, Davies RM, Glavind K (1970). The effect of chlorhexidine mouthrinses on the human oral flora. J Periodontal Res 5:84-89.

Schlagenhauf U, Stellwag P, Fiedler A (1990). Subgingival irrigation in the maintenance phase of periodontal therapy. J Clin Periodontol 17:650-653.

Schlagenhauf U, Horlacher V, Netuschil L, Brecx M (1994). Repeated subgingival oxygen irrigations in untreated periodontal patients. J Clin Periodontol 21:48-50.

Schmid MO, Balmelli OP, Saxer UP (1976). Plaque removing effect of a toothbrush, dental floss and toothpick. J Clin Periodontol 3:157-165.

Schou (1998). Behavioral aspects of dental plaque control measures: An oral health promotion perspective. In: Lang NP, Attström R, Löe H (eds). Proceedings of the European Workshop on Mechanical Plaque Control. Berlin: Quintessence: 156-168.

Schröder U, Granath L (1983). Dietary habits and oral hygiene as predictors of caries in 3-year-old children. Community Dent Oral Epidemiol 11:308-311.

Schroers K (1994). Klinisch kontrollierte Studie zum Nachweis der Wirksamkeit von Bifluorid 12 bei der Behandlung überempfindlicher Zähne. Forschungsbericht.

Schupbach P, Lutz F, Guggenheim B (1992). Human root caries: Histopathology of arrested lesions. Caries Res 26:153-164.

Schwarz JP, Rateitschak-Plüss EM, Guggenheim R, Düggelin M, Rateitschak KH (1993). Effectiveness of open flap root debridement with rubber cups, interdental plastic tips and prophy paste. An SEM study. J Clin Periodontol 20:1-6.

Schwartz M, Gröndahl HG, Pliskin JS, Boffa J (1984). A longitudinal analysis from bite-wing radiographs of the rate of progression of approximal carious lesions through human dental enamel. Arch Oral Biol 29:529-536.

References

Selwitz R, Nowjack-Raymer R, Driscoll W, Li S (1995). Evaluation after 4 years of the combined use of fluoride and dental sealants. Community Dent Oral Epidemiol 23:30–35.

Seppä L, Tuutti H, Louma H (1981). A 2-year report on caries prevention by fluoride varnishes in a community with fluoridated water. Scand J Dent Res 89:143–148.

Seppä L (1982). Fluoride varnishes in caries prevention. Proc Finn Dent Soc 78:1–50.

Seppä L, Luoma H, Hausen H (1982a). Fluoride content in enamel after repeated applications of fluoride varnishes in a community with fluoridated water. Caries Res 16:7–11.

Seppä L, Tuutti H, Luoma H (1982b). 3-year report on caries prevention using fluoride varnishes for caries risk children in a community with fluoridated water. Scand J Dent Res 90:89–94.

Seppä L, Pöllänen L (1987). Caries preventive effect of two fluoride varnishes and a fluoride mouthrinse. Caries Res 21:375–379.

Seppä L, Forss H, Ogaard B (1993). The effect of fluoride application on fluoride release and antibacterial action of glass ionomers. J Dent Res 72:1310–1314.

Seppä L (1994). Fluoride release and effect on enamel softening by fluoride-treated and fluoride-untreated glass ionomer specimens. Caries Res 28:406–408.

Seppä L, Pöllänen L, Hausen H (1994). Caries-preventive effect of fluoride varnish with different fluoride concentrations. Caries Res 28:64–67.

Seppä L, Korhonen A, Nuutinen A (1995). Inhibitory effect on S mutans by fluoride-treated conventional and resin-reinforced glass ionomer cements. Eur J Oral Sci 103:182–185.

Seppä L, Karkkainen S, Hausen H (2000). Caries trends 1992-1998 in two low-fluoride Finnish towns formerly with and without fluoridation. Caries Res 34:462–468.

Seppä L, Hausen H, Karkkainen S, Larmas M (2002). Caries occurrence in a fluoridated and a nonfluoridated town in Finland: A retrospective study using longitudinal data from public dental records. Caries Res 36:308–314.

Serino G, Wennström JL, Lindhe J, Eneroth L (1994). The prevalence and distribution of gingival recession in subjects with a high standard of oral hygiene. J Clin Periodontol 21:57–63.

Shannon IL, West DC (1979). Prevention of decalcification in orthodontic patients by daily self-treatment with 0.4% SnF_2 gel. Pediatr Dent 1:101–102.

Shellis RP, Dibdin GH (1988). Analysis of the buffering systems in dental plaque. J Dent Res 67:438–446.

Shields C, Leverett D, Adair S, et al (1987). Salivary fluoride levels in fluoridated and non-fluoridated communities [abstract 277]. J Dent Res 66(special issue):141.

Sicilia A, Arregui I, Gallego M, Cabezas B, Cuesta S (2002). A systematic review of powered vs. manual toothbrushes in periodontal cause-related therapy. J Clin Periodontol 29(suppl 3):39–54.

Siegrist B, Kornman KS (1982). The effect of supragingival plaque control on the composition of the subgingival microbial flora in ligature-induced periodontitis in the monkey. J Dent Res 7:936–941.

Silness J, Löe H (1964). Periodontal disease in pregnancy. II. Correlation between oral hygiene and periodontal condition. Acta Odontol Scand 22:121–135.

Silverstone LM (1973). Structure of caries enamel including the early lesion. Dent Update 1:101–105.

Silverstone LM (1974). Fissure sealants, laboratory studies. Caries Res 8:2–26.

Silverstone LM (1975). The acid etch technique. *In vitro* studies with special reference to enamel surfaces and the enamel resin interface. In: Proceedings, International Symposium on Acid Etch Technique. St Paul, MN: North Central.

Silverstone LM, Wefel JS, Zimmerman BF, Clarkson BH, Featherstone MJ (1981). Remineralization of natural and artificial lesions in human dental enamel in vitro. Caries Res 15:138–157.

Silverstone LM (1983). Remineralization and enamel caries: Significance of fluoride and effect on crystal diameters. In: Leach SA, Edgar WM (eds). Demineralization and Remineralization of Teeth. Oxford: IRL Press:185–205.

Silverstone LM, Tilliss T, Cross-Poline G, van der Linden E, Stach D, Featherstone J (1992). A 6-week study comparing the efficacy of a rotary electric toothbrush with a conventional toothbrush. Clin Prev Dent 14:29–34.

Simons D, Kidd E, Beighton D, Jones B (1997). The effect of chlorhexidine/xylitol chewing-gum on cariogenic salivary microflora: A clinical trial in elderly patients. Caries Res 31:91–96.

Simons D, Beighton D, Kidd E, Collier F (1999). The effect of xylitol and chlorhexidine acetate/xylitol chewing gums on plaque accumulation and gingival inflammation. J Clin Periodontol 26:388–391.

Simons D, Brailsford S, Kidd EAM, Beighton D (2001). The effect of chlorhexidine acetate/xylitol chewing gum on the plaque and gingival indices of elderly occupants in residential homes. A 1-year clinical trial. J Clin Periodontol 28:1010–1015.

Simonsen RJ (1979). Fissure sealants in primary molars: Retention of coloured sealants with variable etch times at 12 months. J Dent Child 46:382–384.

Simonsen RJ (1980). Preventive resin restorations: 3-year results. J Am Dent Assoc 100:535–539.

Simonsen RJ (1987). Retention and effectiveness of a single application of white sealant after 10 years. J Am Dent Assoc 115:31–36.

Simonsen RJ (1991). Retention and effectiveness of dental sealant after 15 years. J Am Dent Assoc 122:34–42.

Sjögren K, Birkhed D (1993). Factors related to fluoride retention after toothbrushing and possible connection to caries activity. Caries Res 27:474–477.

Sjögren K, Birkhed D, Persson L, Norén J (1993). Salivary fluoride clearance after a single intake of fluoride tablets and chewing gums in children, adults and dry mouth patients. Scand J Dent Res 5:274–278.

Sjögren K, Ekstrand J, Birkhed D (1994). Effect of water rinsing after toothbrushing on fluoride ingestion and absorption. Caries Res 28:455–459.

Sjögren K, Birkhed D, Rangmar B (1995). Effect of a modified toothpaste technique on approximal caries in preschool children. Caries Res 29:435–441.

Sjögren K, Lingström P, Lundberg A, Birkhed D (1997). Salivary fluoride concentration and plaque pH after using a fluoride-containing chewing gum. Caries Res 31:366–372.

Sjögren K, Ruben J, Lingstrom P, Lundberg AB, Birkhed D (2002). Fluoride and urea chewing gums in an intra-oral experimental caries model. Caries Res 36:64–69.

Skartveit L, Tveit AB, Totdal B, Ovrebo R, Raadal M (1990). In vivo fluoride uptake in enamel and dentin from fluoride-containing materials. ASDC J Dent Child 57:97–100.

Skartveit L, Riordan P, al Dallal E (1994). Effect of fluoride in amalgam on secondary caries incidence. Community Dent Oral Epidemiol 22:122–125.

Skjörland K, Gjermo P, Rölla G (1978). Effect of some polyvalent cations on plaque formation in vivo. Scand J Dent Res 86:103–107.

Sköld L, Sundquist B, Eriksson B, Edeland C (1994). Four-year study of caries inhibition of intensive Duraphat application in 11–15-year-old children. Community Dent Oral Epidemiol 22:8–12.

Slots J, Reynolds H, Genco R (1980). *Actinobacillus actinomycetemcomitans* in human periodontal disease: A cross-sectional microbiological investigation. Infect Immun 29:1013–1020.

Smith MC, Lantz EM, Smith HV (1932). The cause of mottled enamel, a defect of human teeth. J Dent Res 12:149–59. [Reprinted from: tech Bull No 32. Tuscon: University of Arizona College of Agriculture 1931].

Smith JF, Blankenship J (1964). Improving oral hygiene in handicapped children by the use of an electric toothbrush. J Dent Child 31:198–203.

Smith A, Moran J, Dangler L, Leight R, Addy M (1996). The efficacy of an antigingivitis chewing gum. J Clin Periodontol 23:19–23.

Smulow JB, Turesky SS, Hill RG (1983). The effect of supragingival plaque removal on anaerobic bacteria in deep periodontal pockets. J Am Dent Assoc 107:737–742.

Socransky SS, Haffajee AD, Cugini M, Smith C, Kent R Jr (1998). Microbial complexes in subgingival plaque. J Clin Periodontol 25:134–144.

Söderholm G (1979). Effect of a Dental Care Program on Dental Health Conditions. A Study of Employees of a Swedish Shipyard [thesis]. Lund, Sweden: Univ of Lund.

Soh L, Newman H, Strahan J (1982). Effects of subgingival chlorhexidine irrigation on periodontal inflammation. J Clin Periodontol 9:66–74.

Sokolne WA, Proskin HM, Stabholz N (2003). Probing depth changes following 2 years of periodontal maintinance therapy including adjunctive controlled release of chlorhexidine. J Periodontol 74:420–427.

Soparkar PM, Newman M, De Paola P (1991). The efficacy of a novel toothbrush design. J Clin Dent 2:107–110.

Sorvari R, Spets-Happonen S, Luoma H (1994). Efficacy of chlorhexidine solution with fluoride varnishing in preventing enamel softening by *Streptococcus mutans* in an artificial mouth. Scand J Dent Res 102:206–209.

Soskolne W, Heasman P, Stabholz A, et al (1997). Sustained local delivery of chlorhexidine in the treatment of periodontitis. A multi-center study. J Periodontol 68:32–38.

Soskolne W, Chajek T, Flashner M, et al (1998). An in vivo study of the chlorhexidine release profile of the PerioChip in the gingival crevicular fluid, plasma and urine. J Clin Periodontol 25:1017–1021.

Soskolne WA, Proskin HM, Stabholz A (2003). Probing depth changes following 2 years of periodontal maintenance therapy including adjunctive controlled release of chlorhexidine. J Periodontol 74:420–427.

Southard GL, Parsons LG, Thomas LG Jr, Boulware RT, Woodall IR, Jones BJB (1987). The relationship of sanguinaria extract concentration and zinc ion to plaque and gingivitis. J Clin Periodontol 14:315–319.

Southard SR, Drisko CL, Killoy WJ, Cobb CM, Tira DE (1989). The effect of 2% chlorhexidine digluconate irrigation on clinical parameters and the level of *Bacteroides gingivalis* in periodontal pockets. J Periodontol 60:302–309.

Sreenivasan P, Gaffar A (2002). Antiplaque biocides and bacterial resistance: A review. J Clin Periodontol 29:965–974.

References

Stabbe K, Tishk M, Overman P, Love J (1988). A comparison of plaque reaccumulation and patient acceptance using a conventional toothbrush and a newly designed toothbrush. Clin Prev Dent 10:10–14.

Stabholz A, Kettering J, Aprecio R, Zimmerman G, Baker PJ, Wikesjö UM (1993). Retention of antimicrobial activity by human root surface after in situ subgingival irrigation with tetracycline HCl or chlorhexidine. J Periodontol 64:134–141.

Stamm JW (1972). Milk fluoridation as a public health measure. J Can Dent Assoc 38:446–451.

Stamm JS, Banting DW, Imrey PB (1990). Adult root caries survey of two similar communities with contrasting natural fluoride levels. J Am Dent Assoc 120:143–149.

Stanley A, Wilson M, Newman H (1989). The in vitro effects of chlorhexidine on subgingival plaque bacteria. J Clin Periodontol 16: 259–264.

Stecksen-Blicks C, Arvidsson S, Holm AK (1985). Dental health, dental care, and dietary habits in children in different parts of Sweden. Acta Odontol Scand 43:59–67.

Stecksen-Blicks C, Gustafsson L (1986). Impact of oral hygiene and use of fluorides on caries increment in children during 1 year. Community Dent Oral Epidemiol 14:185–189.

Steinberg D, Friedman M, Soskolne A, Sela M (1990). A new degradable controlled release device for treatment of periodontal disease: In vitro release study. J Periodontol 61:393–398.

Steinberg LM, Odusola F, Mandel ID (1992). Remineralizing potential, anti plaque and antigingivitis effects of xylitol and sorbitol sweetened chewing gum. Clin Prev Dent 14:31–34.

Stephen KW, Campbell D (1978). Caries reduction and cost benefit after 3 years of sucking fluoride tablets daily at school. A double-blind trial. Br Dent J 144:202–206.

Stephen KW, Boyle IT, Campbell D, McNee S, Boyle P (1984). Five-year fluoridated milk study in Scotland. Community Dent Oral Epidemiol 12:223–229.

Stephen KW, Creanor SL, Russel JI, Burchell CK, Huntington E, Downie CFA (1988). A 3-year oral health dose response study of sodium monofluorophosphate dentifrices with and without zinc citrate: Anticaries results. Community Dent Oral Epidemiol 16:321–325.

Stephen KW, Kay EJ, Tullis JI (1990a). Combined fluoride therapies. A 6-year double-blind school-based preventive dentistry study in Inverness, Scotland. Community Dent Oral Epidemiol 18:244–248.

Stephen KW, Saxton C, Jones C, Ritchie J, Morrison T (1990b). Control of gingivitis and calculus by a dentifrice containing a zinc salt and triclosan. J Periodontol 61:674–679.

Stephen KW, Chestnutt IG, Jacobson APM, et al. (1994). The effect of NaF and SMFP toothpaste on 3-years' caries increments in adolescents. Int Dent J 44:287–295.

Stiles HM, Meyers R, Brunelle JA, Wittig AB (1976). Occurrence of Streptococcus mutans and Streptococcus sanguis in the oral cavity and feces of young children. In: Stiles HM, Loesche WJ, O'Brien TD (eds). Microbial Aspects of Dental Caries: Proceedings of a Workshop on Microbial Aspects of Dental Caries, June 21–24, 1976, St. Simons Island, Georgia. Washington: Information Retrieval:187–199.

Stoltze K, Bay L (1994). Comparison of a manual and a new electric toothbrush for controlling plaque and gingivitis. J Clin Periodontol 21:86–90.

Stookey GK, DePaula PF, Featherstone JDB, et al (1993). A critical review of the relative anticaries efficacy of sodium fluoride and sodium monofluorophosphate dentifrices. Caries Res 27:337–360.

Straffon LH, Dennison JB (1988). Clinical evaluation comparing sealant and amalgam after 7 years: Final report. J Am Dent Assoc 177:751–755.

Strålfors A, Thilander H, Bergenholtz A (1967). Correlation between caries and periodontal disease in the hamster. Arch Oral Biology 12:1213.

Straub A, Salvi G, Lang N (1998). Supragingival plaque formation in the human dentition. In: Lang NP, Attström R, Löe H (eds). Proceedings of the European Workshop on Mechanical Plaque Control. Berlin: Quintessence:72–84.

Strömberg N (1996). Salivens fingeravtryck avslöjar risk för tandlossning. Tandlakartidningen 88:138–141.

Sundberg H (1996). Changes in the prevalence of caries in children and adolescents in Sweden 1985–1994. Eur J Oral Sci 104:470–476.

Sundin B, Birkhed D, Granath L (1983). Is there not a strong relationship nowadays between caries and consumption of sweets? Swed Dent J 7:103–108.

Sundin B, Granath L, Birkhed D (1992). Variation of posterior approximal caries incidence with consumption of sweets with regard to other caries-related factors in 15–18-year-olds. Community Dent Oral Epidemiol 20:76–80.

Suomi JD, Greene JC, Vermillion JR, Doyle J, Chang JJ, Leatherwood EC (1971). The effect of controlled oral hygiene procedures on the progression of periodontal disease in adults: Results after third and final year. J Periodontol 42:152–160.

Sutcliffe P (1977). Caries experience and oral cleanliness of 3- and 4-year-old children from deprived and non-deprived areas in Edinburgh, Scotland. Community Dent Oral Epidemiol 5:213–219.

Svanberg M, Rölla G (1982). *Streptococcus mutans* in plaque and saliva after mouthrinsing with SnF_2. Scand J Dent Res 90: 292–298.

Svanberg M, Westergren G (1983). Effect of SnF_2, administered as mouthrinses or topically applied, on *Streptococcus mutans, Streptococcus sanguis* and *lactobacilli* in dental plaque and saliva. Scand J Dent Res 91:123–129.

Svanberg M (1992). Class II amalgam restorations, glass-ionomer tunnel restorations, and caries development on adjacent tooth surfaces: A 3-year clinical study. Caries Res 26:315–318.

Svatun B, Gjermo P, Eriksen H, Rölla G (1977). A comparison of the plaque inhibiting effect of stannous fluoride and chlorhexidine. Acta Odontol Scand 35:247–250.

Svatun B, Attramadal A (1978). The effect of stannous fluoride on human plaque acidogenicity in situ (Stephan Curve). Acta Odontol Scand 36:211–218.

Svatun B, Saxton CA, van der Ouderaa F, Rölla G (1987). The influence of a dentifrice containing a zinc salt and a nonionic antimicrobial agent on the maintenance of gingival health. J Clin Periodontol 14:457–461.

Svatun B, Saxton C, Rölla G, van der Ouderaa F (1989a). A 1-year study on the maintenance of gingival health by a dentifrice containing a zinc salt and non-anionic antimicrobial agent. J Clin Periodontol 16:75–80.

Svatun B, Saxton C, Rölla G, van der Ouderaa F (1989b). 1-year study of the efficacy of a dentifrice containing zinc citrate and triclosan to maintain gingival health. Scand J Dent Res 97:242–246.

Svatun B, Saxton C, Rölla G (1990). 6-month study of the effect of a dentifrice containing zinc citrate and triclosan on plaque, gingival health and calculus. Scand J Dent Res 98:301–304.

Svatun B, Saxton C, Huntington E, Cummins D (1993a). The effects of a silica dentifrice containing triclosan and zinc citrate on supragingival plaque and calculus formation and the control of gingivitis. Int Dent J 43(suppl 1):431–439.

Svatun B, Saxton C, Huntington E, Cummins D (1993b). The effects of 3 silica dentifrices containing triclosan on supragingival plaque and calculus formation and on gingivitis. Int Dent J 43(suppl 1):441–452.

Swedish Board of Health and Welfare (1979). Caries Prevalence in Swedish Children. Yearly Statistics. Stockholm: Swedish Board of Health and Welfare.

Symons AL, Chu CY, Meyers IA (1996). The effect of fissure morphology and pretreatment of the enamel surface on penetration and adhesion of fissure sealants. J Oral Rehabil 23:791–795.

Takahashi K, Fukazawa M, Motohira H, Ochai K, Nishikawa H, Miyata T (2003). A pilot study of antiplaque effects of mastic chewing gum in the oral cavity. J Periodontol 74:501–505.

Talbott K, Mandel ID, Chilton NW (1977). Reduction of baseline gingivitis scores with repeated prophylaxes. J Prev Dent 4(6):28–29.

Tam LE, Chan GP, Yim D (1997). In vitro caries inhibition effects by conventional and resin-modified glass-ionomer restorations. Oper Dent 22:4–14.

Tatevossian A (1990). Fluoride in dental plaque. J Dent Res 69(special issue):645–652.

Taylor R (1978). Variations of Morphology of Teeth. Springfield: Charles C Thomas.

Teivens A, Mörnstad H, Reventlid M (1996). Individual variation of tooth development in Swedish children. Swed Dent J 20:87–93.

Tellefsen G, Larsen G, Kaligithi R, Zimmerman G, Wikesjö M (1996). Use of chlorhexidine chewing gum significantly reduces dental plaque formation compared to use of similar xylitol and sorbitol products. J Periodontol 67:181–183.

Tempel T, Marcil J, Andre S, Jay S (1975). Comparison of water irrigation and oral rinsing or clearance of soluble and particulate materials from the oral cavity. J Periodontol 46:391–396.

Ten Cate JM, Arends J (1977). Remineralization of artificial enamel lesions in vitro. Caries Res 11:277–286.

Ten Cate JM, Jongebloed WL, Arends J (1981). Remineralization of artificial enamel lesions in vitro. IV. Caries Res 15:60–69.

Ten Cate JM, Duijsters PPE (1983a). Influence of fluoride in solution on tooth demineralization. I. Chemical data. Caries Res 17:193–199.

Ten Cate JM, Duijsters PPE (1983b). Influence of fluoride in solution on tooth demineralization. II. Microradiographic data. Caries Res 17:513–519.

Ten Cate JM (1993). The caries preventive effect of a fluoride dentifrice containing triclosan and zinc citrate, a compilation of in vitro and in situ studies. Int Dent J 43:407–413.

Ten Cate JM, Van Loveren C, Buijs M (1993). Comparison of caries preventive treatments, in a bacterial demineralization model. Caries Res 27:237.

Ten Cate JM (1994). In situ models, physico-chemical aspects. Adv Dent Res 8:125–133.

Ten Cate JM, van Duinen R (1995). Hypermineralization of dentinal lesions adjacent to glass-ionomer cement restorations. J Dent Res 74:1266–1271.

Ten Cate JM, Buijs MJ, Damen JJ (1995). PH-cycling enamel and dentin lesins in the presence of concentrations of fluoride. Eur J Oral Sci 103:362.

Ten Cate JM, Featherstone J (1996). Physicochemical aspects of fluoride-enamel interactions. In: Fejerskov O, Ekstrand J, Burt B (eds). Fluoride in Dentistry. Copenhagen: Munksgaard.

Tenovuo J, Häkkinen P, Paunio P, Emilson C (1992). Effects of chlorhexidine-fluoride gel treatments in mothers on the establishment of mutans streptococci in primary teeth and the development of dental caries in children. Caries Res 26:275–280.

Tenovuo J, Lagerlöf F (1994). Saliva. In: Thylstrup A, Fejerskov O (eds). Textbook of Clinical Cariology, ed 2. Copenhagen: Munksgaard:17–43.

Terezhalmy GT, Gagliardi V, Rybicki L, Kaufman M (1995a). Clinical evaluation of the efficacy and safety of the Ultrasonex toothbrush: A 30-day study. Compend Contin Educ Dent 15:866–874.

Terezhalmy GT, Iffland H, Jelepis C, Waskowski J (1995b). Clinical evaluation of the effect of an ultrasonic toothbrush on plaque, gingivitis and gingival bleeding: A 6-month study. J Prosthet Dent 73:97–103.

Theilade E, Fejerskov O, Horsted M (1976). A transmission electron microscopic study of 7-day old bacterial plaque in human tooth fissures. Arch Oral Biol 21:587–598.

Theilade E, Theilade J (1976). Role of plaque in the etiology of periodontal disease and caries. Oral Sci Rev 9:23.

Thibodeau EA, O'Sullivan DM (1996). Salivary mutans streptococci and dental caries patterns in preschool children. Community Dent Oral Epidemiol 24:164–168.

Thylstrup A, Fejerskov O (1978). Clinical appearance of dental fluorosis in permanent teeth in relation to histological changes. Community Dent Oral Epidemiol 6:315–328.

Thylstrup A, Fejerskov O, Bruun C, Kann J (1979). Enamel changes and dental caries in 7-year-old children given fluoride tablets from shortly after birth. Caries Res 13:265–276.

Thylstrup A, Chironga L, Carvalho JC, Ekstrand KR (1989). The occurrence of dental calculus in occlusal fissures as an indication of caries activity. In: ten Cate JF (ed). Recent Advances in the Study of Dental Calculus. Oxford: IRL Press:211–222.

Thylstrup A, Bruun C, Holmen L (1994). In vivo caries models-mechanisms for caries initiation and arrestment. Adv Dent Res 8: 144–157.

Thylstrup A, Vinther D, Christiansen J (1997). Promoting changes in clinical practice. Treatment time and outcome studies in a Danish public child dental health clinic. Community Dent Oral Epidemiol 25:126–134.

Tinanoff N, Brady J, Gross A (1976). The effect of NaF and SnF$_2$ mouthrinses on bacterial colonization of tooth enamel: TEM and SEM studies. Caries Res 10:415–426.

Tinanoff N, Hock J, Camosci D, Helldén L (1980). Effect of stannous fluoride mouthrinse on dental plaque formation. J Clin Periodontol 7:232–241.

Tinanoff N, Zameck BS (1985). Alteration in salivary and plaque S mutans in adults brushing with 0.4% SnF$_2$ gel once or twice a day. Pediatr Dent 7:180–184.

Tinanoff N, Manwell M, Zameck R, Grasso J (1989). Clinical and radiological effects of daily brushing with either NaF or SnF$_2$ gels in subjects with fixed or removable dental prostheses. J Clin Periodontol 16:284–290.

Tinanoff N (1990). Review of the antimicrobial action of stannous fluoride. J Clin Dent 2(1):21–27.

Toffenetti F (2001). Initial management: To drill or not to drill. In: Wilson NHF, Roulet, JF, Fuzzi MS (eds). Advances in Operative Dentistry. Chicago: Quintessence:89–104.

Torell P, Ericsson Y (1965). 2-year test with different methods of local caries-preventive fluoride application in Swedish schoolchildren. Acta Odontol Scand 23:287–322.

Toumba KJ, Curzon M (1993). Slow-release fluoride. Caries Res 27(suppl):43–46.

Toumba KJ, Curzon MEJ (1994). Plasma fluoride levels following ingestion of a F glass slow release device [abstract 129]. Caries Res 28:217.

Toumba KJ, Curzon MEJ (1996). The prevention of dental caries using fluoride slow-releasing glass devices in children [abstract 121]. Caries Res 30:306.

Toumba KJ, Pollard MA, Curzon MEJ (1996). Fluoride glass slow-release devices and enamel remineralisation in situ [abstract 5]. Caries Res 30:268.

Toumba KJ, Curzon MEJ (1998). Prevention of occlusal caries using fluoride slow-releasing glass devices [abstract 3085]. J Dent Res 77:1017.

Toumba KJ, Arizos S (2000). In vitro and in situ fluoride release and "recharging" capacity of compomers and glass ionomer cements [abstract 89]. Caries Res 34:339.

Toumba KJ (2001). Slow-release devices for fluoride delivery to high-risk individuals. Caries Res 35(suppl 1):10–13.

Triratana T, Tuongratanaphan S, Rustogi K, Petrone M, Volpe A, Petrone D (1993). Plaque/gingivitis efficacy of triclosan dentifrices. J Dent Res 72:334.

Tritten CB, Armitage G (1996). Comparison of a sonic and manual toothbrush for efficacy in supragingival plaque removal and reduction of gingivitis. J Clin Periodontol 23:641–648.

Tucker GJ, Andlaw RJ, Burchell CK (1976). The relationship between oral hygiene and dental caries incidence in 11-year-old children. A 3-year study. Br Dent 141:75–79.

Tveit AB, Espelid I Fjelltveit A (1994). Clinical diagnosis of occlusal dentin caries. Caries Res 28:368–372.

Twetman S, Petersson L, Pakhomov G (1996). Caries incidence in relation to salivary mutans streptococci and fluoride varnish applications in preschool children from low- and optimal-fluoride areas. Caries Res 30:347–353.

Twetman S, Petersson LG (1997a). Effect of different chlorhexidine varnish regimens on mutans streptococci levels in interdental plaque and saliva. Caries Res 31:89–193.

Twetman S, Petersson LG (1997b). Efficacy of a chlorhexidine and chlorhexidine-fluoride varnish mixture to decrease interdental levels of mutans streptococci. Caries Res 31:361–365.

Twetman S, Petersson LG (1998). Comparison of the efficacy of 3 different chlorhexidine preparations in decreasing the levels of mutans streptococci in saliva and interdental plaque. Caries Res 32:113–118.

Twetman S, Grindefjord M (1999). Mutans streptococci suppression by chlorhexidine gel in toddlers. Am J Dent 12:89–91.

Twetman S, Sköld-Larsson K, Modéer T (1999). Fluoride concentration in whole saliva and separate gland secretions after topical treatment with three different fluoride varnishes. Acta Odontol Scand 57:263–266.

Ullsfoss B, Ogaard B, Arends J, Ruben J, Rölla G, Afseth J (1994). Effect of a combined chlorhexidine and NaF mouthrinse: An in vivo human caries model study. Scand J Dent Res 102:109–112.

Vandekerckhove BN, Bollen CM, Dekeyser C, Darius P, Quirynen M (1996). Full- versus partial-mouth disinfection in the treatment of periodontal infections. Long-term clinical observations of a pilot study. J Periodontol 67:1251–1259.

Van der Ouderaa FJ, Cummins D (1989). Delivery systems for agents in supra- and subgingival plaque control. J Dent Res 68(special issue):1617–1624.

Van der Weijden GA, Danser MM, Nijboer A, Timmerman MF, Van der Velden U (1993a). The plaque-removing efficacy of an oscillating/rotating toothbrush. A short-term study. J Clin Periodontol 20:273–278.

Van der Weijden GA, Timmerman MF, Nijboer A, Van der Velden U (1993b). A comparative study of electric toothbrushes for the effectiveness of plaque removal in relation to toothbrushing duration. A timer study. J Clin Periodontol 20:476–481.

Van der Weijden GA, Timmerman MF, Reijerse E, et al (1994). The longterm effect of an oscillating/rotating toothbrush. An 8-month clinical study. J Clin Periodontol 21:139–145.

Van der Weijden GA, Timmerman M, Reijerse E, Mantel M, van der Velden U (1995a). The effectiveness of an electronic toothbrush in the removal of established plaque and treatment of gingivitis. J Clin Periodontol 22:179–182.

Van der Weijden GA, Timmerman MF, Reijerse E, Snoek CM, van der Velden U (1995b). Comparison of two electric toothbrushes in plaque removing ability—Professional and supervised brushing. J Clin Periodontol 22:648–652.

Van der Weijden GA, Timmerman MF, Snoek I, Reijerse E, van der Velden U (1996a). Toothbrushing duration and plaque removing efficacy of electric toothbrushes. Am J Dent 9:31–36.

Van der Weijden GA, Timmerman MF, Reijerse E, Snoek CM, van der Velden U (1996b). Comparison of an oscillating rotating electric toothbrush and a "sonic" toothbrush in plaque-removing ability. J Clin Periodontol 24:1–5.

Van der Weijden GA, Timmerman MF, Reijerse E, Snoek CM, van der Velden U (1996c). Toothbrushing force in relation to plaque removal. J Clin Periodontol 23:724–729.

Van der Weijden GA, Timmerman M, Danser M, van der Velden U (1998). The role of automated toothbrushes: Advantages and limitations of automated toothbrushes. In: Lang NP (ed). Proceedings of the European Workshop on Mechanical Plaque Control. Chicago: Quintessence:138–155.

Van Houte J, Yanover L, Brecher S (1981). Relationship of levels of the bacterium Streptococcus mutans in saliva of children and their parents. Arch Oral Biol 26:381–386.

Van Loveren C, Buijs JF, Buijs MF, ten Cate JM (1996). Protection of bovine enamel and dentine by chlorhexidine and fluoride varnishes in a bacterial demineralization model. Caries Res 30:45–51.

Van Loveren C (2001). Antimicrobial activity of fluoride and its in vivo importance: Identification of research questions. Caries Res 35(suppl 1):65–70.

Van Moer C, Moradi-Sabzevar M, Adriaens P (1996). Anti-gingivitis effect of chlorhexidine chewing gum in periodontal maintenance patients [abstract 1829]. J Dent Res 75(special issue):246.

Van Rijkom HM, Truin GJ, van't Hof MA (1996). A meta-analysis of clinical studies on the caries-inhbiting effect of chlorhexidine treatment. J Dent Res 75:790–795.

Van Rijkom HM, Truin GJ, van't Hof MA (1998). A meta-analysis of clinical studies on the caries-inhibiting effect of fluoride gel treatement. Caries Res 32:83–92.

Van Swol RL, van Scotter DE, Pucher JJ, Dentino AR (1996). Clinical evaluation of an ionic toothbrush in the removal of established plaque and reduction of gingivitis. Quintessence Int 27:389–394.

Van Winkelhoff AJ, van der Velden U, Winkel E, de Graaff J (1986). Black-pigmented *Bacteroides* and motile organisms on oral mucosal surfaces in individuals with and without periodontal breakdown. J Periodontal Res 21:434–439.

Van Winkelhoff AJ, van der Velden U, Clement M, de Graaff J (1988). Intra-oral distribution of black-pigmented *Bacteroides* species in periodontitis patients. Oral Microbiol Immunol 3:83–85.

Vehkalahti M (1989). Occurrence of gingival recession in adults. J Periodontol 60:599–603.

Velu H, Balozet L (1931). Darmous (dystrophic dentaire) du mouton et solubilité du principe actif des phosphates naturels qui le provique. Bull Soc Path Exot 24:848–851.

Verdonschot EH, Bronkhorst EM, Burgersdijk RCW, König KG, Schaeken MJM, Truin GJ (1992). Performance of some diagnostic systems in examinations for small occlusal carious lesions. Caries Res 26:59–64.

Veys R, Barkvoll P, De Boever J, Baert J (1992). Side effects of the use of sodium lauryl sulphate in dentifrice on the oral tissues. J Head Neck Pathol 11:81–84.

Vierrou AM, Manwell MA, Zamek RL, Sachdeva RC, Tinanoff N (1986). Control of *Streptococcus mutans* with topical fluorides in patients undergoing orthodontic treatment. J Am Dent Assoc 113:644–646.

Vogel GL, Carey CM, Chow LC, Gregory TM, Brown WE (1988). Micro-analysis of mineral saturation within enamel during lactic acid demineralization. J Dent Res 67:1172–1180.

Vogel GL, Carey CM, Ekstrand J (1992). Distribution of fluoride in saliva and plaque fluid after a 0.048 mol/L NaF rinse. J Dent Res 71:1553–1557.

Volpe AR, Kapczak LJ, King WJ (1967). In vivo calculus assessment. Part III. Periodontics 5:184–193.

Volpe AR, Mooney R, Zumbrunnen C, Stahl D, Goldman H (1975). A long-term clinical study evaluating the effect of two dentifrices on oral tissues. J Periodontol 46:13–118.

Volpe AR, Petrone M, DeVizio W, Davies R (1993). A review of plaque, gingivitis, calculus and caries clinical efficacy studies with a dentifrice containing triclosan and PVM/MA copolymer. J Clin Dent 4(special issue):31–41.

Volpe AR, Petrone M, De Vizio W, Davies R (1996). A review of plaque, gingivitis, calculus and caries clinical efficacy studies with a fluoride dentifrice containing triclosan and PVM/MA copolymer. J Clin Dent 7:1–14.

Von der Fehr F, Löe J, Theilade E (1970). Experimental caries in man. Caries Res 4:131.

Waaler SM, Rölla G, Skjörland KK, Ogaard B (1993). Effects of oral rinsing with triclosan and sodium lauryl sulphate on dental plaque formation: A pilot study. Scand J Dent Res 101:192–195.

Waerhaug J (1976). The interdental brush and its place in operative and crown and bridge dentistry. J Oral Rehabil 3:107–113.

Waerhaug J (1981). Effect of toothbrushing on subgingival plaque formation. J Periodontol 52:30–34.

Waerhaug M, Gjermo P, Rölla G, Johansen J (1984). Comparison of the effect of chlorhexidine and $CuSO_4$ on plaque formation and development of gingivitis. J Clin Periodontol 11:176–180.

Wahab FK, Shellis RP, Elderton RJ (1993). Effects of low fluoride concentrations on formation of caries-like lesions in human enamel in a sequential-transfer bacterial system. Arch Oral Biol 38:985–995.

Wåler SM, Rölla G (1980). Plaque inhibiting effect of combinations of chlorhexidine and the metal ions zinc and tin. Acta Odontol Scand 38:213–217.

Wåler SM, Rölla G (1982). Comparison between plaque inhibiting effect of chlorhexidine and aqueous solutions of copper- and silver-ions. Scand J Dent Res 90:131–133.

Wåler SM, Rölla G (1985). Importance of teeth and tongue as possible receptor sites for chlorhexidine in relation to its clinical effect. Scand J Dent Res 93:222–226.

Wåler SM (1989). Studies on the Inhibition of Human Dental Plaque [thesis]. Oslo: Univ of Oslo.

Walker CB (1990). Effects of sanguinarine and sanguinaria extract on the microbiota associated with the oral cavity. J Can Dent Assoc 56:13–17.

Wallman C, Krasse B, Birkhed D (1994). Effect of chlorhexidine treatment followed by stannous fluoride gel application on mutans streptococci in margins of restorations. Caries Res 28:435–440.

Wallman C, Birkhed D (2002). Effect of chlorhexidine varnish and gel on mutans streptococci in margins of restorations in adults. Caries Res 36:360–365.

Walls AWG, Murray JJ, McCabe JF (1988). The management of occlusal caries in permanent molars. A clinical trial comparing a minimal composite restoration with an occlusal amalgam restoration. Br Dent J 164:288–292.

Walsh TF, Glenwright H (1984). Relative effectiveness of a rotary and conventional toothbrush in plaque removal. Community Dent Oral Epidemiol 12:160–164.

Wang CW, Corpron R, Lamb W, Strachan D, Kowalski C (1993). In situ remineralization of enamel lesions using continuous vs intermittent fluoride application. Caries Res 27:455–460.

Warren PR, Chater B (1996). An overview of established interdental cleaning methods. J Clin Dent 7:65–69.

Weaks LM, Leschner NB, Barnes CM, Holroyd SV (1984). Clinical evaluation of the Prophy-jet as an instrument for routine removal of tooth stain and plaque. J Periodontol 55:486–488.

Weatherell JA, Deutsch D, Robinson C, Hallsworth AS (1977). Assimilation of fluoride by enamel throughout the life of the tooth. Caries Res 11:85–115.

Weatherell JA, Strong M, Ralph JP, Robinson C (1988). Availability of fluoride at different sites in the buccal sulcus. Caries Res 22:129–133.

Weerheijm KL, de Soet J, van Amerongen W, de Graaff J (1993). The effect of glass-ionomer cement on carious dentine: An in vivo study. Caries Res 27:417–423.

Weerheijm KL, Kreulen CM, de Soet JJ, Groen HJ, van Amerongen WE (1999). Bacterial counts in carious dentine under restorations: 2-year in vivo effect. Caries Res 33:130–134.

Weinstein P, Getz I (1978). Changing Human Behavior: Strategies for Preventive Dentistry. Chicago: Science Research Associates.

Wellock WD, Brudevold F (1963). A study of acidulated fluoride solutions. II. The caries inhibiting effect of single annual topical applications of an acidic fluoride and phosphate solution. A two-year experience. Arch Oral Biol 8:179–182.

Wendt LK, Koch G (1988). Fissure sealant in permanent first molars after 10 years. Swed Dent J 12:181–185.

Wendt LK, Hallonsten A, Koch G, Birkhed D (1994). Oral hygiene in relation to caries development and immigrant status in infants and toddlers. Scand J Dent Res 102:269–273.

Wendt LK, Birkhed D (1995). Dietary habits related to caries development and immigrant status in infants and toddlers living in Sweden. Acta Odontol Scand 53:339–344.

Wendt LK, Koch G, Birkhed D (2001a). On the retention and effectiveness of fissure sealant in permanent molars after 15–20 years: A cohort study. Community Dent Oral Epidemiol 29:302–307.

Wendt LK, Koch G, Birkhed D (2001b). Long-term evaluaiton of a fissure sealing programme in public dental service clinics in Sweden. Swed Dent J 25:61–65.

Wennström J, Lindhe J (1985). Some effects of a sanguinarine-containing mouthrinse on developing plaque and gingivitis. J Clin Periodontol 12:867–872.

Wennström A, Wennström J, Lindhe J (1986). Healing following surgical and nonsurgical treatment of juvenile periodontitis. A 5-year longitudinal study. J Clin Periodontol 13:869–882.

Wennström J, Heijl L, Dahlén G, Gröndahl K (1987a). Periodic subgingival antimicrobial irrigation of periodontal pockets. I. Clinical observations. J Clin Periodontol 14:541–550.

Wennström J, Dahlén G, Gröndahl K, Heijl L (1987b). Periodic subgingival antimicrobial irrigation of periodontal pockets. II. Microbiological and radiographical observations. J Clin Periodontol 14:573–580.

Wennström J, Serino G, Lindhe J, Eneroth L, Tollskog G (1993). Periodontal conditions of adults regular dental care attendants. A 12-year longitudinal study. J Clin Periodontol 20:714–722.

Wennström J (1997). Rinsing, irrigation and sustained local delivery. In: Lang NP, Karring T, Lindhe J (eds). Proccedings of the 2nd European Workshop on Periodontology—Chemicals in Periodontics. Berlin: Quintessence:131–151.

Wescott WB, Starcke EN, Shannon IL (1975). Chemical protection against postirradiation dental caries. Oral Surg Oral Med Oral Pathol 40:709–719.

Wespi HJ (1948). Gedanken zur Fragen der oprimalen Ernährung in der Schwangerschaft. Schweiz Med Wochenschr 7:153–155.

Wespi HJ (1950). Fluoridiertes Kochsalz zur Kariesprophylaxe. Schweiz Med Wochenschr 80:561–564.

Westergaard J, Frandsen A, Slots J (1978). Ultra-structure of the subgingival flora in juvenile periodontitis. Scand J Dent Res 86:421–429.

Westfelt E, Nyman S, Lindhe J, Socransky S (1983a). Use of chlorhexidine as a plaque control measure following surgical treatment of periodontal disease. J Clin Periodontol 10:22–36.

Westfelt E, Nyman S, Lindhe J, Socransky S (1983b). Significance of frequency of professional tooth cleaning for healing following periodontal surgery. J Clin Periodontol 10:148–156.

White DJ (1988). Reactivity of fluoride dentifrices with artificial caries. II. Effects on subsurface lesions: F uptake, F distribution, surface hardening and remineralization. Caries Res 22:27–36.

White CL, Drisko C, Mayberry W, Killoy W, Sackuvich D (1988a). The effect of supervised water irrigation on subgingival microflora [abstract 2298]. J Dent Res 67(special issue):400.

White DJ, Bowman WD, Faller RV, Mobley MJ, Wolfgang RA, Yesinowski JP (1988b). ^{18}F MAS-NMR and solution chemical characterization of the reactions of fluoride with hydroxyapatite and powdered enamel. Acta Odont Scand 46:375–389.

White DJ, Chen WC, Nancollas GH (1988c). Kinetic and physical aspects of enamel remineralization—A constant composition study. Caries Res 22:11–19.

White DJ, Nacollas (1990). Physical and chemical considerations of the role of firmly and loosely bound fluoride in caries prevention. J Dent Res 69:587–594.

White DJ (1991). Processes contributing to the formation of dental calculus. Biofouling 4:209–218.

White DJ (1992). Tartar control dentifrices: Current status and future prospects. In: Embery G, Rölla G (eds). Clinical and Biological Aspects of Dentifrices. Oxford: Oxford UP:277–291.

White DJ, Nelson D, Faller R (1994). Mode of action of fluoride: Application of new techniques and test methods to the examination of the mechanism of action of topical fluoride. Adv Dent Res 8:166–174.

White L (1996). Efficacy of a sonic toothbrush in reducing plaque and gingivitis in adolescent patients. J Clin Orthod 30:85–90.

Whitford GM (1989). The Metabolism and Toxicity of Fluoride. Vol 13: Monographs in Oral Science. Basel: Karger.

Whitford GM (1992). Changing patterns of fluoride intake: Dietary fluoride supplements. Workshop on changing patterns of fluoride intake. Chapel Hill, Univ of North Carolina. J Dent Res 71:1249–1254.

Whitford GM (1996). Fluoride toxicology and health effects. In: Fejerskov O, Ekstrand J, Burt B. Fluoride in Dentistry. Copenhagen: Munksgaard.

Wikesjö U, Reynolds H, Christersson L, Zambon J, Genco R (1989). Effects of subgingival irrigation on *A actinomycetemcomitans*. J Clin Periodontol 16:116–119.

Wiktorsson AM, Martinsson T, Zimmerman M (1991a). Fluoride sources and dental attendance habits among adults in communities with optimal and low water fluoride concentrations. Acta Odontol Scand 49:159–162.

Wiktorsson AM, Martinsson T, Zimmerman M (1991b). Number of remaining teeth among adults in communities with optimal and low water fluoride concentrations. Swed Dent J 15:279–284.

Wiktorsson AM, Martinsson T, Zimmerman M (1992a). Caries prevalence among adults in communities with optimal and low water fluoride concentrations. Community Dent Oral Epidemiol 20:359–363.

Wiktorsson AM, Martinsson T, Zimmerman M (1992b). Salivary levels of lactobacilli, buffer capacity and salivary flow rate related to caries activity among adults in communities with optimal and low fluoride concentrations. Swed Dent J 16:231–237.

Wiktorsson AM (1995). Dental Caries and Dental Fluorosis Among Adults in Two Swedish Communities with Optimal and Low Water Fluoride Concentrations [thesis]. Stockholm, Sweden: Karolinska Institute.

Wilcoxon DB, Ackerman R, Killoy W, Love J, Sakumura J, Tira D (1991). The effectiveness of a counterrotational action power toothbrush on plaque control in orthodontic patients. Am J Orthod Dentofacial Orthop 99:7–14.

Williams B, von Fraunhofer JA (1977). The influence of the time of etching and washing on the bond strength of fissure sealants applied to enamel. J Oral Rehabil 4:139–141.

Williams B, Winter GB (1981). Fissure sealants. Further results at 4 years. Br Dent J 150:183–187.

Williams SA, Curzon M (1990). The interrelationship between caries, oral cleanliness and the use of fluoride toothpaste [abstract 66]. Caries Res 24:413.

Wilson AD, Kent BE (1972). A new translucent treatment cement for dentistry. Br Dent J 132:133–135.

Wilson AD, Groffman DM, Kuhn AT (1985). The release of fluoride and other chemical species from a glass-ionomer cement. Biomaterials 6:431–433.

Wilson AD, McLean JW (1988). Glass-Ionomer Cement. Chicago: Quintessence.

Wilson M (1993). Laboratory assessment of antimicrobial agents for the treatment of chronic periodontitis. Microb Ecol Health Dis 5:143–145.

Wilson S, Levine D, Dequincey G, Killoy W (1993). Effects of two toothbrushes on plaque, gingivitis, gingival abrasion, and recession: A 1-year longitudinal study. Compend Cont Educ Dent 16:569–579.

Wolff LF, Bakdash M, Pihlstrom B, Bandt C, Aeppli D (1989). The effect of professional and home subgingival irrigation with antimicrobial agents on gingivitis and early periodontitis. J Dent Hyg 63:222–225.

World Health Organization (1987). Oral Health Surveys: Basic Methods, ed 3. Geneva: WHO.

World Health Organization (1993). DMFT levels at 12 years 1993. WHO/OHU/DMFT 12. Geneva: WHO.

World Health Organization (1994). WHO Oral Health Country Profile Programme 1994. Geneva: WHO.

World Health Organization (1997). Global Oral Data Bank 1997. Geneva: WHO.

Wright G, Banting D, Feasby W (1979). The Dorchester dental flossing study: Final report. Clin Prev Dent 1:23–26.

Wu CD, Savitt ED (2002). Evaluation of the safety and efficacy of over-the-counter oral hygiene products for the reduction and control of plaque and gingivitis. Periodontol 2000 28:91–105.

Ximénez-Fyvie LA, Haffajee AD, Som S, Thompson M, Torresyap G, Socransky SS (2000). The effect of repeated professional supragingival plaque removal on the composition of the supra- and subgingival microbiota. J Clin Periodontol 27:637–647.

Yankell SL, Emling R, Cohen D, Vanarsdall R (1985). A 4-week evaluation of oral health in orthodontic patients using a new plaque removal device. Compend Cont Educ Dent 6:123–127.

Yankell SL, Edvardsen S, Braaten S, Emling R (1993). Laboratory and clinical evaluations of the Jordan Exact toothbrush. J Clin Dent 4:67–70.

Yankell SL, Emling R (1994). A study of gingival irritation and plaque removal following a three-minute toothbrushing. J Clin Dent 5:1–4.

Yukna RA, Shaklee R (1993). Evaluation of a counter-rotational powered brush in patients in supportive periodontal therapy. J Periodontol 64:859–864.

Zambon JJ, Reynolds H, Chen P, Genco R (1985). Rapid identification of periodontal pathogens in subgingival dental plaque. Comparison of indirect immunofluorescence microscopy with bacterial culture for detection of Bacteroides gingivalis. J Periodontol 56:32–40.

Zero D, van Houte J, Russo J (1986). The intraoral effect on enamel demineralization of extracellular matrix material synthesized from sucrose by streptococcus mutans. J Dent Res 65:918–923.

Zero DT, Fu J, Espeland MA, Featherstone JDB (1988). Comparison of fluoride concentrations in unstimulated whole saliva following the use of a fluoride dentifrice and a fluoride dentifrice and a fluoride rinse. J Dent Res 67:1257–1262.

Zero DT, Raubertas RF, Pedersen AM, Fu J, Hayes AL, Featherstone JDB (1992a). Studies of fluoride retention by oral soft tissue after the application of home-use topical fluorides. J Dent Res 71:1546–1552.

Zero DT, Raubertas RF, Fu J, Pedersen AM, Hayes AL, Featherstone JDB (1992b). Fluoride concentrations in plaque, whole saliva, and ductal saliva after application of homeuse topical fluorides. J Dent Res 71:1768–1775.

Zickert I, Lindvall AM, Axelsson P (1982a). Effect on caries and gingivitis of a preventive program based on oral hygiene measures and fluoride application. Community Dent Oral Epidemiol 10:289–295.

Zickert I, Emilson CG, Krasse B (1982b). The effect of caries preventive measures in children highly infected with the bacterium Streptococcus mutans. Arch Oral Biol 27:861–868.

Zickert I, Emilson C, Krasse B (1987a). Microbial conditions and caries incrment 2 years after discontinuation of controlled antimicrobial measures in Swedish teenagers. Community Dent Oral Epidemiol 15:241–244.

Zickert I, Emilson C, Ekblom K, Krasse B (1987b). Prolonged oral reduction of Streptococcus mutans in humans after chlorhexidine disinfection followed by fluoride treatment. Scand J Dent Res 95:315–319.

Zimmerman A, Flores-de-Jacoby L, Pan P (1993). Gingivitis, plaque accumulation and plaque composition under long-term use of Meridol. J Clin Periodontol 20:346–351.

LIST OF ABBREVIATIONS

A8-F—5-n-octanyl-3-trifluoromethylsalicylanilide

ADA—American Dental Association

AlF_3—aluminum fluoride

Al_2O_3—alumina

AmF—amine fluoride

APITN—Apical Periodontitis Index of Treatment Needs

APF—acidulated phosphate fluoride

BOP—bleeding on probing

$CaCO_3$—calcium carbonate

CaF_2—calcium fluoride

CaGP—calcium glycerophosphate

CCITN—Community Caries Index of Treatment Needs

CEJ—cementoenamel junction

CFU—colony-forming unit

CHX—chlorhexidine

CPITN—Community Periodontal Index of Treatment Needs

DAV—dentin abrasive value

DB—distobuccal

DFPT—decayed and filled primary teeth

DFS—decayed or filled surface

DFT—decayed or filled teeth

DMFS—decayed, missing, or filled surface

DMFT—decayed, missing, or filled teeth

DS—decayed surface

EMRIRF—external modifying risk indicator, risk factor, and prognostic risk factor

F—fluoride

FA—fluorapatite

FDI—Fédération Dentaire Internationale

FHA—fluoridated hydroxyapatite

GBI—Gingival Bleeding Index

GCF—gingival crevicular fluid

GI—Gingival Index

HA—hydroxyapatite

HEMA—2-hydroxyethyl methacrylate

HQ—highest quartile

HR—hazard ratio

Ig—immunoglobulin

IL—interleukin

IMRIRF—internal modifying risk indicator, risk factor, and prognostic risk factor

KF—potassium fluoride

LaF_2—lanthanum fluoride

LQ—lowest quartile

MB—mesiobuccal

MFP—monofluorophosphate

MMP—matrix metalloproteinase

MS—mutans streptococci

Na_3AlF_6—sodium hexafluoroaluminate

Na_2FPO_3—sodium monofluorophosphate

NaF—sodium fluoride

NH_4F—ammonium fluoride

PAL—probing attachment loss
PCPC—professional chemical plaque control
PFRI—Plaque Formation Rate Index
PI—Plaque Index
PMNL—polymorphonuclear leukocyte
PMTC—professional mechanical toothcleaning
PRF—prognostic risk factor
PRP—proline-rich glycoprotein
PRT—periodontal risk test
PTD—probable toxic dose
RDA—radioactive dentin abrasivity
RF—risk factor

RI—risk indicator
SBI—Sulcular Bleeding Index
SiC Index—Significant Caries Index
SiO_2—silicone dioxide
SLS—sodium lauryl sulfate
SMFP—sodium monofluorophosphate
SnF_2—stannous fluoride
SrF_2—strontium fluoride
TF Index—Thylstrup-Fejerskov Index
USPHS—US Public Health Service
WHO—World Health Organization

INDEX